PESTICIDE MICROBIOLOGY

*Microbiological Aspects of Pesticide Behaviour
in the Environment*

Edited by

I. R. HILL

*Imperial Chemical Industries, Plant Protection Division,
Jealott's Hill Research Station, Bracknell, England*

and

S. J. L. WRIGHT

School of Biological Sciences, University of Bath, Bath, England

1978

ACADEMIC PRESS

LONDON   NEW YORK   SAN FRANCISCO

A Subsidiary of Harcourt Brace Jovanovich, Publishers

ACADEMIC PRESS INC. (LONDON) LTD.
24/28 Oval Road,
London NW1

United States Edition published by
ACADEMIC PRESS INC.
111 Fifth Avenue
New York, New York 10003

Library of Congress Catalog Card Number: 77–93204
ISBN: 0–12–348650–5

Printed in Great Britain by
J. W. ARROWSMITH LTD., BRISTOL BS3 2NT

Contributors

ANDERSON, J. R., *Microbiology Unit, National Food Research Institute, Council for Scientific and Industrial Research, P.O. Box 395, Pretoria 0001, South Africa.*

ARNOLD, D. J., *I.C.I. Plant Protection Division, Jealott's Hill Research Station, Bracknell, Berks, RG12 6EY, England.*

BENEZET, H. J., *Department of Entomology, University of Missouri, Columbia, Missouri, U.S.A.*

CRIPPS, R. E., *Shell Research Limited, Woodstock Laboratory, Sittingbourne Research Centre, Sittingbourne, Kent, ME9 8AG, England.*

EDWARDS, C. A., *Rothamsted Experimental Station, Harpenden, Herts., AL5 2JQ, England.*

GRAY, T. R. G., *Department of Biology, University of Essex, Colchester, CO4 3SQ, Essex, England.*

HILL, I. R., *I.C.I. Plant Protection Division, Jealott's Hill Research Station, Bracknell, Berks., RG12 6EY, England.*

JONES, J. G., *Freshwater Biological Association, Windermere Laboratory, The Ferry House, Ambleside, Cumbria, LA22 0LP, England.*

MATSUMURA, F., *Pesticide Research Center, Michigan State University, East Lansing, Michigan, U.S.A.*

NEWMAN, J. F., *I.C.I. Plant Protection Division, Jealott's Hill Research Station, Bracknell, Berks, RG12 6EY, England.*

ROBERTS, T. R., *Shell Research Limited, Woodstock Laboratory, Sittingbourne Research Centre, Sittingbourne, Kent, ME9 8AG, England.*

WOODCOCK, D., *University of Bristol, Research Station, Long Ashton, Bristol, BS18 9AF, England.*

WRIGHT, S. J. L., *School of Biological Sciences, University of Bath, Bath, BA2 7AY, Avon, England.*

v

Foreword

KENNETH MELLANBY

Former Director of Monks Wood Experimental Station, former Head of the Department of Entomology, Rothamsted Experimental Station. Editor, International Journal Environmental Pollution

There is growing concern throughout the world at the way man is affecting, and often damaging, his environment. National and international agencies exist to monitor and control pollution of the air, of water and of the land. There is much popular debate of all environmental topics; this is often marred by error and emotion, but this may be better than complacency and apathy. In fact, where serious environmental damage has been demonstrated, striking improvements in environmental quality have generally occurred subsequently. Thus the air of most cities in developed countries, in Europe, America or Japan, is cleaner today than it was ten or twenty years ago. Though still polluted, many rivers in industrial areas are less filthy than they were in the past. We see that, where the facts are available, and where there is still the will to succeed, civilized man can enjoy the fruits of industrial progress and yet produce a cleaner world for himself and his descendants.

The control of all types of pollution can be expensive, and so it is wise to use resources sensibly to this end. This is not done when the dangers from, for instance, toxic chemicals are overestimated. In the past many useful substances have been denied to mankind because their possible disadvantages have been exaggerated. Control of new chemicals should be firmly based on a sound knowledge of their possible environmental effects. Such substances should not be widely used until their properties, including their toxicity and their stability, are properly understood. But the formulation of unnecessarily stringent precautions may be counterproductive; in some countries they have certainly delayed the introduction of efficient pesticides, and have thus prevented the control of agricultural and medical pests. Control should be based on knowledge, particularly of the sum total of the biological effects, both beneficial and harmful, of substances like pesticides which are likely to enter our environment.

The use of pesticides illustrates well a dilemma which often faces modern man. No one can doubt that their use has made enormous contributions to agriculture and public health. Without pesticides world food supplies would be inadequate for the growing human population—even though this growth has been accelerated by the use of pesticides to conquer many of the major diseases which previously reduced population increase. However, this book is not concerned with this social problem, even though it is of major importance to man's ultimate survival on earth. This book is concerned with the ecological and environmental effects of the pesticides themselves.

Pesticides differ from most other substances which affect our environment in that they are deliberately dispersed by man for what he, the user, thinks are entirely desirable ends. Most other potential pollutants which are widely dispersed are the byproducts of industry and city living which man hopes he can get rid of, harmlessly, by dispersing and diluting them into the air, into rivers or into the sea. This process is usually successful. Many wastes are quickly diluted to harmless levels. However, others are not, hence many major developments in industrial and urban pollution control. These work by containing or destroying the dangerous substances which were previously discharged. But with pesticides the situation is quite different. If they are to be effective, they *must* be introduced into the environment in order to reach the target organisms—the pests—which they are designed to control.

Were a pesticide entirely specific in its action, so that only the target organism was killed, there would be no controversy about its use. Similarly, if it could be used so that the whole amount of the substance discharged reached the pest, and none spilled over to affect other organisms, it would not matter whether or not the poison was specific. But, though great progress has been made in producing more specific (or less unspecific) pesticides, for instance substances much more toxic to insects than to mammals, and though the efficiency of spraying has been greatly improved, there is still much room for improvement. Most pesticides can damage a wide range of organisms, both harmful and beneficial, for broad-spectrum pesticides are generally more use-ful and commercially viable than those suitable for only one pest. Even with the most efficient dispersal methods, more than 99% of most insecticides miss the bodies of the pests at which they are aimed. Herbicides are more effectively applied, but even with them there is substantial loss to the soil or onto non-target organisms. Therefore pesticides must obviously have important ecological effects, other than those intended (that is, the control of the pests themselves) and these must be understood if these substances are to be used safely and efficiently.

Conservationists tend to stress the possible damage which may be done to wildlife by pesticides. In Britain in the late 1950s ornithologists were disturbed to find large numbers of dead birds strewing the countryside in early spring, victims of the use of aldrin and dieldrin seed dressings protecting cereals from attack by the Wheat Bulb Fly. It was shown that these chemicals were passed on to, and concentrated by, predators, which were themselves killed. There was great concern that small populations of slow breeding hawks would be exterminated. This proved to be a world wide problem, brought to public attention by the publication of books like Rachel Carson's *Silent Spring* in 1962, though scientists in many countries were working on the problem many years before this book was published, and have even taken effective action to reduce the harmful effects of these organochlorine insecticides by the voluntary ban of their use as seed dressing on spring-sown cereals in Britain.

The harmful side effects of pesticides on birds, and of herbicides drifting outside crops onto gardens or wild flowers may be easy to observe, and have generally been possible to control. However, many scientists have feared that the wide use of toxic chemicals may have more serious and more permanent effects of a less easily observable kind, by affecting microorganisms. This is one important subject which is discussed in this book. We are increasingly coming to realize that microorganisms, bacteria, fungi, algae, protozoa and small arthropods play important roles in soil, some related to the maintenance of its fertility. The destruction of beneficial bacteria, or shifts in the balance of populations of different species, may have marked effects on agricultural productivity. Soil science is a complex subject at all times, and the introduction of new and powerful chemicals with unknown effects could obviously make it even more complex. So the study of these effects is clearly of great importance.

On the other hand, it has been found that some microorganisms, though destroyed by particular pesticides, may flourish in the presence of others, and even use them as energy sources. The ecological problems relating to the persistence of toxic chemicals in the soil, with potential hazards to wildlife, can obviously be affected by the action of microorganisms which degrade pesticides and so prevent them from having such long-acting effects. This again is an important theme of some of the work described in this book.

Few of us can doubt that, with the growing world population, and the need for greater food production, pesticides will play an increasingly important role in the future. Though we welcome developments in the non-chemical methods of pest control—including the use of parasites and predators, and the use of "genetic engineering" to make pests less viable—it is unlikely that these biological control methods will have anything like universal application. So we must get the greatest benefits and the least damage from the pesticides that are used. This book sets out to show how the knowledge to make this possible may be obtained, and how a better understanding of the relationships of microorganisms to the whole ecology of our environment may be realized. It will be found to be of value to microbiologists, for few can have first hand information on the whole range of topics covered. But it will be particularly useful to other biologists who wish to understand the role of microorganisms in this type of ecological problem which is obviously of such great economic importance in its relation to future problems of food production and public health.

June 1978

Preface

The interest of microbiologists in synthetic pesticidal chemicals extends far beyond the compounds which are aimed at microbial pests.

Why are so many microbiologists around the world now concerned with chemicals which are designed to control other, often totally unrelated, organisms? The answers lie in the ever-increasing global use of pesticides to sustain and enhance man's food supply, together with the remarkable ubiquity, diversity and numbers of the microbes whose collective activities are essential features of life on this planet. Microbes are present in all environmental situations in which pesticides are used and they will therefore encounter such man-made chemicals, however inadvertently, and probably react with them in some way. In simplest terms such interactions can be considered as (i) the action of microbes on pesticides and (ii) the action of pesticides on microbes. In reality the events are unlikely to be so clearly defined or isolated from the influence of other factors controlling the fate and activity of pesticides in the environment. We consider, however, that there is sufficient justification in the area of pesticide science for "Pesticide Microbiology". With the help of several authors we have attempted to produce a book reflecting our belief and which, hopefully, allows pesticide–microbe interactions to be seen in the wider environmental context.

Our intention has been to obtain a book which apart from serving as a reference source for workers already in the field, might also interest those who work in allied areas such as pesticide chemistry, toxicology and microbial ecology. We have also considered the growing number of undergraduate and post-graduate students whose courses and research embrace aspects of environmental microbiology, toxicology and the fate of xenobiotics. For these reasons we have included "introductory" chapters to give basic information on pesticides, microbial ecology and the environmental fate of pesticides. Overlap has been minimized, but retained where necessary for clarity or emphasis.

Of the many aspects of pesticide behaviour in the environment which currently receive scientific attention, there has been a notable acceleration in the microbiological field in industrial, Government-sponsored and academic research institutions globally. The vast amount of information generated is widely dispersed in diverse scientific journals and books, in several languages. This book represents an attempt to bring at least some of this information together between two covers. Investigations in "Pesticide Micro-biology" will increasingly require the skills and collaboration of microbiologists of different disciplines and inclinations. Dare we suggest that the subject can form a convenient bridge between the physiologists and ecologists in microbiology?

We thank the contributing authors, especially those who submitted manuscripts as originally requested in 1975 and have since been very patient whilst we pieced together the rest of the book. We gratefully acknowledge the guidance given by the staff at Academic Press and the work of the typists, especially Mrs. Rita Pratt and Mrs. Lorraine Thorne, who have competently dealt with the manuscripts at various stages. A special word of thanks go to Mrs. Glyn Pearce who so skilfully interpreted our intentions in producing the Figures in Chapter 3 and the design on the book's cover.

Finally, we sincerely thank our families for their great forbearance over the past years in which we have been occupied with the book. In particular we thank our wives, Lyn and Sue, for their considerable encouragement and help with some of the tasks involved. The book is dedicated to them.

I. R. HILL and S. J. L. WRIGHT
June 1978

Contents

CHAPTER 1

Pesticides

J. F. Newman

CHAPTER 2

Microbiological Aspects of the Soil, Plant, Aquatic, Air and Animal Environments

Soil and Plants

T. R. G. Gray

Aquatic Environments

J. G. Jones

The Air and Animals

S. J. L. Wright

CHAPTER 3

The Behaviour and Fate of Pesticides in Microbial Environments

I. R. Hill and S. J. L. Wright

CHAPTER 4

Microbial Transformation of Pesticides

I. R. Hill

CHAPTER 5

Transformations of Pesticides in the Environment – The Experimental Approach

I. R. Hill and D. J. Arnold

CHAPTER 6

Some Methods for Assessing Pesticide Effects on Non-Target Soil Microorganisms and their Activities

J. R. Anderson

CHAPTER 7

Pesticide Effects on Non-Target Soil Microorganisms

J. R. Anderson

CHAPTER 8

Interactions of Pesticides with Micro-Algae

S. J. L. Wright

CHAPTER 9

Pesticides and the Micro-Fauna of Soil and Water

C. A. Edwards

CHAPTER 10

Microbial Degradation of Insecticides

F. Matsumura and H. J. Benezet

CHAPTER 11

Microbial Degradation of Herbicides

R. E. Cripps and T. R. Roberts

CHAPTER 12

Microbial Degradation of Fungicides, Fumigants and Nematocides

D. Woodcock

Chapter 1

Pesticides

JAMES F. NEWMAN

Imperial Chemical Industries, Plant Protection Division,
Jealotts Hill Research Station, Bracknell, Berkshire, England

I. Introduction

Some 10 000 years ago men living in the Tigris/Euphrates valley moved from a hunting and food gathering way of life to a more settled existence in which food plants were sown and harvested. The essential ecological difference between a crop and natural vegetation is one of species diversity, such diversity being ideally absent, certainly much lower, in the crop. The inventors of agriculture no doubt found life easier when substantial pure stands of food plants could be established around them and the labour of hunting and food gathering could be replaced by the more productive labour of cultivation and harvesting.

This easier life was reflected by a population growth rate of between 0·5 and 1·0% per year for agricultural societies (Cipolla, 1965). The settled mode of life, however, can be followed with advantage by other consumers of the crops, whether they be vertebrate, invertebrate, fungal or bacterial, and these forms have reproductive rates often greatly in excess of that of humans. The conditions are clearly favourable for such pests to proliferate. Pests and diseases are those organisms which compete with man for his agricultural land, his growing crops and his harvested produce. The ecological limitations which restricted their numbers in natural diverse vegetation are removed in the artificial crop environment and they emerge as pests, as a logical consequence of man's interference with nature.

The control of pests and diseases is thus a problem as old as agriculture. In earlier times, control upon any large and effective scale was not possible and the shifting ecological balances of pests and other species led to levels of attack which varied widely from year to year. History contains many references to seasons of high pest incidence, from the biblical plagues of Egypt to the failure of the Irish potato crop in the middle of the last century.

II. Types of Pesticides

The earliest attempts to control pests chemically used naturally occurring toxic substances, such as mercury or sulphur, or plant extracts such as nicotine, pyrethrum or derris. The era of modern synthetic pesticides largely dates from 1939 when the insecticidal properties of DDT were discovered. Pesticides today include insecticides for insect control, acaricides for mite control, nematicides for the control of eelworms, rodenticides for rats and mice, fungicides and bactericides for the control of plant diseases and storage rots, and herbicides for the control of weeds. It is convenient also to include the plant growth regulators, which can be used to control the type and amount of plant growth. While these substances are not strictly pesticides, in that they do not kill or regulate numbers of organisms, they are important agricultural chemical implements, applied in a similar way to pesticides.

III. Purpose and Types of Use of Pesticides

Pesticides are used in agriculture for three main purposes – to produce a larger yield of crop, to produce a crop of higher quality and to reduce the input of labour and energy into crop production. Even now, approximately one third of the potential production of the world's crops is destroyed by pests, weeds and diseases, either by direct attack upon the growing crop or by damage and destruction to the stored harvest (Agro-Allied Industries/FAO, 1972). In practice, the appropriate use of pesticides has been responsible for a trebling of cocoa yields in Ghana, an increase of one third in the yield of sugar cane in Pakistan and a doubling of cotton production. Even when pest or disease attacks do not completely destroy crops, a high proportion of the produce may be blemished, so that it demands a lower price in the market and often has a reduced storage life.

The productivity of agriculture, both in terms of output per unit area and output per man employed, has increased greatly during the last century. In Britain in 1850 about 22% of the population were engaged in agriculture but by 1950 this figure had declined to 5%. Comparable figures for the USA are 65% in 1850 and 13% in 1950. This process continues today and is well illustrated by the figures for the number of people supported in food and clothing by the labour of one farm worker in the USA at various times. In 1850 a farm worker produced sufficient for four persons, in 1960 for 25 persons and in 1975 for 44 persons. This increase in efficiency is clearly not solely related to pesticides, the progressive mechanization of farm operations and the development of higher-yielding crop varieties being other important factors. However, pesticide use has undoubtedly played a part in the maintenance of this trend in recent years. The increase in efficiency in terms of output per area per man has been achieved at the cost of a large increase in the input of energy from fossil fuel into agriculture. Horsepower and manpower have been replaced by mechanical power and the fuel input alone to British agriculture has increased by four times between 1938 and 1970 (Blaxter, 1974). The use of the increased rates of fertilizers and pesticides which has done so much to increase agricultural production in recent years also involves a considerable energy input in relation to the extraction and synthesis of the chemicals. There is an area, however, in which pesticide usage can reduce the input of energy to agriculture. The use of chemical weed control as an alternative to mechanical cultivation may effect an overall energy saving. This is particularly so when a heavy and energy-expensive operation such as ploughing and cultivation can be replaced by direct seeding into almost undisturbed ground after the surface weed growth has been destroyed by a herbicide of short biological persistence, such as paraquat. The total energy budget for such a technique has been studied recently (Leach, 1976) and it has been shown that even after allowance has been made for the energy requirement for chemical production there is a saving of about one third in fuel requirement.

The extent to which different types of pesticide are used varies with the differing agricultural and sociological conditions which exist in various parts of the world. Table 1 shows the relative expenditure on different types of pesticides in various areas. In the developed countries of Europe and N. America the greatest expenditure is upon herbicides, reflecting the high cost of labour for other methods of weed control. In the more tropical countries which are in general the less developed areas, the climatic conditions favour insect pests and the greatest expenditure is on insecticides. Some tropical areas show considerable use of fungicides but such use is largely associated with high value plantation crops, often controlled by large commercial organizations elsewhere.

TABLE 1

World pesticide use

Country	% total expenditure on pesticides		
	Herbicides	Insecticides	Fungicides
N. America	61	25	7
W. Europe	49	25	23
Far East	25	45	21
Africa	30	42	28
S. America	30	53	16
Australia, New Zealand	45	38	16
Central America	33	51	15

Apart from normal applications to crops in the field, pesticide usage is important in a number of other situations. Stored food may be attacked by many species of insects, including the larval and adult stages of a wide range of beetles and moths, by mites and by a variety of bacterial and fungal rots and moulds. While appropriate attention to the design of food stores and hygienic handling and carriage of harvested crops is important in reducing the access of pests or in maintaining conditions which do not favour their growth, the use of insecticides and fungicides is essential in both the preservation of stored products and in the disinfestation of storage premises and flour mills.

Pesticides of all types are widely used in forestry. Herbicides are used in nursery beds and in young plantations to prevent the smothering of young trees by weed growth. Insecticides may be required at any stage of tree growth to control attacks by leaf-eating or shoot-boring caterpillars which can otherwise devastate forests over large areas. Timber after felling, during storage and in use requires protection by pesticides from termites, wood-boring beetles and from fungal and bacterial decay. Much timber is used in the production of paper and the paper pulp industry is a large user of fungicides to limit the growth of moulds in paper mills. The industrial use of pesticides includes such applications as the control of weeds on railway tracks, of woody vegetation beneath power lines and of weeds on industrial premises generally. Herbicides are used around storage tanks for oil and other inflammable substances where excessive weed growth might increase fire hazard. Herbicides are used extensively in the control of aquatic weed growth in drains, in irrigation and navigation canals and in water storage reservoirs. The cost of such applications is less than that incurred in mechanical cutting and removal.

One of the most important uses of insecticides is in the field of public health, in the control of the diseases of man and his domestic animals, which are carried by insects and mites. Some of the earliest uses of DDT were in the control of the louse vector of typhus fever and in malarial mosquito control. It is well to remember that this insecticide, now known to have some environmental deficiencies and consequently much criticized, has undoubtedly been responsible for saving more human lives than any other chemical invention. The list of effective insecticide uses in public health is impressive, with massive reductions in the incidence of major insect-borne diseases in many different parts of the world. While much is being done and some success is being achieved in research on alternative biological or other methods of pest control, there is nothing to suggest that the great importance of pesticides in these varied applications is likely to decline in the foreseeable future.

IV. Insecticides

A. ORGANOCHLORINES

DDT is a chemically stable compound having the low vapour pressure of 1.5×10^{-7} mm of mercury at 20°C (Woodwell *et al.*, 1971) and which readily kills insects by contact. In contrast to earlier insecticides it offered the possibility of prolonged control, from a single application, of insects on plants, or, when applied as a residual deposit to the walls of buildings, of the mosquito vectors of malaria. Following its spectacular success in the latter application, DDT went into widespread use in agriculture in the late 1940s. The insecticidal activity of the gamma isomer of hexachlorocyclohexane (gamma BHC or lindane) was discovered in the early 1940s and was used in a similar way to DDT. In the following years chemical research into organochlorine compounds led to the development of a range of insecticides, including aldrin, dieldrin, heptachlor, chlordane and endrin. The direction of development was towards increased insecticidal activity, which was unfortunately often associated with higher mammalian toxicity and increased persistence in the environment. These insecticides all produce their insecticidal action by affecting the functioning of the nervous system. The pesticide molecules interfere with the ionic permeability of nerve cell membranes and so produce an unstable state in which spurious nerve impulses induce uncontrolled activity in the whole organism. The effect appears to depend not upon a chemical interference with any enzyme system but upon a spatial disruption of the cell membrane, related to the shape of the organochlorine molecule.

B. ORGANOPHOSPHORUS COMPOUNDS

The organophosphorus insecticides had their origin in research in Germany in the late 1930s (Schrader, 1963). Compounds of dangerously high mammalian toxicity are included in this group, reflecting the interest in it in relation to chemical weapons during the war years. More recently, much work has been done in this area of chemistry. Far from being dangerously toxic compounds, some of the modern organophosphorus insecticides are among the safest materials available. The oral LD_{50} figure for rats for menazon is $2000\,mg\,kg^{-1}$ body weight, for pirimiphos ethyl $2050\,mg\,kg^{-1}$ and for malathion $2800\,mg\,kg^{-1}$ (Martin and Worthing, 1977). In contrast to the organochlorine insecticides the organophosphorus compounds are essentially chemically reactive materials and persist in the environment for short periods only. Their solubility properties often render them systemic in action, so that they are absorbed into plants and to a varied extent translocated in the sap flow. Effective insect control can thus be obtained with incomplete cover of the foliage. When pronounced, the systemic property may be used to protect the whole plant by a soil application of a granular formulation, as in the use of disulfoton to protect beans from aphid attack. The effect may be more limited, as in the use of a foliar spray of menazon to control aphids on the sides of leaves away from the spray application.

Metabolism of the organophosphorus insecticides is important in various ways and the extensive literature on this matter is well reviewed by O'Brien (1967). Compounds having the thionate structure require activation by oxidation *in vivo* to the thiolate, as in the conversion of parathion to paraoxon. The relative safety to man and other mammals of other organophosphorus insecticides is dependent upon differing metabolic routes in insects and higher animals. Malathion is more toxic to insects than to mammals, having an LD_{50} value of 0.75 mg kg^{-1} to aphids and of about 3000 mg kg^{-1} to rats. This difference is caused by the conversion of malathion in insects to the more toxic malaoxon, while in mammals a carboxyesterase is present which removes one of the terminal ethyl groups to give a metabolite of low activity (Krueger and O'Brien, 1959).

The action of the organophosphorus insecticides depends upon inhibition of the enzyme acetylcholinesterase. The neurotransmitter substance acetylcholine functions in various parts of the nervous system in both insects and mammals and the inhibition of the enzyme responsible for its removal results in disruption of nervous control. The insecticides effect a phosphorylation of the enzyme and this reaction, in contrast to the normal acetylation by acetylcholine, is less readily reversible to regenerate the active enzyme.

C. CARBAMATES

The carbamate insecticides have been developed more recently and include carbaryl, propoxur, aldicarb and pirimicarb, together with the natural plant toxin physostigmine. These materials act in a way comparable with that of the organophosphorus compounds in that they react with and inhibit the action of acetylcholinesterase by effecting, in this case, a carbamylation of the enzyme. Although some carbamate insecticides have a high acute toxicity to mammals, the acute oral LD_{50} of aldicarb to rats being 1.0 mg kg^{-1}, the carbamylation of the enzyme is more readily reversed than is phosphorylation.

The activity of both organophosphorus and carbamate insecticides is dependent upon the fit of the inhibitor molecule to the area of the active site on the enzyme. There is therefore possibility for great variation in the activity of different but related structures. The exploitation of differences in enzyme structures adjacent to the active site in different forms of life gives a range of selective activities to these compounds. The carbamate, pirimicarb, has a high activity against aphids with a relatively low mammalian toxicity, the acute oral LD_{50} to rats being 147 mg kg^{-1} and in practical use it is selectively safe to predatory beetles and honey bees. The chemistry and toxicology of carbamate insecticides is well reviewed by Kuhr and Dorough (1976).

D. PYRETHRINS AND RELATED CHEMICALS

The natural pyrethrins, derived from the flowers of *Pyrethrum cineraefolium*, have long been known as very safe insecticides. They have a powerful contact action, producing a rapid knock-down effect on insects and, together with the chemically and biologically similar synthetic product allethrin, they have been widely used as domestic and public-health fly sprays. These materials are, however, photochemically unstable, have a poor storage life in formulation and lack adequate persistence for use in many pest control applications. Extensive research by Elliot *et al.* (1973) has led to the discovery of related synthetic materials of greater stability and persistence, combined with the low mammalian toxicity of the natural pyrethrins. Permethrin, one of this group of insecticides at present under development, has an acute oral toxicity to mammals in excess of 4000 mg kg^{-1} but is effective against many insect pests at application rates as low as 10 g ha^{-1}.

V. Fungicides

The parasitic fungus has a close relationship with the host plant and the problem of killing the one without damage to the other is clearly greater than in the case of an insect pest upon a plant, where not only is there less intimacy of contact but a greater physiological diversity may be exploited. Fungicides may be divided into protectants, which are used to form a surface deposit on foliage to prevent the germination or establishment of fungal spores; and eradicants which will eliminate an established fungal infection. Even eradicants, however, cannot be expected to repair damage already inflicted by the disease. Some 130 compounds are in use as fungicides and information on their chemistry, biological action and use is provided by Torgeson (1967, 1969).

Prior to the development of synthetic organic fungicides, protection of plants from fungal disease largely depended upon the use of compounds of sulphur or of copper. Elemental sulphur was applied as a dispersion of finely divided particles or as a colloidal solution. Lime sulphur, a mixture of calcium polysulphides, was more soluble in water and so easier to formulate and apply. It subsequently decomposed on the leaf surface to give free sulphur. Bordeaux mixture, introduced in 1885 to control mildew on vines, is largely cupric hydroxide precipitated from a mixture of copper sulphate and calcium hydroxide. Cuprous oxide and oxychloride have also been widely used as protectants. The ultimate fungitoxic action of all the copper compounds is probably a non-specific effect of copper on enzyme proteins.

Mercury compounds have been widely used as fungicides. Organic mercury compounds were introduced in 1915 and have been particularly useful as seed treatments on cereal seeds for the elimination of seed-borne diseases. Modern highly effective organomercurials include PMA (phenylmercury acetate), EMC (ethylmercury chloride) and MMD (methylmercury dicyandiamide) but the high general toxicity of these materials, together with their undesirable environmental accumulation properties, has led to a decline in their use.

Organic sulphur-containing fungicides include thiram, captan, maneb and zineb. These materials are widely used as protectant fungicides on crops and particularly on potatoes for the control of blight (*Phytophthora infestans*). Their persistence on foliage or in the environment is limited and spray applications at intervals of 1–2 weeks through the season are necessary to maintain protection.

Within recent years effective organic fungicides having systemic activity within the plant have been produced. These include benomyl, ethirimol and thiophanate methyl. These fungicides are of low toxicity to mammals, having oral LD_{50} values for the rat between 4000 and 10 000 mg kg^{-1} but they are effective against limited ranges of fungal diseases.

VI. Herbicides

The phenoxy-acid herbicides, MCPA and 2,4-D, were discovered in the early 1940s, arising from research into natural plant growth hormones such as indole-3-acetic acid. They are, in general, more chemically stable than the natural hormones and so produce an excessive and disruptive response in susceptible plants. A range of these materials is now available and their use has had a marked influence upon the pattern of agriculture, particularly in the production of cereals in a much simplified crop rotation. Phenoxy-acids are relatively rapidly metabolized in the microbiological environment of the soil. The phenoxy-acid herbicides are essentially selective in their action, having a much greater effect on broad-leaved plants than on grasses. This selectivity is probably related to a greater ease of translocation to the susceptible meristems in dicotyledons.

The urea herbicides, such as monuron and diuron; and the triazines, such as simazine and atrazine, include many compounds. Physiologically they are total herbicides but can be used selectively by exploiting their low water solubility and low mobility in the soil. They are used as soil-applied herbicides, to be taken up by plant roots, and they may persist in the soil for a long time. A treatment of the surface layers of the soil with simazine, for example, at a rate of 2–4 kg ha^{-1} can keep the area free of shallow-rooted annual weeds for up to one year while not affecting deeper-rooted shrubs or plantation crops in the same area. Simazine is also widely used for weed control in maize, where it acts as a chemically selective material, since the maize plant is able to detoxify the compound to a herbicidally inactive hydroxy derivative.

The bipyridyl herbicides, paraquat and diquat, act as non-persistent total herbicides, killing all green plant tissue with which they come into contact. The chemistry, mode of action and degradation of these compounds has been described by Calderbank (1976). While the bipyridyls are readily decomposed by some microorganisms in pure culture, in practice in the field, the short persistence of their activity is due to their rapid biological inactivation by strong adsorption on to clay minerals in the soil.

VII. The Development of a New Pesticide

A. CHEMICAL ORIGINS

Modern synthetic pesticides have, in general, arisen from the research programmes in the laboratories of the large chemical industries. The discovery of radically new types of pesticide has not so far come about by logical processes of deduction from knowledge of biochemical or biophysical processes in pest and disease organisms but rather from novelty in chemical synthesis followed by biological screening tests to measure activity. The synthetic chemists and biologists concerned with the discovery of new active structures may have ideas on molecular shapes and degrees of reactivity which suggest to them a likelihood of biological activity but the success of prediction in the field of radical new approach has been poor. Massive screening programmes have, however, shown novel biological activity from time to time. Once it has been recognized, the exploration of related structures and the optimization of biological activity can proceed on an increasingly logical basis as knowledge is built up of the structure–activity relationship and of possible biochemical modes of action. Intellectually unsatisfactory though a screening procedure may be, it has produced and continues to produce results. Screening tests for biological activity must cover a wide spectrum of types of activity against a range of organisms. A novel synthesis is likely to produce a small quantity of chemical, so that considerable ingenuity is required in the design of biological tests to use the minimum quantity of material and yet detect novel types of activity. Once activity of interest has been discovered, the demands of the testing procedure may become a little easier, in that larger quantities of chemical are prepared and more limited criteria for activity can be set.

B. BIOLOGICAL EVALUATION

The biological evaluation of a compound emerging from primary screening tests will normally proceed sequentially through laboratory and glasshouse tests to small plot trials in the field and ultimately to larger scale trials in a range of climatic and crop conditions in various parts of the world. The progress of biological evaluation, however, becomes increasingly dependent upon progress in two other fields of investigation which proceed in parallel: the examination and costing of possible manufacturing processes and research upon toxicology and environmental impact. It is clear that an eventual pesticide must not only be effective in controlling the pest or disease in the field conditions in which it occurs but it must do it at a cost which is a little more than repaid in terms of increased crop yield or quality. It must also do it without imposing undue toxic risk to the farmer who uses it, to the consumer of the crop, or to the environment in general. In practice, since research in these various areas is concurrent, the criteria can only be met through a series of decision points as information becomes available. Such decision points would be built into the critical path analysis plan which is an essential part of the overall control of the programme.

C. TOXICOLOGY AND ENVIRONMENTAL IMPACT

The wide range of species of animals and plants in the natural world have all undergone the process of organic evolution and are therefore, to some extent, related and have many biochemical patterns in common. The specificity of toxic substances is thus a relative matter and it is likely that an insecticide, for example, will have some activity to forms of life other than insects. The trend of pesticide evolution over the past thirty years, however, has been towards increased specificity. It has moved away from general poisons, so that apples are no longer sprayed with lead arsenate, weeds sprayed with arsenic compounds or stores fumigated with hydrogen cyanide. The injury of a farm worker by chemical poisoning is today a rare event. The risk of injury from using pesticides on the farm, rates well below the hazards involved in using tractors or even falling off haystacks or imprudently taking liberties with bulls.

The toxicological evaluation of a new chemical starts with the measurement of the acute toxicity when administered by various routes to laboratory animals and is expressed as an LD_{50} value – the dose required to kill 50% of a batch of animals exposed. This is followed by the measurement of chronic toxicity, by the continued administration of small quantities by various routes, as in the food, by application to the skin, or by inhalation of contaminated air. This procedure will include the measurement of the expression of toxic effects in various ways, including carcinogenesis and teratogenesis. Such toxicity measurements ultimately lead to the establishment of no-effect levels, defined as those levels which can be administered continuously without producing any measurable physiological response. The no-effect level is of use in setting safety margins in the allocation of permissible residue levels in foodstuffs and in the environment.

The marketing of pesticides throughout the world is controlled by official registration authorities and a condition of registration is the submission of satisfactory information not only upon toxicology but also upon environmental impact. Under the latter heading, information is required in two areas: the ultimate fate of the chemical in the environment and the effects of the use of the chemical upon the populations of living things which may be exposed to it.

Chemicals may be degraded by spontaneous decomposition, the action of light, or metabolic transformations by living organisms. DDT and the other organochlorine pesticides were put into widespread use with little consideration of possible environmental fate or long-term biological effects. While this is understandable in relation to the time at which DDT was introduced, our knowledge today of possible long-term effects of persistent materials renders such an action unthinkable. Work on DDT (Woodwell *et al.*, 1971) has been directed to a greater understanding of the geochemical cycling of such substances. It was shown that large reservoirs for DDT exist in the atmosphere and in the oceans but that most DDT which has been produced has either been degraded to innocuousness or stored in places where it is not freely available to living things: the biota of the world contains, in total, less than one thirtieth of one year's peak DDT production in the 1960s. This is a very small amount in relation to the total potentially available.

One of the subjects of this book, the metabolic degradation of pesticides by microorganisms, is clearly a most important aspect of the fate of pesticides in soil. Metabolic investigations today, depend heavily upon studies with radio-labelled materials and the full determination of the environmental fate of a chemical may well require the synthesis of a number of samples with the label, at known different locations in the molecule, so that the various fragments may be traced. The metabolites will require precise identification and analytical methods will be required for the determination of major metabolites in addition to the parent pesticide in a range of crops, soils, animals and aquatic environments.

The investigation of the biological impact of residues of pesticides and of their major metabolites also progresses sequentially through the research and development programme. A primary requirement is early recognition of major problems of ecological damage, so that grossly unsuitable materials may be rejected before excessive expenditure upon them has taken place. Many long-term effects in the environment, such as those upon large predatory birds, or accumulation into fish, might not emerge in practical use until large areas of land had been treated for a number of years, so that there is a requirement for laboratory or small plot procedures which will detect such long-term possibilities at an early stage. Environmental testing techniques have been reviewed by Newman (1976).

Persistence in the soil may be studied in small plots, in parallel with laboratory work on microbial metabolism in soil and on physico-chemical interactions with soil constituents. In small plots, of around 6×6 m size, studies may be made on changes in populations of soil invertebrates, such as microarthropods and earthworms and on accumulation of residues into earthworms.

A laboratory study which may also provide useful information at this early stage is the ecological tank described by Metcalf et al. (1971) in which an aquarium tank accommodates both terrestrial and aquatic habitats. Plants, phytophagous insects, decomposer invertebrates, microorganisms and predaceous fish are included in the system. The plants are treated with radio-labelled pesticide and the progress of the pesticide and its degradation products can be followed through the system. While somewhat artificial and simplified, such an arrangement can provide useful information on the likely mobility and degradation of a pesticide and can indicate possibilities for accumulation into fish.

If early work in laboratory and small plot systems does not indicate unacceptable environmental impact, ecological investigations are extended to larger areas to cover population studies on larger forms of life. Terrestrial invertebrates are studied, with particular attention to predatory and parasitic forms in crops and to honey bees. Aquatic work is done to investigate the fate of the chemical in water and to measure effects on aquatic invertebrates and fish; and large area trials are laid down to determine effects in the field on wild birds and mammals.

D. FORMULATION AND APPLICATION

The way in which a chemical is presented in the environment may have a marked effect upon both its initial impact and the duration of its activity. The type of chemical formulation, in solution, mixed in an inert filler, as a granule for soil application, as a stabilized suspension or emulsion, in combination with wetting agents and stickers, or as a seed dressing, can obviously have an important effect upon the localization of the active material, the speed with which it is released, the concentration and the duration of exposure to which organisms are subjected. When highly toxic pesticides are to be applied to the soil, the use of granular formulations, free from dust, is a safer method for the operator than the use of dust or spray applications. Similarly, the use of specialized seed dressings provides an efficient method for applying insecticides or fungicides, in that the active material is placed exactly where it is required to protect the seed and the young plant, while the contamination of the whole body of the soil is limited. From the point of view of microbiological interactions and metabolism, the type of formulation is again important and must be considered in the planning of investigations.

The application method is often related to type of formulation. Placement treatments, such as the use of dressed seed or combine drilling, provide close control of the position of the pesticide in relation to the plant. When field spraying of a crop is required, the droplet size spectrum produced by the machine is important and it is desirable that a minimal proportion of spray should be in droplets small enough to drift laterally. The problem of drift is greatest with aerial applications. These have great advantages in speed of application and freedom from mechanical damage to the crop but need careful control to achieve precise application to the crop area alone.

VIII. Changes Produced by Pesticides

When a biologically active substance is applied to an ecological system it is inevitable that the system will alter in response to the interference. To some extent this is the purpose of the application, to eliminate or at least to reduce the success of the species which we designate disease, pest or weed. The introduction of the hormone herbicides into cereal culture largely eliminated the broad-leaved weeds, permitting the continuous growing of cereals over a greater number of years without a break and so increased the importance of monocotyledonous weeds such as wild oats and couch grass. Populations of microorganisms may also adapt to the presence of a pesticide by changes in species diversity or by adaptation of enzyme systems, so that a pesticide may be more rapidly metabolized by an already adapted microbial population in soil where the same or a related pesticide has been used before. Instances of this type are described by Audus (1969) and by Kearney and Kaufman (1975).

The evolution of genetic resistance in pests and disease-causing organisms to the action of pesticides is a comparable situation. The development of pesticide resistance is well recognized but its expression is very variable in different groups of organisms. Acquired resistance depends upon the selection of genes which are inherited in normal fashion but the origin of these genes is not entirely clear. Possibly they occur naturally in the population at a low rate, until their frequency is increased under the strong selection pressure of residual pesticide treatments, or they may arise anew by mutation. There is much variation in the likelihood of resistance emerging in different groups of pests. Pathogenic fungi seem to be genetically plastic and readily develop strong resistance to fungicides and also to disease resistant crop varieties. Among animal pests, resistance appears rapidly in mites, ticks and to some extent in aphids, yet after 30 years of use, DDT still controls many species of malarial mosquitoes effectively.

While the physiological basis of inherited resistance is varied, the commonest mechanism is an enhanced rate of metabolism of the toxic compound to harmless substances.

References

AGRO-ALLIED INDUSTRIES/FAO (1972). "Pesticides in the Modern World", Newgate Press Ltd, London.

AUDUS, L. J. (1969). "The Physiology and Biochemistry of Herbicides", Academic Press Inc, New York and London.

BLAXTER, K. (1974). Power and agricultural revolution. *New Scientist* **61**, 400–403.

CALDERBANK, A. (1976). Diquat and paraquat. *In* "Herbicides – Chemistry, Degradation and Mode of Action" (P. C. Kearney and D. D. Kaufman, Eds), vol. 2, pp. 501–533, Marcel Dekker Inc, New York.

CIPOLLA, C. M. (1965). "The Economic History of World Population", 3rd edition, Penguin Books, London.

ELLIOT, M., FARNHAM, A. W., JANES, N. F., NEEDHAM, P. H., PULMAN, D. A. and STEVENSON, J. H. (1973). A photostable pyrethroid. *Nature, Lond.* **246**, 169–170.

KEARNEY, P. C. and KAUFMAN, D. D. (1975). "Herbicides – Chemistry, Degradation and Mode of Action", vol. 1, Marcel Dekker Inc, New York and Basel.

KRUEGER, H. R. and O'BRIEN, R. D. (1959). Relationship between metabolism and differential toxicity of malathion in insects and mice. *J. econ. Ent.* **52**, 1063–1067.

KUHR, R. J. and DOROUGH, H. W. (1976). "Carbamate Insecticides: Chemistry, Biochemistry and Toxicology", CRC Press, Cleveland, Ohio.

LEACH, G. (1976). "Energy and Food Production", IPC Science and Technology Press, London.

MARTIN, H. and WORTHING, C. R. (1977). "Pesticide Manual", 5th edition. British Crop Protection Council.

METCALF, R. L., SANGHA, G. K. and KAPOOR, I. P. (1971). Model ecosystem for the evaluation of pesticide biodegradability and ecological magnification. *Environ. Sci. Technol.* **5**, 709–713.

NEWMAN, J. F. (1976). Assessment of the environmental impact of pesticides. *Outlook Agric.* **9**, 9–15.

O'BRIEN, R. D. (1967). "Insecticides – Action and Metabolism", Academic Press Inc, New York and London.

SCHRADER, G. (1963). "Die Entwicklung neur Insektizider Phosphorsaure-Ester", 3rd edition. Verlay Chemie GmbH, Weinheim.

TORGESON, D. C. (1967). "Fungicides: an Advanced Treatise", vol. 1. Academic Press Inc, New York.

TORGESON, D. C. (1969). "Fungicides: an Advanced Treatise", vol. 2. Academic Press Inc, New York.

WOODWELL, G. M., CRAIG, P. P. and JOHNSON, H. A. (1971). DDT in the biosphere: where does it go? *Science, N.Y.* **174**, 1101–1107.

Chapter 2

Microbiological Aspects of the Soil, Plant, Aquatic, Air and Animal Environments

Soil and Plants

TIMOTHY R. G. GRAY

Department of Biology, University of Essex, Colchester, England

Aquatic Environments

J. GWYNFRYN JONES

The Freshwater Biological Association, Ambleside, Cumbria, England

The Air and Animals

S. JOHN L. WRIGHT

School of Biological Sciences, University of Bath, Bath, England

I. General Introduction

This chapter aims to present basic information on the physico-chemical composition and microbiology of the environmental situations in which pesticides are likely to encounter microbes. The Editors decided to include such a chapter in the earnest hope that it would provide a background against which the specific data and opinions contained in subsequent chapters might be viewed.

Whilst primarily intended for those readers who may not be experienced in microbial ecology, it is hoped that it may also be useful to those microbiologists (most?) who are compelled by time and other factors to specialize on certain types of microorganisms or habitats. The editors, moreover, hope to encourage an awareness among pesticide scientists and microbiologists of the microbial habitats which, although regarded as less important than soil and water, have been overlooked unnecessarily. For example, should we not understand how insecticidal aerosols react with the air-spora; how ingested pesticides react with intestinal microorganisms in both "target" and "non-target" animals; how pesticides react with the normal leaf microflora on the plants they are protecting; how topical insecticide applications react with skin microflora on domestic animals; how the skin and gut microflora of fish react to algicides, herbicides and molluscicides applied to their aquatic habitats?

With justification, however, it is the soil and aquatic environments which are most commonly considered in relation to the microbiological consequences of pesticide usage, especially in agriculture. Accordingly, emphasis is given here to these environments through the authoritative accounts written by T. R. G. Gray and J. G. Jones respectively. To maintain a perspective of microbial habitats with regard to the widespread use of pesticides, general descriptive accounts of the atmosphere and animals have been prepared by one of the Editors (S.J.L.W.).

II. The Soil and Plant Environments

A. INTRODUCTION

In order to understand the way in which pesticides are transformed in soil, it is necessary to appreciate the nature and condition of the microorganisms that live there. Information on this has continued to grow, with the result that soil cannot be considered as a single environment but as a complex of micro-environments which usually interact with one another.

These micro-environments are modified by the nature and amount of food substrates for the microbes and the operation of a large number of chemical and physical factors. Food substrates originate from a wide variety of organic materials including all types and parts of dead and dying plants, the roots of living plants, the soil animals and corpses and droppings of the above-ground fauna and the microbes themselves. Much of this material is initially insoluble and present in the form of particles varying in size from microscopic fragments of microorganisms to huge branches and trunks of trees. Table 1, taken from Hill (1967), gives an example of part of the range of such substrates within a forest soil and shows how many alternative sources of food can be found even when the soil is colonized by only one major plant species, *Pinus nigra*. This list is still oversimplified because it gives no idea of the amount of soluble materials present in the soil, resulting from release of material from roots, excretion by animals or leaching from plant remains.

Listing the potential food sources may be misleading if it gives the impression of a static and continuous supply of nutrients for soil microbes. Both long-term seasonal changes and short-term fluctuations in the supply will occur, as well as similar changes in the physical and chemical factors affecting their utilization. These will become evident later, in the treatment of individual environmental factors.

The importance of environmental factors is not easy to assess from a study of the effects of single factors on microbes in laboratory experiments. In natural soils, an alteration in any single environmental factor may set off consequential changes which have a more profound effect on soil metabolism and population balance than might have been anticipated (Stotzky, 1975). Thus a change in the water content

TABLE 1

The occurrence of organic substrates in a pine forest soil[a]

Substrates		% total internal surface area		% total organic surface area	
		F$_2$ layer	H layer	A$_1$ horizon	C horizon
Pine branches	cortex	0·004–0·037	0·005–0·017	0·016–0·17	0·009–0·076
	stele	0·001–0·019	0·002–0·010	0·003–0·07	0·002–0·034
	immature shoots	0·0–0·006	—	—	—
Pine dwarf shoots	cortex	4·6–7·1	—	—	—
	stele	2·2–6·3	—	—	—
Charcoal		0·0–0·02	0·0–0·04	0·0–0·93	0·0–0·9
Bark		0·001–0·023	0·0–0·005	—	—
Foliage leaves		61·3–71·2	—	—	—
Roots	live	—	0·0–0·12	1·0–18·9	1·97–29·0
	dead	—	0·0–0·09	5·3–24·3	7·0–30·3
	mycorrhizal	—	—	45·9–90·0	32·1–61·6
Cones	male	0·0–0·036	—	—	—
	female	0·03–0·20	—	—	—
	partly eaten	0·0–0·007	—	—	—
Seeds		0·0–0·17	—	—	—
Hyphae		0·09–0·18	0·02–0·08	0·27–2·3	0·4–3·0
Sclerotia		—	—	0·0–0·05	0·0–0·08
Resin aggregates		—	0·01–0·06	0·3–0·6	0·31–1·75
Insect exoskeletons		0·0–0·004	0·004–0·02	0·0–0·05	0·0–0·02
Faecal pellets		—	99·6–99·9	—	—
Unidentified		17·0–29·6	99·6–99·9	1·07–4·6	3·8–14·75

[a] After Hill (1967).

will affect aeration which in turn may alter the oxidation–reduction states of inorganic materials. Many reduced compounds are toxic to plants and microbes and so there may be short-term changes in the nature and quality of exudates from plants or long-term changes in the vegetation if these alterations persist.

However, despite the occurrence of apparently unstable micro-environments in soil, it is evident from an examination of the natural environment that if habitats are examined on a large enough scale or over a long enough period of time, the soil as a whole is relatively stable. Thus soil profiles are established in mature plant communities in which horizons with well-defined properties can be characterized. These profiles may persist for very long periods of time, in association with climax vegetation. Indeed, examination of the mean residence time of some organic matter in undisturbed soils has shown that it can persist for thousands of years (Clark and Paul, 1970). However, agri-cultural and forestry practices may cause profound changes in the soil, resulting in the more rapid metabolism of some of this organic matter and a shortening of its mean residence time. These environments, in the process of change, are the principle concern of those interested in pesticides.

B. ORGANIC MATERIAL – QUANTITY AND AVAILABILITY

The net primary productivity of plant material is remarkably constant in terrestrial environments since rates of photosynthesis and rates of plant respiration are usually balanced. Macfadyen (1970) has sugges-ted that the amount of dry matter production approximates 1 kg m^{-2} per year, i.e. $2 \cdot 016 \times 10^8$ joules energy. This material may be eaten by herbivores or it may go directly to the soil upon the death of the plant. The amount of net primary production removed by herbivores, including mammals, molluscs, insects, etc. may be quite large but some of this is eventually returned to the soil on the death of the herbivore. If this fraction is neglected, then it would appear that the percentage of the net primary production decomposed by soil organisms is approximately 93% in a Spartina salt marsh, 57% in a grazed meadow, 30% in a temperate beech wood and 33% in a tropical rain forest (Macfadyen, 1970). It may seem surprising that a smaller proportion of material is decomposed in soil in forest ecosystems than in grasslands but this may only reflect our ignorance of the importance of leaf-feeding insects and the fact that a good deal of the primary production of trees remains in the plant for longer periods of time, increasing the chances of its removal before reaching the soil. In a grazed meadow, the mean standing crop of the above ground parts is about 1 kg m^{-2} while in a beech forest the figure is 15·5 kg m^{-2}.

Leaf material is known to be colonized by microorganisms before reaching the soil, for it has been shown that considerable microbial activity occurs on the external surfaces of living leaves, an environment termed the phylloplane (Kerling, 1958). The environment results from the interaction between the leaf and the atmosphere, the organisms in it are both residents and casual colonizers. Bacteria, yeasts and filamentous fungi are all common (Dickinson, 1965), the bacteria and yeasts occurring mainly along the anticlinal walls of the epidermal cells (Last, 1955). Diurnal variations in the leaf environment are considerable, mainly because of dew formation. Nutrients dissolve in the dew, may be metabolized by microbes and can then be reabsorbed by the plant. Thus a tree having 67 kg of leaves can absorb up to ten litres of water each night (Ruinen, 1961). Inorganic ions, organic acids, sugars (the principle organic components) and amino acids all circulate in this way but some materials appear in the phylloplane as a result of the activity of leaf-sucking insects, e.g. melezitose (Carlisle *et al.*, 1966). Up to 5% of the dry weight of a leaf can be passed out of the leaf each day (Tukey *et al.*, 1957), though some of this is reabsorbed or washed away into the soil by rainfall and not metabolized by microbes on the phylloplane. The occurrence of pesticides in the phylloplane might lead to:

(a) their metabolism
(b) their uptake in dew by the plant
(c) a change in composition of the nutrients in the phylloplane by the elimination of leaf-sucking insects
(d) a change in permeability or wettability of leaf cells and surfaces leading to changed rates of exudation
(e) the modification of nutrient breakdown by inhibitory effects on microbes.

Leaf and branch material arrives on the soil surface at varying times of the year (Bray and Gorham, 1964). In equatorial regions, litter-fall is continuous although most litter falls during the first six months of the year, unless there is a dry season which increases it at other times. In the warm temperate forests of Eastern Australia continuous leaf-falls occur, with maxima corresponding to rises in temperature and precipitation in early spring and summer. However, the most striking seasonal changes occur in cool temperate forests where leaf-fall of deciduous trees is correlated with cooling and decreasing light intensities during the autumn. The amount of litter-fall is variable from year to year and ratios of highest:lowest annual litter-fall range from 2:7 for *Nothofagus* in New Zealand to 1:8 for *Acer saccharum* in the northern hemisphere. Within the litter the proportion of the different components varies although leaves always predominate. The percentage of non-leaf material (fruits, branches, bark, flowers, bud scales, pollen, insects, etc.) is about 30% in forest litter.

Following its deposition on the soil surface, the composition of the litter may change as some of the residues decompose more rapidly than others, even under similar environmental conditions. Thus, coniferous leaves and woods decompose more slowly than those of deciduous trees. Also as stands of trees age, the proportion of harder structures in the litter layer increases while the contribution of herbaceous material decreases (Ovington, 1962). Fast decomposing litter types from deciduous trees include *Fraxinus*, *Betula* and *Tilia* while more slowly decomposing types are *Quercus* and *Fagus*. Thus on a mull soil, after six months, only 5% of *Fraxinus* leaves remain while 85% of *Quercus* leaves are still present (Bocock *et al.*, 1960). Soil type also influences the rate of decomposition (Bocock, 1964) as does the type of leaf, for example sun or shade leaf on a single tree (Heath and Arnold, 1966).

This change in gross composition of the litter leads to a change in the specific materials present in the litter. Thus, during the early stages of decomposition, inorganic ions such as calcium, potassium, sodium, magnesium and phosphate are leached from the leaves. Latter and Cragg (1967) found that the most pronounced losses of ions within the first five months of decomposition of *Juncus* litter were potassium, phosphate, magnesium and calcium. In forests, some of the leached ions are replaced by those washed in from the leaf canopy (Carlisle *et al.*, 1966). There are also changes in the amount of water-soluble organic materials. Burges (1958) recorded a 50% decrease in the first three months of *Casuarina* litter while Sowden and Ivarson (1962) found that the loss of materials was more rapid from deciduous litter than coniferous litter. The rate of loss differs from one species to another, so that in *Quercus petraea*, the content of hot-water-soluble materials, originally at 18%, declined slowly whereas in *Fraxinus excelsior* the initial high values of 32% had dropped to 1·5% after only a month (Gilbert and Bocock, 1960). The slower disappearance of these materials from oak leaves may be because of the relatively high tannin content of the leaves which can inhibit utilization of simpler materials (Basaraba, 1966).

The range of water-soluble organic materials in litter and soils is large and includes organic acids, which occur most frequently in waterlogged soils. Thus acetic, formic, butyric, lactic and succinic acids have all been found in rice paddy soils (Takijima, 1964) while Wang *et al.* (1967) have found several phenolic acids in sugar cane fields. Water-soluble sugars and amino acids are also rarely present in a free state, but occur in polymerized insoluble forms. Putnam and Schmidt (1959) have found 2–387 mg g^{-1} of several amino acids in soil in a free condition. These low values can be explained by the rapidity with which the amino acids are attacked by microbes, for Schmidt *et al.* (1960) found that practically all the amino acids added to soil had been utilized during the first 96 h incubation. Some amino acids are more resistant than others, lysine and tyrosine for example; others, rather than being mineralized become incorporated into microbial cells and humic compounds (Clark and Paul, 1970). Addition of fungicides to soil causes partial sterilization and changes the nature and amounts of free amino acids in the soil. Soil fumigation with a chlorinated propane–propylene mixture increased the amounts of aspartic and glutamic acids (Altman, 1963). Recently, Wainwright and Pugh (1975) showed that high concentrations (250 μg g^{-1} soil) of fungicides such as benomyl and thiram, lowered the concentration of amino acids after a 28 day incubation period, while low concentrations (50 μg g^{-1} soil) resulted in a marked increase in the amount of amino acid–nitrogen extracted. Glycine and threonine were among the compounds that increased in concentration, possibly because they were less susceptible to degradation during the flush of microbial growth following partial sterilization.

All the insoluble organic materials found in plants and animals are likely to find their way into soil. Initially these materials will be in the ratios found in intact plants or animals, but as decomposition proceeds these ratios change. This is partly reflected in changes in the C:N ratio of soil organic matter. The initial value of the C:N ratio of plant material may be as high as 200:1 in woody materials, although values closer to 30:1 are found in some legumes. Where comparatively large amounts of nitrogen are available, the C:N ratio falls during the early stages of decomposition due to the rapid carbon loss, coupled with the less rapid loss of nitrogen from the tissues of the plants and the microbes decomposing them. After the ratio has reached 20:1, the change slows down until a relatively stable ratio of about 12:1 is achieved, similar to that found in microbes (Forbes, 1974).

The effect of the C:N ratio of a substrate can be complicated, as shown by the work of Davey and Papavizas (1963). They found that the saprophytic activity of the pathogenic fungus *Rhizoctonia solani* could be inhibited most strongly by amending soil with cellulose and ammonium nitrate at C:N ratios between 100:1 and 40:1. This inhibition is caused both by nitrogen immobilization and carbon dioxide production but the former is overcome when the C:N ratio has fallen below these levels as mineralized nitrogen becomes available. It appears that at C:N ratios between 30:1 and 15:1, rates of nitrogen immobilization and mineralization are equal, immobilization occurring above these values and mineralization below. In the absence of nitrogen in a substrate, it will be removed from elsewhere in the soil as shown during cellulose decomposition (Tribe, 1960).

The amount of nitrogen in organic materials reaching the soil is lowest in woody material. Allison and Klein (1961) and Allison and Murphy (1962, 1963) found that the nitrogen content of wood and bark ranged from 0·038% to 0·413% with higher figures for bark than for wood. Herbaceous leaf litter contained from 0·5–1·5% nitrogen while legumes contained 1·5–3·0% (Bartholomew, 1965). The values for tree leaf litter are also within this range, with mean values for hardwoods and softwoods being 1·89 and 1·52% respectively (Voigt, 1965).

While the rate at which organic carbon is metabolized in the soil depends upon its nitrogen content, it is possible to recognize at least three fractions (Clark and Paul, 1970). Firstly, there are the decomposing plant residues and their associated microbial biomass which turnover at least once every few years. Thus Jenkinson (1966) showed that 60% of rye grass carbon was lost in six months and 80% in four years, while 30% of the carbon in decomposing bacterial cells was lost as carbon dioxide in ten days. On a larger scale, it is probable that conifer leaves take seven years to decompose and *Fraxinus* leaves less than one year. Secondly, there are microbial metabolites and cell wall constituents that are relatively stable in soil and possess a half-life of 5–25 years. Among the compounds known to stabilize fungal cell walls is melanin (Alexander, 1971) which Bull (1970) has shown to inhibit polysaccharases. Thirdly, there are some very resistant organic matter residues composed of humic components, ranging in age from 250–2500 years (Campbell *et al.*, 1967). Paul *et al.* (1964) also showed that cultivation of the soil reduced the mean residence time of this organic matter from 2200 years to 500 years. The humic materials are derived partly from microbial metabolites and partly from plant remains such as lignin and cellulose. Carbon can be incorporated into humus quite quickly, for Mayaudon and Simonart (1958) have shown that large amounts of ^{14}C from glucose can be found in humic fractions after only seven days. Humic acid probably consists of a core of polymerized polyphenolic units and some nitrogen-containing heterocyclic rings to which are attached side chains including amino acids. Among reasons put forward to explain the resistance of these molecules to decomposition are their polyphenolic structure, the presence of protein complexed with lignin derivatives, the adsorption of the carbonaceous part of the molecules by silicate minerals and the formation of complexes with various toxic, trivalent cations (Alexander, 1965). Swaby and Ladd (1962) have also suggested that the irregular cross-linking within the molecule by a variety of covalent bonds means that degradation can only take place through action of many different extracellular enzymes from a variety of organisms.

The presence in soil of such recalcitrant molecules, the relatively small amounts of fresh organic material reaching the soil each year and the very large microbial populations found in the soil, result in a microflora that spends long periods of time in a resting condition or growing at very slow rates. Gray *et al.* (1974) have shown that in a woodland soil, even if soil animals and fungi were presumed absent and all the organic matter were available to the bacteria, these organisms would still only have a mean generation time of about four days. In micro-environments within the soil, more rapid growth might occur and Hissett and Gray (1976), reviewing instances of rapid growth in soil, concluded that it occurred mostly in response to sudden additions of fresh substrates.

These considerations are relevant if one considers the possible effects of the microflora upon pesticides added to the soil. If a pesticide is readily available to microbes, as are organophosphates, aliphatic acid herbicides and carbamates, then it is likely to be removed rapidly, unless it becomes persistent for some other reason. On the other hand, the rate of addition of pesticides is so low that the effect on the rate of growth of soil microbes through provision of an energy source may not be demonstrable. Thus many pesticides and herbicides are added to soil at rates less than a few kilograms per hectare which is small compared with the total input of organic matter (Gray, 1970). Also energy may be required to initiate decomposition of pesticides which has to be supplied from the soil organic matter. Some pesticides may be metabolized by microbes but not as a source of carbon or energy. This phenomenon of co-metabolism may be caused by the initial breakdown products being toxic to growth or by the inability of the enzymes to attack the products of the initial breakdown (Horvath and Alexander, 1970). The initial transformations may be brought about by resting cells, thus, even in a situation where there is relatively little energy available for growth, pesticide modification could occur.

The toxic effects of added materials may be important, especially when the molecules are recalcitrant and when climatic or edaphic factors prevent decomposition. The evidence for such effects are often contradictory and have been reviewed by Helling *et al.* (1971).

C. THE RHIZOSPHERE EFFECT

In considering the rate of addition of organic matter to soil, the case of the root environment needs special mention. Among the sites of intense microbial activity in the soil are the root surface and the soil adjacent to it, termed the rhizosphere. The root surface is a complicated environment for microbial growth in both a physical and chemical sense. Mosse (1975) reviewed the information on the structure of roots and pointed out that the environments on the root surface include the root hairs, the root cap, the greatly convoluted epidermal and outer cortical cells and the mucilaginous layer or mucigel which sometimes surrounds not only the root cap but also the root surface and the root hairs. Thus, the so-called root surface actually consists of a number of distorted and convoluted surfaces and volumes of material which become less clearly defined on proceeding outwards from the root into the soil. Foster and Rovira (1973) have shown that the colonization of such root surfaces by microbes is not continuous and that, in wheat, while high concentrations of microbes occur in the collapsed cortical and epidermal cells of older roots, colonization of young nodal roots is sparse and is associated with a layer of soil approximately 20nm thick over the whole root surface. In young seminal roots, this layer is absent and most colonization occurs at the junctions of the cortical cells (cf. the phylloplane). In plants possessing a mucilage layer, Greaves and Darbyshire (1972) have shown that soil particles either adhere to or are incorporated in the mucilage, together with many bacteria and fungi. Bacteria were found to stimulate mucilage production as well as increase the numbers of dead or damaged root surface cells. In some cases, a membrane-like boundary to the mucilage was observed which Mosse (1975) suggests has a non-biological origin, for example from surface drying or chemical interaction with the environment, although others have suggested that it represents the epidermal cuticle or is produced by microbial action. If the mucilage retards the outward diffusion of exudates, then the microbes in it would benefit and would also be protected from desiccation. Bacteria in the mucilage have often been observed to have a distinct capsular area around them, which is characteristic of organisms grown in sugar-enriched media in artificial culture.

Material from roots also finds its way into the surrounding soil, either as sloughed-off cells or exudates. The exudates consist of a mixture of amino acids, organic acids, nucleotides, flavonones, sugars and enzymes (Rovira, 1969). However, the amounts and rates of exudation of these compounds are not well established and interpretation of results is often complicated by the fact that measurements are made on plants which are grown in water culture or sand rather than soil, and that microbes on the root may decompose compounds as they are exuded from the root (Brown, 1975). Meshkov (1953) found that peas exuded 0·14–0·23% of their entire dry weight as reducing sugars while corn exuded 0·23–

0·35%. No measurements of amino acids, the predominant component of the exudates, were made. Harmsen and Jager (1963) grew vetch in a synthetic soil and found that 1·6–2·9% of root carbon was released but they did not distinguish between exudates and sloughed off cells. However, Bowen and Rovira (1973) found that 1–2% of the root carbon in wheat was released into field soils of which 0·8–1·6% was insoluble mucilaginous material and only 0·2–0·4% was water-soluble exudate.

Unfortunately, these figures do not tell us the rate of nutrient exudation which is crucial to our understanding of the rate of growth of microbes on the root surface. However, Gray and Williams (1971) used Harmsen and Jager's data to calculate that 3–6% of the root biomass might be lost in six weeks. Barber and Gunn (1974) have suggested that up to 9% of the total increment of the root dry weight is exuded from barley in solid growth media.

Martin (1975) investigated the quantities of ^{14}C-labelled material leached from the rhizosphere of wheat, clover and ryegrass grown in soil and showed that soil type had a marked influence on the measured amount of water-soluble material released from roots. He also suggested that the amount of material released into the rhizosphere as exudates or sloughed-off cells might be much higher than thought earlier. The proportion of ^{14}C fixed by plants found in root-free soils varied from 3·1–5·8% after a seven week growth period, equivalent to 10·4–38·4% of the ^{14}C occurring in the plant roots. How far these figures are affected by leaching the soil at weekly intervals is impossible to say. Griffin et al. (1976) investigated the quantity of organic matter sloughed off roots per gram root dry weight per week and estimated that this represented 95–98% of the total sloughed organic matter plus sugars lost by roots. Thus, the emphasis on sloughed-off cells as opposed to exudates as a cause of the rhizosphere effect is changing. In natural soils, the sloughed-off material would stay very close to the root, blurring the distinction still further between root surface and rhizosphere environments.

The nature of the rhizosphere effects will be varied and the factors affecting both the nature and amount of exudation from plants have been reviewed by Rovira (1969). However, the diffusability of the exudates and their susceptibility to adsorption and decomposition will all be important in determining the distance from the root at which the rhizosphere effect occurs. Generally, the zone of increased microbial activity only extends for 1–2 mm beyond the root surface. Within this zone, changes in other environmental factors such as pH and oxygen concentrations will occur. Thus Nye (1968) has suggested that roots take up more anions than cations and therefore pass out HCO_3^- ions to preserve electrical neutrality. For every equivalent of HCO_3^- secreted,

20 moles of CO_2 are respired and while this may be thought to counteract the effect of the HCO_3^- ions, it should be remembered that CO_2 moves rapidly away from the root via the air-filled pores while HCO_3^- ions are confined to the soil solution around the root. An increase of pH of one unit over a few millimetres around the root should be possible in an acid soil. Greenwood (1970) has also pointed out that root growth is more likely to be retarded by oxygen deficiency than carbon dioxide excess and so the rhizosphere microorganisms may well be ones tolerating or preferring low oxygen tensions (Brown, 1975).

So far we have considered the root environment from the point of view of release of substances and outward flow of materials but it should be remembered that the main purpose of a root is to absorb substances and so there will be a depletion of some nutrients close to the root. The zone of depletion will depend upon the diffusion coefficient for the substance in soil and Nye (1968) has shown that at 10 mm away from a root, potassium ions (diffusion coefficient of 10^{-7} $cm^2 sec^{-1}$) would be almost at normal soil levels, whereas at the root surface they would have dropped to under 40%. The diffusion coefficient for phosphate ions is ten times less than that for potassium and so the extent of nutrient disturbance would be smaller, while that for nitrate ions is ten times greater. In the case of phosphate absorption, the root hairs, projecting from the root surface, will clearly influence the extent of the depletion zone. Thus the nutrient balance in the rhizosphere will be quite different from that in root-free soil and could influence the rate at which pesticides are metabolized.

The composition of exudates in the rhizosphere can also be modified by foliar applications of chemicals and this can influence the occurrence of microbes there. Foliar applications of urea cause bacterial counts to rise and fungal counts to decrease in the rhizosphere (Vrany, 1965). This is due to increased photosynthetic activity of the plant but effects caused by the application of other substances to leaves, for example chloramphenicol, are due to the translocation and exudation of the compounds (Vrany et al., 1962). Vrany (1975) investigated the effects of foliar applications of the herbicide 4-chloro-2-methylphenoxyacetic acid (MCPA) on the rhizosphere microflora of wheat. He observed that this compound, when sprayed onto leaves on its own or in combination with urea, increased the numbers of bacteria in the rhizosphere although it decreased carbon dioxide production from soil when sprayed onto leaves on its own.

Organic substances present in the soil might also affect root exudation and hence the microbial populations. Rovira and Ridge (1973) found that acetic acid could increase exudation by 100% from roots grown in nutrient solutions buffered with acetate. They postulated that this was because of the effect of undissociated acetic acid on cell permeability or on root cell metabolism. The presence of other organic additions to soil–root systems might have similar effects.

D. MOISTURE AND AERATION

As soil dries out, the properties of the soil atmosphere and the pathways for diffusion of gases in the soil change. Griffin (1968) has summarized the factors affecting oxygen diffusion in a steady state system where the rate of oxygen uptake by and the rate of oxygen diffusion to a structure are equal. The rate of diffusion is affected by the oxygen concentration gradients between the surface of the structure, the water in the pores and the atmosphere external to the soil, as well as the geometry of the soil system and the diffusion coefficients of the medium around the structure. Thus rates of oxygen diffusion will fall if the diffusion coefficient or the oxygen concentration at the water/gas interface is lowered or the geometry is altered by rainfall, thus lengthening the diffusion path. To counteract this fall, there would have to be a compensatory reduction in the oxygen concentration at the surface of the structure. However, if this concentration drops below a critical level the oxygen uptake will decrease, since it is likely that the terminal oxidases are governed by Michaelis-Menten kinetics. The K_m values for cytochrome oxidases are about $2 \cdot 5 \times 10^{-8}$ molar though K_m values for oxygen uptake are higher than this (up to 3×10^{-6} molar), partly because the K_m value is influenced by diffusion through the mucilage, cell wall and cytoplasm. Thus oxygen may limit growth of aerobes in soils with high matric potentials where the soil spaces are filled with water and the diffusion rate of oxygen is reduced to $2 \cdot 6 \times 10^{-5}$ cm^2 sec^{-1} (Griffin, 1972).

Since soil particles are often aggregated together to form crumbs, the aeration of soil may be considered in two phases, within the crumbs and in the spaces between them. In a soil crumb, microbial respiration removes oxygen so that a water-saturated crumb with a radius greater than 3 mm would be anaerobic at the centre but aerobic on the outside. However, soil pores are so small that the microbes at the centre of a crumb would soon use up all available substrates (Allison, 1968) and it could be considered, therefore, that most microbial activity would take place at the crumb surfaces. The rates of oxygen diffusion to these surfaces will be greatly influenced by the proportion of the spaces filled with water since this will break the continuity of the gas phase. Rixon and Bridge (1968) have found that when the gas-filled pore space changes from approximately 10 to 20%, the respiratory quotient changes, suggesting the onset of aerobic respiration.

Such changes may also affect the metabolism and movement of pesticides. Thus under anaerobic conditions, reduction of nitro-groups and displacement of chloride groups by hydrogen can take place, often more rapidly than through oxidative pathways occurring in aerobic conditions (Helling et al., 1971). Lignin polyphenols and other soil organic matter components may also act as reducing agents for pesticides, especially in oxygen-poor environments (Crosby, 1970). DDT may be transformed to DDE under aerobic conditions and DDD under anaerobic conditions. The latter transformation is a much more rapid one; after three months, less than 1% of the DDT added to anaerobic soils could be recovered, while in aerobic soils, 75% was still left after six months (Guenzi and Beard, 1968). Similar differences have been reported for lindane (Newland et al., 1969). On the other hand, carbon loss from prometryne, a herbicide, was much faster in aerobic rather than flooded soils (Plimmer et al., 1970).

Pesticide movement may be directly affected by soil moisture. Thus water movement may allow pesticide movement but redistribution and rate of movement are retarded when the water content falls below 40% for clay, 30% for loam and 25% for sand (Griffin, 1972). In the absence of water movement, pesticides might still move by diffusion but the pathways through which they move decrease with decreasing water content. It can be shown that for un-ionized and non-polar molecules, the rate of transfer due to diffusion in a given soil is approximately proportional to the volumetric water content. Volumetric water content increases as bulk density increases, assuming a constant gravimetric water content. This may affect the diffusion coefficients of pesticides as illustrated by the work of Ehlers et al. (1969) on lindane. The biggest effect was on the diffusion coefficient in the vapour phase, so that lindane, a relatively volatile chemical, was affected differently from dimethoate, a non-volatile material. This emphasizes the impossibility of making generalizations about the effect of environment on pesticides.

As soil dries out, many microorganisms are killed by desiccation, whilst others survive as resting propagules, e.g. spores, sclerotia, rhizomorphs, resting hyphae or cells (Warcup, 1957). These effects may be interpreted in terms of the water potential of a soil, the sum of the osmotic and matric potentials (usually expressed in bars, where 1 bar = 10^6 dyne cm^{-2} = 1022 cm water = pF 3·0), which determines the degree of difficulty a microbe will experience in extracting water from a soil. The relationship between soil moisture content and soil water potential for a clay, a loamy soil and a sandy soil is shown in Fig. 1,

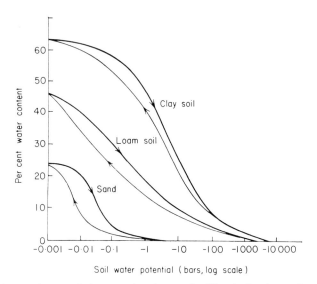

Fig. 1. Moisture characteristic curve for three soils. The drying (upper) and wetting (lower) boundary curves are both shown (after Griffin, 1972). Reproduced by permission of Chapman and Hall, London.

where it can be seen that the relationships for a drying soil and a wetting soil are different.

The effect of water potential on microbial species is varied and has been discussed by Griffin (1972). At potentials less than −145 bar, fungi such as *Aspergillus* and *Penicillium* predominate. The majority of fungi, however, are most frequently isolated at potentials greater than −40 bar. The lower limit for bacterial activity in soil is probably about −80 bar but activity may cease at higher water potentials, even at −5 bar (Cook and Papendick, 1970), because of other effects such as decreased mobility of the organisms. Streptomycetes appear to have a lower limit of about −55 bar, but they can survive as spores in much drier soils. Thus breakdown of materials in dryer soils is likely to be mainly fungal, whereas in wetter soils, bacteria may increase in importance. High activity of bacteria occurred at high water potentials with greatest numbers following amendment of soil with nutrients above −5 bar (Cook and Papendick, 1970) and maximum rates of nitrification between −0·1 and −0·2 bar and sulphur oxidation between −0·03 and −0·06 bar.

E. SOIL REACTION

The acidity or alkalinity of a soil can markedly affect the fate of pesticides directly. Thus, many pesticides decompose rapidly in water under alkaline conditions because of the presence of hydroxyl ions; indeed captan [N-(trichloromethylthio)-3a,4,7,7a-tetrahydrophthalimide)] has a half-life of only about 2·5 h at neutral pH values (Birchfield and Storrs, 1950). It reacts with mercaptides or sulphide ions in solution and the unstable product is hydrolysed to thiophosgene. Hydrolysis of dazomet (tetrahydro-3,5-dimethyl-2H-1,3,5-thiadiazine-2-thione) to methyl isothiocyanate occurs in a few hours in soil (Drechsler and Otto, 1968).

Soil reaction also affects pesticides indirectly through its effects on microbial activity. In general, increased acidity depresses the growth of bacteria more than the growth of fungi. Bacterial activity is usually low when the pH falls below 5·0, except for certain acidophilic forms, whereas fungal activity may continue until the pH has fallen to below 3·0. However, while these relationships can be discerned clearly in aqueous solutions, they are less obvious in soil for a variety of reasons. The presence of clay particles which are negatively charged causes hydrogen ions and other cations to accumulate in excess of anions. Further away from the particle surfaces, the excess of cations decreases until equal proportions of anions and cations are found. In the presence of metallic cations, the hydrogen ions may be displaced from the particle surfaces and thus the nature of the soil and the suspending fluid in which pH measurements are taken is of great importance. Weiss (1963) has shown that negatively charged bacterial cells cause the accumulation of hydrogen ions at their surfaces while Griffin (1972) has pointed out that many fungal hyphae are coated with clay particles which could influence surface pH.

Local variations in pH also occur around particles of organic matter. Thus, Williams and Mayfield (1971) have shown that ammonium ions released from the decomposition of nitrogenous materials added to soil, can accumulate around organic particles raising their surface pH from 4·0 to 7·0. This is sufficient to allow the growth of acid-sensitive streptomycetes and, as shown later by Lowe and Gray (1973), typical soil bacteria such as *Arthrobacter*. Alteration of the pH with calcium carbonate or potassium hydroxide also allowed *Arthrobacter* to grow.

If nitrogenous materials or lime are not added to soil, organic matter becomes increasingly acid as decomposition proceeds. However, Nykvist (1959) has shown that while leaves are initially acid when they fall on the soil surface, leaching of organic acids causes their pH to rise for a few months although this trend is soon reversed and litter becomes very acid after a few years, with a pH below 4·5.

Clay particles, in addition to adsorbing hydrogen ions and other cations, will also adsorb organic compounds, including enzymes, metabolic by-products, microbes, humic substances and pesticides. Organic cations are adsorbed in the same way as metallic cations except that the ionic property of organic cations is pH dependent; they may be adsorbed also by hydrogen bonding, ion dipole and physical forces, such as van der Waal's forces and the organic cation may interact with water (White and Mortland, 1970). The larger the organic cation, the more difficult it is to displace it with metal cations. Organic compounds may also become cationic following adsorption, because there will be a source of protons from the exchangeable hydrogen and other cations at the cation exchange sites.

Different types of clay particles may also affect microorganisms in different ways. Thus the distribution of *Fusarium oxysporum* f. *cubense* is influenced by the occurrence of montmorillonite. This mineral greatly enhances bacterial respiration since it has an efficient buffering action (compared with kaolin or vermiculite), due to its high cation exchange capacity and wide lattice expansion at a wide range of matric potentials which allows free interchange of cations to take place. Increased bacterial growth is presumed to be antagonistic to fungi (Stotzky and Rem, 1966, 1967).

F. TEMPERATURE

Some soils are either permanently hot (those near hot springs) or permanently cold (in polar regions) and the special problems for microbial populations in these areas have been reviewed by Bunt (1971) and Brock (1969). However, within normal soils in temperate and tropical regions, large temperature changes can take place which can affect the activity of microbial populations.

Travleev (1960) reported that while soil temperatures are related to air temperatures, especially at the soil surface, there is a steep temperature gradient within the first few centimetres. Thus in an oak woodland, while the litter had a temperature of 13·8°C, the soil at 10 cm depth was only 5·3°C. However, in winter, when the litter temperature was −1·5°C, the soil at 10 cm depth was +0·2°C. Bleak (1970) showed that when a layer of snow deeper than 60 cm lay on the soil, temperatures at the soil surface ranged from −2·5°C to +1·2°C but even at these temperatures, some microbial activity was evident since large weight losses occurred in leaf litter over the winter period.

Raney (1965) has shown that a close association exists between the moisture content of a soil and its heat absorbing capacity. Only 0·84 joules are required to raise the temperature of 1 g soil by 1°C and if the soil is moist, heat is conducted downwards more efficiently. Thus the temperature gradient in moist soils will be less than that found in dry soils. As the temperature rises, the diffusion rates of gases, including volatilized pesticides, oxygen and dissolved pesticides rises. However, higher rates of microbial respiration may compensate for increased oxygen diffusion and may lead to increased metabolism of pesticides. In mineral soils, decreasing temperatures lead to increased adsorption of pesticides by clay particles, although the reverse may be true for adsorption by organic materials and for particular pesticides such as lindane (Spencer, 1970). This emphasizes further the interdependence of many environmental factors affecting pesticides in soil and underlines the difficulty in establishing cause and effect relationships.

Temperature fluctuations may be responsible for shifts in the microbial population of soils. Thus, Okafor (1966) found that the organisms colonizing chitin in tropical soils where temperatures were always between 28 and 30°C were principally actinomycetes, nematodes and protozoa, whereas in temperate soils where temperatures fluctuated between 2 and 15°C, fungi and bacteria were most important.

G. EFFECTS OF SOME AGRICULTURAL PRACTICES

Russell (1968) has reviewed the changes in the environment that may take place in agricultural soils, following common cultivation practices. He concluded that evidence for major changes in the bacterial flora of soil following crop rotation and liming was lacking, although the fungal flora of the soil and rhizosphere might be changed and some processes stop if the pH fell below 5·0. However, the practice of mulching reduced the rate of drying of the soil surface and the rapidity of temperature change, increased the amount of root tissue and lead to immobilization of nitrogen at the interface between mulch and soil. This technique might, therefore, encourage nitrogen fixation.

Biocidal treatments, in general, increase the amount of water-soluble organic matter in the soil, with heat producing larger effects than irradiation or fumigation and air-drying and freezing causing the least effect (Powlson, 1975). Partial sterilization of soil with steam or chemicals (e.g. for greenhouse use) leads to a flush of ammonium production. The ammonium produced is not nitrified immediately, due to the presence of toxins in the soil and consequently steamed soils are flooded or left for some time before planting. Rovira and Bowen (1966) found that the toxicity could be eliminated by some fungi and bacteria and by leaching with water. Jenkinson (1966) has suggested that partial or complete sterilization of soil increases the rate of mineralization of microbial material since dead or damaged microbes are more easily decomposed than living ones. In contrast, the rate of mineralization of organic carbon in non-microbial material is not generally accelerated, although irradiation and heat treatment can make some soil organic matter more decomposable. These increases in rates of mineralization are accompanied by increased numbers of soil organisms, especially bacteria.

It has also been suggested that larger soil organisms will be killed by ploughing, rotovating or grinding but decreases in populations of soil animals might equally be due to the redistribution of partially decomposed organic matter (Powlson, 1975). Similarly, effects of disturbance on microorganisms are likely to be indirect and due to changes in soil aeration and water holding properties.

It has been shown that prolonged application of herbicides to soil can also affect the microbial population (Voets et al., 1974). Thus, application of the readily degradable atrazine (4 kg ha^{-1} per annum) for 14 years caused the numbers of anaerobic bacteria, spore-forming bacteria and cellulolytic microbes to be reduced permanently, although total numbers of microbes were not changed. Nitrifiers, denitrifiers and amylolytic organisms were temporarily reduced while ammonifiers and proteolytic organisms and azotobacters were temporarily increased in number. The removal of vegetation, which was thought to be responsible for these changes, also leads to a decrease in soil organic matter and a 50% or more reduction in the activities of some soil enzymes. Soils also became more acidic.

Application of manure slurries to soil may cause pollution problems, including surface run-off of organic matter into streams, spread of pathogens and accumulation of toxic metal ions. These problems are particularly acute if more slurry than is needed for fertilizer purposes is added to soil (Skinner, 1975). The addition of slurry may also lead to blockage of soil pores with organic matter carried into the soil in water. Anaerobic conditions may result from this and also from the enhanced respiration of microbes on the added organic matter. Skinner (1975) has also pointed out that the application to soil of liquor from which leaf protein has been extracted results in the proliferation of saccharolytic clostridia and the development of anaerobic conditions. However, subsequently, polysaccharides produced by these organisms stabilized the expanded soil structure formed by gas production after initial application of the liquor.

H. MICROBIAL CONTENT

The degradation of pesticides will clearly be influenced by the nature of the microbial populations present at the time of application, and possibly by the balance between the major groups of organisms which have varying requirements for carbon, nitrogen and other components of pesticides. At present, information on these points, though copious, is of limited value. Most investigators examining the soil microflora have been concerned only with counting the number of propagules of the different microorganisms in soil. Such information on relative numbers of organisms is extremely misleading since the numbers refer to individuals of vastly differing size, degree of metabolic activity and viability. Taking these points in reverse order, it is thought that many of the microbes in soil are dead, principally because of the large discrepancies between counts of intact cells made by direct microscopy and counts of cells capable of forming colonies on relatively non-selective media. It is not unusual to find up to a 1000-fold difference in such counts, especially if some of the more sensitive fluorescent stains are used for the direct microscopic counts. Recent estimates by Anderson and Slinger (1975), using a europium chelate/fluorescent brightener stain suggest viabilities of about 59% but these are likely to be overestimates since the assumption that nucleic acids which pick up the stain disappear from cells quickly after death seems doubtful. Those cells which are viable may also exhibit differing degrees of metabolic activity since the presence of some organisms as spores or slowly respiring vegetative cells cannot be detected by most counting techniques. Thus, Warcup (1957) showed that many fungi that gave rise to colonies on dilution plates originated from spores while Siala et al. (1974) have shown that the majority of *Bacillus subtilis* colonies on dilution plates originate from vegetative cells, at least in acid soils. Finally, the size of different microbes is extremely variable. Even amongst the bacteria, variations in cell weight are large and *Bacillus* species may weigh 10–20 times more than some Gram-negative, rod-shaped bacteria (Gray et al., 1974).

An obvious way of overcoming these difficulties is to determine the relative weights of populations of different organisms but this is fraught with difficulties and requires a large number of assumptions to be made. However, recent measurements made by Dr. R. Hissett, Dr. J. Frankland, Dr. M. J. Swift and myself (unpublished data) suggest that in a woodland soil, while living bacteria represent no less than 37 kg ha^{-1} of woodland, over 110 kg ha^{-1} of living fungal material is present. However, the same calculations based on measurements of living plus dead microbes suggest bacterial weights of 9113 kg ha^{-1} and fungal weights of 566 kg ha^{-1}.

Weights of living material represent the nature of the populations better. Independent evidence on other similar soils has been obtained by Anderson and Domsch (1975) using differential inhibitors in order to partition the contributions of different sectors of the soil microflora from the total soil respiration. They suggest that about 80% of the carbon dioxide evolved is due to fungal respiration. Other respiration measurements on soil animals when subtracted from total soil respiration measurements enable us to make an informed guess of the way in which carbon and energy are used in a natural ecosystem. The fate of the gross primary production in a terrestrial system is shown in Fig. 2.

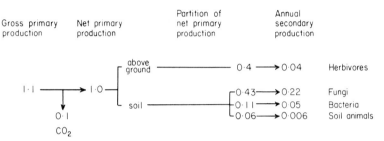

Fig. 2. Partition and utilization of the gross primary production (kg m^{-2}) of plants among animals and microbes. This assumes that animals have a 10% efficiency and microbes a 50% efficiency for conversion of substrate into protoplasm.

Clearly, the figures will vary from one ecosystem to another but the importance of fungi in organic matter decomposition, compared to bacteria and animals, is obvious. Assuming that fungi possess the necessary enzymes, it is these organisms that should be considered in relation to pesticide utilization. Their hyphal pattern of growth will increase the chances of their initiating decomposition, since suitable conditions for growth are more likely to exist somewhere along the length of a filament than around a compact colony of unicells.

III. The Aquatic Environments

A. INTRODUCTION

It would not be possible, in the space available here, to do justice to the vast literature devoted to aquatic microbiology. As a compromise, a brief summary of each environment, the marine, estuarine and fresh water, is given. The reader who requires more basic ecological information is referred to Odum (1971), Hutchinson (1957, 1966) and Hynes (1970). The brief introductory remarks on the physico-chemical nature of the environment are followed by a review of the more recent publications on the micro-biota and its activity and interactions. This is far from comprehensive but it is hoped that it will serve as a backcloth against which the other contents of the volume may be viewed.

1. *Physical and chemical characteristics*

The sea covers approximately 70% of this planet in an interconnected network of oceans which are in continuous circulation. The movement of the sea is largely well defined and governed by wind, Coriolis force, currents caused by temperature and salinity gradients and lunar periodicity. There do not appear to be any regions which are devoid of life but larger populations and activities are recorded in coastal areas. The increased productivity near land margins is due in part to upwelling, caused by movement of surface water from the coast by winds and its replacement by nutrient-rich water from deeper zones (Odum, 1971). Outwelling from nutrient rich estuaries may also exert a similar influence. A factor of major importance to marine biota is the salinity, which averages 3·5% (compared with < 0·05% in fresh water) including 2·7% NaCl. Since circulation of seawater ensures that oxygen depletion rarely occurs, variations in salinity, temperature and depth are among the major factors which govern the distribution of organisms. Apart from the major cations and anions, other ions, including nutrient ions such as phosphate and nitrate, are in such low concentrations that they contribute less than 0·1% of the total salinity, and except in regions of continuous upwelling, may often become limiting.

2. *Microorganisms and their interactions*

The marine environment can be divided into a number of zones, each with characteristic physico-chemical conditions and microbial populations.

The intertidal zone (between high and low tides) is the one with which we are most familiar mainly because of the variety of our seashores and the zonation of plant and animal life on them. The composition of the intertidal microbial population will depend to a large degree on the nature of the substratum (e.g. sand or mud) and its organic matter content (Dale, 1974). Work on marine muds by Smith (1973) has suggested that the bacterial population could account for 30–60% of the total respiration of this community. The habitat is also capable of supporting large numbers of algae (the microscopic species as well as seaweeds) and many benthic diatoms are sufficiently motile to maintain themselves in a suitable light regime in spite of tidal mixing of surface sand and mud. The distribution of blue–green algae, diatoms and bacteria on sand grains was investigated by Meadows and Anderson (1968). Colonies of tens and hundreds of cells were observed and appeared to be more successful on concave surfaces, possibly where abrasion was reduced. The change in the flora with depth of sediment was extremely variable and the only trend noted was a decreasing population towards the high water mark. Munro and Brock (1968) also examined sand populations of bacteria and diatoms and found both in viable condition to depths exceeding 10 cm. Heterotrophic activity in these samples was shown, by means of autoradiography, to be mainly bacterial in origin.

The micro-biota of the neritic zone (that covering the continental shelf) and the oceanic zone (the open ocean beyond the continental shelf) may be conveniently divided into the pelagic (open-water) and benthic (bottom-living) populations. The open water may be further subdivided into (a) the euphotic zone (i.e. to the depth where photosynthesis balances respiration, in general equivalent to the point where light intensity is about 1% of that at the surface) and (b) the aphotic zone (the deeper water). The latter is considerably deeper than its counterpart in the freshwater environment and the euphotic zone is known to vary considerably in depth, being shallower in more turbid coastal waters.

The nanoplankton appear to be important primary producers in the euphotic zone of oceanic and possibly neritic waters, whereas colourless flagellates and sedimenting detritus are suggested as important food sources for zooplankton and benthic consumers in the aphotic zone (Odum, 1971). Marine sediments often consist of an upper oxidized zone below which E_h drops markedly with depth to a region richer in H_2S and Fe^{++} ions. Movement upward of reduced gases from this reduced zone may provide an important source of nutrients for the microorganisms above. If sufficient light is present, algae may be found in the oxidized zone, and photosynthetic, as well as chemolithotrophic, bacteria in the redox discontinuity layer which separates the two zones. Anaerobic bacteria grow and provide food for ciliates in the reduced zone. If the water above the sediments becomes depleted in oxygen, then, as in fresh water, the reduced zone may move upwards into the deeper waters, and bacteria such as sulphide and thiosulphate oxidizers may thrive (Tuttle and Jannasch, 1973).

Bacterial numbers in the sea may vary between less than one litre^{-1} in oceanic water to more than $10^8 \, ml^{-1}$ inshore and between 10^1 and $10^8 \, g^{-1}$ in marine sediments (ZoBell, 1963). It should be stressed however that the count obtained will vary, not only with the site but also with the method used, microbiologists are still unable to quote with confidence the efficiency of counting procedures for aquatic bacteria. Among the species most commonly isolated have been those of *Pseudomonas*, *Vibrio*, *Flavobacterium* and *Achromobacter* but the sea has also acted as the source for studies on many halophilic, psychrophilic, photosynthetic and phosphorescent bacteria.

Some results from recent surveys by Hobbie *et al.* (1972) and Holm-Hansen (1972) are summarized in Fig. 3. Values of the variables are not included since they varied considerably between sites. The general pattern was, however, fairly reproducible and is as illustrated, i.e. higher levels of particulate material, biomass and activity in the euphotic zone. Recorded maxima for particulate organic carbon were around 30 μg C litre^{-1} in oceanic water and 300 μg C litre^{-1} in coastal waters. Phytoplankton and biomass (ATP) carbon values were in the same range. The respiration rates presented were obtained from concentrates of cells with a maximum of 36 μl O$_2$ day^{-1}. The direct counts of bacteria were obtained from similar materials and the peak reading corresponds to a population of 25 cells ml^{-1}. The original publications should be consulted for detailed discussion of the results.

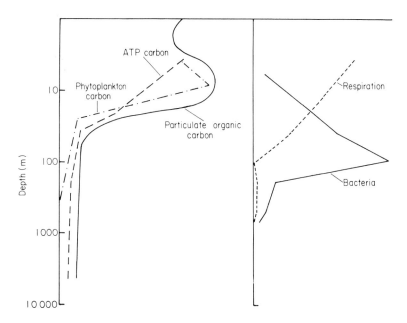

Fig. 3. Variation with depth of a number of variables in coastal and oceanic water. Details of the results are given in the text, and the data are taken from Holm-Hansen (1972) and Hobbie *et al.* (1972).

The location of bacteria in seawater has continued to exercise the minds of marine microbiologists and although Seki (1970) demonstrated a positive correlation between bacterial numbers and particulate material, Wiebe and Pomeroy (1972) concluded that as many as 80% of the bacteria were unattached and were not as common on particles as had previously been supposed. Jannasch (1968) has suggested that slower growing bacteria may be associated with particles, where nutrient concentrations would be higher, whereas the other bacteria could be adapted to grow at low substrate concentrations and therefore compete successfully with other marine microbiota. In a series of studies on the heterotrophic activity of marine microorganisms, Williams (1970) demonstrated that 80% of the recorded activity was associated with organisms which passed through the 8 μm mean pore size filter and 50% with those passing through a 1·2 μm filter. Andrews and Williams (1971) showed that oxidation rates of organic substrates (glucose and amino acids) were higher in the summer than the winter months and may account for organic matter equivalent to 50% of the measured phytoplankton production.

Much of the marine microbiological research has concentrated on the plankton, the algae, zooplankters and bacteria in particular and it is generally thought that fungi and yeasts are far fewer in number, except in areas of high concentration of macrophyte detritus (Odum, 1971) and possibly on the surfaces of fish (Bruce and Morris, 1973). The fact that fewer publications are concerned with benthic microorganisms probably reflects the difficulties of sampling bottom muds, particularly in oceanic regions. Jannasch and Wirsen (1973) reported exceptionally low rates of microbial transformations when substrates were incubated *in situ* for one year. Apart from the zonation already mentioned, very little is known about the interactions of sediment microorganisms, although the attachment of bacteria to dead or moribund algae noted by Oppenheimer and Vance (1960) closely parallels similar observations in fresh water.

C. THE ESTUARINE ENVIRONMENT

1. *Physical and chemical characteristics*

Estuaries may be classified in a number of ways. Among the more common systems of classification are those based on:

(a) Morphology (ranging from the steep sided fjord type to the silted river delta system).

   (b) The degree of water mixing (ranging from negligible where the sea runs as a "salt-wedge" under the river water, to complete mixing by strong tidal action).

   (c) The energetics of the system and the degree of stress to which it is submitted.

In general terms, an estuary is defined as a semi-enclosed coastal body of water which has a free connection with the open sea (Odum, 1971). The system has many unique characteristics and the organisms therein often possess wide tolerances to changes in temperature and salinity. Such a system is particularly susceptible to man-made changes, and its ability to cope with increased organic load may be drastically affected by such factors as the type of estuary, the degree of impoundment and the quantity of toxic material in any effluent present.

2. *Microorganisms and their interactions*

Estuaries tend to be more productive than either of their fresh- or salt-water component parts. This is due, in part, to the presence of macrophytic, benthic and planktonic photosynthetic organisms, which ensure that some level of primary production continues all through the year, with the tidal system ensuring nutrient replenishment and waste-product removal. Estuaries tend to act as nutrient traps, with rapid recycling by the benthos and recovery from deeper mud zones by the combined action of microorganisms and burrowing animals. The stratification of the sediment into oxidized and reduced zones and the associated microbial populations are particularly well developed in estuaries. It should be stressed, however, that pollutants, as well as beneficial substrates, may be caught in the nutrient trap, hence the sensitivity of the estuarine system to the activities of man.

    Comparatively little work has been done on estuarine ecosystems, probably because of the heavy sampling programme that would be required to provide realistic estimates of populations and their activities. Turner and Gray (1962) reported significant variations in the salt tolerance of the bacteria and Crawford *et al.* (1973, 1974) demonstrated considerable patchiness in heterotrophic activity in both time and space. They also suggested that fluctuations in dissolved free amino acids in estuarine water may be related to algal excretion and death, and that particulate production from such heterotrophic sources may be a significant part of the food chain, being equivalent to approximately 10% of the algal production.

    The overall high productivity of estuaries has resulted in the isolation of a wide variety of microorganisms. Hansen and Veldkamp

(1972) isolated a new type of photosynthetic purple bacterium capable of aerobic growth in the dark. Stewart (1965) demonstrated the presence of nitrogen-fixing blue–green algae, and Jones (1974) concluded that, although bacterial nitrogen fixation was present, it was negligible compared with that of the blue–greens. Colwell (1972) reported that yeasts were widely distributed in estuaries, as were bacterial viruses. Nutrient levels are known to affect the phytoplankton crops (Mommaerts, 1969) and although pollution by sewage and dredge spoil tends to cause exponential growth of algae, initial phases of temporary inhibition, possibly due to toxic substances, have been reported (Young and Barber, 1973). The development of tolerance in bacteria to certain pollutants, for example mercury, has also been reported in estuarine environments (Colwell, 1972).

D. THE FRESHWATER ENVIRONMENT

Although it covers a relatively small area of the earth's surface, fresh water is of paramount importance to man as a cheap source of industrial and domestic water and as a convenient means of waste disposal. It can be divided into two main habitats: the running water or lotic habitat; and the still water or lentic habitat.

1. Physical and chemical characteristics of the lotic habitat

Three factors exert a major influence on the biota of streams and rivers. The distribution of animal and plant life is, to a large degree, controlled by

(a) The current strength, which provides us with the simplest system of zonal classification i.e. into rapids and pool sections. The former has a characteristically well scoured bed with colonization by plants and animals capable of attachment to rocks and stones. A pool is defined as a region of slower flow where deposition occurs, particularly of detritus. A silty bottom is thus formed which is particularly suitable for burrowing animals.

(b) Land drainage is important in that it provides both dissolved and particulate substrates for the lotic community. Dissolved inorganic nutrients are consumed to a large degree by the primary producers, namely aquatic macrophytes and the periphyton, particularly attached filamentous green algae and diatoms. The particulate material, including the detritus component, is an important food source for other consumers in the river or stream system.

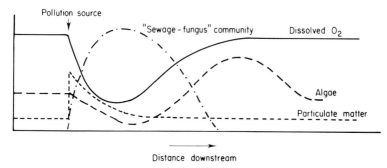

Fig. 4. The effect of organic pollution on a river. (Re-drawn from Hynes, 1960.)

(c) Continuous movement and turbulence often ensures adequate aeration of the water and therefore oxygen depletion is less likely than in a lake or pond, except in cases of extreme pollution (Fig. 4).

Further information on the systems of classification and the general biology of rivers and streams may be found in Hynes (1960, 1970).

2. *Microorganisms of the lotic habitat and their interactions*

Comments on the apparent lack of quantitative information on the microbiology of streams and rivers have been made by several authors (Mann, 1969; Hynes, 1970). An attempt to review the available information on heterotrophic microorganisms and their activity, in more detail than space permits here, was made by Jones (1975) and a summary of factors affecting algal distribution was given by Blum (1960).

Several publications have discussed the relative importance of autochthonous material (formed within the system) and allochthonous material (formed elsewhere) to the lotic habitat. The reader should note that the application of these terms differ slightly from those of Winogradsky (1949) but his "autochthonous" and "zymogenous" microflora could well be responsible for the breakdown of organic matter of internal and external origin respectively. Mann (1969) and Westlake *et al.* (1972) both commented on the importance of the allochthonous input to their systems. Autumn shed leaves are an important source of energy to many streams and are colonized by both fungi and bacteria during the decomposition process (Kaushik and Hynes, 1968) making them more palatable to invertebrate grazers (Peterson and Cummins, 1974). Bacteria and fungi are also thought to be primary colonizers of stone surfaces, followed by diatoms, thus forming a matrix which traps detritus and provides a suitable food source for browsing animals (Madsen, 1972).

Input of organic material, whether natural or man-made, can create a considerable oxygen demand in the river or stream. This is the basis of the well known oxygen-sag curve illustrated in Fig. 4. The oxygen levels are depressed immediately below the point source of pollution; increased numbers of bacteria fungi and yeasts (Cooke, 1961; Woollett and Hendrick, 1970) are recorded and "sewage-fungus" communities may produce unsightly growths (Curtis, 1969). Immediately down stream, the input algal growth may be stimulated and depending on the volume, degree of turbulence and other factors the system may not recover until some considerable distance downstream. The effect of organic materials on rivers has been summarized by Curtis and Harrington (1970).

Enrichment with inorganic nutrients can produce larger standing crops of the important primary producers, namely macrophytes, diatoms and filamentous green algae. The latter in particular may develop to nuisance levels, as exemplified by heavy growths of *Cladophora* species. Examples of extreme inorganic enrichment may be seen in water associated with agricultural drainage channels, particularly after injudicious use of fertilizers. Such additions to our water systems may also affect the plankton but only in slower flowing rivers where production is possible. It is also possible for plankton community respiration to exceed photosynthesis under conditions of extreme enrichment, as demonstrated by Kowalczewski and Lack (1971) on a polluted tributary of the Thames.

The health hazards resulting from pollution of rivers with sewage have been sufficiently well documented for further comment to be unnecessary here, except to note that although growth of coliform bacteria has been recorded in natural waters (Hendricks and Morrison, 1967), self-purification mechanisms (Mitchell, 1971) would probably ensure control of the population.

Two further forms of pollution deserve brief mention since they are receiving increased attention. The action of thiobacilli in mine waters is known to cause a considerable drop in pH, with resulting deleterious effect if these acid wastes enter water courses. More recently, examination of "streamers" at these sites has shown that a community of bacteria, fungi and a few algae develop, which offer the possibility of studies which parallel those on the "sewage-fungus" complex (Dugan *et al.*, 1970). Thermal pollution, especially below power stations, has recently been the cause of some concern. Work by Zeikus and Brock (1972) has shown that, although growth rates were similar above and below the source of pollution, and that each bacterial population was optimally adapted to its temperature regime, larger biomass figures were recorded in the warmer water. It therefore seems possible, under extreme conditions, that a combination of increased bacterial biomass and lower oxygen solubility at higher temperatures, may cause a significant drop in oxygen concentration at some sites.

3. *Physical and chemical characteristics of the lentic habitat*

In the temperate regions of the world a classical pattern of thermal stratification develops in lakes. The top water is warmed in early summer and as a result of density differences the lake is split into three zones which undergo very little intermixing until the autumn overturn induced by lower temperatures, higher winds and increased rainfall. The upper layer (the epilimnion) includes the euphotic zone and is the area of maximum primary production both in the limnetic or open water region, where phytoplankton are the major producers, and the littoral or marginal region, where macrophytes may be the most significant contributors. Their smaller size and shallow basins means that the littoral zone often plays a more significant role in pond systems. Wind-induced turbulence may ensure mixing of the epilimnion but density gradients prevent significant exchange with the cooler deep-water zone called the hypolimnion. The region of rapid temperature change separating the two zones is termed the metalimnion and that of maximum temperature gradient the thermocline (Hutchinson, 1957).

If the light does not penetrate the hypolimnion its physical isolation can result in depletion of oxygen in the deeper water due to a number of biological and chemical oxygen-demanding reactions. The degree of hypolimnetic oxygen depletion will depend on a number of factors. The morphometry of the lake, the volume of the hypolimnion and the area of mud in contact with it are of primary importance. Thus deep lakes, with a large hypolimnion volume/mud area ratio, tend not to become oxygen depleted, whereas shallow lakes where the hypolimnion is often smaller than the epilimnion, may be subject to severe deoxygenation. The degree of productivity or fertility of the lake will also affect the oxygen depletion pattern. If a lake is rich in nutrients then higher production, deposition and decomposition rates result in more severe deoxygenation of the hypolimnion. Further enrichment by the activities of man may exacerbate this problem. Thus the concepts of eutrophy and oligotrophy which have received so much publicity recently may be seen to have morphometric as well as cultural component parts.

Examples of the vertical zonation of a shallow productive lake and a deeper, less productive water body are shown in Fig. 5. The data are

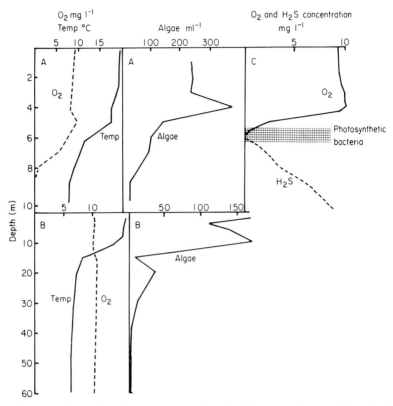

Fig. 5. Vertical profiles of temperature, dissolved O_2 and algae in (A) a shallow productive lake, (B) a deeper less productive lake (data redrawn from Jones, 1972a) and (C) diagrammatic representation of the zone of development of photosynthetic bacteria in a eutrophic lake.

taken from a survey by Jones (1972a) and demonstrate the higher levels of algal standing crops in the euphotic zone, the thermal stratification and the degree of deoxygenation of the hypolimnion. The higher prevalence of anaerobiosis in the hypolimnion in lentic fresh water habitats results in movement upward, into the water column, of the reduced zone of the mud, enriching the water with nutrients previously bound in or under the sediment-water interface. Further details of these processes may be obtained from the classic studies of Mortimer (1941, 1942).

4. *Microorganisms of the lentic habitat and their interactions*

The temperature and resultant oxygen regime of the stratified lake, as described above, is of considerable importance in the determination of the microbiota which develops in lakes.

The heterotrophic bacteria, most of which are Gram negative (Collins, 1960), tend to be the major component of the bacterial flora. Higher numbers are usually recorded in the epilimnion, particularly the oligotrophic water, although recent results from total counts using incident-light fluorescence techniques show a much larger population in the anaerobic, nutrient-rich hypolimnia of eutrophic lakes. Under certain conditions of stratification and rainfall, high counts are obtained in the thermocline region (Collins and Willoughby, 1962). A wide variety of bacterial sizes are observed in any fresh water sample, with a preponderance of very small forms. Although methods have been devised to demonstrate the uptake and mineralization rates of aquatic bacteria (Wright and Hobbie, 1966; Hobbie and Crawford, 1969), it is difficult to ascertain whether all or part of the bacterial population is responsible for the activity (Ramsay, 1974).

Whereas the spatial distribution of aquatic bacteria, particularly with depth, is seen to correlate significantly with a number of variables (Jones, 1972a) temporal changes are more difficult to explain (Jones 1971, 1973). The degree of eutrophication of a lake will also affect the microbial populations. The phytoplankton will respond to concentration of nutrients as well as to the underwater light climate (Talling, 1971) and the other microbiota and their activities are seen to respond to the level of enrichment of the water body (Stewart et al., 1971; Jones, 1972b); indeed, whole lake experimental enrichment (Schindler et al., 1973) has been shown to have profound effects on the microbial population.

Hypolimnetic anaerobiosis and lake enrichment has been shown to affect organisms other than algae and planktonic heterotrophic bacteria. The size of the population of ciliated protozoa in a eutrophic lake has been shown to be affected by the concentration of dissolved oxygen in the water (Goulder, 1974). Zonation of populations occurs in the mud as well as the water and complex interactions involving end-product inhibition and substrate interdependence may occur, as for example between sulphate-reducing and methane-producing bacteria (Cappenberg, 1974a, b; Cappenberg and Prins, 1974).

Photosynthetic bacteria are known to grow in eutrophic lakes particularly in the zone below the metalimnion where oxygen is absent, or barely detectable, but where there is sufficient hydrogen sulphide from the hypolimnion and light from the surface (Fig. 5c). There have been many reports of increased activity in these zones, particularly of bacterial chemosynthesis (Sorokin, 1964, 1968), *Chlorobium* and *Chromatium* spp. (Kuznetsov, 1959), *Pelodictyon aggregatum* (Gorlenko and Lebedeva, 1971) and *Pelochromatium roseum* (Anagnostidis and Overbeck, 1966). The Rhodospirillaceae, on the other hand, appear to develop to a greater degree in littoral muds

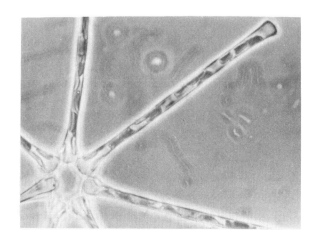

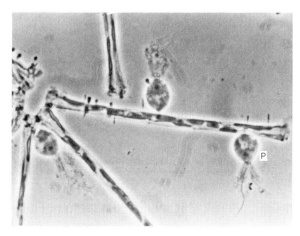

P

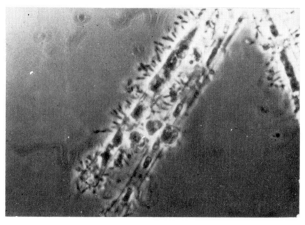

10μm

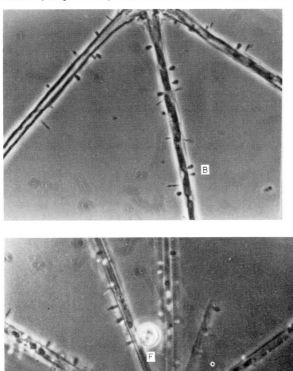

Fig. 6. Stages in the attachment of bacteria (B), fungi (F) and protozoa (P) to a planktonic diatom, *Asterionella formosa* Hass., over a period of one month in Blelham Tarn.

but may be redistributed by turbulence throughout the water column. A more detailed discussion of bacterial distribution patterns was presented by Overbeck (1974).

Microbial interactions are often more evident in the lentic habitat than in rivers and streams. Algae are known to be parasitized by both fungi and protozoa (Canter and Lund, 1953, 1968), often to a degree which results in the decline of the host population. Blue–green algae are also known to be attacked by bacteria some of which resemble myxobacteria (Shilo, 1970; Daft and Stewart, 1971) and others which are spore formers (Reim *et al.*, 1974). Chemotactic responses of bacteria to algae have been reported by Bell and Mitchell (1972) and a summary of some of the interactions involving blue–green algae is given by Whitton (1973). We have, at this laboratory, observed the colonization of algae by fungi, protozoa and bacteria, much of which appears to be of a passive nature and only occurs to a significant degree after the algal population has started to decline. Examples of this phenomenon are shown in Fig. 6.

The role of microorganisms in production and decomposition processes in aquatic environments has, in the past, often been the subject of isolated studies. More recently, however, there has been a tendency to study whole ecosystems, with attempts to produce budgets which include most of the processes discussed above (Wetzel *et al.*, 1972) or to further elucidate the agents involved in a particular process, for example the deoxygenation of the hypolimnion (Burns and Ross, 1972).

IV. The Air

A. INTRODUCTION

The atmosphere of Earth is a gaseous mixture consisting mainly of nitrogen (78%) and oxygen (21%), together with carbon dioxide (0·03%) and traces of inert gases. Although the air, with low availability of water and nutrients, is not a particularly favourable microbial habitat, it is evident that numerous and diverse microorganisms exist in this medium. This fact is only too obvious to the microbiologist who has exposed culture media to the atmosphere, deliberately or otherwise. Especially common in air are the spores of fungi. Although microorganisms are small and light enough to be carried in the air, there is the constant possibility that they will be removed from it again by natural gravitational deposition, impaction, or in the precipitation. Whilst the idea of a true permanent aerial microflora or "aeroplankton" seemed improbable, Gregory (1973) concluded that the aerial environment is not beyond the range of microbial exploitation.

The air affords a convenient medium for the dispersal of some microorganisms, especially those with appropriate adaptations such as spores, cysts, dessication-resistant cells and mechanisms for spore discharge. There are, however, disadvantages to man in the ease with which some microorganisms are dispersed, especially with regard to the spread of some diseases. Notable in this connection are bacterial, viral and fungal diseases of the respiratory tract of man and animals (coupled with allergies) and some fungal diseases of cereal crops. For a full treatise of aerial microbiology the reader should refer in the first instance to Gregory (1973).

Since most pesticides (especially those used in agriculture) at some time during, or even after, application, exist in an aerial phase, there are opportunities for airborne microbes to encounter and react with them. Such encounters may not contribute greatly to the overall fate of pesticides in the environment. It would be wrong, however, to disregard the air as a potential medium for pesticide–microbe interactions.

B. CHARACTERISTICS OF THE ATMOSPHERE

The turbulent tendency of outdoor air generally masks the effects of gravitational acceleration and constant terminal velocity of fall of small particles such as microbes and even small convection currents may be sufficient to keep particles such as spores in aerial suspension.

The atmosphere is characterized by a decrease in barometric pressure, temperature and density with increasing height. The term *troposphere* collectively describes the lower atmospheric layers extending from the ground to approximately 10 km height (Gregory, 1973). Between the troposphere and the stratosphere is a boundary zone, the tropopause. The stratosphere contains meteoric dust from space, whereas the troposphere contains terrestrial dust and organic matter. The troposphere is thus the biologically significant zone and within it are described five separate layers in ascending order (Gregory, 1973):

(a) The laminar boundary layer, describing a thin layer of air with orderly flow lying immediately above the veneer of still air which coats all bodies;

(b) The local eddy layer, where local eddies are caused by small irregularities;

(c) The turbulent boundary layer, created by ground-drag influence on major air currents;

(d) The outer frictional turbulence layer;

(e) The convection layer.

C. MICROBIAL CONTENT

1. Origin and distribution of microorganisms

The air has posed many problems, some of them unique, for the microbiologists who have attempted to analyse it. In order to determine the microbial numbers and types in the air the organisms must somehow be trapped. Many techniques have been devised for this purpose and most involve the deposition of organisms onto a surface upon which they can be examined, either directly or after a period of growth. In the simplest methods, organisms are collected by gravity sedimentation from the air onto glass slides coated with an adhesive (e.g. glycerine jelly), or onto agar media in Petri dishes. Other techniques employ processes such as filtration, centrifugation, impingement or electrostatics in order to capture the microorganisms.

The numbers and types of microorganisms found in air samples vary according to the location, climate, time of day and local environmental factors. The spores of fungi often predominate in samples and may arise from macroscopic as well as microscopic fungi. Non-microbial living particles in the air include the spores of bryophytes and pteridophytes, pollen grains, moss gemmae and lichen propagules. Apart from fungal spores, the microbial forms found in air include myxomycete spores, viruses, protozoan cysts, eukaryotic and prokaryotic micro-algae, and the cells and spores of bacteria. The term "air spora" has been used to describe, rather loosely, the biological particle content of the air.

Whereas the sources of most bacteria in the air are probably the underlying soil and water bodies, the fungal spores are mainly derived from fungi growing on vegetation surfaces or vegetable debris above ground level. Thus, whilst Penicillia and Aspergilli are common in the soil, the predominant forms in air are spores of *Cladosporium* and basidiospores of higher fungi, found on vegetation.

The viruses, bacteria, protozoa and algae are not generally endowed with natural discharge mechanisms. Instead, external agencies such as wind, man and animals release them into the air, usually in aggregates on particles of dust or vegetation. Water droplets contaminated with microorganisms may enter the air from raindrop splashes, wave action, spray, aerosols and bursting bubbles, whilst droplets from coughing and sneezing are an important source of microorganisms in indoor air. Some types of bacteria are, however, well suited to dispersal via the atmosphere. This is particularly evident in the *Streptomyces* group which produce spores on aerial hyphae. The slime-molds (Myxomycetes) also bear fruiting structures raised above the substratum. Adaptations, some of which are remarkably sophisticated, for aerial dispersal of spores are most apparent amongst the fungi. Passive spore liberation may be achieved by the effects of gravity and air currents, pick-up in small droplets of water or mist and raindrop impact on spore-bearing structures. The mechanisms of active spore discharge include rounding-off turgid cells, hygroscopic twitching and ballistic action. Many of the spore release mechanisms are influenced by environmental factors, such as light and humidity.

Diurnal periodicity patterns in aerial fungal spore levels have been established and related to periodicity of spore release. Seasonal variations, especially with respect to airborne bacteria, have been related to some agricultural activities. Bacterial and fungal spores are the most numerous of the air spora near ground level and in open country may average 10 000 m^{-3} (Madelin and Linton, 1971). A mean spore count of 12 500 m^{-3} at 2 m height at Rothamsted in summer was quoted by Gregory (1973).

Whereas in outdoor air there is a general tendency for microbial dispersion and randomization, in the air inside buildings there is a tendency for microorganisms to accumulate, especially where there is inadequate ventilation. The movements and activities of man and domestic animals may significantly influence the numbers of airborne microorganisms inside buildings. Apart from dust particles and moisture droplets, other sources of airborne microorganisms in such situations include the skin scales and hairs which are continuously shed and ejections from the respiratory tract. Microbial contamination of air is usually high in buildings housing animals, due to the presence of hay, straw, fodder, dry excreta and the animals themselves. Particles less than 1 μm in diameter may remain permanently suspended (Madelin and Linton, 1971).

2. *Survival*

A key factor in the survival of aerial microbes is the problem of remaining airborne. Additional stress factors include the low availability of water, temperature fluctuations and radiations. Temperatures in the atmosphere are more likely to be preservative than lethal for most air spora. On the other hand, radiations (especially ultraviolet) are potentially hazardous to the organisms. Pigmented microorganisms are, in general, less sensitive to the damaging effects of u.v. radiation than are non-pigmented types. Although visible light reverses u.v. damage in many microorganisms by a process of photoreactivation, microorganisms which have survived high altitude passage may be expected to show an increased mutation rate on return to Earth (Gregory, 1973). Anderson and Cox (1967), who reviewed data on microbial survival in aerosols, also note that radiation may increase the rate of mutation of bacteria in aerosols.

Some micro-algae may be better suited for aerial survival and dispersal than others. For example, Ehresmann and Hatch (1975) found that a unicellular prokaryotic alga was better able to survive low relative humidities than was a unicellular eukaryotic type. In this respect the prokaryotic alga resembled bacteria in its aerial survival.

3. Microbial types present

Although comparatively low in number, viable micro-algae and protozoa do occur in the air overlying both terrestrial and aquatic
environments (Brown et al., 1964; Schlichting, 1969). Numerous
representatives of different algal types including the green algae, blue–
green algae, diatoms and desmids, have been found in the air. Brown et
al. (1964) recovered viable members of 62 different algal genera from
the atmosphere, whilst Schlichting (1969) calculated that, up to that
date, a total of 187 species of algae and protozoa had been isolated
from air. The algae sampled by Brown et al. were considered to
originate in the soil and the specific composition of the airborne algae
was determined by proximity to soil algal populations and by
meteorological conditions.

Schlichting (1969) found more algae and protozoa in the air over
dry dusty Texas soil than in the air from the forested moist climate of
the Great Lakes or coastal air off Carolina. This suggested that algae
adapted to terrestrial habitats might better resist dehydration stresses
and so be adapted to aerial dispersal or existence. However, Gregory et
al. (1955) had reported that of three contrasting sample sites (city,
countryside and seaside) the highest counts of the unicellular blue–
green alga Gloeocapsa were observed at the seaside site, with an
average of 110 colonies m^{-3} air.

Although algal counts in air are very low in comparison with, for
example, those for fungal spores, they may be high enough to allow
consideration of the air as a dispersal medium for algae with
established survival capacity (Gregory et al., 1955). Actual algal
numbers may reach a few hundred m^{-3} (Gregory, 1973), although
Schlichting (1969) never found more than eight cells m^{-3}. Even so,
such numbers were considered significant in relation to phenomena
such as allergies and aerial radioactivity (Schlichting, 1969). Bacteria
are usually far more numerous than algae and protozoa and are likely
to have been picked up together with dust and soil particles from the
underlying terrain. It would seem that, in general, vegetative bacterial
cells have a fairly transient and perhaps insignificant existence in the
air. The bacteriology of the air, with the exception of those aspects
relating to animal disease transmission, does not appear to have been
systematically examined since the work of the pioneer bacteriologists.
By contrast, there is a vast amount of data on the fungi and plant
pollen of the air spora. The concentration of fungal spores in the air
varies widely according to locality and conditions but the ubiquitous
spores of Cladosporium spp. often predominate. In Britain these are
usually accompanied by Sporobolomyces, Alternaria and various basidiospores and ascospores.

The possible existence of a true air-inhabiting microbial population or "aeroplankton" has been debated. Indeed, there are some grounds for the suggestion, in as much as some essential requirements of life are present in the form of gaseous nitrogen and carbon compounds, together with a variable amount of water vapour. However, the air can be considered as an exacting microbial habitat, probably only tolerated by specialized microorganisms which are somehow maintained in suspension.

Even discounting the existence of an "aeroplankton", the air is an important medium for the dispersal of a vast array of microorganisms including pathogenic bacteria, viruses and fungi. Transport over long distances may be achieved via the atmosphere. This is especially significant in the epidemiology of some fungal diseases of crops. Long distance movement of the uredospores of the wheat stem rust fungus *Puccinia graminis* has been established in outbreaks of the disease in Canada, USA, India and Russia (see Gregory, 1973).

V. Animals

A. INTRODUCTION

A variety of microbial habitats are presented by the diverse free-living animals whose external surfaces and internal tracts are readily accessible to microorganisms. In some cases microbial populations are actually required for the normal functioning of the animal. The relevance of the animal kingdom to the use of pesticides need hardly be stated. The microbes which coexist with animals deserve attention also.

The diverse microbial types found in association with animals include viruses, bacteria, fungi, yeasts, protozoa, mycoplasmas, rickettsiae, actinomycetes and algae; the associations ranging from harmless modes, such as commensalism and symbiosis, to parasitism and pathogenicity. In either case the microbial habitat is at least sheltered from the vagaries of external environments. Some animals also serve as vectors transferring specific microorganisms to new hosts.

B. ANIMALS AS MICROBIAL ENVIRONMENTS

The normal activity of most microorganisms requires quite high water activity (a_w) values, of the order of 0.7 to 0.99. Some microorganisms adopt measures which enable them to survive low a_w extremes by the development of thick cell walls, spores or cysts. On the other hand, some obligate pathogens, for example the spirochaete *Treponema pallidum*, are extremely vulnerable to the effects of drying and die rapidly on exposure to air. In those animals normally exposed to the open air it seems likely that the a_w would be lower on external surfaces, especially in insects with hard, smooth surfaces, than in the various internal and external body secretions or in the alimentary tract. This would apply less to animals living within the soil environment. Apart from affecting the a_w, solute concentrations in body fluids also influence the osmotic pressure to which microorganisms have to adapt.

Animal bodies provide a wide range of habitats in terms of oxygen availability. Aerobic bacteria are typically found on the skin, whilst facultative and obligate anaerobes occur in the low redox potential parts of the alimentary tract. Within the oral cavity are sites of high and low oxygen content and also areas, especially around the teeth, where conditions are rendered anaerobic by the activity of large populations of facultative and microaerophilic organisms.

Light does not appear to be an important factor in determining microbial life in association with animals, except in the case of algal endosymbionts of invertebrates. It is apparent, however, that many of the micrococci and some mycobacteria and yeasts found on the human skin surface are pigmented and may thereby be protected from u.v. wavelengths.

The skin temperature may fluctuate and is sometimes a few degrees lower than the more constant internal temperature. The majority of microorganisms which inhabit animals are mesophiles having temperature optima close to the host's body temperature. There would seem to be little opportunity for the development of thermophiles or psychrophiles, except on the surface of animals continually exposed to temperature extremes.

The pH value may fluctuate from site to site within an animal, possibly to the extent of influencing both the activity and species composition of resident microorganisms, as is evidenced in the human digestive tract.

C. MICROORGANISMS AND VERTEBRATES

The outer surfaces and orifices of vertebrates are continuously exposed to microbial contamination. Some microbes are present under all normal conditions and these, the "normal flora", may have established an equilibrium with the animal. Although the microbial flora of individual vertebrate species may vary according to state of health or environmental factors, it is possible to describe the typical microflora of certain sites of healthy animals. The most comprehensive data, not surprisingly, relate to the microflora of man (see Skinner and Carr, 1974).

1. *The skin*

The secretions of mammalian skin glands contain amino acids, lipids, urea, organic acids, vitamins and mineral salts which are nutritious for heterotrophic microorganisms. The normal flora of the skin is, however, somewhat restricted, possibly due to the anti-bacterial activity of long chain fatty acids and lysozyme secreted by the skin. The skin flora of man consists mainly of Gram-positive coccoid bacteria (staphylococci and non-haemolytic streptococci), coryneforms, mycobacteria and some yeasts. Other organisms may also be present, especially on skin adjacent to orifices.

2. *The respiratory tract*

Microorganisms which enter the upper respiratory tract are trapped in the mucous secretions of the nasal passages. Bacteria commonly found in the nasopharynx of man include staphylococci, viridans-group streptococci, non-pathogenic *Neisseria* spp. and coryneforms. The nasal and bronchial passages filter the inhaled air and trapped particles are expelled in the mucous. Due to this and to secreted immunoglobulins the lung tissues in healthy animals are essentially sterile.

3. *The alimentary tract*

The oral cavity of man, with an overall neutral to slightly alkaline reaction, is potentially a suitable environment for microbial proliferation. Typically found are streptococci, non-pathogenic *Neisseria* spp., lactobacilli, anaerobic spirochaetes, coliform bacteria and yeasts. Approximately two-thirds of the oral microflora may be anaerobic or microaerophilic, the teeth having an important influence. Dental plaque, a heterogeneous material which tenaciously covers the teeth, contains numerous filamentous bacteria, especially of the anaerobic *Fusobacterium* type. Associated with these organisms are populations of lactobacilli and streptococci which ferment carbohydrates to form lactic acid.

The acidic nature of the stomach (approx. pH 2 in man) inhibits most vegetative bacteria, the only organisms favoured being acid-tolerant lactobacilli associated with the stomach lining. Bacterial numbers then increase in the intestines as conditions become more alkaline in the presence of bile secretions. Numbers are high in the large intestine where, due to the activities of facultative anaerobes such as the coliform bacilli and enterococci, a low redox potential develops, allowing the proliferation of strict anaerobes such as *Clostridium* and *Bacteroides* spp. Lactic acid bacteria are prominent in the intestinal flora of pigs, in which under normal husbandry conditions the microbial population contains lactobacilli, streptococci, *Bacteroides* spp. and *Escherichia coli* at all levels of the tract (Kenworthy, 1973).

Smith and Crabb (1961) reported a consistent pattern in the composition of the faecal flora of adult animals of the same species but there were clear differences between animal species. For example, low numbers of *E. coli* and *Clostridium welchii* were found in rabbit, guinea pig, horse and cattle faeces, but high numbers of these bacteria in dog and cat faeces. There were vast differences in the total viable faecal bacteria count between different animal species (Table 2). The

TABLE 2

Numbers of viable bacteria in faeces of adult animals

	Average viable numbers[a] g^{-1} faeces
Cattle	$34\cdot2\times10^4$
Sheep	$37\cdot4\times10^5$
Horses	$20\cdot0\times10^6$
Pigs	$43\cdot1\times10^7$
Rabbits	$50\cdot1\times10^7$
Guinea pigs	$19\cdot9\times10^7$
Mice	$23\cdot7\times10^8$
Humans	$46\cdot2\times10^8$
Dogs	$37\cdot8\times10^7$
Cats	$17\cdot6\times10^8$
Chickens	$56\cdot6\times10^7$

[a] Derived from Smith and Crabb (1961) and Smith (1961).

bacterial content of faeces was similar irrespective of species whilst the animals were young, but became dissimilar between species as the animals aged, possibly due to dietary changes (Smith and Crabb, 1961; Smith, 1961). Changes in the gut microflora of pigs were related to changes in diet and housing conditions (Hill and Kenworthy, 1970; Kenworthy, 1973).

In some cases the external distribution of microorganisms is directly related to the distribution of particular animal species (Lovell, 1957). Thus *Erysipelothrix rhusiopathiae*, an organism primarily associated with pigs, can be found in soil after pigs have been present. Similarly, the incidence in soil of enteric coliform bacteria and the tetanus organism, *Clostridium tetani*, closely parallels the deposition of faeces of domestic animals and the application of animal manure. Some of the fungi which inhabit the dung of herbivorous animals are chance contaminants whilst others are ingested on forage. The coprophilous fungi, including members of the genera *Pilobolus*, *Sordaria* and *Coprinus*, are sometimes characteristic of the type of dung. Spore discharge mechanisms and phototropism are well developed in many of these fungi, thereby increasing the efficiency of spore transfer from dung to the surrounding grass and thence, after animal passage, to fresh dung. It is clear that animal faeces may have a considerable influence on the localized microbial content of environments such as soil and water into which they are voided.

Herbivores are adapted to the digestion of grass and plant material and rely on the cellulolytic enzymic activities of symbiotic gut microorganisms. The alimentary tract in these animals is modified to include enlarged specialized parts. In ruminants (e.g. cow, sheep, camel, goat) the plant polysaccharides are digested in the rumen and reticulum occurring before the stomach. In non-ruminants (e.g. horse, pig, rabbit) the enlargements, the caecum and colon, occur after the stomach.

The rumen is a large organ in which the combined activities of anaerobic bacteria and protozoa uniquely achieve the digestion of plant material, especially cellulose. Since several important farm animals are ruminants the rumen microorganisms assume special significance.

The microbiology of the rumen has been investigated for many years and the reviews of Hungate (1963) and Coleman (1963) respectively concerned its bacterial and protozoan components. Apart from the plant polysaccharide components, ruminants ingest proteins, fats, other organic compounds and minerals. The masticated mixture encounters a vast mixed population of bacteria (approx. 10^{10} cells ml^{-1}) and ciliate protozoa on entering the rumen. With its steady-state conditions of anaerobiosis, pH 5·5 to 7·0, temperature 39° to 40°, regular supply of nutrients and continual removal of products and cells, the rumen has been likened to laboratory continuous culture systems. Whilst the protozoa do not appear to be essential for the rumen fermentations, they undoubtedly assist in some reactions and may control bacterial numbers. Aspects of the biology of rumen ciliate protozoa were described by Eadie (1962, 1967) and Clarke (1964). Hungate (1963) drew attention to the intricate interrelationships of rumen microorganisms and concluded that the rumen microbial population as a whole could be considered as symbiotic. It is therefore usual in microbiological studies of rumen activities to use the natural mixed cultures of rumen fluid. Such studies as have been made in relation to pesticide degradation by rumen microorganisms (see later chapters) testify to this fact. The rumen bacteria are broadly categorized according to their metabolic activities. Cellulose-degrading genera include *Bacteroides*, *Butyrivibrio* and *Ruminococcus*, whilst starch-hydrolysers include *Streptococcus*, *Bacteroides* and *Selenomonas*. Other important genera are *Peptostreptococcus* which ferment lactate and *Veillonella* which decarboxylate succinate.

Sugars released by microbial breakdown of plant polysaccharides in the rumen are fermented to produce volatile fatty acids, carbon dioxide and methane. The fatty acids are utilized by the animal as an energy source and the gases eliminated. The rumen microorganisms also synthesize amino acids and vitamins which can be maintained at a steady level of supply for the animal. Microbial cells are passed from the rumen into the gastrointestinal tract where they are digested by the animal as a source of protein. Although this fate befalls individual rumen bacteria, overall the species involved enjoy a safe and constant ecological niche, thus establishing the mutualistic aspect of the symbiosis.

4. Fish

The flesh of healthy fish, like that of mammals, is considered to be sterile. The slime-coated external surface and the intestines are, however, densely populated by bacteria. Shewan (1962) found $10^2–10^5$ bacteria cm^{-2} on the surface of newly caught healthy fish and $10^3–10^8$ bacteria ml^{-1} of intestinal fluid. One may therefore speculate on the size of the bacterial population associated with a shoal of fish containing several thousand individuals each of surface area approximately $10^3 cm^2$. Evidence suggests that the species composition of the fish microflora is influenced less by the type of fish than it is by external environmental factors (Shewan, 1962) or season (Shewan, 1966). As in the seawater itself, Gram-negative bacteria of the genera *Pseudomonas*, *Achromobacter*, *Acinetobacter*, *Flavobacterium* and *Cytophaga* predominated on marine fish, together with some coryneforms and micrococci (Shewan, 1966). Anaerobic bacteria are not normally present on the fish surface but do occur in the intestines (Shewan, 1962).

5. *Microbial pathogens*

Anti-microbial defence mechanisms of vertebrates are well developed and include primary barriers (skin, mucous membranes and egg shells), phagocytic cells and the sophisticated specific immune response. The use of anti-microbial chemicals in support of the natural processes is widely practised. Antibiotics, in particular, have undoubtedly contributed significantly in the combat of diseases in man and domestic animals. There are, however, some disadvantages in chemotherapy. For example, the killing of pathogens early in the course of infection may preclude the establishment of lasting immunity. There is also the problem of development of microbial resistance to drugs and antibiotics and the possibilities for transfer of the genetic determinants of resistance to previously sensitive pathogens within the same or different animals. The latter suggestion has been considered to be important in relation to dietary supplementation with low levels of antibiotics to promote growth of animals in some methods of intensive rearing. Although the subject is still somewhat controversial, Jukes (1973) considered that concern for the spread and transfer of antibiotic resistance in this way was not substantiated.

D. MICROORGANISMS AND INVERTEBRATES

In many respects the invertebrates have closer relationships with microorganisms than do most vertebrates, with the possible exception of ruminants. Invertebrates, ranging from protozoa to insects, utilize microorganisms directly as food, whilst symbiotic associations of microorganisms and invertebrates are also well established. Furthermore, invertebrates are important agents in the dispersal of some microbial pathogens of plants and animals.

1. *Microbial dispersal by invertebrates*

Apart from the chance transportation of microorganisms which are picked up externally or by ingestion, invertebrates, especially insects, may also act as specific microbial vectors. The invertebrate-dispersed microorganisms include arthropod-borne viruses, Rickettsiae, fungi and protozoa pathogenic to plants or animals.

2. *Nutritional relations of microorganisms and invertebrates*

Microbial breakdown of complex substrates provides nutrients for invertebrate inhabitants of the soil such as nematodes and insect larvae. There are, however, even closer microbe–invertebrate relationships and several symbiotic associations have been described.

The wood-boring ambrosia beetles carry fungal spores which are deposited on the walls of tunnels bored into the wood. The beetles, being unable to utilize the cellulose and lignin components of wood directly, feed off the fungal growth which develops on tunnel walls. The relationship of ambrosia beetles to their associated specific fungi was discussed by Baker (1963). Some species of ants and termites also rely on fungi, even to the extent of cultivating them in subterranean "fungus gardens" off which the larvae feed. The wood-boring termites and cockroaches are also incapable of digesting wood constituents and rely on flagellate protozoa, vast numbers of which are housed in specialized sack-like swellings of the hind gut. In some cases these protozoa have intracellular bacteria which may provide the cellulase activity. In other cases the wood-eating insects rely solely on cellulolytic intestinal bacteria.

The endosymbiotic microorganisms of protozoa include bacteria, fungi, algae and even protozoa. Protozoa and several genera of the larger invertebrates, especially the aquatic coelenterates (e.g. jellyfish and sea anemones), provide most examples of algal endosymbionts (Droop, 1963). Other invertebrate types having endosymbiotic algae include porifera (sponges), platyhelminth worms and molluscs. The symbiont algae usually live inside cells associated with food digestion or transport. The hosts benefit from the photosynthetically derived organic compounds and oxygen, whilst the intracellular environment is constant and nutrient-rich for the algae. Taylor (1973) reviewed the properties of algal symbionts and the physiological aspects of their relationships with the invertebrate hosts.

Microbial endosymbionts are widely found in insects (Brooks, 1963), especially in developmental stages where the diet may lack factors required for normal growth and development. The microorganisms, usually bacteria or yeasts, live in specialized insect cells, mycetocytes, and furnish vitamins and amino acids upon which the hosts are highly dependent. Experimental removal of the microorganisms from some insects has caused retarded growth and development. Furthermore, the insects have developed mechanisms for ensuring continuity of the symbiosis from one generation to the next.

3. Microbial pathogens

A particularly important group of microbial pathogens of insects are varieties of the bacterium *Bacillus thuringiensis* which affect lepidopterous larvae. Toxicity is attributed to protein crystals formed in the bacterial cells concomitant with sporulation and also to a protein in the spores (Somerville and Pockett, 1975). Other important insect pathogenic bacilli include *B. popillae* and *B. lentimorbus* which cause "milky" diseases of some white grubs. The range of microbial pathogens of invertebrates is extended by including the protozoa which attack some insects, molluscs and helminths and also the phycomycete fungi which parasitize aquatic invertebrates. An extensive examination of the microbiological control of insects and mites with bacteria, viruses, fungi, protozoa and nematodes has been compiled by Burges and Hussey (1971).

In preparing this brief review to consider animals as microbial environments, general reference has been made to the following authorities: Brock (1966, 1970); Linton (1971a, b); Madelin *et al.* (1971) and Stanier *et al.* (1971).

References

ALEXANDER, M. (1965). Biodegradation: problems of molecular recalcitrance and microbial fallibility. *Adv. appl. Microbiol.* **7**, 35–80.

ALEXANDER, M. (1971). Biochemical ecology of microorganisms. *A. Rev. Microbiol.* **25**, 361–392.

ALLISON, F. E. (1968). Soil aggregation – some facts and fallacies as seen by a microbiologist. *Soil Sci.* **106**, 136–143.

ALLISON, F. E. and KLEIN, C. J. (1961). Comparative rates of decomposition in soil of wood and bark particles of several soft wood species. *Proc. Soil Sci. Soc. Am.* **25**, 193–196.

ALLISON, F. E. and MURPHY, R. M. (1962). Comparative rates of decomposition in soil of wood and bark particles of several hardwood species. *Proc. Soil Sci. Soc. Am.* **26**, 463–466.

ALLISON, F. E. and MURPHY, R. M. (1963). Comparative rates of decomposition in soil of wood and bark particles of several species of pines. *Proc. Soil Sci. Soc. Am.* **27**, 309–312.

ALTMAN, J. (1963). Increase in nitrogenous compounds in soil following soil fumigation. *Phytopathology* **53**, 870.

ANAGNOSTIDIS, K. and OVERBECK, J. (1966). Methanoxydierer und hypolimnische Schwefel-bakterien. Studien zur okologischen Biocoenotik der Gewarrermikroorganismen. *Ber. dt. bot. Ges.* **79**, 163–174.

ANDERSON, I. D. and COX, C. S. (1967). Microbial survival. *In* "Airborne Microbes", 17th Symp. Soc. gen. Microbiol. (P. H. Gregory and J. L. Monteith, Eds), pp. 203–226. Cambridge University Press, Cambridge.

ANDERSON, J. P. E. and DOMSCH, K. H. (1975). Measurement of bacterial and fungal contributions to respiration of selected agricultural and forest soils. *Can. J. Microbiol.* **21**, 314–322.

ANDERSON, J. R. and SLINGER, J. M. (1975). Europium chelate and fluorescent brightener staining of soil propagules and their photomicrographic counting. II. Efficiency. *Soil Biol. Biochem.* **7** 211–215.

ANDREWS, P. and WILLIAMS, P. J. LeB. (1971). Heterotrophic utilization of dissolved organic compounds in the sea. III. Measurement of the oxidation rates and concentrations of glucose and amino acids in sea water. *J. mar. biol. Ass. U.K.* **51**, 111–125.

BAKER, J. M. (1963). Ambrosia beetles and their fungi, with particular reference to *Platypus cylindrus* Fab. *In* "Symbiotic Associations", 13th Symp. Soc. gen. Microbiol. (P. S. Nutman and Barbara Mosse, Eds), pp. 232–265. Cambridge University Press, Cambridge.

BARBER, D. A. and GUNN, K. B. (1974). The effect of mechanical forces on the exudation of organic substances by the roots of cereal plants grown under sterile conditions. *New Phytol.* **73**, 39–45.

BARTHOLOMEW, W. V. (1965). Mineralization and immobilization of nitrogen in the decomposition of plant and animal residues. *In* "Soil Nitrogen" (W. V. Bartholomew and F. E. Clark, Eds), pp. 285–306. American Soc. Agronomy, Madison.

BASARABA, J. (1966). Effects of vegetable tannins on glucose oxidation by various micro-organisms. *Can. J. Microbiol.* **12**, 787–794.

BELL, W. and MITCHELL, R. (1972). Chemotactic and growth responses of marine bacteria to algal extracellular products. *Biol. Bull. mar. biol. Lab., Woods Hole*, **143**, 265–277.

BIRCHFIELD, H. P. and STORRS, E. E. (1950). Chemical structures and dissociation constants of amino acids, peptides and proteins in relation to their reaction rates with 2,4-dichloro-6-(*o*-chloroanilino)-*s*-triazine. *Contrib. Boyce Thompson Inst.* **18**, 395–418.

BLEAK, A. T. (1970). Disappearance of plant material under a winter snow cover. *Ecology* **51**, 915–917.

BLUM, J. L. (1960). Algal populations in flowing waters. *Spec. publs. Pymatuning Lab. Fld. Biol.* **2**, 11–21.

BOCOCK, K. L. (1964). Changes in the amounts of dry matter, nitrogen, carbon and energy in decomposing woodland leaf litter in relation to the activities of the soil fauna. *J. Ecol.* **52**, 273–284.

BOCOCK, K. L., GILBERT, O. J. W., CAPSTICK, C. K., TWINN, D. C., WAID, J. S. and WOODMAN, M. J. (1960). Changes in leaf litter when placed on the surface of soils with contrasting humus types. I. Losses in dry weight of oak and ash leaf litter. *J. Soil Sci.* **11**, 1–9.

BOWEN, G. D. and ROVIRA, A. D. (1973). Are modelling approaches useful in rhizosphere biology? *In* "Modern Methods in the Study of Microbial Ecology" (T. Rosswall, Ed.), *Bull. Ecol. Res. Comm.* (Stockholm) **17**, 443–450.

BRAY, J. R. and GORHAM, E. (1964). Litter production in forests of the world. *Adv. ecol. Res.* **2**, 101–157.

BROCK, T. D. (1966). "Principles of Microbial Ecology", Prentice-Hall Inc, Englewood Cliffs.

BROCK, T. D. (1969). Microbial growth under extreme conditions. *In* "Microbial Growth", 19th Symp. Soc. gen. Microbiol. (Pauline Meadow and S. J. Pirt, Eds), pp. 15–41. Cambridge University Press, Cambridge.

BROCK, T. D. (1970). "Biology of Microorganisms". Prentice-Hall Inc, Englewood Cliffs.

BROOKS, M. A. (1963). Symbiosis and aposymbiosis in arthropods. *In* "Symbiotic Associations", 13th Symp. Soc. gen. Microbiol. (P. S. Nutman and Barbara Mosse, Eds), pp. 200–231. Cambridge University Press, Cambridge.

BROWN, M. (1975). Rhizosphere micro-organisms – opportunists, bandits or bene-factors. *In* "Soil Microbiology" (N. Walker, Ed.), pp. 21–38. Butterworths, London.

BROWN, R. M., LARSON, D. A. and BOLD, H. C. (1964). Airborne algae: their abundance and heterogeneity. *Science, N.Y.* **143**, 583–585.

BRUCE, J. and MORRIS, E. O. (1973). Psychrophilic yeasts isolated from marine fish. *Antonie van Leeuwenhoek* **39**, 331–339.

BULL, A. T. (1970). Inhibition of polysaccharases by melanin; enzyme inhibition in relation to mycolysis. *Arch. Biochem. Biophys.* **137**, 345–356.

BUNT, J. (1971). Microbial productivity in polar regions. *In* "Microbes and Biological Productivity", 21st Symp. Soc. gen. Microbiol. (D. E. Hughes and A. H. Rose, Eds), pp. 333–354. Cambridge University Press, Cambridge.

BURGES, A. (1958). "Micro-organisms in the Soil". Hutchinson, London.

BURGES, H. D. and HUSSEY, N. W. (Eds) (1971). "Microbial Control of Insects and Mites". Academic Press, London.

BURNS, N. M. and ROSS, C. (1972). "Project Hypo", Canada Centre for Inland Waters Paper No. 6 United States Environmental Protection Agency Tech. Rept. TS-05-71-208-24.

CAMPBELL, C. A., PAUL, E. A., RENNIE, D. A. and McCALLUM, K. J. (1967). Factors affecting the accuracy of the carbon dating method in soil humus studies. *Soil Sci.* **104**, 81–85.

CANTER, H. M. and LUND, J. W. G. (1953). Studies on plankton parasites. II. The parasitism of diatoms with special reference to lakes in the English Lake District. *Trans. Br. mycol. Soc.* **36**, 13–37.

CANTER, H. M. and LUND, J. W. G. (1968). The importance of protozoa in control-ling the abundance of planktonic algae in lakes. *Proc. Linn. Soc. Lond.* **179**, 203–219.

CAPPENBERG, T. E. (1974a). Interrelations between sulfate-reducing and methane producing bacteria in bottom deposits of a freshwater lake. I. Field observations. *Antonie van Leeuwenhoek* **40**, 285–295.

CAPPENBERG, T. E. (1974b). II. Inhibition experiments. *Antonie van Leeuwenhoek* **40**, 297–306.

CAPPENBERG, T. E. and PRINS, R. A. (1974). III. Experiments with ^{14}C-labelled substrates. *Antonie van Leeuwenhoek* **40**, 457–469.

CARLISLE, A. A., BROWN, H. F. and WHITE, E. J. (1966). The organic matter and nutrient elements in the precipitation beneath a sessile oak (*Quercus petraea*) canopy. *J. Ecol.* **54**, 87–98.

CLARK, F. E. and PAUL, E. A. (1970). The microflora of grassland. *Adv. Agron.* **22**, 375–435.

CLARKE, R. T. J. (1964). Ciliates of the rumen of domestic cattle (*Bos taurus* L.). *N.Z. Jl. agric. Res.* **7**, 248–257.

COLEMAN, G. S. (1963). The growth and metabolism of rumen ciliate protozoa. *In* "Symbiotic Associations", 13th Symp. Soc. gen. Microbiol. (P. S. Nutman and Barbara Mosse, Eds), pp. 298–324. Cambridge University Press, Cambridge.

COLLINS, V. G. (1960). The distribution and ecology of Gram-negative organisms other than Enterobacteriaceae in lakes. *J. appl. Bact.* **23**, 510–514.

COLLINS, V. G. and WILLOUGHBY, L. G. (1962). The distribution of bacteria and fungal spores in Blelham Tarn with particular reference to an experimental over-turn. *Arch. Mikrobiol.* **43**, 294–307.

COLWELL, R. R. (1972). Bacteria, yeasts, viruses and related microorganisms of the Chesapeake Bay. *Chesapeake Sci.* **13**, Suppl., 569–570.

COOK, R. J. and PAPENDICK, R. I. (1970). Effect of soil water on microbial antagonism and nutrient availability in relation to soil-borne fungal diseases of plants. *In* "Root Diseases and Soil-borne Pathogens" (T. A. Toussoun, R. V. Bega and P. E. Nelson, Eds), pp. 81–88. University of California Press, Berkely.

COOKE, W. B. (1961). Pollution effects on the fungus population of a stream. *Ecology* **42**, 1–18.

CRAWFORD, C. C., HOBBIE, J. E. and WEBB, K. L. (1973). Utilization of dissolved organic compounds by microorganisms in an estuary. *In* "Estuarine Microbial Ecology" (L. H. Stevenson and R. R. Colwell, Eds), pp. 169–180. University of South Carolina Press, Columbia.

CRAWFORD, C. C., HOBBIE, J. E. and WEBB, K. L. (1974). Utilization of dissolved free amino acids by estuarine micro-organisms. *Ecology* **55**, 551–563.

CROSBY, D. G. (1970). The non-biological degradation of pesticides in soil. *In* "Pesticides in the Soil: Ecology, Degradation and Movement" (G. E. Guyer, Ed.), pp. 86–94. Michigan State University, East Lansing.

CURTIS, E. J. C. (1969). Sewage fungus: its nature and effects. *Wat. Res.* **3**, 289–311.

CURTIS, E. J. C. and HARRINGTON, D. W. (1970). Effects of organic wastes on rivers. *Process Biochemistry* **5**, 44–46.

DAFT, M. J. and STEWART, W. D. P. (1971). Bacterial pathogens of freshwater blue–green algae. *New Phytol.* **70**, 819–829.

DALE, N. G. (1974). Bacteria in intertidal sediments: factors related to their dis-tribution. *Limnol. Oceanogr.* **19**, 509–518.

DAVEY, C. B. and PAPAVIZAS, G. C. (1963). Saprophytic activity of *Rhizoctonia* as affected by the carbon–nitrogen balance of certain organic soil amendments. *Proc. Soil Sci. Soc. Am.* **27**, 164–167.

DICKINSON, C. H. (1965). The mycoflora associated with *Halimione portulacoides*. III. Fungi on green and moribund leaves. *Trans. Br. mycol. Soc.* **48**, 603–610.

DRECHSLER, N. and OTTO, S. (1968). Über den Abbau von Dazomet im Boden. *Residue Rev.* **23**, 49–54.

DROOP, M. R. (1963). Algae and invertebrates in symbiosis. *In* "Symbiotic Associations", 13th Symp. Soc. gen. Microbiol. (P. S. Nutman and Barbara Mosse, Eds), pp. 171–199. Cambridge University Press, Cambridge.

DUGAN, P. R., MACMILLAN, C. B. and PFISTER, R. M. (1970). Aerobic heterotrophic bacteria indigenous to a pH 2·8 acid mine water: predominant slime-producing bacteria in acid streamers. *J. Bact.* **101**, 982–988.

EADIE, J. M. (1962). Inter-relationships between certain rumen ciliate protozoa. *J. gen. Microbiol.* **29**, 579–588.

EADIE, J. M. (1967). Studies on the ecology of certain rumen ciliate protozoa. *J. gen. Microbiol.* **49**, 175–194.

EHLERS, W. W., FARMER, W. J., SPENCER, W. F. and LETEY, J. (1969). Lindane diffusion in soils: II. Water content, bulk density and temperature effects. *Proc. Soil Sci. Soc. Am.* **33**, 505–508.

EHRESMANN, D. W. and HATCH, M. T. (1975). Effect of relative humidity on the survival of airborne unicellular algae. *Appl. Microbiol.* **29**, 352–357.

FORBES, R. S. (1974). Decomposition of agricultural crop debris. *In* "Biology of Plant Litter Decomposition" (C. H. Dickinson and G. J. F. Pugh, Eds), vol. 2, pp. 723–742. Academic Press, London.

FOSTER, R. C. and ROVIRA, A. D. (1973). The rhizosphere of wheat roots studied by electron microscopy of ultra-thin sections. *In* "Modern Methods in the Study of Microbial Ecology", (T. Rosswall, Ed.) *Bull. Ecol. Res. Comm.* (Stockholm) **17**, 92–102.

GILBERT, O. J. W. and BOCOCK, K. L. (1960). Changes in leaf litter when placed on the surface of soils with contrasting humus types. 2. Changes in the nitrogen content of oak and ash leaf litter. *J. Soil Sci.* **11**, 10–19.

GORLENKO, V. M. and LEBEDEVA, E. V. (1971). New green sulphur bacteria with appendages. *Microbiology (Mikrobiologiya)* **40**, 900–903.

GOULDER, R. (1974). The seasonal and spatial distribution of some benthic ciliated protozoa in Esthwaite Water. *Freshwat. Biol.* **4**, 127–147.

GRAY, T. R. G. (1970). Microbial growth in soil. *In* "Pesticides in the Soil: Ecology, Degradation and Movement" (G. E. Guyer, Ed.), pp. 36–41. Michigan State University, East Lansing.

GRAY, T. R. G. and WILLIAMS, S. T. (1971). Microbial productivity in soil. *In* "Microbes and Biological Productivity", 21st Symp. Soc. gen. Microbiol. (D. E. Hughes and A. H. Rose, Eds), pp. 255–286. Cambridge University Press, Cambridge.

GRAY, T. R. G., HISSETT, R. and DUXBURY, T. (1974). Bacterial populations of litter and soil in a deciduous woodland. II. Numbers, biomass and growth rates. *Rev. Ecol. Biol. Sol.* **11**, 15–26.

GREAVES, M. P. and DARBYSHIRE, J. F. (1972). The ultrastructure of the mucilaginous layer on plant roots. *Soil Biol. Biochem.* **4**, 443–449.

GREENWOOD, D. J. (1970). Distribution of carbon dioxide in the aqueous phase of aerobic soils. *J. Soil Sci.* **21**, 314–329.

GREGORY, P. H. (1973). "The Microbiology of the Atmosphere", 2nd edition. Leonard Hill (Intertext), Aylesbury.

GREGORY, P. H., HAMILTON, E. D. and SREERAMULU, T. (1955). Occurrence of the alga *Gloeocapsa* in the air. *Nature, Lond.* **176**, 1270.

GRIFFIN, D. M. (1968). A theoretical study relating the concentration and diffusion of oxygen to the biology of organisms in soil. *New Phytol.* **67**, 561–567.

GRIFFIN, D. M. (1972). "Ecology of soil fungi". Chapman and Hall, London.

GRIFFIN, G. J., HALE, M. G. and SHAY, F. J. (1976). Nature and quantity of sloughed organic matter produced by roots of axenic peanut plants. *Soil Biol. Biochem.* **8**, 29–32.

GUENZI, W. D. and BEARD, W. E. (1968). Anaerobic conversion of DDT to DDD and aerobic stability of DDT in soil. *Proc. Soil Sci. Soc. Am.* **32**, 522–524.

HANSEN, T. A. and VELDKAMP, H. (1972). A new type of photosynthetic purple bacterium. *Antonie van Leeuwenhoek* **38**, 629.

HARMSEN, G. W. and JAGER, G. (1963). Determination of the quantity of carbon and nitrogen in the rhizosphere of young plants. *In* "Soil Organisms" (J. Doeksen and J. van der Drift, Eds), pp. 245–251. North-Holland Publishing Co., Amsterdam.

HEATH, G. W. and ARNOLD, M. K. (1966). Studies in leaf-litter breakdown. II. Breakdown rate of "sun" and "shade" leaves. *Pedobiologia* **6**, 238–243.

HELLING, C. S., KEARNEY, P. C. and ALEXANDER, M. (1971). Behaviour of pesticides in soil. *Adv. Agron.* **23**, 147–240.

HENDRICKS, C. W. and MORRISON, S. M. (1967). Multiplication and growth of selected enteric bacteria in clear mountain stream water. *Wat. Res.* **1**, 567–576.

HILL, I. R. (1967). Application of the fluorescent antibody techniques to an ecological study of bacilli in soil. Ph.D. thesis, University of Liverpool.

HILL, I. R. and KENWORTHY, R. (1970). Microbiology of pigs and their environment in relation to weaning. *J. appl. Bact.* **33**, 299–316.

HISSETT, R. and GRAY, T. R. G. (1976). Microsites and time changes in soil microbe ecology. *In* "The Role of Aquatic and Terrestrial Organisms in Decomposition Processes" (J. Anderson and A. Macfadyen, Eds), pp. 23–39. Blackwell Scientific Publications, Oxford.

HOBBIE, J. E. and CRAWFORD, C. C. (1969). Respiration corrections for bacterial uptake of dissolved organic compounds in natural waters. *Limnol. Oceanogr.* **14**, 528–533.

HOBBIE, J. E., HOLM-HANSEN, O., PACKARD, T. T., POMEROY, L. R., SHELDON, R. W., THOMAS, J. P. and WIEBE, W. J. (1972). A study of the distribution and activity of micro-organisms in ocean water. *Limnol. Oceanogr.* **17**, 544–555.

HOLM-HANSEN, O. (1972). The distribution and chemical composition of particulate material in marine and freshwaters. *Mem. Ist. Ital. Idrobiol.* **29** Suppl., 37–51.

HORVATH, R. S. and ALEXANDER, M. (1970). Cometabolism: a technique for the accumulation of biochemical products. *Can. J. Microbiol.* **16**, 1131–1132.

HUNGATE, R. E. (1963). Symbiotic associations: the rumen bacteria. *In* "Symbiotic Associations", 13th Symp. Soc. gen. Microbiol. (P. S. Nutman and Barbara Mosse, Eds), pp. 266–297. Cambridge University Press, Cambridge.

HUTCHINSON, G. E. (1957). "A Treatise on Limnology, 1. Geography, Physics and Chemistry". John Wiley and Sons, New York.

HUTCHINSON, G. E. (1966). "A Treatise on Limnology, 2. Introduction to Lake Biology and the Limnoplankton". John Wiley and Sons, New York.

HYNES, H. B. N. (1960). "The Biology of Polluted Waters". Liverpool University Press, Liverpool.

HYNES, H. B. N. (1970). "The Ecology of Running Waters". Liverpool University Press, Liverpool.

JANNASCH, H. W. (1968). Growth characteristics of heterotrophic bacteria in sea water. *J. Bact.* **95**, 722–723.

JANNASCH, H. W. and WIRSEN, C. O. (1973). Deep-sea micro-organisms: *in situ* response to nutrient enrichment. *Science, N.Y.* **180**, 1041–1043.

JENKINSON, D. S. (1966). Studies on the decomposition of plant materials in soil. II. Partial sterilization and the soil biomass. *J. Soil Sci.* **17**, 280–302.

JONES, J. G. (1971). Studies on freshwater bacteria: factors which influence the population and its activity. *J. Ecol.* **59**, 593–613.

JONES, J. G. (1972a). Studies on freshwater bacteria: association with algae and alkaline phosphatase activity. *J. Ecol.* **60**, 59–75.

JONES, J. G. (1972b). Studies on freshwater micro-organisms: phosphatase activity in lakes of differing degrees of eutrophication. *J. Ecol.* **60**, 777–791.

JONES, J. G. (1973). Studies on freshwater bacteria: the effect of enclosure in large experimental tubes. *J. appl. Bact.* **36**, 445–456.

JONES, J. G. (1975). Heterotrophic micro-organisms and their activity. *In* "River Ecology" (B. A. Whitton, Ed.), pp. 141–154. Blackwells Scientific Publications, Oxford.

JONES, K. (1974). Nitrogen fixation in a salt marsh. *J. Ecol.* **62**, 553–565.

JUKES, T. H. (1973). Public health significance of feeding low levels of antibiotics to animals. *Adv. appl. Microbiol.* **16**, 1–30.

KAUSHIK, N. K. and HYNES, H. B. N. (1968). Experimental study on the role of autumn-shed leaves in aquatic environments. *J. Ecol.* **56**, 229–243.

KENWORTHY, R. (1973). Intestinal microflora of the pig. *Adv. appl. Microbiol.* **16**, 31–54.

KERLING, L. C. P. (1958). De microflora op het blad van *Beta vulgaris* L. *Tijdschr. Plzieckt.* **64**, 402–410.

KOWALCZEWSKI, A. and LACK, T. J. (1971). Primary production and respiration of the phytoplankton of the Rivers Thames and Kennet at Reading. *Freshwat. Biol.* **1**, 197–212.

KUZNETSOV, S. I. (1959). "Die Rolle der Mikroorganismen im Stoffkreislauf der Seen", VEB Deutscher Verlag der Wissenschaften, Berlin.

LAST, F. T. (1955). Seasonal incidence of *Sporobolomyces* on cereal leaves. *Trans. Br. mycol. Soc.* **38**, 221–239.

LATTER, P. M. and CRAGG, J. D. (1967). The decomposition of *Juncus squarrosus* leaves and microbiological changes in the profile of *Juncus* moor. *J. Ecol.* **55**, 465–482.

LINTON, A. H. (1971a). Micro-organisms and vertebrates: their normal and pathological relationships. *In* "Micro-organisms: Function, Form and Environment" (Lilian E. Hawker and A. H. Linton, Eds), pp. 538–557. Edward Arnold, London.

LINTON, A. H. (1971b). Micro-organisms and vertebrates: host resistance and immunity. *In* "Micro-organisms: Function, Form and Environment" (Lilian E. Hawker and A. H. Linton, Eds), pp. 558–582. Edward Arnold, London.

LOVELL, R. (1957). The biological influences of man and animals on microbial ecology. *In* "Microbial Ecology", 7th Symp. Soc. gen. Microbiol. (R. E. O. Williams and C. C. Spicer, Eds), pp. 315–327. Cambridge University Press, Cambridge.

LOWE, W. E. and GRAY, T. R. G. (1973). Ecological studies on coccoid bacteria in a pine forest soil. II. Growth of bacteria introduced into soil. *Soil Biol. Biochem.* **5**, 449–462.

MACFADYEN, A. (1970). Soil metabolism in relation to ecosystem energy flow and to primary and secondary production. *In* "Methods for the Study of Production and Energy Flow in Soil Communities" (J. Phillipson, Ed.), pp. 167–172. UNESCO, Paris.

MADELIN, M. F. and LINTON, A. H. (1971). Microbiology of air. *In* "Microorganisms: Function, Form and Environment" (Lilian E. Hawker and A. H. Linton, Eds), pp. 529–537. Edward Arnold, London.

MADELIN, M. F., GREENHAM, L. W. and LINTON, A. H. (1971). Micro-organisms and invertebrate animals. *In* "Micro-organisms: Function, Form and Environment" (Lilian E. Hawker and A. H. Linton, Eds), pp. 597–614. Edward Arnold, London.

MADSEN, B. L. (1972). Detritus on stones in small streams. *Mem. Ist. Ital. Idrobiol.* **29**, Suppl., 385–403.

MANN, K. H. (1969). The dynamics of aquatic ecosystems. *In* "Advances in Ecological Research" (J. B. Cragg, Ed.), vol. 6, pp. 1–81. Academic Press, London.

MARTIN, J. K. (1975). ^{14}C-Labelled material leached from the rhizosphere of plants supplied continuously with $^{14}CO_2$. *Soil Biol. Biochem.* **7**, 395–399.

MAYAUDON, J. and SIMONART, P. (1958). Study of the decomposition of organic matter in soil by means of radioactive carbon. II. The decomposition of radioactive glucose in soil and distribution of radioactivity in the humus fractions of soil. *Pl. Soil* **9**, 376–380.

MEADOWS, P. S. and ANDERSON, J. G. (1968). Micro-organisms attached to marine sand grains. *J. mar. biol. Ass. U.K.* **48**, 161–177.

MESHKOV, M. V. (1953) cited by KRASILNIKOV, N. A. (1961). *In* "Soil Microorganisms and Higher Plants". Israel Program for Scientific Translations, Jerusalem.

MITCHELL, R. (1971). Role of predators in the reversal of imbalances in microbial ecosystems. *Nature, Lond.* **230**, 257–258.

MOMMAERTS, J. P. (1969). On the distribution of major nutrients and phytoplankton in the Tamar estuary. *J. mar. biol. Ass. U.K.* **49**, 749–765.

MORTIMER, C. H. (1941). The exchange of dissolved substances between mud and water in lakes. *J. Ecol.* **29**, 280–329.

MORTIMER, C. H. (1942). The exchange of dissolved substances between mud and water in lakes. Parts III and IV. *J. Ecol.* **30**, 147–201.

MOSSE, B. (1975). A microbiologist's view of root anatomy. In "Soil Microbiology" (N. Walker, Ed.), pp. 39–66. Butterworth, London.

MUNRO, A. L. S. and BROCK, T. D. (1968). Distinction between bacterial and algal utilization of soluble substances in the sea. *J. gen. Microbiol.* **51**, 35–42.

NEWLAND, L. W., CHESTERS, G. and LEE, G. B. (1969). Degradation of γ-BHC in simulated lake impoundments as affected by aeration. *J. Wat. Pollut. Control Fed.* **41**, R174–R184.

NYE, P. H. (1968). Processes in the root environment. *J. Soil Sci.* **19**, 205–215.

NYKVIST, N. (1959). Leaching and decomposition of litter. I. Experiments on leaf litter of *Franxinus excelsior*. *Oikos* **10**, 190–211.

ODUM, E. P. (1971). "Fundamentals of Ecology", 3rd edition. W. B. Saunders Co., Philadelphia.

OKAFOR, N. (1966). Ecology of micro-organisms on chitin buried in soil. *J. gen. Microbiol.* **44**, 311–327.

OPPENHEIMER, C. H. and VANCE, M. H. (1960). Attachment of bacteria to the surfaces of living and dead micro-organisms in marine sediments. *Z. allg. Mikrobiol.* **1**, 47–52.

OVERBECK, J. (1974). Microbiology and biochemistry. *Mitt. int. Verein. theor. angew. Limnol.* **20**, 198–228.

OVINGTON, J. D. (1962). Quantitative ecology and the woodland ecosystem concept. *Adv. ecol. Res.* **1**, 103–197.

PAUL, E. A., CAMPBELL, C. A., RENNIE, D. A. and McCALLUM, K. J. (1964). Investigations of the dynamics of soil humus utilizing carbon dating techniques. *Trans. 8th int. Congr. Soil Sci.*, Bucharest, **3**, 201–208.

PETERSON, R. C. and CUMMINS, K. W. (1974). Leaf processing in a woodland stream. *Freshwat. Biol.* **4**, 343–368.

PLIMMER, J. R., KEARNEY, P. C. and CHISAKA, H. (1970). Abstract. *Weed Sci. Soc. Am. Abstr.* p. 167.

POWLSON, D. S. (1975). Effects of biocidal treatments on soil organisms. In "Soil Microbiology" (N. Walker, Ed.), pp. 193–224. Butterworths, London.

PUTNAM, H. D. and SCHMIDT, E. L. (1959). Studies on the free amino acid fraction of soils. *Soil Sci.* **87**, 22–27.

RAMSAY, A. J. (1974). The use of autoradiography to determine the proportion of bacteria metabolising in an aquatic habitat. *J. gen. Microbiol.* **80**, 363–373.

RANEY, W. A. (1965). Physical factors of the soil as they affect micro-organisms. In "Ecology of Soil-borne Plant Pathogens" (K. F. Baker and W. C. Snyder, Eds), pp. 115–118. University of California Press, Berkely.

REIM, R. L., SHANE, M. S. and CANNON, R. E. (1974). The characterization of a *Bacillus* capable of blue–green bactericidal activity. *Can. J. Microbiol.* **20**, 981–986.

RIXON, A. J. and BRIDGE, B. J. (1968). Respiration quotient arising from microbial activity in relation to matric suction and air filled pore space of soil. *Nature, Lond.* **218**, 961–962.

ROVIRA, A. D. (1969). Plant root exudates. *Bot. Rev.* **35**, 35–57.

ROVIRA, A. D. and BOWEN, G. D. (1966). The effects of microorganisms upon plant growth. II. Detoxification of heat sterilised soils by fungi and bacteria. *Pl. Soil* **25**, 129–142.

ROVIRA, A. D. and RIDGE, E. H. (1973). Exudation of ^{14}C-labelled compounds from wheat roots: influence of nutrients, micro-organisms and added organic compounds. *New Phytol.* **72**, 1081–1087.

RUINEN, J. (1961). The phyllosphere. I. An ecologically neglected milieu. *Pl. Soil* **15**, 81–109.

RUSSELL, E. W. (1968). The agricultural environment of soil bacteria. *In* "The Ecology of Soil Bacteria" (T. R. G. Gray and D. Parkinson, Eds), pp. 77–89. Liverpool University Press, Liverpool.

SCHINDLER, D. W., KLING, H., SCHMIDT, R. V., PROCOPOWICH, J., FROST, V. E., REID, R. A. and CAPEL, M. (1973). Eutrophication of lake 227 by addition of phosphate and nitrate: the second, third and fourth years of enrichment, 1970, 1971 and 1972. *J. Fish. Res. Bd Can.* **30**, 1415–1440.

SCHLICHTING, H. E. (1969). The importance of airborne algae and protozoa. *J. Air Pollut. Control Assoc.* **19**, 946–951.

SCHMIDT, E. L., PUTNAM, H. D. and PAUL, E. A. (1960). Behaviour of free amino acids in soil. *Proc. Soil Sci. Soc. Am.* **24**, 107–109.

SEKI, H. (1970). Microbial biomass on particulate organic matter in seawater of the euphotic zone. *Appl. Microbiol.* **19**, 960–962.

SHEWAN, J. M. (1962). The bacteriology of fresh and spoiling fish and some related chemical changes. *In* "Recent Advances in Food Science" (J. Hawthorn and J. Muil Leitch, Eds), pp. 167–193. Butterworths, London.

SHEWAN, J. M. (1966). Some factors affecting the bacterial flora of marine fish. *Suppl. to Medlemsblad. for Den Norske Veterinaerforening* No. 11, 2–12.

SHILO, M. (1970). Lysis of blue–green algae by a myxobacter. *J. Bact.* **104**, 453–461.

SIALA, A., HILL, I. R. and GRAY, T. R. G. (1974). Populations of spore-forming bacteria in an acid forest soil, with special reference to *Bacillus subtilis*. *J. gen. Microbiol.* **81**, 183–190.

SKINNER, F. A. (1975). Anaerobic bacteria and their activities in soil. *In* "Soil Microbiology" (N. Walker, Ed.), pp. 1–19. Butterworths, London.

SKINNER, F. A. and CARR, J. G., Eds (1974). "The Normal Microbial Flora of Man", Academic Press, London.

SMITH, H. WILLIAMS (1961). The development of the bacterial flora of the faeces of animals and man: the changes that occur during ageing. *J. appl. Bact.* **24**, 235–241.

SMITH, H. WILLIAMS and CRABB, W. E. (1961). The bacterial flora of the faeces of animals and man; its development in the young. *J. Path. Bact.* **82**, 53–66.

SMITH, K. L. (1973). Respiration of a sublittoral community. *Ecology* **54**, 1065–1075.

SOMERVILLE, H. J. and POCKETT, H. V. (1975). An insect toxin from spores of *Bacillus thuringiensis* and *Bacillus cereus*. *J. gen. Microbiol.* **87**, 359–369.

SOROKIN, J. I. (1964). On the primary production and bacterial activities in the Black Sea. *J. Cons. Perm. int. Explor. Mer.* **29**, 41–60.

SOROKIN, Yu. I. (1968). Primary production and microbiological processes in Lake Gek-Gel. *Microbiology (Mikrobiologiya)* **37**, 289–296.

SOWDEN, F. J. and IVARSON, K. C. (1962). Decomposition of forest litters. III. Changes in the carbohydrate constituents. *Pl. Soil* **16**, 389–400.

SPENCER, W. F. (1970). Distribution of pesticides between soil, water and air. *In* "Pesticides in the Soil: Ecology, Degradation and Movement" (G. E. Guyer, Ed.), pp. 120–128. Michigan State University, East Lansing.

STANIER, R. Y., DOUDOROFF, M. and ADELBERG, E. A. (1971). "General Microbiology", 3rd edition. Macmillan, London.

STEWART, W. D. P. (1965). Nitrogen turnover in marine and brackish habitats. I. Nitrogen fixation. *Ann. Bot.* **29**, 229–239.

STEWART, W. D. P., MAGUS, T., FITZGERALD, G. P. and BURRIS, R. H. (1971). Nitrogenase activity in Wisconsin lakes of differing degrees of eutrophication. *New Phytol.* **70**, 497–509.

STOTZKY, G. (1975). Microbial metabolism in soil. *Proc. 1st Intersect. Congress IAMS* **2**, 249–261.

STOTZKY, G. and REM, L. T. (1966). Influence of clay minerals on micro-organisms. I. Montmorillonite and kaolinite on bacteria. *Can. J. Microbiol.* **12**, 547–563.

STOTZKY, G. and REM, L. T. (1967). Influence of clay minerals on micro-organisms. II. Montmorillonite and kaolinite on fungi. *Can. J. Microbiol.* **13**, 1535–1550.

SWABY, R. J. and LADD, J. N. (1962). Chemical nature, microbial resistance, and origin of soil humus. *Proc. int. Soils Conf., New Zealand*, pp. 197–202. CSIRO, Adelaide.

TAKIJIMA, Y. (1964). Studies on organic acids in paddy field soils with reference to their inhibitory effects on the growth of rice plants. *Soil Sci. Plant Nutrn.*, Tokyo, **10**, 14–29.

TALLING, J. F. (1971). The underwater light climate as a controlling factor in the production ecology of freshwater phytoplankton. *Mitt. int. Verein. theor. angew. Limnol.* **19**, 214–243.

TAYLOR, D. L. (1973). Algal symbionts of invertebrates. *A. Rev. Microbiol.* **27**, 171–187.

TRAVLEEV, A. P. (1960). The role of forest litter in heat insulation. *Pochvovedenie* **10**, 92–95.

TRIBE, H. T. (1960). Aspects of cellulose decomposition in Canadian soils. II. Nitrate nitrogen levels and carbon dioxide evolution. *Can. J. Microbiol.* **6**, 317–323.

TUKEY, H. B., Jr., WITTWER, S. H. and TUKEY, H. B. (1957). Leaching of carbohydrates from plant foliage as related to light intensity. *Science, N.Y.* **126**, 120–121.

TURNER, M. and GRAY, T. R. G. (1962). Bacteria of a developing salt marsh. *Nature, Lond.* **194**, 559–560.

TUTTLE, J. H. and JANNASCH, H. W. (1973). Sulfide and thiosulfate – oxidising bacteria in anoxic marine basins. *Mar. Biol.* **20**, 64–70.

VOETS, J. P., MEERSCHMAN, P. and VERSTRAETE, W. (1974). Soil microbiological and biochemical effects of long term atrazine application. *Soil Biol. Biochem.* **6**, 149–152.

VOIGT, G. K. (1965). Nitrogen recovery from decomposing tree leaf tissue and forest humus. *Proc. Soil Sci. Soc. Am.* **29**, 756–759.

VRANY, J. (1965). Effect of foliar applications on the rhizosphere microflora. *In* "Plant Microbes Relationship" (J. Macura and V. Vančura, Eds), pp. 84–90. Czech Academy of Sciences, Prague.

VRANY, J. (1975). Biological changes in the rhizosphere of wheat after the foliar application of chlorocholinechloride, urea and 4-chloro-2-methylphenoxyacetic acid. *Fol. Microbiol.* **20**, 402–408.

VRANY, J., VANCURA, V. and MACURA, J. (1962). The effect of foliar applications of some readily metabolized substances, growth regulators and antibiotics on rhizosphere microflora. *Fol. Microbiol.* **7**, 61–70.

WAINWRIGHT, M. and PUGH, G. J. F. (1975). Changes in the free amino acid content of soil following treatment with fungicides. *Soil Biol. Biochem.* **7**, 1–4.

WANG, T. S. C., YANG, T. K. and CHUANG, T. T. (1967). Soil phenolic acids as plant growth inhibitors. *Soil Sci.* **103**, 239–246.

WARCUP, J. H. (1957). Studies on the occurrence and activity of fungi in a wheat field soil. *Trans. Br. mycol. Soc.* **40**, 237–262.

WEISS, L. (1963). The pH value at the surface of *Bacillus subtilis*. *J. gen. Microbiol.* **32**, 331–340.

WESTLAKE, D. F., CASEY, H., DAWSON, H., LADLE, M., MANN, R. H. K. and MARKER, A. F. H. (1972). The chalk stream ecosystem. *In* "Productivity Problems of Freshwaters" (Z. Kajak and A. Hillbricht-Ilbowska, Eds), pp. 615–635. Proc. IBP/UNESCO Symposium. Polish Scientific Publishers, Warsaw–Krakow.

WETZEL, R. G., RICH, P. R., MILLER, M. C. and ALLEN, H. L. (1972). Metabolism of dissolved and particulate detrital carbon in a temperate hard-water lake. *Mem. Ist. Ital. Idrobiol.* **29**, Suppl., 185–243.

WHITE, J. L. and MORTLAND, M. M. (1970). Pesticide retention by soil minerals. *In* "Pesticides in the Soil: Ecology, Degradation and Movement" (G. Guyer, Ed.), pp. 95–100. Michigan State University. East Lansing.

WHITTON, B. A. (1973). Interaction with other organisms. *In* "The Biology of Blue–Green Algae" (N. G. Carr and B. A. Whitton, Eds), pp. 415–443. Blackwell Scientific Publications, Oxford.

WIEBE, W. J. and POMEROY, L. R. (1972). Micro-organisms and their association with aggregates and detritus in the sea: a microscopic study. *Mem. Ist. Ital. Idrobiol.* **29**, Suppl., 325–352.

WILLIAMS, P. J. LeB. (1970). Heterotrophic utilization of dissolved organic compounds in the sea. I. Size distribution of population and relation between respiration and incorporation of growth substrates. *J. mar. biol. Ass. U.K.* **50**, 859–870.

WILLIAMS, S. T. and MAYFIELD, C. I. (1971). Studies on the ecology of actinomycetes in soil. III. The behaviour of neutrophilic streptomycetes in acid soils. *Soil Biol. Biochem.* **3**, 197–208.

WINOGRADSKY, S. (1949). "Microbiologie du Sol: Problèmes et Méthodes", Masson et Cie, Paris.

WOOLLETT, L. L. and HENDRICK, L. R. (1970). Ecology of yeasts in polluted water. *Antonie van Leeuwenhoek* **36**, 427–435.

WRIGHT, R. T. and HOBBIE, J. E. (1966). Use of glucose and acetate by bacteria and algae in aquatic ecosystems. *Ecology* **47**, 447–464.

YOUNG, D. L. K. and BARBER, R. T. (1973). Effects of waste dumping in New York Bight on the growth of natural populations of phytoplankton. *Environ. Pollut.* **5**, 237–252.

ZEIKUS, J. G. and BROCK, T. D. (1972). Effects of thermal additions from the Yellowstone geyser basins on the bacteriology of the Firehole River. *Ecology* **53**, 283–290.

ZoBell, C. E. (1963). Domain of the marine microbiologist. *In* "Symposium on Marine Microbiology" (C. H. Oppenheimer, Ed.), pp. 3–24. Charles C. Thomas, Springfield, Ill.

Chapter 3

The Behaviour and Fate of Pesticides in Microbial Environments

IAN R. HILL[1] and S. JOHN L. WRIGHT[2]

[1] *Imperial Chemical Industries, Plant Protection Division, Jealott's Hill Research Station, Bracknell, Berkshire, England*
[2] *School of Biological Sciences, University of Bath, Bath, England*

I. Introduction

Large populations of microorganisms living in the surface horizons of the soil help to produce and maintain a state of fertility which is required for production of the world's food supply. However, the soil and plants growing in it also support populations of both micro- and macroorganisms which are detrimental to agriculture. Consequently, over the past few decades, increasing and concerted attempts have been made to control these pests with an array of pesticides. Thus it is to the soil or soil-grown crops, that many of the pesticides are applied.

Other components of the total environment also directly receive pesticide applications. One such environment, often neglected in discussions of the interactions between microorganisms and pesticides, is that of the animal, yet both the outer surface and some inner parts (e.g. gastro-intestinal tract) of animals are populated by vast numbers of diverse microorganisms. Furthermore, animals may frequently be invaded by harmful or pathogenic microorganisms. Consequently numerous pesticides, drugs and antibiotics have been developed to combat diseases and are variously formulated for topical, oral, and injection application.

The remaining environments, i.e. bodies of water and the air, receive comparatively small amounts of directly applied pesticides. However, the application of pesticides (e.g. herbicides, algicides, insecticides, molluscicides and piscicides) to aquatic environments is practised to a greater or lesser degree in many countries for the purpose of clearing and sanitizing the water. Pests are not normally present in the atmosphere in large enough quantities at any given time and location to make intra-aerial pesticide application common, although for many years one such usage has been the destruction of swarming locusts.

Whilst pesticides are usually aimed at target organisms within specific components of the environment, the methods of application and the subsequent influences on these chemicals often result in a more widespread distribution. In most cases, entry of a pesticide into one environment, whether soil, air, water, plants or animals, can lead to the redistribution of the chemical and its products into other components of the environment. The fact that pesticides enter natural environments is fundamentally relevant to the theme of this book, since microorganisms inhabit most, if not all, of the situations in which pesticides or their products are likely to occur.

It is therefore appropriate in this chapter to outline how pesticides enter and are redistributed through the environments in which they may interact with, or be influenced by, a variety of physical, chemical and biological agencies. Whilst an appreciation of all factors which may influence pesticide fate in the environment is important in order to place microbe–pesticide interactions into perspective, it is not our purpose to give detailed consideration to these factors except in as much as they influence microbial activities and interactions with pesticides. Many aspects of the environmental fate and behaviour of pesticides have been well covered in reviews of the subject (Crosby and Li, 1969; Graham-Bryce, 1972; Hurtig, 1972; Osgerby, 1973; Edwards, 1974; Haque and Freed, 1974). Haque and Freed (1974) considered pesticide fate in terms of environmental chemodynamics and the dynamics of pesticides in the environment were discussed by Robinson (1973). In a fascinating and lucid account Mellanby (1967) put environmental pollution by pesticides into perspective and also looked towards the future use of pesticides. More recently, this same topic has received further comprehensive examination by Edwards (1973a).

Within the past decade there has been an ever increasing number of reports of the presence of pesticides and their degradation products in unexpected biotic and abiotic situations, albeit often in minute concentrations. This does not necessarily reflect any increase in the dispersal of pesticidal chemicals in the environment, in their ecological effects, or in the accumulation of their residues. Instead, it may be the result of both the continued quest of many ecologists and residue chemists for data concerning the overall environmental impact of pesticides and the improved sensitivity of analytical techniques. Residues are in fact frequently determined in the parts per million (10^{-6}) to parts per trillion (10^{-12}) range.

Many pesticide–environment interactions are widely acknowledged to be of ecological significance and hence merit investigation. Others, however, are of minor importance in that they occur infrequently, in highly localized situations or result in residues so low as to cast considerable doubt upon any environmental significance of the findings. Whilst some "interactions" of this latter type are also described within this chapter it is hoped that the discerning reader will not misinterpret the context in which they are presented or exaggerate their significance.

The introduction and redistribution of pesticides in environments is simply depicted in Fig. 1. This indicates that, regardless of the site or purpose of the application (or the accidental spillage etc.) of a pesticide, the chemical in its original or modified form may be transported within or between environment components, cycled, or even accumulated at various stages in food chains. It therefore seems unlikely that a pesticide would be lost from the biosphere before encountering microorganisms present.

Illustrated in Fig. 1 are the following possible factors relating to movement of pesticides, most of which will be dealt with in later sections of this chapter:

(a) A proportion of the pesticide sprays or dusts applied to soil and aquatic situations may be aerially transported, together with particle-adsorbed chemicals (on soil, spores, plant debris etc.), eventually to be deposited elsewhere.

(b) Conversely, pesticides intentionally applied to aerial environments will be partly deposited on the land or water.

(c) Foliar crop sprays almost always result in deposition of appreciable amounts of pesticide onto the soil by overspray and run-off from leaves and stems.

(d) Death of treated or accidentally contaminated plants and animals can also result in the incorporation of pesticide residues into soil or water.

(e) Pesticides may enter the atmosphere by volatilization from soil, plant and water surfaces.

(f) Soil run-off and wash-off, together with wind-mediated erosion and animal transport, can move pesticides within the soil ecosystem or cause transfer to aquatic systems.

(g) Further losses from soil may occur through leaching of the more mobile pesticides, with the possibility of resultant transfer to bodies of water.

(h) Applications to aquatic environments inevitably result in some of the pesticide being diluted and dispersed without necessarily reaching the target organisms, at least in effective concentrations.

(i) Pesticides may accidentally enter waters and in some cases the soil, by spillage, misuse, cleansing or disposal of containers and from industrial and domestic effluent. Whilst such factors may contribute little to the overall level of environmental contamination by pesticides, they may be very important at times in localized situations.

(j) Pesticide mobility in soil or water may be reduced by adsorption of the chemical to biotic and abiotic surfaces, although in such cases an equilibrium between adsorbed and desorbed states is likely.

(k) Applications to control animal pests, or incidental uptake by non-target animals (e.g. from treated crops) may result in accumulation, metabolism and excretion, or transfer to higher trophic levels in food chains, with possible further accumulation.

In addition to the factors considered above, which primarily relate to movement, pesticides come under a variety of influences which determine their fate. Thus chemical structure, and consequently

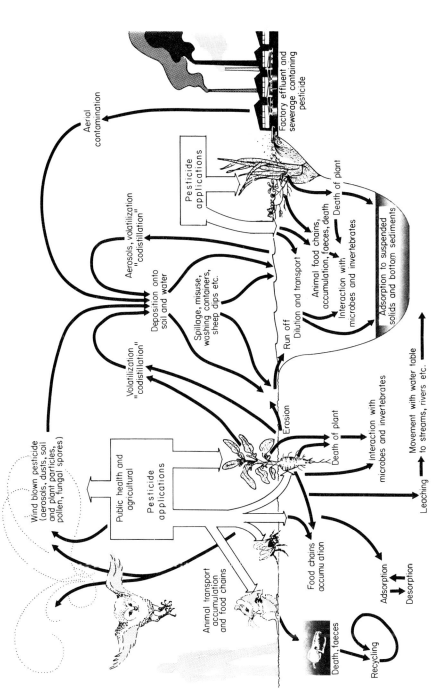

Fig. 1. Entry and movement of pesticides in the ecosystem. (This figure, designed by the authors, was produced at ICI Plant Protection Division and is reproduced with their permission.)

biological activity, may be altered by physical, chemical and biological agencies.

Most pesticides are designed for specific uses but their specificity is rarely absolute. Harmful effects may therefore be exerted on some non-target organisms, especially when high doses are used, deliberately or otherwise. Such effects may not only be caused by the pesticide but also by any transformation products. For ease of application and in order to achieve optimal effectiveness, pesticides are usually formulated with other chemical ingredients. Such additives, although usually present in smaller quantities, are subject to the same environmental influences as the pesticide and should be viewed in a similar way. Some countries specify which additives are acceptable.

The following sections describe in more detail the fate and behaviour of pesticides in individual components of the total environment.

II. Pesticides in the Soil Environment

A. ENTRY

Fig. 2 illustrates many of the direct and indirect routes by which pesticides and their products enter the soil environment. The majority of these chemicals present in soil originate from deliberate applications to the soil or to the foliage of crop plants and weeds when considerable quantities reach the soil by either missing the "target" or by run-off from leaves and stems (Hindin and Bennett, 1970; Cope, 1971). A proportion of the pesticide applied may also eventually become incorporated into the soil as a result of root exudation or death of the plant, particularly after herbicide applications. Further, although mostly small, amounts may remain airborne during application or volatilize from the soil surface and drift away from the site of application to be deposited elsewhere (Akesson and Yates, 1964; Higgins, 1967; Hindin and Bennett, 1970).

Much smaller quantities of pesticides and their transformation products might enter soil in applications of green manure, also in the faeces from treated animals and animals grazing on treated crops and from those receiving pesticides via food chains. Death of such animals could result in addition of further pesticidal residues to the soil.

Pesticides usually enter the soil environment at or below the concentrations recommended for agricultural use, that is mostly in the order of $0.5–5$ kg ha^{-1} and occasionally even lower. If the pesticide is assumed to be evenly distributed in the top 10 cm of soil, the resulting

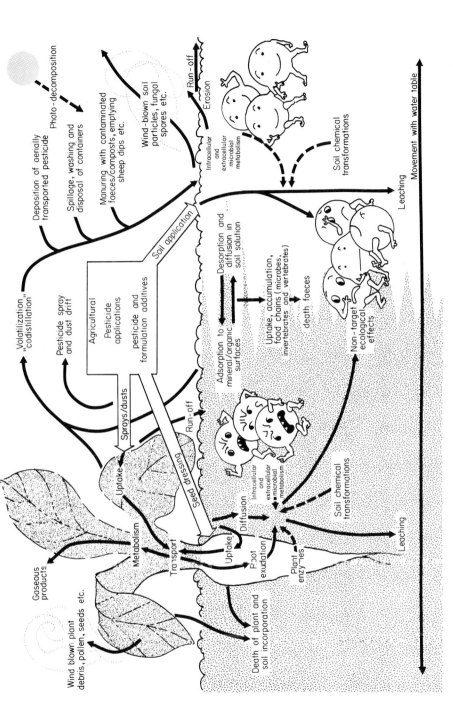

Fig. 2. Pesticides in the soil environment. Key: Movement of pesticide or products →. Agencies affecting the pesticide or products – – →. Rhizosphere – shaded area. Animations – microorganisms. (This figure, designed by the authors, was produced at ICI Plant Protection Division and is reproduced with their permission.)

concentrations will be approximately 0·5–5 ppm. It should thus be apparent that the amounts transferred between and within environments can be exceedingly small and in many instances may be insignificant. That this may be so does not necessarily absolve us from any obligation to quantify these movements and, where practical and desirable, to identify degradation products. However, conclusions drawn from such studies should be kept in perspective and wherever possible be related to data from toxicological evaluations.

Occasionally, spillage or disposal can produce localized regions of high pesticide concentration in soil (Stojanovic *et al.*, 1972).

B. DISPERSAL

Movement of pesticides and their transformation products within the soil environment, or from the soil to other environments, is not only a function of the properties of the pesticide and the soil but also of the prevailing climatic conditions. Pesticide dispersal is thus a complex process. As the transformation of these chemicals in soil may proceed simultaneously with their dispersal, any discussion of pesticide movement should not only relate to the applied chemical but also to any of the products derived from it. The various factors influencing pesticide dispersal within and from soil are described below.

1. *Wind and water erosion*

Both wind and water can move pesticides laterally over soil surfaces (Ashton, 1961; Edwards *et al.*, 1970; Lichtenstein, 1970).

Erosion of soil by water has been shown to move pesticides both in solution (run-off), particularly in the period immediately following application, and adsorbed to soil particles (wash-off) (Barnett *et al.*, 1967; Epstein and Grant, 1968; Ritter *et al.*, 1974). Wind erosion of soil can occur over much greater distances than water erosion and can move pesticides adsorbed to the soil surfaces (Menges, 1964) but in general this phenomenon is poorly documented.

The erosion of pesticide-containing soils is a potential source of contamination of aquatic environments, although the available evidence suggests that in practice such movement is likely to be localized and relatively infrequent (see e.g. Thompson *et al.*, 1970). Studies designed to quantify these effects in agricultural conditions are difficult to effect in practice. The investigations of Barthel *et al.* (1966) failed to indicate any widespread contamination of Mississippi River sediment in regions where the surrounding field crops had received large amounts of pesticide. However, their observations did not include an evaluation of pesticide in solution in run-off water, nor of losses due to desorption or degradation within the aquatic environment. Studies of these phenomena require the establishment of field plots with watersheds leading into specialized collection apparatus, as are presently in operation at US Environmental Protection Agency, Southeast Environmental Research Laboratory, Athens, USA.

Soil surface movements of pesticides are usually accelerated by steep topography, low soil permeability, considerable rainfall and strong adsorption of pesticide to particles on or near the soil surface (LeGrand, 1966; Peach et al., 1973). Several other factors, including the formulation and application rate of the pesticide, method of soil cultivation and the presence of vegetative cover will also affect the redistribution of the pesticide over the soil surface. Smith and Wischmeier (1962) and Chepil and Woodruff (1963) have suggested that vegetative cover may virtually eliminate wind and water erosion in many situations. The selection of pesticides for use in agricultural conditions particularly vulnerable to soil surface wash-off and run-off, e.g. rice fields, requires special consideration.

2. Adsorption

In the soil, pesticides are distributed between the solid surfaces, the soil solution and the soil atmosphere. The equilibria established have been discussed in terms of chemodynamics by Haque and Freed (1974). Adsorption of pesticides to mineral and organic soil particles is mostly a reversible process (Osgerby, 1973) dependent upon the properties of both pesticide and soil (Harris and Warren, 1964; Talbert and Fletchall, 1965). Adsorption, which results in reduced pesticide mobility, is due to bonds both of the Van der Waals type and to those resulting from the electrical charges of the pesticide and the soil surfaces.

Numerous attempts have been made to correlate pesticide adsorption with soil clay or organic matter content, ion exchange capacity and pH. However, most workers agree that the amount of organic matter is the most important soil characteristic affecting the adsorption of the majority of pesticides (see e.g. Lichtenstein and Schultz, 1960; Yaron et al., 1967), although electrolytes and the negatively charged surfaces of clay particles play an important role in adsorbing many cationic pesticides (Knight and Tomlinson, 1967; Hurle and Freed, 1972; White and Mortland, 1970). Dean (1960) found that when organic cations were trapped in monomolecular layers between the expanding lattices of some clay minerals, the molecules were inactivated and rendered inaccessible to microbial attack. Desiccation or drought enhanced this "fixation".

The equilibrium between adsorbed pesticide and that in solution can be altered in many ways. Water, for example, being very polar, competes with the pesticides for adsorption sites (Deming, 1963; Hance, 1965), although it has been proposed that this only becomes an important factor when the moisture content is at or below the wilting point (Edwards, 1974).

Pesticides are mostly formulated to aid application procedures and to increase biological effectiveness. However, adsorption (and leaching) may be influenced by the use of granular formulations and by the choice of appropriate chemical analogues, salts etc. The active ingredients of sprays and dusts usually behave independently of their formulation constituents once in the soil.

Desorption usually occurs readily in the short term, however, Graham-Bryce (1972) has suggested that under field conditions the rate of this process is retarded, becoming particularly slow after cycles of wetting and drying (Hilton and Yuen, 1963; Graham-Bryce, 1967).

The adsorption of pesticides to soil surfaces is of particular environmental importance, in that it can reduce or even eliminate run-off erosion, leaching and volatilization, and may also control both biological activity against the target pest and any undesirable toxicological or ecological effects on non-target organisms. The majority of microorganisms in soil exist on particle surfaces. Pesticide adsorption to these same surfaces may hinder microbial degradation if the chemical is either present in inhibitory concentrations or if the sites for enzyme attack are blocked. For these reasons, laboratory studies in the absence of soil can only rarely demonstrate the characteristics and environmental interactions of a pesticide *in situ*.

3. *Leaching*

Pesticides move in the soil solution both by diffusion and by bulk transfer involving a mass flow of water containing the pesticide. The downward movement of pesticide in solution through the soil profile in the zone above the water table, resulting from rainfall or flooding of land, is termed leaching. Lambert *et al.* (1965) compared this phenomenon to the process of chromatographic separation. The extent of leaching is determined by the solubility, adsorptive properties and rate of degradation of the pesticide, as well as by the natural water movement in, and the physical and chemical characteristics of, the soil. Many pesticides, including the widely used organochlorine insecticides and bipyridyl herbicides are very resistant to leaching (Carter and Stringer, 1970; Edwards and Glass, 1971; Riley *et al.*, 1976). Harris (1967, 1969) and Helling *et al.* (1971) have described the soil mobilities of a wide range of pesticides.

Loss of pesticide by leaching is believed by some authors to be overestimated, due to a lack of consideration for the upward movement of soil water as a result of both evaporation from the soil surface and transpiration by plants (Graham-Bryce and Briggs, 1970). Harris (1964) and Phillips (1964) have shown this upward movement to carry some persistent pesticides back to the root region of crops. Edwards and Glass (1971) and Glass and Edwards (1974) reported much less field-leaching of both methoxychlor and picloram than had been indicated by laboratory data. Thus, although laboratory studies may indicate the potential mobility of pesticides they are also likely to overestimate the true extent of field-leaching.

4. *Volatilization*

Loss from soil to the atmosphere by volatilization has always been expected for many pesticides of relatively high vapour pressure, especially those designed to exhibit fumigant activity. However, in recent years it has become apparent that smaller losses can occur via the same route for numerous pesticides with vapour pressures greater than approximately 10^{-7} mm Hg at 20°C (see e.g. Farmer and Jensen, 1970; Guenzi and Beard, 1970; Harris, 1970; Cliath and Spencer, 1971; Spencer *et al.*, 1973). The losses of pesticides by volatilization are generally at their greatest in the period shortly after application to the soil surface (Gray and Weierich, 1965; Parochetti *et al.*, 1971) but can be reduced by use of granular formulation and subsurface treatments.

Many studies have shown that an increase in soil water content increases pesticide volatilization (Spencer *et al.*, 1969; Spencer, 1970). In the past, this has often been attributed to the phenomenon of co-distillation (Acree *et al.*, 1963; Bowman *et al.*, 1965), a physical or chemical reaction of the pesticide with water resulting in the formation of a product more volatile than the pesticide involved. This term is now considered by some authors to be a misnomer in the present context (Hartley, 1969; Hamaker, 1972; Spencer *et al.*, 1973). The most likely alternative is that these losses are the result of simple displacement of the pesticides from adsorbing sites by water, thereby increasing the pesticide level in the soil solution and soil atmosphere and enhancing the opportunity for losses from the soil by volatilization.

The structure of the chemical and its adsorption to soil surfaces are both important factors in controlling its rate of volatilization (Kearney *et al.*, 1964; Haque *et al.*, 1974). Atmosphere, temperature, wind and relative humidity, plant cover and the formulation and rate of application of the pesticide also exert some influence on volatilization (Danielson and Genter, 1964; Menges, 1964; Spencer *et al.*, 1973; Spencer and Cliath, 1974, 1975). However, few studies other than that of Caro and Taylor (1971) have quantified volatilization under field conditions.

Occasionally, transformation products of pesticides in soil may be more volatile than the parent pesticide, as for example with DDE, a product derived from DDT (Spencer and Cliath, 1972).

5. *Plant uptake*

Pesticides may also be lost from the soil environment by uptake into cultivated and non-cultivated plants (Beall and Nash, 1972; Scott and Phillips, 1973; Hilton *et al.*, 1973). The total amount and the rate of uptake are related to the ability of the plant to absorb the chemical and to the availability of the pesticide to the plant roots (Graham-Bryce and Etheridge, 1970; Graham-Bryce and Coutts, 1971).

Although Edwards (1974) has shown that relatively few of the persistent insecticides are concentrated from the soil into plant tissues, some pesticides have been recovered from crop plants in considerable quantities (Lichtenstein *et al.*, 1967; Onsager *et al.*, 1970; Caro and Taylor, 1971). Fungicidal seed dressings, soil-applied herbicides and other pesticides may have to be absorbed by the plant roots in order to exert their biological activity. However, the amounts taken up are not necessarily large (Graham-Bryce and Coutts, 1971; Walker, 1973). There are relatively few reports in the literature which clearly demonstrate that plants play an important role in the removal of pesticides from soil. Some of the pesticide residues taken up by plant tissues may eventually be returned to the soil, either directly following death of plants or indirectly via food chains and excretion or death of the recipient animals.

Many of the major pesticide-using countries now impose strict legislative controls on chemical manufacturers and distributors, obliging them to demonstrate that the residues in crops are below specified limits.

6. *Animal uptake and transport*

Animals on or in soil can receive pesticides directly by both deliberate application and accidental contact; indirectly by eating treated or contaminated plants or other animals. Soil invertebrates such as earthworms and slugs also ingest pesticides adsorbed onto soil particles (Wheatley and Hardman, 1968) and in some instances may concentrate the chemical in their tissues (Edwards, 1974). A wide variety of soil arthropods have also been found to contain pesticide residues (Davis, 1968; Gish, 1970; Korschgen, 1970). However, data often fail to indicate whether the residues in herbivorous invertebrates are derived from the soil or from ingested plant material. The amount of pesticide and transformation products accumulated from the soil by invertebrates is related to the degree of exposure to the chemicals, to the ability of the animal to absorb and retain them in tissues and to soil and pesticidal factors affecting both mobility and transformation, and thus availability, of the chemical (Wheatley and Hardman, 1968; Davis, 1971).

Invertebrates in soil are only capable of moving pesticides, whether in their tissues or on their outer surfaces, over relatively small distances. Some of this transported pesticide is likely to re-enter the soil via excretion or death of the animals. When invertebrates contaminated with pesticides are eaten by mammals or birds (Jefferies and Davis, 1968; Stickel et al., 1965) the distances over which the chemicals are transported can be increased enormously, even to a global scale (Evans, 1974). Death and excretion may again return pesticide to the soil. However, the amounts of pesticides and their products dispersed by these routes are mostly extremely small and are unlikely to have any ecological significance.

C. ECOLOGICAL EFFECTS

Wherever possible, pesticides and their methods of application are designed to ensure that the chemical exerts its biological activity principally upon the target pest. However, many pesticides are also likely to be adsorbed to and ingested or absorbed by a large number of non-target organisms. Some pesticides exhibit a relatively wide spectrum of biological activity and as a result have the potential to exert undesirable effects on non-target organisms. However, in soil, any such effects may be reduced or prevented by adsorption of the chemical to soil surfaces or by its degradation to less harmful products. Many of the recently developed pesticides are more specific in their expression of biological activity than those marketed earlier and thus have fewer ecological side effects.

Until recently, environmental and ecological studies were mainly concerned with the more persistent pesticides and particularly with their movement and accumulation in food chains. Latterly, however, most pesticides have been subjected to more extensive ecological investigation.

Studies of the effects of pesticides on soil microorganisms have proved extremely difficult, due to the heterogeneity of the environment and its microbial content, to the lack of understanding of many of the microbial processes operative in soil and to the inability of microbiologists to isolate more than a small proportion of the viable soil population. As a result, laboratory studies with pure and mixed microbial cultures have often been used to determine or predict the ecological effects. Such simplified systems are unlikely to reproduce accurately the complex soil environment, where microhabitats do not necessarily reflect the gross natural soil conditions, and in which both pesticides and their transformation products may be in solution or absorbed to soil surfaces and can have direct or indirect effects on the inhabitants. It is quite possible that extrapolation of data from culture studies to soil situations could grossly exaggerate the true effects of pesticides on soil microorganisms *in situ*.

Although there is still a considerable lack of information, it has been shown that undesirable effects of pesticides, applied at normal agricultural rates, on microorganisms and their activities in soil are not common and where they exist are often transient (see Chapter 7). Furthermore, the significance of disturbances that have been observed are often difficult to assess.

D. TRANSFORMATIONS

Although pesticides may be "lost" from the soil environment by leaching, volatilization and via the biota, a proportion of many of these chemicals does remain in this environment. Whilst on or in soil the pesticides are subjected to biological, chemical and photochemical effects capable of causing transformations in their chemical structure. The majority of these transformations are degradative and often remove the biological activity of the pesticide. This ability of the soil and its inhabitants is widespread and only a minority of the pesticides now in use remain unchanged over a long period of time (Alexander, 1973; Edwards, 1973b). The rate of transformation of a pesticide in soil under agricultural conditions is related to its distribution in the soil profile, to its chemical structure and formulation and to the physical, chemical and biological properties of the soil.

In practice, the products of biological, chemical and photochemical activity on pesticides are often the same or similar and it is not always easy to distinguish which of the three agencies has caused a specific transformation, or to establish their relative involvements in a degradative pathway. Many investigators have therefore attempted to isolate the different processes experimentally. However, any method of excluding biological activity is also likely to affect the soil chemistry and although photochemical effects can be avoided in experiments, they cannot themselves be realistically investigated in the absence of both biological and chemical soil activity. Furthermore, the elimination or modification of one or more of these agencies may alter the overall routes, rates and products of pesticide transformation.

1. *Biological transformations*

The rates of transformation of pesticides in sterile and non-sterile soils have often been compared (see e.g. Lichtenstein and Schulz, 1960; Menn *et al.*, 1965) and there is little doubt that many pesticides are extensively metabolized by microorganisms in soil. In several cases, microorganisms which can metabolize the pesticides in laboratory culture media have also been isolated from the soil (Matsumura *et al.*, 1968; Clark and Wright, 1970; Miles *et al.*, 1971; Jagnow and Haider, 1972). The role of the soil microflora, microfauna and microalgae in transforming pesticides is discussed in detail in later chapters of this book.

The rates of transformation of pesticides in soil are related to factors such as the moisture and organic matter content, pH and temperature of the soil as well as to the methods of soil cultivation and irrigation (Lichtenstein and Schulz, 1961; Sheets and Harris, 1965; Guenzi and Beard, 1968; Baker and Applegate, 1970; Bro-Rasmussen *et al.*, 1970; Meggitt, 1970; Parr *et al.*, 1970). Although it is difficult to correlate any single factor with the microbial contribution to pesticide transformations, these factors are all capable of influencing the overall metabolic processes of microorganisms in soil.

In agricultural conditions, considerable quantities of seed-, foliar- and soil-applied pesticides are likely to enter and remain in or pass through the zone of soil in the proximity of and under the influence of the plant roots. In this zone, the rhizosphere (see Chapter 2), the numbers, types and metabolic activity of the microorganisms and the physico-chemical soil conditions are different from those in soil remote from plant roots. Pesticides in the rhizosphere may thus be transformed by routes, and at rates, different from those occurring in the rest of the soil.

The microorganisms most commonly considered to be involved in the metabolism of pesticides in soil are bacteria, fungi and actinomycetes; although this is mainly a reflection of the predominance of these groups in soil, the isolation procedures used and the interests of the investigators. The contribution of the microfauna and microalgae is, however, less clear. Although extracellular enzymes exist in soil (Burns *et al.*, 1972) little is known of their role in the transformation of pesticides.

2. *Chemical transformations*

Less studied and less understood than the biological mechanisms of pesticide transformation are the chemical reactions between soil and pesticides. Non-biological reactions have only been proposed as the main mechanism for the degradation of a few pesticides in soil (Armstrong *et al.*, 1967; Harris, 1967; Skipper *et al.*, 1967; Hance, 1969). However, the importance of chemical reactions should not be underestimated for they may also contribute essential stages to principally biological routes of degradation.

Probably the two most important factors governing chemical transformations in soil are moisture content and pH (Gomaa *et al.*, 1969; Crosby, 1970) although other reagents such as nucleophiles, reducing agents and oxygen may also be involved. The studies of Plimmer and Kearney (1968) have indicated that free radical reactions may degrade pesticides and Steelink and Tollin (1967) have shown the existence of free radicals in soil humic acids. The biota in the soil may in part be responsible for the generation of these free radicals, for example, via hydrogen peroxide produced by microbial extracellular oxidase enzymes.

3. *Photochemical transformations*

The majority of pesticides are susceptible to photodecomposition under experimental conditions, the effects being most pronounced after low wavelength u.v. irradiation. Sunlight reaching the soil surface is usually limited to the 290–450 nm range (Koller, 1965), although Barker (1968) calculated that very small amounts of the lower u.v. wavelengths did reach the Earth's surface. There is sufficient evidence to suggest that photodecomposition occurs under field conditions and that a considerable proportion of some pesticides may be transformed by this mechanism (Slade, 1966; Crosby and Li, 1969; Kearney and Kaufman, 1969). Photochemical effects probably occur mainly prior to soil entry (i.e. in the atmosphere or on plant surfaces), for diffusion and leaching in soil normally ensure that pesticide molecules are rapidly transported to sub-surface sites impenetrable by sunlight. Furthermore, most pesticides are applied to established crops where the foliar cover can drastically reduce the intensity of light reaching the soil surface. Goring *et al.* (1975) expressed considerable doubts as to whether photodecomposition of pesticides in soil has much practical significance.

In situ studies of photochemical transformations in soil can be difficult to interpret and replicate. The factors responsible include variability of natural light intensity, the adsorption of pesticides to soil and plant surfaces with a possible resultant shift in the absorbance spectrum, and also the presence of photochemical sensitizers in the environment (Slade, 1966; Pitts *et al.*, 1969; Ivie and Casida, 1971a, b; Soderquist, 1973). This has mostly resulted in a preference for laboratory investigations, with the inherent difficulty in extrapolating to agricultural conditions.

Whilst the separation of chemical, photochemical and biological mechanisms of pesticide transformations in soil may be a fascinating exercise it serves little purpose to the environmental chemist, who is primarily concerned with what products are formed *in situ*, how much and how fast.

III. Pesticides in Aquatic Environments

Aquatic environments include streams, rivers, ponds, lakes, reservoirs, drainage channels, irrigation systems, paddy fields, estuaries and the sea, none of which are standardized or homogeneous. It is therefore difficult to cover all eventualities regarding pesticide entry and fate in aquatic environments. Thus, Fig. 3 depicts, in a generalized way, the entry and dispersal of pesticides *in aquae*.

A. ENTRY

Pesticide entry into aquatic environments occurs intentionally and unintentionally. The intentional, direct, application of pesticides for controlling vascular weeds and algae, insects and snails is well known. The unintentional, or indirect, introduction of unknown amounts is less readily defined. The entry and quantitative aspects of pesticides in natural aquatic systems have been considered by Cope (1966), Ware and Roan (1970) and Edwards (1973b). The principal direct and intended modes of pesticide entry include: the use of pesticide sprays and granular formulations to control water-inhabiting pests; disposal of domestic and industrial wastes (e.g. from pesticide manufacture and

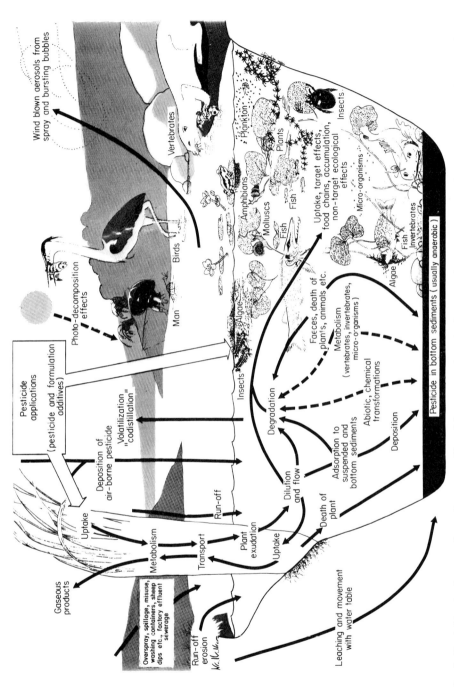

Fig. 3. Pesticides in the aquatic environment. Key: Movement of pesticide or products ↓. Agencies affecting the pesticide or products ---→. (This figure, designed by the authors, was produced at ICI Plant Protection Division and is reproduced with their permission.)

formulation; the food industry; moth proofing); disposal of unused pesticides; on-site cleaning of application and mixing equipment and decontamination procedures; disposal of commodities containing excessive residues; the use of biocides to prevent fouling of industrial cooling waters.

Pesticides may unintentionally and indirectly enter aquatic systems by drift from aerial or ground applications of pesticides and movement via wind, water and soil erosion. Included in this category are run-off, wash-off and leaching from treated land; irrigation water from pesticide target areas; application malpractice and accidents, e.g. spillage, use of the wrong chemical, faulty spraying equipment or technique; accidents involving waterborne pesticide cargo; in domestic and industrial effluents; and via migratory birds.

Generally the quantities of pesticides entering natural waters indirectly are far less significant than those applied directly. Even the organochlorines, the most commonly detected pesticides, are normally present in extremely low concentrations (Hartung, 1975; Holden, 1975).

Aerial drift of pesticides, particularly from aircraft spraying, with resultant contamination of water, is not necessarily restricted to areas adjacent to the site of application. Significant quantities of pesticide-contaminated dust particles may be transported in the atmosphere to be deposited by rainfall great distances away (Cohen and Pinkerton, 1966; Crosby, 1973). Johnson and Ball (1972) considered that airborne pesticide contamination of the Great Lakes could be important in view of their vast surface area.

Run-off from agricultural land was considered as a source of insecticide residues in rivers (Nicholson, 1967). Water-soluble pesticides are more likely to enter in this way than are less soluble types (Graetz et al., 1970). Although the concentrations of herbicide residues reaching rivers via run-off waters are generally low and non-persistent (Frank, 1970; Merkle and Bovey, 1974), run-off is nonetheless an established means of herbicide loss from treated land under conditions of high rainfall (Barnett et al., 1967; Trichell et al., 1968; Bovey et al., 1974; Willis et al., 1975).

Balicka (1974) considered that the extent of infiltration of agricultural pesticides into adjacent surface waters is determined not only by the formulation and application of chemicals but also by climatic conditions, the terrain and physico-chemical properties of the soil. An additional source of pesticide contamination of lakes may be the inflow waters from surrounding watersheds coursing through agricultural land (Rudd and Herman, 1972).

Mullison (1970) found few signs of herbicide contamination or accumulation problems as long as the chemicals were used according to label instructions. Furthermore, Holden (1972) considered that in normal use most pesticides are unlikely to enter fresh waters in biologically significant amounts. However, accidental spillages or careless disposal of surplus concentrates may constitute occasional discharges into streams and rivers which could account for significant fish mortality (Holden, 1972). Stojanovic *et al.* (1972) drew attention to the fact that even after normal emptying of containers of liquid pesticide formulations, as much as 2–3% of the concentrate may remain. There are those who are not too particular about where they jettison such "empty" cans and localized adverse effects may ensue.

B. DISPERSAL, FATE AND ECOLOGICAL EFFECTS

The principle extrinsic factors likely to influence a pesticide's fate include: dilution, chemical and biological degradation, adsorption to living and non-living materials, uptake or absorption by living organisms with consequent possible accumulation, entry into food chains, biological magnification and transport, sedimentation (usually in association with silt particles, plant or animal detritus), volatilization, photodecomposition, effects of water currents and turbulence, aerosols and splashing, and return of the pesticide or metabolic products to the water following death of contaminated biota or by animal excretion. The persistence and fate of a pesticide may be further complicated by the possible concurrent operation of these diverse factors. Water movements pose problems for the study of pesticide persistence and interactions and also enhance the effects of dilution.

1. *Dilution and abiotic movement*

Dilution influences the fate of pesticides in any aquatic system and is likely to be the principal factor in reducing levels of water-soluble herbicides (Frank, 1970). Whilst in small bodies of static water the effects of dilution may be expressed in simple mathematics, the dilution in moving water is far less predictable owing to the hydrodynamics and is complicated by turbulence, weed growth and changing physical features (Frank, 1970, 1972). Turbulence causes resuspension of pesticide-contaminated sediment particulates (Edwards, 1974) and may also promote losses by volatilization (Frank, 1970). It is also possible for pesticides to become concentrated and move in surface slicks (Crosby, 1973). Tatton and Ruzicka (1967), who found organochlorine pesticide residues in a variety of Antarctic birds, concluded that oceanic (and atmospheric) movement may contribute to a worldwide pattern of distribution of such pesticides.

2. *Volatilization*

The volatility of different compounds is a relative value, being ultimately dependent on the vapour pressure, and varies with temperature, water solubility and adsorption characteristics (Parochetti and Warren, 1966; Kenaga, 1972; Spencer *et al.*, 1973).

Volatilization, a process by which even "non-volatile" pesticides are dissipated (Spencer and Cliath, 1972), can be important in the loss of a pesticide from the aquatic environment. For compounds such as acrolein and xylene, used as aquatic herbicides in irrigation channels, volatilization can be the principal factor effecting rapid loss, especially in turbulent water (Frank, 1970, 1972). In a study of the volatilization rates of several organophosphorus and organochlorine insecticides from different substrates, Lichtenstein and Schulz (1970) observed that the most rapid rates were from water, whilst relatively small amounts volatilized from soils.

The amount of pesticide volatilized from water appears to be limited by the compound's water solubility, since volatilization increases with concentration until the approximate maximum water solubility value is reached (Kenaga, 1972). This was illustrated by a study of volatilization of the insecticide dyfonate (*O*-ethyl-*S*-phenyl-ethyl-phosphonodithioate) by Lichtenstein and Schulz (1970). However, although it could be expected that low water solubility and relatively high vapour pressure would enhance volatilization from water, no direct relationship between insecticide solubility and volatilization was found (Lichtenstein and Schulz, 1970).

DDT has high affinity for the air–water interface and low affinity for water, a feature which would suggest ready volatilization of DDT from water. This, however, does not seem to happen in nature, signifying that preferred sorptive surfaces influence volatility (Kenaga, 1972). Such an effect was clearly shown by Lichtenstein and Schulz (1970) who found that in most cases the addition of soil, algal cells, or detergent to water reduced the loss of insecticides by volatilization.

3. *Adsorption*

Pesticides may adsorb to both non-living and animate particles and are often present in suspended and sedimented particles (Crosby, 1973), which may be especially significant in estuarine conditions. Adsorption is very important in consideration of the fate of bipyridyl herbicides which adsorb strongly to plant material and mud (Calderbank, 1971). The comparatively insoluble organochlorine insecticides may be found in higher concentrations in mud and sediments than in the water itself (Edwards, 1974), and it is possible that appreciable insecticide degradation in rivers is attributable to the activity of anaerobic microbes in bottom mud sediments. In contrast, the removal of the herbicide 2,4-D from natural waters by adsorption onto various types of clay particles was insignificant (Aly and Faust, 1964).

Huang (1971) proposed that after adsorption, small amounts of some pesticides are continuously desorbed into the surrounding water, to maintain a dynamic equilibrium. Gerakis and Sficas (1974) concluded that accumulation of pesticides in the bottom sediments of water bodies is an important factor in their disappearance from contaminated water. Furthermore, pesticides may either be directly adsorbed in the sediments or be co-deposited with particulates and algae.

Pesticides may also be adsorbed by the aquatic life, including plankton, invertebrates, vegetation and fish (Wojtalik et al., 1971; Edwards, 1974). The high surface area:volume characteristic of unicellular microorganisms, coupled with the lipophilic nature of many pesticides, may explain the high and often rapid pesticide sorption capacities of microorganisms (Ware and Roan, 1970; Valentine and Bingham, 1974).

4. Absorption or uptake by the biota

It has been suggested (Ware and Roan, 1970) that, partly because of their generally low water solubilities, many pesticides tend to "seek" living organisms. Pesticides can be absorbed by most types of aquatic microbes, the plankton, algae, and higher flora and fauna and thus may enter organisms directly, or indirectly via food chains. Pesticide uptake by aquatic microorganisms would be favoured by their relatively high surface area:volume ratio and enhanced by the selective partitioning of lipophilic compounds (e.g. many insecticides) into cellular lipids, especially at cell surfaces.

The microscopic algae, as primary producers at the base of the aquatic food chain, constitute an important metabolic potential in intimate contact with water. It is possible that these organisms make a significant, somewhat overlooked, contribution to the immediate fate and ultimate breakdown of some pesticides – an aspect which is considered further in Chapter 8.

There are indications (Frank, 1970) that aquatic plants can influence the timing of pesticide residues in water, through absorption by floating parts and subsequent release by submerged parts. Plant growth, especially the massed vegetation of some aquatic weeds, is also effective in removing herbicides and insecticides from water. Sutton *et al.* (1969) found that immersed parrotfeather (*Myriophyllum brasiliense*), which has a strong transpiration stream, was particularly effective in removing simazine from water.

Frank (1970) concluded that in view of the large volume of water compared to the total mass of animal life in natural waters, it is unlikely that the aquatic fauna play a significant role in reducing herbicide levels. However, the chlorinated hydrocarbon insecticides are taken up by invertebrates in fresh and sea water and accumulated to varying degrees, reaching very high levels in oysters and sea squirts (Edwards, 1974). Aquatic animals also concentrate pesticide residues directly from water and evidence suggests that for the organochlorines such direct intake is more important than aquisition of residues via the food or prey (Edwards, 1974; Moriarty, 1975).

5. *Biotic movement*

Following movement from the external water phase into living organisms, further movements include those within individual organisms, movements between different organisms (i.e. along food chains) and transport via an organism, possibly over great distances. The feasibility of long distance biotic pesticide movement is determined by the mobility of the host and the rate of pesticide breakdown within it (Robinson, 1973). There is evidence that contamination of aquatic environments by hydrocarbon insecticides is widespread (Nicholson, 1967; Tatton and Ruzicka, 1967; Robinson, 1973). Sladen *et al.* (1966), who found DDT residues in Antarctic penguins and a seal, considered that whilst local contamination by man was unlikely, long distance transport of DDT residues via air and ocean currents, or migrating whales and birds, was possible.

The transfer of pesticides to bottom sediments through sinking of dead, contaminated biological material is probably significant in view of the biodegradation potential of muds and sediments.

6. *Accumulation, biological magnification, effects on food chains*

Many of the aquatic biota accumulate pesticides from water and apart from the immediate toxic effects which might ensue, pesticide residues may be transferred through one or several trophic levels in food chains, with the possibility that concentration or biological magnification may occur at each succeeding trophic level. The biological impoundment, especially in lipid tissues, and subsequent magnification of the pesticide residues in food chains is well established among organisms of aquatic communities. It has also given rise to the concept of a "lipid pool", an ecosystem storage mechanism preventing rapid degradation or residue loss (Rudd and Herman, 1972).

One of the best known examples of trophic concentration occurs in Clear Lake, California, where the entire lacustrine ecosystem contained chlorinated hydrocarbon insecticide residues (Rudd and Herman, 1972). Concentrations of such insecticides may reach very high, lethal levels in fish-eating birds, so marking the final step in the biological magnification process (Nicholson, 1967). Apart from any other consideration, biological magnification signifies the need to consider aspects beyond the limited (and popular) concept of "acute toxicity" of pesticides (Kawahara and Breidenbach, 1966).

An indication of the possible scale of biological magnification, especially with organochlorine insecticides, was given by Butler (1967). Thus from an estimated 1 ppb DDT in estuarine water, the levels rose to 70 ppb in plankton, 15 ppm in fish and up to 800 ppm in porpoise blubber. Crosby and Tucker (1971) observed that the invertebrate *Daphnia*, typical of the first animal links in the aquatic food chain, could rapidly accumulate DDT in quantities up to 23 000-fold from dilute suspension. Even higher organochlorine insecticide accumulation values were recorded for freshwater invertebrates studied by Johnson et al. (1971), with levels in *Daphnia magna* and *Culex pipiens* 100 000-fold greater than in the water. It was envisaged that foraging invertebrates would contribute through biological magnification to the high levels of DDT found in predaceous salmon, and also that invertebrates might magnify the insecticide degradation products.

The position of fish in the aquatic ecosystem's hierarchy and in relation to predatory birds and man is important when considering the fate of pesticides. Organochlorine insecticides may cause high mortality in fish eggs and fry with drastic effects on reproduction (Holden, 1972) and may even be transmitted from female fish to offspring via the egg (Burdick et al., 1964). Organochlorines were also reported to disrupt the breeding of fish-eating birds (Rudd and Herman, 1972). In contrast, 2,4-D applied to reservoirs for weed control appeared to exert no harmful effect on zooplankton, phytoplankton, invertebrates or fish. Although significant quantities of the herbicide were absorbed by plankton, very little was transmitted along food chains and there was no accumulation in fish (Wojtalik *et al.*, 1971). Similarly, Sears and Meehan (1971) observed no apparent short-term effects on salmonid fishes or aquatic invertebrates when adjacent spruce plantations were sprayed with 2,4-D, although low levels of the pesticide were present in the water. From the limited data available on the effects of herbicides on plankton, Mullison (1970) concluded that there appeared to be no problem of biological magnification of herbicides in food chain organisms in the aquatic environment.

7. *Degradation*

Water is a suitable medium for chemical reactions and is also itself chemically reactive (Crosby, 1973). In alkaline natural aquatic systems many pesticides may undergo alkaline hydrolysis. Other reactions which contribute to abiotic transformation of pesticides in water include reduction, elimination, decarboxylation, oxidation and isomerization (Crosby, 1973).

Detoxification and degradative reactions may occur in higher aquatic plants and animals, also in invertebrates and larvae (Johnson *et al.*, 1971). It seems likely, however, that microbial activity is generally more significant in the biochemical transformation of pesticides in aquatic systems, although the degradation rates may be slower than those achieved by microbes in soil. It is also possible that in natural environments microbes may act synergistically to effect the degradation of pesticides. Some also have remarkable powers of adaptation and genetic transfer whereby degradative capacities might be acquired.

There have been few attempts to distinguish photochemical breakdown from any other processes by which a pesticide may be degraded under natural conditions in aquatic environments: Indeed, it is possible that in some cases, photolytic, chemical and biological forces may all operate to effect net changes in pesticide molecules. On the other hand, laboratory experiments on the photochemical reactions of pesticides are not very relevant to natural situations. Many pesticides contain aromatic moieties and would be expected to absorb u.v. energy, however the available solar u.v. below 290 nm is limited. Aly and Faust (1964) observed that 2,4-D underwent photolysis during u.v. irradiation, but concluded that the u.v. energies of sunlight would be inadequate to decompose 2,4-D. Paraquat is known to be degraded on plant surfaces by the u.v. component of sunlight (Slade, 1966) and this may occur on floating or submerged plants in water. Conversely, paraquat in aqueous solution is not decomposed significantly by sunlight since in this state it exhibits a lower, sharper absorption peak that it does when adsorbed onto surfaces (Slade, 1966).

The extent of pesticide photodegradation in natural waters will be determined by factors such as light quality and quantity, water turbidity and pH, the presence of adsorbing surfaces, chemical reactants and agencies to remove photolysis products, also the presence of photosensitizing substances, a feature which may be particularly important (Crosby and Li, 1969).

8. Toxicity

Apart from the desired toxic effects of pesticides on target species, it is apparent that non-target species may also succumb. Such effects concerning microorganisms are considered in later chapters. An overall picture of pesticide toxicity to aquatic organisms including plankton, invertebrates and fish may be gained from Butler (1965), Kerr and Vass (1973) and Holden (1972, 1973). Ware and Roan (1970), reviewing the toxicity of pesticides to microorganisms, made the important point that apart from the commonly assessed lethal effects, toxicity of pesticidal chemicals may be assessed in terms of other direct effects, for example on growth rate or a specific aspect of metabolism, such as photosynthesis. These authors further drew attention to the fact that since such investigations usually seek and even anticipate population decreases, there may be some unobserved species which flourish under a pesticide regime. For example, Sethunathan and MacRae (1969) found that application of the insecticide diazinon to submerged soil significantly stimulated the actinomycete population and also the algae in the standing water.

9. Indirect effects

Biotic effects of pesticides in aquatic systems other than the direct and food chain–transfer effects have also been observed. The use of herbicides for controlling aquatic macrophyte vegetation would be expected to have an effect on the species composition of the associated invertebrate fauna (Holden, 1972), which in turn could have far reaching consequences for the fish population. Wojtalik et al. (1971) observed that despite a lack of apparent toxicity, there were generic and species changes in macro-invertebrate populations following the use of 2,4-D to control water-milfoil in reservoirs. The changes were attributed to rapid collapse of the milfoil plants which provided the invertebrates with cover from fish predators, a substrate, and epiphytic food in the form of diatoms. Frank (1970) deduced a common pattern from the studies of several workers on the effects of herbicides on phyto-plankton and invertebrate fauna, which also showed a decline in micro-fauna following death of macrophytes. Newbold (1975) showed that herbicide destruction of aquatic macrophyte weeds caused significant changes to the habitats of animals. The effects ranged from oxygen depletion to recolonization by different plant species.

De-oxygenation, especially in static water, results from the death of photosynthetic organisms and the activity of microbes decomposing herbicide-killed vegetation. Such oxygen-depletion may adversely affect and even kill fish populations (Jennings, 1966; Newman and Way, 1966). However, no oxygen deficiency effect on the fish was observed when copper sulphate was applied to control the blue–green alga *Microcystis* in ponds (Crance, 1963). Indeed, the decline in blue–green algae after copper sulphate application was accompanied by an increase in the zooplankton (which would favour the fish population).

There was marked stimulation of indigenous algal growth when the insecticide gamma-BHC was applied at high rates to submerged rice soils (Raghu and MacRae, 1967). The stimulation, particularly apparent with the blue–green algae, was attributed to elimination of small predatory animals.

IV. Pesticides in the Atmosphere

Most studies of the distribution and behaviour of pesticides and microorganisms in the atmosphere have been limited to observations in the troposphere (see Wheatley, 1973 and Chapter 2). Only in recent years has it been fully realized that the atmosphere is important as a reservoir of pesticides, although mostly at extremely low concentrations, and that in it these chemicals can be transported over enormous distances.

A. ENTRY

Agricultural uses are responsible for most of the pesticide entry into the atmosphere. Other sources capable of causing localized aerial distribution of pesticides include public health applications, domestic and industrial hygiene, the fumigation of ships, aircraft and buildings, factories manufacturing, storing or utilizing (e.g. moth proofing) pesticides and the burning of waste organic materials containing pesticide residues. However, air currents and wind will normally result in rapid dilution of pesticides to negligible atmospheric concentrations.

Pesticides can enter the atmosphere in particulate or vapour form during application, by volatilization after deposition, or adsorbed to wind-blown soil or plant particles. The amounts of pesticides entering the atmosphere are related to their vapour pressures as well as to application methods and climatic factors such as air movement, wind speed, temperature and humidity. Pesticides applied to airborne pests, such as locusts and the African armyworm, or from aeroplanes at terrestrial and aquatic targets (Hindin and Bennett, 1970; Rainey, 1974), can result in larger losses to the atmosphere than do terrestrial application techniques. Further factors governing losses by volatilization from soil and water have already been discussed.

Fig. 4 illustrates the most likely possible routes of entry and dispersal of pesticides in the atmosphere.

1. *Entry as particles*

Ludlam (1967) has described how particulate matter of 60 μm diameter or more rarely remains airborne for more than a few metres and how even those particles larger than a few μm can soon return to the earth's surface by gravity. Smaller particles can, however, be transported by air currents, even into the upper layers of the troposphere.

The range of particle sizes produced by pesticide application equipment results in deposition by both impaction and sedimentation, and also the loss into the atmosphere of particles with very low settling velocities (Ebeling, 1963; Akesson and Yates, 1964; Wheatley, 1973). Pesticidal dust particles, many of which are 10 μm diameter or less, are more prone to drifting than are liquid droplets, which are usually considerably larger (Gerhardt and Witt, 1965; Middleton, 1965; MacCollom *et al.*, 1968). However, during application, aqueous spray droplets can evaporate to a smaller size and even further to leave the solid material or a solution of the pesticide in non-volatile formulation components. Such particles will then be more susceptible to drifting. The use of dusts (and even sprays in some circumstances) is declining in favour of granules, and considerable efforts are being made to improve application machinery and chemical formulations to further reduce losses into the air.

Pest control with fog, aerosol and smoke formulations can result in high atmospheric concentrations of pesticides, due to the minute particle sizes involved. However such applications are carried out on a small scale and in enclosed spaces such as buildings and glasshouses.

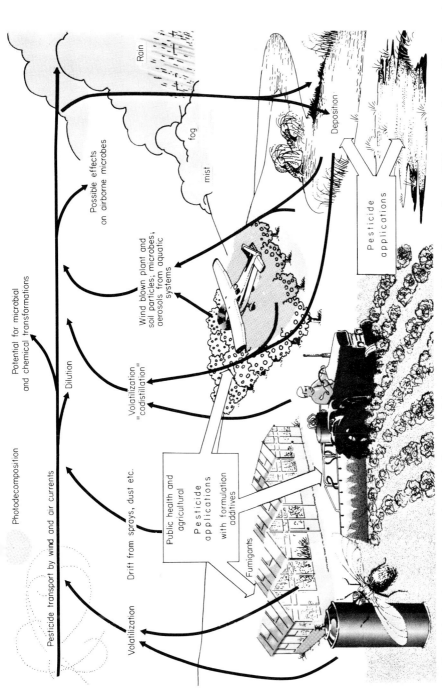

Fig. 4. Pesticides in the atmosphere. Arrows indicate possible movement of pesticides or their products. (This figure, designed by the authors, was produced at ICI Plant Protection Division and is reproduced with their permission.)

2. Entry as vapour during application

Menzer *et al.* (1970) and Kiigemagi and Terriere (1971) have claimed that during the application of some pesticides, losses of up to half of the applied pesticide may occur in the vapour phase. Losses during application have received much less attention than particle drift, due to the problems involved in experimentation. Unfortunately, most of the evidence for such non-particulate losses has been calculated using soil surface recoveries after spraying, from measured and calculated losses due to particle drift and from the vapour pressure of the pesticide. However, the available data suggest that for all but the least volatile of pesticides, a proportion of that applied may enter the atmosphere in the vapour phase (see e.g. Dekker *et al.*, 1950). In many instances the amounts involved are very small and rapidly diluted by air movement.

3. Entry by volatilization from surfaces

Loss of pesticides and their transformation products by volatilization from the surfaces of soil, water, plants and any other biotic or abiotic material receiving applications is probably responsible for a considerable proportion of the total residue in the atmosphere. These losses, reviewed by Spencer *et al.* (1973) are described in other sections of this chapter. Wheatley (1973) has also considered the possibility of continuous contamination of solid and aquatic surfaces by airborne pesticide and its subsequent release back into the atmosphere. He suggests that such a mechanism could smooth-out regions of "high" atmospheric contamination.

4. Entry by wind erosion and water spray

Pesticides adsorbed to the surfaces of soil particles may be distributed into the air by wind erosion, although this is only likely to be extensive with poorly structured or very dry soils and particularly during mechanical cultivation (Cohen and Pinkerton, 1966). Plant material (particularly pollen but also pieces of foliage etc.), and microbial propagules are frequently of sufficiently low mass or size to become wind-blown. Adsorbed or absorbed pesticides on these particles will thus contribute to the total residue in the atmosphere, although the amounts involved will mostly be extremely small.

The atmospheric entry of pesticides from aquatic situations can occur as a result of aerosol formations from wind and wave-spray. Evaporation of the water droplet will leave any pesticide present as suspended solid particles. Although only minute traces of pesticide will enter the atmosphere at any one time, this route may be important in maintaining low background concentrations of some pesticides (Wheatley, 1973).

B. DISPERSAL

Gaseous or finely particulate pesticide and pollen particles with adsorbed or absorbed pesticide, are much more likely to travel long distances in the atmosphere than are chemicals adsorbed to wind-blown soil particles. The practical difficulties of differentiating these forms of pesticide in the air and the problems of studying transport over long distances has resulted in the majority of studies either quantifying the total residues present in the air or examining short distance dispersal.

After the report by West (1964) that low concentrations of pesticides were present in the air over several Californian cities, many other investigators made similar observations in both urban and rural areas, often from analysis of rainfall samples (see e.g. Cohen and Pinkerton, 1966; Tabor, 1966; Stanly et al., 1971; Södergren, 1972).

The finding of pesticides in dust samples from the northeast tradewinds in the air over Barbados and in the snows of Antarctica (Riseborough et al., 1968; Tarrant and Tatton, 1968; Peterle, 1969; Bidleman and Olney, 1974) have implicated air as a global dispersal agent. Tatton and Ruzicka's (1967) observation of organochlorine pesticides in Antarctic wild animals provides less conclusive evidence, for at least part of the residues may have been transferred via food chains or the ocean. Some regions of the world's surface may be more prone than others to the receipt of atmospherically transported pesticide, for example Koeman and Pennings (1970) proposed that organochlorine insecticides entering the atmosphere in the hot arid areas of Africa are mostly maintained in the air over these regions and deposited only after transport to more temperate parts of the globe. Although some pesticides may travel long distances, the amounts involved are extremely small in comparison with the concentrations near to the areas of application and have negligible environmental significance.

Airborne pesticide, in both particulate and vapour phase forms, is deposited mostly by precipitation (wash-out) in rainfall, but also partly by sedimentation or filtering-out (dry-deposition). Studies of deposition are not particularly well advanced and fail to differentiate between the particle and vapour phase origins of pesticide in rainfall.

Seiber *et al.* (1975) have recently reviewed methods available for the determination of pesticides in air.

The speed of dispersal of pesticides and their persistence in the air has received little attention due to the extreme practical problems. Fungal spore clouds have, however, been detected up to 600 miles away from their source 36 hours after release; nuclear explosion debris has been found to travel the earth's circumference in less than one month, whilst being dispersed in a belt of air with a width of 22° of latitude (Wheatley, 1973). Thus it seems inevitable that pesticides entering the atmosphere are similarly transported and rapidly diluted to extremely low concentrations.

Pitter and Baum (1975) described some of the variety of meteorological and pesticide-related factors likely to affect pesticide dispersion in, and subsequent deposition from, the atmosphere, and how such factors influence mathematical models representing dispersion. Crews *et al.* (1975) have comprehensively reviewed diffusion modelling methods available for atmospheric pollutants. Woodwell *et al.* (1971, 1972) suggested by analogy with carbon dioxide, that DDT could persist for up to seven years in the atmosphere but with a mean residence time probably no longer than four years. The amount of global transport and deposition is, however, still in doubt because of the lack of reliable data.

C. ECOLOGICAL EFFECTS

Edwards (1974) has suggested that the amounts of pesticides in the air are not likely to be harmful to humans, particularly as the amounts respired are much smaller than those consumed with food (see, e.g. Stanly *et al.*, 1971). The investigations of Yobs *et al.* (1972) demonstrated that the concentrations of pesticides in the air were well below the threshold limits recommended by the American Conference of Government Industrial Hygienists. Jegier (1969) classified the levels of exposure of populations to atmospheric pesticides into three groups; firstly, high exposure in manufacturing or application locations, secondly, the exposure of communities in the surrounding areas, thirdly, the background concentrations experienced by populations considerable distances from the major sites of usage. The hazards to workers involved in the manufacture of pesticides and to operators applying the pesticides are potentially the greatest. Whilst recognition of this has led to stringently applied safety precautions in many countries, there are still areas of the world where misgivings must be expressed for the apparent lack of concern for people exposed to these high concentrations.

There is little published information on the effects of pesticides in the air on non-target, airborne macroorganisms. Most vertebrates and invertebrates are, however, unlikely to experience hazards greater than result from their normal food intake. During agricultural and public health pesticide applications, some birds and insects may be exposed to relatively high concentrations but only for short periods of time.

There is evidence for the presence of large numbers of micro-organisms in the air (Gregory, 1961; Hirst and Hurst, 1967; Lidwell, 1967). However, the concentrations of pesticides present in the atmosphere are unlikely to have any widespread undesirable micro-biological effects. In localized aerial conditions with both high pesticide concentration and large numbers of microorganisms, as for example occur where non-target fungal spore clouds coincide with, or are caused by, pesticide applications, the ecological effects on the organisms may be considerably greater. Such occurrences will be infrequent and as far as we are aware have not been investigated.

D. TRANSFORMATIONS

Very little is known of the transformations of pesticides in the atmosphere. However, the presence of light, heat, water, oxygen, ozone, free radicals, particles of organic matter, microorganisms, hydrocarbon pollutants and gases such as nitrogen dioxide and sulphur dioxide, provide the potential for microbial, chemical and photochemical reactions (Crosby, 1973; Moilanen et al., 1975).

The spatial distribution of microorganisms in the air makes it unlikely that there will be much contact between them and pesticides in particulate form. Where the pesticides are in vapour form the concentrations are usually low and only relatively small amounts are likely to be absorbed by the airborne cells. The proportion of airborne pesticide transformed by microorganisms is therefore likely to be extremely small.

Whereas few purely chemical reactions of pesticides in the atmosphere have been recorded, the amount of information on photochemical transformations has increased considerably over the past few years. Most photochemical studies (reviewed by Crosby and Li, 1969 and Watkins, 1974) have been conducted on pesticides in solution or as exposed dry deposits. However, more recently, vapour phase photochemistry has progressively received more attention (Klein and Korte, 1971; Korte and Klein, 1971; Moilanen and Crosby, 1972, 1973; Crosby and Moilanen, 1974; Moilanen et al., 1975).

Pesticides in the troposphere, particularly the upper regions, are exposed to light more intense and with lower wavelengths than those on soil and plant surfaces or in aquatic conditions. For pesticides susceptible to photochemical reactions, the transformations during atmospheric transport could therefore be quite extensive. Whilst meteorological factors will influence the rate of these reactions, Nagl and Korte (1972) have also shown that pollutant gases in the atmosphere might affect both the rate and products of the photochemical transformations.

V. Pesticides in Relation to Animals

Whilst plants are essential and rather passive primary producers, animals are by nature more actively involved in the web of food chains ramifying through the total ecosystem. Animals themselves, however, constitute "environments" where microbial interactions with pesticides may occur. Not surprisingly, in view of the concern as to the environmental acceptability of some pesticides, emphasis in recent years has been placed on the analysis and tracing of pesticide residues in animals. This is particularly true of the persistent organochlorine insecticide residues and animals occupying key positions in the food chains, especially those used for human consumption.

It is difficult to depict an animal which does not present too polarized a view of the possible interactions of pesticides with animals and associated microorganisms. However, the use of the cow (Fig. 5) allows the inclusion of a wider range of factors than would have been the case with, for instance, earthworms, insects, protozoa, birds, rodents, or man himself. In relation to animals, pesticide-microorganism interactions may occur essentially, (a) on the outer surface, (b) in the alimentary tract, (c) in the external area influenced by the animal, principally via excretion, and (d) in animal products.

A. ENTRY

1. Direct and intended entry

Pesticides are deposited on or directly enter some animals through the application of chemicals to control animal pests, for example insecticides, rodenticides, acaricides and molluscicides. There is also the use of chemicals, including anti-microbials, applied to domestic animals by various routes and formulations to counteract pests and pathogens. Many of the pests invoked here are themselves members of the animal

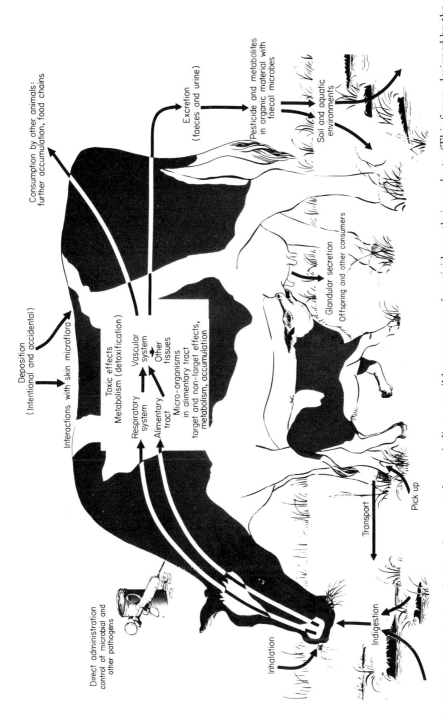

Fig. 5. Pesticides in the animal environment. Arrows indicate possible movement of pesticides or their products. (This figure, designed by the authors, was produced at ICI Plant Protection Division and is reproduced with their permission).

kingdom, for example insects, lice, protozoa and worms. Pesticides for controlling the spread of diseases which have a defined animal-host stage represent another aspect of deliberate application. It has been recognised for some time that the addition of antibiotics to the feed improves the growth performance of animals such as chickens, pigs and calves (Jukes, 1973; see also Hill and Kenworthy, 1970).

2. *Indirect and unintended entry*

The chances of animals being accidentally contaminated by pesticides will to some extent depend on their size, natural distribution, mobility and feeding habits. An animal's surface may be contaminated as it moves through pesticide-treated air, water, soil, or vegetation. Small animals may be inadvertently drenched by pesticide sprays aimed at target plants which provide a habitat. Aerial drift, and to some extent water movement, may extend the area of unintended pesticide–animal contact. Animals may variously ingest pesticide-contaminated water and natural or manufactured food, whilst contaminated air, dust and aerosols may be inhaled.

Wild birds and mammals throughout the world have been found to contain residues and metabolites of organochlorine pesticides, reflecting the use and persistence of these chemicals (Stickel, 1973). Carnivorous birds and mammals tend to accumulate higher residue concentrations than do their herbivorous counterparts. Persistent pesticides may be passed from one animal generation to the offspring via the placenta, milk or eggs (Burdick *et al.*, 1964; Stickel, 1973).

Edwards (1974) reviewed data on the levels of persistent pesticides in animals and animal products. Further information concerning pesticide residues in different animal types may be obtained from the following sources: Kerr and Vass (1973), aquatic invertebrates; Thompson (1973), soil invertebrates; Johnson (1973), fish; Stickel (1973), birds and wild mammals; Davies (1973), man; Foster (1974), poultry.

B. DISPERSAL IN ANIMALS

1. *Fate of pesticides on and within animals*

Physical agencies are likely, in some instances, to influence the dispersal of pesticides from animals. Chemicals applied topically (e.g. sheep dips and dusting powders for insect control on domestic animals) may voltatilize, be displaced by air movement, undergo photochemical change, or be partially removed by contact with other surfaces. Pesticide persistence on the outer surfaces of animals will also be determined by sorptive factors.

It seems reasonable, however, to assume that the overriding factors in the behaviour of pesticides in animals will be of a biological nature and related to the animal's physiology and anatomy. Inhaled pesticide will move to the lungs from where it may either be exhaled or transferred to the circulatory system. The pesticide, or its metabolites, may move along the alimentary tract and be absorbed into the blood and thence relocated in tissues or organs such as the kidneys, liver and brain. The pesticide or metabolites may also be passed out of the animal, via urine, milk, faeces and eggs.

Animals which contain pesticide residues when they die will contribute to the relocation and cycling of pesticides in the ecosystem.

2. Animals as agents of dispersal

In view of their overall capacity for voluntary movement, animals ranging from protozoa and invertebrates to man may be implicated in the translocation of pesticides within and between aerial, terrestrial and aquatic environments. Migrating animals, including birds, may contribute to pesticide dispersal on a small scale to distant environments (Sladen *et al.*, 1966; Robinson, 1973).

Inter-animal transfer of pesticide residues via food chains, with accumulations at higher trophic levels, has been widely reported (Edwards, 1973a, 1974). Such pesticide movement is particularly well documented for chains involving fish and birds. Pesticide accumulation in free-living unicellular protozoa has also been reported. For example, the ciliate *Paramaecium* concentrated DDT several hundred-fold from water (Gregory *et al.*, 1969), indicating that animals feeding on the protozoa could ingest much higher levels than would be obtained directly from the water.

3. Pesticides in foods of animal origin

There is a considerable amount of data on the presence of pesticide residues, especially those of organochlorine insecticides, in foods. Such residues in animal tissues are a likely source of residues in human fat (Edwards, 1974) and several Governments have laid down strict tolerance levels for residues of these pesticides in foods, including those of animal origin. Data relating to many aspects of the presence of pesticides, especially insecticides, in foods, animal tissues, milk and milk products are given by Downey (1972), Duggan and Duggan (1973) and Foster (1974).

There is little evidence that inclusion of antibiotics in the diet of farm animals leads to accumulation of significant residues in the meat (Jukes, 1973). However, organochlorine pesticides have been detected in tissues and eggs of poultry, even without dietary exposure to the chemicals (Stickel, 1973).

C. INTERACTIONS WITH ANIMALS

1. *Detoxication, metabolism and excretion*

Whilst some toxic chemicals may resist metabolism in the animal and be excreted unchanged it is more usual for a degree of transformation to occur. Metabolites may be less, or even more toxic than the original compound (Dorough, 1970).

Most animals appear to have systems capable of metabolism and excretion of xenobiotics. Common reactions in pesticide detoxication and metabolism in animals include hydrolysis, oxidation, hydroxylation and conjugate formation. Enzymes in the liver are often involved. Many pesticides are excreted as conjugates, especially as sulphates and glucuronides and widely different animal types may employ similar degradative routes. Thus, methyl-carbamates were quickly hydrolysed and the products excreted as sulphate and glucuronide conjugates in cows (Dorough, 1970), chicken (Paulson and Zehr, 1971) and in the goat and rat (Paulson *et al.*, 1972). Animal tissues may also transform antibiotics. Thus, chloramphenicol is detoxified in the liver by an acylation to the glucuronide. The glucuronide and the arylamine produced by amide hydrolysis are major urinary metabolites of chloramphenicol (Drasar and Hill, 1974).

2. *Toxic effects and hazards*

Among the effects that have been attributed to the presence of some pesticides, especially organochlorine insecticides, in non-target animals, are toxic and teratogenic actions and reproductive disorders. However, extensive toxicological studies, demonstrating the safety of the product, are now required by many countries prior to their granting permission for sale of the pesticide. The toxic effects of organochlorines on reproductive functions in aquatic animals and in birds are widely documented (e.g. Cope, 1966; Holden, 1972, 1973; Stickel, 1973; Foster, 1974). The development of resistance to pesticides in fish is discussed by Holden (1972).

Herbicides have been regarded generally as less toxic to animals and more readily excreted than insecticides. Although 2,4,5-T is known to be teratogenic at high doses, neither this herbicide nor 2,4-D are considered to affect wildlife or farm animals when applied at normal application rates and there is little evidence for effects on man (Kearney, 1970). There is evidence, however, that some of the commonly used herbicide formulations may be hazardous to farm animals exposed to sprayed vegetation (e.g. Palmer, 1972).

Hocking (1950), considering the dangers of pesticides to honey bees, concluded that whereas insecticides were likely to damage bee populations directly, herbicides might exert an indirect effect by removing bee forage. There have since been a number of conflicting reports on the effects of herbicides on honey bees. More recently Moffett et al. (1972) observed that some herbicides and the oil carriers in which they were formulated were highly toxic.

3. Pesticides in man

Man occupies a unique position with regard to pesticides, for it is he alone who exercises control over their invention, synthesis, formulation, transportation and application, and who may also be accused of permitting their abuse.

Davies (1973) outlined three ways in which humans may be exposed to pesticides:

1. acute exposure (e.g. accidental excessive contamination);
2. chronic exposure (e.g. working in a pesticide "atmosphere");
3. incidental exposure (a consequence of the ubiquity of small amounts of pesticides).

Such information as there is on pesticide residues in humans relates principally to the persistent lipid-soluble organochlorines, especially in adipose tissues, blood and various organs. However, the detection of small quantities of pesticide residues is not in itself evidence of a harmful effect and the public health significance can only be assessed by relating clinical findings to the body burden in poisoned or exposed cases (Durham, 1969).

Concern has also been expressed regarding possible toxic effects of some herbicides to man. Johnson (1971) concluded, however, that the widely used phenoxy-herbicides, including some military defoliants, do not pose a direct threat to man. Similarly, although many agrochemicals could, if used carelessly, adversely affect human health, the hazard is extremely low when weighed against the widescale benefits of such chemicals (Sharratt, 1974).

The incorporation of antibiotics into animal feed has been viewed with concern by some for possible side effects on human public health (Jukes, 1973). This was based on the idea that antibiotic resistance might occur either by episomal transfer from commensal enterobacteria in animals to pathogenic recipient bacteria in man, or more directly, by transfer of resistant pathogens from animals to man. It was also considered that a hazard might arise through ingestion of food containing antibiotic residues. However, Jukes (1973) concluded that in Canada and the US there was insufficient evidence to support these arguments and instead suggested that the responsibility for spreading antibiotic resistance in human pathogens rests primarily with those using antibiotics in clinical practice.

D. INTERACTIONS WITH MICROBES ASSOCIATED WITH ANIMALS

Interactions of pesticides with microbial populations on the surface of animals appear to be likely in circumstances of topical application or accidental deposition of the pesticide. Whereas the ingestion of pesticide-contaminated food signifies the possibility of pesticides initially reacting with the oral microflora, most available data describe pesticide interactions with microbes of the lower alimentary tract.

Evans (1947) mentioned the previous isolation of phenol-utilizing intestinal microorganisms from the faeces of man, horse, cow, sheep and pig. Since many pesticides contain aromatic structures, it may be envisaged that such compounds may also be susceptible to attack by intestinal microorganisms, especially bacteria. Indeed, pesticide-degrading activity need not be the restricted province of individual species and might be more effectively accomplished by the mixed natural populations. Anaerobic and facultatively anaerobic bacteria are provided with a suitable environment in the animal gut and the possibility that the animal may benefit from the association through the activity of pesticide-detoxifying microorganisms is speculative but plausible. The transfer of pesticide-degrading microorganisms to the external environment via the faeces suggests a manner in which animals may influence the microbiological aspects of the fate of pesticides in soil, water and on vegetation.

The reductive dechlorination of pp'DDT to pp'DDD (TDE) was shown to be effected by coliform bacteria isolated from the faeces of untreated rats (Mendel and Walton, 1966). It was concluded that the normal flora of the gastro-intestinal tract were more important as agents for the conversion than the liver, as had previously been suggested (Peterson and Robison, 1964). A similar conclusion was reached by Fries *et al.* (1969) who, like Miskus *et al.* (1965), demonstrated that the conversion of DDT to stable DDD was attributable to the rumen microorganisms in cows. In view of the lower toxicity of DDD isomers, there are important implications in the findings of Miskus *et al.* (1965) and Fries *et al.* (1969) both for the animal and the consumer. Miskus *et al.* (1965) further suggested that the DDD residues detected in milk might be explained by the conversion of DDT-contaminated fodder to DDD by rumen microorganisms, prior to absorption and secretion. Bovine rumen microorganisms also converted organo-phosphorus insecticides to less toxic metabolites which were to some extent secreted in the milk (Ahmed *et al.*, 1958). Examples of pesticide degradation by microbes of the alimentary tract are also seen in the case of some invertebrates. Thus, bacteria isolated from the gut contents of snails, including the schistosome-vector snail, were able to degrade the molluscicide Bayluscide (niclosamide) used for snail control (Etges *et al.*, 1969). The presence of pesticides in milk may antagonize the microorganisms used in the production of cheese and other fermented products (Downey, 1972). However, Ledford and Chen (1969) isolated cheese microorganisms which effected some degradation of *pp'*DDT and *pp'*DDE.

Some antibiotics are metabolized by human intestinal bacteria and with some, such as chloramphenicol and penicillin, enzymic attack on the antibiotic confers resistance on the bacteria. It is interesting to note that bacteria inhabiting mammals such as man can detoxify chloramphenicol by a reaction which differs from that employed by the animal itself (see Section C.1 of this chapter). Enteric bacteria can rapidly inactivate chloramphenicol by acetylation of the hydroxyl groups, whilst slower reactions include nitro-group reduction by nitro-reductase, amidase cleavage of the side chain, and oxidation of the secondary alcohol group.

It has generally been accepted that the shell secreted by birds around the eggs provides a barrier to microbial invasion of the egg contents and that water is necessary for translocation of organisms across the shell (Board, 1966). It can be envisaged that pesticide-induced thinning of egg shells (Stickel, 1973; Foster, 1974) will reduce the effectiveness of the shell barrier, thus increasing the possibility of microbial entry and spoilage. Structural changes in egg shells of DDT-fed quail included a high incidence of large pores, apparently free of cuticle (McFarland *et al.*, 1971), suggesting that the shell's water repellance and barrier effectiveness would be impaired (Board and Halls, 1973).

To some extent it is surprising that the microorganisms which inhabit the external surface and alimentary tract of the diverse fauna have not been more commonly employed in fundamental laboratory investigations of the microbial interactions with pesticides.

VI. Pesticides on Aerial Plant Surfaces

Pesticides are deposited on the aerial surfaces of terrestrial and aquatic plants mostly from the intentional application of dusts and sprays, although small quantities may originate from the volatilized or particulate chemical carried by the air or precipitated in rainwater (Walker, 1966; Beall and Nash, 1971). The concentrations and times of retention of pesticides on plant surfaces will be determined by their chemical properties, by the surface characteristics and rate of uptake by the plant, by environmental losses due to volatilization and wash-off and by factors responsible for any chemical, photochemical and biological degradation (Crafts and Foy, 1962; Ebeling, 1963; Sargent, 1966; Hammerton, 1967; Calderbank, 1968; Hull, 1970). Many of these processes will themselves be influenced by the prevailing environmental conditions, for example rain, radiation, wind, humidity and temperature.

The aerial surfaces of plants (phyllosphere) are known to support populations of microorganisms (Preece and Dickinson, 1971; Diem, 1974) although the often hostile nature of the environment (e.g. low availability of water and the effects of radiation) has led to some controversy over the activity of these organisms. The small amount of evidence from field observations supports the possibility that the saprophytic leaf microflora may have some role to play in the control of pathogenicity (Leben and Daft, 1965; Blakeman, 1971; Blakeman and Fraser, 1971; Crosse, 1971). It is possible that the presence of pesticides (especially fungicides and bactericides) on plant surfaces might not only protect the plant at a particular point in time but could also modify the saprophytic population and its potential plant protective role. There is no clear evidence to support or refute such suggestions and there have been few investigations of the ecological effects of pesticides on plant surfaces. Hislop and Cox (1969), studying the effects of the fungicide captan applied to apples, showed a profound but short-lived inhibitory effect on saprophytes capable of controlling pathogens. Similarly, Stott (1971) found quantitative but not qualitative reductions in the microflora of sugar beet leaves treated with captan and "fungex" and he also observed that within a few weeks of the fungicide applications the leaf flora had returned to normal.

The possible indirect effect of soil-applied pesticides on the microbial populations of the aerial surfaces of plants, via altered patterns of foliar exudation does not appear to have been studied. However, both the amounts and constituents of the leaf exudates can be altered by treatments which alter either the plant nutrient status, as do soil-applied fertilizers and light intensity, or permeability (Last, 1955; Sol, 1967; Blakeman, 1971). It is also possible that the exudates on the leaf surface may themselves modify pesticidal activity (Dunn *et al.*, 1971).

Although the difficulties in studying the plant surface populations of microorganisms and their interactions with pesticides should not be underestimated, it is a task that merits far more attention than it receives at present.

VII. Conclusions

Microorganisms are present in almost all environmental situations in which pesticides are used or enter and it is therefore inevitable that these chemicals will encounter microorganisms in many different circumstances, however inadvertently. In this chapter we have described many of the mechanisms by which pesticides are applied to, otherwise enter and are distributed within, the various components of the environment; together with factors affecting their redistribution and ultimate fate.

Under normal circumstances the amounts of pesticidal chemicals applied in agriculture are very small in comparison with the total volume of the recipient environment. The concentration may be reduced even further by dilution and transformation, and the availability reduced by adsorption etc. However, this should not be taken as a reason for accepting the presence of pesticides without a critical appraisal of their environmental impact. Indeed, over the past decade an ever increasing amount of data has been compiled for both the influence of pesticides on the environment and its constituent biota and vice versa the effect of the environment on these chemicals. Furthermore many Government-controlled agencies scrutinize this information and only when satisfied that any effects are minimal or acceptable will they grant permission for the manufacturing company, or its licensee, to market the chemical.

As a result of both the legislative demands of Governments and the concern of many scientists and environmentalists, the non-target effects of most pesticides (when applied correctly) on the environment are minimal. This is particularly so with the recent generation of pesticides that have received and survived rigorous investigation from a wide range of aspects, including those with a direct bearing on pesticide reactivity with natural environmental biota, including microorganisms.

Any appraisal of the environmental impact of a pesticide should not only consider its behaviour and transformations in the environment and its ecological effects. It must also take into account the toxicological properties and agricultural, economic and public health advantages of the chemical in question.

References

ACREE, F., BEROZA, M. and BOWMAN, M. C. (1963). Codistillation of DDT with water. *J. agric. Fd Chem.* **11**, 278–280.

AHMED, M. K., CASIDA, J. E. and NICHOLS, R. E. (1958). Bovine metabolism of organophosphorus insecticides: significance of rumen fluid with particular reference to parathion. *J. agric. Fd Chem.* **6**, 740–746.

AKESSON, N. B. and YATES, W. E. (1964). Problems relating to application of agricultural chemicals and resulting drift residues. *A. Rev. Entomol.* **9**, 285–318.

ALEXANDER, M. (1973). Nonbiodegradable and other recalcitrant molecules. *Biotechnol. Bioeng.* **15**, 611–647.

ALY, O. M. and FAUST, S. D. (1964). Studies on the fate of 2,4-D and ester derivatives in natural surface waters. *J. agric. Fd Chem.* **12**, 541–546.

ARMSTRONG, D. E., CHESTERS, G. and HARRIS, R. F. (1967). Atrazine hydrolysis in soil. *Proc. Soil Sci. Soc. Am.* **31**, 61–66.

ASHTON, F. M. (1961). Movement of herbicides in soil with simulated furrow irrigation. *Weeds* **9**, 612–619.

BAKER, R. D. and APPLEGATE, H. G. (1970). Effect of temperature and ultraviolet radiation on the persistence of methyl parathion and DDT in soils. *Agron. J.* **62**, 509–512.

BALICKA, N. (1974). Infiltration of pesticides from soil into surface waters. *Pol. Arch. Hydrobiol.* **21**, 173–178.

BARKER, R. E. (1968). The availability of solar radiation below 290 nm and its importance in photomodification of polymers. *Photochem. Photobiol.* **7**, 275–295.

BARNETT, A. P., HAUSER, E. W., WHITE, A. W. and HOLLADAY, J. H. (1967). Loss of 2,4-D in washoff from cultivated fallow land. *Weeds* **15**, 133–137.

BARTHEL, W. F., PARSONS, D. A., MCDOWELL, L. L. and GRISSINGER, E. H. (1966). Potentials for removing pesticides from water. *In* "Pesticides and their Effects on Soils and Water," pp. 128–144. ASA Special Publicn No. 8, Soil Sci. Soc. Am.

BEALL, M. L. and NASH, R. G. (1971). Organochlorine insecticide residues in soybean plant tops: root vs. vapor sorption. *Agron. J.* **63**, 460–464.

BEALL, M. L. and NASH, R. G. (1972). Insecticide depth in soil – effect on soybean uptake in the greenhouse. *J. Environ. Qual.* **1**, 283–288.

BIDLEMAN, T. F. and OLNEY, C. E. (1974). Chlorinated hydrocarbons in the Sargasso Sea atmosphere and surface water. *Science, N.Y.* **183**, 516–518.

BLAKEMAN, J. P. (1971). The chemical environment of the leaf surface in relation to growth of pathogenic fungi. *In* "Ecology of Leaf Surface Micro-organisms," (T. F. Preece and C. H. Dickinson, Eds), pp. 255–268. Academic Press, London.

BLAKEMAN, J. P. and FRASER, A. K. (1971). Inhibition of *Botrytis cinerea* spores by bacteria on the surface of chrysanthemum leaves. *Physiol. Pl. Pathol.* **1**, 45–54.

BOARD, R. G. (1966). The course of microbial infection of the hen's egg. *J. appl. Bact.* **29**, 319–341.

BOARD, R. G. and HALLS, N. A. (1973). The cuticle: a barrier to liquid and particle penetration of the shell of the hen's egg. *Br. Poult. Sci.* **14**, 69–97.

BOVEY, R. W., BURNETT, E., RICHARDSON, C., MERKLE, M. G., BAUR, J. R. and KNISEL, W. G. (1974). Occurrence of 2,4,5-T and picloram in surface runoff water in the Blacklands of Texas. *J. Environ. Qual.* **3**, 61–64.

BOWMAN, M. C., SCHECHTER, M. S. and CARTER, R. L. (1965). Behaviour of chlorinated insecticides in a broad spectrum of soil types. *J. agric. Fd Chem.* **13**, 360–365.

BRO-RASMUSSEN, F., NØDDEGAARD, E. and VOLDUM-CLAUSEN, K. (1970). Comparison of the disappearance of eight organophosphorus insecticides from soil in laboratory soil and in outdoor experiments. *Pestic. Sci.* **1**, 179–182.

BURDICK, G. E., HARRIS, E. J., DEAN, H. J., WALKER, T. M., SKEA, J. and COLBY, D. (1964). The accumulation of DDT in lake trout and the effect on reproduction. *Trans. Am. Fish. Soc.* **93**, 127–136.

BURNS, R. G., EL-SAYED, M. H. and McLAREN, A. D. (1972). Extraction of an urease-active organo-complex from soil. *Soil Biol. Biochem.* **4**, 107–108.

BUTLER, P. A. (1965). Effects of herbicides on estuarine fauna. *Proc. 5th Weed Control Conf.* **18**, 576–580.

BUTLER, P. A. (1967). Pesticides in the estuary. *Proc. Marsh. Estuary Mgmt Symp.* 120–124.

CALDERBANK, A. (1968). The bipyridylium herbicides. *Adv. Pest Control Res.* **8**, 127–235.

CALDERBANK, A. (1971). The fate of paraquat in water. *Outlook on Agric.* **6**, 128–130.

CARO, J. H. and TAYLOR, A. W. (1971). Pathways of loss of dieldrin from soils under field conditions. *J. agric. Fd Chem.* **19**, 379–384.

CARTER, F. L. and STRINGER, C. A. (1970). Residues and degradation products of technical heptachlor in various soil types. *J. econ. Ent.* **63**, 625–628.

CHEPIL, W. S. and WOODRUFF, N. P. (1963). The physics of wind erosion and its control. *Adv. Agron.* **15**, 211–302.

CLARK, C. G. and WRIGHT, S. J. L. (1970). Degradation of the herbicide isopropyl *N*-phenylcarbamate by *Arthrobacter* and *Achromobacter* spp. from soil. *Soil Biol. Biochem.* **2**, 217–226.

CLIATH, M. M. and SPENCER, W. F. (1971). Movement and persistence of dieldrin and lindane in soil as influenced by placement and irrigation. *Proc. Soil Sci. Soc. Am.* **35**, 791–795.

COHEN, J. M. and PINKERTON, C. (1966). Widespread translocation of pesticides by air transport and rain-out. *Adv. Chem. Ser.* **60**, 163–176.

COPE, O. B. (1966). Contamination of the freshwater ecosystem by pesticides. *J. appl. Ecol.* **3** (suppl.), 33–34.

COPE, O. B. (1971). Interactions between pesticides and wildlife. *A. Rev. Entomol.* **16**, 325–364.

CRAFTS, A. S. and FOY, C. L. (1962). The chemical and physical nature of plant surfaces in relation to the use of pesticides and to their residues. *Residue Rev.* **1**, 112–139.

CRANCE, J. H. (1963). The effects of copper sulphate on *Microcystis* and zooplankton in ponds. *Progve Fish Cult.* **25**, 198–202.

CREWS, W. B., BREWER, J. W. and PETERSEN, T. J. (1975). Modeling of atmospheric behavior: A submodel of the dynamics of pesticides. *In* "Environmental Dynamics of Pesticides" (R. Haque and V. H. Freed, Eds), pp. 79–96. Plenum Press, New York.

CROSBY, D. G. (1970). The nonbiological degradation of pesticides in soils. *In* "Pesticides in the Soil: Ecology, Degradation and Movement", pp. 86–94. Michigan State Univ., East Lansing.

CROSBY, D. G. (1973). The fate of pesticides in the environment. *A. Rev. Pl. Physiol.* **24**, 467–492.

CROSBY, D. G. and LI, M.-Y. (1969). Herbicide photodecomposition. *In* "Degradation of Herbicides" (P. C. Kearney and D. D. Kaufman, Eds), pp. 321–363. Marcel Dekker, Inc, New York.

CROSBY, D. G. and MOILANEN, K. W. (1974). Vapor-phase decomposition of aldrin and dieldrin. *Arch. Environ. Contam. Toxicol.* **2**, 62–74.

CROSBY, D. G. and TUCKER, R. K. (1971). Accumulation of DDT by *Daphnia magna*. *Environ. Sci. Technol.* **5**, 714–716.

CROSSE, J. E. (1971). Interactions between saprophyte and pathogenic bacteria in plant disease. *In* "Ecology of Leaf Surface Micro-organisms" (T. F. Preece and C. H. Dickinson, Eds), pp. 283–290. Academic Press, London.

DANIELSON, L. L. and GENTNER, W. A. (1964). Influence of air movement on persistence of EPTC on soil. *Weeds* **12**, 92–94.

DAVIES, J. E. (1973). Pesticide residues in man. *In* "Environmental Pollution by Pesticides" (C. A. Edwards, Ed.), pp. 313–333. Plenum Press, London and New York.

DAVIS, B. N. K. (1968). The soil macrofauna and organochlorine insecticide residues at twelve sites near Huntingdon. *Ann. Appl. Biol.* **61**, 29–45.

DAVIS, B. N. K. (1971). Laboratory studies on the uptake of dieldrin and DDT by earthworms. *Soil Biol. Biochem.* **3**, 221–233.

DEAN, L. A. (1960). Chemistry of pesticides in soils. USDA Agricultural Research Service Symposium, *ARS* **20-9**, 63–69.

DEKKER, G. C., WEINMAN, C. J. and BANN, J. M. (1950). A preliminary report on the rate of insecticide residue loss from treated plants. *J. econ. Ent.* **43**, 919–927.

DEMING, J. M. (1963). Determination of voltatility losses of C^{14}-CDAA from soil surfaces. *Weeds* **11**, 91–96.

DIEM, H. G. (1974). Microorganisms of the leaf surface: estimation of the mycoflora of the barley phytosphere. *J. gen. Microbiol.* **80**, 77–83.

DOROUGH, H. W. (1970). Metabolism of insecticidal methylcarbamates in animals. *J. agric. Fd Chem.* **18**, 1015–1022.

DOWNEY, W. K. (1972). Pesticide residues in milk and milk products in particular the synthetic organic insecticides. *Ann. Bull. Internat. Dairy Fedn.* Part II, pp. 1–51 (plus I to XXXIX).

DRASAR, B. S. and HILL, M. J. (1974). "Human Intestinal Flora", pp. 150–158. Academic Press, London.

DUGGAN, R. E. and DUGGAN, M. B. (1973). Pesticide residues in food. *In* "Environmental Pollution by Pesticides" (C. A. Edwards, Ed.), pp. 334–364. Plenum Press, London and New York.

DUNN, C. L., BEYNON, K. I., BROWN, K. F. and MONTAGNE, J. T. W. (1971). The effect of glucose in leaf exudates upon the biological activity of some fungicides. *In* "Ecology of Leaf Surface Micro-organisms" (T. F. Preece and C. H. Dickinson, Eds), pp. 491–508. Academic Press, London.

DURHAM, W. F. (1969). Body burden of pesticides in man. *Ann. N.Y. Acad. Sci.* **160**, 183–195.

EBELING, W. (1963). Analysis of the basic processes involved in the deposition, degradation, persistence and effectiveness of pesticides. *Residue Rev.* **3**, 35–163.

EDWARDS, C. A. (Ed.) (1973a). "Environmental Pollution by Pesticides", Plenum Press, London and New York.

EDWARDS, C. A. (1973b). Pesticide residues in soil and water. *In* "Environmental Pollution by Pesticides" (C. A. Edwards, Ed.), pp. 409–458. Plenum Press, London and New York.

EDWARDS, C. A. (1974). "Persistent Pesticides in the Environment", 2nd edition, C.R.C. Monoscience Series. Butterworths, London.

EDWARDS, W. M. and GLASS, B. L. (1971). Methoxychlor and 2,4,5-T in lysimeter percolation and runoff water. *Bull. Environ. Contam. Toxicol.* **6**, 81–84.

EDWARDS, C. A., THOMPSON, A. R., BEYNON, K. I. and EDWARDS, M. J. (1970). Movement of dieldrin through soils. I. – From arable soils into ponds. *Pestic. Sci.* **1**, 169–173.

EPSTEIN, E. and GRANT, W. J. (1968). Chlorinated insecticides in runoff water as affected by crop rotation. *Proc. Soil Sci. Soc. Am.* **32**, 423–426.

ETGES, F. T., BELL, E. J. and IVINS, B. E. (1969). A field survey of molluscicide-degrading microorganisms in the Caribbean area. *Am. J. trop. Med. Hyg.* **18**, 472–476.

EVANS, P. R. (1974). Global transport of pesticides by birds. *Chemy Ind.* 197–199.

EVANS, W. C. (1947). Oxidation of phenol and benzoic acid by some soil bacteria. *Biochem. J.* **41**, 373–383.

FARMER, W. J. and JENSEN, C. R. (1970). Diffusion and analysis of carbon-14 labeled dieldrin in soils. *Proc. Soil Sci. Soc. Am.* **34**, 28–31.

FOSTER, T. S. (1974). Physiological and biological effects of pesticide residues in poultry. *Residue Rev.* **51**, 69–121.

FRANK, P. A. (1970). Degradation and effects of herbicides in water. *In* "FAO International Conference on Weed Control", pp. 539–559. Weed Sci. Soc. Am.

FRANK, P. A. (1972). Herbicidal residues in aquatic environments. *Adv. Chem. Ser.* **111**, 135–148.

FRIES, G. R., MARROW, G. S. and GORDON, C. H. (1969). Metabolism of o,p' and p,p'-DDT by rumen microorganisms. *J. agric. Fd Chem.* **17**, 860–862.

GERAKIS, P. A. and SFICAS, A. G. (1974). The presence and cycling of pesticides in the ecosphere. *Residue Rev.* **52**, 69–87.

GERHARDT, P. D. and WITT, J. M. (1965). Pesticide residues and problems resulting from drift from aerial applications of dusts and sprays. *Proc. 12th Int. Congr. Entomol., London*, 1964, 565.

GISH, C. D. (1970). Organochlorine insecticide residues in soils and soil invertebrates from agricultural lands. *Pest. Mon. J.* **3**, 241–252.

GLASS, B. L. and EDWARDS, W. M. (1974). Picloram in lysimeter runoff and percolation water. *Bull. Environ. Contam. Toxicol.* **11**, 109–112.

GOMAA, H. M., SUFFET, I. H. and FAUST, S. D. (1969). Kinetics of hydrolysis of diazinon and diazoxon. *Residue Rev.* **29**, 171–190.

GORING, C. A. I., LASKOWSKI, D. A., HAMAKER, J. W. and MEIKLE, R. W. (1975). Principles of pesticide degradation in soil. *In* "Environmental Dynamics of Pesticides" (R. Haque and V. H. Freed, Eds), pp. 135–172. Plenum Press, New York.

GRAETZ, D. A., CHESTERS, G., DANIEL, T. C., NEWLAND, L. W. and LEE, G. B. (1970). Parathion degradation in lake sediments. *J. Wat. Pollut. Control Fed.* **42**, R76–R94.

GRAHAM-BRYCE, I. J. (1967). Adsorption of disulfoton by soil. *J. Sci. Fd Agric.* **18**, 72–77.

GRAHAM-BRYCE, I. J. (1972). Herbicide movement and availability in soils. *Proc. 11th Br. Weed Control Conf.* 1193–1202.

GRAHAM-BRYCE, I. J. and BRIGGS, G. G. (1970). Pollution of soils. *R.I.C. Reviews* **3**, 87–104.

GRAHAM-BRYCE, I. J. and COUTTS, J. (1971). Interactions of pyrimidine fungicides with soil and their influence on uptake by plants. *Proc. 6th Br. Insectic. Fungic. Conf.* 419–430.

GRAHAM-BRYCE, I. J. and ETHERIDGE, P. (1970). Influence of soil factors on uptake of systemic insecticides by plants. *Proc. VIIth Int. Congr. Plant Protection*, 136–138.

GRAY, R. A. and WEIERICH, A. J. (1965). Factors affecting the vapor loss of EPTC from soils. *Weeds* **13**, 141–147.

GREGORY, P. H. (1961). "The Microbiology of the Atmosphere", Leonard Hill Ltd., London; Interscience Publ. Inc, New York.

GREGORY, W. W., REED, J. K. and PRIESTER, L. E. (1969). Accumulation of parathion and DDT by some algae and protozoa. *J. Protozool.* **16**, 69–71.

GUENZI, W. D. and BEARD, W. E. (1968). Anaerobic conversion of DDT to DDD and aerobic stability of DDT in soil. *Proc. Soil Sci. Soc. Am.* **32**, 522–524.

GUENZI, W. D. and BEARD, W. E. (1970). Volatilization of lindane and DDT from soils. *Proc. Soil Sci. Soc. Am.* **34**, 443–447.

HAMAKER, J. W. (1972). Diffusion and volatilization. *In* "Organic Chemicals in the Soil Environment" (C. A. I. Goring and J. W. Hamaker, Eds), vol. 1, pp. 341–397. Marcel Dekker Inc, New York.

HAMMERTON, J. L. (1967). Environmental factors and susceptibility to herbicides. *Weeds* **15**, 330–336.

HANCE, R. J. (1965). The adsorption of urea and some of its derivatives by a variety of soils. *Weed Res.* **5**, 98–107.

HANCE, R. J. (1969). Further observations of the decomposition of herbicides in soil. *J. Sci. Fd Agric.* **20**, 144–145.

HAQUE, R. and FREED, V. H. (1974). Behavior of pesticides in the environment: "Environmental chemodynamics". *Residue Rev.* **52**, 89–116.

HAQUE, R., SCHMEDDING, D. W. and FREED, V. H. (1974). Aqueous solubility, adsorption and vapor behavior of the polychlorinated biphenyl Aroclor 1254. *Environ. Sci. Technol.* **8**. 139–142.

HARRIS, C. I. (1964). Movement of dicamba and diphenamid in soils. *Weeds* **12**, 112–115.

HARRIS, C. I. (1967). Fate of 2-chloro-*s*-triazine herbicides in soil. *J. agric. Fd Chem.* **15**, 157–162.

HARRIS, C. I. (1969). Movement of pesticides in soil. *J. agric. Fd Chem.* **17**, 80–82.

HARRIS, C. I. and WARREN, G. F. (1964). Adsorption and desorption of herbicides by soil. *Weeds* **12**, 120–126.

HARRIS, C. R. (1970). Laboratory evaluation of candidate materials as potential soil insecticides. III. *J. econ. Ent.* **63**, 782–787.

HARTLEY, G. S. (1969). Evaporation of pesticides. *Adv. Chem. Ser.* **86**, 115–134.

HARTUNG, R. (1975). Accumulation of chemicals in the hydrosphere. *In* "Environmental Dynamics of Pesticides" (R. Haque and V. H. Freed, Eds), pp. 185–198. Plenum Press, New York.

HELLING, C. S., KEARNEY, P. C. and ALEXANDER, M. (1971). Behavior of pesticides in soils. *Adv. Agron.* **23**, 147–240.

HIGGINS, R. E. (1967). Hazards of drift phytotoxicity and hazards to animals. *Proc. W. Weed Control Conf.* **21**, 28–30.

HILL, I. R. and KENWORTHY, R. (1970). Microbiology of pigs and their environment in relation to weaning. *J. appl. Bact.* **33**, 299–316.

HILTON, H. W. and YUEN, Q. H. (1963). Adsorption of several pre-emergence herbicides by Hawaiian sugar cane soils. *J. agric. Fd Chem.* **11**, 230–234.

HILTON, H. W., KAMEDA, S. S. and NOMURA, N. S. (1973). Distribution of picloram residues in sugarcane. *J. agric. Fd Chem.* **21**, 124–126.

HINDIN, E. and BENNETT, P. J. (1970). Occurrence of pesticides in aquatic environments: Part I. Insecticide distribution on an agricultural plot. Tech. Ext. Service, Bull. no 317, Washington State Univ.

HIRST, J. M. and HURST, G. W. (1967). Long-distance spore transport. In "Airborne Microbes" (P. H. Gregory and J. C. Monteith, Eds), pp. 307–344. S.G.M. 17th symp. Cambridge University Press, Cambridge.

HISLOP, E. C. and COX, T. W. (1969). Effects of captan on the non-parasitic microflora of apple leaves. Trans. Br. mycol. Soc. 52, 223–235.

HOCKING, B. (1950). The honeybee and agricultural chemicals. Bee Wld 31, 49–53.

HOLDEN, A. V. (1972). The effects of pesticides on life in fresh waters. Proc. R. Soc. Ser. B, 180, 383–394.

HOLDEN, A. V. (1973). Effects of pesticides on fish. In "Environmental Pollution by Pesticides" (C. A. Edwards, Ed.), pp. 213–253. Plenum Press, London and New York.

HOLDEN, A. V. (1975). Monitoring persistent organic pollutants. In "Organochlorine Insecticides: Persistent Organic Pollutants" (F. Moriarty, Ed.), pp. 1–27. Academic Press, London.

HUANG, J. C. (1971). Effect of selected factors on pesticide sorption and desorption in the aquatic system. J. Wat. Pollut. Control Fed. 43, 1739–1748.

HULL, H. M. (1970). Leaf structure as related to absorption of pesticides and other compounds. Residue Rev. 31, 1–150.

HURLE, K. B. and FREED, V. H. (1972). Effect of electrolytes on the solubility of some 1,3,5-triazines and substituted ureas and their adsorption on soil. Weed Res. 12, 1–10.

HURTIG, H. (1972). Long-distance transport of pesticides. OEPP/EPPO Bull 4, 5–25.

IVIE, G. W. and CASIDA, J. E. (1971a). Sensitized photodecomposition and photosensitizer activity of pesticide chemicals exposed to sunlight on silica gel chromatoplates. J. agric. Fd Chem. 19, 405–409.

IVIE, G. W. and CASIDA, J. E. (1971b). Photosensitizers for the accelerated degradation of chlorinated cyclodienes and other insecticide chemicals exposed to sunlight on bean leaves. J. agric. Fd Chem. 19, 410–416.

JAGNOW, G. and HAIDER, K. (1972). Evolution of $^{14}CO_2$ from soil incubated with dieldrin-^{14}C and the action of soil bacteria on labelled dieldrin. Soil Biol. Biochem. 4, 43–49.

JEFFERIES, D. J. and DAVIS, B. N. K. (1968). Dynamics of dieldrin in soil, earthworms, and song thrushes. J. Wildl. Mgmt 32, 441–456.

JEGIER, Z. (1969). Pesticide residues in the atmosphere. Ann. N.Y. Acad. Sci. 160, 143–154.

JENNINGS, R. C. (1966). Observations on the use of diquat for the control of water weeds. Proc. 8th Br. Weed Control Conf. 2, 586–587.

JOHNSON, B. T., SAUNDERS, C. R., SANDERS, H. O. and CAMPBELL, R. S. (1971). Biological magnification and degradation of DDT and Aldrin by freshwater invertebrates. J. Fish. Res. Bd Can. 28, 705–709.

JOHNSON, D. W. (1973). Pesticide residues in fish. *In* "Environmental Pollution by Pesticides" (C. A. Edwards, Ed.), pp. 181–212. Plenum Press, London and New York.

JOHNSON, H. E. and BALL, R. C. (1972). Organic pesticide pollution in an aquatic environment. *Adv. Chem. Ser.* **111**, 1–10.

JOHNSON, J. E. (1971). The public health implications of widespread use of the phenoxy herbicides and picloram. *Bioscience* **21**, 899–905.

JUKES, T. H. (1973). Public health significance of feeding low levels of antibiotics to animals. *Adv. appl. Mic.* **16**, 1–30.

KAWAHARA, F. K. and BREIDENBACH, A. W. (1966). Potentials for removing pesticides from water. *In* "Pesticides and their Effects on Soils and Water", pp. 122–127. ASA Special Publicn No. 8, Soil Sci. Soc. Am.

KEARNEY, P. C. (1970). Herbicides in the environment. *In* "FAO International Conference on Weed Control, Davis. California, 1970". Weed Sci. Soc. Am.

KEARNEY, P. C. and KAUFMAN, D. D. (Eds) (1969). "Degradation of Herbicides", Marcel Dekker Inc, New York.

KEARNEY, P. C., SHEETS, T. J. and SMITH, J. W. (1964). Volatility of seven *s*-triazines. *Weeds* **12**, 83–87.

KENAGA, E. E. (1972). Guidelines for environmental study of pesticides: Determination of bioconcentration potential. *Residue Rev.* **44**, 73–113.

KERR, S. R. and VASS, W. P. (1973). Pesticide residues in aquatic invertebrates. *In* "Environmental Pollution by Pesticides" (C. A. Edwards, Ed.), pp. 134–180. Plenum Press, London and New York.

KIIGEMAGI, U. and TERRIERE, L. C. (1971). Losses of organophosphorus insecticides during application to the soil. *Bull. Environ. Contam. Toxicol.* **6**, 336–342.

KLEIN, W. and KORTE, F. (1971). Conversion of pesticides under atmospheric conditions and in soil. *In* "Trace Substances in Environmental Health", pp. 71–80. Proc. Univ. Missouri 5th Ann. Conf.

KNIGHT, B. A. G. and TOMLINSON, T. E. (1967). The interaction of paraquat (1 : 1' dimethyl 4 :4'-dipyridylium dichloride) with mineral soils. *J. Soil Sci.* **18**, 233–243.

KOEMAN, J. H. and PENNINGS, J. H. (1970). An orientational survey on the side effects and environmental distribution of insecticides used in tsetse control in Africa. *Bull. Environ. Contam. Toxicol.* **5**, 164–170.

KOLLER, L. R. (1965). *Ultraviolet Radiation*, 2nd edition. Wiley, New York.

KORSCHGEN, L. J. (1970). Soil–food-chain–pesticide wildlife relationships in aldrin-treated fields. *J. Wildl. Mgmt* **34**, 186–199.

KORTE, F. and KLEIN, W. (1971). Recent results on the fate of technical chemicals in the environment on the example of pesticides. *In* "Nuclear Techniques in Environmental Pollution", pp. 535–549. Proc. IAEA Symp. 1970, Vienna.

LAMBERT, S. M., PORTER, P. E. and SCHIEFERSTEIN, R. H. (1965). Movement and sorption of chemicals applied to the soil. *Weeds* **13**, 185–190.

LAST, F. T. (1955). Seasonal incidence of *Sporobolomyces* on cereal leaves. *Trans. Br. mycol. Soc.* **38**, 221–239.

LEBEN, C. and DAFT, G. C. (1965). Influence of an epiphytic bacterium on cucumber anthracnose, early blight of tomato, and northern leaf blight of corn. *Phytopathology* **55**, 760–762.

LEDFORD, R. A. and CHEN, J. H. (1969). Degradation of DDT and DDE by cheese microorganisms. *J. Food Sci.* **34**, 386–388.

LEGRAND, H. E. (1966). Movement of pesticides in the soil. *In* "Pesticides and their Effects on Soils and Water", pp. 71–77. ASA Special Publicn No. 8, Soil Sci. Soc. Am.

LICHTENSTEIN, E. P. (1970). Fate and movement of pesticides in and from soils. *In* "Pesticides in the Soil: Ecology, Degradation and Movement", pp. 101–106. Michigan State University, East Lansing.

LICHTENSTEIN, E. P. and SCHULZ, K. R. (1960). Epoxidation of aldrin and heptachlor in soils as influenced by autoclaving, moisture and soil types. *J. econ. Ent.* **53**, 192–197.

LICHTENSTEIN, E. P. and SCHULZ, K. R. (1961). Effect of soil cultivation, soil surface and water on the persistence of insecticidal residues in soils. *J. econ. Ent.* **54**, 517–522.

LICHTENSTEIN, E. P. and SCHULZ, K. R. (1970). Volatilization of insecticides from various substrates. *J. agric. Fd Chem.* **18**, 814–818.

LICHTENSTEIN, E. P., FUHREMANN, T. W., SCOPES, N. E. A. and SKRENTNY, R. F. (1967). Translocation of insecticides from soils into pea plants. Effects of the detergent LAS on translocation and plant growth. *J. agric. Fd Chem.* **15**, 864–869.

LIDWELL, O. M. (1967). Take-off of bacteria and viruses. *In* "Airborne Microbes" (P. H. Gregory and J. C. Monteith, Eds), pp. 116–137. S.G.M. 17th symp. Cambridge University Press, Cambridge.

LUDLAM, F. H. (1967). The circulation of air, water and particles in the troposphere. *In* "Airborne Microbes" (P. H. Gregory and J. C. Monteith, Eds), pp. 1–7 S.G.M. 17th symp. Cambridge University Press, Cambridge.

MacCOLLOM, G. B., JOHNSTONE, D. B. and PARKER, B. L. (1968). Determination and measurement of insecticide dust particles in atmospheres adjacent to orchards. *Bull. Environ. Contam. Toxicol.* **3**, 368–374.

MATSUMARA, F., BOUSH, G. M. and TAI, A. (1968). Breakdown of dieldrin in the soil by a micro-organism. *Nature, Lond.* **219**, 965–967.

McFARLAND, L. Z., GARRETT, R. L. and NOWELL, J. A. (1971). Normal egg shells and the egg shells caused by organochlorine insecticides viewed by the scanning electron microscope. *In* "Proc. 4th A. Scanning Electron Microsc. Symp.", pp. 377–384. Chicago.

MEGGITT, W. E. (1970). Herbicide activity in relation to soil type. *In* "Pesticides in the Soil: Ecology, Degradation and Movement", pp. 139–141. Michigan State University, East Lansing.

MELLANBY, K. (1967). "Pesticides and Pollution". Collins, London.

MENDEL, J. L. and WALTON, M. S. (1966). Conversion of p,p'-DDT to p,p'-DDD by intestinal flora of the rat. *Science, N.Y.* **151**, 1527–1528.

MENGES, R. M. (1964). Influence of wind on performance of preemergence herbicides. *Weeds* **12**, 236–237.

MENN, J. J., McBAIN, J. B., ADELSON, B. J. and PATCHETT, G. G. (1965). Degradation of *N*-(mercaptomethyl) phthalimide-*s*-(*O*,*O*-dimethyl-phosphorodithioate) (Imidan) in soils. *J. econ. Ent.* **58**, 875–878.

MENZER, R. E., FONTANILLA, E. L. and DITMAN, L. P. (1970). Degradation of disulfoton and phorate in soil influenced by environmental factors and soil type. *Bull. Environ. Contam. Toxicol.* **5**, 1–9.

MERKLE, M. G. and BOVEY, R. W. (1974). Movement of pesticides in surface water. *In* "Pesticides in Soil and Water" (W. D. Guenzi, Ed.), pp. 99–106. Soil Sci. Soc. Am., Madison.

MIDDLETON, J. T. (1965). The presence, persistence and removal of pesticides in air. *In* "Research in Pesticides" (C. O. Chichester, Ed.), pp. 191–198. Academic Press, New York.

MILES, J. R. W., TU, C. M. and HARRIS, C. R. (1971). Degradation of heptachlor epoxide and heptachlor by a mixed culture of soil microorganisms. *J. econ. Ent.* **64**, 839–841.

MISKUS, R. P., BLAIR, D. P. and CASIDA, J. E. (1965). Conversion of DDT to DDD by bovine rumen fluid, lake water and reduced porphyrins. *J. agric. Fd Chem.* **13**, 481–483.

MOFFETT, J. O., MORTON, H. L. and MACDONALD, R. H. (1972). Toxicity of some herbicidal sprays to honey bees. *J. econ. Ent.* **65**, 32–36.

MOILANEN, K. W. and CROSBY, D. G. (1972). Vapor phase photodecomposition of pesticides. *ACS Divn. Pest. Chem. 163rd meeting*, Paper No. 27.

MOILANEN, K. W. and CROSBY, D. G. (1973). Vapor-phase decomposition of *p*,*p*′-DDT and its relatives. *ACS Divn. Pest. Chem. 165th meeting*, Paper No. 21.

MOILANEN, K. W., CROSBY, D. G., SODERQUIST, C. J. and WONG, A. S. (1975). Dynamic aspects of pesticide photodecomposition. *In* "Environmental Dynamics of Pesticides" (R. Haque and V. H. Freed, Eds), pp. 45–60. Plenum Press, New York.

MORIARTY, F. (1975). Exposure and residues. *In* "Organochlorine Insecticides: Persistent Organic Pollutants" (F. Moriarty, Ed.), pp. 29–72. Academic Press, London.

MULLISON, W. R. (1970). Effects of herbicides on water and its inhabitants. *Weed Sci.* **18**, 738–750.

NAGL, H. G. and KORTE, F. (1972). Beitrage zur okologischen chemie – XLVIII Reaktionen von dieldrin mit stickstoffdioxyd und ozon in ultraviolettem licht. *Tetrahedron* **28**, 5445–5458.

NEWBOLD, C. (1975). Herbicides in aquatic systems. *Biol. Conserv.* **7**, 97–118.

NEWMAN, J. F. and WAY, J. M. (1966). Some ecological observations on the use of paraquat and diquat as aquatic herbicides. *Proc. 8th Br. Weed Control Conf.* **2**, 582–585.

NICHOLSON, H. P. (1967). Pesticide pollution control. *Science, N.Y.* **158**, 871–876.

ONSAGER, J. A., RUSK, H. W. and BUTLER, L. I. (1970). Residues of aldrin, dieldrin, and DDT in soil and sugarbeets. *J. econ. Ent.* **63**, 1143–1146.

OSGERBY, J. M. (1973). Processes affecting herbicide action in soil. *Pestic. Sci.* **4**, 247–258.

PALMER, J. S. (1972). Toxicity of a 4-chloro-2-butynyl-*m*-chlorocarbanilate (barban) formulation to cattle, sheep and chickens. *J. Am. vet. med. Ass.* **160**, 338–340.

PAROCHETTI, J. V. and WARREN, G. F. (1966). Vapor losses of IPC and CIPC. *Weeds* **14**, 281–285.

PAROCHETTI, J. V., HEIN, E. R. and COLBY, S. R. (1971). Volatility of dichlobenil. *Weed Sci.* **19**, 28–31.

PARR, J. F., WILLIS, G. H. and SMITH, S. (1970). Soil anaerobiosis: II. Effect of selected environments and energy sources on the degradation of DDT. *Soil Sci.* **110**, 306–312.

PAULSON, G. D. and ZEHR, M. V. (1971). Metabolism of *p*-chlorophenyl *N*-methyl-carbamate in the chicken. *J. agric. Fd Chem.* **19**, 471–474.

PAULSON, G. D., ZEHR, M. V., DOCTER, M. M. and ZAYLSKIE, R. G. (1972). Metabolism of *p*-chlorophenyl *N*-methylcarbamate in the rat and goat. *J. agric. Fd Chem.* **20**, 33–37.

PEACH, M. E., SCHAFFNER, I. P. and STILES, D. A. (1973). Movement of aldrin and heptachlor residues in a sloping field of sandy loam texture. *Can. J. Soil Sci.* **53**, 459–463.

PETERLE, T. J. (1969). DDT in Antarctic snow. *Nature, Lond.* **224**, 620.

PETERSON, J. E. and ROBISON, W. H. (1964). Metabolic products of *p,p'*-DDT in the rat. *Toxic. appl. Pharmac.* **6**, 321–327.

PHILLIPS, F. T. (1964). The aqueous transport of water soluble nematocides through soils. III. – Natural factors modifying the chromatographic leaching of phenol through soil. *J. Sci. Fd Agric.* **15**, 458–463.

PITTER, R. L. and BAUM, E. J. (1975). Chemicals in the air: The atmospheric system and dispersal of chemicals. *In* "Environmental Dynamics of Pesticides" (R. Haque and V. H. Freed, Eds), pp. 5–16. Plenum Press, New York.

PITTS, J. N., KHAN, A. V., SMITH, E. B. and WAYNE, R. P. (1969). Single oxygen in the environmental sciences. Singlet molecular oxygen and photochemical air pollution. *Environ. Sci. Technol.* **3**, 241–247.

PLIMMER, J. R. and KEARNEY, P. C. (1968). Free radical oxidation of pesticides. *Weed Soc. Am. Abstr.* 20.

PREECE, T. F. and DICKINSON, C. H. (Eds) (1971). "Ecology of Leaf Surface Micro-organisms". Academic Press, London.

RAGHU, K. and MacRAE, I. C. (1967). The effect of the gamma-isomer of benzene hexachloride upon the microflora of submerged rice soils. 1. Effect upon algae. *Can. J. Microbiol.* **13**, 173–180.

RAINEY, R. C. (1974). Utilisation of atmospheric processes in the distribution of pesticides. *Chemy Ind.* 199–201.

RILEY, D., TUCKER, B. V. and WILKINSON, W. (1976). Biological unavailability of bound paraquat residues in soil. *In* "Bound and Conjugated Pesticide Residues" (D. D. Kaufman, G. G. Still, G. D. Paulson and S. K. Bandal, Eds), pp. 301–353. ACS Symp. Ser. no. 29.

RISEBOROUGH, R. W., HUGGETT, R. J., GRIFFIN, J. J. and GOLDBERG, E. D. (1968). Pesticides: Transatlantic movements in the north east trades. *Science, N.Y.* **159**, 1233–1236.

RITTER, W. F., JOHNSON, H. P., LOVELY, W. G. and MOLNAU, M. (1974). Atrazine, propachlor and diazinon residues on small agricultural watersheds. *Environ. Sci. Technol.* **8**, 38–42.

ROBINSON, J. (1973). Dynamics of pesticide residues in the environment. *In* "Environmental Pollution by Pesticides" (C. A. Edwards, Ed.), pp. 459–493. Plenum Press, London and New York.

RUDD, R. L. and HERMAN, S. G. (1972). Ecosystem transferal of pesticides residues in an aquatic environment. *In* "Environmental Toxicology of Pesticides" (F. Matsumura, G. M. Boush and T. Misato, Eds), pp. 471–485. Academic Press, New York.

SARGENT, J. A. (1966). The physiology of entry of herbicides into plants in relation to formulation. *Proc. 8th Br. Weed Control Conf.* **2**, 804–809.

SCOTT, H. D. and PHILLIPS, R. E. (1973). Absorption of herbicides by soybean seed. *Weed Sci.* **21**, 71–76.

SEARS, H. S. and MEEHAN, W. R. (1971). Short-term effects of 2,4-D on aquatic organisms in the Nakwasina river watershed, southeastern Alaska. *Pest. Mon. J.* **5**, 213–217.

SEIBER, J. N., WOODROW, J. E., SHAFIK, T. M. and ENOS, H. F. (1975). Determination of pesticides and their transformation products in air. *In* "Environmental Dynamics of Pesticides" (R. Haque and V. H. Freed, Eds), pp. 17–44. Plenum Press, New York.

SETHUNATHAN, N. and MacRAE, I. C. (1969). Some effects of diazinon on the microflora of submerged soils. *Pl. Soil* **30**, 109–112.

SHARRATT, M. (1974). The effects on man of the use of chemicals in agriculture. *In* "Pollution and the Use of Chemicals in Agriculture" (D. E. G. Irvine and B. Knights, Eds), pp. 76–88. Butterworths, London.

SHEETS, T. J. and HARRIS, C. I. (1965). Herbicide residues in soils and their phytotoxicity to crops grown in rotations. *Residue Rev.* **11**, 119–140.

SKIPPER, H. D., GILMOUR, C. M. and FURTICK, W. R. (1967). Microbial versus chemical degradation of atrazine in soils. *Proc. Soil Sci. Soc. Am.* **31**, 653–656.

SLADE, P. (1966). The fate of paraquat applied to plants. *Weed Res.* **6**, 158–167.

SLADEN, W. J. L., MENZIE, C. M. and REICHEL, W. L. (1966). DDT residues in Adelie penguins and a crabeater seal from Antarctica. *Nature, Lond.* **210**, 670–673.

SMITH, D. D. and WISCHMEIER, W. H. (1962). Rainfall erosion. *Adv. Agron.* **14**, 109–148.

SÖDERGREN, A. (1972). Chlorinated hydrocarbon residues in airborne fallout. *Nature, Lond.* **236**, 395–397.

SODERQUIST, C. J. (1973). Dissipation of MCPA in a rice field. M.S. Thesis, Univ. of California, Davis, California.

SOL, H. H. (1967). The influence of different nitrogen sources on (1) the sugars and amino acids leached from the leaves and (2) the susceptibility of *Vicia faba* to attack by *Botrytis fabae*. *Meded Landb-Hogresch. Opzockstns Gent* **32**, 768–775.

SPENCER, W. F. (1970). Distribution of pesticides between soil, water and air. *In* "Pesticides in Soil: Ecology, Degradation and Movement", pp. 120–128. Michigan State University, East Lansing.

SPENCER, W. F. and CLIATH, M. M. (1972). Volatility of DDT and related compounds. *J. agric. Fd Chem.* **20**, 645–649.

SPENCER, W. F. and CLIATH, M. M. (1974). Factors affecting vapor loss of trifluralin from soil. *J. agric. Fd Chem.* **22**, 987–991.

SPENCER, W. F. and CLIATH, M. M. (1975). Vaporization of chemicals. *In* "Environmental Dynamics of Pesticides" (R. Haque and V. H. Freed, Eds), pp. 61–78. Plenum Press, New York.

SPENCER, W. F., CLIATH, M. M. and FARMER, W. J. (1969). Vapor density of soil-applied dieldrin as related to soil-water content, temperature and dieldrin concentration. *Proc. Soil. Sci. Soc. Am.* **33**, 509–511.

SPENCER, W. F., FARMER, W. J. and CLIATH, M. M. (1973). Pesticide volatilization. *Residue Rev.* **49**, 1–47.

STANLY, C. W., BARNEY, J. E., HELTON, M. R. and YOBS, A. R. (1971). Measurement of atmospheric levels of pesticides. *Environ. Sci. Technol.* **5**, 430–435.

STEELINK, C. and TOLLIN, G. (1967). Free radicals in soil. *In* "Soil Biochemistry" (A. D. McLaren and G. H. Peterson, Eds), vol. 1, pp. 147–169. Marcel Dekker Inc., New York.

STICKEL, L. F. (1973). Pesticide residues in birds and mammals. *In* "Environmental Pollution by Pesticides" (C. A. Edwards, Ed.), pp. 254–312. Plenum Press, London and New York.

STICKEL, W. H., HAYNE, D. W. and STICKEL, L. F. (1965). Effects of heptachlor-contaminated earthworms on woodcocks. *J. Wildl. Mgmt* **29**, 132–155.

STOJANOVIC, B. J., KENNEDY, M. V. and SHUMAN, F. L. (1972). Edaphic aspects of the disposal of unused pesticides, pesticide wastes and pesticide containers. *J. Environ. Qual.* **1**, 54–62.

STOTT, M. A. (1971). Studies on the physiology of some leaf saprophytes. *In* "Ecology of Leaf Surface Micro-organisms" (T. F. Preece and C. H. Dickinson, Eds), pp. 203–210. Academic Press, London.

SUTTON, D. L., DURHAM, D. A., BINGHAM, S. W. and FOY, C. L. (1969). Influence of simazine on apparent photosynthesis of aquatic plants and herbicide residue removal from water. *Weed Sci.* **17**, 56–59.

TABOR, E. C. (1966). Contamination of urban air through the use of insecticides. *Trans. N.Y. Acad. Sci.*, Ser. 2, **28**, 569–578.

TALBERT, R. E. and FLETCHALL, O. H. (1965). The adsorption of some *s*-triazines in soils. *Weeds* **13**, 46–52.

TARRANT, K. R. and TATTON, J. O'G. (1968). Organochlorine pesticides in rainwater in the British Isles. *Nature, Lond.* **219**, 725–727.

TATTON, J. O'G. and RUZICKA, J. H. A. (1967). Organochlorine pesticides in Antarctica. *Nature, Lond.* **215**, 346–348.

THOMPSON, A. R. (1973). Pesticide residues in soil invertebrates. *In* "Environmental Pollution by Pesticides" (C. A. Edwards, Ed.), pp. 87–133. Plenum Press, London and New York.

THOMPSON, A. R., EDWARDS, C. A., EDWARDS, M. J. and BEYNON, K. I. (1970). Movement of dieldrin through soils. II. – In sloping troughs and soil columns. *Pestic. Sci.* **1**, 174–178.

TRICHELL, D. W., MORTON, H. L. and MERKLE, M. G. (1968). Loss of herbicides in runoff water. *Weed Sci.* **16**, 447–449.

VALENTINE, J. P. and BINGHAM, S. W. (1974). Influence of several algae on 2,4-D residues in water. *Weed Sci.* **22**, 358–363.

WALKER, A. (1973). Availability of linuron to plants in different soils. *Pestic. Sci.* **4**, 665–675.

WALKER, C. H. (1966). Some chemical aspects of residue studies with DDT. *J. appl. Ecol.* **3** (Suppl.), 213–222.

WARE, G. W. and ROAN, C. C. (1970). Interaction of pesticides with aquatic microorganisms and plankton. *Residue Rev.* **33**, 15–45.

WATKINS, D. A. M. (1974). Some implications of the photochemical decomposition of pesticides. *Chemy Ind.* 185–190.

WEST, I. (1964). Pesticides as contaminants. *Arch. Environ. Health* **9**, 626–633.

WHEATLEY, G. A. (1973). Pesticides in the atmosphere. *In* "Environmental Pollution by Pesticides" (C. A. Edwards, Ed.), pp. 365–408. Plenum Press, London and New York.

WHEATLEY, G. A. and HARDMAN, J. A. (1968). Organochlorine insecticide residues in earthworms from arable soils. *J. Sci. Fd Agric.* **19**, 219–225.

WHITE, J. L. and MORTLAND, M. M. (1970). Pesticide retention by soil minerals. *In* "Pesticides in the Soil: Ecology, Degradation and Movement", pp. 95–100. Michigan State University, East Lansing.

WILLIS, G. H., ROGERS, R. L. and SOUTHWICK, L. M. (1975). Losses of diuron, linuron, fenac and trifluralin in surface drainage water. *J. Environ. Qual.* **4**, 399–402.

WOJTALIK, T. A., HALL, T. F. and HILL, L. O. (1971). Monitoring ecological conditions associated with wide-scale applications of DMA 2,4-D to aquatic environments. *Pest. Mon. J.* **4**, 184–203.

WOODWELL, G. M., CRAIG, P. P. and JOHNSON, H. A. (1971). DDT in the biosphere: Where does it go? *Science, N.Y.* **74**, 1101.

WOODWELL, G. M., CRAIG, P. P. and JOHNSON, H. A. (1972). Atmospheric circulation of DDT. *Science, N.Y.* **177**, 725.

YARON, B., SWOBODA, A. R. and THOMAS, G. W. (1967). Aldrin adsorption by soils and clays. *J. agric. Fd Chem.* **15**, 671–675.

YOBS, A. R., HANAN, J. A., STEVENSON, B. L., BOLAND, J. J. and ENOS, H. F. (1972). Levels of selected pesticides in ambient air of the United States. *ACS Divn. Pest. Chem. 163rd meeting*, Paper no. 24.

Chapter 4

Microbial Transformation of Pesticides

IAN R. HILL

Imperial Chemical Industries, Plant Protection Division,
Jealott's Hill Research Station, Bracknell, Berkshire, England

I. Introduction

There is often a fundamental conflict between the need for a sustained level of biological activity of a pesticide in the environment and the desirability of degradation of the chemical to less toxic and ecologically less damaging molecules. Where the pesticide is applied directly to soil or water, persistence *per se* may be particularly important for maximum agricultural and economic benefit. However, where the intended biological activity of the pesticide is expressed on or in plant foliage there is no requirement for persistence of the excess chemical reaching the soil or water and its rapid degradation in these environments becomes wholly desirable.

Most of the crop protection pesticides developed in the early part of this century were inorganic and very stable in the environment. With the advent and replacement of the inorganic, by organic pesticides, in the 1940s, the problem of persistence was thought to have been solved. It was believed that biological and non-biological degradative agencies, together with volatilization, would soon dispose of these new products. Furthermore, pest control was considerably improved and application rates were reduced. The fallacy of these beliefs was only realized with the discovery that some pesticides, in particular the chlorinated hydrocarbon insecticides, were not being degraded but were accumulating in the environment (see reviews by Alexander, 1973 and Edwards, 1973). As a result of these observations, studies of the persistence of pesticides in soil became quite common, assessments being made of the amount of the parent chemical and/or biological activity remaining at various periods after application. This rather unitary approach in a medium such as soil, with its heterogeneity of biological, chemical and physical phenomena, was extremely limited in the type of information it produced.

Microorganisms possess the enzymic capacity to degrade a large proportion of naturally occurring chemicals. Although most synthetic pesticides do not share an overall structural resemblance with natural products, both of these groups of chemicals have many substituents and bond arrangements in common. Consequently, many of the man-made pesticides introduced into the environment are microbially degraded, mostly by enzymes evolved in response to the presence of natural substrates. The properties of the products of pesticide degradation are usually considerably different to those of the parent, not only in biological activity, but also in toxicological, volatility and absorptive characteristics. It is thus important not only to know how much parent compound or total biological activity remains in soil or water, at intervals after pesticide application, but also to identify and quantify the products of degradation and, where necessary, to investigate the properties of these products.

The manufacturer of a pesticide has an ethical and a legal obligation to study the fate of his product in the environment, the latter obligation being in response to Government requirements preceding sales of the chemical (see Chapters 1 and 5). Knowledge of the fate, distribution and activity of the pesticide in soil or water can also provide information on which to base modifications in the pesticidally active ingredient or its formulation additives.

Further discussion in this chapter is mainly restricted to microbial transformations of pesticides. It is, however, essential to appreciate that in the environment these reactions cannot be viewed entirely in isolation but are frequently inseparable from the chemical, photochemical and other biological, reactions, which may provide alternative routes or stages within an essentially degradative pathway.

II. Pesticide Transformations

A. DEFINITIONS OF TERMINOLOGY

Many of the terms used to describe the chemical changes undergone by pesticides are often poorly defined and wrongly used in the literature.

Frehse (1976) has recently discussed and defined *persistent* and *persistence*. These are behavioural terms defining the period for which a pesticide remains unaltered. No strict definitions can be applied as they are qualified by subjective judgements of the observer. *Recalcitrant* is used by some authors to indicate substances which can persist unchanged in the environment for many years.

The term *degradation* is frequently used to denote any change in molecular structure of the pesticide or its products but perhaps should more correctly be restricted to only "destructive" changes of the chemical to simpler components. Some authors have invoked an even more rigid definition, referring to the capacity of the molecule to be completely mineralized, that is, broken down aerobically to products such as carbon dioxide, water, sulphate, nitrate, ammonia, dinitrogen, etc., and anaerobically to methane and hydrogen sulphide also. Terms such as *decomposition* and *breakdown* are generally considered as analogous to degradation. *Disappearance* and *dissipation*, regularly used in the literature, are far less specific and usually only refer to the "apparent loss" (within the constraints of the techniques applied) of the parent pesticide or named transformation products. Processes such as conjugation, epoxidation and condensation are classified as *additive*, where the addition of a chemical group to the pesticide or its products results in a new product with some degree of stability. Additive reactions do not include those where a transient increase in molecular size occurs as an integral part of a degradative process. The term *transformation* encompasses any change in the chemical under study, whether degradative or additive.

Many terms are used to relate chemical changes to alterations in the biological activity of the pesticide and its products against target and non-target organisms. *Toxification* (or *toxication*) describes reactions in which the toxicity of the molecule, to defined target or non-target organisms, is enhanced. Brooks (1972) has, for example, described six types of enzyme-catalysed reactions of organophosphorus insecticides which can result in products of increased cholinesterase inhibitory activity. *Detoxification* (or *detoxication*) implies the converse reaction, the toxicity of the chemical being decreased. *Activation* is usually restricted to transformations whereby an inactive or insufficiently active chemical attains the desired pesticidal activity against the target organism after entry into the environment. It is also used in some contexts to indicate changes in the chemical resulting in an altered spectrum of biological activity. Considerable numbers of activation reactions have been reported and many have been attributed to microorganisms.

Biological involvement in a pesticide transformation may be indicated by the use of the prefix *bio-*, as in *bio-degradation*, *bio-transformation*, etc. Active biological participation, involving enzymes, justifies use of the term *metabolism*. Prior to use of this term, proof of enzymic participation should be available from studies using isolated, purified enzyme preparations.

Most of the remaining terminology encountered is either self-explanatory or too specific to discuss here.

B. MICROBIAL POTENTIAL

In the environments to which pesticides are applied there is generally a wealth of different types of microorganisms. The characteristics of the indigenous populations are, at least in part, determined by the available nutrient sources but also by the physico-chemical properties of their surroundings. In soil, the most striking example of this is shown by the enhanced numbers and metabolic activities of microbial populations in the region of plant roots. In aquatic conditions, the nature of the microflora may also be influenced by the surrounding soils and by the industrial or sewage effluents with which the water is polluted. These considerations have been discussed in Chapter 2.

Microorganisms are able to degrade a wide variety of chemicals, from simple polysaccharides, amino acids, proteins, lipids, etc. to the more complex materials such as plant residues, crude oils, waxes and rubbers. Indeed the numbers and types of chemicals metabolized are enormous, and it was this potential that led to the hypothesis of microbial infallibility. This proposed that somewhere some type of organism exists which, under suitable conditions, can degrade any chemical theoretically capable of being degraded. The proposal itself is impossible to confirm or refute. There are, however, some naturally occurring materials (e.g. humic acids, surface components of pollen grains, chitin and gaseous and liquid hydrocarbons) that have remained almost unaltered in the environment for at least many hundreds of years (Alexander, 1973). Whilst some of these are suitable substrates for microbial metabolism in culture media, indicating that environmental factors rather than microbial fallibility are responsible for their persistence in situ, others such as humic acids are little altered even under laboratory conditions highly conducive to microbial activity.

The majority of pesticides are degraded fairly rapidly in the environment and even under conditions of repeated annual applications are unlikely to accumulate. The enzymes involved in the metabolism of pesticides may be constitutive (Kaufman, 1965) or may require induction, either by the pesticide or an alternative chemically related inducer (Alexander 1965a, b). Tu et al. (1968) observed that whilst a thermoactinomycete possessed constitutive enzymes for the conversion of aldrin to dieldrin, a Fusarium sp. required a period of adaptation. Whiteside and Alexander (1960), studying the degradation of 2,4-D in soil, showed that the pesticide was degraded after a lag phase. Following a second application of 2,4-D, the herbicide was degraded more rapidly and without a lag period; degradation was even faster after a third application. Numerous investigators have reported similar observations with other pesticides (see e.g. Kaufman, 1964; McGrath, 1976). Audus (1960) suggested that microorganisms could develop the ability to degrade pesticides either by chance mutation or enzyme adaptation. Where a lag phase is observed in the degradation of a pesticide by natural microbial populations, it is difficult to establish whether this results from enzyme induction in cell communities already present and the multiplication of such cells or from proliferation of mutants capable of degrading the compound by constitutive or adaptive enzymes. The absence of a lag phase does not necessarily indicate the presence of constitutive enzymes but may be due to co-metabolic transformation of the pesticide, the alternative substrate from which the organism obtains its growth and energy source already being present in the environment.

Only a small proportion of pesticides are extremely persistent in the environment with considerable residues detectable many years after

application (Alexander, 1973; Edwards, 1973). Many of these pesticides provided valuable crop protection at economic rates and thus
became extensively used throughout the world. Alexander (1966b)
proposed several possible explanations for the prolonged persistence
of such compounds in the environment. They were:

(a) Organisms present may lack any enzymatic or biological potential for degrading the pesticide.

(b) The pesticide may be unable to penetrate the cell.

(c) The concentration or physical nature of the pesticide in the
environment may inhibit organisms or their enzymes.

(d) The pesticide molecule may be inaccessible due to adsorption
or to being coated with relatively impenetrable compounds.

(e) The steric configuration of the pesticide molecule prevents or
hinders enzymic attack.

(f) The environment is toxic or deficient in some factor essential
for growth of microorganisms able to degrade the pesticide.

However, microorganisms have been isolated from soil, aquatic and
animal sources that can, in culture media, degrade many of the
environmentally persistent pesticides (IRRI, 1966; Patil et al., 1970;
Matsumura and Boush, 1971). Furthermore, whilst many organochlorine insecticides (e.g. γ-BHC, DDT, methoxychlor) are persistent
in non-flooded soil they have been found to degrade under flooded,
anaerobic conditions (Hill and McCarty, 1967; Newland et al., 1969;
Sethunathan, 1973; Watanabe, 1973). These findings indicate that in
many cases environmental, or a combination of environmental and
microbial factors, and not microbial fallibility are responsible for
prolonging persistence.

From studies in nutrient media it has become evident that many
microorganisms cannot use pesticides as a carbon or energy source but
will metabolize them in the presence of alternative substrates
(Matsumura and Boush, 1971; Focht, 1972; Buswell and Cain, 1973).
This phenomenon has been termed co-metabolism, an extension of
the original co-oxidative mechanism reported by Leadbetter and
Foster (1959). Reviews by Raymond et al. (1971) and Horvath (1972)
have adequately described the occurrence of this process with both
pesticidal and non-pesticidal chemicals, and have presented evidence
for its existence in natural environments. It is therefore possible that
the stability of some pesticides in the environment may in part be due
to a lack of suitable co-metabolic sources of carbon and/or energy.

Some pesticides which are degraded in vitro may persist in the
environment because the organisms responsible for their degradation
lack competitive ability and are consequently uncommon in that

environment. Hegeman (1971), reviewing the evolution of metabolic pathways in bacteria, suggested that in many cases organisms have not yet "come to grips with the class of refractory substrates in question". He added that there is some hope that the metabolic apparatus necessary to transform the man-made compounds now accumulating in the soil and water will evolve in the microorganisms of these habitats; but stressed that even if the desired metabolic sequences could be attained by mutation under laboratory conditions, it is still impossible to gauge the likely competitive ability of any such organism on placing it in the environment.

C. FACTORS AFFECTING PESTICIDE TRANSFORMATIONS IN THE ENVIRONMENT

The routes and/or rates of transformations of pesticides in the environment are influenced by environmental factors, agricultural techniques and the properties of the pesticidal product. Many of these factors were discussed in Chapter 3.

Most of the available data on the factors affecting pesticide transformations are from soil studies; aquatic environments and the air are relatively unexplored. Ware and Roan (1970) stated that many of the organisms found in soil are often present in aquatic conditions. They proposed that in such bodies of water the metabolism of the pesticide may be qualitatively, although not necessarily quantitatively, similar to that in soil. However, the biological, physical and chemical properties of agricultural soils and bottom sediments (often anaerobic) of ponds, rivers, etc. are often so distinctly different that extrapolation of soil data to aqueous environments should be viewed with considerable caution.

Experimentally, it is usually only possible to investigate the effects of one or a small number of factors influencing transformation, at any one time. However, both soil and water are heterogeneous environments in which the fate of a pesticide is controlled by the totality of direct interactions between the pesticide and its surroundings; although indirect effects may be equally important. Most abiotic environmental factors have some effect on the volatilization and/or mobility of the pesticide, thus influencing its availability to degradative agencies. Furthermore, both abiotic (environmental and pesticidal) and biotic (e.g. plants) factors can modify the types and numbers of microorganisms present, including those that are able to metabolize the chemical.

1. Environmental factors

In soil and water, pesticide transformations are affected by climate (temperature, precipitation, wind and sunlight), soil properties (aerobic/anaerobic state, organic matter, pH, mineral surfaces, available nutrients) and the biota (flora and fauna; macro- and micro-). In soil, methods of cultivation and irrigation also play a role, whilst in some bodies of water, pressure may indirectly influence transformations.

Most of the environmental factors have been adequately discussed in Chapter 3 and by Burns (1975). However, relatively little attention has been paid to pesticide transformations in either flooded agricultural soils or aquatic environments (see reviews by Kuwatsuka, 1972; Sethunathan, 1973; Watanabe, 1973). There is insufficient data available on which to base predictions of the relative rates (or routes) of degradation of pesticides in flooded and non-flooded soils. Experimental results do however, suggest that whilst hydrolysis, reductive dechlorination and nitro-reduction may be enhanced in the flooded state, dehydrochlorination and ring cleavage are less favoured. Studies in the author's laboratory with the pyrimidine pesticides ethirimol, pirimiphos-ethyl, pirimiphos-methyl and pirimicarb have shown that hydrolytic cleavage of both carbamate and phosphorothioate substituents occurs in flooded and non-flooded soil but heterocyclic ring cleavage is considerably reduced by flooding.

Few studies have investigated the effect of the degree of anaerobiosis in flooded soil, on pesticide degradation. Sidderamappa and Sethunathan (1975) found that the lower the redox potential, the more rapidly were BHC isomers degraded. High temperatures and large amounts of organic matter stimulate microbial activity and in flooded soils will cause a fall in the redox potential. Burge (1971) found that the microbial degradation of DDT, an insecticide persistent in aerobic but not in anaerobic soil, was inhibited by 2% oxygen.

Bacteria and fungi are considered to be mainly responsible for pesticide metabolism in soil. The contribution of the microfauna and microalgae is uncertain and has received relatively little attention. Similarly, the role of extracellular enzymes in transforming pesticides is far from clear. In aquatic or flooded soil situations even less is known of the relative involvement of the different macro- and microorganisms and any extracellular enzymes.

2. Pesticide factors

The chemical structure of the pesticide molecule often has a marked effect upon the rates and routes of both biotic and abiotic transformations. The relationships between structure and biodegradability are more fully discussed in Section III.

The types of formulation components (e.g. solvents, emulsifiers, inert fillers, wetters, activators, adsorbents, buffers, foaming agents, adjuvants and synergists) entering the environment with the pesticidal active ingredient have been described by Ebeling (1963) and Fay and Melton (1973). These additives and any impurities in the product are potentially capable of affecting the soil biota and the soil physical and chemical properties; thus they could indirectly influence the transformation of the pesticide. The type of formulation can also affect volatilization and mobility, consequently modifying the amount of pesticide available for transformation and photodecomposition.

Once in soil or water, many formulation components behave independently of the pesticide, particularly as a result of their different water solubilities and adsorptive properties. However, in static water, in soil where the water movement is upwards and with formulations where the pesticide and the additives remain in close proximity, the conditions will be most conducive to the expression of any effect that the additives may exert on the processes of pesticide transformation.

Attempts have been made to reduce microbial metabolism of pesticides in soil, by incorporating biocidal materials in the formulation (Osgerby, 1973). The large amounts necessary are, however, unlikely to be viewed favourably by environmentalists or by Governments from whom permission for marketing is required.

Granules and other controlled-release formulations, although often manufactured to prolong or control the target activity of the active ingredient are also likely to affect the rate of transformation of the pesticide. Similar, but less pronounced effects may result from the use of different particle sizes of water-insoluble pesticides. Booth (1975, unpublished presentation, ACS Symp. Bound and Conjugated Pesticide Residues; Vail, USA) reported a more rapid degradation of difluronyl added to soil in a finely particulate form, than when applied in a solvent. He concluded that the reduced rate of degradation in the latter case was due to the presence of large particles of the pesticide formed on evaporation of the solvent.

The active ingredient of some pesticides may be manipulated by the manufacturer to control water solubility, as in the case of 2,4-D where the esters are less soluble than the salts. The solubility of the product can influence both its mobility and degradation.

3. Pesticide combinations

Considerable numbers of pesticides are applied to plants, soil and aquatic sites as part of multiple pesticide programmes. Combinations of pesticides in the environment, though commonly resulting from successive applications of different chemicals, may also occur from the simultaneous use of mixtures of different products. The mixture of pesticides may either be formulated by the manufacturer and marketed as a single product, or it may be prepared as a mix by the operator.

Kaufman (1972) has reviewed the relatively small number of studies of transformation of pesticide combinations in soil. He concluded that although the combined presence of different pesticides in soil often had no effect on the behaviour of the individual components (see for example Hubbell et al. 1973), three effects can occur:

(a) Increased degradation rate. Many workers have reported that enzymes capable of transforming pesticides may be induced by the presence of other, chemically related substrates (Steenson and Walker, 1957; Audus, 1960; Miyazaki et al., 1969). The serial application of pesticides with analogous structures could result in similar effects.

(b) Increased persistence has been reported with several pesticide combinations (Kaufman et al., 1970; Kaufman et al., 1971; Wagenbreth and Kluge, 1971; Priest and Stephens, 1975). Lichtenstein (1966) observed similar effects when alkyl benzene sulphonate detergents were added to soils containing the organophosphorus insecticides parathion and diazinon. As yet, such effects have only been reported for the parent pesticide. Kaufman (1972) has pointed out that similar reactions may occur with the pesticide transformation products. The mechanisms responsible, although little studied, could involve either inhibition of microorganisms involved in the transformation processes or competitive enzyme inhibition.

(c) The third effect considered by Kaufman (1972) is the formation of complexes from interaction of the pesticides or their residues. Miller and Lukens (1966) observed a loss of biological activity of metham (methyldithiocarbamic acid) when in the presence of halogenated nematocides and attributed this to the esterification of the metham by the nematocides. Asymmetrical azobenzenes may result from condensation of aniline molecules derived from different aniline-containing herbicides in soil (Bartha, 1969). The importance and frequency of such interactions in the environment is presently unknown and merits further investigation.

4. Agricultural techniques

Application methods, soil cultivation techniques and the nature of the crop cover can all affect the persistence of a pesticide in soil. As with formulation components, these factors mostly exert their effects indirectly by controlling mobility and soil surface adsorption of the pesticide. The amount of pesticide available for transformation in soil or water will be reduced as a result of "losses" by volatilization. Volatilization is related to both soil factors, in particular soil moisture, and chemical characteristics of the pesticide. However, the method of application to soil (e.g. surface or sub-surface), subsequent cultivation techniques and the amount of crop cover can also play a large part in controlling pesticide loss from the soil by this route.

Plants can influence the fate of pesticides both indirectly and directly. The amount of a pesticide and/or its transformation products in soil may be reduced by root uptake. However, this material may re-enter the soil environment in a metabolized form, via root exudation. Similarly, pesticides applied to foliage may enter soil via translocation and root exudation in their original or altered forms, providing further substrates for soil transformations. The rate and route of transformation of pesticides in soil may be modified in the presence of plant roots and the associated microflora. At ICI Plant Protection Division we have shown enhanced rates of degradation of ethirimol (Hill and Arnold, 1973) and a thiazolidin-oxime carbamate (Hill, unpublished data) in soil when barley and maize, respectively, were present.

The rate and frequency of agricultural applications of pesticides have also been shown to affect their rates of transformation in soil. Where soils have been treated with different amounts of pesticide, the higher dose levels almost always result in larger amounts of transformation products (Lichtenstein et al., 1971), although the products formed at the different rates are usually the same. When a pesticide is degraded by inducible enzymes, each successive application is degraded faster until a maximum rate is reached.

III. Microbial Transformations of Pesticides

A. INTRODUCTION

It has proved difficult to make direct determinations of the reactions and routes by which pesticides are transformed by microorganisms in the environment. Chemical, microbial and other biotic reactions cannot be distinguished readily and predictions of the biological reaction pathways have not always been possible because of the "foreign" nature of many of these chemicals to the organisms. As a result, only the broad biological (generally assumed to be microbiological) contribution can be assessed *in situ*, using reaction kinetics or sterilized and non-sterilized soils, whilst the specific processes of metabolism must be studied *in vitro* with the isolated components of the natural microbial populations. The observations from the latter studies are often extrapolated to the field situation and the isolated products of pesticide transformation *in situ* compared to those from *in vitro* microbial metabolism. The dangers of superimposing laboratory observations, from pure culture microbial studies, on environmental situations are discussed in Chapter 5.

Microorganisms differ widely in their ability to metabolize "foreign" chemicals. Whilst a large number of reactions of pesticides have been observed in the presence of microorganisms, many workers have failed to differentiate enzymatic processes from those resulting from the extra- or intracellular conditions of redox and pH, or from the presence of heavy metals, etc.

Although details of many enzyme reactions with pesticides are still lacking, the volume of knowledge is steadily increasing. Studies of the oxygenases, for example, are well advanced and many of these enzymes have been isolated, purified and prepared in crystalline form (Hayaishi and Nozaki, 1969). Investigation of the requirements, often complex, for enzyme activity have mostly progressed simultaneously. *Pseudomonas putida*, for example, can hydroxylate camphor in the presence of NADH, molecular oxygen, a flavoprotein and the iron sulphur protein putidaredoxin (Katagiri *et al.*, 1968).

The principal reactions involved in pesticide metabolism, indeed in the metabolism of most organic compounds include β-oxidation, oxidative dealkylation, thioether oxidation, phosphorothionate oxidation, epoxidation of carbon–carbon double bonds, hydroxylation, aromatic ring cleavage, hydrolysis, dehalogenation, nitro-reduction, condensation and conjugate formation. The number of reactions is few compared to the number of compounds metabolized, reflecting the relatively small number of chemical types. It is no surprise that there have been many observed correlations between the chemical structure of different pesticides and their routes of metabolism. The numbers and types of substituents can all influence the rate of metabolism (Sheets, 1958; Burger *et al.*, 1962; MacRae and Alexander, 1965; Alexander and Lustigman, 1966). The positions of linkage of side chains onto aromatic rings can also affect the reaction rate (Alexander, 1965a).

The pesticide may serve as a carbon, nitrogen or energy source for the microorganism, or may only be a substrate for co-metabolism. There is now considerable evidence that the microbial metabolism of some pesticides involves co-oxidative stages (Horvath, 1970a, 1970b, 1972). However, there is little direct proof of co-oxidation occurring under field conditions.

The site (chromosomal or plasmid) of genetic information for coding the enzymes responsible for pesticide metabolism has received little attention. The occurrence of bacterial drug resistance factors composed of extrachromosomal infectious plasmids and non-infectious resistance determinants is well documented. Waid (1972) has reviewed the circumstantial evidence for the presence of bacterial transfer factors able to confer specific pesticidal degradative ability upon a microbial recipient. He suggests that these factors may carry genetic information controlling the enzymes that are responsible for the degradation of pesticides and may effectively preserve the ability of natural microbial populations to metabolize these chemicals. Johnston and Cain (1973) reported experiments showing a strict correlation between the ability of bacteria to degrade benzene-sulphonate and the presence of a piece of extrachromosomal covalently-closed circular DNA but they were unable to demonstrate any transmission of the factor. Chakrabarty (1972, 1974) has, however, characterized different infectious plasmids, in pseudomonads, specifying enzymes capable of degrading n-octane and salycilate and a further plasmid conferring tolerance to high concentrations of mercury salts. Novick and Roth (1968) characterized similarly carried genes in *Staphylococcus aureus* effecting resistance to metallic ions at high concentrations. The evidence for the extrachromosomal nature of these plasmids is wholly genetic, their isolation not yet having proved possible.

The remainder of this section is devoted to a review of many of the microbial processes relevant to pesticide transformations. Much of the data stems from laboratory studies with microbes grown in controlled culture conditions and from studies of the isolated and purified enzyme systems from these organisms. Information for microbially mediated pesticide reactions under natural environmental conditions has, of necessity, been mostly extracted from the literature on soil studies. Excellent and far more comprehensive reviews have been presented for selected hydrocarbon groups by McKenna (1972), Dagley (1971), Chapman (1972), Gibson (1972), Markovetz (1972) and Callely (1974).

B. β-OXIDATION

Although few pesticides are aliphatic fatty acids, or capable of being transformed to fatty acids, many aromatic pesticides have fatty acid side chains, or substituents readily converted to this form. These units are microbially metabolized by β-oxidation (Fig. 1), a stepwise cleavage of C-2 fragments requiring ATP and coenzyme A and resulting in the formation of activated acetic acid units.

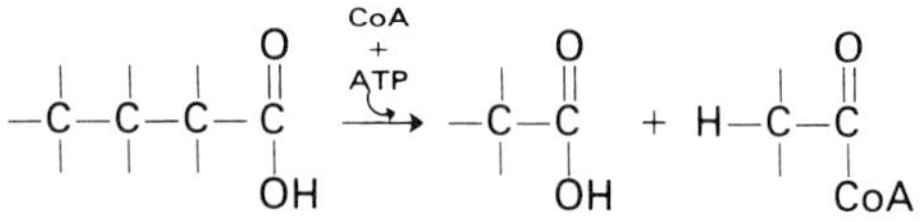

$$
\begin{array}{c}
\overset{\displaystyle O}{\underset{\displaystyle OH}{-C-C-C-C}} \quad \overset{\overset{\text{CoA}}{\underset{\text{ATP}}{+}}}{\longrightarrow} \quad \overset{\displaystyle O}{\underset{\displaystyle OH}{-C-C}} \;+\; \overset{\displaystyle O}{\underset{\displaystyle CoA}{H-C-C}}
\end{array}
$$

Fig. 1. β-Oxidation.

The bio-degradability of n-fatty acid groups, by β-oxidation, has been well documented. Hegeman (1971) suggested that the n-alkyl forms are metabolized following initial oxidation of the α- or β-carbons and Meikle (1972) has described the reaction pathway. Far less is known of branched fatty acids and although simple branched chains may undergo β-oxidation (Kennedy, 1957), mechanisms involving α-oxidation and ω-hydroxylation have also been demonstrated (Jones, 1968; Yano et al., 1971). As β-oxidation requires two protons on the α- and the β-carbons, extensive branching or terminal substitution rendering the α- and β-positions inaccessible to initial attack can impede metabolism. Substituents away from the α- and β-carbons may merely hinder oxidation. Single methyl substituted aliphatic compounds, for example, are more resistant to microbial attack than the corresponding unsubstituted molecule (van der Linden and Thijsse, 1963) whilst dimethyl substitution can render the compound even more refractile (Hammond and Alexander, 1972).

β-oxidation of the ω-phenoxyalkanoate herbicides with more than two carbon atoms in their side chains has been particularly well studied, both in microbial cultures (Byrde and Woodcock, 1957; Webley et al., 1957; Taylor and Wain, 1962; Kearney, 1966; Loos, 1969) and in soil (Gutenman and Lisk, 1964; Gutenman et al., 1964). The alkanoic moieties with large even numbers of carbon atoms are oxidized to products with smaller even numbers, whilst those with odd numbers of carbons may form valerate and propionate derivatives. Further oxidation of the propionate results in the formation of an unstable phenoxyformate and consequent release of the phenol. β-oxidation does not necessarily destroy the biological activity of the pesticide and in some cases may operate as an activation mechanism.

C. OXIDATIVE DEALKYLATION

Alkyl groups of pesticides are commonly attached to nitrogen and oxygen atoms, less frequently to the carbons and only rarely to sulphur and other atoms.

Although N- and O-linked alkyl groups are readily dealkylated by microorganisms it would appear that those directly bonded to a carbon skeleton are often more resistant to metabolism (Gaunt and Evans, 1961; Kearney and Helling, 1969; Probst and Tepe, 1969). The limited number of observations of S-dealkylation may be due to the greater tendency for oxidation of the sulphur to the sulphone and sulphoxide.

These dealkylation reactions are mostly non-specific oxidations requiring mixed-function oxidases and a hydrogen donor such as NADPH.

1. N-*dealkylation*

The microbially mediated removal of an alkyl group from a nitrogen atom (Fig. 2) has been recorded for numerous pesticides and apparently proceeds via unstable hydroxy-methyl intermediates (Geissbühler, 1969; Kaufman and Kearney, 1970). The relative stability of the hydroxy-methyl derivatives of N-methylcarbamates has aided their isolation from microbial cultures grown in the presence of these pesticides (Liu and Bollag, 1971; Bollag and Liu, 1972). Fish *et al.* (1956) from studies of schradan (*bis-NNN'N'*-tetramethylphosphorodiamidic anhydride) also proposed the formation of a transitory N-oxide (Fig. 2) and subsequent rearrangement to the N-hydroxymethyl isomer. The presence of this stage remains unestablished and the complete sequence of intermediates has yet to be isolated in a single dealkylation reaction.

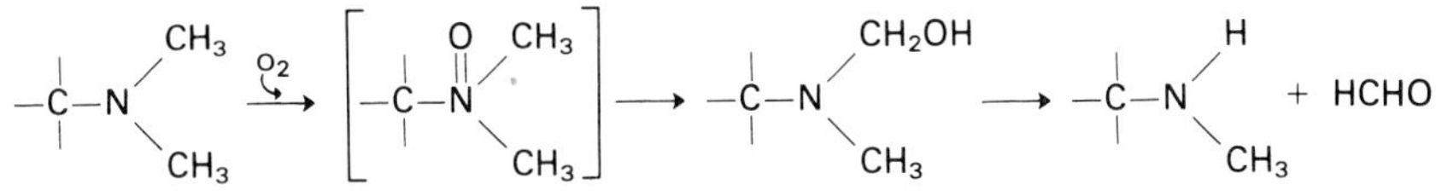

Fig. 2. N-dealkylation.

In mammals, the mixed-function oxidases are tightly bound to membranes and have been difficult to isolate as active preparations. Similar problems have been experienced in preparing cell-free extracts of these enzymes from microorganisms. Laanio *et al.* (1973) have, however, prepared a cell-free extract of *Aspergillus fumigatus* able to dealkylate dinitramine ($N'N'$-diethyl-2,6-dinitro-4-trifluoromethyl-m-phenylenediamine) in the presence of ferrous ions and NADPH.

N-dealkylation is of considerable importance as a stage (often the first) in the metabolism of many pesticides, for example s-triazines, phenyl ureas, alkylcarbamates, alkylamides, dinitrotoluidines and bipyridyls. The presence of an N-dialkyl group usually results in a stepwise reaction via the monoalkyl to the completely dealkylated product (Geissbühler *et al.*, 1963; Geissbühler, 1969). Observations in my laboratory, however, suggest that for some *N*-dialkylated pesticides, the desalkyl derivative is formed much more readily than the didesalkyl product. Whilst *N*-dealkylation has been reported in soil, strong adsorption of the pesticide, as can occur with dimethyl phenyl ureas (Kearney and Plimmer, 1970), may reduce their availability for bio-degradation. Studies with several herbicidal ureas have shown that adsorption increases (fenuron < monuron < diuron < neburon) as degradation in soil decreases (Sheets, 1958; Boerner, 1965; Frank, 1966; Geissbühler, 1969).

The dealkylation of pesticides in animals may, in part, be due to intestinal microorganisms. Dealkylated products of trifluralin (2,6-dinitro-*NN*-dipropyl-4-trifluoromethylaniline) have been isolated from the faeces or urine of ruminants and rats (Emmerson and Anderson, 1966; Golab *et al.*, 1969). The same products were also formed in artificial rumen fluid (Golab *et al.*, 1969) and by cultures of rumen bacteria (Williams and Feil, 1971).

N-dealkylation does not completely detoxify all pesticides (Geissbühler *et al.*, 1963; Dalton *et al.*, 1966; Kaufman and Kearney, 1970) and can even result in an increase in biological activity (Klowzowski, 1965; Kaufman and Kearney, 1970). The target activity of insecticides is, however, mostly lost after *N*-dealkylation.

There have been relatively few observations of *N*-dealkoxylations, although Tweedy *et al.* (1970b) observed loss of the alkoxy group from metobromuron (3-[4-bromophenyl]-1-methoxy-1-methyl urea) in fungal culture. The methyl groups on the quaternary bipyridyl herbicides diquat (1,1'-ethylene-2,2'-bipyridylium ion) and paraquat (1,1'-dimethyl-4,4'-bipyridylium ion) can be microbially *N*-dealkylated (Funderburk and Bozarth, 1967). Meikle (1972) has suggested that this mechanism may also involve mixed-function oxidases. There have been few studies of the ease with which different size alkyl groups are cleaved from pesticides. Kaufman and Blake (1970) did, however, observe that fungal isolates had specific preferences for cleavage of either the *N*-ethyl and *N*-isopropyl substituents of atrazine (2-chloro-4-ethylamino-6-isopropylamino-1,3,5-triazine).

2. O-*dealkylation*

Enzyme mediated *O*-demethylation has long been considered important in the degradation of plant residues incorporated in soil.

Cleavage of the alkoxylated moieties on aromatic compounds (Fig. 3) has been the subject of extensive investigation. Cartwright and

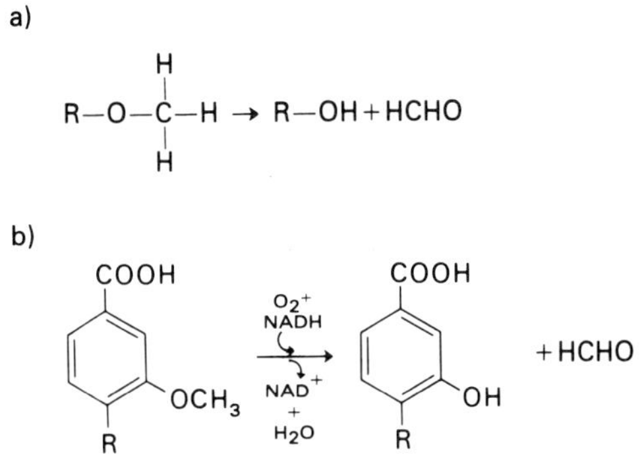

a)

$$R-O-\overset{\displaystyle H}{\underset{\displaystyle H}{C}}-H \rightarrow R-OH+HCHO$$

b)

Fig. 3. O-dealkylation. (a. generalized b. vanillate, m-methoxybenzoate and veratrate (Ribbons, 1970)).

Smith (1967) prepared cell-free preparations that, in the presence of oxygen, reduced glutathione and NADPH, demethylated the methoxyl group of vanillic acid. More recently, from studies of methoxybenzoate O-demethylation, Buswell and Mahmood (1972), have demonstrated a requirement for reduced pyridine nucleotides only. It has since been shown that these reactions are catalysed by mono-oxygenases that introduce hydroxyl groups (Ribbons, 1970; Toms and Wood, 1970), forming hemiacetals which are hydrolysed to leave the phenol and an aldehyde. These enzymes do not appear to be specific to alkyl groups of differing size (Cartwright et al., 1971). The complete pathway still needs to be elucidated clearly.

It is well established that microorganisms can cleave the O-alkyl ether linkages on pesticides; and has been observed in soil (Freed, 1966; Stenersen, 1969; Kapoor et al., 1970), in microbial cultures, their cell free extracts and enzyme preparations (Bollag et al., 1967; Tiedje and Alexander, 1969; Gamar and Gaunt, 1971) and also in rumen fluid (Cook, 1957).

Konrad et al. (1969) showed that in an aqueous medium inoculated with an extract of soil, malathion (S-(1,2-di[ethoxycarbonyl]ethyl)-dimethyl phosphorothiolothionate) was O-dealkylated following a lag phase. In soil, dealkylation may occur but malathion degradation proceeds mainly by chemical, rather than biological activity, with the formation of the thiomalic ester, the free acid and dimethyl phosphate. Pesticides with more than one alkoxyl group may either be multiply-dealkylated, as was observed by Stenersen (1969) when the insecticide bromophos (O-(4-bromo-2,5-dichlorophenyl)OO-dimethyl phosphorothioate) was incubated in the presence of several fungi, or alternatively may lose only one of these substituents. Booth-royd et al. (1961) found that different microorganisms could demethoxylate griseofulvin (7-chloro-4,6,2′-trimethoxy-6′-methylgris-2′-en-3,4′-dione) at different positions (Fig. 4). Henderson (1957), from studies with veratric acid, proposed that aromatic molecules with more than one substituted methoxyl group may be preferentially attacked at the *p*-position. However, studies of the degradation of di- and trihydric derivatives of phenolic acids by rat caecal micro-organisms indicated that the *m*-methoxyl groups were more labile than *p*-methoxyls (Scheline and Longberg, 1965; Scheline, 1968; Meyer and Scheline, 1972).

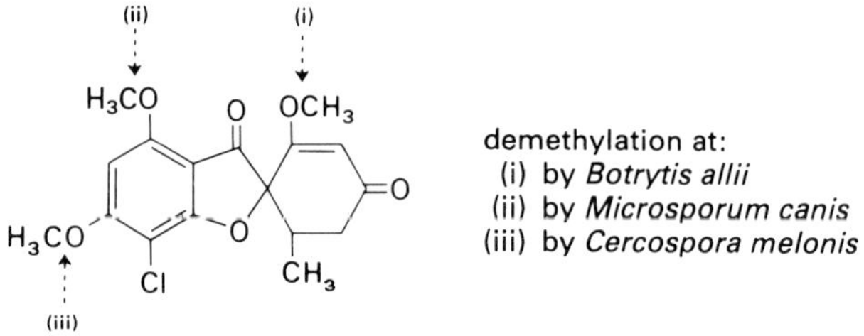
(ii)
(i)
H3CO
O
OCH3
O
O
H3CO
Cl
CH3
(iii)
demethylation at:
(i) by Botrytis allii
(ii) by Microsporum canis
(iii) by Cercospora melonis

Fig. 4. *O*-dealkylation of griseofulvin (Boothroyd *et al.*, 1961).

Whilst *O*-dealkylation has been observed in soil and microbial cultures and attributed to biological activity, Plimmer *et al.* (1968) have shown that similar reactions can occur by free radical reactions in model systems. Free radicals do exist in soil but as yet the transformations in which they participate are poorly understood.

ω-Phenoxyalkanoate herbicides also possess substituents bonded to the aromatic nucleus by an ether linkage. Whilst β-oxidation can progressively remove C-2 fragments from the alkanoate, cleavage can also occur at the ether oxygen (Fig. 5), the latter reaction invariably resulting in a loss of biological activity. Cleavage of the ether bridge occurs by the mechanism already described for *O*-dealkylation (Tiedje and Alexander, 1969; Gamar and Gaunt, 1971) and was demonstrated by Helling *et al.* (1968) to involve a cleavage between the ether-oxygen and the aliphatic side chain. This reaction has been observed in soil

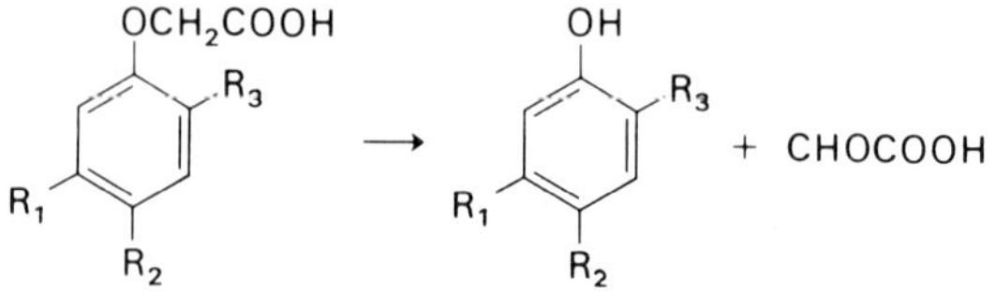
OCH2COOH
R3
R1
R2
OH
R3
R1
R2
+ CHOCOOH

Fig. 5. Oxidative ether cleavage of ω-phenoxyalkanoate herbicides.

(Kearney *et al.*, 1967; Loos, 1969) and in microbial cultures and their cell free extracts (Helling *et al.*, 1968; Loos, 1969).

MacRae and Alexander (1963) and Tiedje and Alexander (1969) have shown that a reductive cleavage can also occur, at the ether-oxygens with alkanoate groups containing more than two carbon atoms in the backbone of the aliphatic chain (Fig. 6).

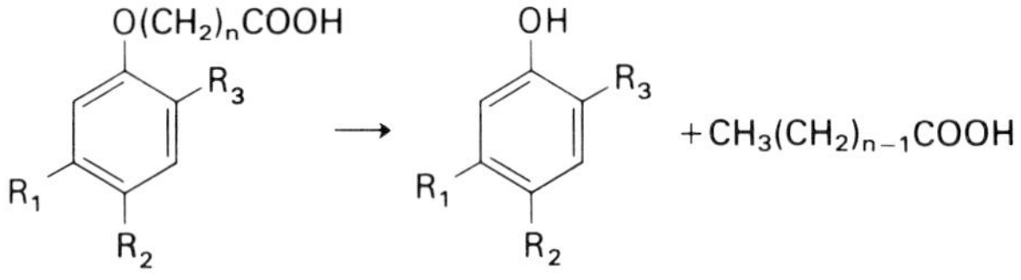

Fig. 6. Reductive ether cleavage of ω-phenoxyalkanoate herbicides.

O-dealkylation is an important mechanism for removal of the biological activity of pesticides, in particular the organophosphorus insecticides and the phenoxyalkanoate herbicides.

3. C-*dealkylation*

The fate of alkyl groups directly bonded to the carbon skeletons of aromatic and heterocyclic rings is less predictable than those attached to oxygen and nitrogen atoms. Cresols, for example, are metabolized by pseudomonads to methyl catechols, without loss of the alkyl substituent (Bayly *et al.*, 1966; Ribbons, 1966), and to protocatechuic acid after microbial dealkylation (Dagley and Patel, 1957). Harrison and Ribbons (1971) observed that other alkyl substituents, such as butyl and propyl, may also remain attached to the ring during catechol formation.

C-dealkylation, initiated by mono-oxygenases, again results in the formation of hydroxymethyl derivatives (Fig. 7). Further metabolism occurs by successive dehydrogenation, via an aldehyde, to form a carboxylic acid (Dagley and Patel, 1957). Nozaka and Kusunose (1968) prepared a microbial cell-free extract that dealkylated the methyl substituents on both xylene and toluene, prior to hydroxylation and ring cleavage. NADH and FAD were required for maximal activity. Esser (1970) suggested that these reactions are most commonly encountered where direct ring hydroxylation is hindered.

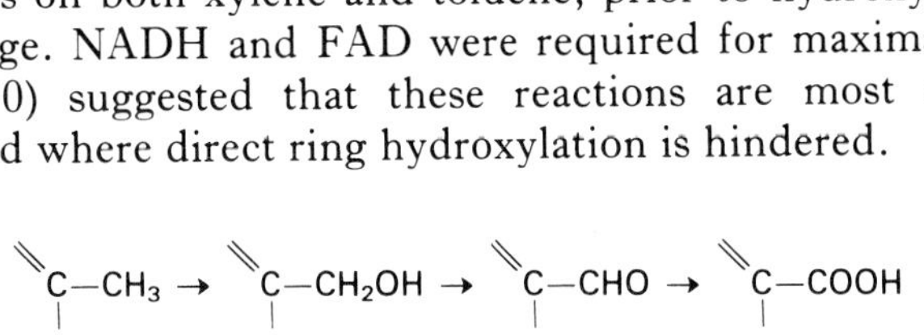

Fig. 7. C-dealkylation.

The oxidative metabolism of ring-substituted methyl groups has been recorded for many aromatic pesticides in biological systems, the benzylic alcohol intermediate in many cases having been isolated. In microorganisms, there have been few such observations reported. Wallnöfer *et al.* (1972) showed that the methyl substituents of the fungicide furcarbanil (2,5-dimethyl-3-furanilide) were oxidized to hydroxy-methyl groups in fungal culture but they were unable to isolate the carboxylic acid derivative. Kapoor *et al.* (1972) have reported microbial *C*-dealkylation of the organochlorine insecticide methoxychlor (1,1,1-trichloro-2,2-di[4-methoxyphenyl]ethane). Conversely, the methyl group of MCPA (4-chloro-2-methylphenoxyacetic acid) was retained past the muconic acid stage when the herbicide was metabolized by a soil bacterium (Gaunt and Evans, 1961, 1971a, b).

Dealkylation of large alkyl groups attached directly to the aromatic rings of pesticides has also been recorded. Both bufencarb (3-(1-methylbutyl)phenylmethyl carbamate and 3-(1-ethylpropyl)phenyl methylcarbamate) and diazinon (*OO*-diethyl-*O*-(2-isopropyl-6-methyl-4-pyrimidinyl) phosphorothioate) (Fig. 8) are oxidized at the alkyl side chain, in soil and in microbial culture (Getzin and Rosefield, 1966; Gunner *et al.*, 1966; Tucker and Pack, 1972).

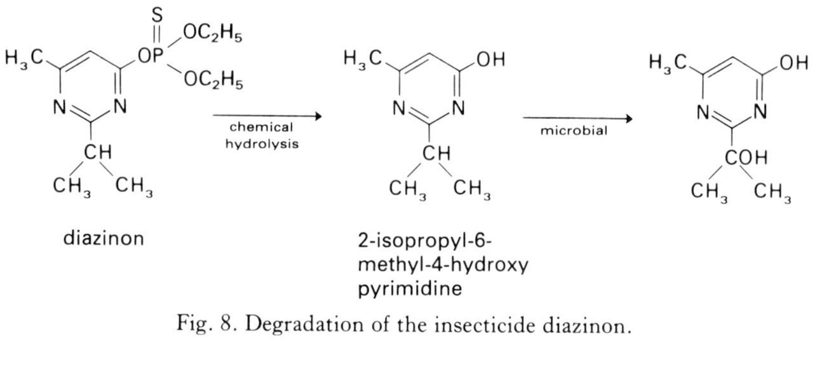

Fig. 8. Degradation of the insecticide diazinon.

D. THIOETHER OXIDATION

The divalent thioether sulphur moiety present in many pesticides is readily oxidizable to the sulphoxide but the reaction then ceases or only proceeds slowly to form the sulphone (Fig. 9), possibly due to the increased polarity of the sulphoxide.

Thioether oxidation has been observed in soil (Menn *et al.*, 1960; Bull, 1968; Chin *et al.*, 1970; Andrawes *et al.*, 1971), in model ecosystems (Kapoor *et al.*, 1970) and in microbial cultures (Ahmed and

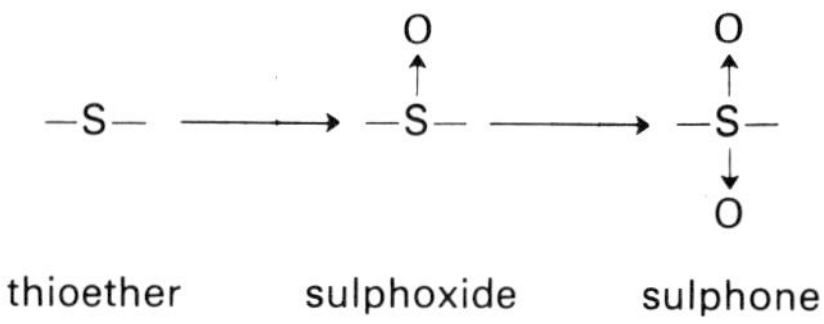

$$\text{—S—} \longrightarrow \overset{\overset{\textstyle O}{\uparrow}}{\text{—S—}} \longrightarrow \overset{\overset{\textstyle O}{\uparrow}}{\underset{\underset{\textstyle O}{\downarrow}}{\text{—S—}}}$$

thioether   sulphoxide   sulphone

Fig. 9. Thioether oxidation.

Casida, 1958). Menn *et al.* (1960) attributed the oxidation of carbophenothion (*S*-(4-chlorophenylthiomethyl)*OO*-diethyl phosphorodithioate) in soil to microbial activity because of the increased persistence of the pesticide under sterile conditions. There have been similar reports for prometryne(2,4-bis-isopropylamino-6-methylthio-1,3,5-triazine) (Gysin, 1962; Plimmer and Kearney, 1969) and aldicarb (2-methyl-2-(methylthio)propionaldehyde *O*-[methylcarbamoyl] oxime) (Coppedge *et al.*, 1967). However, relatively few thioether transformations have been conclusively attributed to microbial activity. The studies reported by Ahmed and Casida (1958) suggest that different organisms might be responsible for formation of the sulphone than for the sulphoxide. Chin *et al.* (1970) observed degradation of carboxin (2,3-dihydro-6-methyl-5-phenylcarbamoyl-1,4-oxathiin) to the sulphoxide in sterilized soil and proposed that microbial activity was not essential for the conversion.

Whilst the sulphur oxidation products commonly retain the desired biological activity of the parent pesticide, they may also possess increased hydrolytic stability (Coppedge *et al.*, 1967; Bull, 1968).

E. DECARBOXYLATION

Although aliphatic carboxyl groups of pesticides are potentially susceptible to microbial oxidation, reduction and conjugation, the majority of observations have been of β-oxidation (Section IIIB). The ability of microorganisms to decarboxylate (Fig. 10) has, however, long been known (see Wallen *et al.*, 1959). For decarboxylation of aliphatic acids, the presence of a suitably activated carboxyl, such as an α-carbonyl group, is normally required (Gunter, 1953; Miyazaki *et al.*, 1970; Wedermeyer, 1967).

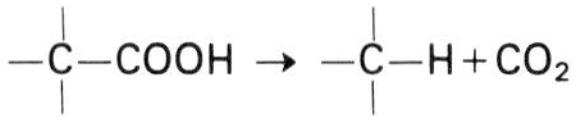

$$-\overset{|}{\underset{|}{C}}-COOH \rightarrow -\overset{|}{\underset{|}{C}}-H + CO_2$$

Fig. 10. Decarboxylation.

The literature contains many examples of the microbial decarboxylation of carboxyl groups on aromatic rings. However, several different mechanisms exist. True decarboxylation, the replacement of the carboxyl by hydrogen, and evolution of carbon dioxide, was demonstrated by Rao *et al.* (1969). They also isolated the carboxylase enzyme responsible for the decarboxylation of *O*-pyrocatechuic acid. In a review of the metabolic processes of gastro-intestinal microorganisms, Scheline (1973) proposed that decarboxylation is primarily a reaction occurring with *p*-hydroxylated compounds. Further ring substitution inhibited the reaction.

More common than the decarboxylation of carboxyl groups on the aromatic ring is their replacement by a hydroxyl group. The formation of 2,5-dihydroxypyridine from nicotinic acid, by C-6 hydroxylation and oxidative decarboxylation, was reported by Behrman and Stanier (1957) (Fig. 11).

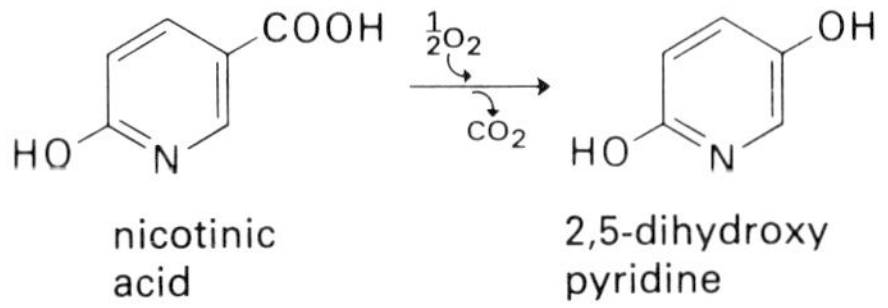

nicotinic   2,5-dihydroxy
acid   pyridine

Fig. 11. Replacement of an aromatic carboxyl by a hydroxyl group (Behrman and Stanier, 1957).

F. EPOXIDATION

Microbially catalysed oxidation of unsaturated carbon–carbon double bonds results in the formation of an epoxide (Fig. 12).

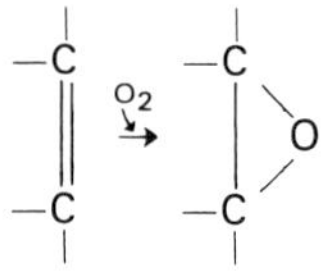

Fig. 12. Epoxidation.

Epoxidation of the chlorinated cyclodiene insecticides was originally observed in soil and from studies using sterile and non-sterile soils (Lichtenstein and Schulz, 1960) was attributed at least in part to microbial activity. Subsequent studies with microbial isolates in laboratory culture has confirmed their ability to convert aldrin (1, 2, 3, 4, 10, 10-hexachloro-1,4,4a,5,8,8a-hexahydro-*exo*-1,4-*endo*-5,8-

dimethanonaphthalene) to the epoxide dieldrin (Patil *et al.*, 1970, 1972; Tu *et al.*, 1968) and heptachlor (1,4,5,6,7,8,8-heptachloro-3a,4,7,7a-tetrahydro-4,7-methanoindene) to its epoxide (Bourquin *et al.*, 1972; Miles *et al.*, 1969, 1971). May and Abbott (1973) have isolated an enzyme system, from a pseudomonad, capable of converting alkenes to their corresponding epoxide in the presence of NADH and molecular oxygen.

G. AROMATIC HYDROXYLATION

Microbial metabolism of many aromatic pesticides is initiated by hydroxylation, a reaction catalysed by mixed function oxygenases and requiring the presence of molecular oxygen and a hydrogen donor (NADP or NADPH). Hydroxylated substituents also occur as intermediates in other major reaction pathways (for example in *N*-dealkylation, Section IIIC).

The hydroxylation of aromatic compounds (Fig. 13), achieved by many microorganisms, introduces more polar groups into the ring as a prerequisite to ring cleavage and use of the ring carbon for microbial growth. The final hydroxylated products of these reactions are stable in comparison with many of the hydroxylated aliphatic derivatives (e.g., the hydroxymethyl intermediates of *N*-dealkylation) and can be readily isolated. Chapman (1972) and Dagley (1972) have reviewed the evidence for the existence of two distinct routes by which microbial oxidative hydroxylation of homocyclic and heterocyclic aromatic rings occurs. In the first, the incorporation of one atom of molecular oxygen, by a mono-oxygenase, introduces a single hydroxyl group into the molecule. The remaining oxygen atom is reduced to water by the pyridine nucleotides NADH and NADPH, although transfer of reducing power to the site of hydroxylation may involve intermediaries such as pteridines (Guroff and Rhoads, 1967) and putidaredoxin (Katagiri *et al.*, 1968). In the second reaction route, using a molecule of oxygen, di-oxygenases catalyse the formation of peroxides which are then dehydrogenated to leave a *cis*-dihydrodiol. Relatively little is known of the mechanisms by which the oxygen is incorporated, although studies of the relative stereochemistry of the hydroxyl groups is beginning to provide some insight into this mechanism (Jeffrey *et al.*, 1975).

Both heterocyclic and non-heterocyclic aromatic rings can be microbially hydroxylated under aerobic and anaerobic conditions without the involvement of molecular oxygen. Pyridine, *N*-methyl-isonicotinic acid and nicotinic acid have been shown to be

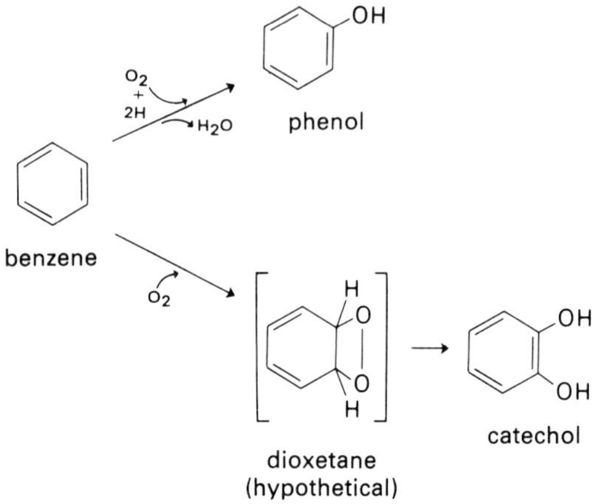

Fig. 13. Aromatic ring hydroxylation (hydroxyl oxygen derived from molecular oxygen).

hydroxylated by an oxygen atom derived from water (Hayaishi and Kornberg, 1952; Dagley and Johnson, 1963; Hirschberg and Ensign, 1971; Orpin *et al.*, 1972b) (Fig. 14).

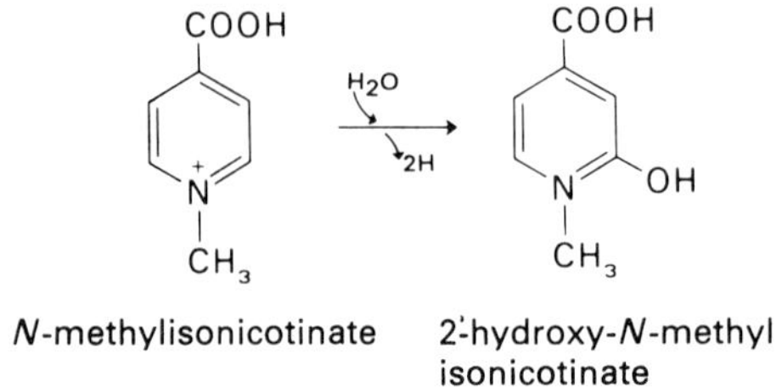

N-methylisonicotinate   2-hydroxy-*N*-methyl
                           isonicotinate

Fig. 14. *N*-methylisonicotinate hydroxylation (hydroxyl oxygen derived from water) (Orpin *et al.*, 1972b).

Dutton and Evans (1969) demonstrated an alternative anaerobic pathway in the photosynthetic autotroph *Rhodopseudomonas palustris*. The anaerobic photometabolism of benzoate involved an initial reduction of the substrate to the cyclic aliphatic acid, cyclohexene carboxylate, followed by hydroxylation with water supplying the hydroxyl group (Fig. 15). Relatively little is known of the environmental significance of hydroxylations involving water.

The position of hydroxyl group insertion on the aromatic ring (and subsequent ring metabolism) depends upon the type, size, number and position of the substituent groups already present on the molecule. Mono-substituted benzene rings are almost always hydroxylated at the

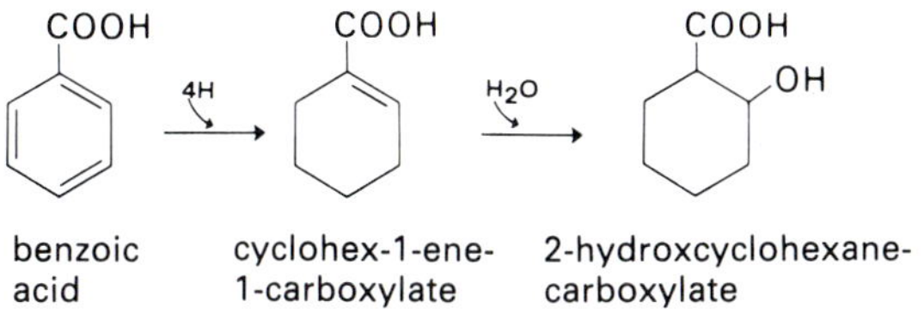

Fig. 15. Anaerobic hydroxylation of benzoate by *Rhodopseudomonas palustris* (Dutton and Evans, 1969).

ortho- and *para-* positions, relative to the substituent. Occupation of these positions by a second substituent can cause an alternative place-ment of the hydroxyl group. For example, 2,4-disubstitution results in hydroxylation at C-5, which is in fact *ortho-* and *para-* to the two substituent positions. Substituted aromatic molecules may be hydrox-ylated at alternative positions by different microorganisms. 4-Hydrox-yphenyl acetic acid may be metabolized to either homogentisic acid (Blakely, 1972) or to homoprotocatechuic acid (Sparnins *et al.*, 1974). Similarly, Davey and Gibson (1974) and Gibson *et al.* (1974) have demonstrated that different pseudomonads metabolize *p-* and *m-*xylene by either an initial oxidation of one of the two methyl groups or by direct oxidation of the aromatic ring.

Woodcock and his colleagues (Byrde and Woodcock, 1957; Clifford and Woodcock, 1964; Faulkner and Woodcock, 1961, 1965) have studied the influence of halogen substituents on microbial hydrox-ylation of the aromatic ring. *Aspergillus niger* hydroxylated phenoxy-acetic acid and related compounds predominately in the *para-* position but to some extent at the *ortho-*carbon. *Meta-*hydroxylation was not observed. Chlorinated phenoxyacetates were, however, less specifically hydroxylated, 2-chlorophenoxy-acetic acid forming all possible hydroxy-derivatives. When the herbicide 2,4-D (2,4-dichloro-phenoxyacetic acid) was provided as the substrate, 2,5-dichloro-4-hydroxyphenoxy acetic acid was isolated, indicating a transfer of a chlorine atom to an alternative position on the ring. Such intramolec-ular migration, induced by hydroxylation and termed "NIH shift", is mainly restricted to *para-*hydroxylation (Daly *et al.*, 1968). Halogens can either be liberated by direct replacement with a hydroxyl (Faulk-ner and Woodcock, 1961, 1965; Johnston *et al.*, 1972), or retained on the ring through aromatic hydroxylation and ring cleavage. The larger the halogen atom the more it can interfere with and reduce aromatic ring oxidation.

The substituents on the aromatic ring also have a considerable influence on the rate of enzymic ring hydroxylation. Alexander and

Lustigman (1966) demonstrated that substituents on a benzene ring increased resistance to biodegradation in the order carboxy- < hydroxy- < amino- < methoxy- < sulphonate- < nitro-. Phenol and benzoic acid are very rapidly metabolized but further substitution of these compounds reduces the rate at which they are metabolized. The observations of Kameda *et al.* (1957) illustrated the effects of substituent position on the degradation of hydroxy-, methoxy-, nitro- and amino- benzoic acids by pseudomonads isolated from soil. Whilst *ortho-* and *para-* forms were metabolized by some of the cultures, the *meta-* isomer was particularly resistant. Cain and Cartwright (1959) obtained similar results with a *Nocardia* sp. grown on isomers of nitrobenzoic acid. Although the results of many such studies are expressed in relation to the rate of ring cleavage it is generally recognized that the observed effects are a reflection of the relative ease with which hydroxylation proceeds.

The majority of investigations of enzymic hydroxylations have been with non-pesticidal chemicals. Whilst far less is known of the processes and intermediates of aromatic pesticide hydroxylation, the evidence from studies of both pesticides and model compounds and from identification of their hydroxylated products suggests that the mechanisms are the same as those demonstrated for their non-pesticidal counterparts. Studies with pesticides have as yet only demonstrated the incorporation of molecular oxygen. Bollag *et al.* (1968) and Bollag *et al.* (1967) have partially purified a microbial enzyme that can hydroxylate 2,4-dichlorophenol, a degradation product of several phenoxyalkanoate herbicides. The enzyme, requiring molecular oxygen and NADPH, also converted other phenols to their corresponding catechols. Some pesticides possess alkane or alkane-forming groups directly bonded on to a carbon atom of the aromatic nucleus (e.g. chlorfenac; 2,3,6-trichlorophenylacetic acid). It has long been known that both the side chain and aromatic ring of non-pesticidal phenylalkanes with an odd number of carbons in the alkane moiety can be degraded. It is only relatively recently, however, that Sariaslani *et al.* (1974) have demonstrated hydroxylation and ring cleavage of a phenylalkane (1-phenyldodecane) with an even number of carbons in the side chain. The alkane underwent β-oxidation to leave the acetic derivative, which remained on the ring through hydroxylation and cleavage.

Microorganisms can also dehydroxylate C- and N- hydroxyl groups. The dehydroxylation of ^{14}C-protocatechuic and ^{14}C-homoprotocatechuic acids in conventional, antibiotic treated and germ-free animals (Dacre *et al.*, 1968; Dacre and Williams, 1962, 1968) was attributed to the enzymic activity of bacteria. *In vitro* microbial studies have shown the dehydroxylation of caffeic and dihydrocaffeic acid by *Pseudomonas* sp. (Perez-Silva and Rodrigues, 1967).

H. AROMATIC, NON-HETEROCYCLIC RING CLEAVAGE

Cleavage of an aromatic ring, by oxygenase enzymes, usually initiates the final stages of biodegradation of the chemical, its utilization as a carbon and energy source and ultimately the release of carbon dioxide and water. Many investigations of the ring cleavage mechanism have been on chemicals with structures simpler than those of most pesticides.

For cleavage of the benzene nucleus at least two hydroxyl groups must be present, and in either an *ortho-* or a *para-* relationship. *Ortho*-dihydric phenols (as e.g. catechol and protocatechuic acid) undergo ring-fission by two distinct mechanisms (Fig. 16) utilizing molecular oxygen and dioxygenase enzymes.

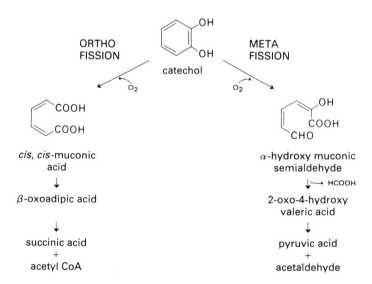

Fig. 16. Ring fission of catechol.

(a) The first involves cleavage between the hydroxyl groups to form a dicarboxylic acid (as for example *cis,cis*-muconic acid from catechol and β-carboxy-*cis,cis*-muconic acid from protocatechuic acid). This reaction is commonly referred to as *ortho* ring fission (Ornston and Stanier, 1966). Further metabolism of dicarboxylic acids by bacteria proceeds via lactones to an oxoadipate which can enter the tri-carboxylic acid cycle through acetyl CoA and a succinate. Cain *et al.* (1968a) suggested that the intermediates of fungal metabolism by the *ortho* route might be different to those described above.

(b) The second pathway, *meta* ring fission (Dagley and Gibson, 1965; Sala-Trepat and Evans, 1971), involves cleavage between a carbon with a hydroxyl substituent and an adjacent non-substituted carbon. The resultant aldehydo- or keto-acid (for example the α-hydroxymuconic semialdehydes from catechol and protocatechuic acid) may be metabolized further by either hydrolytic elimination of formic acid or by dehydrogenation.

Meta-dihydric phenols require the introduction of a third hydroxyl group, to form a 1,2,4-trihydric phenol, before ring fission can proceed via the *ortho* or the *meta* fission pathways (Chapman, 1972).

A third mechanism of ring cleavage, the "gentisate" pathway (Fig. 17), can occur when the ring is 1,4-hydroxylated (i.e. *para*-dihydric). Fission of gentisic acid and related compounds, also catalysed by dioxygenases, is between the carboxylated carbon and the adjacent hydroxyl group (Chapman, 1972) to form a maleyl pyruvate. Further metabolism by isomerization and hydrolysis will form fumarate and pyruvate which can enter the tricarboxylic acid cycle.

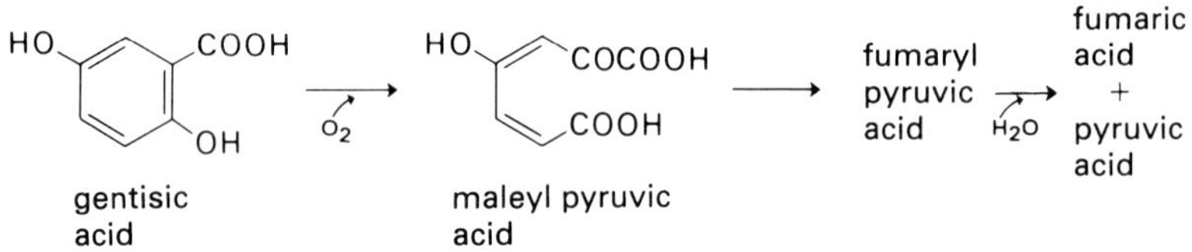

Fig. 17. Ring fission of gentisic acid.

A fourth ring cleavage route, of which very little is known, follows reductive hydroxylation of benzoate (Dutton and Evans, 1969). The hydroxylated cyclohexane carboxylate undergoes dehydrogenation and isomerization before being cleaved by a hydration reaction, to form a pimelic acid (Fig. 18).

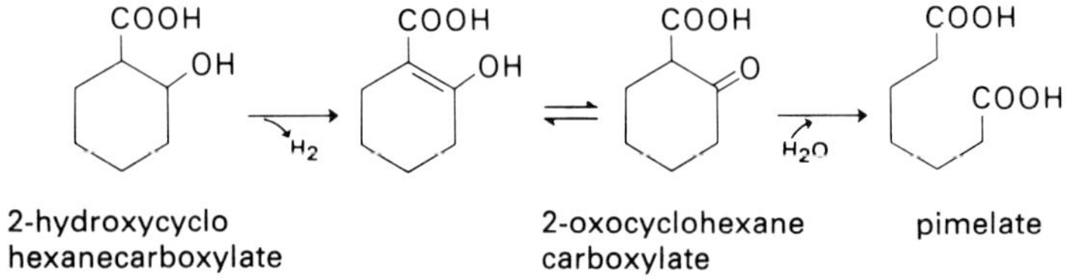

Fig. 18. Ring fission of hydroxycyclohexanecarboxylate by *Rhodopseudomonas palustris* (Dutton and Evans, 1969).

Microbial metabolism of polynuclear aromatics usually follows a stepwise pattern, involving sequential hydroxylation and ring cleavage, until a single ring of the basic catechol, protocatechuic acid or gentisic acid structure remains.

Whether *ortho*-dihydric phenols are metabolized by the *ortho* or *meta* route depends upon the microorganism involved, the structure of the hydroxylated phenol (or its precursors) and, where co-oxidative degradation is evident, upon the growth substrate.

Many aromatic pesticides have ring substituents which may still be present after ring hydroxylation. Schemes for the metabolism of substituted 1,2-dihydric phenols (Chapman, 1972; Meikle, 1972) are valuable aids to the prediction of intermediates in the degradation of similar chemicals, which are pesticidal. Dagley (1971) has suggested that the *meta* fission pathway serves as a general pathway for bacterial metabolism of alkyl substituted aromatic compounds. The electron doning (*ortho-* and *para*-directing) and electron withdrawing (*meta-* directing) effects of substituents can also influence the ring fission route (Meikle, 1972), the former activating the enzymes of the *meta* pathway and the latter those of the *ortho* route. The hydroxylated product need not be an inducer of either route, as for example with catechol where the original substrate (before hydroxylation) apparently induces the appropriate pathway (Hegeman, 1971).

Whilst there are numerous reports of the microbial fission of single and multiple ringed aromatic pesticides in soil and aquatic environments and in culture, there have been relatively few studies in which the route and intermediates have been elucidated. Where detail is available, the metabolic pathways appear to be closely related to the *ortho*, *meta* and "gentisic" acid schemes described. Tiedje *et al.* (1969) using an *Arthrobacter* sp. cell-free extract and Evans *et al.* (1971) with pseudomonads have isolated typical *ortho* fission products from the hydroxylated derivatives of 2,4-D. The chlorine atoms were retained on the ring through ring cleavage. Other phenoxyalkanoate herbicides may be similarly metabolized (Loos, 1969; Menzie, 1969; Gaunt and Evans, 1971a, b), although Horvath's findings (1970b) showed that *meta* fission cannot be excluded from consideration. The bicyclic ring of methylcarbamate insecticides such as carbaryl (1-naphthyl methyl-carbamate) appears to be metabolized via α-naphthol. That is, by a mechanism similar to that reported for naphthalene (Fernley and Evans, 1958; Kaufman, 1970).

Most of the laboratory studies of pesticide metabolism by isolated microorganisms have shown that although numerous bacteria can cleave aromatic rings there are remarkably few similar observations for fungi (and even fewer for algae). This is somewhat surprising for not only can some fungi attack and cleave simple benzyl substrates (Cain *et al.*, 1968a) but Henderson and Farmer (1955) found that ring fission of lignin-related aromatics occurred in the presence of a large range of fungal isolates.

I. AROMATIC, HETEROCYCLIC RING CLEAVAGE

Microorganisms can metabolize heterocyclic compounds by dis-placement or modification of substituent groups and by ring hydroxylation or hydrolysis and cleavage, but in general they are considerably less studied than their homocyclic equivalents. Furthermore, the variety of different types and numbers of the hetero-atoms (Fig. 19), exerting their own specific influences on metabolic reactions, makes predictions of metabolic routes extremely difficult. Meikle (1972) and Callely (1974) have presented comprehensive reviews of the microbial metabolism of pesticidal and non-pesticidal heterocyclic compounds. Although many heterocyclic molecules are rapidly cleaved, others are remarkably stable both in environmental situations and in microbial cultures.

The heterocyclic rings of pesticides are mostly aromatic and 5- or 6-membered, with S, N and O as the hetero atoms (Fig. 19). Whilst the pyridines, pyrimidines and triazines are single rings, a large number of multiple-ringed pesticidal heterocyclics are also manu-factured.

Studies of transformations of these pesticides in environmental situations, as well as in the presence of microbial cultures, have often demonstrated ring cleavage (by evolution of $^{14}CO_2$ from ^{14}C-radiolabelled ring carbon atoms), but the intermediate metabolites and the degradation routes have rarely been demonstrated. Many natural compounds in the environment also contain 5- and 6-membered heterocyclic rings and these have attracted some investigation.

1. 5-Membered rings

5-Membered rings with a pyrrole, furan, thiophen or imidazole struc-ture are all microbially cleaved (Sjaastad and Kay, 1970; Callely, 1974), although ring substituents may retard or inhibit the process. Where 5-membered rings are attached to heterocyclic or non-hetero-cyclic 6-membered rings, e.g. indole, tryptophan and nicotine, the former are usually cleaved first (Stanier *et al.*, 1951; Gherna and Rittenberg, 1962; Gherna *et al.*, 1965; Fujioka and Wada, 1968), with or without initial hydroxylation (Fig. 20).

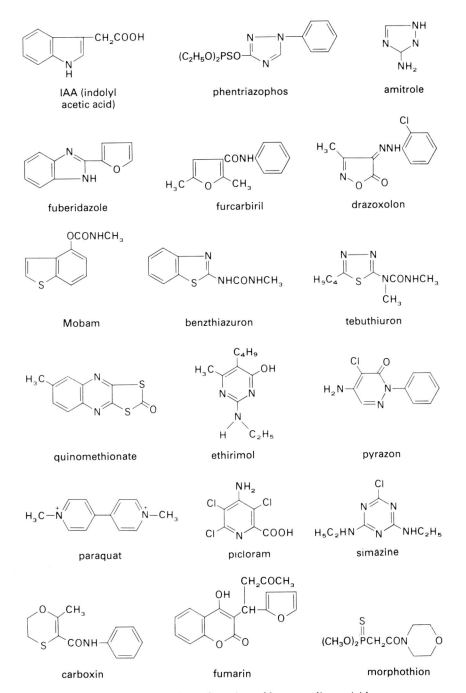

Fig. 19. Some of the wide variety of heterocyclic pesticides.

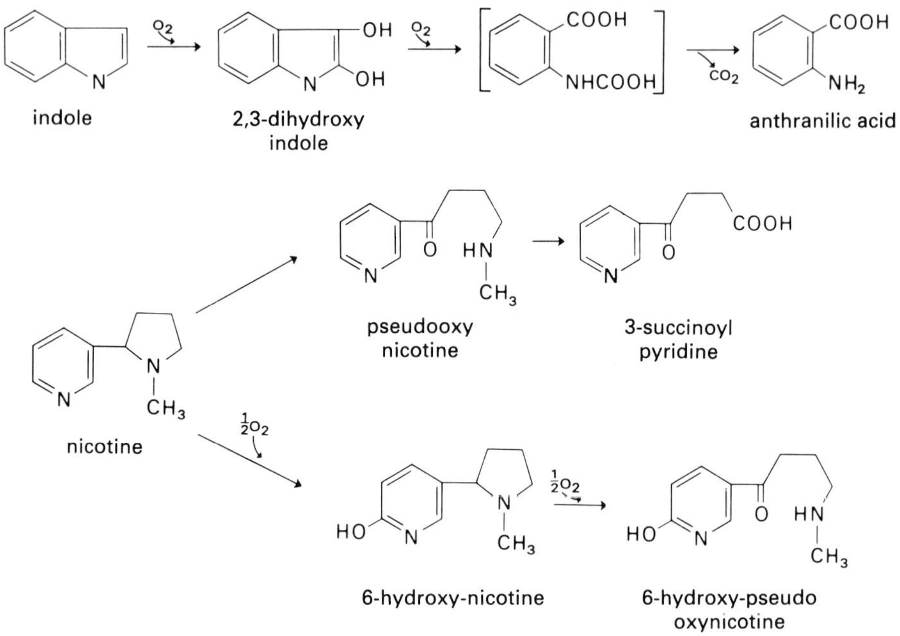

Fig. 20. Cleavage of the 5-membered rings of indole (Fujioka and Wada, 1968) and nicotine (Gherna and Rittenberg, 1962; Gherna et al., 1965).

Pesticides with 5-membered rings undergo reactions similar to their non-pesticidal equivalents. For example Cripps (1973) has shown 1,2-thiophen-2-carboxylic acid to be metabolized by a route similar to that of 2-furoic acid (Trudgill, 1969). There appear to be no reports of microbial ring fission of triazoles, the disappearance of the herbicide amitrole in soil being attributed to a chemical rather than a biological mechanism (Kaufman et al., 1968).

2. 6-Membered rings

The 6-membered heterocyclic rings of pesticides are mostly the nitrogen-containing pyridines, pyrimidines and triazines, although other structures including S and O atoms are evident in a few pesticides (Fig. 19).

(a) Pyridines. Possibly more is known of the fission of pyridines than of any other heterocycle-containing pesticide. Pyridines are naturally present in the environment (for example in coenzymes and bacterial spores) but may also be unintentionally introduced in the effluents from industrial users of pyridine solvents. Microorganisms have been isolated that are able to use pyridine and its derivatives as a sole source of carbon and energy (Houghton and Cain, 1972; Shukla, 1973). Only actinomycetes have been isolated that can use these substrates as a nitrogen source (Ensign and Rittenberg, 1963). Many of the isolates show a high degree of substrate specificity. Although the pyridine-N can be recovered as NH_3, all attempts to prepare pyridine-degrading cell-free extracts for detailed metabolism studies have failed. Atkinson (Microbiological Research Establishment, Porton Down, pers. comm., 1977) has prepared cell-free extracts of several mesophilic bacteria that in the presence of DTT and FAD can degrade paraquat with the formation of a product of higher E600 nm than the herbicide. Watson and Cain (1975) using radiolabelled substrates, inhibitors and microbial mutants, have successfully isolated several intermediates in the degradation of pyridine. They presented evidence for the presence of two metabolic routes of biodegradation, one by a *Bacillus* sp. and the other by a *Nocardia* sp. The *Bacillus* appeared, initially, to open the ring between C2 and C3, to form a semialdehyde which was further metabolized to formamide and succinic semialdehyde by C6-N cleavage. The mechanism of C2–C3 ring opening was not defined but oxygenase-catalysed hydroxylation was considered unlikely. The interpretation of the data from studies with the *Nocardia* sp. was that the ring was opened by hydrolysis at C2-N, and after a further cleavage at C6-N released glutaric dialdehyde and NH_3.

Studies of nicotinic acid (Behrman and Stanier, 1957; Ensign and Rittenberg, 1964), 2-, 3-, and 4-hydroxypyridine and 3,4-dihydroxypyridine (Watson *et al.*, 1974a, b; Cain *et al.*, 1974) have all

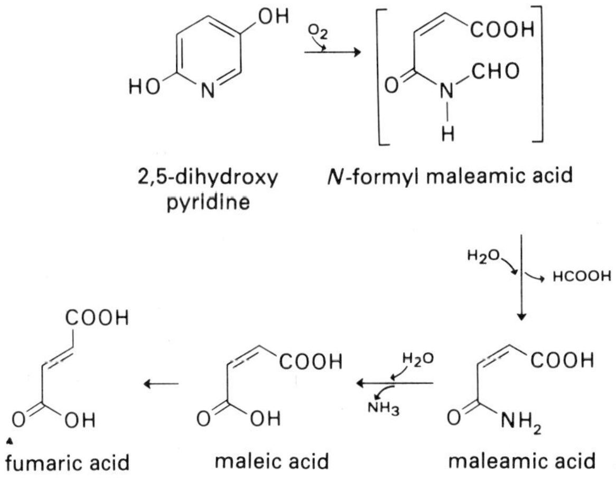

2,5-dihydroxy   *N*-formyl maleamic acid
pyridine

fumaric acid   maleic acid   maleamic acid

Fig. 21. Maleamate pathway of pyridine metabolism (Behrman and Stanier, 1957).

demonstrated hydroxylation and C2–C3 ring cleavage by oxygenase enzymes. The presence of an unsubstituted C6 can also result in 6-hydroxylation. Whilst most of these pyridines are metabolized via the maleamate pathway (Fig. 21), 3,4-dihydroxypyridine, formed from 4-hydroxypyridine, exhibited a *meta* type fission (probably oxygenase-catalysed) with subsequent hydrolysis and deamination to form ammonia, pyruvate and formate (Fig. 22).

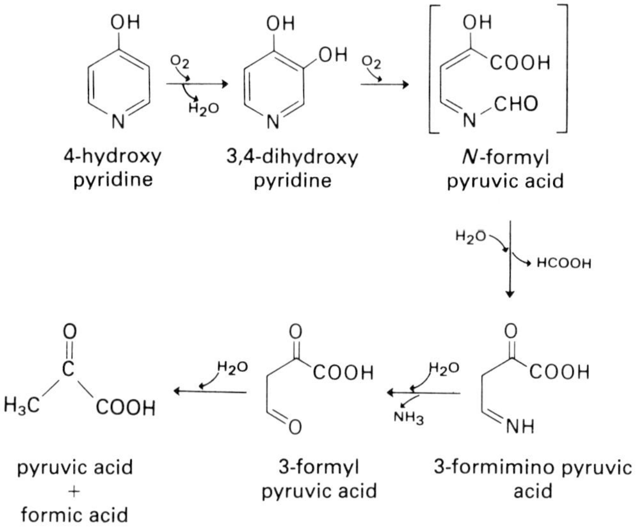

Fig. 22. 4-Hydroxypyridine metabolism (Watson *et al.*, 1974a, b).

Houghton and Cain (1972) have suggested that the oxidative formation of *para*-diols from pyridines is more commonly encountered than the *ortho*-diol pathway, this being the reverse of the situation with monocyclic aromatic molecules.

Studies with pyridoxic acids (Sparrow *et al.*, 1969) have indicated that the pyridine ring may be cleaved by dioxygenases (possibly involving cyclic peroxide formation) between a ring carbon with a methyl substituent and an adjacent unsubstituted carbon.

Many pyridine pesticides are microbially metabolized with cleavage of the ring. Although the bypyridyl herbicides paraquat and diquat are both metabolized in culture (Baldwin *et al.*, 1966; Funderburk, 1969; Anderson and Drew, 1972) they are not transformed in soil due to their strong absorption to mineral surfaces (Dixon *et al.*, 1970; Riley *et al.*, 1976). Picolinamide and *N*-methylisonicotinic acid, photolytic products of diquat and paraquat, respectively, are both microbially cleaved between the C2 and C3 ring carbons (Cain *et al.*, 1970; Orpin *et al.*, 1972a, b). Orpin's group found both these products to be degraded via the maleamate pathway with the *N*-methylisonicotinate being *N*-dealkylated before ring cleavage. Cain and his colleagues delineated an alternative route for *N*-methylisonicotinate, in which the methyl group was retained through ring opening and the products of the reaction were methylamine and succinic acid. Meikle *et al.* (1966) noted that ring ^{14}C-picloram was degraded in soil with evolution of $^{14}CO_2$ but the only intermediate isolated was the 6-hydroxy derivative (Redemann *et al.*, 1968). Meikle (1972) has suggested that this product, never present at greater than 1% of the applied pesticide, is not on the main degradation pathway. Naik *et al.* (1972) attempted to correlate the chemical structure of a large number of pyridines, including the herbicides picloram and pyriclor, with their degradation in soil suspensions. Although they obtained a rough correlation between increasing numbers of substituents and increasing persistence and demonstrated the influence of substituent position and type on degradation, their method of analysis took no account of soil-surface adsorption. In their system, 2- and 3-hydroxypyridines and 2,3-dihydroxypyridine disappeared relatively slowly.

(b) *Pyrimidines.* The ability of microorganisms to cleave rings containing more than one nitrogen atom has long been known from studies of bio-degradation of the pyrimidine components of nucleic acids in culture and in soil (Di Carlo *et al.*, 1951; Hayaishi and

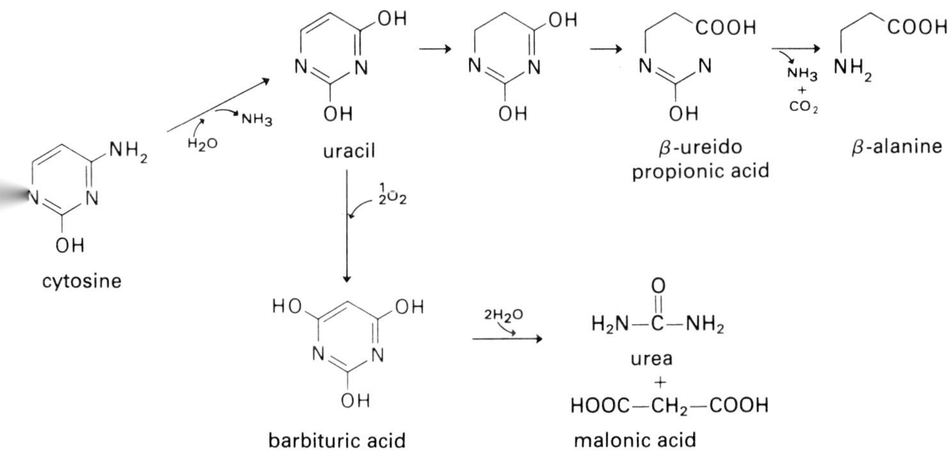

Fig. 23. Cytosine metabolism.

Kornberg, 1952; LaRue and Spencer, 1968; Drobnikova, 1971; Drobnikova and Pospisilova, 1972). The pyrimidine cytosine is metabolized by two routes (Fig. 23); to form either β-alanine, ammonia and carbon dioxide or urea and malonic acid.

Although the pyrimidine herbicides bromacil and terbacil (Gardiner *et al.*, 1969), the insecticides diazinon (Gunner and Zuckerman, 1968), pirimicarb, pirimiphos-ethyl and pirimiphosmethyl (Hill and Arnold, unpublished data) and the fungicide ethirimol (Hill and Arnold, 1973) all undergo microbially catalysed ring cleavage in soil, and in the case of diazinon in microbial culture also, no ring fission intermediates could be isolated and identified. The pyrimidine ring is cleaved under aerobic soil conditions, but in anaerobic environments, as may occur within aggregates of soil, with increasing soil depth and with flooding, the reaction is retarded or completely inhibited (Sethunathan and Yoshida, 1969; Hill, unpublished data).

(c) *Triazines.* No natural equivalent of the triazine ring exists in the environment. The herbicidal *s*-triazine nucleus is relatively resistant to ring cleavage. There are, however, data to suggest that the ring may be destroyed slowly in soil (Skipper *et al.*, 1967; Kaufman and Kearney, 1970) and in cultures of soil organisms (Roeth *et al.*, 1969; Kaufman and Kearney, 1970).

(d) *Other heterocycles.* Studies of pyridazine herbicides have shown that they are susceptible to microbial attack in soil (Rahn and Zimdahl, 1973), but isolation of products was not attempted. DeFrenne *et al.* (1973) have, however, successfully prepared a bacterial cell-free extract capable of degrading pyrazon (5-amino-4-chloro-2-phenyl-3-pyridazinone) and have isolated and characterized several intermediates. Whilst dihydroxylation and *ortho* cleavage of the homocyclic ring was observed, with concomitant release of the heterocyclic moiety, the pyridazine ring was not metabolized. A carboxylated pyrone ring was also formed from the homocyclic ring but as it could not be metabolized further was assumed to have been the product of an enzymatic (lactonization) side reaction.

Callely (1974) reported that microorganisms have been isolated which degrade the morpholine ring containing N and O hetero-atoms.

J. HYDROLYSIS

Biologically, the incorporation of a molecule of water into a substrate (hydrolysis) is catalysed by a large group of universally distributed enzymes, including esterases, amidases, nitrilases, phosphatases and chitinases.

Numerous pesticides contain ester linkages which may undergo hydrolysis to leave the corresponding acid together with an alcohol or an amine (Fig. 24).

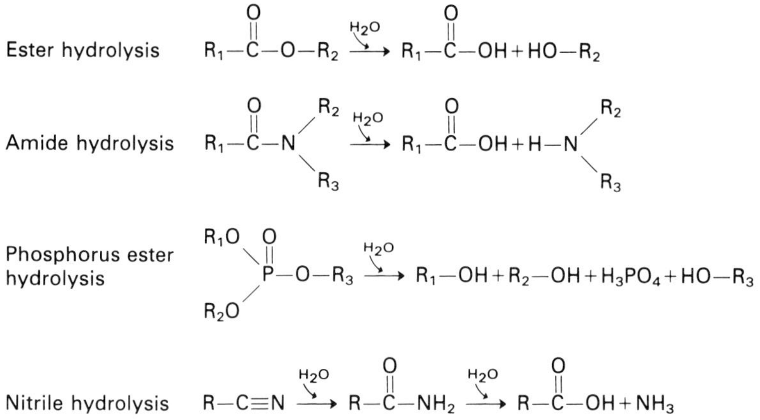

Ester hydrolysis
$$R_1-\overset{\overset{\displaystyle O}{\|}}{C}-O-R_2 \xrightarrow{H_2O} R_1-\overset{\overset{\displaystyle O}{\|}}{C}-OH + HO-R_2$$

Amide hydrolysis
$$R_1-\overset{\overset{\displaystyle O}{\|}}{C}-N\overset{R_2}{\underset{R_3}{\diagup}} \xrightarrow{H_2O} R_1-\overset{\overset{\displaystyle O}{\|}}{C}-OH + H-N\overset{R_2}{\underset{R_3}{\diagup}}$$

Phosphorus ester hydrolysis
$$\overset{R_1O}{\underset{R_2O}{\diagdown}}\!\!P-O-R_3 \xrightarrow{H_2O} R_1-OH + R_2-OH + H_3PO_4 + HO-R_3$$

Nitrile hydrolysis
$$R-C\equiv N \xrightarrow{H_2O} R-\overset{\overset{\displaystyle O}{\|}}{C}-NH_2 \xrightarrow{H_2O} R-\overset{\overset{\displaystyle O}{\|}}{C}-OH + NH_3$$

Fig. 24. Hydrolyses.

The lipophilic substrates are mostly converted into more hydrophilic products. The products of hydrolysis may therefore be more widely distributed by leaching than is the parent pesticide. Hydrolysis usually results in a loss of pesticidal activity. This reaction can also occur chemically under both acid and alkaline conditions and may be catalysed in soil by the presence of iron salts, nitrogenous compounds and heavy metal ions, making the separate biological and chemical contributions of the environment difficult to quantify. Even in microbial culture interpretation can be difficult. Chemical hydrolyses may occur both in the medium and within the microbial cells. However, hydrolytic enzymes may also operate inside the cell, on the outer surface of the cytoplasmic membrane (Rose, 1965), or even extracellularly in the culture medium (Lu and Liska, 1969). Soil can also possess a considerable extracellular complement of such enzymes (Skujins, 1967). Getzin and Rosefield (1971) suggested that malathion (S-(1,2-di(ethoxycarbonyl)ethyl)-OO-dimethyl phosphorodithioate) hydrolysis in soil might be due to such enzymes.

The comparison of hydrolytic cleavage of pesticides in soils of differing pH to that in buffered solutions (Konrad et al., 1967) should be viewed with some caution. In the soil environment the physical and chemical characteristics of the microhabitats can be very different from the values obtained by bulk measurements. Assessment of biological hydrolytic activity in soil is not always possible, since the sterilization techniques necessary to eliminate the biological contribution may also destroy micro-habitats and alter the soil chemistry. However, from the wealth of data available it would appear that in the environment both biological and chemical reactivity play a part in the hydrolytic cleavage of both pesticidal and non-pesticidal substrates. The ability of microorganisms to enzymatically hydrolyse pesticides has been proven by the isolation and partial or complete purification of active enzyme systems (Kearney, 1965, 1967; Matsumura and Boush, 1966; Sharabi and Bordeleau, 1969; Lanzilotta and Pramer, 1970a, b; Wallnöfer and Bader, 1970; Engelhardt *et al.*, 1971).

Of the numerous pesticides possessing potential sites for hydrolytic enzyme activity, the carbamate and organophosphorous compounds have probably received the greatest attention.

(a) *The carbamates.* Whether alkyl, aryl or oxime the carbamates are esters and amides (Fig. 25) which according to their structure should undergo hydrolysis to produce carbamic or carbonic acid esters which are unstable and degrade, releasing carbon dioxide. Where *N*-alkyl groups are present enzymatic *N*-dealkylation may precede hydrolysis. Although hydrolysis of the carbamate could occur via either the amide or ester linkage the latter is considered more probable (Dittert and Higuchi, 1963; Coppedge *et al.*, 1967; Bartha *et al.*, 1968; Kaufman, 1970).

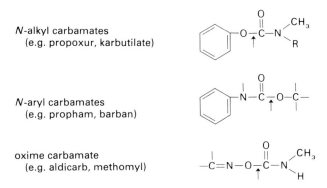

N-alkyl carbamates
(e.g. propoxur, karbutilate)

N-aryl carbamates
(e.g. propham, barban)

oxime carbamate
(e.g. aldicarb, methomyl)

Fig. 25. Sites of hydrolysis of carbamate pesticides (arrows indicate probable site of cleavage).

(b) *The organophosphate pesticides.* These exist as phosphorothioates and phosphates. Conversion of the former to the latter appears to be chemically and not microbially mediated (Meikle, 1972). Natural organophosphates are commonly encountered (e.g. nucleic acids, nucleotides and phospho-proteins) and many microorganisms possess phosphatases capable of hydrolytically releasing the inorganic phosphate from these compounds (Mounter and Tuck, 1956; Ahmed and Casida, 1958; Matsumura and Boush, 1968). Many of the phosphate and phosphorothioate pesticides have been reported to be microbially hydrolysed in soil (Lichenstein and Schulz, 1964; Sethunathan and Yoshida, 1969, 1973a, b; Beynon and Wright, 1969), but so chemically reactive are these groups that it is only occasionally that such conclusions can be made with conviction. The phosphonate (P-C) moiety present in a few pesticides and not particularly common in natural products (Kittredge and Roberts, 1969), is enzymatically cleaved (Alam and Bishop, 1969) but not by phosphatases.

(c) *Pesticides with other ester groupings.* Hydrolytic cleavage of for example, the $-\overset{|}{N}-CO-\overset{|}{N}-$ of the urea herbicides, $-\overset{|}{\underset{|}{C}}-\overset{|}{N}-CO-\overset{|}{\underset{|}{C}}-$ of anilide herbicides, and the $-C\equiv N$ nitrile linkage occurs in much the same way and has been attributed to enzymatic activity in soil and in microbial cultures (Cartwright and Cain, 1959a, b; Beynon and Wright, 1968; Lanzilotta and Pramer, 1970b; Bartha, 1971; Engelhardt et al., 1971, 1972).

The hydrolysis of various other esters have been reported. For example using conventional and germ-free rat caecal content Eriksson and Gustafsson (1970) demonstrated sulphate ester hydrolysis of sodium estrone sulphate and attributed this to microbial activity. Microorganisms have also been implicated in the sulphamate hydrolysis of cyclamate (Drasar et al., 1972; Goldberg et al., 1969). Pesticides, non-pesticides and their products may be conjugated, in plants and animals, to form glycosides. The conjugates may be released into soil or aquatic environments. Although there have been few studies of glycoside hydrolysis by soil and water microorganisms, many bacteria from the alimentary tracts of animals have been shown to possess glucuronidase (Barber et al., 1948; Hawksworth et al., 1971; Soleim and Scheline, 1972) and β-glucosidase (Hawksworth et al., 1971) activity.

Studies of pesticides have illustrated the electronic and steric effects of substituents on the stability of ester bonds. The more electrophilic the substituents the more they make an ester bond susceptible to nucleophilic attack by the hydroxyl ions. Alkyl and quaternary groups attached via oxygen to phosphorus atoms increase stability, possibly due to steric blocking of the ester bond (O'Brien, 1967). The type and position of substituents on a "nearby" phenyl ring can have similar effects (Kearney, 1967; Kearney and Helling, 1969; Sharabi and Bordeleau, 1969), although Cripps (1971) suggested that such groups are too far away from the reactive site to exert steric hindrance. Instead, Cripps proposed that such interference with the hydrolysis of the carbamate moiety may be related to the inductive or electron-withdrawing effects of the substituents and their effect on the acidity of the proton on the carbamate nitrogen.

Some of the insecticidal cholinesterase-inhibiting organophosphates and carbamates reduce the rate of degradation of propham (iso-propyl phenylcarbamate) and chlorpropham (isopropyl 3-chloro-phenylcarbamate) in soil and in microbial cultures (Westmacott and Wright, 1975; Kaufman et al., 1970; Priest and Stephens, 1975) thereby lengthening the duration of effective herbicidal activity. This ability was attributed to competitive inhibition of the phenyl-carbamate hydrolysing enzyme. Brookes (1972) has suggested that the sites on propham, for hydrolysis, have factors in common with the active sites for cholinesterase inhibition in animals.

K. HALOGEN REACTIONS

The enzymatic removal or transfer of halogen substituents is often an important stage in the removal of biological activity and the degradation of halogenated pesticides. Microorganisms are not usually quite so adept at dealing with halogen atoms as they are with the majority of other substituents on pesticides. This is reflected in the prolonged environmental persistence of many halogenated insecticides (Edwards, 1973). Halogens on aliphatic groups are, however, generally more reactive than their aromatic equivalents.

Four major reaction routes have been described by which halogens are either removed or transferred to alternative sites on pesticide molecules. These are hydroxyl replacement; dehydrohalogenation; reductive dehalogenation; and halogen migration ("NIH shift"). Microbial enzymes have been implicated in all of these processes, although in the environment chemical reactivity may sometimes be wholly or partly responsible (Skipper et al., 1967).

The prediction of routes and rates of microbial metabolism of halogenated pesticides has been relatively unsuccessful, for the reactions of both aliphatic and aromatic molecules can be considerably influenced by the type, number and position of halogen atoms, as well as by the presence of other substituents (Hirsch and Alexander, 1960; Alexander, 1965b; Kearney and Plimmer, 1970).

1. *Hydroxyl replacement*

Hydrolytic dehalogenation (Fig. 26), in which a halogen is replaced with a hydroxyl group, occurs in both aliphatic and aromatic molecules, but is catalysed by different enzyme systems. In aliphatic compounds the hydroxyl group is supplied by water. Goldman (1972), reviewing the enzymology of carbon–halogen bonds has discussed this reaction at length. Halidohydrolase enzymes have been isolated and characterized (Davis and Evans, 1962; Goldman, 1965; Goldman *et al.*, 1968) and found to be active on a variety of halogenated fatty acids, releasing the corresponding hydroxy- or keto-acid for utilization as a carbon and energy source.

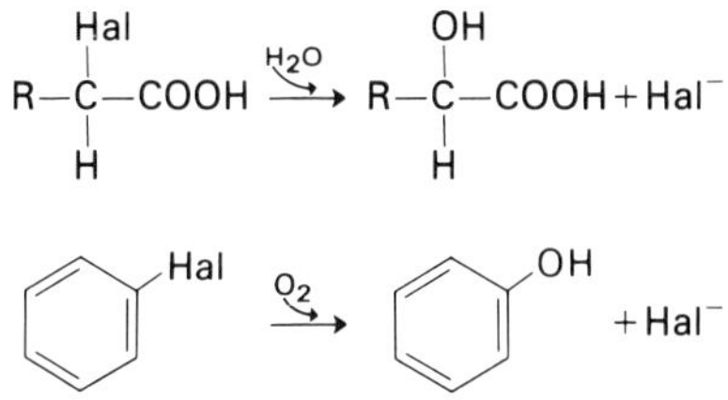

Fig. 26. Hydrolytic dehalogenation of aliphatic and aromatic compounds.

By contrast, the removal of halogens attached to an aromatic compound involves replacement by a hydroxyl group with the oxygen atom derived from molecular oxygen (Guroff *et al.*, 1966; Hayaishi and Nozaki, 1969). Meikle (1972) has suggested that this displacement, catalysed by mono-oxygenases, is incidental to aromatic ring hydroxylation and occurs particularly if the halogen is at a susceptible position on the ring.

Mono-, di- and tri-haloacetic acids, including the herbicides TCA (trichloroacetic acid) and dalapon (2,2-dichloropropionic acid), are dehalogenated by microbial hydroxyl replacement (Kearney *et al.*, 1964; Goldman, 1965; Kearney *et al.*, 1965; Kearney *et al.*, 1969). Engst and Kujawa (1967) have shown *pp*-DDT to undergo complete hydrolytic dehalogenation of the alkyl group with formation of an acetic acid derivative. The aliphatic chlorinated group of nitrapyrin (2-chloro-6-trichloromethyl pyridine), a nitrification inhibitor, undergoes the same reaction (Meikle and Hamilton, 1964).

There are relatively few examples of the hydrolytic dehalogenation of aromatic pesticides. Products of this reaction have, however, been isolated from microbial cultures incubated with halogenated phenoxyacetate and chlorbenzoate groups similar to those present in many herbicides (Faulkner and Woodcock, 1961; Johnston *et al.*, 1972).

It has also been demonstrated that halogen substituents may be retained during hydroxylation and displaced either immediately after ring cleavage or following the formation of halogenated succinates and acetates (Cain *et al.*, 1968b; Tiedge *et al.*, 1969; Duxbury *et al.*, 1970).

The participation of "neighbouring" groups in hydrolytic dehalogenation has also been described. Goldman *et al.* (1967) and Milne *et al.* (1968) have demonstrated the simultaneous dehalogenation and decarboxylation of 2-fluorobenzoate by a pseudomonad during a single hydroxylation reaction involving molecular oxygen.

2. *Halogen migration*

An alternative process to the enzymatic replacement of substituents on aromatic nuclei is their migration to an adjacent site on the ring (Fig. 27). This phenomenon, induced by hydroxylation, has been termed "NIH shift" (Guroff *et al.*, 1967). Meikle (1972) has suggested that this reaction is least likely to occur with compounds such as phenols and anilines, but is favoured by molecules of the alkylbenzene, tryptophan, halobenzene and anisole type. Although various substituents may undergo "NIH shift" reactions, few have been encountered in studies of pesticide transformations.

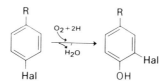

Fig. 27. Halogen migration.

Faulkner and Woodcock (1964, 1965) isolated a transformation product of the herbicide 2,4-D that demonstrated the occurrence of halogen migration. Guroff *et al.* (1966), from studies with bacterial phenylalanine hydroxylases and *p*-halogenated phenylalanines, found the predominant reaction product to be *m*-chlorotyrosine. The available data indicate that "NIH shift" reactions are limited to *p*-hydroxylations, and that the incoming hydroxyl group is mostly *ortho* to the migrating substituent. Meikle (1972) has reviewed the orientation of hydroxyl groups resulting from aromatic ring microbial oxidation.

3. Reductive dehalogenation

Many of the chlorinated pesticides are relatively resistant to enzymatic dehalogenation. Yet in recent years it has become evident that some of these compounds can be degraded slowly in both aerobic and anaerobic situations, and also in microbial cultures in the laboratory. The degradative reactions mostly involve dehydrohalogenation or reductive dehalogenation.

Reductive dehalogenation of aliphatic halides (Fig. 28) is a major degradation route, favoured by anaerobic conditions and probably involving carriers such as reduced cytochrome oxidase and FAD (Wedemeyer, 1966; French and Hoopingarner, 1970).

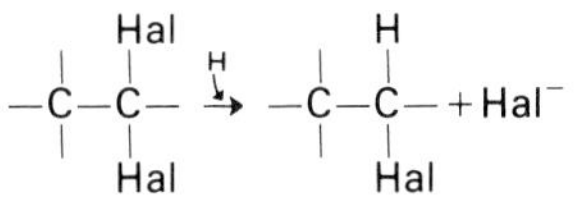

$$-\overset{|}{\underset{|}{C}}-\overset{\overset{\displaystyle Hal}{|}}{\underset{\underset{\displaystyle Hal}{|}}{C}}- \overset{H}{\longrightarrow} -\overset{|}{\underset{|}{C}}-\overset{\overset{\displaystyle H}{|}}{\underset{\underset{\displaystyle Hal}{|}}{C}}- + Hal^-$$

Fig. 28. Reductive dehalogenation.

p,p'-DDT (1,1,1-trichloro-2,2-di(4-chlorophenyl)ethane) is converted to TDE (1,1-dichloro-2,2-di(4-chlorophenyl)ethane) (Fig. 29) in anaerobic soil (Duffy and Wong, 1967; Guenzi and Beard, 1967; Bhuiya and Rothwell, 1973), in lake water (Miskus et al., 1965), by the intestinal microorganisms of the northern anchovy and the rainbow trout (Wedemeyer, 1968; Malone, 1970), in anaerobic sludge inoculated with methane-producing and sulphate-reducing bacteria (Hill and McCarty, 1967), by a wide variety of microorganisms in culture (Kallman and Andrews, 1963; Chacko et al., 1966; Johnson et al., 1967) and in a disrupted cell suspension of *Aerobacter aerogenes* (Wedemeyer, 1967).

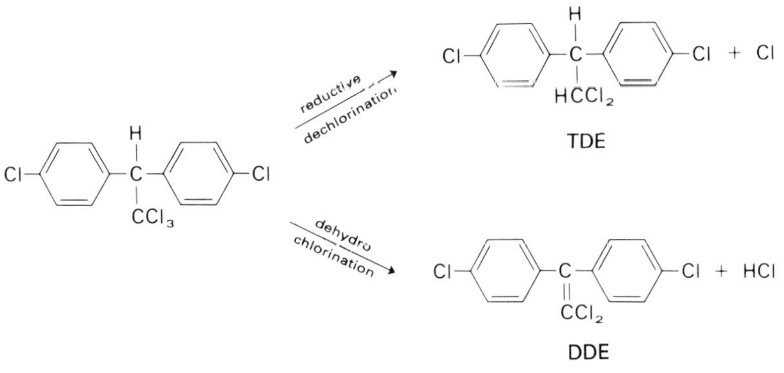

Fig. 29. Halogen reactions of p,p'-DDT.

When BHC (1,2,3,4,5,6-hexachlorocyclohexane) is incubated anaerobically, chloride ions are released (MacRae *et al.*, 1969; Sethunathan *et al.*, 1969). An unidentified product was considered to have possibly been pentachlorocyclohexane, from reductive dechlorination. The halogen nematocide, ethylene dibromide and related aliphatic halides are also reductively dehalogenated (Castro and Belser, 1968), although a dehydrochlorination reaction may be responsible for removal of the second halogen atom. Bartnicki and Castro (1969) isolated an enzyme from a *Flavobacterium* sp. that dehalogenated halohydrins by epoxide formation and elimination of HCl or HBr.

4. *Dehydrohalogenation*

Dehydrohalogenation (Fig. 30) has been observed with both aromatic and aliphatic halides (although relatively infrequently for the latter) and, like reductive dehalogenation, has been reported for many of the chlorinated pesticides. Dehydrohalogenation involves the removal of adjacent halogen and hydrogen atoms and the formation of a C–C double bond.

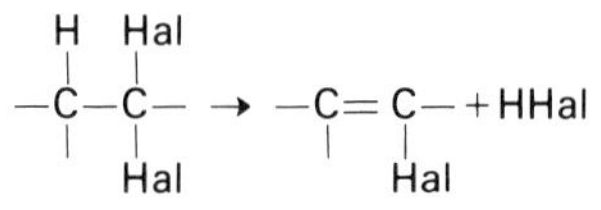

$$
\begin{array}{ccc}
\text{H} & \text{Hal} & \\
| & | & \\
-\text{C}-\text{C}- & \rightarrow & -\text{C}=\text{C}- + \text{HHal} \\
| & | & \qquad | \quad | \\
& \text{Hal} & \quad \text{Hal}
\end{array}
$$

Fig. 30. Dehydrohalogenation.

In aerobically incubated soil and in microbial cultures an aliphatic chlorine on p,p'-DDT is dehydrohalogenated forming DDE (1,1-dichloro-2,2-di(4-chlorophenyl)ethylene) (Fig. 29) (Guenzi and Beard, 1968; Matsumura and Boush, 1968; Fries *et al.*, 1969; Miyazaki and Thorsteinson, 1972). Small amounts of the same product are also formed under anaerobic conditions (Stenerson, 1965; Guenzi and Beard, 1967).

γ-Pentachlorocyclohex-1-ene, formed by dehydrochlorination of lindane (γ-BHC) in moist soil (Yule *et al.*, 1967), was attributed to microbial activity. From studies involving simulated cattle dips, Allan (1955) isolated strains of *Escherichia coli* and *Clostridium sporogenes* that dehalogenated lindane with the formation of both benzene and monochlorobenzene, but only in small amounts.

Dehalogenation of chlorocatechols may occur immediately following ring cleavage, also possibly by a dehydrochlorination mechanism during formation of the lactone bond (Bollag *et al.*, 1968; Tiedje *et al.*, 1969; Duxbury *et al.*, 1970).

Dalapon was dechlorinated by a partially purified enzyme preparation of an *Arthrobacter* sp. (Kearney *et al.*, 1964), leaving pyruvic acid. Although this reaction was originally believed to be a hydrolytic dehalogenation it is more likely that the initial reaction is a dehydrochlorination.

L. NITRO-REDUCTION

Aromatic, but not aliphatic, nitro compounds are microbially reduced to amines, probably via nitroso- and hydroxylamino- intermediates (Yamashina *et al.*, 1954; Cartwright and Cain, 1959a, b) (Fig. 31).

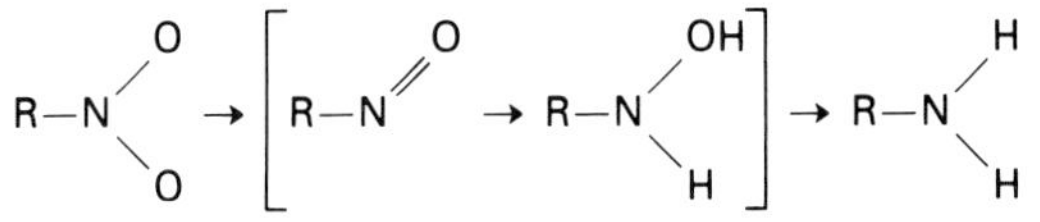

$$R-N\begin{matrix}O\\\\O\end{matrix} \rightarrow \left[R-N\begin{matrix}O\\\\\end{matrix} \rightarrow R-N\begin{matrix}OH\\\\H\end{matrix}\right] \rightarrow R-N\begin{matrix}H\\\\H\end{matrix}$$

Fig. 31. Reduction of aromatic nitro groups.

Nitro-reductase systems have been noted in many microorganisms (Madosingh, 1961; Chacko *et al.*, 1966; Matsumura *et al.*, 1968; Hamdi and Tewfik, 1970), and many pesticides have been shown to be susceptible to their effects. The insecticide parathion (diethyl-4-nitrophenyl phosphorothionate), for example, was converted to its more persistent amino analogue under both aerobic and anaerobic conditions, by microorganisms isolated from a lake sediment (Graetz *et al.*, 1970).

The reduction of several nitro pesticides has been observed in soil, and the transformation attributed to microbial activity (Lichenstein and Schulz, 1964; Probst and Tepe, 1969; Ko and Farley, 1969; Smith *et al.*, 1973; Parr and Smith, 1973). *In situ* and *in vitro* studies have also implicated intestinal microorganisms in the nitro-reduction of trifluralin (2,6-dinitro-*NN*-dipropyl-4-trifluoromethylaniline) (Golab *et al.*, 1969; Williams and Feil, 1971). It seems very unlikely that these reactions are due to chemical activity. Loss of the nitro group from pesticides usually results in a loss of biological activity.

The amines formed by nitro-reduction may be further metabolized by *N*-dealkylation, oxidative deamination, carbonyl and hydroxyl replacement reactions, amide formation or condensation to form azo compounds.

M. MISCELLANEOUS REACTIONS

Several reactions, which have been recorded for only a few pesticides, have thus far been omitted from discussion. One major reaction of pesticides in plants and animals is conjugation, the biological coupling of a pesticide or its products with non-pesticidal molecules. Whilst microorganisms have frequently been shown to carry out conjugation reactions such as methylation, formylation, acetylation and glycosilation, there have been relatively few reports of pesticides or their model compounds undergoing these reactions. Examples of pesticide conjugations involving microorganisms have been documented by Sijpesteijn *et al.* (1962), Loos *et al.* (1967b), Tweedy *et al.* (1970a, b), Curtis *et al.* (1972), Kearney and Plimmer (1972), Cserjesi and Johnson (1972), Wiese and Vargas, (1973). Unfortunately, proof of active microbial enzyme participation is rare. Similar reactions can also occur spontaneously or as a result of chemical reactivity in the "medium". Some heavy metal pesticides are known to chemically complex with amino acids (Zentmayer and Rich, 1956). Thiol groups in fungal cells can react with some fungicides, for example, captan (*N*-(trichloromethylthio)-3a,4,7,7a-tetrahydrophthalimide) (Richmond and Somers, 1966, 1968), indeed such fungicides may rely on reactions with thiol, amino and other nucleophilic groups in microorganisms or soil organic matter for expression of their biological activity. Gause *et al.* (1967) showed the initial step in the reaction of thiols with fungicidal quinones such as chloranil (2,3,5,6-tetra-chloro-1,4-benzoquinone), to be a "redox" reaction.

Condensation reactions of pesticides have been observed with the aniline derivatives of the chloroaniline herbicides. Although aniline itself does not form a condensation product, various mono- and dichlorinated forms produce symmetrical and asymmetrical azobenzenes in the presence of microorganisms (Bordeleau and Bartha, 1970; Bartha, 1971; Bordeleau *et al.*, 1972; Lay and Ilnicki, 1974). Bordeleau *et al.* (1972) proposed that the condensation occurs spontaneously after microbial peroxidase enzymes have produced labile intermediates. Plimmer *et al.* (1970) have demonstrated that nitrite can interact with two molecules of a chlorinated aniline to form a triazene. However only very small amounts of the azobenzenes and triazenes appear to be formed in soil (Sprott and Corke, 1971; Burge, 1972), possibly because their aniline precursors are degraded or react with or are rapidly bound to soil humic constituents.

Apart from studies of methylation reactions, the heavy metal pesticides have not received much attention. The C–Hg bond of the fungicide PMA (phenylmercuric acetate) has, however, been shown to be susceptible to cleavage by pseudomonads (Furakawa *et al.*, 1969). Tonomura and Kanzaki (1969) isolated cell-free preparations capable of this reaction in the presence of NADH and sulphydryl groups. Matsumura *et al.* (1971a) found no evidence of methylmercury compounds in soil or aquatic situations treated with PMA but did find a diphenylmercury product, together with free benzene and mercury. Von Endt *et al.* (1968) isolated soil organisms capable of cleaving the C–As bond of MSMA (monosodium methane-arsonate), a methylated arsenical herbicide, with the formation of inorganic arsenate.

Patil *et al.* (1970) and Matsumura *et al.* (1971b) have reported the microbial conversion of dieldrin to photodieldrin and endrin to keto-endrin. This isomerization reaction results in environmentally more stable products, but does not provide the microorganism with any utilizable source of energy and carbon.

It is possible that reactions of pesticides may occur in the environment that have so far escaped attention. Soil binding occurs with many pesticides and/or their transformation products, but because of our lack of understanding of the structure and properties of organic matter and, except in a few instances (e.g. Hill, 1976), because of our inability to extract and characterize many of these bound pesticide residues, we remain ignorant of the mechanisms of their formation.

IV. Conclusions

Microorganisms have the enzymic potential to metabolize the majority of pesticides currently in use. Many of the reactions are well understood and the necessary conditions delineated.

Studies of the transformation of pesticides in soil and water, environments in which microorganisms play a considerable role, contribute significantly to the extremely high costs of preparing environmental impact data for registration authorities. One of the few ways by which such costs might be reduced would be to decrease the practical work now necessary. However, our ability to predict either the routes or rates of transformation of a pesticide in the environment is still too rudimentary for this approach to be viable at present. It is to this end that microbiologists and biochemists should strive.

References

AHMED, M. K. and CASIDA, J. E. (1958). Metabolism of some organo-phosphorus insecticides by microorganisms. *J. econ. Ent.* **51**, 59–63.

ALAM, A. V. and BISHOP, S. H. (1969). Growth of *Escherichia coli* on some organophosphonic acids. *Can. J. Microbiol.* **15**, 1043–1046.

ALEXANDER, M. (1965a). Biodegradation: problems of molecular recalcitrance and microbial fallibility. *Adv. appl. Microbiol.* **7**, 35–80.

ALEXANDER, M. (1965b). Persistence and biological reactions of pesticides in soils. *Proc. Soil Sci. Soc. Am.* **29**, 1–7.

ALEXANDER, M. (1966). Biodegradation of pesticides. *In* "Pesticides and their Effects on Soils and Water", pp. 78–84, ASA Special Publicn No. 8. Soil Sci. Soc. Am., Madison.

ALEXANDER, M. (1973). Nonbiodegradable and other recalcitrant molecules. *Biotechnol. Bioeng.* **15**, 611–647.

ALEXANDER, M. and LUSTIGMAN, B. K. (1966). Effect of chemical structure on microbial degradation of substituted benzenes. *J. agric. Fd Chem.* **14**, 410–413.

ALLAN, J. (1955). Loss of biological efficiency of cattle-dipping wash containing benzene hexachloride. *Nature, Lond.* **175**, 1131–1132.

ANDERSON, J. R. and DREW, E. A. (1972). Growth characteristics of a species of *Lipomyces* and its degradation of paraquat. *J. gen. Microbiol.* **70**, 43–58.

ANDRAWES, N. R., BAGLEY, W. P. and HERRETT, R. A. (1971). Fate and carry over properties of Temik aldicarb pesticide [2-methyl-2-(methylthio)propionaldehyde O-(methylcarbamoyl)oxime] in soil. *J. agric. Fd Chem.* **19**, 727–730.

AUDUS, L. J. (1960). Microbiological breakdown of herbicides in soils. *In* "Herbicides and the Soil" (E. K. Woodford and G. R. Sagar, Eds), pp. 1–19. Blackwell Scientific Publications, Oxford.

BALDWIN, B. C., BRAY, M. F. and GEOGHEGAN, M. J. (1966). The microbial decomposition of paraquat. *Biochem J.* **101**, 15P.

BARBER, M., BROOKSBANK, B. W. L. and HASLEWOOD, G. A. D. (1948). Destruction of urinary glucuronide by bacteria. *Nature, Lond.* **162**, 701–702.

BARTHA, R. (1969). Pesticide interaction creates hybrid residue. *Science, N.Y.* **166**, 1299–1300.

BARTHA, R. (1971). Fate of herbicide-derived chloroanilines in soil. *J. agric. Fd Chem.* **19**, 385–387.

BARTHA, R., LINKE, H. A. B. and PRAMER, D. (1968). Pesticide transformations: production of chloroazobenzenes from chloroanilines. *Science, N.Y.* **161**, 582–583.

BARTNICKI, E. W. and CASTRO, C. E. (1969). Biodehalogenation. The pathway for transhalogenation and the stereochemistry of epoxide formation from halohydrins. *Biochemistry* **8**, 4677–4680.

BAYLY, R. C., DAGLEY, S. and GIBSON, D. T. (1966). The metabolism of cresols by species of *Pseudomonas*. *Biochem. J.* **101**, 293–301.

BEHRMAN, E. J. and STANIER, R. Y. (1957). The bacterial oxidation of nicotinic acid. *J. biol. Chem.* **228**, 923–945.

BEYNON, K. I. and WRIGHT, A. N. (1968). Persistence, penetration and breakdown of chlorthiamid and dichlobenil herbicides in field soils of different types. *J. Sci. Fd Agric.* **19**, 718–722.

BEYNON, K. I. and WRIGHT, A. N. (1969). Breakdown of the insecticide Gardona, on plants and in soils. *J. Sci. Fd Agric.* **20**, 250–256.

BHUIYA, Z. H. and ROTHWELL, D. F. (1973). Conversion of DDT to DDD in flooded soil. *Pl. Soil* **39**, 193–196.

BLAKELY, E. R. (1972). Microbial conversion of *p*-hydroxyphenylacetic acid to homogentisic acid. *Can. J. Microbiol.* **18**, 1247–1255.

BOERNER, H. (1965). The decomposition of afalon [*N*-(3,4-dichlorophenyl)-*N'*-methoxy-*N'*-methyl-urea] and aresin [*N*-(*p*-chlorophenyl)-*N'*-methoxy-*N'*-methylurea] in the soil. *Z. PflKrankh. PflPath. PflSchutz.* **72**, 516–531.

BOLLAG, J. M. and LIU, S. Y. (1972). Hydroxylations of carbaryl by soil fungi. *Nature, Lond.* **236**, 177–178.

BOLLAG, J. M., HELLING, C. S. and ALEXANDER, M. (1967). Metabolism of 4-chloro-2-methylphenoxyacetic acid by soil bacteria. *Appl. Microbiol.* **15**, 1393–1398.

BOLLAG, J. M., BRIGGS, G. G., DAWSON, J. E. and ALEXANDER, M. (1968). Enzymatic degradation of chlorocatechols. *J. agric. Fd Chem.* **16**, 829–833.

BOOTHROYD, B., NAPIER, E. J. and SOMERFIELD, G. A. (1961). The demethylation of griseofulvin by fungi. *Biochem. J.* **80**, 34–37.

BORDELEAU, L. M. and BARTHA, R. (1970). Azobenzene residues from aniline-based herbicides; evidence for labile intermediates. *Bull. Environ. Contam. Toxicol.* **5**, 34–37.

BORDELEAU, L. M., ROSEN, J. D. and BARTHA, R. (1972). Herbicide-derived chloroazobenzene residues: pathway of formation. *J. agric. Fd Chem.* **20**, 573–578.

BOURQUIN, A. W., ALEXANDER, S. K., SPEIDEL, H. K., MANN, J. E. and FAIR, J. F. (1972). Microbial interactions with cyclodiene pesticides. *Develop. Ind. Microbiol.* **13**, 264–278.

BROOKS, G. T. (1972). Pathways of enzymatic degradation of pesticides. *In* "Environmental Quality and Safety" (F. Coulston and F. Korte, Eds), vol. 1, pp. 106–164. Academic Press, New York.

BULL, D. L. (1968). Metabolism of UC-21149 (2-methyl-2 (methylthio) pro-pionaldehyde-*O*-(methylcarbamoyl) oxime) in cotton plants and soil in the field. *J. econ. Ent.* **61**, 1598–1602.

BURGE, W. D. (1971). Anaerobic decomposition of DDT in soil. Acceleration by volatile components of alfalfa. *J. agric. Fd Chem.* **19**, 375–378.

BURGE, W. D. (1972). Microbial populations hydrolysing propanil and accumulation of 3,4-dichloroaniline and 3,3',4,4'-tetrachloroazobenzene in soils. *Soil Biol. Biochem.* **4**, 379–386.

BURGER, K., MacRAE, I. C. and ALEXANDER, M. (1962). Decomposition of phenoxyalkyl carboxylic acids. *Proc. Soil. Sci. Soc. Am.* **26**, 243–246.

BURNS, R. G. (1975). Factors affecting pesticide loss from soil. *In* "Soil Biochemistry" (E. A. Paul and A. D. McLaren, Eds), vol. 4, pp. 103–141. Marcel Dekker Inc, New York.

BUSWELL, J. A. and CAIN, R. B. (1973). Microbial degradation of piperonylic acid by *Pseudomonas fluorescens*. *FEBS Lett.* **29**, 297–300.

BUSWELL, J. A. and MAHMOOD, A. (1972). Bacterial degradation of p-methoxybenzoic acid. *Arch. Mikrobiol.* **84**, 275–286.

BYRDE, R. J. W. and WOODCOCK, D. (1957). Fungal detoxication. 2. The metabolism of some phenoxy-n-alkyl carboxylic acids by *Aspergillus niger*. *Biochem. J.* **65**, 682–686.

CAIN, R. B. and CARTWRIGHT, N. J. (1960). Properties of some aromatic ring-opening enzymes of species of the genus *Nocardia*. *Biochim. biophys. Acta* **37**, 197–213.

CAIN, R. B., BILTON, R. F. and DARRAH, J. A. (1968a). The metabolism of aromatic acids by micro-organisms. Metabolic pathways in the fungi. *Biochem J.* **108**, 797–828.

CAIN, R. B., TRANTER, E. K. and DARRAH, J. A. (1968b). The utilisation of some halogenated aromatic acids by *Nocardia*. Oxidation and metabolism. *Biochem. J.* **106**, 211–227.

CAIN, R. B., WRIGHT, K. A. and HOUGHTON, C. (1970). Microbial metabolism of the pyridine ring: bacterial degradation of 1-methyl-4-carboxypyridinium chloride, a photolytic product of paraquat. *Meded. Fac. Landb. Rijksuniv. Gent* **35**, 785–798.

CAIN, R. B., HOUGHTON, C. and WRIGHT, K. A. (1974). Microbial metabolism of the pyridine ring. Metabolism of 2- and 3-hydroxypyridines by the maleamate pathway in *Achromobacter* sp. *Biochem. J.* **140**, 293–300.

CALLELY, A. G. (1974). The microbial breakdown of heterocyclic compounds. *In* "Industrial Aspects of Biochemistry" (B. Spencer, Ed.), pp. 515–532. Spec. Meet. Fed. Eur. Biochem. Soc. Dublin.

CARTWRIGHT, N. J. and CAIN, R. B. (1959a). Bacterial degradation of the nitrobenzoic acids. *Biochem. J.* **71**, 248–261.

CARTWRIGHT, N. J. and CAIN, R. B. (1959b). Bacterial degradation of the nitrobenzoic acids. 2. Reduction of the nitro group. *Biochem. J.* **73**, 305–314.

CARTWRIGHT, N. J. and SMITH, A. R. W. (1967). Bacterial attack on phenolic ethers. An enzyme system demethylating vanillic acid. *Biochem. J.* **102**, 826–841.

CARTWRIGHT, N. J., HOLDOM, K. and BROADBENT, D. A. (1971). Bacterial attack on phenolic ethers. Dealkylation of higher ethers and further observations on O-demethylases. *Microbios* **3**, 113–130.

CASTRO, C. E. and BELSER, N. O. (1968). Biodehalogenation. Reductive dehalogenation of the biocides ethylene dibromide, 1,2-dibromo-3-chloropropane and 2,3-dibromobutane in soil. *Environ. Sci. Technol.* **2**, 779–783.

CHACKO, C. I., LOCKWOOD, J. I. and ZABIK, M. (1966). Chlorinated hydrocarbon pesticides: degradation by microbes. *Science, N.Y.* **154**, 893–895.

CHAKRABARTY, A. M. (1972). Genetic basis of the biodegradation of salicylate in *Pseudomonas*. *J. Bact.* **112**, 815–823.

CHAKRABARTY, A. M. (1974). Dissociation of a degradative plasmid aggregate in *Pseudomonas*. *J. Bact.* **118**, 815–820.

CHAPMAN, P. J. (1972). An outline of reaction sequences used for the bacterial degradation of phenolic compounds. *In* "Degradation of Synthetic Organic Molecules in the Biosphere", pp. 17–55. National Academy of Sciences. Washington D.C.

CHIN, W. T., STONE, G. M. and SMITH, A. E. (1970). Degradation of carboxin (Vitavax) in water and soil. *J. agric. Fd Chem.* **18**, 731–732.

CLIFFORD, D. R. and WOODCOCK, D. (1964). Metabolism of phenoxyacetic acid by *Aspergillus niger* van Tiegh. *Nature, Lond.* **203**, 763.

COOK, J. W. (1957). *In vitro* destruction of some organophosphate pesticides by bovine rumen fluid. *J. agric. Fd Chem.* **5**, 859–863.

COPPEDGE, J. R., LINDQUIST, D. A., BULL, D. L. and DOROUGH, H. W. (1967). Fate of 2-methyl-2-(methylthio) propionaldehyde O-(methylcarbamoyl)oxime (Temik) in cotton plants and soil. *J. agric. Fd Chem.* **15**, 902–910.

CRIPPS, R. E. (1971). The microbial breakdown of pesticides. *In* "Microbial Aspects of Pollution" (G. Sykes and F. A. Skinner, Eds), pp. 255–266. Academic Press, London.

CRIPPS, R. E. (1973). The microbial metabolism of thiophene-2-carboxylate. *Biochem. J.* **134**, 353–366.

CSERJESI, A. J. and JOHNSON, E. I. (1972). Methylation of pentachlorophenol by *Trichoderma virgatum. Can. J. Microbiol.* **18**, 45–49.

CURTIS, R. F., LAND, D. G., GRIFFITHS, N. M., GEE, M. G., ROBINSON, D., PEEL, J. L., DENNIS, C. and GEE, J. M. (1972). 2,3,4,6,-Tetrachloro-anisole: association with musty taint in chickens and microbiological formation. *Nature, Lond.* **235**, 223–224.

DACRE, J. C. and WILLIAMS, R. T. (1962). Dehydroxylation of (^{14}C) protocatechuic acid in the rat. *Biochem. J.* **84**, 81P.

DACRE, J. C. and WILLIAMS, R. T. (1968). The role of the tissues and gut micro-organisms in the metabolism of (^{14}C) protocatechuic acid in the rat. Aromatic dehydroxylation. *J. Pharm. Pharmacol.* **20**, 610–618.

DACRE, J. C., SCHELINE, R. R. and WILLIAMS, R. T. (1968). The role of the tissues and gut flora in the metabolism of (^{14}C)-homoprotocatechuic acid in the rat and rabbit. *J. Pharm. Pharmacol.* **20**, 619–625.

DAGLEY, S. (1971). Catabolism of aromatic compounds by microorganisms. *In* "Advances in Microbial Physiology" (A. H. Rose and J. F. Wilkinson, Eds), vol. 6, pp. 1–46. Academic Press, London.

DAGLEY, S. (1972). Microbial degradation of stable chemical structures: general features of metabolic pathways. *In* "Degradation of Synthetic Organic Molecules in the Biosphere", pp. 1–16. National Academy of Sciences, Washington D.C.

DAGLEY, S. and GIBSON, D. T. (1965). The bacterial degradation of catechol. *Biochem. J.* **95**, 466–474.

DAGLEY, S. and JOHNSON, P. A. (1963). Microbial oxidation of kynurenic, xanthurenic and picolinic acids. *Biochim. biophys. Acta* **78**, 577–587.

DAGLEY, S. and PATEL, M. D. (1957). Oxidation of p-cresol and related compounds by a *Pseudomonas. Biochem J.* **66**, 227–233.

DALTON, R. L., EVANS, A. W. and RHODES, R. C. (1966). Disappearance of diuron from cotton field soils. *Weeds* **14**, 31–33.

DALY, J., GUROFF, G., JERINA, D., UDENFRIEND, S. and WITKOP, B. (1968). Intramolecular migrations during hydroxylation of aromatic compounds. The NIH shift. *Adv. Chem. Ser.* **77**, 279–289.

DAVEY, J. F. and GIBSON, D. T. (1974). Bacterial metabolism of *para-* and *meta-* xylene: oxidation of a methyl substituent. *J. Bact.* **119**, 923–929.

DAVIES, J. I. and EVANS, W. C. (1962). The elimination of halide ions from aliphatic halogen substituted organic acids by an enzyme preparation of *Pseudomonas dehalogenans*. *Biochem. J.*, **82**, 50P.

de FRENNE, E., EBERSPACHER, J. and LINGENS, F. (1973). The bacterial degradation of 5-amino-4-chloro-2-phenyl-3(2H)-pyridazinone. *Eur. J. Biochem.* **33**, 357–363.

di CARLO, F. J., SCHULTZ, A. S. and McMANUS, D. K. (1951). The assimilation of nucleic acid derivatives and related compounds by yeasts. *J. biol. Chem.* **189**, 151–157.

DITTERT, L. C. and HIGUCHI, T. (1963). Rates of hydrolysis of carbamate and carbonate esters in alkaline solution. *J. Pharm. Sci.* **52**, 852–857.

DIXON, J. B., MOORE, D. E., AGNIHOTRI, N. P. and LEWIS, D. E. (1970). Exchange of diquat in soil clays, vermiculite, and smectite. *Proc. Soil Sci. Soc. Am.* **34**, 805–808.

DRASAR, B. S., RENWICK, A. G. and WILLIAMS, R. T. (1972). The role of the gut flora in the metabolism of cyclamate. *Biochem. J.* **129**, 881–890.

DROBNIKOVA, V. (1971). Oxidative patterns of purines and pyrimidines in soil after specific and non-specific preincubation. *Zbl. Bakt.* Abt. II **126**, 735–745.

DROBNIKOVA, V. and POSPISILOVA, V. (1972). Decomposition of pyrimidine derivatives in soil. The effect of specific microflora and specific preincubation. *Zbl. Bakt.* Abt. II **127**, 98–112.

DUFFY, J. R. and WONG, N. (1967). Residues of organochlorine insecticides and their metabolites in soils in atlantic provinces of Canada. *J. agric. Fd Chem.* **15**, 457–464.

DUTTON, P. L. and EVANS, W. C. (1969). The metabolism of aromatic compounds by *Rhodopseudomonas palustris*. A new reductive method of aromatic ring metabolism. *Biochem. J.* **113**, 525–536.

DUXBURY, J. M., TIEDJE, J. M., ALEXANDER, M. and DAWSON, J. E. (1970). 2,4-D metabolism: enzymatic conversion of chloromaleylacetic acid to succinic acid. *J. agric. Fd Chem.* **18**, 199–201.

EBELING, W. (1963). Analysis of the basic processes involved in the decomposition, degradation, persistence and effectiveness of pesticides. *Residue Rev.* **3**, 35–163.

EDWARDS, C. A. (1973). "Persistent Pesticides in the Environment", 2nd edition. CRC Press, Cleveland.

EMMERSON, J. L. and ANDERSON, R. C. (1966). Metabolism of trifluralin in the rat and the dog. *Toxicol. Appl. Pharmacol.* **9**, 84–97.

ENGELHARDT, G., WALLNÖFER, P. R. and PLAPP, R. (1971). Degradation of linuron and some other herbicides and fungicides by a linuron-inducible enzyme obtained from *Bacillus sphaericus*. *Appl. Microbiol.* **22**, 284–288.

ENGELHARDT, G., WALLNÖFER, P. R. and PLAPP, R. (1972). Identification of *N,O*-dimethylhydroxylamine as a microbial degradation product of the herbicide linuron. *Appl. Microbiol.* **23**, 664–666.

ENGST, R. and KUJAWA, M. (1967). Enzymatischer Abbau des DDT durch Schimmelpilze. 2. Mitt. Reaktionsverlauf des enzymatischen DDT-Abbaues. *Nahrung* **11**, 751–760.

ENSIGN, J. C. and RITTENBERG, S. C. (1963). A crystalline pigment produced from 2-hydroxypyridine by *Arthrobacter crystallopoietes* n. sp. *Arch. Mikrobiol.* **47**, 137–153.

ENSIGN, J. C. and RITTENBERG, S. C. (1964). The pathway of oxidation of nicotinic acid by a *Bacillus* species. *J. biol. Chem.* **239**, 2285–2291.

ERICKSSON, H. and GUSTAFSSON, J. A. (1970). Steroids in germ-free and conventional rats. Sulpho- and glucuronohydrolase activities of caecal contents from conventional rats. *Eur. J. Biochem.* **13**, 198–202.

ESSER, H. O. (1970). The biodegradation of pesticides in the soil. *Meded. Fac. Landb. Rijksuniv. Gent* **35**, 753–783.

EVANS, W. C., SMITH, B. S. W., FERNLEY, H. N. and DAVIES, J. I. (1971). Bacterial metabolism of 2,4-dichlorophenoxyacetate. *Biochem. J.* **122**, 543–551.

FAULKNER, J. K. and WOODCOCK, D. (1961). Fungal detoxication. Pt V. Metabolism of *o*- and *p*-chlorophenoxyacetic acids by *Aspergillus niger*. *J. Chem. Soc.* 1961, Part IV, 5397–5400.

FAULKNER, J. K. and WOODCOCK, D. (1964). Metabolism of 2,4-dichlorophenoxyacetic acid (2,4-D) by *Aspergillus niger*. *Nature, Lond.* **203**, 865.

FAULKNER, J. K. and WOODCOCK, D. (1965). Fungal detoxication. Pt VII. Metabolism of 2,4-dichlorophenoxyacetic and 4-chloro-2-methylphenoxyacetic acids by *Aspergillus niger*. *J. Chem. Soc.* 1965, Part I, 1187–1191.

FAY, B. F. and MELTON, J. J. (1973). Auxilliary chemicals used with pesticides. *Proc. W. Soc. Weed Sci.* **26**, 18–24.

FERNLEY, H. N. and EVANS, W. C. (1958). Oxidative metabolism of polycyclic hydrocarbons by soil pseudomonads. *Nature, Lond.* **182**, 373–375.

FISH, M. S., JOHNSON, N. M. and HORNING, E. C. (1956). *t*-Amine oxide rearrangements. *N,N*-dimethyl-tryptamine oxide. *J. am. Chem. Soc.* **78**, 3668–3671.

FOCHT, D. D. (1972). Microbial degradation of DDT metabolites to carbon dioxide, water and chloride. *Bull. Environ. Contam. Taxicol.* **7**, 52–56.

FRANK, P. A. (1966). Persistence and distribution of monuron and neburon in an aquatic environment. *Weeds* **14**, 219–222.

FREED, V. H. (1966). *In* "Pesticides and their Effects on Soils and Water", ASA Spec. Publicn. No. 8, pp. 25–43. Soil Science Society of America, Madison.

FREHSE, H. (1976). The perspective of persistence. *In* "Proc. Symp. Persistence of Insectic. Herbic", pp. 1–39, BCPC Monograph no. 17.

FRENCH, A. L. and HOOPINFARNER, R. A. (1970). Dechlorination of DDT by membranes isolated from *Escherichia coli*. *J. econ. Ent.* **63**, 756–759.

FRIES, G. R., MARROW, G. S. and GORDON, C. H. (1969). Metabolism of *o,p'*- and *p,p'*-DDT by rumen microorganisms. *J. agric. Fd Chem.* **17**, 860–862.

FUJIOKA, M. and WADA, H. (1968). The bacterial oxidation of indole. *Biochim. biophys. Acta* **158**, 70–78.

FUNDERBURK, H. H. (1969). Diquat and paraquat. *In* "Degradation of Herbicides" (P. C. Kearney and D. D. Kaufman, Eds), pp. 283–297. Marcel Dekker Inc, New York.

FUNDERBURK, H. H., Jr. and BOZARTH, G. A. (1967). Review of the metabolism and decomposition of diquat and paraquat. *J. agric. Fd Chem.* **15**, 563–567.

FURAKAWA, K., SUSUKI, T. and TONOMURA, K. (1969). Decomposition of organic mercurial compounds by mercury-resistant bacteria. *Agric. biol. Chem.* **33**, 128–130.

GAMAR, Y. and GAUNT, J. K. (1971). Bacterial metabolism of 4-chloro-2-methyl-phenoxyacetate. Formation of glyoxylate by side-chain cleavage. *Biochem. J.* **122**, 527–531.

GARDINER, J. A., RHODES, R. C., ADAMS, J. B. and SOBOCZENSKI, E. J. (1969). Synthesis and studies with 2-C^{14}-labelled bromacil and terbacil. *J. agric. Fd Chem.* **17**, 980–986.

GAUNT, J. K. and EVANS, W. C. (1961). Metabolism of 4-chloro-2-methyl-phenoxyacetic acids by a soil microorganism. *Biochem. J.* **79**, 25P–26P.

GAUNT, J. K. and EVANS, W. C. (1971a). Metabolism of 4-chloro-2-methyl-phenoxyacetate by a soil pseudomonad. Preliminary evidence for the metabolic pathway. *Biochem. J.* **122**, 519–526.

GAUNT, J. K. and EVANS, W. C. (1971b). Metabolism of 4-chloro-2-methyl-phenoxyacetate by a soil pseudomonad. Ring-fission, lactonising and delactonising enzymes. *Biochem. J.* **122**, 533–542.

GAUSE, E. M., MONTALVO, D. A. and ROWLANDS, J. R. (1967). Electron spin resonance studies of the chloranil–cysteine, dichlone–cysteine and dichlone–glutathione reactions. *Biochim. biophys. Acta.* **141**, 217–219.

GEISSBÜHLER, H. (1969). The substituted ureas. *In* "Degradation of Herbicides" (P. C. Kearney and D. D. Kaufman, Eds), pp. 79–111. Marcel Dekker Inc, New York.

GEISSBÜHLER, H., HASELBACH, C., AEBI, H. and EBNER, L. (1963). The fate of N'-(4-chlorophenoxy)-phenyl-N,N-dimethyl urea (C-1983) in soils and plants. III. Breakdown in soils and plants. *Weed Res.* **3**, 277–297.

GETZIN, L. W. and ROSEFIELD, I. (1966). Persistence of diazinon and Zinophos in soils. *J. econ. Ent.* **59**, 512–516.

GETZIN, L. W. and ROSEFIELD, I. (1971). Partial purification and properties of a soil enzyme that degrades the insecticide malathion. *Biochim. biophys. Acta* **235**, 442–453.

GHERNA, R. L. and RITTENBERG, S. C. (1962). Alternate pathways in nicotine degradation. *Bact. Proc.* 107.

GHERNA, R. L., RICHARDSON, S. H. and RITTENBERG, S. C. (1965). The bacterial oxidation of nicotine. VI. The metabolism of 2,6-dihydroxypseudooxynicotine. *J. biol. Chem.* **240**, 3669–3674.

GIBSON, D. T. (1972). The microbial oxidation of aromatic hydrocarbons. *Critical Rev. Microbiol.* **1**, 199–223.

GIBSON, D. T., MAHADEVAN, V. and DAVEY, J. F. (1974). Bacterial metabolism of *para*- and *meta*-xylene: oxidation of the aromatic ring. *J. Bact.* **119**, 930–936.

GOLAB, T., HERBERG, R. J., DAY, E. W., RAUN, A. P., HOLZER, F. J. and PROBST, G. W. (1969). Fate of carbon-14 trifluralin in artificial rumen fluid and in ruminant animals. *J. agric. Fd Chem.* **17**, 576–580.

GOLDBERG, L., PAREKH, C., PATTI, A. and SOIKE, K. (1969). Cyclamate degradation in mammals and *in vitro. Toxicol. Appl. Pharmacol.* **14**, 654.

GOLDMAN, P. (1965). The enzymatic cleavage of the carbon-fluorine bond in fluoroacetate *J. biol. Chem.* **240**, 3434–3438.

GOLDMAN, P. (1972). Enzymology of carbon–halogen bonds. *In* "Degradation of Synthetic Organic Molecules in the Biosphere", pp. 147–165. National Academy of Sciences, Washington D.C.

GOLDMAN, P., MILNE, G. W. A. and PIGNATARO, M. T. (1967). Fluorine containing metabolites from 2-fluoro benzoic acid by *Pseudomonas* species. *Archs Biochem. Biophys.* **118**, 178–184.

GOLDMAN, P., MILNE, G. W. A. and KEISTER, D. B. (1968). Carbon–halogen bond cleavage. III. Studies on bacterial halidohydrolases. *J. biol. Chem.* **243**, 428–434.

GRAETZ, D. A., CHESTERS, G., DANIEL, T. C., NEWLAND, L. W. and LEE, C. B. (1970). Parathion degradation in lake sediments. *J. Wat. Pollut. Control Fed.* **12**, 76–94.

GUENZI, W. D. and BEARD, W. E. (1967). Anaerobic biodegradation of DDT to DDD in soil. *Science, N.Y.* **156**, 1116–1117.

GUENZI, W. D. and BEARD, W. E. (1968). Anaerobic conversion of DDT to DDD and aerobic stability of DDT in soil. *Proc. Soil Sci. Soc. Am.* **32**, 522–524.

GUNNER, H. B. and ZUCKERMAN, B. N. (1968). Degradation of diazinon by synergistic microbial action. *Nature, Lond.* **217**, 1183–1184.

GUNNER, H. B., ZUCKERMAN, B. M., WALKER, R. W., MILLER, C. W., DEUBERT, K. H. and LONGLEY, R. E. (1966). The distribution and persistence of diazinon applied to plant and soil and its influence on rhizosphere and soil microflora. *Pl. Soil* **25**, 249–264.

GUNTER, S. E. (1953). The enzymatic oxidation of *p*-hydroxy-mandelic acid to *p*-hydroxybenzoic acid. *J. Bact.* **66**, 341–346.

GUROFF, G., and RHOADS, C. A. (1967). Phenylalanine hydroxylase from *Pseudomonas* species (ATCC 11299a): purification of the enzyme and activation by various metal ions. *J. biol. Chem.* **242**, 3641–3645.

GUROFF, G., KONDO, K. and DALY, J. (1966). The production of *meta*-chlorotyrosine from *para*-chlorophenylalanine by phenylalanine hydroxylase. *Biochem. biophys. Res. Commun.* **25**, 622–628.

GUROFF, G., DALY, J. W., JERINA, D. M., RENSON, J., WITKOP, B. and UDENFRIEND, S. (1967). Hydroxylation induced migration: The NIH shift. *Science, N.Y.* **157**, 1524–1530.

GUTENMAN, W. H. and LISK, D. J. (1964). Conversion of 4-(2,4-DB) to 2,4-dichlorophenoxycrotonic acid (2,4-DC) and production of 2,4-D from 2,4-DC in soil. *J. agric. Fd Chem.* **12**, 322–325.

GUTENMAN, W. H., LOOS, M. A., ALEXANDER, M. and LISK, D. J. (1964). Betaoxidation of phenoxyalkanoic acids in soil. *Proc. Soil Sci. Soc. Am.* **28**, 205–207.

GYSIN, H. (1962). Triazine herbicides. Their chemistry, biological properties and mode of action. *Chemy Ind.* 1193–1400.

HAMDI, Y. A. and TEWFIK, M. S. (1970). Degradation of 3,5-dinitrocresol by *Rhizobium* and *Arthrobacter* spp. *Soil Biol. Biochem.* **2**, 163–166.

HAMMOND, M. W. and ALEXANDER, M. (1972). Effect of chemical structure on microbial degradation of methyl-substituted aliphatic acids. *Environ. Sci. Technol.* **6**, 732–735.

HARRISON, J. E. and RIBBONS, D. W. (1971). Bacterial oxidation of methylenedioxy compounds. *Bact. Proc.* 110.

HAWKSWORTH, G., DRASAR, B. S. and HILL, M. J. (1971). Intestinal bacteria and the hydrolysis of glycosidic bonds. *J. med. Microbiol.* **4**, 451–459.

HAYAISHI, O. and KORNBERG, A. (1952). Metabolism of cytosine, thymine, uracil and barbituric acid by bacterial enzymes. *J. biol. Chem.* **197**, 717–732.

HAYAISHI, O. and NOZAKI, M. (1969). Nature and mechanisms of oxygenases. *Science, N.Y.* **164**, 389–396.

HEGEMAN, G. D. (1971). The evolution of metabolic pathways in bacteria. *In* "Degradation of Synthetic Organic Molecules in the Biosphere", pp. 56–72. National Academy of Sciences, Washington D.C.

HELLING, C. S., BOLLAG, J. M. and DAWSON, J. E. (1968). Cleavage of ether-oxygen bond in phenoxyacetic acid by an *Arthrobacter* species. *J. agric. Fd Chem.* **16**, 538–539.

HENDERSON, M. E. K. (1957). Metabolism of methoxylated aromatic compounds by soil fungi. *J. gen. Microbiol.* **16**, 686–695.

HENDERSON, M. E. K. and FARMER, V. C. (1955). Utilisation by soil fungi of *p*-hydroxybenzaldehyde, ferulic acid, syringaldehyde and vanillin. *J. gen. Microbiol.* **12**, 37–46.

HILL, D. W. and McCARTY, P. L. (1967). Anaerobic degradation of selected chlorinated hydrocarbon pesticides. *J. Wat. Pollut. Control Fed.* **39**, 1259–1277.

HILL, I. R. (1976). Degradation of the insecticide pirimicarb in soil – characterisation of "bound" residues. *In* "Bound and Conjugated Pesticide Residues" (D. D. Kaufman, G. G. Still, G. D. Paulson and S. K. Bandal, Eds), pp. 358–361. ACS Symp. Ser. No. 29.

HILL, I. R. and ARNOLD, D. J. (1973). Degradation of ethirimol in soil. *Proc. 7th Brit. Insectic. Fungic. Conf.* 47–55.

HIRSCH, P. and ALEXANDER, M. (1960). Microbial decomposition of halogenated propionic and acetic acids. *Can. J. Microbiol.* **6**, 241–249.

HIRSCHBERG, R. and ENSIGN, J. C. (1971). Oxidation of nicotinic acid by a *Bacillus* species: source of oxygen atoms for the hydroxylation of nicotinic acid and 6-hydroxynicotinic acid. *J. Bact.* **108**, 757–759.

HORVATH, R. S. (1970a). Microbial cometabolism of 2,4,5-trichlorophenoxyacetic acid. *Bull. Environ. Contam. Toxicol.* **5**, 537–541.

HORVATH, R. S. (1970b). Cometabolism of methyl- and chloro-substituted catechols by an *Achromobacter* sp. possessing a new meta-cleaving oxygenase. *Biochem. J.* **119**, 871–876.

HORVATH, R. S. (1972). Microbial co-metabolism and the degradation of organic compounds in nature. *Bact. Rev.* **36**, 146–155.

HOUGHTON, C. and CAIN, R. B. (1972). Microbial metabolism of the pyridine ring. Formation of pyridinediols (dihydroxy-pyridines) as intermediates in the degradation of pyridine compounds by microorganisms. *Biochem. J.* **130**, 879–893.

HUBBELL, D. H., ROTHWELL, D. F., WHEELER, W. B., TAPPAN, W. B. and RHOADS, F. M. (1973). Microbiological effects and persistence of some pesticide combinations in soil. *J. Environ. Qual.* **2**, 96–99.

IRRI (1966). Annual Rept. The International Rice Research Institute, Laguna, Philippines.

JEFFREY, A. M., YEH, H. J. C., JERINA, D. M., PATEL, T. R., DAVEY, J. F. and GIBSON, D. T. (1975). Initial reactions in the oxidation of naphthalene by *Pseudomonas putida. Biochemistry* **14**, 575–584.

JOHNSON, B. T., GOODMAN, R. N. and GOLDBERG, H. S. (1967). Conversion of DDT to DDD by pathogenic and saprophytic bacteria associated with plants. *Science, N.Y.* **157**, 560–561.

JOHNSTON, H. W., BRIGGS, G. G. and ALEXANDER, M. (1972). Metabolism of 3-chlorobenzoic acid by a pseudomonad. *Soil Biol. Biochem.* **4**, 187–190.

JOHNSTON, J. B. and CAIN, R. B. (1973). The control of benzene-sulphonate degradation in bacteria. *Spec. Meet. Fed. Eur. Biochem. Soc.* Abs. no. 137.

JONES, D. F. (1968). Microbial oxidation of long-chain aliphatic compounds. II. Branched-chain alkanes. *J. Chem. Soc.* **1968**, C, 2809–2815.

KALLMAN, B. J. and ANDREWS, A. K. (1963). Reductive dechlorination of DDT to DDD by yeast. *Science, N.Y.* **141**, 1050–1051.

KAMEDA, Y., TOYOURA, E. and KIMURA, Y. (1957). Metabolic activities of soil bacteria towards derivatives of benzoic acid, amino acids and acylamino acids. *Kanazawa Daigaku Yakugakubu Kenkyu Nempo* **7**, 37: *Chem. Abstr.* **52**, 4081.

KAPOOR, I. P., METCALF, R. L., NYSTROM, R. F. and SANGHA, G. K. (1970). Comparative metabolism of methoxychlor, methiochlor and DDT in mouse, insects and in a model ecosystem. *J. agric. Fd Chem.* **18**, 1145–1152.

KAPOOR, I. P., METCALF, R. L., HIRWE, A. S., LU, P. Y., COATS, J. R. and NYSTROM, R. F. (1972). Comparative metabolism of DDT, methylchlor, and ethoxychlor in mouse, insects and in a model ecosystem. *J. agric. Fd Chem.* **20**, 1–6.

KATAGIRI, M., GANGULI, B. N. and GUNSALUS, I. C. (1968). A soluble cytochrome P-450 functional in methylene hydroxylation. *J. biol. Chem.* **243**, 3543–3546.

KAUFMAN, D. D. (1964). Microbial degradation of 2,2-dichloropropionic acid in five soils. *Can. J. Microbiol.* **10**, 843–852.

KAUFMAN, D. D. (1965). Degradation of carbamate herbicides in soil. *J. agric. Fd Chem.* **15**, 582–591.

KAUFMAN, D. D. (1970). Pesticide metabolism. *In* "Pesticides in the Soil: Ecology, Degradation, Movement", pp. 73–86. Michigan State Univ., East Lansing.

KAUFMAN, D. D. (1972). Degradation of pesticide combinations. *Pest. Chem. Proc. Int. Cong. Pestic. Chem.* **6**, 175–204.

KAUFMAN, D. D. and BLAKE, J. (1970). Degradation of atrazine by soil fungi. *Soil Biol. Biochem.* **2**, 73–80.

KAUFMAN, D. D. and KEARNEY, P. C. (1970). Microbial degradation of *s*-triazine herbicides. *Residue Rev.* **32**, 235–265.

KAUFMAN, D. D., PLIMMER, J. R., KEARNEY, P. C., BLAKE, J. and GUARDIA, F. S. (1968). Chemical vs. microbial decomposition of amitrole in soil. *Weed Sci.* **16**, 266–272.

KAUFMAN, D. D., KEARNEY, P. C., von ENDT, D. W. and MILLER, D. E. (1970). Methylcarbamate inhibition of phenylcarbamate metabolism in soil. *J. agric. Fd Chem.* **18**, 513–519.

KAUFMAN, D. D., BLAKE, J. and MILLER, D. E. (1971). Methylcarbamates affect acylanilide herbicide residues in soil. *J. agric. Fd Chem.* **19**, 204–206.

KEARNEY, P. C. (1965). Purification and properties of an enzyme responsible for hydrolysing phenylcarbamates. *J. agric. Fd Chem.* **13**, 561–564.

KEARNEY, P. C. (1966). Metabolism of herbicides in soils. *Adv. Chem. Ser.* **60**, 250–262.

KEARNEY, P. C. (1967). Influence of physicochemical properties on biodegradability of phenylcarbamate herbicides. *J. agric. Fd Chem.* **15**, 568–571.

KEARNEY, P. C. and HELLING, C. S. (1969). Reactions of pesticides in soils. *Residue Rev.* **25**, 25–44.

KEARNEY, P. C. and PLIMMER, J. R. (1970). Relation of structure to pesticide decomposition. *In* "Pesticides in the Soil: Ecology, Degradation and Movement", pp. 65–72. Michigan State Univ., East Lansing.

KEARNEY, P. C. and PLIMMER, J. R. (1972). Metabolism of 3,4-dichloroaniline in soils. *J. agric. Fd Chem.* **20**, 584–585.

KEARNEY, P. C., KAUFMAN, D. D. and BEALL, M. L. (1964). Enzymatic dehalogenation of 2,2-dichloropropionate. *Biochem. Biophys. Res. Commun.* **14**, 29–33.

KEARNEY, P. C., HARRIS, C. I., KAUFMAN, D. D. and SHEETS, T. J. (1965). Behaviour and fate of chlorinated aliphatic acids in soils. *Adv. Pest. Control Res.* **6**, 1–30.

KEARNEY, P. C., KAUFMAN, D. D. and ALEXANDER, M. (1967). Biochemistry of herbicide decomposition in soils. *In* "Soil Biochemistry" (A. D. McLaren and G. H. Peterson, Eds), vol. 1, pp. 318–342. Marcel Dekker Inc, New York.

KEARNEY, P. C., KAUFMAN, D. D., von ENDT, D. W. and GUARDIA, F. S. (1969). TCA metabolism by soil micro-organisms. *J. agric. Fd Chem.* **17**, 581–584.

KENNEDY, E. P. (1957). Metabolism of lipides. *Ann. Rev. Biochem.* **26**, 119–148.

KITTREDGE, J. S. and ROBERTS, E. (1969). A carbon-phosphorus bond in nature. *Science, N.Y.* **164**, 37–42.

KLOWZOWSKI, T. T. (1965). Variable toxicity of triazine herbicides. *Nature, Lond.* **205**, 104–105.

KO, W. H. and FARLEY, J. D. (1969). Conversion of pentachloronitrobenzene to pentachloroaniline in soil and the effect of these compounds on soil microflora. *Phytopathology* **59**, 64–67.

KONRAD, J. G., ARMSTRONG, D. E. and CHESTERS, G. (1967). Soil degradation of diazinon, a phosphorothionate insecticide. *Agron. J.* **59**, 591–594.

KONRAD, J. G., CHESTERS, G. and ARMSTRONG, D. E. (1969). Soil degradation of malathion, phosphorodithioate insecticide. *Proc. Soil Sci. Soc. Am.* **33**, 259–262.

KUWATSUKA, S. (1972). Degradation of several herbicides in soils under different conditions. *In* "Environmental Toxicology of Pesticides" (F. Matsumura, G. M. Boush and T. Misato, Eds), pp. 385–400. Academic Press, New York.

LAANIO, T. L., KEARNEY, P. C. and KAUFMAN, D. D. (1973). Microbial metabolism of dinitramine. *Pestic. Biochem. Physiol.* **3**, 271–277.

LANZILOTTA, R. P. and PRAMER, D. (1970a). Herbicide transformation I. Studies with whole cells of *Fusarium solani*. *Appl. Microbiol.* **19**, 301–306.

LANZILOTTA, R. P. and PRAMER, D. (1970b). Herbicide transformation II. Studies with an acylamidase of *Fusarium solani*. *Appl. Microbiol.* **19**, 307–313.

LA RUE, T. A. and SPENCER, J. F. T. (1968). The utilisation of purines and pyrimidines by yeasts. *Can. J. Microbiol.* **14**, 79–86.

LAY, M. M. and ILNICKI, R. D. (1974). Peroxidase activity and propanil degradation in soil. *Weed Res.* **14**, 111–113.

LEADBETTER, E. R. and FOSTER, J. W. (1959). Oxidation products formed from gaseous alkanes by the bacterium *Pseudomonas methanica*. *Archs Biochem. Biophys.* **82**, 491–492.

LICHTENSTEIN, E. P. (1966). Increase of persistence and toxicity of parathion and diazinon in soils with detergents. *J. econ. Ent.* **59**, 985–993.

LICHTENSTEIN, E. P. and SCHULZ, S. R. (1960). Epoxidation of aldrin and heptachlor in soils as influenced by autoclaving, moisture, and soil types. *J. econ. Ent.* **53**, 192–197.

LICHTENSTEIN, E. P. and SCHULZ, S. R. (1964). The effects of moisture and microorganisms on the persistence and metabolism of some organophosphorus insecticides in soils, with special emphasis on parathion. *J. econ. Ent.* **57**, 618–627.

LICHTENSTEIN, E. P., FUHREMANN, T. W. and SCHULZ, K. R. (1971). Persistence and vertical distribution of DDT, lindane, and aldrin residues 10 and 15 years after a single soil application. *J. agric. Fd Chem.* **19**, 718–721.

LIU, S. Y. and BOLLAG, J. M. (1971). Carbaryl decomposition to 1-naphthyl carbamate by *Aspergillus terreus*. *Pestic. Biochem. Physiol.* **1**, 366–372.

LOOS, M. A. (1969). Phenoxyalkanoic acids. *In* "Degradation of Herbicides" (P. C. Kearney and D. D. Kaufman, Eds), pp. 1–49. Marcel Dekker Inc, New York.

LOOS, M. A., ROBERTS, R. N. and ALEXANDER, M. (1967a). Phenols as intermediates in the decomposition of phenoxyacetates by an *Arthrobacter species*. *Can. J. Microbiol.* **13**, 679–690.

LOOS, M. A., ROBERTS, R. N. and ALEXANDER, M. (1967b). Formation of 2,4-dichlorophenol and 2,4-dichloroanisole from 2,4-dichlorophenoxyacetate by *Arthrobacter* sp. *Can. J. Microbiol.* **13**, 691–699.

LU, J. Y. and LISKA, B. J. (1969). Lipase from *Pseudomonas fragi* II. Properties of the enzyme. *Appl. Microbiol.* **18**, 108–113.

MacRAE, I. C. and ALEXANDER, M. (1963). Metabolism of phenoxyalkyl carboxylic acids by a *Flavobacterium* species. *J. Bact.* **86**, 1231–1235.

MacRAE, I. C. and ALEXANDER, M. (1965). Microbial degradation of selected herbicides in soil. *J. agric. Fd Chem.* **13**, 72–76.

MacRAE, I. C., RAGHU, K. and BAUTISTA, E. M. (1969). Anaerobic degradation of the insecticide lindane by *Clostridium* sp. *Nature, Lond.* **221**, 859–860.

MADHOSINGH, C. (1961). The metabolic detoxication of 2,4-dinitrophenol by *Fusarium oxysporum. Can. J. Microbiol.* **7**, 553–567.

MALONE, T. C. (1970). *In vitro* conversion of DDT to DDD by the intestinal microflora of the northern anchovy, *Engraulis mordax. Nature, Lond.* **227**, 848–849.

MARKOVETZ, A. J. (1972). Subterminal oxidation of aliphatic hydrocarbons by microorganisms. *Critical Rev. Microbiol.* **1**, 225–237.

MATSUMURA, F. and BOUSH, G. M. (1966). Malathion degradation by *Trichoderma viride* and a *Pseudomonas* species. *Science, N.Y.* **153**, 1278–1280.

MATSUMURA, F. and BOUSH, G. M. (1968). Degradation of insecticides by a soil fungus, *Trichoderma viride. J. econ. Ent.* **61**, 610–612.

MATSUMURA, F. and BOUSH, G. M. (1971). Metabolism of insecticides by microorganisms. *In* "Soil Biochemistry" (A. D. McLaren and J. Skujins, Eds), vol. 2, pp. 320–336. Marcel Dekker Inc, New York.

MATSUMURA, F., BOUSH, G. M. and TAI, A. (1968). Breakdown of dieldrin in the soil by a microorganism. *Nature, Lond.* **219**, 965–967.

MATSUMURA, F., GOTOH, Y. and BOUSH, G. M. (1971a). Phenyl mercuric acetate metabolic conversion by microorganisms. *Science, N.Y.* **173**, 49–51.

MATSUMURA, F., KHANVILKAR, V. G., PATIL, K. C. and BOUSH, G. M. (1971b). Degradation pathways of endrin by certain soil microorganisms. *J. agric. Fd Chem.* **19**, 27–32.

MAY, S. W. and ABBOTT, B. J. (1973). Enzymic epoxidation II. Comparison between the epoxidation and hydroxylation reactions catalysed by the ω-hydroxylation system of *Pseudomonas oleovorans. J. biol. Chem.* **248**, 1725–1730.

McGRATH, D. (1976). Factors that influence the persistence of TCA in soil. *Weed Res.* **16**, 131–137.

McKENNA, E. J. (1972). Microbial metabolism of normal and branched chain alkanes. *In* "Degradation of Synthetic Organic Molecules in the Biosphere", pp. 73–97. National Academy of Sciences, Washington D.C.

MEIKLE, R. W. (1972). Decomposition: qualitative relationships. *In* "Organic Chemicals in the Soil Environment" (C. A. I. Goring and J. W. Hamaker, Eds), pp. 145–251. Marcel Dekker Inc, New York.

MEIKLE, R. W. and HAMILTON, P. M. (1964). Nutrient-conserving agents. Loss of 2-chloro-6-(trichloro-methyl)-pyridine from soil. *J. agric. Fd Chem.* **12**, 207–218.

MEIKLE, R. W., WILLIAMS, E. A. and REDEMANN, C. T. (1966). Metabolism of Tordon herbicide (4-amino-3,5,6-trichloropicolinic acid) in cotton and decomposition in soil. *J. agric. Fd Chem.* **14**, 383–387.

MENN, J. J., PATCHETT, G. G. and BATCHELDER, G. H. (1960). The persistence of trithion, an organophosphorus insecticide in soil. *J. econ. Ent.* **53**, 1080–1082.

MENZIE, C. M. (1969). *Metabolism of pesticides.* Special Scientific Report, Wildlife No. 27. Bureau of Sport, Fisheries and Wildlife, Washington.

MEYER, T. and SCHELINE, R. R. (1972). 3,4,5-trimethoxy-cinnamic acid and related compounds. I. Metabolism by the rat intestinal microflora. *Xenobiotica* **2**, 383–390.

MILES, J. R. W., TU, C. M. and HARRIS, C. R. (1969). Metabolism of heptachlor and its degradation products by soil microorganisms. *J. econ. Ent.* **62**, 1334–1337.

MILES, J. R. W., TU, C. M. and HARRIS, C. R. (1971). Degradation of heptachlor epoxide and heptachlor by a mixed culture of soil microorganisms. *J. econ. Ent.* **64**, 839–841.

MILLER, P. M. and LUKENS, R. J. (1966). Deactivation of sodium N-methyl-dithiocarbamate in soil by nematocides containing halogenated hydrocarbons. *Phytopathology* **56**, 967–970.

MILNE, G. W. A., GOLDMAN, P. and HOLTZMAN, J. L. (1968). The metabolism of 2-fluorobenzoic acid. II. Studies with $^{18}O_2$. *J. biol. Chem.* **243**, 5374–5376.

MISKUS, R. P., BLAIR, D. P. and CASIDA, J. E. (1965). Conversion of DDT to DDD by bovine rumen fluid, lake water, and reduced porphyrins. *J. agric. Fd Chem.* **13**, 481–483.

MIYAZAKI, S. and THORSTEINSON, A. J. (1972). Metabolism of DDT by fresh water diatoms. *Bull. Environ. Contam. Toxicol.* **8**, 81–83.

MIYAZAKI, S., BOUSH, G. M. and MATSUMURA, F. (1969). Metabolism of ^{14}C-chlorobenzilate and ^{14}C-chloropropylate by *Rhodotorula gracilis*. *Appl. Microbiol.* **18**, 972–976.

MIYAZAKI, S., BOUSH, G. M. and MATSUMURA, F. (1970). Microbial degradation of chlorobenzilate (ethyl 4,4′-dichlorobenzilate). *J. agric. Fd Chem.* **18**, 87–91.

MOUNTER, L. A. and TUCK, K. D. (1956). Dialkylfluorophosphatases of micro-organisms, II. Substrate specificity studies. *J. biol. Chem.* **221**, 537–541.

NAIK, M. N., JACKSON, R. B., STOKES, J. and SWABY, R. J. (1972). Microbial degradation and phytotoxicity of picloram and other substituted pyridines. *Soil Biol. Biochem.* **4**, 313–323.

NEWLAND, L. W., CHESTERS, G. and LEE, G. B. (1969). Degradation of γ-BHC in simulated lake impoundments as affected by aeration. *J. Wat. Pollut. Control Fed.* **41**, R174–R188.

NOVICK, R. P. and ROTH, C. (1968). Plasmid-linked resistance to inorganic salts in *Staphylococcus aureus*. *J. Bact.* **94**, 1335–1342.

NOZAKA, J. and KUSUNOSE, M. (1968). Metabolism of hydrocarbons in micro-organisms. Part I. Oxidation of p-xylene and toluene by cell-free enzyme preparations of *Pseudomonas aeruginosa*. *Agric. biol. Chem.* **32**, 1033–1039.

O'BRIEN, R. D. (1967). "Insecticides, Action and Metabolism", Academic Press, New York.

ORNSTON, L. N. and STANIER, R. Y. (1966). The conversion of catechol and protocatechuate to β-ketoadipate by *Pseudomonas putida*. 1. Biochemistry. *J. biol. Chem.* **241**, 3776–3786.

ORPIN, C. G., KNIGHT, M. and EVANS, W. C. (1972a). The bacterial oxidation of N-methylisonicotinate, a photolytic product of paraquat. *Biochem. J.* **127**, 833–844.

ORPIN, C. G., KNIGHT, M. and EVANS, W. C. (1972b). The bacterial oxidation of picolinamide, a photolytic product of diquat. *Biochem. J.* **127**, 819–831.

OSGERBY, J. M. (1973). Processes affecting herbicide action in soil. *Pestic. Sci.* **4**, 247–258.

PARR, J. F. and SMITH, S. (1973). Degradation of trifluralin under laboratory conditions and soil anaerobiosis. *Soil Sci.* **115**, 55–63.

PATIL, K. C., MATSUMURA, F. and BOUSH, G. M. (1970). Degradation of endrin, aldrin and DDT by soil microorganisms. *Appl. Microbiol.* **19**, 879–881.

PATIL, K. C., MATSUMURA, F. and BOUSH, G. M. (1972). Metabolic transformation of DDT, dieldrin, aldrin, and endrin by marine microorganisms. *Environ. Sci. Technol.* **6**, 629–632.

PEREZ-SILVA, G. and RODRIGUES, D. (1967). Deshidroxilación del ácido cafeico por varias especies de pseudomonas. *Boln. R. Soc. esp. Hist. nat. (Biol).* **65**, 401–405.

PLIMMER, J. R. and KEARNEY, P. C. (1969). Prometryne degradation in soils and light. *Abstr. Am. Chem. Soc.*, 158th meeting New York, Sept 8–12, A30.

PLIMMER, J. R., KEARNEY, P. C. and ROWLANDS, J. R. (1968). Free radical oxidation of pesticides. *Weed Sci. Soc. Am. Abstr.*, p. 20.

PLIMMER, J. R., KEARNEY, P. C., CHISAKA, H., YOUNT, J. B. and KLINGEBIEL, U. I. (1970). 1,3-bis(3,4-dichlorophenyl)triazene from propanil in soils. *J. agric. Fd Chem.* **18**, 859–861.

PRIEST, B. and STEPHENS, R. J. (1975). Studies on the breakdown of *p*-chlorophenyl methylcarbamate I. In soil. *Pestic. Sci.* **6**, 53–59.

PROBST, G. W. and TEPE, J. B. (1969). Trifluralin and related compounds. *In* "Degradation of Herbicides" (P. C. Kearney and D. D. Kaufman, Eds), pp. 255–282. Marcel Dekker Inc, New York.

RAHN, P. R. and ZIMDAHL, R. L. (1973). Soil degradation of two phenyl pyridazinone herbicides. *Weed Sci.* **21**, 314–317.

RAO, P. V. S., MOORE, K. and TOWERS, G. H. N. (1969). *O*-pyrocatechuic acid carboxy-lyase from *Aspergillus niger*. *Archs Biochem. Biophys.* **122**, 466–473.

RAYMOND, R. L., JAMISON, V. W. and HUDSON, J. O. (1971). Hydrocarbon cooxidation in microbial systems. *Lipids* **6**, 453–457.

REDEMANN, C. T., MEIKLE, R. W., HAMILTON, P., BANKS, V. S. and YOUNGSON, C. R. (1968). The fate of 4-amino-3,5,6-trichloropicolinic acid in spring wheat and soil. *Bull. Environ. Contam. Toxicol.* **3**, 80–96.

RIBBONS, D. W. (1966). Metabolism of *o*-cresol by *Pseudomonas aeruginosa* T_1. *J. gen. Microbiol.* **44**, 221–231.

RIBBONS, D. W. (1970). Stoichiometry of *O*-demethylase activity in *Pseudomonas aeruginosa*. *FEBS Lett.* **8**, 101–104.

RICHMOND, D. V. and SOMERS, E. (1966). Studies of the fungitoxicity of captan. IV. Reactions of captan with cell thiols. *Ann. appl. Biol.* **57**, 231–240.

RICHMOND, D. V. and SOMERS, E. (1968). Studies of the fungitoxicity of captan. VI. Decomposition of ^{35}S-labelled captan by *Neurospora crassa* conidia. *Ann. appl. Biol.* **62**, 35–43.

RILEY, D., TUCKER, B. V. and WILKINSON, W. (1976). Biological unavailability of bound paraquat residues in soil. *In* "Bound and Conjugated Pesticide Residues" (D. D. Kaufman, G. G. Still, G. D. Paulson and S. K. Bandal, Eds), pp. 301–353. ACS Symp. Ser. No. 29.

ROETH, F. W., LAVY, T. L. and BURNSIDE, O. C. (1969). Atrazine degradation in two soil profiles. *Weed Sci.* **17**, 202–205.

ROSE, A. H. (1965). "Chemical Microbiology". Butterworths, London.

SALA-TREPAT, J. M. and EVANS, W. C. (1971). The meta cleavage of catechol by *Azotobacter* species. *Eur. J. Biochem.* **20**, 400–413.

SARIASLANI, F. S., HARPER, D. B. and HIGGINS, I. J. (1974). Microbial degradation of hydrocarbons. Catabolism of 1-phenylalkanes by *Nocardia salmonicolor. Biochem. J.* **140**, 31–45.

SCHELINE, R. R. (1968). Metabolism of phenolic acids by the rat intestinal microflora. *Acta Pharmacol. Toxicol.* **26**, 189–205.

SCHELINE, R. R. (1973). Metabolism of foreign compounds by gastrointestinal microorganisms. *Pharmacol. Rev.* **25**, 451–523.

SCHELINE, R. R. and LONGBERG, B. (1965). The absorption, metabolism and excretion of the sulphonated azo dye, acid yellow, by rats. *Acta Pharmacol. Toxicol.* **23**, 1–14.

SETHUNATHAN, N. (1973). Microbial degradation of insecticides in flooded soil and in anaerobic cultures. *Residue Rev.* **47**, 143–165.

SETHUNATHAN, N. and YOSHIDA, T. (1969). Fate of diazinon in submerged soil, accumulation of hydrolysis product. *J. agric. Fd Chem.* **17**, 1192–1195.

SETHUNATHAN, N. and YOSHIDA, T. (1973a). A *Flavobacterium* sp. that degrades diazinon and parathion. *Can. J. Microbiol.* **19**, 873–875.

SETHUNATHAN, N. and YOSHIDA, T. (1973b). Parathion degradation in submerged rice soils in the Philippines. *J. agric. Fd Chem.* **21**, 504–506.

SETHUNATHAN, N., BAUTISTA, E. M. and YOSHIDA, T. (1969). Degradation of benzene hexachloride by a soil bacterium. *Can. J. Microbiol.* **15**, 1349–1354.

SHARABI, N. E. and BORDELEAU, L. M. (1969). Biochemical decomposition of the herbicide *N*-(3,4-dichlorophenyl)-2-methylpentanamide and related compounds. *Appl. Microbiol.* **18**, 369–375.

SHEETS, T. J. (1958). The comparative toxicities of four phenyl urea herbicides in several soil types. *Weeds* **6**, 413–424.

SHUKLA, O. P. (1973). Microbial decomposition of pyridine. *Ind. J. Exp. Biol.* **11**, 463–465.

SIDDERAMAPPA, R. and SETHUNATHAN, N. (1975). Persistence of gamma-BHC and beta-BHC in Indian rice soils under flooded conditions. *Pestic. Sci.* **6**, 395–403.

SIJPESTEIJN, A. K., KASLANDER, J. and von der KECH, G. J. M. (1962). On the conversion of sodium dimethyldithiocarbamate into its α-amino-butyric acid derivative by microorganisms. *Biochim. biophys. Acta* **62**, 587–589.

SJAASTAD, O. V. and KAY, R. N. B. (1970). *In vivo* rupture of the imidazole ring of histamine. *Experientia* **26**, 1197–1198.

SKIPPER, H. D., GILMOUR, C. M. and FURTICK, W. R. (1967). Microbial versus chemical degradation of atrazine in soils. *Proc. Soil Sci. Soc. Am.* **31**, 653–656.

SKUJIŅŠ, J. J. (1967). Enzymes in soil. *In* "Soil Biochemistry" (A. D. McLaren and G. H. Peterson, Eds), vol. 1, pp. 371–414. Marcel Dekker Inc, New York.

SMITH, R. A., BELLES, W. S., SHEN, K. W. and WOODS, W. G. (1973). The degradation of dinitramine (N^3,N^3-diethyl 2,4-dinitro-6-trifluoromethyl-*m*-phenylendiamine) in soil. *Pestic. Biochem. Physiol.* **3**, 278–288.

SOLEIM, H. A. and SCHELINE, R. R. (1972). Metabolism of xenobiotics by strains of intestinal bacteria. *Acta. Pharmacol. Toxicol.* **31**, 471–480.

SPARNINS, V. L., CHAPMAN, P. J. and DAGLEY, S. (1974). Bacterial degradation of 4-hydroxyphenyl acetic acid and homoprotocatechuic acid. *J. Bact.* **120**, 159–167.

SPARROW, L. G., HO, P. P. K., SUNDARAM, T. K., ZACH, D., NYMS, E. J. and SNELL, E. E. (1969). The bacterial oxidation of vitamin B_6 VII. Purification, properties and mechanism of action of an oxygenase which cleaves the 3-hydroxypyridine ring. *J. biol. Chem.* **244**, 2590–2600.

SPROTT, G. D. and CORKE, C. T. (1971). Formation of 3,3',4,4'-tetrachloroazobenzene from 3,4-dichloroaniline in Ontario soils. *Can. J. Microbiol.* **17**, 235–240.

STANIER, R. Y., HAYAISHI, O. and TSUCHIDA, M. (1951). The bacterial oxidation of tryptophan. I. A general survey of the pathways. *J. Bact.* **62**, 355–366.

STEENSON, T. I. and WALKER, N. (1957). The pathway of breakdown of 2,4-dichloro- and 4-chloro-2-methyl-phenoxyacetic acids by bacteria. *J. gen. Microbiol.* **16**, 146–155.

STENERSON, J. H. V. (1965). DDT metabolism in resistant and susceptible stable flies and in bacteria. *Nature, Lond.* **207**, 660–661.

STENERSEN, J. (1969). Degradation of P^{32}-bromophos by microorganisms and seedlings. *Bull. Environ. Contam. Toxicol.* **4**, 104–112.

TAYLOR, H. F. and WAIN, R. L. (1962). Side chain degradation of certain ω-phenoxyalkane carboxylic acids by *Nocardia coeliaci* and other micro-organisms isolated from soil. *Proc. R. Soc.*, Ser. B **156**, 172.

TIEDJE, J. M. and ALEXANDER, M. (1969). Enzymatic cleavage of the ether bond of 2,4-dichlorophenoxyacetate. *J. agric. Fd Chem.* **17**, 1080–1084.

TIEDJE, J. M., DUXBURY, J. M., ALEXANDER, M. and DAWSON, J. W. (1969). 2,4-D metabolism: Pathway of degradation of chlorocatechols by *Arthrobacter* sp. *J. agric. Fd Chem.* **17**, 1021–1026.

TOMS, A. and WOOD, J. M. (1970). The degradation of *trans*-ferulic acid by *Pseudomonas acidovorans*. *Biochemistry* **9**, 337–343.

TONOMURA, K. and KANZAKI, F. (1969). The reductive decomposition of organic mercurials by cell-free extract of a mercury resistant pseudomonad. *Biochim. biophys. Acta* **184**, 227–229.

TRUDGILL, P. W. (1969). The metabolism of 2-furoic acid by *Pseudomonas* F-2. *Biochem. J.* **113**, 577–587.

TU, C. M., MILES, J. R. W. and HARRIS, C. R. (1968). Soil microbial degradation of aldrin. *Life Sci.* **7**, 311–322.

TUCKER, B. V. and PACK, D. E. (1972). Bux insecticide soil metabolism. *J. agric. Fd Chem.* **20**, 412–416.

TWEEDY, B. G., LOEPPKY, C. and ROSS, J. A. (1970a). Metobromuron: acetylation of the aniline moiety as a detoxification mechanism. *Science, N.Y.* **168**, 482–483.

TWEEDY, B. G., LOEPPKY, C. and ROSS, J. A. (1970b). Metabolism of 3-(*p*-bromophenyl)-1-methoxy-1-methyl urea (metobromuron) by selected soil microorganisms. *J. agric. Fd Chem.* **18**, 851–853.

van der LINDEN, A. C. and THIJSSE, G. J. E. (1963). Degradation of 2-methylhexane by a *Pseudomonas*. *In* "Symposium on Marine Microbiology" (C. H. Oppenheimer, Ed.), pp. 475–480. C. C. Thomas, Illinois.

von ENDT, D. W., KEARNEY, P. C. and KAUFMAN, D. D. (1968). Degradation of monosodium methanearsenic by soil microorganisms. *J. agric. Fd Chem.* **16**, 17–20.

WAGENBRETH, D. and KLUGE, E. (1971). Action of fungicides (Thiuram, Vapam) on the degradation of triazine herbicides in soil. *Arch. Pflanz.* **7**, 451–459.

WAID, J. S. (1972). The possible importance of transfer factors in the bacterial degradation of herbicides in natural ecosystems. *Residue Rev.* **44**, 65–71.

WALLEN, L. L., STODOLA, F. H. and JACKSON, R. W. (1959). "Type Reactions in Fermentation Chemistry", US Dept Agric., Agric. Res. Service.

WALLNÖFER, P. R. and BADER, J. (1970). Degradation of urea herbicides by cell-free extracts of *Bacillus sphaericus*. *Appl. Microbiol.* **19**, 714–717.

WALLNÖFER, P. R., KONIGER, M., SAFE, S. and HUTZINGER, O. (1972). Metabolism of the systemic fungicide 2,5-dimethyl-3-furane carboxylic acid anilide (BAS 3191) by *Rhizopus japonicus* and related fungi. *J. agric. Fd Chem.* **20**, 20–22.

WARE, G. W. and ROAN, C. C. (1970). Interaction of pesticides with aquatic microorganisms and plankton. *Residue Rev.* **33**, 15–45.

WATANABE, I. (1973). Decomposition of pesticides by soil microorganisms – Special emphasis on the flooded soil condition. *Jap. Agric. Res. Quart.* **7**, 15–18.

WATSON, G. K. and CAIN, R. B. (1975). Microbial metabolism of the pyridine ring. Metabolic pathways of pyridine biodegradation by soil bacteria. *Biochem J.* **146**, 157–172.

WATSON, G. K., HOUGHTON, C. and CAIN, R. B. (1974a). Microbial metabolism of the pyridine ring. The hydroxylation of 4-hydroxypyridine to pyridine-3,4-diol (3,4-dihydroxypyridine) by 4-hydroxypyridine 3-hydroxylase. *Biochem. J.* **140**, 265–276.

WATSON, G. K., HOUGHTON, C. and CAIN, R. B. (1974b). Microbial metabolism of the pyridine ring. The metabolism of pyridine-3,4-diol (3,4-dihydroxypyridine) by *Agrobacterium* sp. *Biochem. J.* **140**, 277–292.

WEBLEY, D. M., DUFF, R. B. and FARMER, V. C. (1957). Formation of a β-hydroxy acid as an intermediate in the microbial conversion of monochlorophenoxybutyric acids to the corresponding substituted acetic acids. *Nature, Lond.* **179**, 1130–1131.

WEDERMEYER, G. (1966). Dechlorination of DDT by *Aerobacter aerogenes*. *Science, N.Y.* **152**, 647.

WEDERMEYER, G. (1967). Dechlorination of 1,1,1-trichloro-2,2-bis (p-chloro-phenyl)ethane by *Aerobacter aerogenes*. I. Metabolic products. *Appl. Microbiol.* **15**, 569–574.

WEDERMEYER, G. (1968). Role of intestinal microflora in the degradation of DDT by rainbow trout (*Salmo gairdneri*). *Life Sci.* **7**, 219–223.

WIESE, M. V. and VARGAS, J. M. (1973). Interconversion of chloroneb and 2,5-dichloro-4-methoxyphenol by soil microorganisms. *Pestic. Biochem. Physiol.* **3**, 214–222.

WESTMACOTT, D. and WRIGHT, S. J. L. (1975). Studies on the breakdown of p-chlorophenyl methylcarbamate II. In cultures of a soil *Arthrobacter* sp. *Pestic. Sci.* **6**, 61–68.

WHITESIDE, J. S. and ALEXANDER, M. (1960). Measurement of microbiological effects of herbicides. *Weeds* **8**, 204–213.

WILLIAMS, P. P. and FEIL, V. J. (1971). Identification of trifluralin metabolites from rumen microbial cultures. Effect of trifluralin on bacteria and protozoa. *J. agric. Fd Chem.* **19**, 1198–1204

YAMASHINA, I., SHIKATA, S. and EGAMI, F. (1954). Enzymic reduction of aromatic nitro, nitroso and hydroxyl amino compounds. *Bull. Soc. Chem., Japan* **27**, 42–45.

YANO, I., FURUKAWA, Y. and KUSUNOSE, M. (1971). α-Oxidation of long-chain fatty acids in cell-free extracts of *Arthrobacter simplex*. *Biochim. biophys. Acta* **239**, 513–516.

YULE, W. N., CHIBA, M. and MORLEY, H. V. (1967). Fate of insecticide residues. Decomposition of lindane in soil. *J. agric. Fd Chem.* **15**, 1000–1004.

ZENTMAYER, G. A. and RICH, S. (1956). Reversal of fungitoxicity of 8-quinolinol and copper-8-quinolinolate by other chelators. *Phytopathology* **46**, 33.

Chapter 5

Transformations of Pesticides in the Environment – The Experimental Approach

IAN R. HILL and DAVID J. ARNOLD

Imperial Chemical Industries, Plant Protection Division, Jealott's Hill Research Station, Bracknell, Berkshire, England

I. Introduction

Natural environments, in particular soil and water, are not easy milieu in which to investigate the transformation of pesticides. The wealth and heterogeneity of their biological, chemical and physical attributes has deterred many investigators from attempting studies in these environments. Instead these workers have been content to examine the microbial metabolism of pesticides in nutrient media and to attempt to extrapolate their data to the natural situation. Indeed, until relatively recently the scientific literature on pesticide degradation was dominated by results from *in vitro* studies with pure and mixed cultures of microorganisms and with cell-free extracts and purified enzyme systems. Such studies can be of considerable scientific value, but there has been an unfortunate and undesirable tendency to assume that they also reflect the fate of the pesticide in the environment.

This assumption is unjustified because transformations of pesticides in soil and natural water may result from chemical and photochemical reactivity in the environment as well as from the activity of cellular or extracellular components of the biota (microorganisms, algae, higher plants, animals). Also, the natural processes of adsorption, volatilization, uptake by biota, leaching, etc. can all influence the availability of the chemical to microorganisms. For example, paraquat is extensively metabolized by *Lipomyces starkeyi*, a yeast, in culture media (Anderson and Drew, 1972) but is not degraded in soil because of its strong adsorption to clay surfaces (Riley *et al.*, 1976). Furthermore, the metabolic activity of any microorganism in laboratory media may not be reflected in nature, because of either poor competitive ability of the organism, or adverse environmental conditions.

Over the past few years the inadequacies of *in vitro* pesticide degradation studies have largely, although not entirely, been recognized. As a consequence of the increasing number of soil studies, considerable progress has been made towards a better understanding of the fate of pesticides in this environment. Aquatic environments have, however, received relatively little attention, a reflection of both the practical problems involved and the relatively small number of pesticides directly applied to bodies of water.

The impetus towards "environmentally relevant" studies was stimulated not only by the endeavour of many scientists and the public concern over the introduction and possible accumulation of man-made chemicals in the environment, but also by Government-introduced controls over the sale and use of pesticides. In many countries the sale of a pesticide, by a manufacturer or licensee, cannot proceed without the sanction of a government-controlled agency. Since 1972, with the passing of amendments to the Federal Insecticide, Fungicide and Rodenticide Act, the United States of America has increasingly imposed more and more stringent regulations on the pesticide industry. The Environmental Protection Agency (EPA) is responsible for instituting these regulations, designed to ensure effectiveness of the pesticide and to protect human and environmental health, and has issued "guidelines" (EPA, 1975a) as to the types of information required for an assessment of the "environmental impact" of a product. These include an extensive evaluation of the degradation of the pesticide in the environment. In more recent years many other countries have followed suit, introducing regulations of varying degrees of complexity.

II. *In Vitro* Microbiological Studies

A. ISOLATION OF MICROORGANISMS

Laboratory studies of the microbial metabolism of pesticides have utilized organisms either obtained from culture collections or isolated from soil, aquatic environments, sewerage, rumen content, etc. The isolation of organisms able to degrade organic compounds has often followed enrichment of the source with the product under study or with a structurally related chemical (Henderson, 1965; Horvath, 1972).

Whilst isolation is often carried out under field or laboratory-simulated field conditions, some workers have favoured the use of soil suspensions (Naik *et al.*, 1972) or perfusion techniques (Smith, 1962; Morrill and Dawson, 1964; Sharp and Taylor, 1969). Isolation of cultures of these organisms is mostly carried out by dilution and plating techniques (Gray and Williams, 1971) on appropriate synthetic media. The methods used rarely isolate more than a small proportion of the viable microbial population and in particular are likely to miss auxotrophic organisms, many non-sporing fungi and organisms with highly specific nutrient or physical requirements. Furthermore, enrichment techniques tend to favour organisms which can metabolize the pesticide as a carbon and energy source and may exclude those capable of co-metabolic transformation. Until recently few workers have considered the micro-algal and micro-faunal populations as able to contribute significantly to pesticide transformations *in situ*. Yet, both these groups are known to accumulate pesticides (Wheeler, 1970; Södergren, 1971) and the micro-algae have been shown to metabolize these chemicals (Sweeney, 1969; Zuckerman *et al.*, 1970). In some aquatic environments these organisms may be present in extremely large numbers, which makes their apparent rejection in these studies all the more surprising. A final disadvantage of enrichment procedures is that the chemical is often applied at high concentrations, which may inhibit organisms capable of metabolizing lower concentrations. There is a very good case for isolating organisms, for studies of pesticide metabolism, from both enriched and non-amended sources.

B. PESTICIDE METABOLISM

During or following isolation, microorganisms are selected that are capable of metabolizing the pesticide in solid or liquid culture media. The method of assessment differs widely from one pesticide to another. In a few instances "active" cultures have been distinguished by observations of specific reactions in the medium (Kaufman and Kearney, 1965; Clark and Wright, 1970a). Metabolic studies, however, have mainly followed traditional methodology, the organisms being incubated in a variety of media containing the pesticide in the presence or absence of other sources of carbon, nitrogen and energy. The fastidious requirements of some microbes can be accommodated by the inclusion of, for example, amino acids and growth factors.

The disappearance of pesticide from a culture supernatant cannot itself be taken as a satisfactory indication of metabolism. Losses of unchanged pesticide from the medium can occur by volatilization, intracellular accumulation, and by adsorption to cell surfaces (Gregory et al., 1969) and glassware. The detection of pesticide products is thus an essential criterion for confirming transformation. The use of a radio-labelled substrate not only facilitates analysis, but when used in "enclosed" incubation systems allows the collection of volatilized products evolved from the culture medium into the air and enables the preparation of a "balance-sheet" for recovery of the pesticide and its transformation products.

Observation of the presence of pesticide transformation products in microbial cultures and their absence in appropriate controls is not necessarily an indication of microbial metabolism. Changes in the chemical structure of the pesticide can be caused by the physico-chemical properties of the medium following microbial growth or of the intracellular constituents of the organism. Proof of metabolism is only conclusively obtained following the isolation and, where possible, the purification of active enzyme systems.

C. RELATIONSHIP TO IN SITU TRANSFORMATIONS

It has already been stated that observations in culture can rarely be extrapolated to field circumstances and that organisms able to metabolize the pesticide *in situ* are not necessarily active *in vitro*. One possible exception was reported by Gunner *et al.* (1966). They found that diazinon in soil resulted in a selective enrichment of an *Arthrobacter* sp. and a *Streptomyces* sp. In cultural conditions neither organism could cleave the pyrimidine ring, whereas in a mixed culture of both organisms 15–20% of the radio-labelled ring carbon was evolved as $^{14}CO_2$ within 18 h (Gunner and Zuckerman, 1968). There was then good, although circumstantial, evidence that these two microorganisms were stimulated in soil because of their synergistic ability to degrade the pesticide. Similar synergistic relationships and enhanced degradative abilities of mixed cultures have been observed by other authors (Bollag and Liu, 1971; Bordeleau and Bartha, 1971). If such synergistic interactions are common, and there is every possibility that they are, they may be responsible for the inability of many investigators to demonstrate the metabolism in pure culture of pesticides that are degraded under field conditions. Bounds *et al.* (1969) observed an "associative microbial action" whereby a *Micrococcus* sp. reversed the inhibitory effects of dalapon on an *Agarbacterium* sp. in cultural conditions. Although in this case neither organism metabolized the pesticide, similar reactions could possibly confer a similar resistance on microbes also able to metabolize such a chemical. In many studies of pesticides in soil no significant increase in the numbers of any microorganism occurs, thus the chances of obtaining the correct permutation and concentration of synergistic organisms (should such organisms exist) under conducive cultural conditions are slight.

Senior *et al.* (1976) using a continuous-flow enrichment culture technique isolated a stable, seven-membered microbial community capable of degrading the herbicide dalapon by "cooperative metabolism". The technique is of considerable value and goes some way toward bridging the gap between conventional mono- or mixed-culture studies and field studies. However, extrapolation of the results to field situations remains speculative.

Whilst pesticide formulation components are sometimes included in soil studies, their use in culture media should be viewed with caution. In soils and aquatic environments the additives may be dispersed (leached, adsorbed, etc.) at different rates from the pesticide, thus even if potentially capable of influencing pesticide transformations they may not do so. In a culture vessel the continued presence of the additives may affect the rate or processes of metabolism of the pesticide component, as has been demonstrated with the petrochemical formulation additives of propham (Clark and Wright, 1970b).

D. MAINTENANCE OF MICROBIAL CULTURES

The maintenance of laboratory cultures of microorganisms able to metabolize pesticides has not always proved easy. Many workers have noted the progressive loss of metabolic ability of many organisms when maintained by regular subculture and growth on or in nutrient media, even when the medium contains the pesticide (Reid, 1960; Johnson and Cain, 1973; Hill, unpubl. data). Johnson and Cain (1973) suggested that such instability might be due to loss during the growth phase of the genetic elements essential for processes such as pesticide metabolism. In order to avoid or reduce these losses of activity and to prevent the fall in viability which commonly occurs with some soil organisms maintained on laboratory nutrient media (Wellman and Walden, 1964; Cox, 1968; Cunningham, 1973) microbial cultures can be stored either lyophilized (freeze-dried) or at extremely low temperatures (e.g. under liquid nitrogen; at $-196°$).

The use of lyophilization is well established (Harris, 1954) and has been found to preserve the viability of many biological specimens almost indefinitely (Davis, 1963). The method relies on the preparation and storage of a number of ampoules of each organism, for which considerable space is required. Even more disadvantageous is the fact that many mycelial forms are difficult to lyophilize (Fennel *et al.*, 1950) and many fungi prepared from spore suspensions fail to survive (Hwrang, 1960). Within the last two decades ultra low temperature storage of cultures, mostly in vials immersed in liquid nitrogen, has attracted considerable attention (Cox, 1968; Cunningham, 1973; Daily and Higgins, 1973). The technique requires relatively little storage space, and although some cells suffer damage or death the majority of cultures show a reasonable proportion of survivors. Mazur (1962) reported that often a higher proportion of cells survived after freezing than after lyophilization. Ashwood-Smith (1965) found no indications of genetic changes in bacteria subjected to the freezing, storage and thawing procedures. The principal disadvantage of "cold storage" is that the laboratory operating the technique must be ensured of regular supplies of liquid nitrogen or a similar refrigerant.

E. THE VALUE OF MICROBIAL STUDIES

The use of microorganisms for laboratory culture studies of pesticide metabolism, both *per se* and as an aid to the interpretation of field data, is well established. They may also be used, however, as a tool for the production of transformation products of potential value in chromatographic identification of unknown products in extracts of soil.

The solvents necessary for the efficient extraction of pesticides and their products from soil and aquatic sediments also extract a large amount of extraneous non-pesticidal materials. It is often the case that these natural soil constituents prevent or make difficult the isolation and purification of sufficient quantities of the pesticidal products for identification by u.v., mass spectrometry, infra-red spectrometry, etc. Preliminary identification by co-chromatography may then be necessary. Should speculative synthesis fail to produce reference markers which co-chromatograph with the extracted unknowns, it may be necessary to find alternative sources of reference compounds. Such sources should ideally be "simple" enough to enable the isolation and purification of sufficient quantities of any pesticide transformation products formed. They include aqueous or organic solutions subjected to irradiation; samples from natural aqueous environments (with viable microbial populations); culture media incubated with pure or mixed microbial cultures; and eluates from perfusion columns or from simulated soil systems (e.g. using glass microbeads).

III. Investigations in Natural Environments

The complexity and heterogeneity of natural environments have always presented numerous, and sometimes insuperable, methodological problems to research workers investigating the *in situ* metabolic activities of the endogenous microflora. The value of any study of the degradation of a synthetic organic molecule in these environments, where biological, chemical and physical dynamic equilibria may be disturbed by experimental techniques and where many of the biological and chemical components are still largely uncharacterized, relies heavily upon the methodology employed.

Studies of the transformation of a pesticide in soil can be carried out either in the laboratory or under field conditions. It is, however, desirable that such studies examine the effects of specific environmental parameters on transformation, are reproducible and that the resulting data are suitable for comparison with that from other investigations carried out at different periods of time and perhaps in different parts of the world. As both the rates and routes of dispersal and transformation of pesticides in soil may be influenced by climatic conditions, it is necessary to study these processes under controlled laboratory conditions. The experimental and analytical advantages gained from the use of radio-labelled pesticides further justify the use of laboratory studies, for application of radio-chemicals to several field soil sites generates a variety of "handling" and site-control problems.

Critics of laboratory studies point out that both soil sampling and storage can induce alterations in the natural microbial populations (Stotzky et al., 1962; Jensen, 1963; Grossbard and Hall, 1964), possibly due to both physical (pH, moisture content, temperature, etc.) and chemical (alteration in availability of soil organic matter, etc.) changes. It is worth noting, however, that storage can be avoided and that agricultural soils in situ are themselves regularly disturbed by cultivation and crop harvesting techniques.

Pramer and Bartha (1972) have expressed concern over the lack of consideration of soil as a viable dynamic entity and have outlined procedures for minimizing the effects of disturbance during soil collection, preparation, treatment and incubation. Storage of soil prior to use is usually undesirable, as also is drying of the soil except in so far as to facilitate sieving. Despite evidence in the literature indicating the adverse effects on soil properties and the microflora of air drying soil (Stevenson, 1956; Soulides and Allison, 1961; Robinson et al., 1965; Robinson and Chase, 1967), and the observation that the processes of pesticide transformation may also be altered (Bartha, 1971), studies are still reported using this unnecessary pretreatment stage. Some workers have used methods for "activating" the microbial populations in soil prior to application of the pesticide (Chisaka and Kearney, 1970; Guth et al., 1976). However, the relative merits of soil storage and reactivation, versus minimal storage have not been adequately investigated.

A. LABORATORY SOIL INCUBATION METHODS

The methods of incubating pesticide-treated soils have often been partly determined by the "analytical" procedures used to identify and quantify the parent chemical and its soil degradation products. The analytical techniques are more fully discussed in Section IV. However, the use of radio-labelled pesticides allows the greatest degree of methodological flexibility and enables the detection of both non-volatile and volatile pesticide and products.

Pesticide treated soil can be incubated in open or enclosed containers, the latter enabling the collection of volatile products. In the former, losses of pesticide or products evolved into the air can only be quantified by inference. Furthermore, in open vessels the soil may dry out rapidly, although Tu (1970) suggests that thin polyethylene film closures reduce soil moisture loss without impeding gaseous exchange.

In the simplest enclosed system, soils are incubated in sealed vessels such as plastic bags, glass jars or purpose designed flasks (Fig. 1)

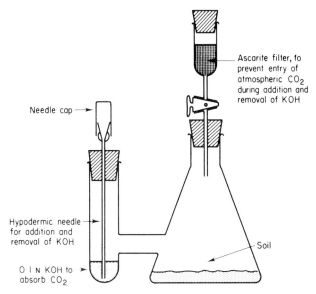

Ascarite filter, to prevent entry of atmospheric CO_2 during addition and removal of KOH

Needle cap →

Hypodermic needle → for addition and removal of KOH

0·1 N KOH to ► absorb CO_2

Soil

Fig. 1. Incubation flask for the measurement of the persistence of pesticides in soil (Bartha and Pramer, 1965). (Figure by courtesy of The Williams and Wilkins Co., Baltimore, USA.)

(Bartha and Pramer, 1965; Nommik, 1971; Balasubramanian and Kanehiro, 1974), these usually being opened at intervals to allow gas transfer. However, the partial pressure of oxygen does not necessarily remain at a constant level and fluctuations in microbial populations or their metabolic activities can result. Pramer and Bartha (1972) reported that 1·0 g of an "average" soil could exhaust the oxygen in 1·0 cc of air in 24 h, and therefore recommended daily aeration of closed vessels where the ratios of vessel volume (ml) to sample weight (g) are 10 or less. Also enclosed but more satisfactory than sealed vessels are those with a continuous flow of air or oxygen-free gas (Bartholomew and Broadbent, 1950; Stotzky, 1960; Parr and Smith, 1969) (Fig. 2); see also Section V, B. However, some authors have reported difficulty in maintaining soil moisture levels in such systems. Specified levels of humidity can be maintained (Winston and Bates, 1960) but it is generally adequate to saturate the air stream with water

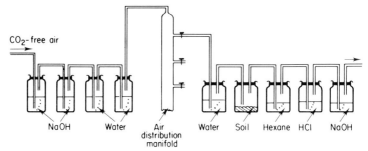

Fig. 2. Apparatus for the measurement of pesticide degradation in soil incubated in a continuous flow of air (Parr and Smith, 1969). (Figure by courtesy of The Williams and Wilkins Co., Baltimore, USA.)

and at regular intervals to compensate the soil, for any weight losses, by the addition of water. Using radio-labelled compounds, these enclosed systems can be adapted with ease to the collection of labelled volatile products by incorporating a series of "trapping" solutions in the effluent gas stream. Anderson (1975) described the use of a vessel, without an air flow, that avoided alterations in the partial pressures of oxygen and carbon dioxide.

The EPA guidelines (EPA, 1975b) indicate that only data from enclosed systems (whether sealed or with a gas flow) will be acceptable for registration submissions. The West German registration authority, however, specify their requirements more definitively (BBA, 1973). As they do not require data for volatile pesticide products from degradation studies, they specify the use of flasks sealed with a cotton wool bung. Most other registration authorities are at present more flexible and have not issued detailed guidelines.

B. SOIL BIOLOGICAL VERSUS CHEMICAL TRANSFORMATION

In order to demonstrate the relative biological and chemical involvement in the transformation of pesticides in soil it may be necessary to manipulate the soil conditions to exclude biological activity.

The biological contribution of soil to pesticide transformation has frequently been implied from observation of the kinetics of the transformation processes. Where initial "losses" of pesticide are slow or absent (lag phase) but increase progressively with time and reach a steady state (usually a logarithmic rate of change) for a period of time, it is deemed likely that the observed effects result from "enrichment" of the soil with organisms able to transform the pesticide (Whiteside and Alexander, 1960; Raghu and MacRae, 1966). Further applications of the pesticide may be transformed more rapidly, either without or with a reduced lag phase. The absence of a lag phase does not, however, necessarily imply a wholly chemical route. Microorganisms may metabolize pesticides using constitutive enzymes and also by co-metabolic reactions; neither necessarily involving a lag phase. Yaron et al. (1974) studying the degradation of the acaricide/insecticide azinphosmethyl in non-sterile and γ-irradiated soils proposed that a lag phase could even occur in the absence of viable microorganisms. Soil was irradiated with 3 Mrad, a dose which in our experience does not necessarily ensure a sterile medium. Unfortunately, Yaron et al. (1974) do not record whether their soil was screened for viable organisms either before or during experimentation, or whether incubation procedures were designed to maintain the soil in a sterile state.

An alternative approach to determining the biological involvement of soil has been to inhibit the microorganisms in the soil by chemical means. Audus (1951) employed sodium azide and concluded that degradation of 2,4-D was entirely a result of biological activity. Subsequently various workers have used mercuric salts, phenolics, antibiotics, organic solvents, soil sterilants, etc. (Munnecke, 1958; Riepma, 1962; Spanis et al., 1962; Ashton, 1963; Rahn and Zimdahl, 1973). In order to overcome the large "losses" of inhibitor due to adsorption, considerable amounts may need to be added to the soil. The addition of such quantities can, however, affect the chemical reactivity and adsorptive properties of the soil. Skipper and Westermann (1973) suggested that the use of the gaseous sterilant propylene oxide was preferable to either sodium azide or autoclaving for preparing sterile soils for studies of chemical transformations. However, their propylene oxide-treated soils were slightly phytotoxic.

The most popular methods of controlling the biological reactivity of soil organisms are steam sterilization (autoclaving) and irradiation. High energy ionizing radiation may be provided by an electron beam, hard X-rays or gamma (γ)-rays, the latter having been most commonly used for sterilizing soil samples. The high temperatures and pressures of steam sterilization may alter soil surface adsorption characteristics (Furmidge and Osgerby, 1967) and cause some organic matter degradation (Warcup, 1957). γ-Irradiation does not seem to alter the cation exchange capacity, particle size distribution or moisture desorption properties of soil (Eno and Popenoe, 1964) as does autoclaving, but both methods result in an increased release into the soil solution and an increased extractability, of nitrogen, phosphorus, sulphur and manganese (Bowen and Cawse, 1964; Eno and Popenoe, 1964). γ-Irradiation does, however, seem to result in a greater release of ethylene (and other hydrocarbons) into the soil and air than does autoclaving (Rovira and Vendrell, 1972). Both the organic matter content of the soil (Eno and Popenoe, 1964) and its moisture content can affect the relative merits of the different sterilization methods. For example, Salonius *et al.* (1967) found that with moist soil, γ-irradiation produced a greater release of ammonium-nitrogen, soluble carbohydrates and soluble ninhydrin-positive compounds than did autoclaving, whereas in dry soils the reverse was true. This could be a reflection of enzyme activity in the presence of soil moisture and lack of it in dry soil.

γ-Irradiation does have the particular disadvantage that it fails to remove all biological activity; enzymic activity derived from microbes and/or plants being retained by the sterilized soil (McLaren *et al.*, 1957; Vela and Wyss, 1962) and often for prolonged periods. Comparisons of pesticide degradation in autoclaved and γ-irradiated soils have shown more rapid degradation in the latter (Getzin and Rosenfield, 1968; Hill and Arnold, unpubl. data), possibly indicating the presence of enzyme activity in the irradiated soils.

Using sterilized soils, many workers have demonstrated that degradation of pesticides is often mediated by the soil biota or their extracellular enzymes (Holstun and Loomis, 1956; Lichtenstein and Schulz, 1960; Ragab and McCollum, 1961). However, chemical reactivity has been shown to contribute both to discrete stages in degradative pathways and to the sum total of degradation, by causing the same transformations as do the biota.

C. LABORATORY SOIL/PLANT SYSTEMS

Many workers have investigated the "loss" of parent pesticide from the soil in field situations, in the presence of growing crops. Few, however, have attempted to quantify the relative contributions to pesticide transformations, of the soil and its biota "away from the plant roots" and that of the chemical and biological activity in the soil around the roots (i.e. in the rhizosphere). The ability of plant roots to create this unique environment around themselves has already been discussed in Chapters 2 and 3. In the field, with growing crops, a considerable proportion of the pesticide applied to soil is likely to enter and either remain in, or pass through, rhizosphere zones. Any pesticide in the rhizosphere may be transformed at different rates or even by different routes from that in non-rhizosphere soil as a result of the altered soil conditions (e.g. nutrients, oxygen, carbon dioxide, water), the presence of exuded root enzymes and the metabolic activity of the highly specialized rhizosphere microflora.

Laboratory studies of pesticide degradation in soil do not normally include the use of soil containing viable plant roots and thus may fail to reflect the true fate of the pesticide in field conditions where crops are planted and weeds exist. However, in the literature there are reports that illustrate the feasibility of studying the transformation of pesticides in soil in which plants are growing. Such techniques are more complex than those including soil alone and if used to replace the latter would either reduce the number of soil types and physical conditions that could be investigated, or would entail a considerable increase in expenditure. We do not believe that plant/soil studies should replace those in soil alone but instead should complement these studies.

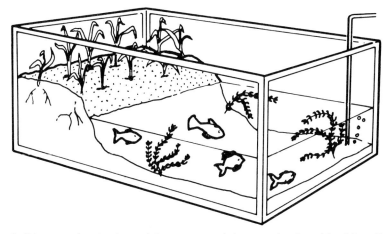

Fig. 3. Diagram of a simple model ecosystem of the type developed by Metcalf and co-workers (see Metcalf, 1974).

Model ecosystems (Fig. 3) containing sand, water, plants and animals have been extensively used for studies of the movement, accumulation and transformation of pesticides (see e.g. Metcalf, 1974). However, the physical conditions in such systems bear little likeness to that in the field. Thus, whilst valuable as indicators of potential pesticide transfer and accumulation along food chains, the technique provides little useful information on either physical movement or transformation of the chemical in the environment. More recently, similar studies incorporating soil (Isensee 1975; Isensee and Jones, 1975; Isensee *et al.*, 1976) have increased the scope and value of the technique.

Lichtenstein *et al.* (1974) described a model system (Fig. 4) in which water percolated through the soil/root system and was collected for

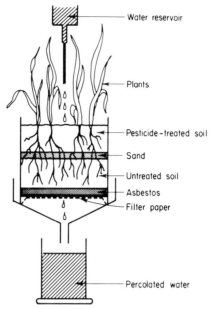

Fig. 4. Model system for studies of the degradation of pesticides in soil in the presence of plants (Lichtenstein *et al.*, 1974). (Figure by courtesy of the American Chemical Society, Washington, D.C.)

analysis. Harvey and Reiser (1973) incubated plants in soil treated with radio-labelled pesticide and, using an enclosed system with air flow, recovered radio-labelled volatile material from the air stream. Best and Weber (1974) carried out a similar study but also incorporated the facility for collection of leached products (Fig. 5).

Beall and Nash (1971), concerned about the foliar sorption of pesticide volatilized by soil, developed a method for growing plants in soil in which the foliar and soil/root environments were separated by a barrier. We have designed and used a slightly different apparatus which also includes the provision for separate recovery of the volatile products from the soil/root and foliar environments (Section V, C).

Beall *et al.* (1976) have described the use of an enclosed "agro-ecosystem", in which a number of plants may be grown in soil and pesticide degradation studied under differing regimes of rainfall, wind velocity and light.

D. FIELD SOILS

Though clearly desirable, studies of pesticide transformations in "undisturbed" field situations are difficult to carry out and to inter-

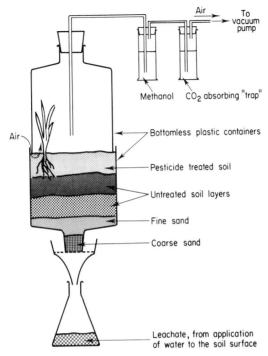

Fig. 5. Diagram of the soil–plant incubation apparatus used by Best and Weber (1974) for studies of the degradation and movement of s-triazines in soil. (Figure by courtesy of the Weed Science Society of America, Champaign.)

pret, particularly due to the variability of the climate. Furthermore, losses by leaching and volatilization are not always considered and are more difficult to quantify in the field than in laboratory studies.

Some workers have sampled the atmosphere after field applications of pesticides in order to estimate losses by volatilization (Caro *et al.*, 1971). McGarity and Rajaratnam (1973) and Dowdell *et al.* (1972) have described methods for measuring volatilization of nitrogen and ethylene, respectively, in field soils but their techniques have not as yet been used in pesticide studies. Leaching experiments have been carried out under field conditions by sampling soil from various depths (Baur *et al.*, 1972; La Fleur *et al.*, 1973), by the use of lysimeters (Glass and Edwards, 1974) or even by investigating surrounding watersheds (Haas *et al.*, 1971), although in the latter type of study the effects of leaching and run-off or wash-off may be difficult to distinguish. It is generally considered that laboratory studies of the leaching of pesticides and their naturally formed transformation products using soil columns are a satisfactory and representative method of determining their soil mobility.

The use of radio-labelled pesticides, enabling a more detailed investigation of *in situ* transformations under a variety of field conditions, is not always feasible because of restrictions in their usage at dispersed sites and the difficulty of maintenance and monitoring at situations distant from the laboratory. However, it is possible to overcome this problem by preparing experimental plots using soils transported from their natural situations. Such soils must be left for a considerable period of time before use in order to re-establish equilibria. Beynon *et al.* (1974) described the metabolism of dimethylvinphos in outdoor tanks of established soil. Führ *et al* (1976) have reported similar investigations.

Studies have been reported in which radio-labelled pesticides have been applied either to soil or plants in pot experiments (Buckland *et al.*, 1973), but possibly of more value are studies carried out *in situ* under field conditions (Andrawes *et al.*, 1971a, b; Baude, *et al.*, 1974; Weisgerber *et al.*, 1974; Golab *et al.*, 1975; Hill and Arnold, Section V, D).

The limitations placed on field studies by the necessity for controls on radio-chemicals, could be removed by the use of the ^{13}C-enriched pesticides. Roberts (1976) recently described the advantages of the use of this and other stable isotopes.

E. AQUATIC STUDIES

The majority of studies with pesticides in aquatic environments have been concerned with either the toxicity to and accumulation of the chemical in the biota, or with the persistence of the parent molecule. There have been relatively few studies of the pathways and products of pesticide transformation in this environment. Consequently there is little information on methodology for studying pesticide degradation in aquatic conditions. Frehse (1976) recently stated that "flowing waters cannot be simulated by a model" and that "storage of 'biological' model water is impossible". Whilst not underestimating the difficulties, we believe that insufficient attempts have been made, as yet, to examine the problems and possibilities.

Pure cultures (Paris and Lewis, 1973) and mixed cultures (Paris *et al.*, 1975) have been used to investigate the degradative capacity of microorganisms of aquatic origin. Pesticides have also been incubated statically in lake (Miskus *et al.*, 1965) and river water (Dorn *et al.*, 1976) and with shaking in river water (Eichelberger and Lichtenberg, 1971) to investigate degradation. In the latter study the authors considered that the results were only valid for the model system used and were not suitable for extrapolation to an *in situ*, naturally flowing stream. Water and sediment systems have also been used, both with shaking to give an aerobic suspension (Miyazaki *et al.*, 1975; Johnson and Lulves, 1975) and under nitrogen to give anaerobic conditions (Johnson and Lulves, 1975). Model ecosystems have been employed to study degradation (Sanborn, 1974) but, for reasons outlined in Section IIIC, the experimental conditions bear little resemblance to natural conditions.

The reported degradability of many non-pesticide organic molecules in natural waters has often been derived from BOD (biochemical oxygen demand) or COD (chemical oxygen demand) measurements, mostly recorded under laboratory conditions. The data from such studies are of limited value, reflecting only metabolism or chemical transformations involving oxygen uptake. No details of adsorptive processes, or of the routes and products of transformations, can be deduced from the data, and molecules transformed by co-metabolic processes may not increase the oxygen demand above that of controls. Many other methods reported in the literature such as loss of foaming properties, changes in surface activity and colorimetric analyses have similar deficiencies.

The detergent industry, with its products more often entering aqueous environments than do those of the pesticide industry, has in recent years seen the introduction of National and EEC (European Economic Community)-recommended biodegradability test procedures. These include "die-away" and activated sludge tests (Swisher, 1970; OECD, 1971), representing natural water and sewage plant conditions respectively. Whereas the activated sludge tests can be realistically compared to commercial sewage units, the "die-away" test is not representative of any natural aquatic environment. In most studies using these methods, only persistence of the parent molecule has been measured, although with the use of radio-labelled substrates the techniques can be adapted to provide information on degradative pathways and products. Nooi *et al.* (1970) have studied the degradative pathway of fatty alcohol/ethylene oxide condensates in activated sludge, using substrates ^{14}C-labelled in the alkyl chain and the ethoxylate chain. As far as we are aware there are no published reports in which pesticide transformations have been intensively studied using activated sludge systems.

In a satisfactory study of the degradation of a pesticide in water, the complete fate of the chemical should be elucidated using radio-labelled substrate and an enclosed incubation system to recover volatile products. It would be extremely difficult and costly to carry this out in a field study, thus laboratory models are necessary; despite the problem of preventing gross alterations in the biota and the physical and chemical conditions of the water and sediment during incubation.

The sampling sites, for laboratory investigations, could be ponds, lakes or rivers; and at best should be unpolluted and available for sampling at different times of the year.

F. OTHER MICROBIAL ENVIRONMENTS

1. *Plants*

The surfaces of plants exposed to the atmosphere support populations of microorganisms (see Chapter 3), and may receive considerable doses of pesticides. There have been very few attempts to study pesticide transformations on these surfaces *in situ*. Whilst it is not difficult to devise methods for such investigations it would not be easy to distinguish transformations mediated by plant exudates (chemical and enzymatic) from those involving microorganisms.

2. *Animals*

Although the metabolism of pesticides, antibiotics, etc. administered to, accidentally deposited on, or eaten by animals has received considerable attention over the past few years, there are many difficulties in demonstrating that, *in situ*, microorganisms participate in the process. Many microorganisms isolated from animal sources have been shown to metabolize pesticides (Peterson and Robison, 1964; Mendel and Walton, 1966; Fries *et al.*, 1969; Etges *et al.*, 1969). Attempts have been made to simulate rumen conditions in the laboratory and observations have also been made using isolated samples of rumen and intestinal fluids (Miskus *et al.*, 1965; Fries *et al.*, 1969).

3. *Air*

There have been no reported studies of microbial transformations of pesticides in the atmosphere, due partly to the difficulties of methodology, but also due to the relative unimportance of these reactions.

IV. Analysis of Pesticides and Products

A variety of analytical techniques and aids have been developed for the quantification and identification of pesticides and their transformation products. The methods available for analysis have often partly determined the systems in which the pesticide-treated soil or water are incubated. Three broad categories of analytical technique can be defined:

1. Bioassay methods

Bioassays (Guth et al., 1969; Walker, 1970) are most frequently developed to assess target and non-target activity of the pesticide. Data from such analyses are sometimes related by the investigator to persistence of the parent molecule. However, bioassays measure the biological activity of both the pesticide and transformation products. Furthermore, where the activity is lost or reduced they fail to distinguish between transformations which remove biological activity, inactivation due to adsorption, or losses due to leaching and volatilization.

2. Indirect methods

Various indirect methods have been developed which indicate, but do not necessarily quantify or characterize, changes in the chemical structure of the pesticide. The loss of chlorine, as chloride, can be demonstrated from organochlorines in soil (Jensen, 1957; Castro and Belser, 1966) but conversely the absence of chlorine release does not necessarily indicate a lack of degradation or of halogen reactivity. For example, the "NIH Shift" reaction, discussed in Chapter 4, is the transfer of a chlorine atom from one position on the aromatic nucleus to an adjacent site without loss to the molecule. In addition, a halogen substituent may be retained in metabolites well down a degradative pathway.

Increases in the evolution of carbon dioxide from soil respiration have also been used as an indicator of pesticide transformations, the source of carbon for the increase being attributed to the pesticide molecule (Tu, 1970; Stojanovic et al., 1972). Unless a radio-labelled pesticide is used and $^{14}CO_2$ is evolved there seems little reason for accepting the validity of this technique. Death of some organisms and enhanced respiration of others using nutrients released from lysed cells could be responsible, as also could disturbances in population balances resulting in enhanced numbers of organisms able to more rapidly metabolize available natural substrates.

3. Direct methods

A variety of analytical techniques have been developed that can distinguish residues of a pesticide and its transformation products, either in the presence of other extraneous material extracted from soil or after "clean-up" procedures to remove substances interfering with analytical procedures. The basic direct methods of analysis are: functional group; chromatographic; spectroscopic; and radiochemical; used either alone or in combinations. The development and use of some of these techniques depends upon a prior knowledge of the structure of the product being analysed.

The most satisfactory method of identifying and quantifying the parent chemical and any volatile and non-volatile products involves the use of pesticides into which radioactive atoms have been introduced. Any products of pesticide transformations, retaining the radio-label can be analysed by suitable radio-chemical techniques. The most commonly used radioactive isotope is ^{14}C, although studies using ^{3}H, ^{32}P, ^{35}S and ^{36}Cl have also been reported. Wherever possible the radio-label should be in a "meaningful" position in the molecule, i.e. in stable (e.g. aromatic or heterocyclic rings) or environmentally important (e.g. carbamate moieties) parts of the molecule. However, synthetic routes and the availability of radio-labelled starter or inter-mediary materials necessary for synthesis can influence the site and choice of the radio-isotope. Where different isomers are present in a commercially marketed pesticide, the different isomeric components should be studied separately. Determination of the fate of molecular fragments of complex molecules may be further aided by preparations separately radio-labelled in more than one position, or by multi-isotope labelling.

V. Soil Studies in the Authors' Laboratory

The format of any experimental study of pesticide transformations carried out by a pesticide manufacturing company is considerably influenced by the registration requirements of the countries in which sales of the product are anticipated. The following discussion and descriptions apply to the studies of the transformations of pesticides in soil as carried out by ICI Plant Protection Division and relate to investigations of the rates, products and, where possible, the routes of transformation. Studies of pesticide residues in soil, in which the parent pesticide and any major or biologically active transformation products are quantified from numerous "actual use" field soil samples, are carried out separately and are not discussed here.

A. THE EXPERIMENTAL REQUIREMENTS

Some of the constraints concerning the methodology of studies have already been outlined in Section III. It was evident that the prime criteria in investigations of the fate of pesticides in soil were the definition and reproducibility of experimental conditions, the ability to quantitatively recover all of the applied chemical ("balance" study), the maintenance of the soil in a condition as natural as possible and the incorporation of a wide range of simulated environmental conditions. To fulfil these requirements it is necessary to conduct experiments under laboratory conditions using radio-labelled pesticides and enclosed incubation systems with a flow-through of air.

At ICI Plant Protection Division most, but not all, of the studies of pesticide transformations in soil, for registration purposes, are carried out under laboratory conditions in the absence of plants. In order to ascertain the effect of viable plant roots and their associated microflora on the rate and products of pesticide transformations in soil, supplementary laboratory studies are carried out with plants grown in pesticide-treated soils. A small number of field studies are also undertaken in the presence of plants, to confirm the validity of the data from laboratory studies.

B. LABORATORY STUDIES IN THE ABSENCE OF PLANTS

Experimentation with any one pesticide is not necessarily restricted to the methods and conditions described below, nor are all of the incubation conditions always used. Instead, the requirement for information is evaluated independently for each pesticide and the programme of work tailored to its anticipated agricultural uses. The major experimental variables considered are shown in Fig. 6.

1. *Sampling and preparation of soil*

When, in practice, the pesticide is likely to be applied to or accidentally deposited on a wide range of agricultural soils, the laboratory studies are carried out with at least four distinct soil types. The EPA require that these soils must not have been treated with pesticide during the previous three years.

The soil is collected from the field and transported in thin-walled polythene bags, folded over but not sealed, to help preserve moist aerobic conditions and reduce any tendency for fluctuations in microbial populations. Thin polythene allows the passage of air but

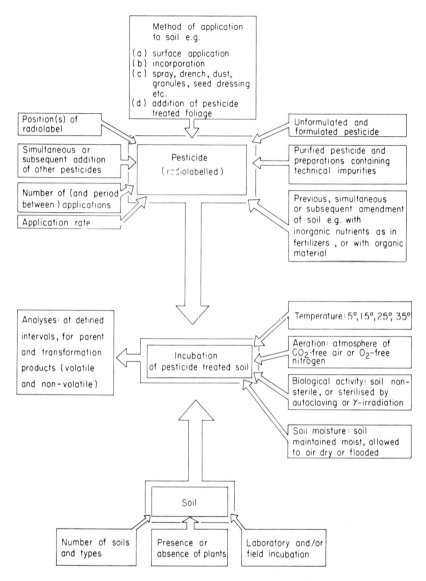

Fig. 6. Experimental variables considered during the establishment of studies of pesticide transformation in soil.

not water vapour (Stotzky *et al.*, 1962). For practical purposes of dispensing and treating "homogenous" samples, the soil is passed through a 2 mm sieve (to exclude roots, stones, etc.). Any soil too wet to sieve is partially air dried, at laboratory temperature, to a condition whereby sieving is facilitated. The moisture content of the sieved soil is determined by drying at 105°C to constant weight, and a rapid determination of moisture holding capacity (MHC) made using Keen–Raczkowski boxes (Keen and Raczkowski, 1921; Coutts, 1930). Sufficient deionized water is mixed into the bulk sieved soil so that after pesticide application the soil moisture content is 40% of its MHC (30% for peat soils).

Soils for sterilization are similarly prepared and either autoclaved at 121°C for 30 min on each of two successive days, or γ-irradiated at 5·0 Mrad in sealed polystyrene containers.

2. Preparation of pesticide

Radio-labelled preparations are used in all our studies of pesticide transformations in soil. The radio-labelled chemical and unlabelled preparations, where required for dilution of the specific activity, are purified before application to soil, e.g. by chromatographic and recrystallization techniques. The pure preparations are membrane-filtered to remove any insoluble contaminants resulting from the purification procedures, e.g. colloidal silica.

Soil studies are usually carried out using non-formulated pesticides, as commercial formulations can vary for different crops or types of application and from time to time may be modified to improve plant uptake, biological activity, etc. Furthermore, the EPA issue a list of approved formulants which, if used, do not require further study. Where agricultural practice involves the use of aqueous preparations the non-formulated pesticide is, wherever possible, applied to the soil as a solution or suspension in water. Despite the effects of adding organic solvents to soil, the low water-solubility of some non-formulated pesticides often necessitates the use of small amounts of organic solvents in the aqueous preparation. Alternatively, the pesticide might be added to the soil as a solid.

3. Application of pesticide to soil

Soils are treated with non-formulated and formulated pesticides by the methods recommended for normal agricultural practice, for example:

(a) Soil surface application of sprays, drenches, dusts, granules, etc., as may occur by direct application or from overspray and "run-off" during foliar treatments.

(b) Incorporation into the soil as granules, dusts, liquids etc.

(c) Where seed dressing is recommended the concentrations of the pesticide in plant-free soil may be simulated using pesticide coated glass beads.

(d) Addition of pesticide-treated foliage to soil, to simulate deposition from the decay of foliage following herbicide applications.

The recommended application rates for organic pesticides in most crop/soil situations do not often exceed $1-2$ kg ha^{-1} of active ingredient. It may be considered desirable to treat soil both at this rate and also at the very high rates which may arise from spillage of the pesticide, misuse, washing out of containers, etc. Many pesticides are applied annually but some are applied at repeated intervals during the growing season. Laboratory studies are designed to incorporate such treatments within the experimental programme.

Pesticide combinations in soil result from both simultaneous and successive applications of more than one pesticide. Where the manufactured product contains more than one pesticide it may be necessary to study the transformations of each component in the presence of the others. However this is not practical for other agriculturally used combinations, due to the large number possible.

The addition of nutrients to soil may occur in agriculture with, after or before pesticide application, in the form of inorganic fertilizers and organic manures. Agricultural amendment of soils with nutrients is effected in many differing ways, thus it is again impractical to make routine studies.

4. Incubation of pesticide-treated soils

Incubation of English soils with pesticide is always commenced within three days of field sampling. For foreign soils a further two days are allowed for transportation.

Samples of pesticide-treated soil (20–30 g) are incubated in glass crystallizing dishes (soil pots), 4 cm diam $\times$ 3 or 6 cm high, supported on siliconized wire racks inside cylindrical glass incubation units (Fig. 7). Each cylinder constitutes a single incubation unit and all the soil pots in any one are replicates containing the same soil type and receiving the same pesticide treatment. The number of pots per unit reflects the projected number of soil analytical time-points during the incubation period. Pots of soil may be included for further additions of pesticide, alterations in environmental conditions (e.g. waterlogging), removal for studies of leaching of the pesticide and its transformation products, etc.

Studies at 40% and 30% MHC are carried out in 3 cm high pots containing 30 g soil (20 g peat soil). Soil samples for flooding are

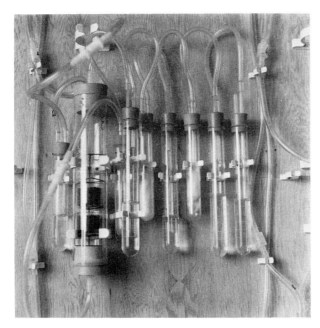

Fig. 7. Apparatus for the incubation of pesticide-treated soil in a continuous flow of air. (Air flows from left to right. Tubes contain, from the left, water, pots of treated soil, glass wool, $0.1N$ H_2SO_4, glass wool, an organic solvent, glass wool, and two of ethanolamine (to trap CO_2); however, different solutions may be used in studies of the degradation of different pesticides.)

dispensed into 6 cm high pots and flooded with sterile distilled water to a depth of 2.5 cm above the soil surface. Incubation units containing soils sterilized by autoclaving or by γ-radiation are assembled inside laminar flow cabinets, using pre-sterilized components and sterile pesticide (usually membrane filtered).

Each soil unit is incubated as an enclosed system, through which is passed a continuous stream of CO_2-free air or O_2-free nitrogen for maintenance of aerobic and anaerobic conditions, respectively. Flooded soils are incubated in an air stream. Sterility of autoclaved and irradiated soils is maintained using miniature air line filters (Microflow, Fleet, England) at the entry and exit of the incubation unit. To prevent rapid drying out of the soil, the air and nitrogen are saturated with water by bubbling through deionized water. Each incubation unit is supplied with approximately 150 ml of the appropriate gas per hour per pot of soil, estimated by a "bubbler" tube containing water, in line immediately before the soil unit (Fig. 7).

The effluent air or nitrogen from each soil incubation unit is bubbled through a series of "trapping" solutions (Fig. 7), designed to collect any voltatilized pesticide and volatile products of degradation. Kearney and Kontson (1976) have reported the use of polyurethane foam to trap some volatile materials whilst Anderson (1975) used oil-coated glasswool. We have not yet evaluated these techniques.

The moisture levels of the soils are maintained by the addition of sterile deionized water to each pot, approximately every eight weeks, to restore the soil to its original weight. Some samples of soil are allowed to air dry (in a stream of dry CO_2-free air) during incubation.

The number of experimental variables and replicates involved in the study of each pesticide necessitates a considerable number of incubation units. It is then for economy of space that the soil pots are incubated in vertical glass cylinders, which are supported from the walls of a constant temperature room (Fig. 8), at $25° \pm 1°$.

Fig. 8. Constant temperature (25°C) room, in which soils are incubated. (The apparatus [*] between groups of soil incubation units are for the production and distribution of CO_2-free, water-saturated air.)

5. Sampling soils for analysis

Complete pots of pesticide-treated soil are removed for analysis at "zero-time" and at regular intervals thereafter. Pots of sterile soil are removed from their incubation units inside laminar flow cabinets, using sterile instruments and aseptic techniques. The sterility of the sampled soil is checked by removing approximately 0.1 g soil, shaking it in a small volume of diluent and adding aliquots of the suspension to a range of broth and agar media. Soils incubated anaerobically are sampled inside an inflatable "isolator" (Fig. 9) containing an atmosphere of oxygen-free nitrogen, to prevent changes in the atmosphere causing fluctuations in soil microbial and chemical activity.

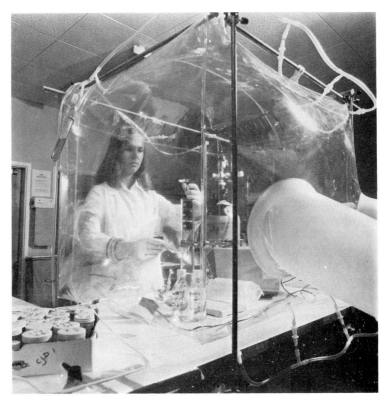

Fig. 9. Nitrogen gas-filled isolator for sampling or watering soils incubated in a stream of nitrogen gas.

6. *Pesticide and degradation product analysis*

All the pesticides studied are radio-labelled and all quantitative measurements of the parent pesticide and any transformation products are made using conventional radio-chemical techniques. Analyses of the "trapping" solutions for radio-activity are made at regular intervals throughout the period of incubation. "Trapping" solutions containing radioactivity are further studied to identify the evolved radio-labelled products. The soil is normally analysed immediately after sampling, whole pots of soil being extracted with appropriate solvents. The pesticide and its products are initially extracted from soil with non-polar and polar solvents that are unlikely to alter chemically, the pesticide residues. Preliminary studies are always carried out to determine the efficiency and effects of extraction procedures on the parent chemical and any available potential trans-formation products, and to ensure that losses due to volatilization do not occur during the extraction and subsequent analytical procedures.

It is often difficult to isolate pesticide transformation products, from soil extracts, sufficiently pure for identification by u.v. spectra, mass spectra, infra-red spectrometry and other similar techniques. In such cases it may be necessary to characterize the pesticide products by co-chromatography and by isotope dilution with pure, authenticated reference compounds.

The radio-labelled material remaining in soil after polar and non-polar solvent extraction is generally referred to as unextractable or "bound". This material may be transformation products or parent pesticide, adsorbed to mineral or organic surfaces and not released by the extraction techniques; trapped within the matrix of the organic matter; or even chemically incorporated within the structure of the soil organic matter. Characterization of "bound" residues is attempted after further extraction of the soil with acid, alkali and salt solutions, with ion exchange resins or by isotope exchange. Studies of "bound" pesticide residues are severely limited by a lack of understanding of the structure and reactivity of soil organic matter and by an inability to satisfactorily isolate and fractionate its components without altering their chemical structures. Determination of the nature and site of "binding" of pesticide residues may sometimes be possible by acid-alkali and chromatographic fractionation of the extracts, by isotope dilution and by molecular weight estimations using gel permeation and ultrafiltration techniques.

At the end of the incubation period the apparatus (glassware, tub-ing, etc.) is solvent-washed and the washings analysed for the presence of radioactivity.

C. LABORATORY STUDIES IN THE PRESENCE OF PLANTS

To investigate effectively the influence of viable plant roots and their associated microflora on the transformations of pesticides in soil it is again essential to use radio-labelled pesticides and enclosed systems.

However, in a system in which there is no separation of the foliar and soil-root environments data may be difficult to interpret. Volatile products may be either from the plant, via root uptake and plant metabolism, or from the soil-root zone. Furthermore $^{14}CO_2$ evolved from the soil may in part be taken up and fixed by plant foliage. It is therefore important to isolate the foliar environment from that enclosing the soil and roots.

1. Incubation of plants and soil

At ICI Plant Protection Division glass/perspex units have been designed and built to fulfil the above requirements. They are complex by comparison with the studies involving soil alone, additional analyses are involved and, consequently extensive use is limited. However they are used to investigate rates of pesticide transformation, and the products formed in the presence of roots, in comparison to soil without plants.

Fig. 10a. Soil-plant incubation unit containing a field bean plant.

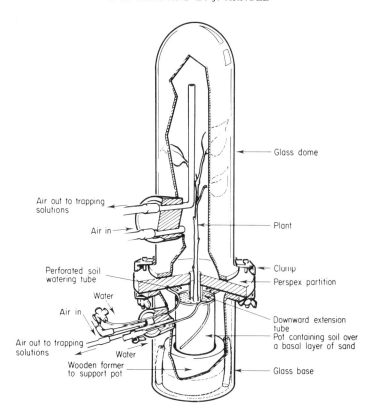

Fig. 10b. Soil-plant incubation unit. Cut-away diagram.

Fig. 10a and b shows an individual unit and Fig. 11 the horizontal perspex partition used to separate the foliar and soil-root environments. Approximately 250 g soil at 40% MHC, on a basal layer of 50 g sand, is contained in a 250 ml glass beaker (9·5 × 7·0 cm diam.) held in place in the lower of two glass chambers by wooden formers. The horizontal partition, with rubber "O" rings, is placed in position so that the central, downward extension tube touches the soil surface. The upper glass chamber is placed in position and the unit clamped together. With the ports sealed the unit is then tested, under pressure, for air leaks.

Plants grown inside the units, from seed or from transplanted seedlings, are allowed to grow up through the hole in the partition into the upper chamber. After a short period of incubation a 2 mm thick layer of flexible silicone sealant (Silastic 9161 RTV and hardener catalyst N 9162; Hopkin and Williams, Chadwell Heath, England) (Stotzky *et al.*, 1961) is applied to the portion of soil surface enclosed

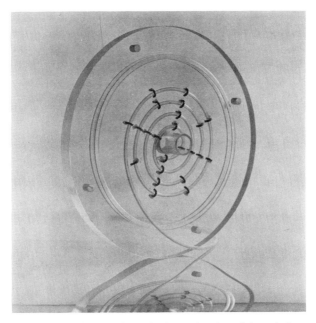

Fig. 11. Perspex partition from soil incubation unit, viewed from below to show the coiled polyethylene tubing, used to water the soil surface, and the downward extension tube.

by the downward extension tube. This effectively isolates the atmospheres in the soil-root and foliar chambers whilst leaving the majority of the soil surface exposed to the air and available for gaseous transfer. As the plant grows, the silicone seal allows expansion of the stem. Any leakage of air past the silicone is rectified by addition of further 1 mm layers of the sealant.

Application of the pesticide to soil may be by seed dressing or by application onto or into the soil during or after assembly of the unit. Pesticide may also be applied to the plant foliage.

Air is supplied to both chambers separately. The effluent air from each chamber is passed through separate sets of "trapping" solutions for recovery of radio-labelled gaseous or volatile products. Water loss from the soil, estimated by weighing the complete unit and making an allowance for the increasing plant weight, is made good by addition of equal volumes of deionized water to the base of the soil pot (into the sand layer) and to the soil surface (via a coil of perforated polyethylene tube attached to the underside of the perspex partition). The water is "injected", by syringe, through stopcocks and hypodermic needles in the bung sealing the port in the lower chamber.

Sets of five such units (each unit normally containing the same pesticide/soil/plant combination) are installed on the upper platform of incubation trolleys (Fig. 12). The lighting system, built into the trolleys contains the facility for use of fluorescent tubes and tungsten filament bulbs. Light reaching the soil surface can be reduced, where necessary, by inserting a segmented black perspex disc around the stem on the upper surface of the perspex partition. Pots of soil without plants, as controls, are incubated on the bottom platform of the trolley and are treated and analysed simultaneously with the soil/plant incubation units. The lower compartment of the soil/plant units and the complete control pots are normally enclosed by light-proof covers.

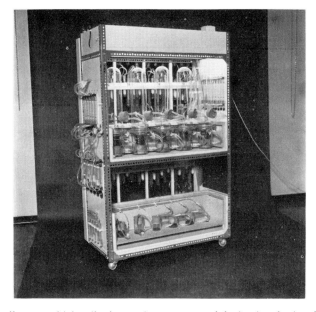

Fig. 12. Trolleys on which soil–plant units are mounted during incubation. The upper box contains fluorescent and tungsten lamps. The lower compartment of the soil–plant units and the complete control pots on the lowermost shelf are normally enclosed by a cover.

2. Analysis of plants, soil and trapping solutions

At regular intervals during incubation complete soil/plant units and control units are removed for analysis. The plant foliar and root systems and the soil and sand are all carefully separated and each analysed independently. Inevitably, small roots and root hairs remain in the soil and sand, and some of the soil particles remain adhering to the root. This method is, however, considered preferable to root washing or ultrasonication techniques which may wash the pesticide out of root surface tissues. The sand and soil are analysed as previously described. The plant material is either combusted for total radioactive residues, or extracted and pesticide products present quantitatively and qualitatively analysed, where possible.

D. FIELD EXPERIMENTS

As has already been emphasized, field data are less easy to obtain than laboratory data and "balance studies" are difficult to establish under natural field conditions. However, a small number of field experiments are done to assess the validity of the observations from laboratory studies.

1. *Incubation of soil and plants*

"Balance studies" are not carried out because of the practical difficulties and time required to prepare and establish lysimeters for the recovery of all leached products. Also, the desirability that soil and plant be watered naturally by rainfall, conflicts with the need to collect volatile radio-labelled products. In our studies, products leached below 40 cm depth of soil are not accounted for. Volatile products are, however, collected but only during a 24 hour period (midday to midday) every 48 hours.

The plants are grown in, and the radio-labelled pesticide applied to, soil within polypropylene pipes (7·5 cm × 46 cm) sunk vertically to a depth of approximately 40 cm in the soil. Groups of three pipes are covered by 60 cm high bell jars (Fig. 13) mounted on stainless steel rings (Fig. 14) embedded 5 cm into the soil surface. The bell jar fits into the channels around the steel ring which is filled with water to provide a seal against loss of volatile products. The water is regularly sampled for radioactivity. The air is sucked out of the bell jar and through "trapping" solutions, for collection of volatile radio-labelled products. The use of a midday to midday, 24 h on/24 h off cycle reduces the heating effect inside the jar and mostly allows a proportion of the rainwater to fall on the treated soil. Where necessary the effect of the presence of plants on soil pesticide transformations can be examined by parallel studies without plants present.

2. *Analyses of plants, soil and trapping solutions*

Single pipes with soil and plants (where present) are levered from the soil at regular intervals during incubation. Treatments for "zero-time"

Fig. 13. Apparatus used in field studies of the degradation of pesticides in soil.

Fig. 14. Stainless steel ring on which field bell-jars (see Fig. 13) are mounted. The ring is pushed down into the soil and the jar fits into the outer channel.

analyses are effected in pipes adjacent to the area to be covered by the bell jar. A final analysis is made either when the plant has reached maturity or immediately prior to the time at which a new crop might be sown. The plant foliar material is removed and the pipe sawn into 4×10 cm lengths. The soil from each section is separated from the root material and the plant and soil samples analysed, as previously described.

VI. Conclusions

The manufacturers of plant protection chemicals study the environmental fate of products both as an ethical responsibility and as a commercial necessity; the latter in response to the registration requirements of many countries. The studies should be reproducible and the methodology should facilitate quantitative and qualitative analyses of volatile and non-volatile pesticide and products. These conditions can only be truly satisfied in laboratory studies using radio-labelled pesticides.

It is essential that these studies include a realistic evaluation of pesticide transformation in the environment under investigation. For example, studies of the fate of a pesticide in soil must be carried out using soil in as near to its field condition as is possible. Microbial cultures grown in nutrient media, perfusion columns and soil suspensions do not provide data of relevance to agricultural conditions. The acceptance of the validity of laboratory studies, and the availability of radio-labelled chemicals enables balance-sheet studies (recovery of all the applied radio-label) of pesticide transformations in soil to be implemented under a wide range of experimental conditions. These may include studies of the effects on transformation of environmental factors such as soil type, temperature, moisture content, aerobiosis, anaerobiosis and flooding; and where necessary of pesticide-related factors such as formulation components, impurities in commercial preparations, safteners, synergists, nutrient amendments, pesticide combinations and serial applications.

The majority of such investigations are executed in the absence of plants. The known effects of plants on soil microbial populations and their activity, and on the physical and chemical characteristics of the soil surrounding the roots, emphasizes the desirability of some studies of pesticide transformations in soil being done in the presence of plants. The relative complexity of such investigations whether in the laboratory or the field prevents their extensive use. However soil/plant studies should be used to supplement and confirm or refute observations made in soil alone.

The amount of data required by registration agencies is still in a state of flux, regularly being changed, often becoming increasingly stringent and demanding, usually retrospectively applied. Experimental programmes to provide registration data for the fate of pesticides in soil often extend over one to two years, or more. It has thus become necessary to anticipate the changing requirements and to continuously investigate the use of new or modified techniques that enable a better, more comprehensive and more relevant understanding of the fate of pesticides in the soil environment.

References

ANDERSON, J. P. E. (1975). Influence of temperature and moisture on volatilisation, degradation and binding of diallate in soil. *Z. PflKrankh. PflSchutz*, Sdh. VII, 141–146.

ANDERSON, J. R. and DREW, E. A. (1972). Growth characteristics of a species of *Lipomyces* and its degradation of paraquat. *J. gen. Microbiol.* **70**, 43–58.

ANDRAWES, N. R., BAGLEY, W. P. and HERRETT, R. A. (1971a). Fate and carryover properties of Temik aldicarb pesticide [2-methyl-2-(methylthio)propionaldehyde-O-(methylcarbamoyl)oxime] in soil. *J. agric. Fd Chem.* **19**, 727–730.

ANDRAWES, N. R., BAGLEY, W. P. and HERRETT, R. A. (1971b). Metabolism of 2-methyl-2(methylthio)propionaldehyde-O-(methylcarbamoyl)-oxime (Temik aldicarb pesticide) in potato plants. *J. agric. Fd Chem.* **19**, 731–737.

ASHTON, F. M. (1963). Fate of amitrole in soil. *Weeds* **11**, 167–170.

ASHWOOD-SMITH, M. J. (1965). On the genetic stability of bacteria to freezing and thawing. *Cryobiology* **2**, 39–43.

AUDUS, L. J. (1951). The biological detoxification of hormone herbicides in soil. *Pl. Soil* **3**, 170–192.

BALASUBRAMANIAN, V. and KANEHIRO, Y. (1974). An improved soil incubation apparatus. *Comm. Soil Sci. Plant Analys.* **5**, 283–290.

BARTHA, R. (1971). Altered propanil biodegradation in temporarily air-dried soil. *J. agric. Fd Chem.* **19**, 394–395.

BARTHA, R. and PRAMER, D. (1965). Features of a flask and method for measuring the persistence and biological effects of pesticides in soil. *Soil Sci.* **100**, 68–70.

BARTHOLOMEW, W. V. and BROADBENT, F. E. (1950). Apparatus for control of moisture, temperature and air composition in microbiological respiration experiments. *Proc. Soil Sci. Soc. Am.* **14**, 156–160.

BAUDE, F. J., PEASE, H. L. and HOLT, R. F. (1974). Fate of benomyl on field soil and turf. *J. agric. Fd Chem.* **22**, 413–418.

BAUR, J. R., BAKER, R. D., BOVEY, R. W. and SMITH, J. D. (1972). Concentration of picloram in the soil profile. *Weed Sci.* **20**, 305–309.

BBA (1973). "Unterlagen zum Verhalten von Pflanzenschultzmitteln im Boden Rahmen des Zulassungsverfahrens". Merkblatt no. 36.

BEALL, M. L. and NASH, R. G. (1971). Insecticide residues in soybean plant tops: root vs. vapour sorption. *Agron. J.* **63**, 460–464.

BEALL, M. L., NASH, R. G. and KEARNEY, P. C. (1976). Agroecosystem. A laboratory model ecosystem to simulate agricultural field conditions for monitoring pesticides. Proc. EPA Conference on Modeling and Simulation, Environ. Protect. Agency, pp. 790–793.

BEST, J. A. and WEBER, J. B. (1974). Disappearance of *s*-triazines as affected by soil pH using a balance-sheet approach. *Weed Sci.* **22**, 364–373.

BEYNON, K. I., ROBERTS, T. R. and WRIGHT, A. N. (1974). The degradation of the herbicide benzoylprop-ethyl following its application to wheat. *Pestic. Sci.* **5**, 429–442.

BOLLAG, J. M. and LIU, S. Y. (1971). Hydroxylations of carbaryl by soil fungi. *Nature, Lond.* **236**, 177–178.

BORDELEAU, L. M. and BARTHA, R. (1971). Ecology of a herbicide transformation: synergism of two soil fungi. *Soil Biol. Biochem.* **3**, 281–284.

BOUNDS, H. C., MAGEE, L. A. and COLMER, A. R. (1969). The reversal of dalapon toxicity by microbial action. *Can. J. Microbiol.* **15**, 1121–1124.

BOWEN, H. J. M. and CAWSE, P. A. (1964). Some effects of gamma radiation on the composition of the soil solution and soil organic matter. *Soil Sci.* **98**, 358–361.

BUCKLAND, J. L., COLLINS, R. F., HENDERSON, M. A. and PULLIN, E. M. (1973). Radiochemical distribution and decline studies with bromoxynil octanoate in wheat. *Pestic. Sci.* **4**, 689–700.

CARO, J. H., TAYLOR, A. W. and LEMON, E. R. (1971). Measurement of pesticide concentrations in air overlying in treated field. *Proc. Internat. Symp. Identif'n Measure. of Environ. Pollutants*, Ottawa, 72–75.

CASTRO, C. E. and BELSER, N. O. (1966). Biodehalogenation. Reductive dehalogenation of the biocides ethylene dibromide, 1,2-dibromo-3-chloropropane and 2,3-dibromobutane in soil. *Environ. Sci. Technol.* **2**, 779–783.

CHISAKA, H. and KEARNEY, P. C. (1970). Metabolism of propanil in soils. *J. agric. Fd Chem.* **18**, 854–858.

CLARK, C. G. and WRIGHT, S. J. L. (1970a). Detoxication of isopropyl *N*-phenylcarbamate (IPC) and isopropyl *N*-3-chlorophenylcarbamate (CIPC) in soil, and isolation of IPC-metabolizing bacteria. *Soil Biol. Biochem.* **2**, 19–26.

CLARK, C. G. and WRIGHT, S. J. L. (1970b). Degradation of the herbicide isopropyl N-phenylcarbamate by *Arthrobacter* and *Achromobacter* spp. from soil. *Soil Biol. Biochem.* **2**, 217–226.

COUTTS, J. R. H. (1930). "Single value" soil properties. A study of the significance of certain soil constants. III Note on the technique of the Keen–Raczkowski box experiment. *J. agric. Sci., Camb.* **20**, 407–413.

COX, C. S. (1968). Method for the routine preservation of micro-organisms. *Nature, Lond.* **220**, 1139.

CUNNINGHAM, J. L. (1973). Preservation of rust fungi in liquid nitrogen. *Cryobiology* **10**, 361–363.

DAILY, W. A. and HIGGINS, C. E. (1973). Preservation and storage of microorganisms in the gas phase of liquid nitrogen. *Cryobiology* **10**, 364–367.

DAVIS, R. J. (1963). Viability and behaviour of lyophilized cultures after storage for twenty-one years. *J. Bact.* **85**, 486–487.

DORN, S., OESTERHELT, G., SUCHY, M., TRAUTMANN, K. H. and WIPF, H. K. (1976). Environmental degradation of the insect growth regulator 6,7-epoxy-1-(p-ethylphenoxy)-3-ethyl-7-methylnonane (Ro 10-3108) in polluted water. *J. agric. Fd Chem.* **24**, 637–639.

DOWDELL, R. J., SMITH, K. A., CREES, R. and RESTALL, S. W. F. (1972). Field studies of ethylene in the soil atmosphere – equipment and preliminary results. *Soil Biol. Biochem.* **4**, 325–331.

EICHELBERGER, J. W. and LICHTENBERG, J. J. (1971). Persistence of pesticides in river water. *Environ. Sci. Technol.* **5**, 541–544.

ENO, C. F. and POPENOE, H. (1964). Gamma radiation compared with steam and methyl bromide as a soil sterilizing agent. *Proc. Soil Sci. Soc. Am.* **28**, 533–535.

EPA (1975a). Environmental Protection Agency pesticide program. Guidelines for registering pesticides in the United States. *Federal Register* **40**, 26802–26928.

EPA (1975b). Environmental Protection Agency pesticide program. Guidelines for registering pesticides in the United States. *Federal Register* **40**, 26891–26896.

ETGES, F. T., BELL, E. J. and IVINS, B. E. (1969). A field survey of molluscicide-degrading microorganisms in the Caribbean area. *Am. J. trop. Med. Hyg.* **18**, 472–476.

FENNEL, D. I., ROPER, K. B. and FLICKINGER, M. H. (1950). Further investigations on the preservation of mold cultures. *Mycologia* **42**, 135–147.

FREHSE, H. (1976). The perspective of persistence. *Proc. BCPC Symp: Persistence of Insect. Herbic.* Monograph no. **17**, 1–39.

FRIES, G. R., MARROW, G. S. and GORDON, C. H. (1969). Metabolism of o,p'- and p,p'-DDT by rumen microorganisms. *J. agric. Fd Chem.* **17**, 860–862.

FUHR, F., CHENG, H. H. and MITTELSTAEDT, W. (1976). Pesticide balance and metabolism studies with standardized lysimeters. *Landw. Forsch. Sond.* **32**, 272–278.

FURMIDGE, C. G. L. and OSGERBY, J. M. (1967). Persistence of herbicides in soil. *J. Sci. Fd Agric.* **18**, 269–273.

GETZIN, L. W. and ROSENFIELD, I. (1968). Organophosphorus insecticide degradation by heat labile substances in soil. *J. agric. Fd Chem.* **16**, 598–601.

GLASS, B. L. and EDWARDS, W. M. (1974). Picloram in lysimeter runoff and percolation water. *Bull. Environ. Contam. Toxicol.* **11**, 109–112.

GOLAB, T., BISHOP, C. E., DONOHO, A. L., MANTHEY, J. A. and ZORNES, L. L. (1975). Behaviour of ^{14}C oryzalin in soil and plants. *Pestic. Biochem. Physiol.* **5**, 196–204.

GRAY, T. R. G. and WILLIAMS, S. T. (1971). "Soil Micro-organisms". Oliver and Boyd, Edinburgh.

GREGORY, W. W., REED, J. K. and PRIESTER, L. E. (1969). Accumulation of parathion and DDT by some algae and protozoa. *J. Protozool.* **16**, 69–71.

GROSSBARD, E. and HALL, D. M. (1964). An investigation into the possible changes in the microbial population of soils stored at −15°C. *Pl. Soil* **21**, 317–332.

GUNNER, H. B. and ZUCKERMAN, B. N. (1968). Degradation of diazinon by synergistic microbial action. *Nature, Lond.* **217**, 1183–1184.

GUNNER, H. B., ZUCKERMAN, B. N., WALKER, R. W., MILLER, C. W., DEUBERT, K. H. and LONGLEY, R. E. (1966). The distribution and persistence of diazinon applied to plant and soil and its influence on rhizosphere and soil microflora. *Pl. Soil* **25**, 249–264.

GUTH, J. A., GEISSBUEHLER, H. and EBNER, L. (1969). Dissipation of urea herbicides in soil. *Meded. Fac. Llandb. Rijksuniv. Gent.* **34**, 1027–1037.

GUTH, J. A., BURKHARD, N. and EBERLE, D. O. (1976). Experimental models for studying the persistence of pesticides in soils. *Proc. BCPC Symp: Persistence of Insectic. Herbic.*, BCPC Monograph no. **17**, 137–157.

HAAS, R. H., SCIFRES, C. J., MERKLE, M. G., HAHN, R. R. and HOFFMAN, G. O. (1971). Occurrence and persistence of picloram in grassland water courses. *Weed Res.* **11**, 54–62.

HARRIS, R. J. C. (1954). "Biological Applications of Freezing and Drying". Academic Press Inc., New York.

HARVEY, J. and REISER, R. W. (1973). Metabolism of methomyl in tobacco, corn and cabbage. *J. agric. Fd Chem.* **21**, 775–783.

HENDERSON, M. E. K. (1965). Enrichment in soil of fungi which utilise aromatic compounds. *Pl. Soil* **23**, 339–350.

HOLSTUN, J. T. and LOOMIS, W. E. (1956). Leaching and decomposition of 2,2-dichloropropionic acid in several Iowa soils. *Weeds* **4**, 205–217.

HORVATH, R. S. (1972). Microbial co-metabolism and the degradation of organic compounds in nature. *Bact. Rev.* **36**, 146–155.

HWRANG, S. W. (1960). Effects of ultra-low temperatures on the viability of selected fungus strains. *Mycologia* **52**, 527–529.

ISENSEE, A. R. (1975). Progress and status report on fresh water ecosystem methodology development. In *Subst. Chem. Program – The First Year of Progress, III.* Environ. Protect. Agency Symp., pp. 45 52.

ISENSEE, A. R. and JONES, G. E. (1975). Distribution of 2,3,7,8-tetrachlorodibenzo-*p*-dioxin (TCDD) in aquatic model ecosystem. *Environ. Sci. Technol.* **9**, 668–672.

ISENSEE, A. R., HOLDEN, E. R., WOOLSON, E. A. and JONES, G. E. (1976). Soil persistence and aquatic bioaccumulation potential of hexachlorobenzene (HCB). *J. agric. Fd Chem.* **24**, 1210–1217.

JENSEN, H. L. (1957). Decomposition of chloro-substituted aliphatic acids by soil bacteria. *Can. J. Microbiol.* **3**, 151–164.

JENSEN, V. (1963). Studies on the microflora of Danish beech forest soils 1. The dilution plate count technique for the enumeration of bacteria and fungi in the soil. *Zbl. Bakt.* Abt. II **116**, 13–32.

JOHNSON, B. T. and LULVES, W. (1975). Biodegradation of di-*n*-butyl phthalate and di-2-ethylhexyl phthalate in freshwater hydrosoil. *J. Fish Res. Bd Can.* **32**, 333–339.

JOHNSON, J. B. and CAIN, R. D. (1973). The control of benzene-sulphonate degradation in bacteria. *Abs. Commun. Spec. Meet. Fed. Eur. Biochem. Soc.* Abs no. 137.

KAUFMAN, D. D. and KEARNEY, P. C. (1965). Microbial degradation of isopropyl-N-3-chlorophenylcarbamate and 2-chloroethyl-*N*-3-chlorophenylcarbamate. *Appl. Microbiol.* **13**, 443–446.

KEARNEY, P. C. and KONTSON, A. (1976). A simple system to simultaneously measure volatilisation and metabolism of pesticides from soils. *J. agric. Fd Chem.* **24**, 424–426.

KEEN, B. A. and RACZKOWSKI, M. (1921). The relation between the clay content and certain physical properties of soil. *J. agric. Sci., Camb.* **11**, 441–449.

LA FLEUR, K. S., WOJECK, G. A. and McCASKILL, W. R. (1973). Movement of toxaphene and fluometuron through Dunbar soil to underlying ground water. *J. Environ. Quality* **2**, 515–518.

LICHTENSTEIN, E. P. and SCHULZ, K. R. (1960). Epoxidation of aldrin and heptachlor in soils as influenced by autoclaving, moisture and soil type. *J. econ. Ent.* **53**, 192–197.

LICHTENSTEIN, E. P., FUHREMANN, T. W. and SCHULZ, K. R. (1974). Translocation and metabolism of [^{14}C]-phorate as affected by percolating water in a model soil–plant ecosystem. *J. agric. Fd Chem.* **22**, 991–996.

MAZUR, P. (1962). Mechanisms of injury in frozen and frozen dried cells. *In* "Culture Collections: Perspectives and Problems" (S. M. Martin, ed.), pp. 59–70. Univ. Toronto Press, Toronto.

McGARITY, J. W. and RAJARATNAM, J. A. (1973). Apparatus for the measurement of losses of nitrogen as gas from the field and simulated field environments. *Soil Biol. Biochem.* **5**, 121–131.

McLAREN, A. D., LUSE, R. A. and SKUJIŅS, J. J. (1957). Sterilisation of soil by irradiation and some further observations on soil enzyme activity. *Proc. Soil Sci. Soc. Am.* **26**, 371–377.

MENDEL, J. L. and WALTON, M. S. (1966). Conversion of *pp'*-DDT to *pp'*-DDD by intestinal flora of the rat. *Science, N.Y.* **151**, 1527–1528.

METCALF, R. L. (1974). A laboratory model ecosystem to evaluate compounds producing biological magnification. In *Essays in Toxicology* **5**, 17–38.

MISKUS, R. P., BLAIR, D. P. and CASIDA, J. E. (1965). Conversion of DDT to DDD by bovine rumen fluid, lake water and reduced porphyrins. *J. agric. Fd Chem.* **13**, 481–483.

MIYAZAKI, S., SIKKA, H. C. and LYNCH, R. S. (1975). Metabolism of dichlobenil by microorganisms in the aquatic environment. *J. agric. Fd Chem.* **23**, 365–368.

MORRILL, L. G. and DAWSON, J. E. (1964). An improved percolation system. *Proc. Soil Sci. Soc. Am.* **28**, 710–711.

MUNNECKE, D. E. (1958). The persistence of non-volatile diffusible fungicides in soil. *Phytopathology* **48**, 581–585.

NAIK, M. N., JACKSON, R. B., STOKES, J. and SWABY, R. J. (1972). Microbial degradation and phytotoxicity of picloram and other substituted pyridines. *Soil Biol. Biochem.* **4**, 313–323.

NOMMIK, H. (1971). A technique for determining mineralisation of carbon in soils during incubation. *Soil Sci.* **112**, 131–136.

NOOI, J. R., TESTA, M. C. and WILLEMSE, S. (1970). Biodegradation mechanism of fatty alcohol non-ionics. Experiments with some ^{14}C-labelled stearyl alcohol/ethylene oxide condensates. *Tenside* **7**, 61–65.

OECD (1971). "Pollution by Detergents". Organisation Econ. Co-op. Develop., Paris.

PARIS, D. F. and LEWIS, D. L. (1973). Chemical and microbial degradation of ten selected pesticides in aquatic systems. *Residue Rev.* **45**, 95–124.

PARIS, D. F., LEWIS, D. L., BARNETT, J. T. and BAUGHMANN, G. L. (1975). Microbial degradation and accumulation of pesticides in aquatic systems. *USA Environmental Protection Agency*, Report EPA-660/3-75-007.

PARR, J. F. and SMITH, S. (1969). A multipurpose manifold assembly: Use in evaluating microbiological effects of pesticides. *Soil Sci.* **107**, 271–276.

PETERSON, J. E. and ROBISON, W. H. (1964). Metabolic products of p,p'-DDT in the rat. *Toxic. appl. Pharmac.* **6**, 321–327.

PRAMER, D. and BARTHA, R. (1972). Preparation and processing of soil samples for biodegradation studies. *Environ. Letters* **2**, 217–224.

RAGAB, M. T. H. and McCOLLUM, J. P. (1961). Degradation of C^{14} labelled simazine by plants and soil microorganisms. *Weeds* **9**, 72–84.

RAGHU, K. and MacRAE, I. C. (1966). Biodegradation of the gamma isomer of benzene hexachloride in submerged soils. *Science, N.Y.* **154**, 263–264.

RAHN, P. R. and ZIMDAHL, R. L. (1973). Soil degradation of two phenyl pyridazinone herbicides. *Weed Sci.* **21**, 314–317.

REID, J. J. (1960). Bacterial decomposition of herbicides. *Proc. N.E. Weed Contr. Conf.* **14**, 19–30.

RIEPMA, P. (1962). Preliminary observations on the breakdown of 3-amino-1,2,4-triazole in soil. *Weed Res.* **2**, 41–50.

RILEY, D., TUCKER, B. V. and WILKINSON, W. (1976). Biological unavailability of bound paraquat residues in soil. *In* "Bound and Conjugated Pesticide Residues" (D. D. Kaufman, G. G. Still, G. D. Paulson and S. K. Bandal, Eds), pp. 301–353. ACS Symp. Ser. no. 29, American Chemical Society, Washington D.C.

ROBERTS, T. R. (1976). Experimental models for studying the fate of pesticides in plants. *Proc. BCPC Symp: Persistence of Insectic. Herbic.*, BCPC Monograph no. 17, 159–168.

ROBINSON, J. B., SALONIUS, P. O. and CHASE, F. E. (1965). A note on the differential response of *Arthrobacter* spp. and *Pseudomonas* spp. to drying in soil. *Can. J. Microbiol.* **11**, 746–748.

ROVIRA, A. D. and VENDRELL, M. (1972). Ethylene in sterilised soil: its significance in studies of interactions between microorganisms and plants. *Soil. Biol. Biochem.* **4**, 63–69.

SALONIUS, P. O., ROBINSON, J. B. and CHASE, F. E. (1967). A comparison of autoclaved and gamma-irradiated soils as media for microbial colonisation experiments. *Pl. Soil* **27**, 239–248.

SANBORN, J. R. (1974). The fate of 10 selected pesticides in the aquatic environment. *USA Environmental Protection Agency*, Report EPA-660/3-74-025.

SENIOR, E., BULL, A. T. and SLATER, J. H. (1976). Enzyme evolution in a microbial community growing on the herbicide dalapon. *Nature, Lond.* **263**, 476–479.

SHARP, R. F. and TAYLOR, B. F. (1969). A new soil percolator for the elective culture of soil organisms. *Soil Biol. Biochem.* **1**, 191–194.

SKIPPER, H. D. and WESTERMANN, D. T. (1973). Comparative effects of propylene oxide, sodium azide, and autoclaving on selected soil properties. *Soil Biol. Biochem.* **5**, 409–414.

SMITH, W. K. (1962). A mechanical soil percolator. *J. appl. Bact.* **25**, 83–86.

SÖDERGREN, A. (1971). Accumulation and distribution of chlorinated hydrocarbons in cultures of *Chlorella pyrenoidosa* (Chlorophyceae). *Oikos* **22**, 215–220.

SOULIDES, D. A. and ALLISON, F. E. (1961). Effect of drying and freezing soils on carbon dioxide production, available mineral nutrients, aggregation and bacterial population. *Soil Sci.* **91**, 291–298.

SPANIS, W. C., MUNNECKE, D. E. and SOLBERG, R. A. (1962). Biological breakdown of two organic mercurial fungicides. *Phytopathology* **52**, 455–462.

STEVENSON, I. L. (1956). Some observations on the microbial activity in remoistened air-dried soils. *Pl. Soil* **8**, 170–182.

STOJANOVIC, B. J., KENNEDY, M. V. and SHUMAN, F. L. (1972). Edaphic aspects of the disposal of unused pesticides, pesticide wastes and pesticide containers. *J. Environ. Quality* **1**, 54–62.

STOTZKY, G. (1960). A simple method for the determination of the respiratory quotient of soils. *Can. J. Microbiol.* **6**, 439–452.

STOTZKY, G., CULBRETH, W. and MISH, L. B. (1961). A sealing compound for use in biological work. *Nature, Lond.* **191**, 410.

STOTZKY, G., GOOS, R. D. and TIMONIN, M. I. (1962). Microbial changes occurring in soil as a result of storage. *Pl. Soil* **16**, 1–18.

SWEENEY, R. A. (1969). Metabolism of lindane by unicellular algae. *Proc. 12th Conf. G. Lakes Res.* 98–102.

SWISHER, R. D. (1970). "Surfactant biodegradation". Marcel Dekker Inc. New York.

TU, C. M. (1970). Effect of four organophosphorus insecticides on microbial activities in soil. *Appl. Microbiol.* **19**, 479–484.

VELA, G. R. and WYSS, O. (1962). The effect of gamma radiation on nitrogen transformation in soils. *Bact. Proc.* 24.

WALKER, A. (1970). Persistence of pronamide in soil. *Pestic. Sci.* **1**, 237–239.

WARCUP, J. H. (1957). Chemical and biological aspects of soil sterilisation. *Soils Fertil.* **20**, 1–5.

WEISGERBER, I., KOHLI, J., KAUL, R., KLEIN, W. and KORTE, F. (1974). Fate of aldrin-^{14}C in maize, wheat and soils under outdoor conditions. *J. agric. Fd Chem.* **22**, 609–612.

WELLMAN, A. M. and WALDEN, D. B. (1964). Qualitative and quantitative estimates of viability for some fungi after periods of storage in liquid nitrogen. *Can. J. Microbiol.* **10**, 585–593.

WHEELER, W. B. C. (1970). Experimental absorption of dieldrin by *Chlorella. J. agric. Fd Chem.* **18**, 416–419.

WHITESIDE, J. S. and ALEXANDER, M. (1960). Measurement of microbiological effects of herbicides. *Weeds* **8**, 204–213.

WINSTON, P. W. and BATES, D. H. (1960). Saturated solutions for the control of humidity in biological research. *Ecology* **41**, 232–237.

YARON, B., HEUER, B. and BIRK, Y. (1974). Kinetics of azinphosmethyl losses in the soil environment. *J. agric. Fd Chem.* **22**, 439–441.

ZUCKERMAN, B. M., DEUBERT, K., MACKIEWICZ, M. and GUNNER, H. (1970). Studies on the biodegradation of parathion. *Pl. Soil* **33**, 273–281.

Chapter 6

Some Methods for Assessing Pesticide Effects on Non-Target Soil Microorganisms and their Activities

JOHN R. ANDERSON

Microbiology Unit,
National Food Research Institute, C.S.I.R. Pretoria, South Africa

I. Introduction

The fact that millions of tons of food are wasted annually to the ravages of pests and diseases and that many thousands of people die for want of sufficient food is justification enough for the continued controlled use of chemical pesticides. Until biological (natural) control methods are shown to be practical on a large scale, the present ends justify the present means, provided legislation exists to ensure that such chemicals as may be toxic to man himself are never present in his food. Most countries operate on this principle. Legislation exists concerning the correct labelling, handling and application of pesticides as well as minimal tolerable levels of residues in plant material, animal, bird and fish tissues, soil, water and air. One dereliction on the part of those who legislate is the lack of defined legislation concerning the possible effects of pesticides on those microorganisms in soil and water which are of constant and invaluable assistance to mankind and which vastly outnumber those that are pathogens.

These valuable microorganisms, the non-target organisms, are subjected to herbicides, insecticides, nematocides, acaricides, plant growth stimulators or retarders, algicides, fungicides and bactericides, which are applied increasingly to soil. Of these chemicals, only the last three are aimed specifically against microorganisms which are undesirable but which are, unfortunately, intimate relatives of the non-target microorganisms. The herbicides and insecticides, although not aimed at microorganisms, may nonetheless prove detrimental to the non-target organisms by virtue of their chemical nature which in most cases is designed to interfere with a biochemical pathway or function all too frequently common to both target and non-target species.

Tests to establish the possible ill-effects of pesticides on non-target organisms are of immense importance to agricultural and non-agricultural interests. The inhibition of nitrifying bacteria could cause a decrease in the rate of nitrate formation or an increase in nitrite accumulation. Although nitrate leaches from soil (agronomically undesirable) and may increase eutrophication in lakes (undesirable ecologically and aesthetically) it is an important chemical in soil and water and essential to denitrifying bacteria, while nitrites are known to be toxic to plants and animals. Inhibition of organic matter-degrading microorganisms would affect the rate of release of nutrient materials available for plant growth; conversely their stimulation may rapidly deplete both carbon and nitrogen reserves in soils with an inherently low fertility status. Inhibition of *Rhizobium* species or mycorrhizal fungi could result in the failure of green-cropping and necessitate applications of chemical fertilizers where they need not have been applied – a costly exercise in modern agriculture.

Inhibition of the natural antagonists of plant pathogenic fungi and nematodes could lead to an increase in the incidence of plant disease and root-knot nematode infestations.

These examples serve to underline the need for a system of monitoring effects of pesticides on non-target organisms. As yet, no legislation exists, nor is it likely until such times as soil and aquatic microbiologists, soil zoologists, biochemists, geneticists and plant pathologists can agree on what parameters are important in indicating a potential hazard and, more important still, the methods to be used which are acceptable for determining the potential hazard. All of these disciplines are involved because of the complicated interrelationships that already exist within and between microorganisms and macroorganisms.

A formidable task confronts present and future generations of soil and aquatic microbiologists. Far from being an established and routine task, monitoring the effects of pesticides on non-target organisms is still in its infancy and suffers from policy disagreements which are largely based on personality differences and individual choice.

It is current practice in many cultivars to apply mixtures of pesticides in order to minimize the costs associated with treating the crop for individual pests and diseases. Annually more and more pesticides are released by chemical companies for use in agriculture and therefore the importance of an agreed system of monitoring becomes manifestly urgent.

In this chapter are outlined some methods, past and present, which are available for monitoring effects of pesticides on soil microorganisms, together with some new ideas.

II. Inherent Difficulties Associated With Test Methods

Since microorganisms coexist in soil with other species it is illogical to presume that the study of a single species in pure culture will demonstrate the probable effect of a pesticide on that organism in the soil environment.

Whereas simple chemical and physical factors that influence soil microbial activity can be easily listed, and their effects described in general terms, it is far more difficult to describe their combined effects on even a single species. It is still more difficult to grasp the effects that a change in activity of one species will have on other species. Furthermore, since chemicals can interact with inorganic and organic components of soil, and because commercial pesticides are mixtures of chemicals (active ingredient plus formulation additives), it becomes obvious that the study of pesticide effects in pure culture can never provide dependable information on the likely behaviour of a pesticide in soil. Even if the culture is rendered non-viable within 60 seconds of adding pesticide, the only permissible conclusion is that the compound is toxic to the organism in pure culture.

The advent of the bipyridylium herbicides, with their strong adsorption characteristics in the presence of clay minerals, has underlined both the futility of extrapolation from pure cultures to soil and the need for studies to be conducted in the presence of soil or in soil. Paraquat dichloride is probably a classic example. Burns and Audus (1970) demonstrated conclusively that the ions move rapidly from solution onto the surface of clay minerals contained in isolation within a dialysis membrane. Other studies (Knight and Denny, 1970) have shown that the surface-adsorbed paraquat ions are capable of progressively stronger adsorption, eventually becoming trapped within the expanded clay lattice from whence their removal can only be effected by violent physical or chemical means. Despite these facts concerning the behaviour of the herbicide in soil, or in the presence of clay, several workers have demonstrated the toxicity of paraquat to several species of soil organisms. Wallnöfer (1968) showed paraquat ions to be inhibitory to aerobic bacteria such as *Azotobacter* sp. and *Eschericia coli* in pure culture. Similarly, Kokke (1970) demonstrated that the presence of paraquat in agar media inhibited the development of certain types of soil bacteria and Szegi (1970) found that growth and activity of some micro-fungi was adversely affected by paraquat. The results obtained by these and other workers are not necessarily invalidated by the absence of soil. The effects noted represent what could happen to the various microbial species if soils of very low clay content become saturated with paraquat ions to the extent that all available clay adsorption sites are occupied and free paraquat exists in the soil solution.

Since most agronomic soils possess sufficient clay minerals to absorb the paraquat applied over a period of several hundreds of years, the results of pure culture tests cannot be considered to be an accurate assessment of what will occur in soil. Burns and Audus (1970) demonstrated that the presence of soil minerals prevented the degradation of paraquat ions by the soil yeast *Lipomyces starkeyi*, and Anderson and Drew (1976) showed that paraquat at normal and ten times normal field concentrations had no effect on numbers of *Azotobacter* spp. in soil.

The lesson is manifestly apparent. Pesticides may behave differently towards soil organisms in the absence of soil, whilst in soil the organisms may not be able to act upon pesticides as they would do in culture. Quite apart from pesticides, the effect of clay minerals on the organisms themselves has been clearly demonstrated (Stotzky, 1966; Stotzky and Rem, 1967).

It is important, however, to stress that despite the negative aspects of testing for pesticide effects and degradation in pure culture, the value of axenic culture cannot be disregarded. For example, Wallnöfer (1968) has demonstrated, in pure culture, the probable site of action of paraquat ions on *Azotobacter* sp. which implies that, should the organism and chemical compound ever occur in situations in which clay minerals are either absent, or sufficiently far removed as to permit paraquat ions to exert an effect, an inhibition of this organism may occur.

It is equally important to emphasize that currently the availability of precise and reproducible methods for analysis of a large number of interrelated biological activities leaves much to be desired, and researched, in soil microbiology. This is by no means the fault of the researcher; on the contrary, it is due to the difficult medium in which they must work. Neither does it apply to all methods of analysis for biological activity, but it does unfortunately embrace a large number of test methods as is evidenced by their high coefficients of variability and standard errors.

Thus, in attempting to evaluate effects of pesticides on non-target organisms, an anomaly is presented to the researcher, namely to have the confidence of the analyses offered by pure cultures or the uncertainty of results offered by *in situ* soil experimentation. Compromising with semi-artificial media may also compromise the degree of doubt attached to each system. The only satisfactory solution to this anomaly appears to lie in the use of both pure cultures and *in situ* soil tests. The question then is whether soil testing should precede pure culture for confirmation, or vice versa.

Another major problem in choosing a system for testing pesticide effects concerns the use of field plots or laboratory-incubated soil. With herbicides in particular, it is not possible at present to compare treated and untreated field plots since the untreated (control) plot will retain its full vegetative cover and can never act as a true control. The herbicide-killed vegetation will die *in situ*, its aerial and subterranean parts adding to the existing organic matter which serves as the carbon and nitrogen sources for microorganisms. Control plots retain their cover and the roots in particular continue to support a healthy rhizosphere population of microorganisms. Thus, valid comparisons of the microflora cannot be made. Removal of plants from control plots will not suffice even if the uprooted vegetation is left on the surface. Vegetation burnt-off from control plots is also not strictly comparable with herbicide-killed plots.

The only possible alternatives to field plots in this case are:

(a) Removal of soil from the field, sieving to remove the vegetative cover's roots and then treating the sieved soil with herbicide leaving some untreated as the control.

(b) As in (a) but treating the sieved-off vegetative cover's roots and leaves with herbicide and incorporating it into the sieved soil.

The control in this case has untreated roots and leaves incorporated into it.

(c) The use of an internationally agreed herbicide with a proven nil-effect on soil microorganisms and their activities as the control plot herbicide treatment in the field.

In the case of foliar sprays, especially those designed for use in orchards, it is extremely difficult to calculate the quantity of active ingredient of the pesticide that will reach the soil as run-off following spraying, as a result of wash-off by rain and from leaf abscisson (assuming that the leaves retain some pesticide). It is probably better to assume that all the foliar spray will reach the soil and to calculate the concentration that should be applied to soil based on the concentration that will reach the tree. Alternatively, the concentration reaching soil as run-off can be ignored and instead treated leaves can be mixed in with sieved soil and studied in the laboratory. This method, however, makes sampling of the soil for microbial populations and activity very difficult since soil in contact with leaf material may be very different from that which is not in contact.

In addition, there are the problems of moisture content, aerobiosis and temperature of the soil. Since we are attempting to judge effects of pesticides applied to field soils, how do we reconcile the age-old practice of controlled incubation conditions with the fact that, in field soil, microorganisms are subjected to continuing changes of temperature, moisture content and aeration? Perhaps in good scientific tradition it should be argued that by controlling these variables we are better able to monitor changes in population and activity. Can it not be argued that in soil a microorganism may not be inhibited or stimulated by a pesticide at a temperature 5°C lower or higher, by a moisture content 10 to 20% lower or higher and by a different ratio of air to carbon dioxide? Perhaps this is one of the contributions that pure cultures can make in verifying suspected pesticide effects *in situ*. Since the implementation of controlled incubation conditions is currently favoured, perhaps more attention should be given to the choice of temperature and moisture content in particular. Many researchers will assert that sampling from field treated plots is the best approach since it deals with *in situ* conditions. These workers will assiduously record the moisture content of the sample, ostensibly and quite rightly, to facilitate calculations on a dry weight basis; but how many will record the temperature of the soil or its gaseous composition at the same time? It probably does not matter if they do, since once the soil core is taken its gaseous composition and temperature will undergo changes anyway. This poses the question of why we should bother with field samples since these changes must affect the viability, relative numbers and activity of many of the soil's microorganisms. Anomaly chases anomaly.

One of the major differences between the microbial ecologist studying natural changes in the soil populace or the distribution of a particular species of organism, and those who monitor the effects of pesticides on non-target organisms, is that it is almost certain the latter will attribute changes in the populace or its activity, directly to the pesticide and its metabolites. In all other forms of ecology such changes will be considered due to competition for space, oxygen and nutrients or due to antibiotic effects, temperature, parasitism or predation. In most assessments of the effects of pesticides on non-target organisms the decrease in numbers of a species is seldom if ever attributed to possible stimulation of protozoa, *Bdellovibrio* sp., bacterio- or actinophage, the release of toxic organic acids as a result of a chemical reaction between pesticide and organic matter, clathrate effects or chelations. Whilst any of these secondary (or even tertiary) effects could be due to the presence of the pesticide they are not direct effects and thus may never be demonstrated in pure culture systems. Most researchers simply will not have time nor, possibly, an appropriate technique available, to ascertain whether or not the effect is secondary or tertiary.

The final hurdle in what amounts to a complex series of decisions is the analysis and interpretation of the experimental data. The method of analysis and indeed the design of the entire experiment, should be the task of a statistician who must be given to understand the complexities of soil biological research. No less will suffice.

III. Methods

... every theory, no matter how fantastic, is innocent of the crime of invalidity until proven guilty. TOM GODWIN.

The complexity of the soil microbiological habitat is reflected in the many methods that have been devised to study it. No current method can be said to be so truly direct for studying soil organisms *in situ* as to preclude all possible disruptions to the environment that surrounds and supports them. The choice of methods is therefore understandably wide and varied in the approach to a seemingly insurmountable problem.

A. ENUMERATION OF SOIL MICROORGANISMS

A pre-requisite in the study of any ecosystem is the enumeration of its component species per unit area or volume. In the case of plants or animals, numbers per unit area are fairly easily assessed without disruption of the ecosystem's integrity. The nature of soil and the size of microorganisms precludes such simplicity and inevitably leads to a disruption of their habitat in the process of enumeration. Thus, in the case of soil microorganisms, enumerating the populace can only provide information on numbers and not as yet, on their spatial relationships to each other or to the inorganic and organic components of the soil.

Whilst numbers of soil bacteria, fungi, actinomycetes and algae may not equate with activity, it is nonetheless important to establish whether changes in activity reflect a transient inhibition of the particular enzymic activity or a reduction in numbers of organisms capable of exerting that activity. It should be borne in mind, however, that all types of soil organisms have not yet been definitively assessed and that any one type of enzymic activity cannot be assigned with assurance only to those types which have been identified.

1. Dilution agar plate counts

This method is one of the oldest known to soil microbiologists. In principle, the soil organisms are serially diluted to provide a discernable number of colonies per agar plate. It is assumed that the nutrients in the medium will support the growth of cells into colonies which act as the marker for the presence of viable individuals in the soil. Factors which deprive the method of accuracy are:

(a) The agar media generally used (there are a large number of formulations to choose from) only permit the development of the more aerobic heterotrophic bacteria, certain yeasts and fungi and a limited number of types of actinomycetes. It is generally agreed that any three media for the isolation of bacteria, fungi and actinomycetes respectively, only permit the development of 1 to 10% of the numbers determined by direct microscopy. Nikitin and Makarieva (1970) and Aristovskaya and Zavarzin (1971) have suggested that the estimates given by dilution agar plate counts may be in the region of only 0·001% to 0·01% of the total population as determined by microscopy. The inability of laboratory media to support growth of more members of the microbial populace has been ascribed to nutritional fastidiousness, dormancy, inhibitions of some organisms by others, and to profound differences in physico-chemical properties between the laboratory and the soil environments, such as the concentrations of oxygen and carbon dioxide, which are known to influence the growth of some fungi in particular (Stotzky and Goos, 1965). It is also known that some fungi are sensitive to the presence of volatile metabolites produced by bacteria in the same agar plate (Moore-Landecker and Stotzky, 1972).

(b) The procedures for sampling, diluting and agitating the diluent, to separate organisms from each other and the soil's particulate matter, are many and varied. Agitation may take the form of hand or mechanical shaking, maceration, or ultrasonication at low amplitudes, all with or without "wetting" agents such as tetrasodium pyrophosphate. The time taken to resuspend the soil and the temperature at which the agar medium is held prior to receiving aliquots of the dilutions, are added variables. All of these factors can disrupt the original ratio of species. Since the majority of fungi developing on agar plates are those that produce abundant spores (Warcup, 1960), enumeration of the fungal and probably also the actinomycete component of the soil microflora is very misleading. Soil washing (Williams *et al.*, 1965) to remove spores prior to agar plating has not proved entirely successful, and micromanipulation to isolate hyphal fragments (Warcup, 1960) is an involved and tedious alternative to conventional plating (Stotzky, 1972).

(c) The presence of pesticide material in the soil, or added to the agar medium, presents different problems in the estimation of numbers. In the first instance, the diluent used to obtain the required dilutions of soil organisms may remove the effect of the pesticide. In the second case, although the addition of untreated soil dilutions to agar media containing pesticide may simulate the addition of pesticide to soil, the absence of clay mineral and other soil particles could exaggerate the effects on the microorganisms.

(d) The use of wetting agents and ultrasonication to disperse microorganisms in pesticide-treated soil may result in chemical interactions between wetting agent and pesticide. Clark and Wright (1970) found that the herbicide propham could be dissolved to a small extent by ultrasonication in an aqueous medium. Such a process might result in an artificially high effect of the pesticide on soil organisms plated in agar media.

(e) The usual periods of incubation of dilution agar plates and the carbon content of most of the media used are both conducive to the more rapidly growing types of soil microorganisms. Casida (1965) showed the existence of a coccoid organism in a soil dilution of 10^{-9}, indicating its presence in great numbers. The organism was fastidious and slow growing and easily swamped by the faster growing species.

An organism, common to many soil types, but which prefers carbon-free media was reported by Orenski *et al.* (1966). Nikitin and Makar'yeva (1970) and Aristovskaya (1973) have demonstrated the existence in soil of unusual shaped organisms which do not grow on conventional laboratory media. Some of these organisms are thought to be present in large numbers and some have been shown to possess simple hold-fasts which may prevent their release from soil by conventional methods of mixing and serial dilution.

Despite the stated limitations of the dilution agar plate technique it remains as a useful tool in the study of effects of pesticides on non-target organisms. Using the technique, it is possible to determine whether or not a species shift occurs as a result of pesticide treatment, provided the organisms concerned in the shift are capable of growth on the media used. In addition, by the use of selective media or media with different pH values, specific antibiotics and other inhibitors, it is possible to study the distribution of some individual species of soil organisms.

2. Extinction dilution counts

This method for enumerating organisms is more appropriate to unicellular organisms than to mycelial forms and is usually confined to highly selective media.

In principle the method is useful in so far as numbers of any substrate-specific organisms can be estimated on the basis of mathematical tables of probability. However, the method suffers from the assumption that the organisms will be logarithmically distributed in the 10^{-1} to 10^{-10} serial dilutions used for inoculating each of ten replicate tubes of medium per dilution.

The method differs from the dilution agar plate method in that all ten dilutions are used to inoculate the medium, but as with dilution agar plates, consideration must be given to the effects of dilution on the activity of the pesticide or to the likely exaggerated effects of pesticide in the medium on very small numbers of cells at the higher dilutions.

The method has a notoriously high coefficient of variation and has not been extensively used in soil microbiology for this reason; it is less accurate than the dilution agar plate method (Taylor, 1962). The extinction dilution method generally requires more laboratory glassware and incubation space and can be very tedious if confirmation of the presence of a particular organism has to be made by microscopy.

Possible uses for the method are the determination of the lowest dilution of a pesticide that will begin to affect a particular organism's activity or morphology, the highest dilution required to remove the inhibitory effects of a pesticide adsorbed onto microbial cell walls, or the effects of a pesticide on the sporulation of mycelial forms of soil microorganisms. In the latter possibility, Anderson (unpublished data) found that some substituted pyrimidine fungicides and insecticides increased the sporulation of actinomycetes in soil. The extinction dilution method may permit a quantitative estimate to be made of the extent of such increased sporulation.

3. *Direct microscopy of soil and soil suspensions*

The spatial relationship of soil microorganisms to each other and to micro-invertebrates, plant roots and mineral or organic particles, is an important aspect of soil microbial ecology but present methodology is largely inadequate for its elucidation. Electron microscopes, both scanning and transmission, have given some information on spatial relationships and morphology of organisms in soil (Jenny and Grossenbacher, 1963; Nikitin, 1964; Gray, 1967). Uncertainties with the method are the small sample sizes used (which could be overcome by using many more samples), evacuation and shadowing which are thought to result in artefacts, and difficulties in identifying objects on the electron-micrographs. There can be no doubts that when the difficulties involved with electron microscopy are resolved it will provide a great deal of information on spatial relationships and possibly the effects of pesticides on morphological changes in the soil microorganisms. Furthermore, it is difficult at present to envisage any other technique that will provide convincing information on spatial relationships because the discontinuity of soil, its opacity and the difficulties of obtaining adequate focal depths are the major limitations in developing a technique based on conventional light microscopy.

4. *Examination of soil surfaces and sections*

Kubiena (Waksman, 1952) described a technique for direct observation of organisms in undisturbed soil surfaces or profiles by which incident-type illuminators permitted observations of the larger soil organisms. The use of micromanipulators in conjunction with this method permits isolation principally of fungi but the lack of in-depth focus and high magnifications preclude the method as a means of studying bacterial associations with organic matter or soil particles, other than those immediately on the surface (Waksman, 1952).

Casida (1969) developed a technique using an inverted stage microscope and unamended soil contained in glass vessels for the study of micro-colonies. Casida was able to follow the growth and movement of soil organisms by sequential photography. The method could be of use in studying pesticide-induced changes in the micro-colonization of particles adjoining the inner surface of the glass vessel but lack of focal depth within the soil sample is a major limitation of the method.

Some workers have attempted to show spatial relationships and distribution of microorganisms by means of soil sections. The principle of sectioning involves use of soil cores (Alexander and Jackson, 1955; Burges and Nicholas, 1961; Nicholas *et al.*, 1965) or leaf litter (Minderman, 1956) which are deep frozen prior to impregnation and sectioning. Soil cores frozen in liquid nitrogen and then freeze-dried can be fixed and stained prior to impregnating with resin, but staining and fixing has been found to result in unwanted disturbances of the soil and poorer preparations (Burges and Nicholas, 1961). Impregnation with resin immediately after freeze-drying is followed by a lengthy period of grinding to reduce thickness and produce a flat surface, followed by polishing with carborundum. Sections of 50 μm thickness were found by Burges and Nicholas (1961) to be convenient for studying hyphae.

Increasing gelatin concentrations can be used for impregnating soil or litter samples followed by fixing in 10% formalin and hardening in 80–90% methyl alcohol. The procedure is lengthy, requiring 10 to 14 days before sections of *c*. 8–10 μm can be microtomed from the cores or blocks.

Neither of these variations of the technique permit adequate depth of focus at high magnifications nor the observation of bacteria or other micro-propagules. Only fungal hyphae can be observed and even then it is not possible to observe hyphal growth and distribution in the same micro-habitat over a period of time, a factor which mediates against its use as a method for monitoring effects of pesticides on non-target organisms.

5. *Examination by contact or impression slides*

Conn, Rossi and Cholodny (Waksman, 1952) developed similar methods in which uncoated glass slides are brought into contact with the soil surface or inserted into the soil and allowed to remain for various periods of time. After removal, soil particles and organisms on one side of the slide are fixed, stained and examined. Whilst a considerable amount of information has accrued from the earlier use of the contact slide principle (Waksman, 1952) it has also been criticized in that the counts obtained represent that of an area of soil, in most cases disturbed, rather than from a volume of soil. Furthermore, the condensation of soil moisture at the interface of soil and slide permits the rapid growth of some species, among them predaceous and sessile forms, and does not represent a true picture of the microbial population at that point in the soil.

Rossi *et al.* (1936) varied the procedure by using agar-coated or uncoated glass slides which were pressed firmly against the soil surface, removed, fixed and stained. Here again the method only reflects the area of soil in contact with the slide and in sandy soils in particular does not retain much material. A more recent variation involves the use of nitro-cellulose in amyl acetate containing 5% castor oil to coat the glass slide. This adhesive is applied to the slide just before it is pressed against the soil surface. The adhesive plus soil particles and organisms is allowed to dry, then stained and examined. The method can be repeated on the same area, or at a slightly greater depth within the same area, in order to obtain a three-dimensional picture of the organisms in the soil profile (Brown, 1958). This method may be of some value since it is possible that, having applied a pesticide to soil, successive films taken at increasing depths or distances from the point of application, could reveal changes in numbers, types and morphology of organisms, as the zone of influence of the pesticide is reached or surpassed. A disadvantage of the method is that organisms within organic matter and larger soil aggregates may not be observed.

Lingappa and Lockwood (1963) devised a method suitable for studying microbial changes in soil under laboratory conditions, employing the impression slide principle. In this method, soil is spread uniformly in a suitable vessel. Aqueous phenolic rose bengal solution is applied to the soil to stain the microorganisms, followed by a few drops of collodion which are applied to the soil and allowed to spread. After drying, the collodion film is peeled off, mounted in mineral oil and examined. Here again, successive films of the same surface can be taken at intervals. Questionable points on this technique are whether the phenolic rose bengal will adequately differentiate living and dead organisms, and will the stain and pesticide interact thereby confusing possible true pesticide effects.

6. *Examination of soil particles, suspensions and smears*

Waksman (1952) described a method in which soil is suspended in 0·015% gelatin solution as fixative. A few drops of the suspension are spread and allowed to dry on glass slides followed by staining with an acid dye and examination under high magnification. This method is simple, rapid and quantitative and lends itself to frequent examination of pesticide-treated soil. Only the method of fixing and staining have been altered in more recent variations of this basic technique.

Another simple, effective, method consists of mixing a known amount of indigotin with a known volume of soil and establishing the ratio of erythrosin-stained cells to indigotin particles (Thornton and Gray, 1934). For laboratories that do not possess sophisticated ultraviolet light microscopes, the method has much to recommend it and suffers from no more or less criticism than more recent methods.

Cholodny (1934) introduced the "soil chamber" method in which fine particles of soil are placed on a moist slide, incubated in a moist chamber and examined periodically for any organisms growing out of the particles. With or without the presence of a cover slip, the variation in size of the soil particles imposes a serious limitation on the depth of focus, preventing observations of the smaller particles between larger ones.

Gray et al. (1968) described a variation in which the soil particles are pressed into a film of agar, slightly deeper than the largest particles, by means of a cover slip which can be cemented into place. The methods of Cholodny (1934) and Gray et al. (1968) suffer from the lack of staining of organisms which, being on the surface of soil particles, are extremely difficult to discern. To overcome this Parkinson et al. (1971) suggested staining by means of a glass cylinder covered at one end by nylon mesh which acts to retain the soil particles when the cylinder is placed inside a staining vessel of slightly larger diameter. This apparatus was designed by Hill and Gray (1967) for the fluorescent-antibody staining of soil bacteria. As cautioned by Parkinson et al. (1971), the process of staining organisms on soil particles may result in the loss of microorganisms by dislodging or displacement in the staining solution.

The technique of Jones and Mollison (1948) overcomes problems of variable particle size and staining of organisms on the surface of particles. The original methodology calls for grinding of soil in a pestle and mortar through several changes of sterile water to liberate the organisms, but ultrasonication and dilution are at least as efficient and somewhat quicker. The resulting suspension is added to molten agar at 50°–60°C, thoroughly mixed and allowed to settle for 5–10 seconds. Small aliquots of the soil–agar suspension are introduced under the cover slip of a haemocytometer of known depth. After gelling, the agar film is floated free in sterile water and onto a glass slide. Once the film has dried on the slide it can be stained with phenolic aniline blue (Jones and Mollison, 1948) or examined by phase contrast microscopy (Parkinson et al., 1971).

Numbers of bacteria and lengths of fungal hyphae can be recorded by examining the mounted preparations under oil immersion. The phenolic aniline blue stain does not allow distinction between living and dead organisms and furthermore the relationship between microorganisms and mineral or organic matter particles is difficult to discern. However, the method does allow accurate, quantitative assessments of the amount of total microbial biomass in soil samples.

Soil smears, as an alternative to soil-agar films, are quicker to prepare and can be quantified by spreading a known volume of the soil suspension (prepared as for the Jones and Mollison method) over a 1 cm^2 area of a grease-free slide. After heat fixing over a steam generator, the smears can be stained with a variety of stains, mounted and examined for bacteria and hyphal strands. Thoroughly de-greased slides are essential for the technique to provide a uniform distribution of particles and microorganisms. This method is subject to the same criticisms as the soil-agar film method but is considerably quicker, especially when many soil samples are to be examined. Soil smears also have a much shallower focal depth than soil–agar films, which minimizes the extent of in-depth focusing required during counting.

Strugger and Hilbrich (1942) introduced the practical use of fluorescence microscopy. Rouschal and Strugger (1943) and Strugger (1948) described a technique for the fluorescent microscope examination of microorganisms in the soil *in vivo* using soil suspensions stained with acridine orange. This stain is supposed to differentiate living (green fluorescing) from dead (red fluorescing) cells but the differentiation is dependent upon cell structure and exact concentration of dye.

Acridine orange tends to stain both soil particles and organic matter and the resulting background of fluorescence makes observation of green or orange cells more difficult and tedious for the microscopist. Zvyagintsev (1963) noted that substances such as sodium pyrophosphate, Ziels' carbol fuchsin or diazine green had a quenching effect on the excess acridine orange in soil smears and improved the clarity and brilliance of fluorescing cells.

Babiuk and Paul (1970) described a method which is more suitable for the study of soil smears. The method combines the quantitative accuracy and rapid preparation of the soil smear with a quick staining method using freshly prepared fluorescein isothiocyanate. After staining and mounting, the 1 cm^2 area of soil smear is examined under incident ultraviolet lighting. Soil organisms, by virtue of a protein layer around the cell, fluoresce an apple green colour. The major criticism of the method is that the entire preparation tends to fluoresce in shades of green. The fact that soil particles and organic matter may also possess a protein layer (Esterman *et al.*, 1959) results in fluorescence of soil organic particles which may confuse counting of the smaller soil organisms in particular. Differentiation of cells on or in organic fractions thus becomes difficult with this method.

Darken (1962) described the uptake and fluorescence of the disodium salt of 4,4′-bis[4-anilino-6-bis (2-hydroxyethyl)amino-*s*-triazin-2-ylamino]2,2′-stilbene disulphonic acid in soil fungi and bacteria. Hyphal strands in particular continued to fluoresce green during the continued growth and elongation of the strands. The major disadvantage of this and many other fluorescent stains is that the emission wavelength is frequently too close to the excitation wavelength and observation of the fluorescent cells is consequently more difficult.

The wider separation of emission and excitation energies of europium thenoyltrifluoroacetonate was noted by Scaff *et al.* (1969) who used this stain to differentiate between living (red fluorescent) and dead or dying (pink to no fluorescence) bacterial cells.

By combining the stains of Darken (1962) and Scaff *et al.* (1969) and using the technique of Jones and Mollison (1948) for soil–agar films, Anderson and Westmoreland (1971) showed that it was possible to differentiate living (red fluorescent) from dead (green fluorescent) hyphal strands. Counts of microorganisms obtained were initially about the same as those from soil–agar films stained with phenolic aniline blue. The intensity of staining and the ease of observation of organisms in soil–agar films was considerably improved by increasing the concentration of europium thenoyltrifluoroacetonate and using incident ultraviolet illumination (Anderson and Slinger, 1975a).

The use of the two fluorescent stains with soil smears, prepared by the method of Babiuk and Paul (1970), or soil–agar films, enabled photomicrographic recording of the requisite number of microscope fields for statistical evaluation. After processing, the photographic films can be projected and the total number of red and green fluorescing organisms recorded. This method enables observation of fluorescent cells within organic matter or on the surface of soil particles in either soil–agar films or soil smears.

Disadvantages of the use of soil–agar films with the combined fluorescent stains are that a long staining period is required to enable the europium salt to chelate with the nucleic acids and other vital constituents of viable cells. Furthermore, the depth of the agar film ($c.\,20\,\mu$m) imposes the need for in-depth focusing during counting. The depth of film produced by most haemocytometers is such that photomicrography cannot record the presence of fluorescing cells at depth within the film. To overcome the latter disadvantage, special haemocytometers of considerably reduced sample chamber depth were used for producing soil–agar films of $c.$ 10 μm. The use of such films, or the use of soil smears which are quicker to prepare and require a greatly reduced staining time, enabled more accurate photomicrographic recording of the fluorescent cells (Anderson and Slinger, 1975b).

Any method of fluorescent staining which enhances the differentiation of living and dead cells and allows observation of the colonization of organic matter and soil particles is to be preferred over non-fluorescent stains. The method should also be quantitative and rapid to enable many samples of soil to be processed. Such methods are particularly useful in studying effects of pesticides on non-target organisms, since the ratio of living to dead cells at regular sampling intervals after treatment will indicate the toxic nature of the pesticide.

7. Other methods of direct microscopic examination

The methods outlined above do not allow identification of the stained microbiomass members nor is it possible to differentiate between viable active cells and those which, although technically viable, have an impaired enzymic function due to inhibition by a pesticide.

By means of fluorescent antisera it is possible to identify specific microbes in pesticide-treated and untreated soils and to follow increases or decreases in the numerical frequency of the organisms with time. The method involves the production of antibodies, usually by rabbits injected at frequent intervals with the particular organism (attenuated or killed). The precipitated γ-globulin fraction from blood samples is dialysed, buffered, and adjusted to the required protein content which is then conjugated with crystalline fluorescein isothiocyanate. The conjugated protein-dye is molecular seived to remove uncombined dye and protein and collected as individual fractions. The appropriate specific fraction of the conjugate is used for detecting the presence in soil of the particular microorganisms which originally served as antigen.

Schmidt and Bankole (1962) used the technique for detecting *Aspergillus flavus* in soil. Species of nematode-trapping fungi were detected in soil by Eren and Pramer (1966) while Hill and Gray (1967) applied the technique to the detection of *Bacillus subtilis*. Fluorescent antisera have been produced for several species of organisms and the method is now potentially valuable for microbial ecology studies. Like all methods, it is not without criticism. In this case the method proves problematic when applied to soils with a high clay content but Bohlool and Schmidt (1968) described a technique which helps to overcome much of the non-specific adsorption of dye by the clay minerals. The method involves pre-treatment of the soil with a gelatin–rhodamine conjugate which prevents non-specific staining and acts as a counter-stain for the background.

Another criticism of the fluorescent antisera method is that relatively few organisms can be studied because it is not currently possible to produce specific antisera for all soil microorganisms. However, it is possible that in time antisera for a great number of individual species will come into existence and be held in central collections.

The contact slide method discussed earlier is more accurately described as the buried slide or immersion slide method. Many variations of immersion methods have come into sporadic use in soil microbiology and have served to provide some information on types of organisms present, their distribution within the soil profile and their rate of colonization of the immersed object. Examples of immersion methods include media-filled tubes (Chesters, 1940), nylon mesh (Waid and Woodman, 1957), immersion plates (Thornton, 1952) and various baits. All of these are intended to attract or trap micro-organisms which may then be evaluated by direct microscopy or by isolation onto specific agar media.

Immersion methods are highly selective and can only be used to determine the presence of specific types of soil organisms or their colonization ability. The usually abnormally high nutrient status of such media may result in rapid overgrowth by organisms with a short generation time or a constitutive enzyme system capable of utilizing the media in advance of the desired forms or slower growing forms. Furthermore, the use of tubes or plates will probably create an altered gaseous regime thereby imposing a further selective or restrictive factor. Large objects such as tubes or plates will necessitate the removal and replacement of soil thereby interfering with the natural spatial relationships of the organisms present. Whatever the object, only a single observation per immersion will be possible because of the disturbance to the soil upon its removal.

A modification of the immersion method that has elicited much interest is the capillary, or pedoscope, technique developed by Aristovskaya and Parinkina (1961) and Perfil'ev and Gabe (1969). Small glass capillaries (pedoscopes) with square or oblong cross-section are buried in soil with the capillaries arranged in the direction of the moisture flow. The inside of the capillaries may contain a coating of nutrients and can be buried in this form or partially filled with water. Generally an equilibration of from one to three months is allowed before disinterring the capillaries, which can then be examined microscopically *per se* or after staining (Aristovskaya, 1973). It is claimed that the capillaries can be re-buried after examination but the possibility of some disturbance of the soil must be considered.

The simplicity of the capillary technique lends itself to the study, for example, of the effects of pesticidal seed dressings. Pedoscopes placed at increasing distances from a seed, in one or several directions, could be of value in determining the effectiveness of the pesticide on plant pathogens in the soil, or in determining the effects of the pesticide on non-target organisms some of which will ultimately be responsible for forming the seedling's rhizosphere microbiomass. It may even be possible to arrange pedoscopes around seeds in a manner that will facilitate smaller roots entering the capillaries or passing very close to them, thus providing some information on the rhizosphere microbes themselves.

8. *Enumeration of soil algae*

The enumeration of algae is all too frequently overlooked in soil microbiological investigations yet the contribution of algae to soil fertility is unmistakable. It is known that algae contribute organic matter which enable other organisms to colonize arid regions or volcanic ash of recent origin. They also aid in preventing soil erosion by forming thin crusts which bind soil particles, prevent leaching and provide a reserve supply of essential nutrients for plant growth (Allison *et al.*, 1937).

Tchan (1952) described a method of enumerating algae using a haemocytometer and a standard light microscope modified to provide the required wavelength for chlorophyll excitation. Anderson (unpublished data) modified the Tchan method by using a high pressure mercury vapour lamp as source of ultraviolet light in conjunction with incident light optics. Soil samples (*c.* 1 g) are incubated in glass or plastic containers under a battery of fluorescent lights. Since the algal population of most agricultural soils is low, being approximately $2 \cdot 7 \times 10^4$ cells g^{-1} in non-rhizosphere soil (Rouatt *et al.*, 1960), the soil may be augmented with a known number of cells of one or several species of algae and incubated for one to two weeks. At intervals during this period, 1 g samples of soil are removed and shaken vigorously or sonicated at low frequency to disperse the cell aggregates into 9 ml of water. Aliquots of the suspension are introduced into the 0·05 mm deep sample chamber of specially manufactured, Thoma ruled, haemocytometers (Hawksley, "Cristalite", Lansing, Sussex, UK). The counting grid is viewed under incident ultraviolet light using BG12 or BG38 excitation filters and a yellow gelatin (Kodak No. 12) barrier filter. The red fluorescing algal cells are easily visible in the shallower depth of soil suspension in the modified haemocytometer. The remaining soil in containers under fluorescent lamps can be treated with pesticides of different concentrations and samples examined periodically after treatment to determine the survival potential of algae in the presence of pesticides compared to that in untreated soil. Twenty five known species of soil algae have been examined in this way with no indications of weak or uncertain fluorescence. Prolonged exposure of a single field to the incident ultraviolet light does, however, result in "bleaching" of the cells with a loss of red fluorescence. Consequently enumerations must be conducted rapidly.

B. ESTIMATIONS OF THE SOIL MICROBIAL POPULATION
BY CHEMICAL ANALYSES

All living cells possess certain biochemicals in common but not necessarily in the same quantities. Thus, while it is true that a thorough extraction of soil for a certain cellular chemical can indicate the relative density of microorganisms in soil, it can never provide data on exact numbers or types of microorganisms present.

Some of the more common biochemicals to be found in all microorganisms are: adenosine triphosphate (ATP), nicotinamide-adenine dinucleotide (NAD), nicotinamide-adenine dinucleotide phosphate (NADP), flavin-adenine dinucleotide (FAD), flavin mononucleotide (FMN), deoxyribonucleic acid (DNA), ribonucleic acid (RNA) and cytochromes of various kinds. Since these compounds will also be found in insects, plant and animal tissue, it is obvious that extraction of soil for these compounds as an indication of microbial biomass only is not possible, especially under conditions where plant material has been freshly added to soil or exists as a natural vegetative cover.

Whilst much is known of the melting point and base composition of nucleic acids the separation of nucleic material extracted from soil into groups of microbial, plant or animal origin remains a formidable task especially if hydrolysis of the nucleic material has occurred to different extents within the extract.

1. *ATP determination*

Despite the known and presumed difficulties, attempts have been made to estimate the soil's microbial biomass by extraction of ATP. Doxtader (1969) described a method in which microbial cells are lysed, their ATP partitioned in buffered alcohol and then introduced into a luciferin–luciferase solution contained in a scintillation vial. The emission of a quantum of light from the reaction of enzyme-substrate with ATP was measured in a scintillation counter. Ausmus (1973) described a more elaborate method for soil which involved the use of a computer-linked photomultiplier tube mounted above a centrifugal rotor assembly. Sample (ATP) and enzyme–substrate, placed in adjoining chambers in the rotor, are forced into mixing chambers under centrifugal force and the light emission repeatedly measured by the photomultiplier tube. The emission signals are integrated by computer into μg of ATP per μl of sample on the basis of supplied standards.

Anderson and Davies (1973) described a method based on the use of the Luminescence Biometer (du Pont de Nemours) which consists of a light-proof sample chamber, a photomultiplier tube and an electronic recording circuit. The extracted ATP is rapidly injected through a rubber septum into the enzyme-substrate mixture contained in glass vials within the sample chamber. The peak emission of light from the reaction is automatically recorded and presented either as a trace on a strip-recorder or as a digital display.

Some difference of opinion exists concerning the most efficient and reliable method of lysing microbial cells for the extraction of ATP. Cold sulphuric acid gives good extraction (Ausmus, 1971) but must be suspect since at acid pH values ATP may rapidly hydrolyse. It is claimed, however, that cold sulphuric acid extraction followed by rapid neutralization and filtration through a cation exchange resin has no adverse affect on the ATP molecule. Butanol extraction is consistent and safer but less efficient and involves more steps in the extraction procedure each of which should, ideally, be conducted at temperatures of $c.$ 5°–8° C.

In addition to the problems of extraction, ATP is prone to adsorption on clay minerals. Anderson and Davies (1973) found that oven-sterilized soils with a high clay content retained significant amounts of pure ATP and that dilution of the soil was required to reduce the clay content and to overcome adsorption. It was found necessary when estimating microbial ATP in soil to screen soil suspensions through 20 μm nylon mesh, to remove macroorganisms and plant debris, and to use oven-sterilized controls to which were added known amounts of ATP as a means of measuring the potential adsorption of the soil in question. Ideally, ATP recovery values should be obtained from sterile controls for each soil investigated and ATP values from non-sterile soil samples adjusted accordingly.

Ausmus (1973) assigned "biomass ratios" for several species of algae, bacteria, fungi and actinomycetes and concluded that ATP concentrations were greater when cells were in logarithmic growth phase than in lag or stationary phases. A knowledge of the relative ATP content of cells and also the cell concentrations at various stages of their growth cycle is of value in subsequent routine studies on the growth of an organism in pure culture. In soil, however, ATP values are of little use due to the extreme variability, both in size and metabolic state, of the members of the soil's microbial populace, unless they can be separated into specific groups or ratios of one type to other types. Greaves *et al.* (1973) investigated the ATP content of the bacterial biomass of a basin peat and concluded that the assigning of an ATP content to individual cells in order to calculate numbers of bacteria per gram of soil is not practical. Lee *et al.* (1971) suggested that the ATP content of soil or water represents the "energy flux" of the biomass at the time of sampling.

Thus ATP measurements *per se* are currently of no value in estimating the size of the biomass in any environment since it is known that activity (or energy potential) does not necessarily equate with numbers. It remains as a further useful gauge of changes occurring within soil in response to an applied parameter such as the addition of pesticides. Considerably more research is required on ATP extractions and their significance, before this method attains wider acceptance for estimating the soil's microbial activity.

2. *Photosynthetic pigments*

Algae and other photosynthetic microorganisms may be estimated in soil or water by extraction and quantification of photosynthetic pigments. Tchan (1959) used organic solvents to extract soil for chlorophyll and related the concentration, measured spectrophotometrically, to the abundance of algae. El-Din Sharabi and Pramer (1973) compared extinction dilution (MPN), direct counts by fluorescence microscopy, and a novel method which included partial purification of chlorophyll extracted from soil and its estimation by means of spectrophotofluorometric analysis. Of the three methods, the latter was by far the most accurate, the most reproducible and the least time-consuming. It is conceivable that this approach could be applied to other biochemicals of common cellular occurrence as a means of estimating the microbial biomass of soil.

3. *Other microbial components and products*

Since muramic acid exists in the cell walls of both Gram-positive and Gram-negative bacteria, as well as in their spores, Millar and Casida (1970) used the release of D-lactic acid (the hydrolysis product of muramic acid) from alkaline hydrolysis as a measure of the numbers of bacteria and bacterial spores in soil. Average levels of muramic acid in pure cultures of bacteria were found by these workers to be 4×10^{-3} μg per million Gram-positive bacteria, 5×10^{-4} μg per million Gram-negative bacteria and $6 \cdot 6 \times 10^{-2}$ μg per million aerobic bacterial spores. However, levels in soil were found to be much higher than could be accounted for by these results, suggesting that there are many more muramic acid-containing organisms in soil, or that dead cells contributed to the figures obtained. The inability to distinguish between muramic acid of viable and dead cells prevents the adoption of this chemical method for estimating the bacterial biomass of pesticide-treated soil unless a method of isotope labelling of natural muramic acid, for a defined period, can be devised. The amount of labelled D-lactic acid would then be a fraction of the total obtained from all the muramic acid in soil and would relate to the size of the bacterial population at the time of labelling.

Since partial or complete sterilization of soil renders a small fraction of the organic matter decomposible, it has been assumed that this fraction is largely composed of microbial organic matter. Jenkinson (1966) suggested that the size of the soil microbial biomass can be "roughly estimated" from the size of the flush of carbon dioxide liberated by soil organisms re-inoculated into the sterilized soil. The several assumptions made in this chemical evaluation method make it no more or less than is claimed – a rough measure of the soil's microbial biomass.

Electron micro-probe analysis (Todd *et al.*, 1973) has shown that bacteria and fungi inhabiting mixed deciduous litter are able to concentrate essential plant nutrients. The method combines scanning electron microscopy with X-ray diffraction spectrophotometry and hence, in addition to a visual display on a cathode-ray tube, a second image is produced quantitatively to the X-ray emission which enables estimates to be made of the calcium, magnesium and potassium content of the microbiomass. This advanced chemical analysis method for estimating biomass is as yet insufficiently tested and may suffer from the same drawbacks as electron microscopy in that only small samples are used, necessitating many preparations in order to obtain reliable data.

C. METHODS FOR MEASURING GROWTH RATES AND DISTRIBUTION OF INDIVIDUAL MICROBIAL SPECIES IN SOIL

Another important aspect of soil microbial ecology studies in relation to pesticides is the measurement of rates of growth and the distribution of individual species of non-target organisms. Such measurements are difficult to make because of the very nature of soil. Ideally, the method used should permit the recovery of marker species over a given time span with the minimum disturbance of soil, simulate soil conditions, provide precise data on the rate of growth of the organism under study and be rapid, reproducible and flexible in its application.

In one method (McLennan, 1928), a known number of fungal spores is added to sterile soil in flasks held at 9°C. Agar plates are inoculated with spores of the fungus and incubated at 9°C, or at a temperature more favourable for growth. After seven days, a dilution agar plate count is made to determine numbers of spores in the original soil–spore mixture or the number of colonies produced as a result of growth of the fungus in the soil. This result is compared to the growth rate of the fungus on agar. While the method is not without criticism it could be used to determine survival of spores in pesticide-treated sterile soil.

Blair (1945) placed slides vertically and radially in soil contained in pots and a disc of fungus next to each slide at a particular marked point on the slide. The slides were covered with soil and at intervals some removed, fixed and stained. The extent of hyphal growth on the slide was measured in relation to the starting point.

A modification by Chinn (1953) used slides which are dipped into a liquid agar suspension containing fungal spores (at $c.$ 50°C). After the film of agar has set, the slides are placed in soil adjusted to 50% of its moisture holding capacity, and allowed to incubate for various intervals. Slides removed from the soil are cleaned on one side and then stained with cotton blue in lactophenol. Spores in the stained agar are examined for germination, elongation and lysis.

Both of these methods are adaptable to pesticide-treated soil but the question of whether permeation of pesticide into the agar film does or does not occur would have to be established by prior experimentation. The contribution of the agar to rapid growth of the fungus could mask any inhibitory effect and possibly invalidate the technique.

Various modifications of the glass tube method of Fawcett (1925) have been used in growth rate studies. Some of these modifications, such as that by Evans (1955), used soil in place of agar media. A glass tube with several side arms is filled with sterile or non-sterile soil and inoculated at one end with the test fungus. Samples are taken at intervals through the side arms and examined for the fungus, thus giving an indication of its rate of growth with time. The method could be used in comparing rates of growth through untreated and pesticide-treated soils.

None of the above methods meet all the requirements stated for a convincing demonstration of growth rates in soil. However, the replica plating method of Stotzky (1965a) does provide the researcher with more control of these requirements. This method involves the precise placement of inocula and possibly pesticides (or other amendments) at specific points in soil contained in Petri dishes. A replicator, consisting of wire needles protruding from an acrylic plastic disc in a pattern which matches exactly the holes in a template, also of acrylic plastic, are used for precise alignment of the inocula onto the soil in the Petri dishes. Periodically the replicator and template are used for transferring organisms from their allocated positions in the soil plates to appropriate agar media in Petri dishes. Permanent marks on the replicator, template, soil and agar plates ensure that the geometric integrity of the pattern of inoculations is maintained and hence the spread and survival of organisms with time can be plotted. The system is simple and ingenious and is especially suitable for studying the growth rates and survival of organisms (not only fungi as with other methods) in pesticide-treated soils. A valid criticism of the method is that, of necessity, disturbed soil is used and, furthermore, it is sterilized soil.

It may be possible to modify the replica plate method by using a large diameter soil corer that would provide broad, flat cores of relatively undisturbed soil which could be immediately placed in Petri dishes (or specially designed vessels to permit deeper cores to be taken). The Petri dishes or vessels could then be used *per se* or subjected to γ-radiation sterilization in a moist state to retain the integrity of the core. The coring of soil to different profile depths in the same sample hole could also be of value in determining the ability of the inocula to survive in the different profiles, with or without pesticides. It is possible to use the replica plate method in conjunction with fluorescent antibody staining of isolates of the original inocula. The use of ^{14}C-labelled organisms or specific genetic markers such as auxotrophs or lysogenic strains is also possible (Stotzky, 1972).

Several methods have been described which permit the study of the ability of soil organisms to colonize certain substrates. Garrett (1956) defined this "competitive saprophytic ability" as the summation of physiological characteristics that contribute to success in competitive colonization of dead organic substrates. While this concept has been applied most frequently to comparing saprophytic behaviour of root-infecting fungi it can also be applied to obligate saprophytes and could be of great value in determining the effects of pesticides, especially herbicides, on the ability of soil organisms to colonize treated plant material.

The highly important interrelationship between pathogenic organisms and soil organisms which parasitize them could conceivably be investigated utilizing the principle of Stotzky's replica plating method or the methods of Butler (1953), Wastie (1961) or Garrett (1963). The principle in these latter cases involves mixing inocula of a particular organism with sieved soil in various ratios. The particular organic substrate whose colonization is to be studied is introduced into each of the soil–inoculum mixtures. Possibly, pesticide-treated wheat straw or pathogen-infected portions of plant material with and without various concentrations of pesticide could be used. After incubation for a suitable period of time, the ability of the test organism to colonize the substrate in competition with other organisms, or in the presence of pesticides, is assessed in a number of ways. Pieces of substrate are removed from the soil–inoculum mixture, surface-sterilized, washed in sterile water and plated on appropriate agar medium to determine the extent of invasion of the substrate fragments by the test organism. Alternatively, seeds of a plant which can be parasitized by the test organism, or the pathogen on plant material which was incubated in the soil–inoculum mixture, are planted and the degree of infection resulting from each soil–inoculum mixture assessed.

Direct microscopic examination of the substrate incubated in the soil–inoculum mixtures is also possible and competitive ability could be quantified by the use of fluorescent antibodies against the pathogen in question.

D. METHODS FOR ASSESSING ACTIVITY OF THE SOIL MICROBIAL BIOMASS

Jensen (1939) presented evidence that, at low temperatures, an inverse relationship exists between numbers of soil microorganisms and their activity. Despite this, and despite the teachings of our mentors that numbers do not necessarily equate with activity, there persists an either conscious or subconscious assumption that a direct correlation exists between numbers and activity.

In terms of microbiological investigations of soils with different temperature, pH, moisture, tillage or organic matter conditions, the assertion that numbers and activity are not necessarily directly correlated is valid. Whether this is necessarily so in the case of pesticide applications to soil remains to be proven, for despite considerable literature relating to the effects of pesticides on soil microflora no definitive work has yet emerged to confirm or deny this concept. Nor does it follow that an intensive investigation with one or several pesticides, to demonstrate the authenticity of the original concept, will be valid for all other pesticides, present and future.

It is therefore important that, in the case of pesticides, the concept of numbers of organisms not necessarily correlating with their activity be approached with caution. Some pesticides may follow the concept, some may increase numbers and activity while some may be found to increase or decrease numbers disproportionately to activity. As has already been pointed out, one of the pre-requisites for ecological studies is that changes in the composition and activities of populations be followed over a period of time in, ideally, an undisturbed environment (Stotzky, 1972).

1. *Respirometry*

The most widely employed methods for studying activity of the soil microbial biomass are based on their metabolic activities. Respiratory measurements are probably the most accepted and implemented of all activity measurements. Respiration is usually well correlated with other indices of microbiological activity such as carbon, nitrogen, phosphorus and sulphur transformations, accumulation of metabolic intermediates, pH changes and organic matter content (Stotzky, 1965b).

(a) *Carbon dioxide evolution* : Although carbon dioxide (CO_2) evolution is generally considered as indicative of an actively metabolizing biomass, some may arise by chemical decarboxylation, from cell-free enzymes or by the action of added amendments, including commercially formulated pesticides, on soil carbonates. Oxygen may be utilized during purely chemical reactions (Bunt and Rovira, 1955) and both O_2 and CO_2 may be absorbed in soil–water interfaces (Runkles *et al.*, 1958). The contribution of other organisms (e.g. nematodes, protozoa and soil insects) to O_2 consumption and CO_2 evolution must also be considered in overall respiratory measurements in soil.

Ideally, the more undisturbed the soil, the more accurately the results will reflect its true metabolic activity. Thus, methods have been devised whereby soil respiration is measured in the field. Wallis and Wilde (1957) extracted the soil's gaseous phase through metal cylinders or glass funnels embedded in the soil. The gaseous phase was drawn through N NaOH for two hours and the Na_2CO_3 formed was then determined by the barium chloride method. Witkamp (1966) utilized round boxes inverted over dishes containing a small volume of 0·1 N KOH. Intimate contact between the rim of the boxes and the soil is required to exclude atmospheric CO_2. The amount of CO_2

absorbed after various periods of incubation is measured by titration with 0.179 N HCl as recommended by Conway (1950).

A more sophisticated method for *in situ* measurements was devised by Reiners (1968). A steel sleeve measuring $20 \times 50 \times 20$ cm is inserted into soil until the top is only slightly above ground level or the litter layer and covered with a tightly-fitting plastic lid containing inlet and outlet tubes. Ambient air, pumped into the system at rates of 19–25 litres min^{-1}, is distributed across the 20 cm width of enclosed soil by means of a manifold. Air, passing through the outlet tube, as well as the incoming air, are both analysed for their CO_2 content by means of an infra-red gas analyser, the difference representing respired CO_2.

There are many problems in measuring respiration in the field, for the sake of using undisturbed soil. The conditions imposed by the apparatus itself may cause deviations from normal biological activity. Overestimates of CO_2 evolution by the soil's microbiomass may result from plant roots, macroorganisms and soil animals. Many workers feel loath, or do not have the facilities, to leave expensive apparatus in the field or to establish studies on undisturbed soil for some experiments (e.g. respiration) whilst investigating other parameters on disturbed soil samples in the laboratory. In short, the laboratory is a far more congenial atmosphere, and less Spartan, for conducting respiration studies.

Laboratory methods for studying soil respiration usually involve prepared soil, although undisturbed cores (Howard, 1968) can be used. Simple systems usually involve glass jars to contain the soil and a system for trapping CO_2. This may involve an inner well contained within a sealed system. The usual limitation in such systems is that oxygen becomes limiting, depending on the size of the container.

Other simplified systems involve the use of soil in glass jars and either slight positive air pressure, or a suction system, to provide a flow-through of CO_2-free air. The incoming air must first be dried and then passed through a chemical system to remove CO_2 so that accurate titrimetric analyses for respired CO_2 can be done. Failure to dry the air usually results in blockages in the CO_2-absorbing system due to the hygroscopic nature of the absorbent. The CO_2-free air emerging from such systems requires re-wetting if rapid dehydration of the soil samples is to be avoided. After passage through the soil samples, the air is passed through 1–2 N KOH in separate containers to remove respired CO_2 which is then determined titrimetrically. It is generally necessary to employ two KOH traps in series, the second to prevent a suck-back of atmospheric CO_2 in the event of a decreased flow rate through the system. Flow rates of air in such systems should be slow, in the order of 1–2 litres h^{-1}.

A more elegant and reproducible system of forced aeration has been described by Stotzky (1965b). CO_2-free air is used to flush residual air from a 1-gallon master glass jar which contains one large sample of soil, or several samples in smaller glass jars (2–3 per master jar) if soil samples are required at intervals during respiration studies. The outlet tube is connected to a bubble tower containing glass beads, to disperse the gas bubbles, and a known amount of standard NaOH or KOH.

Forced-aeration, or "flow-through" systems are generally criticized for their inability to remove all CO_2 from the soil. Most operators allow the incoming air to flow across the surface of the soil samples, but it is relatively simple to arrange the inlet tube such that the CO_2-free air is forced to permeate upwards through the soil. Another criticism is that the use of CO_2-free air for purging the soil will probably limit growth of autotrophs. It is also important in considering overall respiration in such systems to use soil samples sufficiently large to be representative of the many forms of life inhabiting the soil under investigation. Flow rates and efficient trapping of CO_2 are vital aspects in such systems.

The amendment of soil, in flow-through systems, with sugars or finely chopped grass cuttings (with and without pesticides) at various rates of application, will provide only basic information on the overall respiration of the soil's microorganisms. Even so, some CO_2 will inevitably arise from independent inorganic and organic chemical reactions and also from plant enzymes and macroorganisms. It is especially important to obtain information on the biodegradation of the pesticide under investigation to ascertain the likely contribution from degradation of the pesticide molecule to the CO_2 evolved in the system. At normal or recommended field rates, the contribution from the pesticide may be small but at exaggerated high rates of application the contribution could be significant, particularly from a pesticide with a very short half life in soil.

In addition to titrimetry, several other methods for measuring CO_2 from soil respiration should be noted. These are the Haldane gas analyser (Baver, 1948), infra-red absorption (Miller, 1953), gas chromatography (Keulemans, 1959) and mass spectrometry (Dunning, 1955).

(b) *Oxygen uptake* : Oxygen uptake as an alternative means of measuring soil respiration has been practised for some decades but is not favoured as much as CO_2 evolution for the following reasons advanced by Stotzky (1965b). (i) In the manometric determination of oxygen uptake, gases other than CO_2 may be evolved as a result of microbial activity which may interfere with the manometric measurement.

(ii) In using oxygen uptake measurements to assess microbial activity a Respiratory Quotient of 1 is assumed, a condition which is seldom fulfilled in soil. (iii) Oxygen uptake measurements underestimate the level of respiratory activity because anaerobic soil conditions often exist in which CO_2 is evolved without oxygen uptake. (iv) As with CO_2, non-biological factors may exaggerate the oxygen uptake measurements.

Other criticisms of Warburg manometric measurements of soil respiration are that the size of the soil sample and the volume of KOH are generally too small, especially if extended periods of study are to be used, and that measurements of O_2-uptake are made in an atmosphere devoid of freely available CO_2 which must induce modifications to the soil micro-flora (e.g. autotrophs). Despite these criticisms, Warburg manometric measurements of oxygen uptake remain popular with some researchers. One of the main advantages of Warburg manometry is that it allows the accurate measurement of O_2-uptake from small replicate soil samples over short periods of time amounting to hours instead of days. The technique permits studies on the effects of inhibitors of specific components of the soil microbial population, effects of specific substrates on the potential biochemical activities of the soil microbiomass and the effects of various environmental conditions on soil microbial activity, since factors such as temperature, moisture and CO_2 concentrations within the flask can be controlled.

Parkinson and Coups (1963) described modifications to the Warburg technique which, it was claimed, have the advantage that larger samples can be dealt with. This produces more consistent results in samples from the same soil horizon or samples exposed to the same treatment. Furthermore, the KOH can be readily replaced in the annular well via the side arm, facilitating long-term experimentation; the soil sample is easily placed in and removed from the flask, permitting chemical or biological analyses of the soil after its respiratory activity has been measured.

The Warburg technique has been adequately described, in various modifications, by Katznelson and Rouatt (1957), Parkinson and Coups (1963) and Howard (1968). The latter authority has also described a method for undisturbed cores of soil and litter, in which weighed amounts of soil are placed in glass tubes and closed at one end with glass wool. Known amounts of fragmented litter are placed on top of, or incorporated into, the soil and kept moist. The tubes are placed in an upright box in the field and protected from rain by means of a roof over the box. At various intervals the tubes are removed from the field for oxygen uptake studies in the laboratory using a Dixon manometer and a special glass adaptor (Howard, 1968). The advantages of this technique are that the same soil and litter can be used for repeated measurements and that movement of the tubes from the field does not involve disturbance of the soil samples. Further advantages are that the disappearance of litter from the tubes can only occur by decomposition and that the environmental conditions influencing the samples closely resemble those in the field. However, it is highly probable that the latter claim is open to similar criticisms as the contact or buried slide techniques mentioned earlier, relating to rapid growth of some species in contact with the glass surface.

Another form of oxygen uptake manometry is that described by Birch and Friend (1956). This ingenious method involves the use of soil in a closed system with NaOH or KOH to remove CO_2 and a side arm, or ancillary outlet, which is connected to a glass tube. The glass tube is partly immersed in electrolyte ($1 \text{ N } H_2SO_4$) contained in a beaker in which there is an inverted gas-tube or burette filled and in continuous contact with the electrolyte in the beaker. Both the gas tube and the glass tube are fitted with platinum electrodes. Direct current from a 6 or 12 volt accumulator is supplied to the electrodes. As oxygen is consumed by the soil and CO_2 from the soil absorbed by the alkali, a partial vacuum is formed in the flask which draws the electrolyte up the glass tube. When contact is made with the electrode in the glass tube, oxygen is generated which replaces that consumed by the soil. At the same time hydrogen is generated in the gas tube and accumulates, by displacement of the electrolyte, in the top of this tube. Measurement of the volume of hydrogen produced represents twice the volume of oxygen consumed.

Criticisms of this method are similar to those of the Warburg technique, namely that the soil microflora is deprived of CO_2 and furthermore the atmosphere above the soil is virtually pure oxygen. The effects of barometric pressure will be compensated for by the control unit containing an equivalent volume of glass beads and expansion of hydrogen gas within the gas-tube, with increases in room temperature, can be compensated for by deliberately producing an equivalent volume of hydrogen gas within the control unit's gas-tube. Birch's respirometer has, more recently, been modified considerably to provide greater accuracy and compensating mechanisms for temperature and barometric pressure (Birch and Melville, 1969).

Regardless of the method selected, measurements of respiratory activity in soil can, at best, only provide information on the gross metabolic activity of the soil's microorganisms together with contributions from macroorganisms, cell-free enzymes and chemical reactions. Not only is it virtually impossible to separate the various contributors, but, even more frustrating, it is impossible with conventional methods to ascertain which section of the microbiomass is contributing most to the respiratory activity.

In a hypothetical situation, a pesticide causes a 50% reduction in the respiratory activity and numerical frequency of a group of organisms. At the same time another group's metabolic activity is stimulated by 50% but no increase in their numerical frequency occurs. The nett result shown by respiration studies and dilution agar plate counts is no effect and a negative (or inhibitory) effect, respectively. The question thus arises as to whether conventional respirometry is worth the effort. Stotzky (1972) suggests that respiration studies may be more meaningful if substrates are added that can only be utilized by some species. Under these conditions respiratory measurements can provide much information on the development of these species and the effect on them of various environmental variables. Such substrate-orientation of respirometry should receive greater consideration.

(c) *Radiorespirometry* : The use of ^{14}C-labelled compounds, simple or complex, can provide valuable information on the metabolic activity of groups of soil microorganisms but cannot, for the moment, completely replace conventional measurements of total respiration. This is because the introduction of a single ^{14}C-labelled compound will favour the growth and hence measurement of respiratory activity of only that fraction of the soil's microbial biomass that is capable of utilizing the compound or its degradation products.

On the other hand, the liberation of $^{14}CO_2$ from uniformly labelled plant residue added to soil may be considered as having a closer relationship to the overall respiration of soil since the mineralization of this material will represent the activities of the greater proportion of types of soil microorganisms. Unless the compound used is particularly prone to purely chemical decomposition, the researcher will be more assured that the $^{14}CO_2$ evolved is of biological origin even though cell-free enzymic activity and intracellular activity may remain inseparable.

The use of ^{14}C-labelled organic compounds for studying the effects of pesticides on the respiratory activity of soil can provide useful information on inhibitions or stimulations of certain groups of soil organisms. Uniformly labelled ^{14}C-plant residue can be treated with the pesticide to be studied and incubated in soil for considerable periods of time. $^{14}CO_2$ evolved as a result of the degradation of the plant residue can be collected in alkali traps hourly, daily or weekly to gauge the effect of the pesticide on the mineralization of the plant residue. Closed systems whose $^{14}CO_2$ traps are changed at regular intervals, or flow-through systems using CO_2-free air can be used.

The mineralization of ^{14}C-labelled organic matter, whilst providing information on the rate of this process in the presence of pesticides, probably represents microbiological activity with a higher degree of certainty than conventional respirometric methods. However, it still only depicts the gross reaction and provides no data on possible inhibitions of organisms associated with the step-wise or sequential degradation of the organic matter. In the absence of precise data concerning the exact sequence of organic matter degraders for all soil types and conditions, and in the absence of adequate knowledge of the sequence of intermediates produced during decomposition of organic matter, more detailed studies in relation to the effects of pesticides by this technique are severely limited.

It is, however, possible to utilize commercially available ^{14}C-labelled compounds representative of some of the known intermediates of organic matter degradation. ^{14}C-Labelled mono-, di- and polysaccharides, as well as protein (of plant or algal origin) and a variety of amino- and other organic acids are available commercially and can be added to pesticide-treated soils to study possible effects of pesticides on the ability of organisms to degrade these intermediates. In this case it is useful to pre-treat a large number of individual soil samples with the pesticide, and then incubate with batches of soil without pesticide. At intervals during incubation, soil samples are removed and treated with a representative radioactive compound such as starch or glucose. Since these compounds are likely to be rapidly metabolized it will be necessary to discard the samples after two to seven days and replace them with another batch of the pesticide-treated samples. The $^{14}CO_2$ evolved from each batch is determined independently. The advantage of this method is that the effects of prolonged exposure of the soil to the pesticide can be studied and batches of soil incubated without pesticide serve as a control for fluctuations in the starch or glucose-utilizing microflora.

Another approach, described by Hubbard (1973) makes use of $^{14}CO_2$ applied in the gaseous environment of incubating soil samples. The assimilation of $^{14}CO_2$ by chemo- and photosynthetic autotrophs (and to a lesser extent by some heterotrophs) is detected by measuring the ^{14}C in the organic fraction liberated after pyrolysis of the soil. Hubbard also points out that quantification of $^{14}CO_2$ evolved during short incubation periods of soil treated with ^{14}C-labelled glucose and amino acids permits the measurement of metabolic processes without considering the complications of cell proliferation.

The methods of Mayaudon and Simonart (1958, 1959a, 1959b, 1960) using ^{14}C-labelled organic matter, cellulose and lignin, could be usefully applied to studying effects of pesticides on degradation of organic matter fractions.

Domsch *et al.* (1973) have described a method for the simultaneous measurement of radioactive and inactive CO_2 in which the time-course of the priming action, following the addition of ^{14}C-glucose to soil, can be monitored and also the differentiation of respiration peaks resulting from simultaneous degradation of ^{14}C- and inactive carbon compounds. The method is of particular interest since it can be readily adapted to studying the effects of pesticides on ^{14}C-compound degradations.

2. *Microcalorimetry*

Virtually every process, whether physical, chemical or biological is accompanied by either the evolution or absorption of heat which is a function of the extent of the process concerned. Calorimetry may therefore be regarded as an analytical tool. Provided the calorimeter is sufficiently sensitive, biological processes can be recorded even in soil and aquatic regimes where the numbers of organisms are not normally as great as in pure culture systems.

Newman and Norman (1943) used air-dry soil samples, spread out in a thin layer and exposed to an atmosphere of high relative humidity for seven days to minimize non-biological thermal changes due to wetting. In this method, a sample of the soil (50 to 350 g) is placed in a Dewar flask and an appropriate volume of water, or soluble substrate, is added and the flask kept in a constant temperature environment. Changes in temperature in the flask are monitored with a sensitive galvanometer, readings on which are a measure of the differences in e.m.f. between two thermocouples, one of which is placed in soil in the Dewar flask and the other in a constant temperature environment outside the sample flask (e.g. a second, sealed, Dewar flask containing water as the reference liquid).

A more sensitive, but elaborate, method was described by Clark *et al.* (1962) and calls for 48 temperature sensing thermistors inserted in soil samples in insulated containers, a 52-point stepping switch, clock mechanisms, a bridge-type strip recorder and a constant temperature room.

A variety of designs for microcalorimeters, some of which are available commercially, has been described by Wadsö (1970) and the applications to which these instruments can be put have been described by Mortensen *et al.* (1973).

E. STUDY OF RATE OF SUBSTRATE CHANGE

The increasing use of radioactive isotopes of C, N, P, S and other elements has revolutionized the study of microbiological transformations in soil and aquatic regimes. However, not all laboratories are fortunate to possess this potential. It is therefore pertinent to mention less sophisticated (instrumentally) methods and to reassure those who are unable to use isotopes that these methods, whilst frequently more time consuming, are still valid.

1. *Molecular nitrogen-fixation*

Despite its antiquity the soil plate technique for *Azotobacter*, devised by Winogradsky (1926), remains a quick and reliable method for demonstrating the presence of these aerobic, free-living, nitrogen-fixing bacteria. The method consists of mixing a carbon source (preferably mannitol) into soil moistened to a paste, which is then placed in Petri dishes and the surface of the soil carefully smoothed over. After incubation, *Azotobacter* colonies can be seen on the surface of the soil. For a large number of soil samples, the method can be used to obtain an approximate idea of the numbers of viable *Azotobacter* cells existing in the soil. It is preferable to other culture methods, especially for examining the effects of pesticides on *Azotobacter*, since the test is conducted in soil.

Dilution agar-plate counts using media to favour the growth of azotobacters have been used, but this method is subject to the same difficulties and open to the same criticisms as are "total" microbial counts. A less accurate but more practical method involves the use of selective saline liquid media and the extinction dilution of soil samples using replicate tubes for each dilution (Augier, 1956). The presence of *Azotobacter* in a tube after incubation is indicated by a characteristic film on the surface of the liquid medium. It is frequently necessary to confirm these observations by microscopy. Although more practical, this method is not ideal for studying the effects of pesticides since the process of dilution may remove non-lethal inhibitory effects of the pesticide. Incorporating pesticide into the medium such that all tubes contain the same amount of pesticide cannot be recommended, since as dilutions increase the soil and numbers of cells decrease so that the effects of the pesticide are exaggerated relative to undiluted soil.

The enumeration of free-living nitrogen-fixing microorganisms in soil, whilst being of some use in evaluating the effects of pesticides on their numerical distribution, cannot be used as a direct measure of their nitrogen-fixing capabilities. The extent of nitrogen-fixation in soil can only be measured accurately by means of ^{15}N-isotope studies or by the acetylene reduction method. The former method requires expensive analytical equipment, namely a mass spectrometer, preferably one which has not been used for analysis of other elements since these interfere with the ^{15}N analysis. The acetylene reduction technique (Dilworth, 1970) brings accurate, quantitative assessment of N_2-fixation within the grasp of most laboratories. The technique involves flushing of assay vials, containing cell suspensions or soil, with an argon–oxygen mixture to remove atmospheric nitrogen prior to introducing a known volume of acetylene. After incubation, the gas phase in the assay vials is analysed for ethylene by gas chromatography. To obviate the necessity of flushing each vial separately, Mackenzie and MacRae (1971) described a modification using a plastic bag with glove apertures, a gas inlet/outlet and an entry port for the addition of materials, which facilitates the flushing-out of atmospheric nitrogen from the assay vials and ensures a nitrogen-free atmosphere for the subsequent stages of introducing acetylene gas into the vials.

Knowles *et al.* (1973) investigating the kinetics of $^{15}N_2$-fixation and acetylene reduction in soil, and the effects of both oxygen and acetylene on nitrogenase activity, found no significant effect of N_2 on acetylene reduction but an inhibition of N_2-fixation by acetylene, suggesting that replacement of the soil atmosphere with argon may not be necessary.

Whilst widely used, some workers have cautioned that the acetylene reduction technique may not be as reliable for soil samples as for pure culture systems. Bergerson (1970), working with detached soyabean nodules and also with pure cultures of *Azotobacter vinelandii* and *Klebsiella aerogenes* concluded that caution was necessary in applying the technique since major errors are likely to result when the conditions in the assays are not carefully matched with the conditions under which nitrogen fixation is occurring. Knowles *et al.* (1973) found that the presence of 0·1 atmospheres of acetylene in a soil system prevented the increase in the rate of CO_2 evolution normally observed after addition of glucose to the soil, and also inhibited the proliferation of anaerobic nitrogen-fixers. The implication is that the acetylene reduction technique may only provide data on the N_2-fixing activity of aerobic microorganisms.

The use of this technique for N_2-fixation in soil may thus be in a state of flux until more information is available or until the methodology is perfected. A recent and welcome adjunct to the technique has been the development of a colorimetric method for determining ethylene (La Rue and Kurz, 1973). While gas chromatography is undoubtedly more accurate, the colorimetric method enables those laboratories which do not possess such equipment to undertake N_2-fixation studies.

2. Nitrification and denitrification

The simplest and most convenient method for studying the effects of pesticides on nitrification is by incubation of soil in containers with the addition of known amounts of ammonium ions. At set intervals of time, replicate flasks of each treatment are removed and the soil analysed for residual ammonium ions as well as nitrite and nitrate ions.

Lees and Quastel (1946) described a system for the perfusion of soil with an aerated liquid and since then many variations and modifications of the original Lees–Quastel apparatus have been described. The original device utilized a column of soil connected to a reservoir containing a solution of ammonium ions. Droplets of the solution were induced to flow to the soil column under slight positive air pressure. After percolating through the soil the liquid returned to the reservoir. A sampling port in the reservoir permitted removal of aliquots for analysis of NH_4^+, NO_2^- and NO_3^- ions.

Lees and Quastel (1946) listed advantages and disadvantages of their method. Among the advantages is the fact that the soil is kept at a constant and evenly distributed water content; temperature variations in the soil are minimized; soil is undisturbed during the experiment and its gassing (by air or any other atmosphere) is guaranteed; sampling and the addition of inhibitors, if required, is facilitated; and the soil and perfusate can be analysed individually.

The disadvantages of the system are that removal of perfusate for analysis decreases the total volume of perfusate without a compensating decrease in the amount of soil and also that the soil will always be near to waterlogged. The latter disadvantage is a very real hazard as any worker who has used the system knows. The apparatus requires careful construction and maintenance otherwise liquid accumulates above the soil or does not percolate at all. The use of peristaltic pumps does permit a greater control over rate of perfusion without reliance upon compressed air systems (Wright and Clark, 1969).

Denitrification in soil is usually closely coupled to other phases of the nitrogen cycle such as nitrogen-fixation, ammonification and nitrification and is difficult to evaluate in the presence of other gas-evolving processes such as respiration. One method for evaluating dissimilatory nitrate reduction is to inoculate two sets of tubes containing a liquid medium of nitrate and glucose which serve as the nitrogen and carbon sources, respectively. The tubes of media are inoculated with aliquots of a serial dilution of soil and tests are made daily to check for the disappearance of nitrates and the appearance and disappearance of nitrites. Using the most probable number (extinction dilution) system it is possible to calculate the number of micro-organisms responsible and to construct an activity curve (de Barjac, 1952) but Todd and Nuner (1973) found that this approach gave inconsistent results. Analyses for total nitrogen remaining at the end of the experiment can be used to evaluate the losses in gaseous form (Treccani, 1961).

Wheeler (1963) used the Lees–Quastel type of soil perfusion apparatus to study the relation between ammonification, nitrification and denitrification, while McGarity (1961) used a Warburg apparatus to trace the release of carbon dioxide, nitrogen and nitrogen oxide in an argon atmosphere. A sophisticated apparatus was used by McGarity et al. (1958) for continuous monitoring of the soil's humidity and oxygen content, the rate of oxygen consumed and release of gases from the soil. Wijler and Delwiche (1954) used a ^{15}N-enriched atmos-phere to study nitrogen-fixation and denitrification. The isotopic dis-tribution in the total nitrogen present as nitrate, nitrite, ammonia, atmospheric N_2 and amide-nitrogen was determined.

Payne (1973) used a method for separating the gaseous mixtures of N_2, N_2O and NO by chromatography on polyaromatic beads. This method is probably the best available for the quantitative assessment of the rate of dissimilatory nitrate reduction. The system was used for studying the time course of nitrogen released from nitrate by the marine organism *Pseudomonas perfectomarinus* in pure culture, but would be easily adaptable to the study of denitrification in soil. Helium was used as the sparging agent to maintain an anaerobic environment and thermal conductivity used to detect the presence of the various nitrogenous gases produced. Payne suggested that the unique *c*-type cytochromes involved in reduction of NO_2, NO and N_2O might be used as indicators of denitrification in marine ecosys-tems.

3. Proteolysis and ammonification

The distinction between proteolysis and ammonification is somewhat arbitrary. Proteolysis may be defined as the breaking down of large protein molecules into increasingly smaller molecules by a whole series of enzymes. The final stage of the deamination may be defined as ammonification but since proteins and amino acids are not the only compounds possessing an amide or amine group, the definition is somewhat tenuous.

Few techniques have been developed for evaluating the microflora responsible for proteolysis. Pochon and Chalvignac (1952) mixed particles of sterilized animal tissues with soil, isolated the organisms developing on them and tested each for its ability to lyse protein. Coagulated blood serum in Petri dishes, inoculated with aliquots of a serial dilution of soil, and isolation of the colonies which caused liquefaction, was the method used by Chalvignac (1953). Lajudie and Chalvignac (1956) used a selective gelatin culture medium and the most probable number method for determining numbers of protein-liquefying microorganisms.

While the above methods enable both isolation and computation of the numbers of organisms, the best method for evaluating the activity of proteolytic organisms in pesticide-treated soil is by incubating soil with uniformly ^{14}C-labelled protein and collecting any $^{14}CO_2$ evolved. The extent and rate of breakdown of the protein will be reflected by the amount of $^{14}CO_2$ liberated, irrespective of the number of steps or types of microorganisms involved.

Ammonification measurements in soil may easily be confounded by absorption of the ammonium ions onto clay particles. Thus, Kauffmann and Chalvignac (1950) mixed a variety of nitrogenous substrates into sand and inoculated the sand with microflora extracted from soil, to obviate the absorption of ammonium ions onto clay.

Greenwood and Lees (1956) modified the Lees–Quastel apparatus to measure the extent of oxygen utilization during studies on ammonification and nitrification of amino acids or casein hydrolysate. Since absorption and nitrification can contribute a significant error in ammonification studies by such methods, liquid elective media containing an amino acid may be used in conjunction with the most probable number method to minimize error. The disappearance of the amino acid and appearance of ammonia in each of the dilutions of soil can be recorded as a function of time and dilution. However, preclusion of errors due to absorption and nitrification is probably best accomplished by using a ^{14}C-labelled amino acid and monitoring the $^{14}CO_2$ evolved. Losses of ^{14}C-amino acid by absorption, unchanged, into soil organisms can be examined by pyrolysis of the soil afterwards.

4. *Transformation of other substrates*

Pochon (1957) has described a method for determining the potential activity of the microbial biomass as indicated by changes detectable in cultures. Provided the nutrients and/or metabolites can be identified by simple chemical means, the method has wide scope for studying the disappearance of the substrate or the appearance of metabolites. Basically, the method is that of the most probable number (extinction dilution) technique but the results are expressed differently. A soil suspension is prepared in water and then diluted stepwise in 10-fold dilutions to 10^{-10}. Each dilution is inoculated into triplicate tubes of the desired selective medium and incubated at the required temperature. Aliquots for analysis are removed daily for six days and thereafter every other day for a period of 14 days. Changes in the medium are recorded and plotted against time. Curves for different soils or rates of pesticide treatment may be compared and at the end of the 14-day period, the most probable numbers of organisms capable of using the chosen selective medium can be determined. The method is extremely useful in cases where the extraction of residual substrate from soil would be problematic or where the destruction of metabolites (required as proof of utilization of the substrate) by other soil organisms is likely to be very rapid.

The similarity to the normal extinction dilution method, however, also means that the method is liable to the same criticisms, chief among which is the possibility of a diluent effect on the pesticide which would nullify its effect on microbial cells.

Wherever it is possible to use ^{14}C-labelled substrates, this should be done in preference to extinction dilution or other forms of pure and semi-pure culture. The difficulty frequently encountered, however, is the non-availability of labelled material. In such cases, the researcher must either devise a new method or resort to extinction dilution methods, preferably of the type described by Pochon (1957). Examples of compounds which cannot normally be bought commercially as ^{14}C-labelled material are pectin and other hemicelluloses, lignin, chitin and nucleic acids. In most of these cases it is known to be difficult to adequately label the compound or, that once labelled, the material is liable to rapid deterioration due to instability of the molecule.

Hemicelluloses are a poorly defined group of compounds of two distinct types: those consisting of repeating units of sugars and uronic acid (the polyuronides) and those which consist solely of sugars (the cellulosans). Among the commonest types are the pectins, pentosans, mannans, xylans and galactans. They are difficult to purify, being closely associated with other chemicals, and their separation often results in changes in their chemical structure.

Methods for studying the activity of the three enzymes known to be involved in pectinolysis are usually based on extinction dilution of soil into: (a) A liquid medium containing carrot cubes, for protopectinase studies (Lambina, 1957). (b) A calcium pectate gel seeded with the soil dilutions prior to gelling of the medium, for polygalacturonidase activity. (c) Apple pectin, for pectin methylesterase activity which results in acidification of the medium (Kaiser and Prévot, 1958).

Studies on other hemicelluloses have involved the use of complex substrate materials, mostly characterized by their method of extraction and purification from oak sawdust (Pochon and Augier, 1954). The only method of a quantitative nature involves the extinction dilution of soil into liquid media containing the substrate (Pochon and Augier, 1956) but the interpretation of results remains nebulous.

Chitin is another biological residue which is often present in soil in significantly high quantities, being the main material in the exoskeletons or carapaces of Arthropods. Clarke and Tracey (1956) and Skerman (1967) described methods for preparing purified chitin which can be used in pure culture systems to study chitin hydrolysis, although this form is not particularly suitable for studies in soil. The study of chitinoclastic activity in pesticide-treated soil is not readily accomplished with such preparations (unless ^{14}C-labelled material is used) and consideration should be given to the use of chitin strips. Chitin breakdown in soil is very slow in the unpurified state (Veldkamp, 1955) and therefore the use of purified chitin, which degrades more rapidly, can only be used as a gauge of the chitinoclastic potential in pesticide-treated soils. A possible method for purified chitin strips could be based upon gravimetric analysis rather than chemical detection of the N-acetylglucosamine or N,N'-diacetylchitobiose decomposition products of chitin, both of which are liable to be produced in small, virtually undetectable amounts in soil and are themselves liable to further microbial degradation. By weighing chitin discs or strips before and after extended incubation in soil it may be possible to determine the rate of chitinoclastic activity with time.

Lipids and nucleic acids are organic residues which must be present in soil in considerable quantities. Whereas some fats can be obtained in ^{14}C-labelled form and can therefore be studied in soil quite conveniently, nucleic acids appear to be difficult to obtain labelled in a form which has a reasonably long period of stability (Radiochemical Centre, Amersham, England – personal communication). Studies on the degradation of nucleic acids can be accomplished by incorporating purified DNA into agar media inoculated with soil dilutions. A convenient medium for such a study is that described by Jacobs et al. (1964) in which DNA is suspended in nutrient agar. The hydrolysis of DNA is detected by precipitation of the unhydrolysed DNA with dilute hydrochloric acid and observation for zones of hydrolysis around individual colonies. The method, applied to soil, is liable to the same criticisms and precautions as all other methods based on the use of soil dilutions and artificial media.

The importance of studying the degradation of particular fractions of tissues in soil is now recognized, more especially in view of the increasing use of systemic pesticides. The uptake of systemic pesticides by plants, or the ingestion of such materials by insects, and the incorporation of the pesticide molecules into the plant or insect's cells instead of simple surface depositions, means that particulate components of the tissues may be rendered less amenable to biodegradation in soil and hence may accumulate, thereby altering the sequence in which key members of the soil's organisms can attack the intact residue.

F. SOIL ENZYMIC ACTIVITY

Skujins (1967) reviewed the techniques available for assay of many enzymes and indicated that very rarely can enzymic activity in soils be correlated with soil fertility and microbial activity. This is because measurement of soil enzymic activity is a manifestation of the free enzymes absorbed on soil particles, those released from lysed organisms or still active within dead, but unlysed, cells, those released from plant roots and soil animals, as well as the contributions from living roots, microbial cells and soil animals. The measurement, therefore, of a single contributing source of enzymic activity in soil is not possible, neither is it necessarily desirable.

In the case of pesticide-treated soils it is possible that one or several pesticides may inactivate a group of organisms normally responsible for the degradation of a particular substrate in soil. Such inactivation may be either short-lived or prolonged and in the latter case it is important to know whether the substrate's delayed transformation will affect those organisms that would normally benefit from the transformation. As a purely hypothetical case, the pesticidal inhibition of all active ureolytic organisms in soil would increase the already high competition for nitrogen among non-ureolytic soil organisms dependent upon ammoniacal nitrogen, unless cell-free urease, together with that from plant tissues, could continue to hydrolyse urea molecules.

Whilst the separation of enzymic contributors to the conversion of a substrate in soil remains a difficult task, some segregation may be possible using physical or chemical methods to eliminate or inhibit contributions from other sources. It may, for example, be possible to filter a soil suspension in water or buffer and test the resulting cell-free filtrate for enzymic activity. However, the enzyme in this case will probably be present in an extremely dilute form and there is no guarantee that some of the enzyme does not arise from the lysis of cells during filtration. To further concentrate the enzyme, use could be made of the "hollow fibre" system of de-watering. Enzymes bound to clay minerals may possibly be released by elution with increasing concentrations of ammonium sulphate. Whilst some of these techniques may have been tried, there is currently no recommended or standardized procedure for the recovery of soil enzymes. This area of soil biological research is as yet in its infancy, requiring much applied research before its significance can be gauged.

Many soil "enzymic" systems have been investigated but in general the question as to whether the systems do produce a true cell-free enzymic preparation are debatable. Typically, these methods involve the pre-treatment of soil suspensions with toluene (before substrate is added) to limit the proliferation of cells possessing the enzymic activity in question. While such treatment may prevent the synthesis of further enzyme, it does not separate that which is present as truly cell-free enzyme from that present in cells as a constitutive enzyme at low (maintenance) levels and does not take into account the source of enzyme (microbial cells, plant roots or seeds). After toluene-treatment, the substrate is added, the soil solution is incubated for a short period of time and chemical or spectrophotometric analysis made for disappearance of substrate or appearance of intermediates and end-products. Hofmann and co-workers (Hofmann and Seegerer, 1951; Hofmann and Schmidt, 1953; Hofmann and Hoffmann, 1953a,b; Hofmann and Hoffmann, 1954) have done much in establishing techniques based on this concept, which has been used for saccharase (Seegerer, 1953), α- and β-glucosidases (Drobnik, 1955; Hofmann and Hoffmann, 1953b), α- and β-galactosidases (Hofmann and Hoffmann, 1954; Hofmann and Hoffmann, 1955), protease (Hofmann and Teicher, 1957) and urease (Hofmann and Schmidt, 1953).

Other workers (McLaren et al., 1957; McLaren, 1963) have suggested the use of radiation-sterilized soil. The use of irradiation, whilst not recommended for phosphatases, which are partially inactivated (Pochon et al., 1969), has been used to study urease activity (McLaren et al., 1957).

Of several methods available for studying the ill-defined phosphatases of soil, the method of Tabatabai and Bremner (1969) is the most reproducible and accurate. For peroxidase activity the method of Bartha and Bordeleau (1969) can be recommended.

An enzyme test requiring no sterilization of the soil and one which is of practical importance and significance is the so-called dehydrogenase test. Since the oxidation of organic substrates is coupled with the reduction of molecular oxygen, with the transfer of hydrogen by dehydrogenase enzymes, it is possible to measure the microbial activity by measuring dehydrogenase activity. Several variations of Lenhard's (1956) original method for dehydrogenase have been suggested and attention to soil type, pH, organic matter content and the solvent system for extraction have been fully investigated by a number of researchers (see, Glathé and Thalmann, 1970). The basis of the test lies in incubating 10–20 g of soil, at 90% of its moisture holding capacity, with 200 mg of calcium carbonate and 2 ml of 1% 2,3,5-triphenyltetrazolium chloride solution for 24 hours. In the absence of oxygen, this salt serves as a terminal acceptor for the hydrogen removed from the oxidizable carbon substrate. The salt, reduced to the corresponding formazan, is extracted by solvents (methanol or acetone) and determined by spectrophotometry at 485 nm.

Stevenson (1959) has shown that dehydrogenase activity is significantly correlated to oxygen uptake but is not necessarily correlated with numbers of organisms. The test, therefore, is either an extremely useful adjunct to other respiration studies, or can stand alone, and is both simple and quick requiring the minimum of apparatus and space.

G. RHIZOSPHERE AND PHYLLOSPHERE

Most soil microbiologists regard the subject of rhizosphere studies as the realm of other soil microbiologists. If blame is to be layed somewhere, then it must be at the door of the non-rhizosphere soil microbiologist himself for segregating (if only subconsciously) the realm of soil microbiology into rhizosphere and non-rhizosphere. There can be no doubt that studies on the rhizosphere are more difficult than non-rhizosphere studies since the plant root imposes additional parameters which influence the distribution, type and response of the soil's microorganisms. Dilution agar plate counts become complicated by the need to remove organisms from the root surface as well as the soil adhering to it. Direct and electron microscopy cannot be utilized with thin layers of soil for enumerating components of the system and must now include sectioning techniques which add further uncertainty to the true distribution of microorganisms on or in the system. Respiration and enzyme studies must now include contributions from yet another source, the plant. Yet, the relationship between plant and microbe remains the most important practical aspect of soil microbiological studies since our ability to survive off the land is contingent, in part, upon a healthy relationship between the two.

The phyllosphere, or more correctly the phylloplane, is a subject of study which only in the last decade has achieved a status of its own. Formerly the domain of the plant pathologist, the study of the plant leaf surface is now recognized as vital to a better understanding of factors influencing the fate of animate and inanimate entities such as spores, viruses, pollutants and pesticides. As with rhizosphere studies, the increasing use of systemic pesticides must impart to phylloplane studies an even greater degree of importance in the future. An alteration, for example, to the saprophytic microflora of the phylloplane by a systemic pesticide may have undesirable effects on the subsequent storage of leaf material, with possibly dire consequences if the alteration predisposes the stored material to myco- or other toxin formation.

1. *Rhizosphere*

Root-microbe associations have been summarized by Gray and Williams (1971) as follows:

  (i) Associations in which the microbes are saprophytic, in soil around roots or on root surfaces. These are unspecific, involving many different microorganisms, and healthy root tissues are not invaded to any great extent.

 (ii) Associations in which the microbes are saprophytic for much of their life but can invade living root tissue in certain circumstances, for example the soil-inhabiting pathogens and certain mycorrhizal fungi.

(iii) Association in which the microbes are active mainly, or only, within living roots. These are specific associations often between one microbial species and one plant species and include the root-inhabiting fungi, some mycorrhizal fungi and *Rhizobium* spp.

All three groups are subject to the influence of pesticides. The most commonly investigated association is that of group (iii) and of these the genus *Rhizobium* has received the most attention. The susceptibility of *Rhizobium* spp. to herbicides has been extensively investigated by a variety of methods ranging from colony counts on specific agar media to determination of leghaemoglobin content and ability of *Rhizobium* spp. to infect appropriate legumes. In areas of the world where leguminous crops are of agronomic necessity and, especially in times of nitrogenous fertilizer shortages or intensification of agriculture, the effects of pesticides on the nodulation capacity of *Rhizobium* becomes a meaningful and practical area for study.

The effects of pesticides, especially fumigants, on the incidence of soil-borne pathogens such as *Armillaria*, *Pythium* and *Rhizoctonia*, all of which may invade plant root systems, are particularly important. Fumigants may affect these rhizosphere pathogens either by direct action, or indirectly by stimulating the growth of natural antagonists such as strains of *Trichoderma*.

Apart from considerations of *Rhizobium* or root pathogens, there are the, as yet, unclarified associations of ectotrophic, endotrophic and vesicular–arbuscular mycorrhizal fungi with plant roots and also the associations of algae and bacteria with non-leguminous plants. These associations are all thought to be symbiotic to some degree and are probably susceptible to pesticide applications to the soil or plant.

Estimates of microbial numbers in rhizosphere soil can be made by dilution agar plate counts using a wide variety of media. Numbers of organisms developing on the plates can then be related to the weight of rhizosphere soil, determined by the method of Timonin (1940) or by the volume displacement method of Reyes and Mitchell (1962). The chief criticisms of the dilution agar plate method for rhizosphere soil are the same as for non-rhizosphere soil, discussed earlier. A possible advantage of the method is that the amount of pesticide present in the rhizosphere soil can be determined by chemical analyses of the soil and would represent a more intimate association between pesticide and an ecosystem than would the analysis for pesticide in non-rhizosphere soil.

The soil-plate technique of Parkinson and Thomas (1965) permits the isolation of fungi from the soil around individual roots, or the total roots of a plant, or on a zonal basis; that is, from particles of soil taken from defined parts of the root system. These authors also suggest the use of a soil washing technique (Parkinson and Thomas, 1965) in which plant roots are shaken in sterile water to remove the rhizosphere soil which is then placed in a soil-washing box (Watson, 1960; Williams *et al.*, 1965) and treated by the soil-washing procedure. This permits separation of fungal spores from hyphae since the latter tend to remain attached to soil particles. The particles are then plated in agar media. Neither of these techniques permit studies on the rhizosphere bacteria but are useful for specialized studies on the effect of pesticides on rhizosphere fungi.

Dilution agar plate counts can also be used for rhizoplane studies. Louw and Webley (1959) advocated a similar procedure to that suggested by Timonin (1940) for rhizosphere soil. After removing the rhizosphere soil by shaking in sterile water, the roots are removed and resuspended in fresh sterile water with glass beads. After vigorous shaking, aliquots of the water are plated in agar media and the numbers of organisms developing are related to the wet weight of root material.

Cook and Lochead (1959) used a root maceration technique in which rhizosphere soil is first removed by shaking roots in water, after which the roots are removed and shaken in three changes of sterile water. The roots are then weighed and resuspended in sterile water in which they are macerated. Dilutions are prepared from the blended suspension and plated in agar media. Numbers of organisms are related to the weight of macerated root tissue. Apart from the criticisms of dilution agar plate counts, a further disadvantage of this technique may be the inhibition of rhizoplane organisms by toxins such as phenols, alkaloids and glycosides released from the plant and which Wood (1967) found to be active against various organisms. Variations of the root maceration technique are described by Stover and Waite (1953) and also by Singh (1965).

Serial root washing in 30 or more changes of sterile water was advocated by Harley and Waid (1955) for the complete removal of fungal hyphae and spores from the surface of roots. An advantage of this method is that slower growing fungi, normally overgrown by more rapidly developing forms, can be isolated. Waid (1956) also suggests the use of a root dissection procedure in which roots, washed free of rhizosphere soil, are cut into small segments (c. 2mm long) and placed in sterile water in Petri dishes. Under a dissecting microscope the root fragments are separated into outer cortex and stele plus some inner cortex. Each part is placed in separate dishes and may be further dissected if necessary. The dissected fragments are then placed in acidified Czapek–Dox agar and incubated. This method permits a study of the spatial relationship of different fungal species to be made on and within root tissues.

Apart from dilution-agar plate counts or agar plate isolation methods for studying rhizosphere and rhizoplane organisms, several methods, originally designed for non-rhizosphere soil studies, have been adapted to studies of plant root microorganisms. Thus, Starkey (1938) adapted the Rossi and Cholodny buried slide method to study rhizosphere organisms. In this method microscope slides are buried vertically or horizontally at different depths in soil. Seeds are planted above the slides and after root development has occurred, the slides are removed, dried and stained in phenolic rose bengal and examined microscopically. Parkinson (1957) evolved an impression slide technique to study mycelial development in the rhizosphere. Roots, shaken to remove excess soil, are cut into appropriate lengths and placed on slides previously coated with a thin layer of nitro-cellulose in amyl acetate which provides a surface to which the rhizosphere soil adheres when the roots are removed. After drying, the slides are stained in phenolic aniline blue and examined. Direct microscopy of root surfaces has also been attempted by Parkinson and Clarke (1961), Moreau and van Effenterre (1961) and Johnen and Anderson (unpublished data). In the first of these, thoroughly rinsed roots are cut into sections and stored in formalin-acetic-alcohol preservative until required. Staining is achieved by lactophenol and cotton blue treatment after which the segments are mounted and examined. The Moreau and van Effenterre (1961) method uses washed roots which are immersed in a varnish of nitrocellulose with chlorinated solvents and a plasticizer. After 12 h, the roots are removed and allowed to dry after which the hardened varnish films are dissected free from the root, mounted on slides and examined. Fluorescent stains of the europium salt and fluorescent brightener type were used by Johnen and Anderson to stain smaller diameter roots which were then observed under incident-ultraviolet light. The combined stain (Anderson and Slinger, 1975a) permitted differentiation of tissues and organisms on the root surface.

Parkinson's (1957) soil box is a versatile device lending itself to isolations of organisms of the rhizosphere or rhizoplane as well as to direct microscopy of growing roots and impression slides of root surfaces. The box, of wood or perspex, can be fitted on one side with removable perspex or glass microscope slide panels. The box is filled with soil in which seeds are planted, and is turned so that the plant roots grow down onto the perspex or glass panels. The box is then re-orientated so that the roots grow down the profile of panels. Panels can be removed at any depth to permit insertion of a perspex panel with perforations through which small tubes containing agar can be introduced such that the agar comes into contact with the root but not the soil. The tubes of agar can be removed at intervals, the agar core extruded, segmented, and the segments plated in nutrient medium. If microscope slides are used in place of perforated perspex panels, these can be removed after marking the outline of the root on the back (outside) of the slide and stained, mounted and examined microscopically for organisms adhering to the portion of slide formerly in contact with the root. If the slides are not removed, fungal (and possibly) bacterial cells adjacent to the root can be observed through the glass slide by using a form of incident lighting (Johnson and Curl, 1972).

In general, dilution agar plate counts and the use of selective isolation media can only provide limited information on the microbiomass of roots. The inability of plate count media to support all forms of the microbiomass is already known and selective media can only impart information on specific activity groups. Similarly, conventional microscopy of any type can at best only provide information on total numbers and possibly the spatial relationships of organisms, but can never identify the organisms. Fluorescent antibody techniques, whilst capable of demonstrating the distribution of specific types of organisms on the root surface, are probably limited by the number and specificity of antibodies available. Electron microscopy has shown the physical relationship between organisms and root tissues but is limited in both the area which it can cover and its inability to identify specific functional groups of organisms. Nevertheless, electron microscopy has been a valuable tool for elucidating some of the mysteries of the rhizoplane.

Gnotobiotic plants and artificial rhizospheres have been used in attempts to understand the interaction between root exudates and the microbiomass. Exudates can be obtained in a number of ways and are of particular importance in view of the increasing use of systemic pesticides.

Flentje et al. (1963) suggested collecting exudates by growing uncontaminated seedlings in sterile water in Petri dishes for a few days in the dark. The aqueous solution is concentrated, filter-sterilized and stored in the lyophilized state. Rovira (1956a) grew selected, uncontaminated seedlings in large test tubes containing acid-washed quartz sand and a plant nutrient solution. After a period of growth the plants were removed and the roots rinsed. The rinsate and sand-nutrient solution were retained, the sand washed and the washings combined with the nutrient solution and root rinsate which were then filtered or centrifuged. The supernatant was concentrated under vacuum and stored.

Ayers and Thornton (1968) used sand towers of washed, ignited, quartz sand to grow seedlings for a period after which the roots were rinsed in situ by flushing with three 100 ml volumes of distilled water which were then bulked, evaporated to a 50-fold concentration and stored at low temperature.

Buxton (1960) collected exudates from seedlings growing in sterile sand perfused with sterile nutrient solution. This perfusion system permitted samples to be withdrawn at different stages of growth of the plant, evaporated in vacuo and stored frozen. A modification of this system (Buxton, 1962) enables the growth of larger plants, such as banana, in a separate non-sterile unit with a single root diverted to grow into the sterile sand of the perfusion apparatus.

The root exudates can be added to single or mixed microbial cultures (Chan and Katznelson, 1961, 1962) or allowed to dialyse into the soil from within membranes (Timonin, 1941; Riviére, 1958).

Timonin's collodion membrane acts as an artificial rhizosphere, providing information on the types of organisms which are stimulated by the diffusing root exudates contained within the collodion sack, and may be of particular interest in the study of pesticides in root exudates and their effect on certain groups of organisms.

Buxton (1962) compared the germination of fungal spores in the presence and absence of root exudates while Rovira (1956b) tested the growth response to exudates by adding bacterial or mycelial suspensions to sterile liquid medium in test tubes which contained small amounts of exudate.

Another approach to the influence of root exudates on the growth of the rhizosphere microorganisms is described by Rovira (1963). Surface sterilized seeds, inoculated with single or mixed cultures of organisms, are allowed to grow in sterile sand or liquid media. Development of particular organisms in relation to actual roots and their exudates is therefore possible but the system is devoid of influence from non-sterile soil and interactions with other forms of rhizosphere microorganisms. This method may not be suitable for studying root–fungal associations because it is well known that many root-invading fungi are more destructive in their attack upon susceptible plants grown in a sterile environment than to plants exposed to a mixed microflora (Clark, 1969). The system has, however, been widely used to gain a better understanding of the influence of soil organisms on the plant.

None of the methods outlined above appear to provide a convenient means whereby the rhizoplane organisms can be effectively removed from the surface of the root. In this respect, ultrasonication of roots in ultrasonication cleaning tanks, as opposed to single probe units, presents a possible method of stripping larger areas of root surface more quickly, with the minimum volume of liquid. Zak (1962) used this method for isolating fungal symbionts from pine mycorrhizae. After stripping, it may yet become possible to effect a rapid concentration of cells free from root tissue and soil particles by means of density gradient centrifugation, which Mishustina (1973) used for separating microorganisms from soil and mud deposits. Another future possibility may be the use of countercurrent electrophoresis for separating organisms into different sized groups after they have been stripped from the surface of the root.

In posing the question whether modelling systems might be useful for rhizosphere studies, Bowen and Rovira (1973) point out that few quantitative attempts have been made to integrate the mass of information on microbiology and chemistry of the rhizosphere, root exudates and growth rates of organisms and plant roots. They suggest that a modelling framework could assist in a fundamental understanding of the dynamics of root colonization and growth of microorganisms in the rhizosphere.

2. Phyllosphere and phylloplane

Studies of microbial colonization of leaf surfaces are analogous to work on the rhizosphere in that both are complicated by the process of plant elongation. Unlike the rhizosphere, however, variations between leaves of different plants and even on the same plant are much greater and therefore cannot be generalized, thus adding a further parameter to be considered in phylloplane studies.

Methods commonly used in phylloplane studies include culturing of organisms washed from the surface of leaves. Such procedures, like soil and rhizosphere studies, are confounded by the adherence of cells, particularly hyphae, to the leaf surface (Dickinson, 1971). Unlike soil, with its more intimate association of clay minerals and microbial cells, the bacteria, fungi and yeasts of the phylloplane exist in a relatively clean environment and are consequently more readily separated from the leaf surface. Agar plate cultures, whilst capable of providing a means of enumerating those species capable of growth on the media employed, do not give a true reflection of the fungi since many of these exist as hyphal forms which may or may not produce an abundance of spores. Agar plate counts for phylloplane fungi, therefore, mainly reflect the density of spores produced by these fungi and to some extent the numbers of spores *per se* which have impinged upon the leaf surface from airborne spora. Studies by Dickinson (1967) and Beech and Davenport (1971) have led to the conclusion that washing leaf surfaces followed by agar plate counts does not accurately distinguish various microbial forms but does provide an indication of population trends (Dickinson, 1971). Surfactants can help to release organisms from the phylloplane but must be employed carefully since many have fungitoxic properties (Steiner and Watson, 1965). The presence of pesticide spray deposits on leaves may also complicate leaf-washing methods since the pesticide will be uniformly dispersed in the washwater used for plating into agar media with the result that all the organisms may now be in contact with the pesticide whereas on the leaf surface some organisms may have existed between droplet deposits or in areas untouched by the spray. Dickinson (1971) suggests washing on membrane filters to remove all traces of the pesticide from the organisms collected in leaf washings.

According to Dickinson (1971) the selection of suitable general purpose agar media for isolation of phylloplane organisms does present some problems, but not as many as in the selection of media for enumerating soil microbes. The provision of light during incubation is an important parameter not normally encountered in soil microbiology, it being necessary to ensure the development of diagnostic sporing structures.

In contrast to the use of leaf washings for isolation of phylloplane organisms, Dickinson (1971) describes a method of culture from squares or discs of washed leaf material. These are placed in nutritionally poor media such as tap water agar or soil extract agar to ensure that colony development around each leaf piece is sparse, thus enabling the edges and surface to be examined microscopically for sporing structures of fungi which may not form extensive colonies.

Leaf impression methods have a considerable advantage over other methods for removing organisms from the leaf surface, namely that they can be carried out *in situ* thus avoiding the handling and transportation of leaves. The main criticism of leaf impression methods centres on the problem of ascertaining which organisms are removed from the leaf by the technique (Dickinson, 1971). Beech and Davenport (1971) outlined the methods for taking leaf impressions, which vary from the use of a coating of nail varnish which is subsequently peeled-off and examined (Masurovsky and Jordan, 1960), or acetate adhesive tapes (Edwards and Hartman, 1952), to the use of molten agar poured directly onto the leaf surface (Angelotti and Foter, 1958). Adhesive tapes, applied at prescribed pressures by a lever-operated applicator (Woodworth and Newgard, 1963), may then be examined by direct microscopy or laid on agar media, incubated in moist conditions and stained with cotton blue prior to microscopic examination.

Leaves may also be pressed directly onto the surface of agar media and removed prior to incubating. This method gives the distribution patterns of organisms on the leaves, but unfortunately moulds tend to overgrow other organisms. Agar "sausages" (Hodges and Wildman, 1947) are a variation of the plate method in that the exposed end of an agar "sausage" is pressed against the leaf, the slice of agar is cut-off, incubated in moist conditions and examined periodically for microbial growth.

Swabbing techniques or successive rinsing of leaves in separate aliquots of sterile water, like leaf impression methods, do not remove all organisms and furthermore do not provide information on the sequence or pattern of distribution on the leaf surface. As suggested by Beech and Davenport (1971), the use of ultrasonication and vortex mixing may yet provide a more quantitative method for removing organisms from leaf surfaces. Unfortunately, as for soil, such methods can only provide quantitative data and not necessarily information on ecological associations. In all probability direct microscopic examination of leaf segments coupled with impression techniques and enumeration of specific genera by "total count" methods will have to be resorted to in order to obtain the maximum information on the distribution of microorganisms on leaf surfaces. In preliminary studies, Anderson (unpublished data) stained microorganisms on leaf surfaces using a europium salt and a fluorescent brightener and observed the surface by incident u.v. light microscopy. Whilst incapable of identifying specific genera, this method may be of some use in elucidating viable and non-viable counts and associations between organisms.

Maceration of leaves, according to Dickinson (1971), presents two major problems, namely the release of leaf chemicals which may interfere with the growth or survival of some species of organisms. Wood (1967) has indicated that many angiosperm toxins, such as phenols, alkaloids and glycosides, which arise during maceration, may affect growth of various organisms. Furthermore, the fragmentation of some hyphal forms may lead to artificially higher estimates of fungal members of the leaf microflora.

Other methods of isolating or identifying organisms on leaf surfaces include stimulation of some species *in situ* on the leaves by altering the environment in their favour, or by inducing leaf senescence to favour growth of specific saprophytes; spore fall methods, in which leaves are attached to the inside lid of a Petri dish, and species which forcibly discharge their spores (e.g. Sporobolomycetaceae) are allowed to eject the spores directly onto agar medium in the lower half of the dish; baiting methods (Sparrow, 1960; Gaertner, 1965) and the use of selective media (Dickinson, 1971).

IV. Conclusions

Science is dedicated to investigating the unknown and to improving upon methods of investigation such that the unknown becomes less so. It follows therefore that methods are of paramount importance in any investigation. Reviewing existing methods is not a difficult undertaking. The real difficulty comes in recommending the best method for a particular line of investigation and in this context it would be naive to suggest to others which methods are best. If one assumes that all soil microbiology laboratories have recourse to unlimited capital and staff then the recommendations would be more easily made but such idyllic resources are not common.

The use of modern methods and equipment will provide the investigator with an assurance that the data obtained are likely to be more accurate and will also possibly reduce the time involved, but where it is not possible to use modernizations older methods should be equally acceptable provided they have been critically evaluated for reliability. A case in point is the use of flow-through respirometer studies using standard glassware. Carried out in a strictly controlled manner, results from this method are just as acceptable as those from some of the more recent, convenient but often expensive techniques such as the infra-red gas detection system.

Methods should always be considered in the context of suitability, acceptability and reproducibility. Those that are convenient but which have inherently high standard errors should not be used where there exists an alternative and statistically valid method, even if it does take longer. Standard error can frequently be minimized by replication although the plot of error versus replication does show a maximum value beyond which replications make little difference.

It is apparent that in several cases the use of an alternative method brings its own limitations. Replacing the agar plate count by another method for estimating population size is probably the most difficult problem to resolve. The agar plate count provides some information on numbers as well as species since it permits isolations of species to be made, although the method is recognized to have severe limitations in providing true estimates of microbial population size. The alternatives are direct microscopy and chemical analyses neither of which, at the moment, permit identifications of species. Chemical analyses, despite the advantages of having the precision and speed of chemistry whilst requiring minimal sample size and replications, are currently unacceptable for determining population size because distribution of the chemical in question may be uneven among different genera of soil organisms.

There appears, at present, to be no immediate answer to the methodological impasse in determining population size other than a compromise, namely, the use of agar plate counts, direct counts and suitable chemical analyses together, to provide a mass of information from which population size can be gauged, though only for comparative purposes. In some cases more precise information on numbers and distribution of particular species or groups of organisms may best be determined by means of fluorescent antibodies but it is doubtful whether any one method will ever completely replace all others.

Respiration and substrate utilization require little additional comment. Obviously, the use of gas detectors and radioactive substrates can widen the scope of studies and cannot be considered as merely convenience tools since they provide greater accuracy and a more finite interpretation of microbial activities within soil.

Studies of enzymic activity in soil as an integral part of the ecosystem requires greater definition. Simply measuring total enzymic activity of a soil will not suffice. The contributions from different species of microorganisms, plants and cell-free sources will have to be separated if the studies are to be more meaningful. This appears to be more a matter of technique rather than the absence of methods for detecting enzyme activity, although too much dependence has been placed upon existing methods designed for single species in conditions unlike soil.

By far the greater and most important deficiency in modern soil microbiology knowledge remains the interrelationships that exist between microorganisms *per se*, microorganisms and plants, and microorganisms and macroorganisms other than plants. Whereas it is agreed that in order to understand the relationship between individuals it is first necessary to understand the individuals concerned, we are surely at a stage in soil microbiology where greater effort and emphasis should be placed on studying these vital associations and the physical, chemical and biological factors that can promote or inhibit them. Annually we receive a barrage of information, much of it repetitious, on the effects of such-and-such on this organism or that activity but seldom do we see the effects of such-and-such on the interrelationships between various component members of the ecosystem. Even in rhizosphere studies it is the numbers and types of organisms present that receive all the attention, rather than the relationships between the various members of the population or indeed between the organisms and the plant.

Commendable progress has been made by some researchers. The effects of pesticides on *Bdellovibrio* infections of bacteria, the effect of herbicides and other pesticides on natural antagonists of some plant pathogens and on the change in epidemiology of certain diseases are notable but are only scratchings on the surface of a deep and relatively unknown field of study. Increased endeavour in the study of the relations between species will undoubtedly bring to light more methods and techniques to delight those wishing to be involved and perhaps eventually help reviewers.

References

ALEXANDER, F. E. S. and JACKSON, R. M. (1955). Preparation of sections for study of soil micro-organisms. *In* "Soil Zoology" (D. K. McE. Kevan, Ed.), pp. 433–441. Butterworths, London.

ALLISON, F. E., HOOVER, S. R. and MORRIS, H. J. (1937). Physiological studies with the nitrogen-fixing alga, *Nostoc muscorum. Bot. Gaz.* **98**, 433–463.

ANDERSON, J. R. and DAVIES, P. I. (1973). Investigations on the extraction of adenosine triphosphate from soils. In "Modern Methods in the Study of Microbial Ecology" (T. Rosswall, Ed.), Bull. Ecol. Res. Comm. (Stockholm) 17, 271–273.

ANDERSON, J. R. and DREW, E. A. (1976). Effects of pure paraquat dichloride, "Gramoxone W" and formulation additives on soil microbiological activities. III. Estimations of soil microflora and enzyme activity in field treated soil. Zbl. Bakt. Abt. II, 131, 125–135.

ANDERSON, J. R. and SLINGER, J. M. (1975a). Europium chelate and fluorescent brightener staining of soil propagules and their photomicrographic counting. 1. Methods. Soil Biol. Biochem. 7, 205–209.

ANDERSON, J. R. and SLINGER, J. M. (1975b). Europium chelate and fluorescent brightener staining of soil propagules and their photomicrographic counting. II. Efficiency. Soil Biol. Biochem. 7, 210–216.

ANDERSON, J. R. and WESTMORELAND, D. (1971). Direct counts of soil organisms using a fluorescent brightener and a europium chelate. Soil Biol. Biochem. 3, 85–87.

ANGELOTTI, R. and FOTER, M. J. (1958). A direct surface agar plate laboratory method for quantitatively detecting bacterial contamination on non-porous surfaces. Fd Res. 23, 170–174.

ARISTOVSKAYA, T. V. (1973). The use of capillary techniques in ecological studies of microorganisms. In "Modern Methods in the Study of Microbial Ecology" (T. Rosswall, Ed.), Bull. Ecol. Res. Comm. (Stockholm) 17, 47–52.

ARISTOVSKAYA, T. V. and PARINKINA, O. M. (1961). New methods of studying soil microorganism associations. Soviet Soil Sci. 12–20.

ARISTOVSKAYA, T. V. and ZAVARZIN, G. A. (1971). Biochemistry of iron in soil. In "Soil Biochemistry" (A. D. McLaren and J. Skujinš, Eds), vol. 2, pp. 385–408. Marcel Dekker, New York.

AUGIER, J. (1956). A propos de la numeration des Azotobacter en milieu liquide. Annls Inst. Pasteur, Paris 91, 759–765.

AUSMUS, B. S. (1971). Adenosine triphosphate – a measure of active microbial biomass. Biol. Sol. 14, 8–9.

AUSMUS, B. S. (1973). The use of the ATP assay in terrestrial decomposition studies. In "Modern Methods in the Study of Microbial Ecology" (T. Rosswall, Ed.), Bull. Ecol. Res. Comm. (Stockholm) 17, 223–234.

AYERS, W. A. and THORNTON, R. H. (1968). Exudation of amino acids by intact and damaged roots of wheat and peas. Pl. Soil 28, 193–207.

BABIUK, L. A. and PAUL, E. A. (1970). The use of fluorescein isothiocyanate in the determination of the bacterial biomass of grassland soil. Can. J. Microbiol. 16, 57–62.

BARTHA, R. and BORDELEAU, L. (1969). Cell-free peroxidases in soil. Soil Biol. Biochem. 1, 139–143.

BAVER, L. D. (1948). "Soil Physics", 2nd edition. Wiley, New York.

BEECH, F. W. and DAVENPORT, R. R. (1971). A survey of methods for the quantitative examination of the yeast flora of apple and grape leaves. In "Ecology of Leaf

Surface Micro-organisms" (T. F. Preece and C. H. Dickinson, Eds), pp. 139–157. Academic Press, London.

BERGERSEN, F. J. (1970). The quantitative relationship between nitrogen fixation and the acetylene-reduction assay. *Aust. J. biol. Sci.* **23**, 1015–1025.

BIRCH, H. F. and FRIEND, M. T. (1956). Humus decomposition in East African soils. *Nature, Lond.* **178**, 500–501.

BIRCH, H. F. and MELVILLE, M. (1969). An electrolytic respirometer for measuring oxygen uptake in soils. *J. Soil Sci.* **20**, 101–110.

BLAIR, L. D. (1945). Techniques for soil fungus studies. *N.Z. Jl Sci. Technol.* (Feb.), 258–271.

BOHLOOL, B. B. and SCHMIDT, E. L. (1968). Non-specific staining: its control in immunofluorescence examination of soil. *Science, N.Y.* **162**, 1012–1014.

BOWEN, G. D. and ROVIRA, A. D. (1973). Are modelling approaches useful in rhizosphere biology? *In* "Modern Methods in the Study of Microbial Ecology" (T. Rosswall, Ed.), *Bull. Ecol. Res. Comm.* (Stockholm) **17**, 443–450.

BROWN, J. C. (1958). Soil fungi of some British sand dunes in relation to soil type and succession. *J. Ecol.* **46**, 641–664.

BUNT, J. S. and ROVIRA, A. D. (1955). The effect of temperature and heat treatment on soil metabolism. *J. Soil Sci.* **6**, 129–136.

BURGES, A. and NICHOLAS, D. P. (1961). Use of soil sections in studying amounts of fungal hyphae in soil. *Soil Sci.* **92**, 25–29.

BURNS, R. G. and AUDUS, L. J. (1970). Distribution and breakdown of paraquat in soil. *Weed Res.* **10**, 49–58.

BUTLER, F. C. (1953). Saprophytic behaviour of some cereal root fungi. I. Saprophytic colonization of wheat straw. *Ann. appl. Biol.* **40**, 284–297.

BUXTON, E. W. (1960). Effects of pea root exudate on the antagonism of some rhizosphere micro-organisms towards *Fusarium oxysporum* f. *pisi. J. gen. Microbiol.* **22**, 678–689.

BUXTON, E. W. (1962). Root exudates from banana and their relationship to strains of the *Fusarium* causing Panama wilt. *Ann. appl. Biol.* **50**, 269–282.

CASIDA, L. E. (1965). An abundant microorganism in soil. *Appl. Microbiol.* **13**, 327–334.

CASIDA, L. E. (1969). Observations of microorganisms in soil and other natural habitats. *Appl. Microbiol.* **18**, 1065–1071.

CHALVIGNAC, M. A. (1953). Mesure des pouvoirs amylolytique et proteolytique des terres en aerobiose. *Annls Inst. Pasteur, Paris* **84**, 816–818.

CHAN, E. C. S. and KATZNELSON, H. (1961). Growth of *Arthrobacter globiformis* and *Pseudomonas* sp. in relation to the rhizosphere effect. *Can. J. Microbiol.* **7**, 759–767.

CHAN, E. C. S. and KATZNELSON, H. (1962). Bacterial interactions in relation to the rhizosphere effect. *Proc. 8th Int. Congr. Microbiol.*, Sec. B., 53.

CHESTERS, C. G. C. (1940). A method for isolating soil fungi. *Trans. Br. mycol. Soc.* **24**, 352–355.

CHINN, S. H. F. (1953). A slide technique for the study of fungi and actinomycetes in soil with special reference to *Helminthosporium sativum*. *Can. J. Bot.* **31**, 718–724.

CHOLODNY, N. G. (1934). A soil chamber as a method for the microscopic study of the soil microflora. *Arch. Mikrobiol.* **5**, 148.

CLARK, C. G. and WRIGHT, S. J. L. (1970). Detoxication of isopropyl *N*-phenyl-carbamate (IPC) and isopropyl *N*-3-chlorophenylcarbamate (CIPC) in soil, and isolation of IPC-metabolizing bacteria. *Soil Biol. Biochem.* **2**, 19–26.

CLARK, F. E. (1969). Ecological associations among soil micro-organisms. *In* "Soil Biology: Reviews of Research", pp. 125–161. UNESCO, Paris.

CLARK, F. E., JACKSON, R. D. and GARDNER, H. R. (1962). Measurement of microbial thermogenesis in soil. *Soil Sci. Soc. Amer. Proc.* **26**, 155–160.

CLARKE, P. H. and TRACEY, M. V. (1956). The occurrence of chitinase in some bacteria. *J. gen. Microbiol.* **14**, 188–196.

COOK, F. D. and LOCHEAD, A. G. (1959). Growth factor relationships of soil microorganisms as affected by proximity to the plant root. *Can. J. Microbiol.* **5**, 323–334.

CONWAY, E. J. (1950). "Microdiffusion and Volumetric Error". Lockwood, London.

DARKEN, M. A. (1962). Absorption and transport of fluorescent brighteners by microorganisms. *Appl. Microbiol.* **10**, 387–393.

DE BARJAC, H. (1952). La puissance denitrificante du sol. Mise au point d'une technique d'evaluation. *Annls Inst. Pasteur, Paris* **83**, 207–212.

DICKINSON, C. H. (1967). Fungal colonization of *Pisum* leaves. *Can. J. Bot.* **45**, 915–927.

DICKINSON, C. H. (1971). Cultural studies of leaf saprophytes. *In* "Ecology of Leaf Surface Micro-organisms" (T. F. Preece and C. H. Dickinson, Eds), pp. 129–137. Academic Press, London.

DILWORTH, M. J. (1970). The acetylene reduction method for measuring biological nitrogen fixation. *Rhizobium Newsletter* **15**, 7–15.

DOMSCH, K. H., ANDERSON, J. P. E. and AHLERS, R. (1973). Method for simultaneous measurement of radioactive and inactive carbon dioxide evolved from soil samples during incubation with labelled substrates. *Appl. Microbiol.* **25**, 819–824.

DOXTADER, K. G. (1969). Estimation of microbial biomass in the soil on the basis of adenosine triphosphate measurements. Seminar at the 69th Ann. Meeting of the Am. Soc. Microbiol., Miami, USA.

DROBNIK, J. (1955). Spaltung der Stärke durch den enzymatischen Komplex der Böden. *Folia Biol., Praha* **1**, 29.

DUNNING, W. J. (1955). The application of mass spectroscopy to problems of chemical analysis. *Q. Rev. chem. Soc.* **9**, 23 50.

EDWARDS, R. W. and HARTMAN, E. (1952). A simple technique for collecting fungus specimens from infected surfaces. *Lloydia* **15**, 39.

EL-DIN SHARABI, N. and PRAMER, D. (1973). A spectrophotofluorometric method for studying algae in soil. *In* "Modern Methods in the Study of Microbial Ecology" (T. Rosswall, Ed.), *Bull. Ecol. Res. Comm.* (Stockholm) **17**, 77–84.

EREN, J. and PRAMER, D. (1966). Application of immunofluorescent staining to studies of the ecology of soil micro-organisms. *Soil Sci.* **101**, 39–45.

ESTERMAN, E. F., PETERSON, G. H. and McLAREN, A. D. (1959). Digestion of clay-protein, lignin-protein and silica-protein complexes by enzymes and bacteria. *Proc. Soil Sci. Soc. Am.* **23**, 31–36.

EVANS, E. (1955). Survival and re-colonization by fungi in soil treated with formalin or carbon disulphide. *Trans. Br. mycol. Soc.* **38**, 335–346.

FAWCETT, H. S. (1925). Maintained growth rates in fungus cultures of long duration. *Ann. appl. Biol.* **12**, 191–198.

FLENTJE, N. T., DODMAN, R. L. and KERR, A. (1963). The mechanism of host penetration by *Thanatephorus cucumeris. Aust. J. biol. Sci.* **16**, 784–799.

GAERTNER, A. (1965). Köderverfahren zur isolierung niederer phycomyceten. *Zbl. Bakt.* Abt I, 450–461.

GARRETT, S. D. (1956). "Biology of Root Infecting Fungi". Cambridge University Press, Cambridge.

GARRETT, S. D. (1963). "Soil Fungi and Soil Fertility". Pergamon Press, Oxford.

GLATHÉ, H. and THALMANN, A. (1970). Über die mikrobielle Aktivitat und ihre Beziehungen zu Fruchtbarkeitsmerkmalen einiger Ackerboden unter besonderer Berucksichtigung der Dehydrogenaseaktivitat (TTC-Reduktion). *Zbl. Bakt.* Abt. II, **124**, 1–23.

GRAY, T. R. G. (1967). Stereoscan electron microscopy of soil microorganisms. *Science, N.Y.* **155**, 1668–1670.

GRAY, T. R. G. and WILLIAMS, S. T. (1971). "Soil Microorganisms". Oliver and Boyd, Edinburgh.

GRAY, T. R. G., BAXBY, P., HILL, I. R. and GOODFELLOW, M. (1968). Direct observation of bacteria in soil. *In* "The Ecology of Soil Bacteria" (T. R. G. Gray and D. Parkinson, Eds), pp. 171–192. Liverpool University Press, Liverpool.

GREAVES, M. P., WHEATLEY, R. E., SHEPHERD, H. and KNIGHT, A. H. (1973). Relationships between microbial populations and adenosine triphosphate in a basin peat. *Soil Biol. Biochem.* **5**, 685–687.

GREENWOOD, D. J. and LEES, H. (1956). Studies on the decomposition of amino acids in soils. *Pl. Soil* **7**, 253–268.

HARLEY, J. L. and WAID, J. S. (1955). A method of studying active mycelia on living roots and other surfaces in the soil. *Trans. Br. mycol. Soc.* **38**, 104–118.

HILL, I. R. and GRAY, T. R. G. (1967) Application of the fluorescent antibody technique to an ecological study of bacteria in soil. *J. Bact.* **93**, 1888–1896.

HODGES, F. A. and WILDMAN, J. D. (1947). A contact culture method for detecting molds on surfaces. *Science, N.Y.* **106**, 425–426.

HOFMANN, E. and HOFFMANN, G. (1953a). Über das Verkommen von α- und β-glykosidasen im Boden. *Naturwissenschaften* **40**, 511.

HOFMANN, E. and HOFFMANN, G. (1953b). Über das Enzymsystem unserer Kulturböden. 4. Die β-Glucosidase. *Biochem. Z.* **324**, 397–400.

HOFMANN, E. and HOFFMANN, G. (1954). Über das Enzymsystem unserer Kulturböden. 5. α- und β-Galactosidase. *Biochem. Z.* **325**, 329–332.

HOFMANN, E. and HOFFMANN, G. (1955). Über das Enzymsystem unserer Kultur-böden. 6. Amylase. *Z. PflErnähr. Düng. Bodenk.* **70**, 97–104.

HOFMANN, E. and SCHMIDT, W. (1953). Über das Enzymsystem unserer Kultur-böden. 2. Urease. *Biochem. Z.* **324**, 125–127.

HOFMANN, E. and SEEGERER, A. (1951). Über das Enzymsystem unserer Kulturböden. I. Saccharase. *Biochem. Z.* **322**, 174–179.

HOFMANN, E. and TEICHER, K. (1957). Das Enzymsystem unserer Kulturböden. 7. Protease. *Z. PflErnähr. Düng. Bodenk.* **77**, 243–251.

HOWARD, P. J. A. (1968). The use of Dixon and Gilson respirometers in soil and litter respiration studies. Merlewood Research & Development Paper No. 5. Nature Conservancy, U.K.

HUBBARD, J. S. (1973). Radiorespirometric methods in measurement of metabolic activities in soil. *In* "Modern Methods in the Study of Microbial Ecology" (T. Rosswall, Ed.), *Bull. Ecol. Res. Comm.* (Stockholm) **17**, 199–206.

JACOBS, S. I., WILLIS, A. T. and GOODBURN, G. M. (1964). Pigment production and enzymatic activity of Staphylococci: the differentiation of pathogens from com-mensals. *J. Path. Bact.* **87**, 151–156.

JENKINSON, D. S. (1966). Studies on the decomposition of plant material in soil. II. Partial sterilisation and the soil biomass. *J. Soil Sci.* **17**, 280–302.

JENNY, H. and GROSSENBACHER, K. (1963). Root–soil boundary zones as seen by the electron microscope. *Soil Sci. Soc. Amer. Proc.* **27**, 273–277.

JENSEN, H. L. (1939). Contributions to the microbiology of Australian soils. V. *Proc. Linn. Soc. N.S.W.* **64**, 601–608.

JOHNSON, L. F. and CURL, E. A. (1972). "Methods for the Research on the Ecology of Soil-Borne Plant Pathogens". Burgess Publishing Co., Minnesota.

JONES, P. C. T. and MOLLISON, J. E. (1948). A technique for the quantitative estimation of soil micro-organisms. *J. gen. Microbiol.* **2**, 54–69.

KAISER, P. and PRÉVOT, A. R. (1958). Methode amelioree d'etude quantitative de la pectinolyse dans le sol. *C.r. hebd. Séanc. Acad. Sci. Paris* **247**, 1065–1067.

KATZNELSON, H. and ROUATT, J. W. (1957). Manometric studies with rhizosphere and non-rhizosphere soil. *Can. J. Microbiol.* **3**, 673–678.

KAUFFMANN, J. and CHALVIGNAC, M. A. (1950). Recherches sur les methodes de mesure du pouvoir ammonificateur d'une terre. *Annls Inst. Pasteur, Paris* **79**, 228–232.

KEULEMANS, A. I. M. (1959). "Gas Chromatography", 2nd edition. Reinhold, New York.

KNIGHT, B. A. G. and DENNY, P. J. (1970). The interaction of paraquat with soil: adsorption by an expanding lattice clay mineral. *Weed Res.* **10**, 40–48.

KNOWLES, R., BROUZES, R. and O'TOOLE, P. (1973). Kinetics of nitrogen fixation and acetylene reduction, and effects of oxygen and of acetylene on these processes in a soil system. *In* "Modern Methods in the Study of Microbial Ecology" (T. Rosswall, Ed.), *Bull. Ecol. Res. Comm.* (Stockholm) **17**, 255–262.

KOKKE, R. (1970). Pesticide and herbicide interaction with microbial ecosystems. *Antonie van Leeuwenhoek* **36**, 580–581.

LAJUDIE, J. and CHALVIGNAC, M. A. (1956). Appreciation de l'activite proteolytique de la microflore du sol. *Annls Inst. Pasteur, Paris* **90**, 359–361.

LAMBINA, V. A. (1957). Distribution of soil bacteria fermenting plant protopectin in fields under grassland crop rotation. *Microbiology (Mikrobiologiya)* **26**, 71–78.

LaRUE, T. A. and KURZ, W. G. W. (1973). Estimation of nitrogenase using a colorimetric determination for ethylene. *Pl. Physiol.* **51**, 1074–1075.

LEE, C. C., HARRIS, R. F., WILLIAMS, J. D., SYERS, J. K. and ARMSTRONG, D. E. (1971). Adenosine triphosphate in lake sediments: II. Origin and significance. *Proc. Soil Sci. Soc. Am.* **35**, 86–91.

LEES, H. and QUASTEL, J. H. (1946). Biochemistry of nitrification in soil. I. Kinetics of, and the effect of poisons on, soil nitrification as studied by a soil perfusion technique. *Biochem. J.* **40**, 803–815.

LENHARD, G. (1956). Die Dehydrogenaseaktivitat des Bodens als Mass fur die Mikroorganismentategkeit im Boden. *Z. PflErnähr. Düng. Bodenk.* **73**, 1–11.

LINGAPPA, B. T. and LOCKWOOD, J. L. (1963). Direct assay of soils for fungistasis. *Phytopathology* **53**, 529–531.

LOUW, H. A. and WEBLEY, D. M. (1959). The bacteriology of the root region of the oat plant grown under controlled pot culture conditions. *J. appl. Bact.* **22**, 216–226.

MACKENZIE, K. A. and MacRAE, I. C. (1971). Apparatus for the manipulation of gas mixtures used in the acetylene reduction technique. *Soil Biol. Int. News Bull.* No. 14, 12–15.

MASUROVSKY, E. B. and JORDAN, W. K. (1960). A plastic replica-embedding and staining technique for studying the behaviour of microorganisms on food contact surfaces. *J. Dairy Sci.* **43**, 1000–1003.

MAYAUDON, J. and SIMONART, P. (1958). Study of the decomposition of organic matter in soil by means of radioactive carbon. II. The decomposition of radioactive glucose in soil and distribution of radioactivity in the humus fractions of soil. *Pl. Soil* **9**, 376–380.

MAYAUDON, J. and SIMONART, P. (1959a). Etude de la decomposition de la matiere organique dans le sol au moyen de carbone radioactif. IV. Decomposition de pigments fol. *Pl. Soil* **11**, 176–180.

MAYAUDON, J. and SIMONART, P. (1959b). Etude de la decomposition de la matiere organique dans le sol au moyen de carbone radioactif. V. Decomposition de cellulose et de lignine. *Pl. Soil* **11**, 181–191.

MAYAUDON, J. and SIMONART, P. (1960). Decomposition of cellulose C^{14} and lignin C^{14} in the soil. *In* "The Ecology of Soil Fungi" (D. Parkinson and J. S. Waid, Eds), pp. 257–259. Liverpool University Press, Liverpool.

McGARITY, J. W. (1961). Denitrification studies on some south Australian soils. *Pl. Soil* **14**, 1–21.

McGARITY, J. W., GILMOUR, C. M. and BOLLEN, W. B. (1958). Use of an electrolytic respirometer to study denitrification in soil. *Can. J. Microbiol.* **4**, 303–316.

McLAREN, A. D. (1963). Enzyme activity in soils sterilized by ionising radiation and some comments on micro-environments in nature. *In* "Recent Progress in Microbiology" (N. E. Gibbons, Ed.), pp. 221–229. Univ. Toronto Press, Toronto.

McLAREN, A. D., RESHETKO, L. and HUBER, W. (1957). Sterilization of soil by irradiation with an electron beam and some observations on soil enzyme activity. *Soil Sci.* **83**, 497.

McLENNAN, E. (1928). The growth of fungi in the soil. *Ann. appl. Biol.* **15**, 95–109.

MILLAR, W. N. and CASIDA, L. E. (1970). Evidence for muramic acid in soil. *Can. J. Microbiol.* **16**, 299–304.

MILLER, F. A. (1953). Applications of infrared and ultraviolet spectra to organic chemistry. *In* "Organic Chemistry: An Advanced Treatise III" (H. Gilman, Ed.), pp. 122–177. Wiley, New York.

MINDERMAN, G. (1956). The preparation of microtome sections of unaltered soil for the study of soil organisms *in situ*. *Pl. Soil* **8**, 42–48.

MISHUSTINA, I. (1973). Submicroscopic forms in marine muds isolated by a density gradient method and studied by electronmicroscopy. *In* "Modern Methods in the Study of Microbial Ecology" (T. Rosswall, Ed.), *Bull. Ecol. Res. Comm.* (Stockholm) **17**, 143–149.

MOORE-LANDECKER, E. and STOTZKY, G. (1972). Inhibition of fungal growth and sporulation by volatile metabolites from bacteria. *Can. J. Microbiol.* **18**, 957–965.

MOREAU, R. and van EFFENTERRE, A. (1961). La technique des repliques et l'observation directe des elements fongiques de la rhizosphere. *Annls. Inst. Pasteur, Paris* **101**, 619–625.

MORTENSEN, U., NORÉN, B. and WADSÖ, I. (1973). Microcalorimetry in the study of the activity of microorganisms. *In* "Modern Methods in the Study of Microbial Ecology" (T. Rosswall, Ed.), *Bull. Ecol. Res. Comm.* (Stockholm) **17**, 189–197.

NEWMAN, A. S. and NORMAN, A. G. (1943). An examination of thermal methods for following microbiological activity in soil. *Proc. Soil Sci. Soc. Am.* **8**, 250–253.

NICHOLAS, D. P., PARKINSON, D. and BURGES, N. A. (1965). Studies of fungi in a podzol. II. Application of the soil sectioning technique to the study of amounts of fungal mycelium in the soil. *J. Soil Sci.* **16**, 258–269.

NIKITIN, D. I. (1964). Use of electron microscopy in the study of soil suspensions and cultures of microorganisms. *Soviet Soil Sci.* 1964, 636–641.

NIKITIN, D. I. and MAKAR'YEVA, E. D. (1970). Use of the electron microscope for quantitative estimation of microorganisms in soil suspensions. *Soviet Soil Sci.* **1970**, 608–612.

ORENSKI, S. W., BYSTRICKY, V. and MARAMOROSCH, K. (1966). The occurrence of microbial forms of unusual morphology in European and Asian soils. *Can. J. Microbiol.* **12**, 1291–1298.

PARKINSON, D. (1957). New methods for the qualitative and quantitative study of fungi in the rhizosphere. *Pédologie, Gand.* **7**, 146–154.

PARKINSON, D. and CLARKE, J. H. (1961). Fungi associated with the seedling roots of *Allum porrum* L. *Pl. Soil* **13**, 384–390.

PARKINSON, D. and COUPS, E. (1963). Microbial activity in a podzol. *In* "Soil Organisms" (J. Doeksen and J. van der Drift, Eds), pp. 167–175. North Holland, Amsterdam.

PARKINSON, D. and THOMAS, A. (1965). A comparison of methods for the isolation of fungi from rhizosphere. *Can. J. Microbiol.* **11**, 1001–1007.

PARKINSON, D., GRAY, T. R. G. and WILLIAMS, S. T. (1971). "Methods for Studying the Ecology of Soil Micro-organisms". IBP Handbook No. 19. Blackwell, Oxford.

PAYNE, W. J. (1973). The use of gas chromatography for studies of denitrification in ecosystems. *In* "Modern Methods in the Study of Microbial Ecology" (T. Rosswall, Ed.), *Bull. Ecol. Res. Comm.* (Stockholm) **17**, 263–268.

PERFIL'EV, B. V. and GABE, D. R. (1969). "Capillary methods of studying microorganisms". Oliver and Boyd, Edinburgh.

POCHON, J. (1957). Principe et application d'une methodologie quantitative. *Pédologie, Gand* **7**, 56–61.

POCHON, J. and AUGIER, J. (1954). Premières recherches sur l'attaque des hémicelluloses dans le sol. *Trans. 5th Int. Congr. Soil Sci.*, Léopoldville **3**, 43–45.

POCHON, J. and AUGIER, J. (1956). Recherches sur la numération des hémicellulolytiques telluriques. *Trans. 6th Int. Congr. Soil Sci.*, Paris. **3**, 277–280.

POCHON, J. and CHALVIGNAC, M. A. (1952). Recherches sur la proteolyse bacterienne au sein du sol. *Annls. Inst. Pasteur, Paris* **82**, 690–695.

POCHON, J., TARDIEUX, P. and D'AGUILAR, J. (1969). Methodological problems in soil biology. *In* "Soil Biology: Reviews of Research", pp. 13–63. UNESCO, Paris.

REINERS, W. A. (1968). Carbon dioxide evolution from the floor of three Minnesota forests. *Ecology* **49**, 471–483.

REYES, A. A. and MITCHELL, J. E. (1962). Growth response of several isolates of *Fusarium* in rhizospheres of host and non-host plants. *Phytopathology* **52**, 1196–1200.

RIVIÉRE, J. (1958). Etude microbiologique de la rhizosphere du blé. II. Realisation et analyse d'une rhizosphere artificielle. *Annls. Inst. Pasteur, Paris* **95**, 231–234.

ROSSI, G. M., RICCARDO, S., GESUÉ, G., STANGANELLA, M. and WANG, T. K. (1936). Direct microscopic and bacteriological examination of the soil. *Soil Sci.* **41**, 53–66.

ROUATT, J. W., KATZNELSON, H. and PAYNE, T. M. B. (1960). Statistical evaluation of the rhizosphere effect. *Proc. Soil Sci. Soc. Am.* **24**, 271–273.

ROUSCHAL, C. and STRUGGER, S. (1943). A new method for observation of microorganisms in the soil *in vivo*. *Naturwissenschaften* **31**, 300.

ROVIRA, A. D. (1956a). Plant root excretions in relation to the rhizosphere effect. I. The nature of root exudate from oats and peas. *Pl. Soil* **7**, 178–217.

ROVIRA, A. D. (1956b). Plant root excretions in relation to the rhizosphere effect. II. A study of the properties of root exudate and its effects on the growth of microorganisms isolated from the rhizosphere and control soil. *Pl. Soil* **7**, 195–208.

ROVIRA, A. D. (1963). Microbial inoculation of plants. I. Establishment of free-living nitrogen-fixing bacteria in the rhizosphere and their effects on maize, tomato and wheat. *Pl. Soil* **19**, 304–314.

RUNKLES, J. R., SCOTT, A. D. and NAKAYAMA, F. S. (1958). Oxygen sorption by moist soils and vermiculite. *Proc. Soil Sci. Soc. Am.* **22**, 15–18.

SCAFF, W. L., DYER, D. L. and MORI, K. (1969). Fluorescent europium chelate stain. *J. Bact.* **98**, 246–248.

SCHMIDT, E. L. and BANKOLE, R. O. (1962). Detection of *Aspergillus flavus* in soil by immunofluorescent staining. *Science, N.Y.* **136**, 776–777.

SEEGERER, A. (1953). Der Saccharasegehalt des Bodens als Masstab seiner biologischen Aktivitat. *Z. PflErnähr. Düng. Bodenk.* **61**, 251–260.

SINGH, K. G. (1965). Comparison of techniques for the isolation of root-infecting fungi. *Nature, Lond.* **206**, 1169–1170.

SKERMAN, V. B. D. (1967). "A Guide to the Identification of the Genera of Bacteria", 2nd edition. Williams and Wilkins, Baltimore.

SKUJINŠ, J. J. (1967). Enzymes in Soil. *In* "Soil Biochemistry" (A. D. McLaren and G. H. Peterson, Eds), pp. 371–414. Marcel Dekker, New York.

SPARROW, F. K. (1960). "Aquatic Phycomycetes". University of Michigan, Ann Arbor.

STARKEY, R. L. (1938). Some influences of the development of higher plants upon the microorganisms in the soil. VI. Microscopic examination of the rhizosphere. *Soil Sci.* **45**, 207–249.

STEINER, G. W. and WATSON, R. D. (1965). The effect of surfactants on growth of fungi. *Phytopathology* **55**, 1009–1012.

STEVENSON, I. L. (1959). Dehydrogenase activity in soils. *Can. J. Microbiol.* **5**, 229–235.

STOTZKY, G. (1965a). Replica plating techniques for studying microbial interactions in soil. *Can. J. Microbiol.* **11**, 629–636.

STOTZKY, G. (1965b). Microbial Respiration. *In* "Methods of Soil Analysis. Part 2. Chemical and Microbiological Properties" (C. A. Black *et al.*, Eds), pp. 1550–1572. American Soc. Agron., Madison.

STOTZKY, G. (1966). Influence of clay minerals on microorganisms. III. Effect of particle size, cation exchange capacity and surface area on bacteria. *Can. J. Microbiol.* **12**, 1235–1242.

STOTZKY, G. (1972). Activity, ecology, and population dynamics of microorganisms in soil. *Cr. Rev. Microbiol.* **2** (1), 59–137.

STOTZKY, G. and GOOS, R. D. (1965). Effect of high CO_2 and low O_2 tensions on the soil microbiota. *Can. J. Microbiol.* **11**, 853–868.

STOTZKY, G. and REM, L. T. (1967). Influence of clay minerals on microorganisms. IV. Montmorillonite and kaolinite on fungi. *Can. J. Microbiol.* **13**, 1535–1550.

STOVER, R. H. and WAITE, B. H. (1953). An improved method for isolating *Fusarium* spp. from plant tissue. *Phytopathology* **43**, 700–701.

STRUGGER, S. (1948). Fluorescence microscope examination of bacteria in soil. *Can. J. Res.* **26**(B), 188–193.

STRUGGER, S. and HILBRICH, P. (1942). Fluorescence microscopic differentiation between live and dead bacteria with the aid of acridine orange staining. *Deut. Tierarztl. Wochenschr.* **50**, 121–130.

SZEGI, J. (1970). Effect of some herbicides on the growth of cellulose-decomposing microscopic fungi. *Meded. Fac. Llandb. Rijksuniv. Gent* **35**, 559–561.

TABATABAI, M. A. and BREMNER, J. M. (1969). Use of p-nitrophenylphosphate for the assay of soil phosphatase activity. *Soil Biol. Biochem.* **1**, 301–307.

TAYLOR, J. (1962). The estimation of numbers of bacteria by tenfold dilution series. *J. appl. Bact.* **25**, 54–61.

TCHAN, Y. T. (1952). Counting soil algae by direct fluorescence microscopy. *Nature, Lond.* **170**, 328–329.

TCHAN, Y. T. (1959). Study of soil algae. III. Bioassay of soil fertility by algae. *Pl. Soil* **10**, 220–231.

THORNTON, R. H. (1952). The screened immersion plate. A method for isolating soil microorganisms. *Research, Lond.* **5**, 190–191.

THORNTON, R. H. and GRAY, P. H. H. (1934). The numbers of bacterial cells in field soils, as estimated by the ratio method. *Proc. R. Soc.* Ser B, **115**, 522–543.

TIMONIN, M. I. (1940). The interaction of higher plants and soil microorganisms. I. Microbial population of rhizosphere of seedlings of certain cultivated plants. *Can. J. Res.* **18**(C) 307–317.

TIMONIN, M. I. (1941). The interaction of higher plants and soil microorganisms. III. Effect of by-products of plant growth on activity of fungi and actinomycetes. *Soil Sci.* **52**, 395–413.

TODD, R. L. and NUNER, J. H. (1973). Comparison of two techniques for assessing denitrification in terrestrial ecosystems. *In* "Modern Methods in the Study of Microbial Ecology" (T. Rosswall, Ed.), *Bull. Ecol. Res. Comm.* (Stockholm) **17**, 277–278.

TODD, R. L., CROMACK, K. and KNUTSON, R. M. (1973). Scanning electron microscopy in the study of terrestrial microbial ecology. *In* "Modern Methods in the Study of Microbial Ecology" (T. Rosswall, Ed.), *Bull. Ecol. Res. Comm.* (Stockholm) **17**, 109–118.

TRECCANI, V. (1961). Richerche sul potere denitrificante del terreno e proposta di un metodo quantitativo di valutazione. *Annali Microbiol.* **11**, 15–18.

VELDKAMP, H. (1955). A study of the aerobic decomposition of chitin by microorganisms. *Meded. LandbHoogesch. Wageningen* **55**, 127–174.

WADSÖ, I. (1970). Microcalorimeters. *Q. Rev. Biophys.* **3**, 383–427.

WAID, J. S. (1956). Root dissection: A method of studying the distribution of active mycelia within root tissue. *Nature, Lond.* **178**, 1477–1478.

WAID, J. S. and WOODMAN, M. J. (1957). A method of estimating hyphal activity in soil. *Symp. Methods d'Etudes Mikrobiologiques du Sol* **3**, 1–6.

WAKSMAN, S. A. (1952). "Soil Microbiology". Wiley, New York.

WALLIS, G. W. and WILDE, S. A. (1957). Rapid determination of carbon dioxide evolved from forest soils. *Ecology* **38**, 359–361.

WALLNÖFER, P. (1968). Untersuchengen über die antimikrobielle wirkung von diquat und paraquat. *Z. Pflanzenkr. Pflanzenschutz* **75**, 218–224.

WARCUP, J. H. (1960). Methods for isolation and estimation of activity of fungi in soil. *In* "The Ecology of Soil Fungi" (D. Parkinson and J. S. Waid, Eds), pp. 3–21. Liverpool University Press, Liverpool.

WASTIE, R. L. (1961). Factors affecting competitive saprophytic colonization of the agar plate by various root-infecting fungi. *Trans. Br. mycol. Soc.* **44**, 145–149.

WATSON, R. D. (1960). Soil washing improves the value of the soil dilution and the plate count method of estimating populations of soil fungi. *Phytopathology* **50**, 792–794.

WHEELER, B. E. J. (1963). Denitrification in percolated soils. *Pl. Soil* **19**, 219–232.

WIJLER, J. and DELWICHE, C. C. (1954). Investigations on the denitrifying process in soil. *Pl. Soil* **5**, 155–169.

WILLIAMS, S. T., PARKINSON, D. and BURGES, N. A. (1965). An examination of the soil washing technique by its application to several soils. *Pl. Soil* **22**, 167–186.

WINOGRADSKY, S. N. (1926). Etudes sur la microbiologie du sol. *Annls Inst. Pasteur, Paris* **40**, 455.

WITKAMP, M. (1966). Decomposition of leaf litter in relation to environment, microflora and microbial respiration. *Ecology* **47**, 194–201.

WOOD, R. K. S. (1967). "Physiological Plant Pathology". Blackwell Scientific Publications, Oxford.

WOODWORTH, H. H. and NEWGARD, P. M. (1963). Measuring cleaning effectiveness. *Res. Ind., Stanford* **15**, 8–9.

WRIGHT, S. J. L. and CLARK, C. G. (1969). Controlled soil perfusion with a multi-channel peristaltic pump. *Weed Res.* **9**, 65–68.

ZAK, B. (1962). Direct isolation of fungal symbionts from pine mycorrhizae. *Phytopathology* **52**, 34.

ZVYAGINTSEV, D. G. (1963). Use of quenchers in the investigation of soil microorganisms by fluorescence microscopy. *Microbiology (Mikrobiologiya)* **32**, 622–625.

Chapter 7

Pesticide Effects on Non-Target Soil Microorganisms

JOHN R. ANDERSON

*Mircobiology Unit, National Food Research Institute, C.S.I.R.
Pretoria, South Africa.*

I. Introduction

It is not intended in this chapter to condemn, condone, or propound arguments regarding particular pesticides according to their effects on non-target microorganisms, since the available evidence will not permit it. At the most, generalizations can be made and trends identified. In some cases the results obtained by independent researchers are conflicting. This merely underlines our present inadequate knowledge, the differences between soil types and climatological areas and the inconsistent methods used in determining effects of pesticides on non-target organisms.

Whilst pesticide structure–activity relationship studies have been advanced considerably in recent years, such knowledge is largely concerned with effects of molecular arrangements on the target organisms and has not yet been extended to the non-target microorganisms, interesting, if difficult, this approach would be.

This review mainly covers work published between c. 1965 and 1976. The chemical names of pesticides herein referred to by common names are given in the pesticide list at the end of the book.

II. Factors Influencing Pesticide Effects on Non-Target Microorganisms

A. SOIL VARIABILITY

The texture, mineral clay and organic matter (O.M.) contents, pH, conductivity and moisture holding capacities (M.H.C.) of soils are known to vary considerably with, and even within, geographical regions. Furthermore, a given soil-type may exhibit minor localized variations within short distances and whereas these small differences may lack significance to the soil chemist they may be sufficient to determine whether or not a specific microbiological activity will be present.

When the above minor differences within one soil type are further influenced by soil temperature variability, profile characteristics and drainage, it becomes obvious to the soil microbiologist that the potential variations in population type and activities may be considerable. The differences in soil microbial biomass composition as a direct result of soil variation or inconsistency may therefore determine the extent of effect imparted by the pesticide and will also be important to its chemical reactivity.

B. AGRONOMIC PRACTICE AND PLANT COVER

In soils of the same chemical or geological class, the type of plant cover may influence the composition of part of the soil microbial biomass. Soil under grass will have a somewhat different gross biomass compared to the same soil under grass–clover leys, wheat, maize, citrus etc. The differences will stem mainly from differences in the composition of the attendant rhizosphere populations which are themselves determined largely by the nature of root exudates and the composition of the organic matter which each form of plant cover produces. In addition, the type of plant cover may also affect soil pH and moisture. Different soil types under identical plant cover may also produce variations in type and quantity of root exudate as a result of differences in rates of plant growth or in the soil nutrient levels.

Considerable differences exist between cropped and uncropped soil and this is undoubtedly due to the presence of an active rhizosphere in the cropped soil. However, in tropical and sub-tropical regions the difference may also result from rapid oxidation of organic matter leading to a decrease in oxidizable carbon in the fallow soil whereas the cropped soil receives constant replenishment from roots and leaf material. The condition of the fallow soil will also determine the type of soil microbial biomass and its activity. Ploughing in crop residues generally accelerates microbial decomposition of the organic matter whereas leaving residues to accumulate favours slower oxidation. Thus the type, mass and activity of microorganisms under these two extremes of agronomic practice can differ considerably. Further profound differences in microbial mass and composition can be expected as a result of irrigation, form of tillage, type and frequency of fertilizer application, the cropping system and the nature of organic material reaching the soil.

In essence, the variations in microbial strain and species composition which are determined by soil type and conditions, agronomic practice and plant populations may be important in determining whether or not a pesticide will affect population size and activity of the microbial biomass.

C. PESTICIDE FORMULATION

The form in which the pesticide is added to soil is especially important. If the commercial formulation is used, the investigator can never be certain whether or not the formulation additives are responsible for any observed changes in the microbial activity. If the pure active pesticide ingredient only is used the investigator will not be certain that the effects noted would not have been exacerbated by the presence of the formulation additives.

On this basis it would appear that investigations should be conducted separately with (i) the pure active ingredient, (ii) the formulation additives and (iii) the combined active ingredient and formulation additives. For examples of the effects of individual components, see Wilkinson and Lucas (1969a, b, c), Anderson and Drew (1976a, b, c) and Quilt et al. (in press).

Even so-called inert ingredients (e.g. pure kaolinite) may affect the soil microorganisms. Whilst rates of application per hectare may not be great for the inert ingredients, it is the method of application that could render their presence important. Thus, applications of pelleted pesticide in the planting furrow or around and between plants within rows will result in a higher localized concentration of the pesticide and the inert carrier than the equivalent rate spread uniformly over one hectare as a dust or spray. A localized increase of the carrier in a very sandy soil, for example, could increase microbial activity in the vicinity of the pesticide. Stotzky (1966) demonstrated that clay minerals can affect the activities of soil microorganisms, whilst Novakova (1974) showed that kaolinite-amended nutrient salts medium enhanced mineralization of peptone in the medium.

D. TYPE OF PESTICIDE AND METHOD OF APPLICATION

The difference between non-systemic and systemic pesticides may be of considerable importance in determining the outcome of investigations on their effects on soil microorganisms. Those tissues not in direct contact with a non-systemic pesticide will probably be microbially degraded in the usual manner, whereas a systemic pesticide may influence the rate of degradation of all component tissues that have absorbed it. This raises the question as to whether investigations on non-systemic pesticides should be executed with only those organic matter fractions that are actually in contact with the pesticide (e.g. leaf epidermis or insect exoskeleton) rather than intact wheat straws, leaves, etc. Finely divided plant material, intimately mixed with the non-systemic pesticide, would approach field conditions, but it could be argued that under natural conditions finely divided plant material takes some time to form through the action of soil macroflora and that microbial degradation normally commences much earlier with the intact plant material. With systemic pesticides it is possible to spray or root-drench plants with the pesticide and obtain plant material with the pesticide incorporated in the tissues for organic matter decomposition studies. These considerations are highly pertinent to simulating degradation in the laboratory and should be considered carefully by the investigator.

The methods of application of pesticides are not always duplicated accurately under laboratory conditions. The incorporation of pesticides into soil to be used for respiration, enumeration or organic matter degradation studies seldom approaches the actual conditions of field applications and consequently the effects recorded in the laboratory may often be remote from natural conditions. In a grass ley, for example, very little of a fungicide or herbicide may actually contact the soil surface, the bulk of the material remaining on the leaves. This material may become incorporated into soil but significant changes may have occurred to the chemical in the interim. These changes might include photodegradation and also metabolism by phyllosphere organisms and the plant itself. Thus, soaking grass cuttings in solutions of the pesticide and incorporating this material into the soil may approach the field conditions more accurately than mixing soil and pesticide and then adding organic matter in order to study its rate of degradation or invasion by microorganisms.

Foliar applications of sprays to large plants and trees are a particular problem for the investigator since the extent and pattern of pesticide run-off on to the underlying soil (with or without plant cover) is very difficult to duplicate in the laboratory. In this case it is probably better to mix soil samples intimately with the pesticide, stipulating that the effects noted will apply only to soil which receives run-off from the leaves and that soil which does not receive run-off may serve as a re-inoculation zone for the affected areas.

The above discussion has dealt with the simulation of pesticide application to soil and plant material in the laboratory. There is much to be said for applying the pesticide to field plots and removing samples of the "naturally" treated soil and plant material to the laboratory in order to conduct investigations. However, in many cases the investigator is forced to process these samples (by sieving, bulking or diluting) in order to conduct the experiments. These physical disruptions to the sample may totally negate the principle of using treated field samples by redistributing the pesticide into soil and plant material that would normally not be influenced by it. Hence, the soil microbiologist is faced with yet another dilemma which could seriously affect the conclusions arrived at regarding the effects of a pesticide on non-target organisms.

Finally, there is the problem of rates of application of the pesticide. Many investigators consider that at least two rates, namely the normal recommended field rate and 10 to 100 times this rate, are sufficient for an investigation into effects of a pesticide. Bearing in mind the necessity for replications of both treatment rates and the analyses, this approach is reasonable, particularly if the active ingredient, the formulation chemicals and the formulated pesticide are to be investigated individually. There is also merit in including a sub-recommended rate of, for example, one-tenth the normal field rate.

E. LABORATORY TECHNIQUE

Methods available for studying the effects of pesticides on non-target organisms have been outlined in Chapter 6. Subtle variations in an otherwise standard method may, and frequently do, lead to results sufficiently different to those obtained by other workers. Thus, ultrasonication of soil suspensions, whilst providing greater dispersion of soil particles and propagules, is seldom adequately described for other investigators to duplicate the work. The performance of two separate models of ultrasonicator may be sufficiently different as to affect the degree of dispersion of organisms in identical soil types.

Other variables in soil microbiological technique which may affect results obtained in different laboratories include the number of times a dilution agar plate is rotated to disperse the organisms, the length of time between diluting the soil and pouring the plates, the conditions of incubation (e.g. in light or dark), randomized positioning of the plates in the incubator, or the provision for ventilation of the chamber. Perfusion experiments may differ considerably in the rate and size of droplet delivery to the soil sample. Moisture content of soils, determined by three well known methods, may provide different answers on the same sample and direct microscopy counts of soil smears, using aniline dyes or fluorescent staining, can give widely different results even between two operators using the same preparation and method of staining.

Perfection in duplicating soil microbiological experiments is very difficult to achieve. Jensen (1968) stated that: "By collaboration between several laboratories it was demonstrated that plate counts on the same soil samples made in different laboratories under conditions as strictly as possible still show significant differences. It was not possible to establish the cause of these differences exactly, but the plate count method involves many operations which are difficult to specify in detail. One of the most important sources of error was believed to be the variation in methods of preparation of the plating medium (soil extract)". Jensen points out that the numerous publications relating to the plate count technique makes it virtually impossible to standardize. Many inexplicable divergences and disagreements between different investigators do not permit standardization of the technique and therefore, by inference, reproducibility of results. This applies especially in determining the effects of pesticides on the greater number of non-target organisms that exist in soil as opposed to the few involved in bio-assay techniques for determining effectiveness of a compound.

One of the greatest sources of divergent results obtained for a particular pesticide continues to be the use of pure culture experiments, the results of which are extrapolated to soil conditions. Whilst pure cultures undoubtedly have a vital role in providing information on the mode of action of pesticides on non-target organisms in particular, their use in determining whether the pesticide will have an effect in soil is rapidly losing favour. If such methods must be used, then the addition of pure clay minerals would greatly improve the credibility of the results.

The majority of factors listed above constantly attend the efforts of soil microbiologists in their attempts to establish the probable effects of pesticides, *in situ*, on the soil's microbial biomass. There can be few sciences with as many variables and pitfalls as this. Many of the difficulties are within our ability to rectify but this will only be achieved by international agreement on methodology and technique for monitoring pesticide effects. I once witnessed an irate soil scientist at an international symposium on methodology loudly denounce any form of agreement because "nobody was going to tell *him* how to conduct his investigations". *Quo vadit studium mikrobium in solo?*

III. Effects of Pesticides on Non-Target Soil Microorganisms

In the following account of the effects of pesticides on microorganisms, four general categories of pesticides have been recognized; herbicides, fungicides, insecticides and others (mainly fumigants). In some cases it is possible to provide only the barest essential details. This is due to the fact that a high proportion of the international literature is not available in translated form, in which case only that information appearing in abstracting journals can be presented. Also, all too frequently, details of the type of formulation used, the nature of the soil type, its pH and organic content are not given. This not only applies to abstracted or summarized data since even the more widely circulated international journals are unfortunately often guilty of allowing authors to omit such vital details.

Notes on tabulated data and abbreviations

1. In tables and text, organic matter and moisture holding capacity have been abbreviated to O.M. (o.m.) and MHC, respectively. Active ingredient of pesticide has been abbreviated to a.i.

2. Rates of application of pesticides to media have been standardized to parts per million (ppm), and to soil as kilograms (kg) or litres per hectare (ha). In the case of soil, all conversions from other units to kg or litres ha^{-1} have been based upon 10^8 grams of soil ha^{-1} to a depth of 1 cm, assuming an average bulk density for soil of 1·0. All conversions have been made in this manner unless a specific depth of incorporation of the pesticide into the soil has been given by the original author(s).

3. Rates designated in the tables as e.g. 1·0; 5; 10 indicate three separate rates were used. Rates designated as 1·0–20 indicate a range of concentrations from 1·0 to 20, inclusive.

4. Where soil-type or pesticide concentration has not been given, the lack of information has been denoted by (−) in the tables.

5. In some cases, data in abstracting journals refer to the use of "recommended field rates". In such cases, the recommended field rates given in the Pesticide Manual, 4th edition (Martin and Worthing, 1974), have been inserted in the tabulated information and denoted * to indicate that the rates are approximate since field rates may vary from country to country.

6. No attempt has been made to convert quoted rates of commercially formulated pesticides into a.i. rates unless the percentage a.i. in the pesticide is given by the original author(s). Rates given in the following tables are, therefore, shown as kg or litres ha^{-1} or as kg a.i. or litres a.i. ha^{-1}, depending on the information available.

7. Where specific salts of a pesticide have been used, this information has been included in the appropriate column of the tables or in the text. Unless otherwise stated, the form of the compound given can be assumed to be the most common form used or available.

Notes on ratios of effects of pesticides

This concept is one in which an attempt is made to obtain an overall guide to, or summary of, the general effects that a particular group of pesticides may have on certain soil microbiological processes. It cannot be considered to be unequivocally accurate for individual pesticidal formulations. All stimulatory effects and instances where there is no effect are added and designated as positive. Similarly, all inhibitions are added together and designated as negative. The ratio of positive to negative effects thus describes the *effect ratio*. The number of soil-types and pesticides used and whether or not increases followed by decreases of population numbers or activities occur have been taken into account in assessing the overall effect ratio of pesticides on a particular soil microbiological process.

A. HERBICIDES

1. *Effects on soil bacteria and their activities*

(a) *Bacterial numbers.* In very general terms, few herbicides have any great or prolonged adverse effect on the total bacterial component of soils. Individual bacterial species are frequently adversely affected to some degree but the effects are seldom permanent. The usual pattern is one of an initial decrease in total numbers followed by a return to normal or even an increase in numbers. Such effects may be due to a disruption of the rhizosphere and non-rhizosphere bacteria imposed by the killing of the vegetative cover, followed by utilization of the dead plant material which allows for increased populations. The question many investigators have asked is "does the increase due to the killed vegetation mask inhibitions of non-zymogenous organisms"? Investigation of individual functional groups of bacteria has indicated that this may, indeed, occur.

The nutritional status of soils may have a significant impact on the type of response obtained. In the poorer type of soil the destruction of vegetation will provide additional oxidizable carbon compounds and increases in the population would therefore be expected. Soils richer in organic and inorganic nutrients may also be expected to show increases. The difference between the two soil types in response to a particular herbicide may, however, only become apparent after the initial surge of activity induced by the killed vegetation, suggesting a need for longer periods of monitoring.

Evidence for the role of the growth environment's nutrient status comes from the work of Voderberg (1961) who showed that inhibition of soil bacterial growth by simazine occurred on potato broth agar but not with peptone-glucose agar. Balicka and Bilodub-Pantera (1964) also demonstrated the effects of nutrients; their results indicating that the presence of organic nutrients was correlated to ability of bacteria to overcome the toxic effects of atrazine. Pantera (1966) found 20% of soil isolates to be sensitive to atrazine and 40% to be tolerant. The sensitive species were especially sensitive in media of low nutrient status but tolerant species were not affected by the type of medium. Whilst it is not possible to extrapolate these findings to the soil environment in terms of degree of inhibition, the results suggest that herbicide effects on numbers of microorganisms in treated soil could depend on the soil's nutrient status.

Other effects of herbicides on soil bacteria in culture media are shown in Table 1, from which the overall effect ratio is 0·32, i.e. largely inhibitory. In contrast, the effects on bacteria in soil are generally far less severe, with an effect ratio of $1 \cdot 2$, and in many cases recovery of depressed populations occurs (Table 2). The difference between the two effect ratios is probably sufficient commentary against the use of pure cultures for determining pesticide effects on soil microorganisms; unless for the purposes of studying mode of action and the physiology of bacteria in the presence of herbicides.

(b) *Nitrification*. Nitrification by chemosynthetic autotrophic bacteria is a process best studied *in situ* rather than in separate pure cultures of the organisms responsible. This is because in soil it is a complete process (NH_4^+ to NO_3^-) whereas in cultures it is incomplete, giving NH_4^+ to NO_2^- by *Nitrosomonas* spp. and NO_2^- to NO_3^- by *Nitrobacter* spp. Conditions in soil which favour the process as a whole are therefore important to a successful oxidation of NH_4^+ to NO_3^- and any factor likely to disrupt the process, such as the presence of pesticides, is best studied *in situ* in soil. Only when the mode of action of the pesticide on one or other of the participants is sought, should pure cultures be resorted to.

Whereas it has been established that some heterotrophic organisms are capable of producing NO_2^- and NO_3^- from NH_4^+ and organic nitrogen in pure cultures (Eylar and Schmidt, 1959; Gunner, 1963; Doxtader and Alexander, 1966) this form of nitrification in soil has rarely been reported. However, Ishaque *et al.* (1971) and Ishaque and Cornfield (1976) reported that soils of pH $4 \cdot 2$ converted ammoniacal nitrogen to nitrate nitrogen in the absence of isolatable chemosynthetic autotrophic bacteria and suggested that heterotrophic nitrification may contribute significantly to nitrification of ammoniacal nitrogen in some soils.

Whatever the organisms concerned, nitrification remains one of the most extensively studied soil processes and the accumulation of data, particularly in respect of pesticide effects, makes it increasingly difficult to generalize on the effects of these chemicals. To date, the most inhibitory chemicals, excluding those specifically designed to inhibit nitrification, are sodium chlorate and calcium cyanamide. The former can inhibit nitrifiers for up to a year whilst the latter has been reported to almost completely eliminate nitrification (Audus, 1970a).

In culture studies, Winely and San Clemente (1968), using EPTC and chlorpropham at rates equivalent to field rates, found that NO_2^- oxidation by *Nitrobacter* sp. was inhibited and that higher rates stopped the process completely. Further studies (Winely and San Clemente, 1969) led to the conclusion that inhibition occurred either by the action of EPTC on the $NADH_2$-oxidase system or by the action of chlorpropham on "other important enzyme systems" of the organism. Farmer *et al.* (1965) found that simazine at 6 ppm inhibited *Nitrobacter agilis* but not *Nitrosomonas europaea*. Simazine at 5, 10 and 100 ppm was found by Kulinska (1967b) to have no effect on pure cultures of nitrifiers. Linuron, MCPA, 2,4,5-T and simazine at 100 ppm inhibited nitrifiers in culture (Torstensson, 1974). The differences in effects for simazine may be due to differences in the strains concerned.

In soil studies, Debona and Audus (1970), using perfusion columns, established an increasing order of inhibitory effectiveness for ten herbicides, all at high concentrations, against nitrification. The order was: dichlobenil < paraquat < picloram < 2,3,6-TBA = chlorthiamid < bromoxynil < chlorflurazole < ioxynil < propanil. Endothal stimulated nitrification at rates up to 1000 ppm. Proliferation of nitrifiers was not affected by paraquat, endothal or 2,3,6-TBA. Chandra and Grover (1967) using loam soil treated with 22 herbicides at rates equivalent to normal and twice normal field rates found no correlation between nitrification activity and counts of fungi and bacteria; implying that the process was affected independently of effects on other organisms and population levels. Thorneburg and Tweedy (1973) found that atrazine and simazine at 1·2 and 5·0 kg ha^{-1}, simazine plus atrazine each at 1·2 kg ha^{-1} and paraquat, trifluralin, picloram and alachlor at 0·6 and 1·2 kg ha^{-1}, had no effect on nitrification in silt loam soil (pH 6·0; 2% O.M.). Other effects of herbicides on soil nitrification *in situ* are shown in Table 3.

In considering the overall effects listed for nitrification an effect ratio of 1·4 is obtained. In the individual groups of herbicides, most of the benzoic and phenol-type herbicides listed were inhibitory whereas the triazines are almost equally divided between no effect, stimulatory and inhibitory effects, thereby giving an effect ratio of 2. The ratio for urea-type herbicides is 0·95 but carbamate and carbanilate herbicides, on the other hand, show a ratio of 5·5.

An interesting approach to studies on the effects of herbicides on soil nitrification is that of Domsch and Paul (1974) who tested the effects of 35 herbicides by means of experimentation and mathematical modelling. They concluded that most of the herbicides had a negligible effect at normal field rates and were only effective against nitrification when the soil pH was below 7·0. Oxidation of NO_2^- to NO_3^- appeared to be more sensitive than oxidation of NH_4^+ to NO_2^-.

(c) *Denitrification*. Denitrification, or dissimilatory nitrate or nitrite reduction, is a common and important activity in soil, yet it remains one of the least investigated. In unpublished work, Anderson and Drew noted that columns of soil (calcareous loam, pH 8·0) perfused with a paraquat solution in what appeared to be a well-aerated, well-drained system, turned a bright purple–mauve in the centre of the columns. Paraquat ions, when reduced to their radical will show this change and since there was no evidence for microbial or chemical breakdown of paraquat, the only tenable explanation was that the "well-aerated" system had at least become micro-aerophilic. Under such conditions denitrifiers can survive at the expense of bound oxygen, as is found in nitrate ions. The reduction of nitrate to nitrous oxide and nitrogen gas thus represents a nett loss of nitrogen to microorganisms and plants. The major factor dissuading investigation of denitrification is its accurate and convenient measurement, and because it is not frequently reported, this has led to the assumption by some that it is not a common soil reaction and that it can be of no great consequence.

Hart and Larson (1966) examined the effect of 2,4-D at 0·221 to 1·105 ppm on the growth of three denitrifying bacteria and found that aerobic growth was less affected than anaerobic growth with NO_3^- as electron acceptor. *Pseudomonas denitrificans* was less sensitive to 2,4-D than was *Achromobacter nephridii* or *Ps. stutzeri*. Manometric experiments to compare the effects of 2,4-D on the reduction of NO_3^- to NO_2^- by resting cells of *Ps. denitrificans* showed that when NO_3^- acted as the electron (e^-) acceptor, the rate of gas formation was decreased by 50 and 100% with *c.* 0·77 and 1·56 ppm 2,4-D, respectively; when NO_2^- was used as the e^- acceptor, gas production still occurred even with *c.* 2·3 ppm 2,4-D but the rate of production was decreased. The data indicate that different bacteria show varying sensitivity to this herbicide and that some correlation exists between the response to 2,4-D and the enzyme content and physiology of microorganisms (Hart and Larson, 1966).

Karki *et al.* (1972) found that sodium chlorate in nitrate–glucose broth inhibited denitrification of the nitrate; whilst in a study of the denitrification potential of a soil bacterium in culture, Bollag and Nash (1974) concluded that inhibition by phenylurea and aniline herbicides depended upon the number of halogen substituents on the aromatic ring. Two substituents per molecule inhibited whereas a single chloro- or bromo-substituent did not. Gizbullin and Solov'eva (1968) found that Alipur, murbetol and dalapon, at the equivalent of field rates, were not toxic to denitrification in culture.

The effects of a number of herbicides on denitrification in soils is shown in Table 4. The overall effects of herbicides on denitrification, from studies listed, show a ratio of 1·82 with no particular group of herbicide emerging as predominantly inhibitory or stimulatory.

(d) *Rhizobia and legume nodulation.* Ten years ago a leading agriculturalist was heard to proclaim that "legumes will grow just as well with applications of nitrogenous fertilizer in place of nodule bacteria and therefore nodulation is really not of major importance". Since then the price of nitrogenous fertilizers have virtually doubled and they are in short supply in many parts of the world. Thus the maligned agent of symbiotic nitrogen-fixation emerges triumphant in a world of increasingly diminishing food supplies. The importance of determining whether herbicides (and other pesticides) affect nodulation now assumes a greater economic role than hitherto, yet relatively few microbiologists investigate this aspect of pesticide effects in soil.

In culture, normal field rate equivalents of six herbicides were investigated in relation to the respiration of *Rhizobium japonicum* by Dunigan *et al.* (1970) who found that only trifluralin slightly reduced respiratory activity. Grossbard (1970) demonstrated that *Rhizobium trifolii*, pre-treated with 25 to 50 ppm asulam could cause damage to the host plant but if the bacterial cells were washed to remove residually adsorbed herbicide, there was no effect on the host or nodulation of its roots. Kapusta and Rouwenhorst (1973) found that six strains of *R. japonicum* exhibited differential sensitivity to chlorpropham applied at 3 to 48 ppm. Pyrazon and DNOC were the most toxic of several herbicides to *Rhizobium meliloti, R. trifolii* and *R. leguminosarum* (Jensen, 1969). Pyrazon has also been shown to damage the host plant if present on *R. trifolii* cells (Grossbard, 1970) and to inhibit the growth of fast-growing strains of *R. meliloti, R. trifolii* and *R. leguminosarum* when present at concentrations of more than 1000 ppm, or slower-growing strains (*R. lupini, R. japonicum*) at rates of 100 to 500 ppm (Kaszubiak, 1966). The latter author also found that the halogenated aliphatics, TCA and dalapon, at rates up to 20 000 ppm, were not toxic to the strains of *Rhizobium* spp. tested (Kaszubiak, 1966).

Resistance to TCA and dalapon was greater for *R. leguminosarum* than for the host plant, *Pisum sativum* (Sud *et al.*, 1973), while Jensen (1969) demonstrated that both these herbicides were the least toxic in cultures of *R. meliloti, R. trifolii* and *R. leguminosarum*. A number of *Rhizobium* strains tolerated 300 ppm TCA (Krasil'nikov, 1967).

Namdeo and Dube (1973b) found that dalapon and paraquat, albeit at the elevated rates of 500 and 1000 ppm, respectively, induced mutagenic effects in rhizobia. Whilst Namdeo and Dube (1973b) draw attention to the useful possibilities of herbicide-induced mutants, such as selection of more efficient strains, potentially harmful mutagenic effects may also occur. These workers (Namdeo and Dube, 1973b) found that dalapon and paraquat at up to 100 ppm were tolerated by rhizobia isolated from either herbicide-treated or untreated grassland soils, but both herbicides showed mutagenic activity at these concentrations. Gillberg (1971) found that dinoseb and MCPA also gave rise to rhizobia mutants whose ability to infect legumes was not affected and that in MCPA-induced mutants, nitrogen-fixing ability may have been improved. Some of the mutants withstood higher than normal rates of the two herbicides, which Gillberg points out may be of use in combating possible harmful effects of herbicides on wild-type strains in soils regularly treated with herbicides.

The phenoxy-acids 2,4-DB and MCPB inhibited fast-growing strains of *R. meliloti*, *R. trifolii* and *R. leguminosarum* at concentrations exceeding 1000 ppm and slow-growing strains of *R. lupini* and *R. japonicum* at 100 to 500 ppm (Kaszubiak, 1966); while 2,4,5-T at 200, 600 and 1000 ppm inhibited *R. leguminosarum* with increasing concentration, the highest rate giving complete inhibition (Foldesy *et al.*, 1972). *R. leguminosarum* showed decreased growth at 50 ppm 2,4-D but 250 ppm was needed for significant decreases in growth of *R. meliloti* and *R. japonicum* (Gaur and Misra, 1972). Some rhizobia tolerated 2,4-D at 300 ppm (Krasil'nikov, 1967) and growth of *R. meliloti*, *R. trifolii*, *R. leguminosarum* and *R. phaseoli* was inhibited by 5 to 300 ppm of MCPA, MCPB or 2,4-D (Mickovski, 1966). Temporary suppression of growth of rhizobia was obtained with rates of 50 and 100 ppm 2,4-DB but more than 500 ppm was necessary for complete inhibition (Jordan and Garcia, 1969). When the 2,4-DB was added to cells in the log-phase, there was no effect on growth but when added during the lag-phase, inhibition occurred. The pre-incubation of rhizobia in up to 10 ppm 2,4-DB did not affect their nodulation capabilities (Jordan and Garcia, 1969).

Discrepancies in the rates of application of phenoxy-acids needed for inhibition of growth in culture may be explained by the suggestion of Makawi and Abdel-Ghaffar (1970) that the sensitivity of rhizobia to herbicides may be more a property of the strain and not the species of nodule bacteria. An example supporting this premise is the work of Kaszubiak (1966) who found that *Rhizobium* strains which grew faster than others were generally more resistant to herbicides. Whereas the triazines, simazine and prometryne, were not toxic to the strains tested, linuron, chloroxuron and fluometuron were toxic to different extents at rates of 20 to 20 000 ppm. Strain differences to both concentration and time of exposure of cells to herbicides was also found. The three most toxic herbicides were diuron, linuron, dinoseb acetate and the mixture of propham + diuron. These were toxic at several hundred ppm to fast-growing strains and to slow-growing strains at 100 ppm (Kaszubiak, 1966, 1968a). Neither atrazine nor simazine, at normal field rate equivalents, were found by Avrov (1966) to be toxic to rhizobia but Krasil'nikov (1967) found that only some strains could tolerate simazine at 300 ppm. Atrazine (25 and 50 ppm), absorbed onto cells of *R. trifolii*, could cause damage to the host plant but if the cells were washed after exposure to atrazine, no damage occurred (Grossbard, 1970). Gramoxone (0·16 and 16 ppm) inhibited growth of eight strains of rhizobia (Manninger *et al.*, 1972).

The effects of some herbicides on rhizobia and nodulation of legumes in soil is shown in Table 5. The most toxic herbicides at relatively higher rates than normally recommended were found by Jensen (1969) to be substituted phenols and pyrazon while the least toxic were dalapon, simazine and prometryne. Szegi *et al.* (1974) have shown that, in sand culture, the recommended field rate of Gramoxone decreased the yield of peas inoculated with rhizobia. Whereas sand culture, as opposed to soil, cannot be regarded as a fair test medium for bipyridyls there are agricultural soils that are known to contain a high proportion of sand and very little clay. Bipyridyls of the paraquat type are not recommended for such soils.

The importance of time of application of a herbicide to soils cropped with legumes was demonstrated by Hamdi and Tewfik (1969). These workers found that nodulation and growth of cowpea was inhibited by trifluralin at recommended field rates when applied on the day of planting but were stimulated when the herbicide was applied 27 days before planting.

From the data listed in Table 5, the overall impression is that herbicides can harm rhizobia and nodulation of legumes. The ratio of positive:negative effects is 0·94, indicating a small nett negative effect. The anilines, trifluralin and nitralin appear to be largely negative towards rhizobia and nodulation, as also are the carbamates listed. Most of the triazines appear (Table 5) to be inhibitory even at recommended field rates. Whilst the data cannot be said to be a complete listing of all results found for herbicides, there are indications that herbicide applications to legumes or to soil designated for legume growth, should be carefully evaluated prior to recommending their use.

(e) *Free-living, nitrogen fixation.* The disruptive effect of herbicides on the vegetative cover may be particularly important to aerobic and anaerobic N_2-fixing bacteria. The death of plants will result in an increased level of oxidizable organic matter from leaves, stems and roots. The consequent increase in bacterial activity will result in a greater utilization of the soil's inorganic and organic nitrogen. Without these forms of nitrogen, free-living N_2-fixing organisms will require to fix atmospheric nitrogen if they are to survive. Provided that a suitable form of oxidizable carbon is available from the degrading plant debris, N_2-fixing organisms should be capable of survival unless the herbicide has had a direct inhibitory effect on them.

In general, free-living N_2-fixing organisms are not greatly affected by herbicides except at high rates of application. In culture, DNOC and prometryne at 10 000 ppm were toxic to *Azotobacter chroococcum* whereas linuron, diphenamid and prometryne at less than 10 000 ppm were not toxic (Rankov *et al.*, 1966). In a later report Rankov and Genchev (1969) established that all strains of *A. chroococcum* were stimulated in the presence of 1% mannitol and 20 000 ppm diphenamid, but when the concentration of mannitol was decreased to 0·1%, morphological changes occurred in cells. N_2-fixing ability decreased with increased diphenamid concentration and decreasing mannitol content of the medium.

Babak (1968) investigated ten *Azotobacter* strains and found halophilic strains to be inhibited by 1000 ppm 2,4-DES and 2,4-DEP but mesophilic strains required 2000 to 3000 ppm of these herbicides for inhibition to occur. Similarly, whilst 5000 to 6000 ppm chlorazine, DCNB or diphenamid inhibited halophilic *Azotobacter* sp. strains, mesophilic strains needed 6000 to 7000 ppm. The same author also reported herbicide-induced pigmentation effects in azotobacters.

Goguadze (1968) found that pure cultures of *Beijerinckia* sp. were sensitive to 0·005 ppm of diuron but were resistant to simazine. *Azotobacter* sp. with simazine and diuron at 0·02 ppm at pH 7·0 produced populations of $1·5 \times 10^6$ and $1·2 \times 10^6$ for the two herbicides, respectively, but at pH 5·0 simazine depressed growth and diuron arrested it completely. Kulinska (1967b), however, found that rates of 5, 10 and 100 ppm simazine had no effect on glucose utilization, respiration or nitrogen uptake by *Azotobacter* sp. Pyrazon at up to 500 ppm stimulated, whereas >1000 ppm TCA retarded *Azotobacter* sp. (Zharasov, 1969). Misra and Gaur (1971) found that 50 to 2000 ppm simazine slightly stimulated *Azotobacter* spp. but at the same rates 2,4-D had no effect.

Langkramer (1970) found that Gramoxone at the equivalent of ten times field rates inhibited *A. chroococcum* in culture and Szegi *et al.* (1974) showed that low concentrations of Gramoxone inhibited *A. chroococcum* and that this organism's sensitivity to the herbicide was increased in media containing inorganic nitrogen compounds.

Electron microscope studies of an *Azotobacter* sp. grown *in vitro* with 5000 ppm amitrole revealed alterations to structural components, such as plasma membranes, and an increase in filamentous forms (Kretschmar *et al.*, 1970). Wegrzyn (1971) found that 1 and 10 ppm eptam stimulated growth of *A. chroococcum* but that 100 and

1000 ppm chloroxuron, chlorpropham, atrazine, aresin and eptam were inhibitory; while at 100 ppm linuron and chlorpropham inhibited growth and nitrogen-fixation. At 1 to 1000 ppm, dalapon, Alipur and Murbetol inhibited growth of *Azotobacter* sp. (Gizbullin and Solov'eva, 1968) but *Clostridium pasteurianum* was not affected. Concentrations of 200 to 1000 ppm of 2,4,5-T had increasingly toxic effects on *A. chroococcum*, but some indication of adaptation to these rates was noticed by Foldesy *et al.* (1972).

A culture solution containing 3000 ppm prometryne stimulated growth of *Azotobacter* (Szabo, 1964). Growth of *A. chroococcum* was suppressed by siduron at 10 ppm and also 10 ppm of its degradation product 2-methylcyclo-hexylamine (Fields and Hemphill, 1968), whereas 5000 ppm of diuron was necessary to inhibit *A. chroococcum*.

Under soil conditions many herbicides at normal field rates have little to no inhibitory effect on free-living microbial N_2-fixation (Table 6). The total number of positive effects outweighs the number of negative effects, producing a ratio of 1·65. No single group of herbicides appears to have a greater effect than any other.

2. *Effects on soil fungi and actinomycetes*

Fungal and actinomycete colonies appearing on dilution agar plates are frequently used as a means of enumerating these microbial forms, but the method leaves considerable doubt as to the true numbers in soil. Thus, use of the technique to measure effects of herbicides on mycelial forms is not considered to provide sufficiently accurate data. Accordingly, many investigators have opted to study effects of chemicals on these organisms using isolates growing on specific natural or artificial media.

(a) *Growth and activity*. Many soil fungi exhibit a high degree of tolerance to herbicides but this should not imply that they are totally unaffected. Valaskova (1968) found that soil fungi grown *in vitro* were less affected by systemic than contact herbicides. This worker found that the minimum inhibitory concentration of PCP and DNOC was *c*. 10 ppm, whereas for dalapon it was 15 000 to 80 000 ppm, for a number of soil fungi (Valaskova, 1968, 1969).

The triazines and substituted urea herbicides have been investigated more extensively than any other group. Uhlig (1966) found that simazine at up to 1000 ppm increased the growth, in culture, of the mycorrhizal fungus *Scleroderma vulgare* but inhibited growth of *Hebeloma crustiliniforme. Tricholoma pessundatum* and *T. saponaceum* were not affected. At 100 ppm, 70 to 75% of the herbicide was absorbed by mycelia of *S. vulgare* and *T. pessundatum*. However, 100 ppm simazine and 40 ppm monolinuron weakly inhibited the growth of another mycorrhizal fungus, *Suillus variegatus* and metobromuron (40 ppm) was strongly inhibitory (Sobotka, 1970). At normal field rate equivalents (*c.* 40 ppm) in culture, simazine had no adverse effect and sometimes stimulated growth of mycorrhizal fungi, but the mixture Alipur at *c.* 60 ppm was toxic to these fungi (Andruszewska, 1973). Concentrations of from 100 to 1000 ppm of prometryne, atrazine, linuron and metobromuron inhibited growth and sporulation of some soil fungi; but of these compounds, prometryne had the least effect on sporulation of *Fusarium* sp. while metobromuron at 100 ppm stimulated conidial formation in *Alternaria* sp. and inhibited this process at 1000 ppm (Plamadeala, 1972). Growth of *Mucor spinosum* was inhibited, but growth of *Aspergillus niger* and *Penicillium tardum* were increased by atrazine and simazine at 4, 10 and 100 ppm (Nepomiluev and Kuzyakina, 1972). *Paecilomyces varioti*, *A. niger*, *A. flavus*, *A. tamarii* and *P. funiculosum* in liquid culture were found (Bakalivinov, 1972) to be inhibited by fluometuron at *c.* 10–20 ppm, but respiration of *P. funiculosum* was stimulated by linuron (*c.* 5–20 ppm), whilst atrazine (*c.* 10–40 ppm) and prometryne (*c.* 5–15 ppm) reduced urease, respiratory and catalase activity in fungi.

Aspergillus fumigatus, *A. niger*, *A. ustus*, *A. flavus*, *A. oryzae* and *A. paradoxus* were all resistant to propazine, diuron, linuron (and dalapon) at 100 ppm in culture; fluometuron, however, at 0·1 ppm inhibited *A. fumigatus* (Venkataraman and Rajyalakshmi, 1971). Atrazine at 4·4 to 140 ppm in potato dextrose agar was tolerated by more saprophytic fungi than phytopathogens (Richardson, 1970) and protein synthesis in *Neurospora crassa* increased in the presence of *c.* 216 ppm atrazine (Schröder *et al.*, 1970). In cultures of *Fusarium* spp. and *Monolinia fructicola*, atrazine at 10 ppm inhibited the hexose monophosphate shunt whilst atrazine, simazine and fluometuron at 10 ppm decreased glucose catabolism in *M. fructicola* (Tweedy and Loeppky, 1968). The triazines simazine and prometryne and the ureas linuron, monolinuron and chloroxuron (as well as the carbamate, chlorpropham) all at rates of 0·2 to 1000 ppm, inhibited soil yeasts (Balicka *et al.*, 1969). Growth of *A. versicolor* and *A. fischeri* was suppressed by 10 ppm siduron in culture medium but that of *A. ochraceus*, *A. candidus*, *A. nidulans*, *A. flavus*, *Botrytis cinerea* and *Rhizopus nigricans* was not affected. None of these fungi were affected by 2-methylcyclohexylamine, a decomposition product of siduron (Fields and Hemphill, 1968). Diuron (10 ppm) inhibited the growth of many fungal isolates but few were affected by monuron or fenuron at 100 ppm (Doxtader, 1968).

Phenoxyacetic and phenoxybutyric compounds at 10^{-6} M stimulated *A. niger* growth but at 10^{-4} M inhibited growth and endogenous and mitochondrial respiration, the site of inhibition in the respiratory chain apparently involving ubiquinone (Smith and Shennan, 1966). At 400 ppm, 2,4-D inhibited *Penicillium herquei*, *Fusarium nivale*, *Theilavia terricola* and *Cunninghamella echinulata* but lower rates had no effect (Leelavathy, 1969). Mycelial growth of *Paecilomyces varioti*, *A. niger*, *A. flavus*, *A. tamarii* and *Penicillium funiculosum* in liquid culture was stimulated by 2,4-D (*c.* 5–20 ppm) and MCPA (*c.* 5–20 ppm) and the latter compound increased catalase activity in *A. flavus* and *A. niger* and urease activity in *A. tamarii* (Bakalivinov, 1972). Nucleic acid synthesis was increased by 140% when *N. crassa* was grown in media containing 220 ppm 2,4-D, but with a similar concentration of the carbamate herbicide chlorpropham, growth was completely inhibited (Schröder *et al.*, 1970). At rates of 100 to 1000 ppm, 2,4-D inhibited growth and spore formation of *Alternaria* sp. and *Fusarium* sp. but at 100 ppm conidial formation of the *Alternaria* sp. was increased (Plamadeala, 1972). The effects of 2,4-D and Gramoxone on 20 fungi in culture was examined by Szegi (1970) who found that 2,4-D at 30 and 60 ppm frequently stimulated growth of the fungi, which were only inhibited by 120 and 250 ppm. Paraquat (as Gramoxone), however, completely inhibited some of the fungi at 25 ppm. In a further investigation, Szegi and Gulyás (1971) found that fungi from Gramoxone-treated soil grew well in agar media containing 2000 ppm Gramoxone but fungi from untreated soil were inhibited by this rate. Mycorrhizal fungi were also found to be sensitive to Gramoxone applied at 40 to 2000 ppm in the growth medium (Andruszewska, 1973).

Cultures of *Suillus variegatus* were inhibited by PCP (80 ppm), pyrazon (50 ppm) and paraquat (as Gramoxone) at 75 ppm but in the latter case submerged growth was normal and only at 100 ppm did Gramoxone completely inhibit the mycorrhizal fungus (Sobotka, 1970). Pyrazon at 20 ppm was found by Zharasov (1969) to be toxic to many soil fungi while at 10 ppm, pyrazon and also dinoseb acetate caused morphological changes in soil fungal isolates (Valaskova, 1967). Bromacil at rates up to 2000 ppm had no effect on *A. tamarii*, *A. flavus*, *A. oryzae*, *Penicillium brevicompactum*, *P. funiculosum*, *Cunninghamella lunata* or *M. verrucaria*, whilst inhibition of *A. niger* by less than 1500 ppm was compensated by increasing the nitrogen level in the medium (Pancholy and Lynd, 1969). Lenacil, molinate and cycloate adversely affected soil fungi at rates of 4000 to 10 000 ppm (Chulakov and Zharasov, 1973). Picloram was tolerated at 1000 ppm and at 0·1, 1·0, 10·0 and 100 ppm increased the dry weight and red pigmentation of *A. tamarii* (Rieck and McCalla, 1969). Also at 1000 ppm, growth of *A. terreus, Fusarium oxysporum* var. *vasinfectum, Penicillium digitatum, Pythium ultimum, Trichoderma lignorum* and *Verticillium albo-atrum* was not affected by picloram (as Tordon) (Goring *et al.*, 1967).

Fields *et al.* (1967) and Hemphill and Fields (1967) found that chlorthal dimethyl (750 ppm) stimulated the growth of fungi, whilst Khikmatov *et al.* (1969) showed that *p*-chloroformanilide (up to 33 ppm) had no effect on soil fungal isolates. Additions of up to 445 ppm trifluralin to nutrient media favoured the growth of penicillia and aspergilli (Rankov, 1971a) and cacodylic acid or MSMA at 10 000 ppm had no effect on *Penicillium* spp., *Aspergillus nidulans* and *Trichoderma viride* (Bollen *et al.*, 1974). Ustinex GL, a mixture of diuron, bromacil, aminotriazole and 2,4-D, inhibited mycelial extension of fungi on agar containing 50–200 ppm (Teuteberg, 1970).

From the above it is seen that, with few exceptions, herbicides applied to growth media at rates approximating the recommended field rates of the herbicides are not severely inhibitory to fungal isolates. With some of the exceptions, such as bipyridyls, strong adsorption to clay minerals in soil would probably negate any inhibitory effect found in culture.

Actinomycetes in culture are generally unaffected, with few exceptions, by the equivalent of recommended field rates of herbicides in culture. Thus, alachlor increased antibiotic production of soil actinomycetes at low concentrations but higher rates inhibited them (Gusterov *et al.*, 1972). *p*-Chloroformanilide at 6·6, 16·6 and 33 ppm had no effect on actinomycetes (Khikmatov *et al.*, 1969). The carbamate herbicides pebulate and EPTC at 1–100 ppm had little effect on growth, morphology, pigmentation or antibiotic production by *Streptomyces* spp. (Krezel and Kosinkiewicz, 1972), but chlorpropham at 5 ppm inhibited antibiotic production of *S. griseus* (Krezel and Leszczynska, 1970). Cycloate and molinate, however, only inhibited actinomycetes at elevated rates of 4000 to 10 000 ppm (Chulakov and Zharasov, 1973).

Triazines such as desmetryne inhibited 50% of actinomycetes in culture at concentrations of 8–100 ppm (Gusterov *et al.*, 1972) but other triazines (e.g. atrazine and simazine) at rates of 1–100 ppm had no effect on actinomycetes (Krezel and Kosinkiewicz, 1972). However, prometryne at 10 ppm inhibited antibiotic production of *S. griseus* (Krezel and Leszczynska, 1970). Simazine (50 ppm) caused genetic changes in *S. globisporus*, altering spore formation, gelatin liquefaction and antibiotic formation without affecting growth (Kissné and Kiss,

1966). Simazine and prometryne at 100 ppm were weakly inhibitory towards streptomycetes (Balicka *et al.*, 1969).

The substituted ureas monolinuron and chloroxuron were inhibitory to streptomycetes at 100 ppm; linuron was exceptionally inhibitory at this concentration (Balicka *et al.*, 1969) and inhibited antibiotic production in *S. griseus* at 10 ppm (Krezel and Leszczynska, 1970). With rates exceeding 25 ppm, linuron reduced growth and induced changes in morphology, pigmentation and antibiotic production in streptomycetes (Krezel and Kosinkiewicz, 1972). Diuron (10 ppm) inhibited growth of many actinomycetes but few were affected by monuron or fenuron at rates of 100 ppm (Doxtader, 1968).

Changes in spore-formation, gelatin liquefaction and antibiotic formation in *S. globisporus* occurred with 2,4-D and DNOC, both at 50 ppm (Kissné and Kiss, 1966). Trifluralin (8–100 ppm) inhibited 50% of actinomycetes on agar media (Gusterov *et al.*, 1972), whilst pyrazon (Zharasov, 1969) was not toxic to actinomycetes when added to media at normal field rates (*c.* 17–34 ppm). Goring *et al.* (1967) found that 1000 ppm of picloram (as Tordon) in media had no effect on *Streptomyces scabies*, whilst trifluralin at rates up to 445 ppm favoured growth of actinomycetes with colourless or grey aerial mycelium (Rankov, 1971a). Whereas no adverse effects were found with DMPA incorporated into media, the degradation products 2,4-dichlorophenol and *O*-methyl-isopropyl-phosphoramidothioate, respectively, suppressed and stimulated the growth of many actinomycete species (Hemphill and Fields, 1967). This latter example of herbicide effects on soil organisms exemplifies the need to obtain degradation data on parent compounds and to include these products in studies of the parent compound to obtain a more meaningful profile of possible effects.

Effects of herbicides on soil fungi and actinomycetes *in situ* are given in Table 7. The aniline-type herbicides show a predominantly positive effect. Carbamates, diazines and halogenated aliphatics are generally negative whilst the phenolics, triazines and substituted ureas show roughly equal positive and negative effects. The overall effect ratio for all the herbicides listed in Table 7 is 1·09.

(b) *Pathogenic fungi and their antagonists.* The effect of pesticides on plant pathogenic fungi and on their natural antagonists is one of the most important aspects of the study of effects of agricultural chemicals on non-target, as well as target, organisms. Since the whole premise of pesticide use is to overthrow pests which decrease crop-potential, any factor pertaining to their demise, with or without the aid of natural antagonists, is pertinent. In so doing, however, it is important to safeguard those natural antagonists for which there are currently no efficient substitutes and to ensure that in eliminating one pest we do not also eliminate antagonists of others and so create additional, possibly greater, problems in another direction.

In some cases predisposition to disease, or resistance to disease, is increased as a result of herbicide applications. In other cases, herbicides may favour survival of phytopathogens while others favour their natural antagonists. Wilkinson and Lucas (1969a, b) found that linuron and Gramoxone were more fungitoxic than MCPA or simazine but that fungi (especially *Helminthosporium* and *Fusarium*) differed in their sensitivity. *Trichoderma viride*, a natural antagonist of *Sclerotium rolfsii* and other pathogenic fungi, was especially sensitive to the paraquat ions in Gramoxone. The latter also affected the spores of *Rhizopus stolonifer* at concentrations as low as 10 ppm, preventing them from producing mycelia on agar media. Concentrations of 1 and 5 ppm reduced the growth rate by 40% and germ-tube deformities were also observed. In further experiments, Wilkinson and Lucas (1969c) obtained evidence for a synergistic inhibitory effect between formulation additives of Gramoxone and paraquat ions on the respiration of *F. culmorum* and *T. viride*, especially with young cultures.

Additional evidence for antagonistic effects of bipyridyls on fungi was found by Wilkinson (1969) who showed that diquat ions could reduce the parasitization of *F. culmorum* by *T. viride* when both fungi were inoculated on to bean leaves sprayed with 1560 or 3120 ppm of diquat solution (as Reglone). Diquat ions (1000 ppm) also completely inhibited *R. stolonifer* but permitted growth of *Aspergillus niger* on wheat chaff or potato haulm.

The Rodriguez-Kabana and Curl school have investigated several aspects of the effects of pesticides on phytopathogenic fungi and their antagonists. Paraquat at 12·5 to 1000 ppm in culture media or 1·25 to 100 kg ha^{-1} in sandy loam, severely reduced growth rate, glucose utilization and carbon dioxide evolution by the pathogen *Sclerotium rolfsii* (Rodriguez-Kabana et al., 1966a, 1967b; Curl et al., 1967) but at concentrations in sandy loam of 0·25 to 2·0 kg ha^{-1} neither paraquat nor prometryne had much effect on sclerotial production in *S. rolfsii* or *Sclerotina trifoliorum* (Curl and Rodriguez-Kabana, 1971). The effects of atrazine, EPTC, diuron and paraquat on *Rhizoctonia solani* in liquid culture were examined in terms of growth rate and mycelial weight. Both parameters were inhibited by all rates of atrazine (10 to 70 ppm), in relation to concentration. Diuron (0·02 to 2·0 ppm) and EPTC (0·8 to 40 ppm) had little effect. Paraquat (6·25 to 100 ppm) inhibited growth rate, decreased mycelial weight and induced a marked lag-phase (except at the lowest rates) in the growth of *R. solani* (Rodriguez-Kabana *et al.*, 1966b).

Populations of *Rhizopus* and *Curvularia* spp. in a rice paddy soil were decreased by propanil at 2, 3 or 4 kg ha^{-1}, nitrofen at 2, 4 or 6 kg ha^{-1} and by prometryne at 0·25, 0·5 and 1·0 kg ha^{-1}, but later population numbers of the pathogens increased again (Chowdhury *et al.*, 1972). In cinnamon alluvial and skeletal soils, *Trichoderma* spp. were not significantly affected by prometryne at 5–7 kg ha^{-1}, Isocil at 3–5 kg ha^{-1} or fluometuron at 8–10 kg ha^{-1} (Goguadze and Gogiya, 1971) whilst in a pale chestnut soil *Trichoderma* spp. were found to predominate after treatments of 4·5 kg ha^{-1} EPTC or pyrazon and 12–18 kg ha^{-1} TCA (Zharasov *et al.*, 1972). Other examples of the effects of herbicides on phytopathogenic fungi and their antagonists are shown in Table 8.

The effect ratio for herbicides on phytopathogenic fungi and their antagonists in soil is 0·81, indicating that both groups of organisms may be susceptible to herbicides. The ratio for phytopathogens alone is 0·76, suggesting that herbicides may aid in controlling some phytopathogens. Insufficient data are available on antagonists of pathogens to enable an effect ratio to be derived, but as a general observation at least some antagonists appear not to be adversely affected by herbicides and may in fact be among the first organisms to re-colonize soil in which fungal populations have been adversely affected by herbicides.

3. *Effects on soil algae*

Despite the importance of algae in soil, comparatively few studies on their reactions to pesticides have been undertaken. This may be due to the fact that, in general, they are not easily enumerated other than by cultural methods and even then relatively few types emerge on isolation media. Direct microscopy for algae (see Chapter 6) is an alternative, but the relatively low numbers of algal cells in most agricultural soils give this method uncertain value for routine counts. The recent technique of spectrophotofluorimetry for algae in soil and aqueous environments (Sharabi and Pramer, 1973) may prove useful in algal enumeration. Several examples of effects of herbicides on micro-algae *in vitro* are given in Chapter 8, and only those investigations not cited in that chapter will be mentioned here.

Amitrole, atrazine, monuron and diuron, at 18 to 50 ppm, inhibited soil algae more readily in culture than in soil (Kiss, 1966, 1967). Only high rates of simazine inhibited cultures of soil algae isolated from the rhizosphere and rhizoplane of maize (Kaiser and Reber, 1966) and propazine had variable effect on several strains of soil algae. Methabenzthiazuron (4 ppm) retarded algal growth (Hugé, 1970), whilst fluometuron (0·06 and 16 ppm) reduced growth of *Chlorella pyrenoidosa* by 19·6% and 98·9%, respectively, (Blythe and Frans, 1972). In the presence of 1220 ppm 2,4-D, *C. pyrenoidosa* developed vesiculation of the plasma membrane, mitochondrial swelling with disorganization of cristae and disruption of the chloroplast lamellar system (Bertagnolli and Nadakavukaren, 1970).

Chlorella vulgaris was sensitive to 1–5 ppm dinitramine (Belles, 1972) and mixed cultures of soil algae were inhibited by 1 to 1000 ppm chlorthal dimethyl (Fields *et al.*, 1967; Hemphill and Fields, 1967). Pure paraquat dichloride (11·4 and 114 ppm), Gramoxone W and formulation additives (57 and 570 ppm), added to a culture medium seven days after inoculation with a mixed culture of unicellular algae, inhibited growth of the algae (Anderson and Drew, 1976a).

The effects of herbicides on soil algae *in situ* can be quite severe but generally not as marked as in cultures. Some herbicide effects on algae *in situ* are given in Table 9, from which an effect ratio of 0·45 is obtained. This indicates, as would be expected, that in general herbicides are very inhibitory towards algae.

4. *Effects on combined activities of the soil microbial biomass*

Since functions such as cellulolysis, respiration, enzyme secretion or cell-free enzymic activity and ammonification are not solely due to any single group of soil organisms and are difficult to separate, it is expedient to group such activities into general headings.

(a) *Cellulolytic activity and organic matter degradation.* Pure culture experiments for studying effects of herbicides on cellulolysis and organic matter degradation are limited by the nature of the medium that can be employed. Some information will become available through the use of pure cellulose, pectin, chitin and lignin media, but with the wide differences between the organisms capable of utilizing these substrates, pure culture evaluations for cellulolysis and organic matter degraders are inevitably cumbersome and lengthy.

Szegi and Gulyás (1971) grew cellulolytic fungi, isolated from Gramoxone-treated soil, on media containing 2000 ppm Gramoxone and found good growth on the cellulose medium, whereas cellulolytic fungi from untreated soil were inhibited by the herbicide. Szegi (1972) found that 2,4-D at 0·03 to 0·25 ppm affected few species of cellulolytic organisms and usually only at the higher rate, but with paraquat at 0·003 to 0·03 ppm, most species were inhibited even at the low concentration. Penicillia and aspergilli were the most sensitive to paraquat while *Fusarium* spp. were most resistant.

Sobieszczanski (1968) used herbicide-containing agar medium on which were placed filter paper discs inoculated with suspensions of bacteria or fungi. It was found that 10 ppm simazine, atrazine and prometryne had little effect on the growth of five species of cellulolytic bacteria, except in the case of *Cellfalcicula* sp. which was slightly inhibited by simazine and prometryne and *Cytophaga* sp. which was similarly inhibited by simazine. At 10 ppm, there were both minor inhibitions and stimulations of five species of cellulolytic fungi by each triazine. With 158 ppm, however, the three triazines severely inhibited the bacteria and in some cases no growth was recorded. Fungal inhibitions were more marked at the higher concentration.

Atrazine and simazine, even at high concentrations, had no effect on *Sporocytophaga myxococcoides* and neither did diuron (Lembeck and Colmer, 1967). Other substituted urea herbicides such as linuron and monolinuron, at 10–158 ppm, slightly inhibited cellulolytic fungi and bacteria at the low rate (especially with monolinuron) but were totally inhibitory to most organisms at the high rate (Sobieszczanski, 1968).

CIPC, chloroxuron, pebulate and EPTC at rates of 10–368 ppm varied in their effect on five strains of cellulolytic bacteria (Sobiesz-czanski, 1968). CIPC inhibited most of the strains, chloroxuron inhibited two strains but pebulate and EPTC had virtually no effect. At high rates CIPC totally inhibited all five strains of bacteria as did chloroxuron, but pebulate and EPTC were not as severe. The five strains of cellulolytic fungi subjected to the same herbicides were inhibited at low rates with CIPC and chloroxuron, but pebulate and EPTC exhibited very slight inhibition and in some cases a stimulation. High rates of CIPC and chloroxuron again severely inhibited most fungi but pebulate and EPTC were less toxic.

Lembeck and Colmer (1967) showed that 2,3,6-TBA, amiben, dicamba and tricamba, at 1000 ppm or more, totally inhibited cellu-lolysis by *S. myxococcoides*. The most toxic herbicide tested was DMPA which gave a 20% inhibition at only 2 ppm.

Effects of herbicides on cellulose and/or organic matter degradation in soil are shown in Table 10. Since herbicides kill vegetation, treatment with these chemicals results in an increase in organic matter both on the surface and within the soil. Such increased availability of material for decomposition by the many types of organisms capable of degrading organic matter would, logically, imply that the population numbers and activity would increase following herbicide treatment, provided that the chemical itself was not inhibitory. The ratio of positive to negative effects obtained from the data cited is 1·31, indicating that, in general, the premise stated above is correct, although the value is not as high as might be expected indicating that some herbicides are inhibitory towards cellulolytic and organic matter-degrading microorganisms. Acylamides, anilines and benzoic-type herbicides, are largely inhibitory, whilst carbamates, diazines, halogenated aliphatics and phenolics on average have little or no effect. The phenoxy-acids and quaternaries are usually found to be inhibitory whilst triazines and substituted ureas are approximately equal in positive and negative effects on these organisms.

(b) *Respiration*. A reasonable idea of the overall metabolic activity of the soil's microbes can be obtained by measuring respiration, although the microbial biomass is not, of course, the sole contributor to soil respiration. It is pertinent to obtain some idea of the likely rate of degradation of the chemical beforehand, since this may contribute significantly to CO_2-evolution (or O_2-uptake), especially where high rates of the compound are applied to the soil.

In culture, prometryne (2·41 ppm) and linuron (2·49 ppm) stimulated O_2-uptake and CO_2-evolution, while chlorpropham (2·14 ppm) stimulated O_2-uptake only in bacteria isolated from the rhizosphere of maize (Karpiak and Iwanowski, 1969). Monolinuron (5 ppm) stimulated respiration in strains of *Agrobacterium tumefaciens* (Kulinska and Romanov, 1970), and fluometuron (1–20 ppm) increased the CO_2-evolution from *Aspergillus flavus* (Chopra et al., 1973). Respiratory activity in *Escherichia coli*, *Aerobacter aerogenes* and *Micrococcus lysodeiktikus* was inhibited by 2·21 ppm 2,4-D while *Pseudomonas fluorescens* and *Staphylococcus aureus* required 3·31 and 0·33 ppm, respectively (Lamartinere et al., 1969).

From the data listed in Table 11, herbicides in general have a slightly inhibitory effect on respiratory activity in soil since the effect ratio is 0·91. Trifluralin is consistently inhibitory and many of the halogenated aliphatics, phenoxy-acids and substituted ureas are also inhibitory. Of the triazines listed, simazine was more often than not inhibitory, more so than other triazines. The effect ratio, indicating a nett inhibitory effect, does not, however, correlate with the findings for organic matter and cellulose degradation; a fact which is not easily explained.

(c) *Other soil-enzymic activity*. Interest in, and realization of, the importance of soil enzymic activity has increased rapidly over the last decade but as yet the reliability and reproducibility of many of the analytical methods remains uncertain despite the efforts of a few dedicated researchers. Much of the problem lies in interference from soil clay minerals and salts and also in the fact that cell-free enzymes in soil may be present only in very small quantities despite the known protective actions of humic material and adsorption of the proteins to clay surfaces.

Whilst dehydrogenase activity is known to be closely correlated to soil respiratory activity, it is included here since many workers regard it as a separate enzymic activity.

Diphenamid, at rates of 1000 to 20 000 ppm, increased catalase, polyphenol oxidase and peroxidase activity of *Azotobacter chroococcum* cultures but decreased the cytochrome oxidase and dehydrase activity (Rankov and Genchev, 1969). 2,4-D (20–40 000 ppm) suppressed pectolytic enzymes of *Aspergillus niger*, the effect increasing with concentration (Gaponenkov and Nogumanova, 1969). MCPA, atrazine, prometryne and mixtures of simazine and aminotriazole or atrazine, mecoprop and 2,4,5-T were applied to fungal cultures at the equivalent of field rates (Bakalivinov, 1972). MCPA increased catalase activity in *Aspergillus flavus* and *A. niger* and urease in *A. tamarii*. The mixture of simazine and aminotriazole also increased urease in *A. tamarii* but prometryne and the atrazine, mecoprop and 2,4,5-T mixture decreased urease in *A. flavus*. Atrazine decreased urease in all species examined and atrazine, prometryne and the atrazine, mecoprop and 2,4,5-T mixture inhibited catalase in *A. flavus*, *A. niger* and *Paecilomyces varioti* (Bakalivinov, 1972). Cultures of bacteria from rhizosphere soil showed lower protease, urease and dehydrogenase activity in the presence of linuron (10 ppm) and chlorpropham (100 ppm), but simazine and atrazine, also at 100 ppm, had less inhibitory effect (Krezel and Musial, 1969).

Both picloram and 2,4-D at 5, 50, 250, 500 and 1000 ppm affected pectolytic enzyme formation by the fungus *Ceratocystis ulmi*; maximum activity of polygalacturonase (PGU) and pectin methylesterase (PME) was obtained with 250 ppm picloram, but polymethylgalacturonase (PMGU) was found with 50 and 250 ppm (Sekhon *et al.*, 1974). PMGU activity was inhibited by 1000 ppm of either picloram or 2,4-D, and PME activity was decreased by 500 ppm or more of 2,4-D. Activity of all three enzymes was stimulated by 250 ppm or less of either herbicide (Sekhon *et al.*, 1974).

The effects in soil of herbicides on some enzymic activities is shown in Table 12. The nett positive effects are greater than the nett negative effects, giving a ratio of 1·7, indicating that in general herbicides may increase enzymic activities. The effects of individual herbicide-groups are difficult to assess in this case because of differences in the types of enzymes evaluated.

(d) *Ammonification.* Ammonification of organic material, resulting in the release of NH_4^+ ions, is a key process in soil supplying readily available nitrogen to plants and microorganisms. Without ammonification nitrification would be severely limited and NO_3^- ions, the preferred form of nitrogen for some plants and microorganisms, would become of limited availability. The fact that, in many cases, ammonification is stimulated by herbicide applications is not surprising since the death of plants and possibly much of their attendant rhizosphere and phyllosphere populations, will result in an increase in proteins available for transformation.

Balicka and Sobieszczanski (1969b) used peptone media with herbicides added at 2–1000 ppm. Only the highest concentrations of herbicides inhibited ammonification and concentrations of 100–250 ppm had no effect, with the exception of linuron which was toxic at lower rates. Applications of 100–250 ppm in media is equivalent to 10 kg ha^{-1} to a depth of 1 cm, or 1·0 kg ha^{-1} to a depth of 10 cm, neither of which rates can be considered excessive and both of which are within the usual herbicide rates applied under field conditions. The effects of some herbicides on ammonifying microorganisms and the ammonification process in soil are shown in Table 13.

The ratio of positive to negative effects derived from Table 13 is 1·74, suggesting that ammonifying microorganisms benefit from the increase in soil organic matter resulting from the killing effect of herbicides on weeds and grasses, and that few herbicides have an inhibitory effect on these microorganisms. The three largest groups of herbicides, the phenoxy-acids, triazines and urea-type compounds, are mainly positive in their effect on these organisms.

5. *Miscellaneous effects*

2,4-D, atrazine and simazine, applied at field rates, increased the nicotinic acid content in cropped but not in fallow chernozem and atrazine also increased the biotin content of cropped soil (Mereshko, 1969). In a medium podzol (7% organic matter), atrazine (10 and 20 kg ha^{-1}) increased the amino acid content of the soil (Prozorova, 1967). The amino acid content of a light loam (1·6% humus) initially declined and later rose after treatment with atrazine, simazine or propazine at 3 kg ha^{-1} (Stepanova, 1967).

At less than 5 ppm, fluometuron (as Cotoran) increased the survival of bacteriophages in nutrient broth (Trull, 1969); whilst Sansing and Cho (1970) found that 1,3,5-triazines did not affect the biological activity of DNA in *Bacillus subtilis*.

At low rates (<10 ppm), DNOC suppressed the paraffin-decomposing ability of eight *Nocardia* spp. whilst atrazine, simazine and 2,4-D generally stimulated this activity, except at high concentrations of atrazine and simazine (Gergely, 1973). *Verticillium dahliae* survived treatment with 50 000 ppm chloramben but gave rise to a wide variety of mutants on sub-culturing (Hubbeling and Basu Chaudhary, 1970). Linuron (10 ppm), chlorpropham (5 ppm) and prometryne (10 ppm) inhibited antibiotic production by *Streptomyces griseus* (Krezel and Leszczynska, 1970) while propanil and its degradation products, 3,4-dichloroaniline and 3,3',4,4'-tetrachloroazobenzene were found by Prasad (1970) to be mutagenic towards the *meth₃* locus in *Aspergillus nidulans*. Leszczynska (1970) found that atrazine, prometryne, linuron, monolinuron, eptam (EPTC), tillam (pebulate) and chlorpropham varied in their effects on soil yeasts depending on the type and concentration of herbicide and strain of yeast. Most of these herbicides stimulated excretion of vitamins by *Rhodotorula* spp. but inhibited this in *Cryptococcus* spp.

Stimulation of growth of the photosynthetic bacterium *Rhodospirillum rubrum* was found (Kulinska and Romanov, 1970) to be inversely related to herbicide concentration. Some increase in carotenoid concentration was observed which seemed to be related to the effects on growth. Production of bacteriochlorophyll was inhibited, especially in ten day old cultures, by rates of 1 and 5 ppm of simazine and 5 ppm of monolinuron.

B. FUNGICIDES

1. *Effects on soil bacteria and their activities*

Unless bacteria are directly inhibited by fungicides, the effects of fungicides on them may be expected to be indirectly stimulatory in providing additional organic substrate (as killed fungi) for their growth and also, where applicable, by reducing antibiotic production by fungi. Decreases in fungal activity as a result of fungicide action in soil may further reduce competition for available nutrients, oxygen, space and water, but in other cases may decrease availability of certain types of nutrients by inhibiting fungal degradation of organic materials, such as lignin.

(a) *Bacterial numbers.* Three soil bacterial species were used by Chinn (1973) to bioassay eight fungicides applied to soil. Methylmercury dicyandiamide (MMDD) was the only fungicide inhibiting the bacteria at 1 ppm. MMDD and pyrazophos were the most inhibitory, followed by thiram and maneb in decreasing order; captan showed little or no activity. When benomyl (*c.* 50 ppm) was added to agar media, numbers of bacterial colonies appearing on the plates increased with increasing benomyl concentration in the soil from which the dilutions were made (van Faassen, 1974).

In a study using captan and thiram, both at 20 kg ha^{-1} of loam soil, Naumann (1971c) found that the bacterial population decreased to 60% of that in untreated soil during the first week after treatment with captan and then increased over the next three weeks to 180% of the control before finally returning to the control level. With thiram, bacterial numbers increased steadily to 180% of the untreated soil's population. Houseworth and Tweedy (1973) found that captan and thiram, at 0·1 or 1·0 kg ha^{-1} in a silt loam, caused fluctuations in bacterial numbers which were inversely related to the fungal population, and in combination with atrazine neither fungicide exhibited synergistic or antagonistic effects on the population. In an unspecified soil type (pH 5·3 and 3·1% O.M.), captan (9 kg ha^{-1}), thiram (6·7 or 13·4 kg ha^{-1}) and quintozene (5·6 or 11·2 kg ha^{-1}), significantly increased numbers of heterotrophic soil bacteria while benomyl (4 or 20 kg ha^{-1}) had little effect (Wainwright and Pugh, 1974). In further studies with captan (0·5, 2·5, 5·0 and 10 kg ha^{-1}) added to an unspecified soil type (pH 6·7 and 2·9% O.M.), taken from beneath cultivated grass, the fungicide generally increased bacterial numbers (Wainwright and Pugh, 1975a). Similar effects of captan were noticed by Agnihotri (1971), but the fungicide has also been implicated in adverse effects on bacterial numbers in soil (Mahmoud et al., 1972).

Benomyl at 2·2 to 89·6 kg a.i. ha^{-1} in silt loam, fine sand and sandy loam soils had no effect on total numbers of bacteria (Peeples, 1974). In sandy soil not previously treated with benomyl, numbers of bacteria increased with increasing rate (2·0, 5·0, 10 kg ha^{-1}) of this fungicide, but in similar soils which had previously received up to 2 kg ha^{-1} of benomyl, an application of 2 kg ha^{-1} under laboratory conditions decreased the total count 200 to 1000 times (van Faassen, 1974). Whilst soils with different humus contents reacted differently to benomyl (3·0–30 kg ha^{-1}) in a humus sand it had little effect on numbers of soil bacteria (Hofer et al., 1971).

In loam soils, quintozene at rates up to 20 kg ha^{-1} had no effect on soil microorganisms or their metabolic activities in the absence of added energy substrates (Farley and Lockwood, 1968; Lockwood, 1970). However, in the presence of added glucose, quintozene increased bacterial numbers (Farley and Lockwood, 1969; Ko and Farley, 1969). Since quintozene can be used as a fungicide, insecticide or herbicide it is important to note that the compound is relatively innocuous towards soil bacteria at normal rates of application. The monosodium salt of hexachlorophene (Isobac), at rates equivalent to $0·1, 0·2, 0·4, 0·8$ and $1·6 \, kg \, ha^{-1}$ in soil, altered the bacterial composition and favoured the growth of a bacterium which was antagonistic to fungal pathogens (Pinckard, 1970).

Anilazine (Dyrene) and maneb, at $1·5, 6·0, 24$ and $96 \, kg \, ha^{-1}$ on an acid lateritic clay (pH $5·0$) and an alluvial loam (pH $7·7$), increased bacterial populations, but the effect dissipated ten months after treatment, except with the highest rate of maneb (Dubey and Rodriguez, 1974). Fenaminosulf (Dexon) added at c. $1·1$ to $4·4 \, kg \, ha^{-1}$ to a soil (pH $6·4$), had no effect on the numbers of bacteria, whilst at higher rates ($8·8$ to $35 \, kg \, ha^{-1}$) there was no consistent effect on spore-forming and non-sporing bacteria (Agnihotri, 1973).

From the above data on soil-applied fungicides, positive effects far outweigh the negative effects, giving a ratio of $3·5$. This indicates that in general the fungicides favour an increased bacterial population in soil or at least have no harmful effect upon them. This is possibly due to the removal of competition and to the provision of additional substrate material in the form of killed fungi.

(b) *Nitrification*. Fungicides may severely suppress nitrification in soil. Audus (1970b) concluded that strong suppression of nitrification, followed by slow recovery, is shown by the following fungicides (recovery time in days in parenthesis): nabam (60), thiram (60), ferbam (28), maneb (25), zineb (17).

Captan, thiram and Verdasan either stimulated or had no effect on nitrification in grass soil when applied at low concentrations (captan $0·5–10$, thiram $0·5–5$ and Verdasan $0·1–0·5 \, kg \, ha^{-1}$). At higher rates, however, captan ($25 \, kg \, ha^{-1}$), thiram (10 and $25 \, kg \, ha^{-1}$) and Verdasan ($1·0–5·0 \, kg \, ha^{-1}$) inhibited the process (Wainwright and Pugh, 1973). In further experiments on an unspecified soil type, Wainwright and Pugh (1974) showed that captan ($9 \, kg \, ha^{-1}$), thiram ($6·7 \, kg \, ha^{-1}$), dicloran ($2 \, kg \, ha^{-1}$) and formalin (500 litres ha^{-1}) inhibited nitrification markedly whereas quintozene ($5·6 \, kg \, ha^{-1}$) had only a slight effect. Mahmoud *et al.* (1972) also found that captan adversely affected soil nitrification.

Maneb, zineb and tribasic copper, applied repeatedly to soil, resulted in decreased nitrification and nitrogen mineralization (Dubey, 1970). In acid lateritic clay (pH $5·0$) and alluvial loam (pH $7·7$), anilazine (Dyrene) and maneb at $1·5$ to $96 \, kg \, ha^{-1}$ inhibited nitrification; inhibition was less in the rapidly-nitrifying loam than in the slower-nitrifying clay and in both cases the fungicides inhibited *Nitrosomonas* spp. but not *Nitrobacter* spp., maneb being the more toxic (Dubey and Rodriguez, 1970).

Benomyl at $1 \cdot 5$ to $30 \, \mathrm{kg \, ha^{-1}}$ in a humus sand decreased nitrification after four weeks incubation (Hofer *et al.*, 1971) whereas in sandy soils receiving the equivalent of $1 \cdot 0$, $2 \cdot 5$ and $10 \, \mathrm{kg \, ha^{-1}}$, van Faassen (1974) found no effect for the first six weeks incubation but then significantly more nitrification in treated soil than in untreated soil. In mixed cultures of *Nitrosomonas* sp. and *Nitrobacter* sp., however, van Faassen (1974) found that 20 ppm benomyl inhibited oxidation of NO_2^- to NO_3^- and that 200 ppm delayed the oxidation of NH_4^+ to NO_2^-, as well as the oxidation of NO_2^- to NO_3^-.

1-Chloro-2-nitropropane (Lanstan) at a concentration equivalent to $1 \cdot 0 \, \mathrm{kg \, ha^{-1}}$ and dicloran (Botran) at $1 \cdot 0$ to $10 \, \mathrm{kg \, ha^{-1}}$ inhibited nitrification in a fine sandy loam but quintozene and tecnazene at up to $100 \, \mathrm{kg \, ha^{-1}}$ had only a slight effect (Caseley and Broadbent, 1968). Fentin acetate ($1 \cdot 0 \, \mathrm{kg \, ha^{-1}}$) added to a loam soil initially decreased and then increased nitrification in one experiment but had no effect in a similar experiment (Barnes *et al.*, 1973), while fenaminosulf (Dexon) in an unspecified soil (pH $6 \cdot 4$; $3 \cdot 5\%$ O.M.) inhibited nitrification at $35 \, \mathrm{kg \, ha^{-1}}$ but not at lower rates of $1 \cdot 1$ to $1 \cdot 4 \, \mathrm{kg \, ha^{-1}}$ (Agnihotri, 1973).

From the limited data available, the effect ratio of $0 \cdot 54$ indicates that fungicides can be very inhibitory towards soil nitrification. Whilst this may be regarded as agronomically desirable for some crops, there are cases, for example lettuce, where decreased NO_3^- content can be harmful (Maw and Kempton, 1973). In addition, the inhibition of nitrification is a disruption of a natural process in soil and represents an altered ecological system that may create stress within the entire soil ecosystem.

(c) *Denitrification.* In reviewing earlier work, Audus (1970b) noted that most dithiocarbamates are inhibitory whilst alkyl-diamino-dithiocarbamates, such as nabam and maneb, are more toxic than their dialkylamino counterparts ferbam, thiram or ziram. In this latter group the inhibitory effect is proportional to the number of dithio-carbamate radicals. Thus, ferbam is more toxic than thiram and the latter more so than ziram (Audus, 1970b).

(d) *Rhizobia and legume nodulation.* Many fungicides show some degree of toxicity towards rhizobia and/or nodulation. Toxicity is often associated more with some strains of rhizobia than others. Thus, the strains associated with *Trifolium*, *Phaseolus* and *Pisum* spp. may be highly susceptible, whereas those associated with *Melilotus* spp. may be either susceptible or resistant. Quinonoxime benzoylhydrazine has been found to be highly toxic to strains from *Trifolium* and *Phaseolus* but can also stimulate growth in some strains from *Melilotus* (Brakel, 1963 – cited by Audus, 1970b). Much of the work reported on the effects of fungicides deals with *in vitro* susceptibility and in many cases the "soil" investigations are confined to sand-culture of legumes.

Afifi *et al.* (1969) found that at the lowest concentration (0·03 ppm) captan, thiram, Ceresan, dichlone and methylarsinic sulphide (Rhizoctole) were toxic to some, but not all, rhizobia strains, whereas in the range 300 to 30 000 ppm most strains were inhibited. The two mercurial fungicides MMDD and Ceresan were respectively strongly toxic and lethal to seven strains of *Rhizobium leguminosarum* but thiram, captan, dichlone and chloranil were less inhibitory (Kecskés and Vincent, 1969a). According to Daitloff (1970a) the *in vitro* toxicity of fungicides to rhizobia followed the order Ceresan > thiram > captan > chloranil, while Gillberg (1971) found that *Rhizobium meliloti* strains generally tolerated higher rates of captan and a mixture of benzimidazole and carbamylic acid (Neo-voronite) *in vitro* than did strains of *Rhizobium trifolii* or *R. leguminosarum*. All strains were inhibited by 50 ppm captan but spontaneous mutants resistant to this concentration were isolated. Passage through captan-containing media failed to induce resistant mutants in those strains that were not resistant to this fungicide but the method did invoke resistant mutants when Neo-voronite was used at 100 ppm. The mutants were able to infect host plants, suggesting that the use of rhizobia mutants may be a means of ensuring that pesticides will not adversely affect these bacteria in practice.

In-vitro experiments with 33 fungicides and 21 strains of *R. leguminosarum* showed (Kecskés, 1970) that captan, thiram, dichlone and chloranil were moderately inhibitory whereas Ceresan and MMDD were strongly inhibitory. Makawi and Abdel-Ghaffar (1970) found that captan, Ceresan, Agrosan and benomyl were bactericidal towards a number of strains of rhizobia, in contrast to herbicides and insecticides which had only slight effects on these organisms. Investigations with thiram and its degradation product, sodium dimethyl dithiocarbamate, showed that some rhizobia were sensitive but the degree of sensitivity depended upon the strain of *Rhizobium* spp. and the pH (Sud and Gupta, 1972). Fisher and Clifton (1976) found that benomyl, triforine and tridemorph had little effect on strains of *R. leguminosarum* and *R. meliloti* with concentrations of up to 200 ppm in agar medium of pH 5·5 to 7·5. Using purest grade unformulated fungicides, Fisher (1976) showed that growth of *R. trifolii* in agar with 10, 50, 100 and 200 ppm each of captan, dodine and oxycarboxin was markedly decreased. An octylphenol/ethylene oxide condensate (Ethylan CP) also decreased growth but benomyl, carboxin, dimethirimol, ethirimol, thiram, tridemorph and thiophanate-methyl had little effect even at the highest concentrations. Triforine at 100 or 200 ppm was slightly inhibitory, but not at 10 or 50 ppm. In liquid culture at rates of 10 to 1000 ppm the inhibitory effects of captan, dodine and oxycarboxin were confirmed (Fisher, 1976).

Staphorst and Strijdom (1976) investigated the effects *in vitro* of 13 fungicides on rhizobia strains capable of nodulating *Vigna anguiculata*. The fungicides could be divided into three broad groups on the basis of their toxicity to each of two strains. The least toxic included quintozene, benomyl, fenaminosulf and 2-hydroxypropyl methane-thiosulphonate (HPMTS). A mixture of dicloran and captafol, captan, and a mixture of captan with phenylmercuryacetate were intermediate in their effect, whilst carboxin, thiram, mancozeb, maneb and 2-(thiocyanomethylthio)benzothiazole (TCMTB) were the most toxic.

The effects of pesticides on rhizobia in soil or in sand-culture are frequently measured in terms of degree of nodulation, nodule weight, leghaemoglobin content or nitrogen-fixing efficiency. It is often the case that results obtained by these criteria do not wholly agree with the indications obtained from the growth *in vitro* of rhizobia in the presence of the same fungicides. Thus, whilst thiram was found to inhibit cultures of *R. leguminosarum* (Kecskés and Vincent, 1969a) it had no effect on the nodulation of vetch in sand-culture or in various soils under greenhouse and light-chamber conditions (Kecskés and Vincent, 1969b). In the latter case, Ceresan strongly inhibited nodulation, MMDD was found to be unstable and harmful to vetch seeds and chloranil, dichlone, captan and copper oxychloride had an intermediate effect on nodulation (Kecskés and Vincent, 1969b).

Whereas captan has been found by other workers to slightly inhibit nodulation, Kratka and Ujevic (1970) found that, at rates equivalent to $0 \cdot 3 \, \text{kg ha}^{-1}$, it had no effect on nodulation of vetch. Further inconsistencies between *in-vitro* and *in-vivo* tests are demonstrated in the work of Staphorst and Strijdom (1976) who showed that whilst quintozene was among the least toxic *in vitro* and *in vivo*, thiram, which was among the most toxic *in vitro*, was relatively without effect on nodulation or on dry mass of entire plants. Furthermore, whilst captan had an intermediate effect *in vitro*, it inhibited nodulation for the first 25 days after planting treated seed but then showed no further harmful effects. Benomyl allowed nodulation of plants by one strain of *Rhizobium* but prevented nodulation by another strain, whereas tests *in vitro* had shown it to be among the least toxic. A characteristic lack of tap-root nodulation in all plants from fungicide-treated seeds $(2 \, \text{g kg}^{-1})$ was noted. Benomyl, fenaminosulf, HPMTS, dicloran plus captafol, captan, Saadtan, carboxin, mancozeb, maneb and TCMTB all inhibited nodulation to some extent, Saadtan and maneb totally (Staphorst and Strijdom, 1976).

Using sterile and non-sterile soil, Fisher (1976) found that with root drenches of dodine, captan, carbendazim, Ethylan CP, thiram and ethirimol, at rates equivalent to 25 and 50 kg ha^{-1} in each case, the dry weight and nitrogen content of *R. trifolii* – inoculated white clover plants were not affected. Only three of the fungicides had any effect on N_2-fixation in excised nodules. Ethylan CP at 25 kg ha^{-1} increased fixation but at 50 kg ha^{-1} had no effect. Thiram at 50 kg ha^{-1} increased N_2-fixation significantly. Oxycarboxin at rates equivalent to 2·5, 5·0 and 15 kg ha^{-1} significantly decreased N_2-fixation, whereas ethirimol did not.

Fungicides appear to differ in their effects on growth of rhizobia *in vitro* or *in vivo* and there are differences in effects in practice depending on the strain of rhizobia involved. There is nonetheless a very real danger that legume nodulation and N_2-fixation may be affected directly by seed-dressings, fungicides applied during the growing season or residual fungicides from previous applications. Daitloff (1970b) has suggested that toxic effects of fungicides may be prevented by coating legume seed with a layer of polyvinyl-acetate after fungicide-treatment but before inoculation with rhizobia.

The data obtained from the available literature suggest that fungicides have a variable effect on rhizobia, nodulation and the N_2-fixing efficiency of nodulated plants. There is certainly very little evidence to condone testing of fungicides *in vitro* as a guide to possible effects *in vivo*. Substantial differences in strain tolerance also exist, making the assessment of fungicide effects even more difficult. The data presented here give a positive to negative effect ratio of 1·0, suggesting that fungicides have a probability of 0·5 of adversely affecting rhizobia and nodulation in soil. Some workers have even suggested that in the absence of a suitable fungicide with no effects on rhizobia or nodulation whatsoever, those which give good seed protection but slightly delay or otherwise limit nodulation are perfectly acceptable, especially if strains of rhizobia which are tolerant to the particular fungicide can be grown and maintained as an inoculum.

(e) *Free-living nitrogen fixation.* With the finding of a semi-symbiotic enhancement of growth of *Azotobacter* spp. in the roots of certain grass species (Dobereiner, 1968), the possible influence of residual fungicides in soil may become particularly important to the nitrogen economy of grassland agriculture, or where these important grass species are grown in rotation with crops.

Nine pesticides, at 4, 40 and 400 ppm, had no effect on *Azotobacter chroococcum* cultures, but copper oxychloride at 40 and 400 ppm completely inhibited growth (Gil Diaz-Ordoñez *et al.*, 1970). The sensitivity of rhizosphere isolates to thiram and its degradation product, sodium dimethyl dithiocarbamate, was studied at pH values of 5–9 (Sud and Gupta, 1972); *A. chroococcum* was more sensitive than other isolates, especially at pH 7, and isolates from roots of different plants varied in their susceptibility to these chemicals. Pure cultures of *A. chroococcum* were inhibited by zineb (5000 and 50 000 ppm), captan (3000 and 30 000 ppm), ferbam (3000 and 30 000 ppm), thiram (5000 ppm), folpet (30 000 ppm) and maneb (30 000 ppm), whilst an antibiotic mixture, Fungicidin, had no effect at 5000 ppm (Langkramer, 1970). Benomyl (2 kg ha^{-1}) had no effect on *Azotobacter* spp. in sandy soil (van Faassen, 1974) but captan applied at field rates to cotton cultivars inhibited *Azotobacter* spp. in the rhizosphere soil (Mahmoud *et al.*, 1972).

2. Effects on soil fungi and actinomycetes

Fungicides are designed to kill undesirable fungi. Not unexpectedly, other fungi are also affected since built-in selectivity of the type that would preclude harmful effects to non-pathogenic fungi and antagonists of pathogenic fungi is, as yet, seemingly impossible to achieve. Recovery of fungi from the effects of fungicides is usually quite rapid but, unfortunately, this may also apply to pathogenic fungi. As a general rule, all species are immediately suppressed to some extent and recovery (by re-invasion, germination of protected or resistant spores, or mutation) may be relatively rapid or slow, depending on prevailing soil conditions. The spectrum of species in the soil after treatment may be considerably altered ("species shift") and may persist for long periods. Alterations in the competition for substrate material and resulting changes in end-product metabolism are probably responsible for the rapid increase in numbers of a particular species or group of fungi.

(a) *Growth*. MMDD was active at only 1 ppm against fungal and actinomycete cultures, whereas thiram, maneb, and especially captan, were less toxic (Chinn, 1973). In experiments using thin layer chromatography plates spread with soil, Helling *et al.* (1974) found that bioassaying with soil fungal species provided information not only on probable movement of the fungicides in soil but also on relative sensitivities of fungi. Of the ten fungi used for bioassay, *Trichoderma viride* was sensitive to the greatest number of fungicides.

Fenaminosulf (as Dexon) in potato dextrose agar at rates of up to 300 ppm had no effect on *Mortierella* spp. (Agnihotri, 1973), whilst benomyl at 25 ppm had a fungistatic effect on fungal propagules from a humus sand (Hofer *et al.*, 1971). Benomyl (4 and 20 kg ha^{-1}), captan (9 kg ha^{-1}), quintozene (5·6 and 11·2 kg ha^{-1}) and thiram (6·7 or 13·4 kg ha^{-1}) applied to field plots of a soil (pH 5·3; O.M. 3·1%) caused increases in fungal propagules 28 days after treatment (Wainwright and Pugh, 1974).

Quintozene, at rates up to 20 kg ha^{-1} in loam soils, produced no changes in numbers of fungi or actinomycetes and no obvious species shifts; but in soil amended with various energy sources, chitin in particular, it strongly suppressed numbers of actinomycetes. With glucose-amended soil, fungi usually increased but some decreases were also noted with some notable qualitative changes; *Fusarium* spp. in particular increased in the presence of quintozene. The species shifts correlated with the relative sensitivities of organisms to quintozene in agar culture (Farley and Lockwood, 1968, 1969; Ko and Farley, 1969). Selective stimulation of quintozene – tolerant *Helminthosporium victoriae* conidia on soil surfaces, with and without added energy sources, was investigated by Farley and Lockwood (1969). The results suggest that quintozene suppressed competition from other fungi and actinomycetes and that the enhanced availability of nutrients promoted germination of *H. victoriae* conidia and subsequent elongation. Katan and Lockwood (1970) investigating quintozene effects in a loam to which alfalfa was added, found that the pesticide (1·0 kg ha^{-1}) affected the rate of colonization of the alfalfa particles. Numbers of fungal propagules increased with time for 18 days and then decreased, resulting in an overall decrease of 20 to 80%. Qualitative changes in the fungi were also noted. Lockwood (1970) suggests that the accumulation of quintozene by organic residues may mean that colonization of, for example, chitin and cellulose, could become "fungicide directed" inasmuch as quintozene appears to affect only microorganisms in an active metabolic condition and which are sensitive to the compound. Accumulation of quintozene within mycelium of tolerant species is significant since upon death of the mycelium, decomposition of the cell-material may be affected by its presence. Quintozene accumulation in *R. solani* was found to reach 300 μg g^{-1} of moist mycelium (Ko and Lockwood, 1968).

Other effects of fungicides on soil fungi and actinomycetes are summarized in Table 14. The ratio of overall positive to negative effects, from Table 14 and the preceding text, is *c.* 0·5, indicating, as expected, a predominantly inhibitory effect.

Audus (1970b) has listed the general response of a number of fungi to a range of fungicides. Phycomycetes such as *Mucor, Rhizopus* and *Mortierella* spp. show sensitivity to thiram, nabam, mercurials, captan, metham-sodium, dazomet and allyl alcohol, whereas the Ascomycetes (various *Aspergillus, Penicillium, Gliocladium, Chaetomium, Chaetomidium* and *Melanospora* spp.) generally recover rapidly. The ubiquitous *Fungi Imperfecti* show a varied pattern of recovery/susceptibility, with the Sphaeropsidales being very resilient. *Trichoderma viride* appears to be uniformly resistant to many fungicides in addition to those mentioned above, whereas the Fusaria are generally highly sensitive.

A very important aspect of the effects of fungicides on soil fungi concerns the phylloplane microflora. Given that most saprophytic phylloplane fungi have their origin in soil, the inhibition/stimulation effect of a foliar-applied fungicide on an organism may be highly significant in subsequent events when the leaf eventually becomes incorporated into the soil. Ethirimol ($1 \cdot 5$ kg a.i. ha^{-1}) did not affect populations of saprophytic filamentous fungi or ballistosporic yeasts whilst zineb ($0 \cdot 35$ kg a.i. ha^{-1}) slightly reduced populations of fungi but had no effect on *Sporobolomyces* spp. (Dickinson, 1973). Tridemorph ($0 \cdot 5$ kg a.i. ha^{-1}) was less effective as a general antifungal compound than was fentin acetate ($0 \cdot 1$ kg ha^{-1}) plus maneb ($0 \cdot 9$ kg ha^{-1}), which decreased numbers of yeasts on leaves. Dickinson (1973) indicates that these studies suggest that a significant proportion of leaf microbes are susceptible to fungicides, especially those known to have a wide spectrum of activity against pathogens. Systemic fungicides, it is postulated, could have less effect. Whilst Dickinson notes that the consequences of fungicide-reduced populations on leaves could be twofold, namely (i) the killing of saprophytes which compete with pathogens, leading to the establishment of spray-tolerant pathogens and (ii) the extension of leaf life by delaying fungal-induced senescence, it is also highly probable that alterations to the phylloplane population could significantly affect microbial decomposition of the leaf tissues in soil.

(b) *Pathogenic fungi and their antagonists.* Of increasing importance and concern is the possible increase in pathogenic fungi and/or the increase in the incidence of relatively "minor" diseases of plants that have been treated with fungicides aimed at the control of "more important" diseases. An example of the interrelationship between pathogenic fungi and fungicides is reported by Backman *et al.* (1975) who noted that fungicidal treatment of groundnuts succeeded in controlling *Cercospora* leaf spot but in unsprayed plots *Sclerotium rolfsii* damage was lower than in sprayed plots. Benomyl-sprayed plots had a higher incidence of *S. rolfsii* due to a direct effect of the fungicide on this pathogen or due to an indirect effect, by killing its antagonist *Trichoderma viride*. The highest infestations of *S. rolfsii* were associated with plots receiving fungicides which had no effect on *S. rolfsii* but which were toxic to *T. viride*.

Various fungicides were tested for their effects on the frequency of re-isolation of *Rhizoctonia solani* from soil seeded with this pathogen (Popov and Zdrozhevskaya, 1972). Quintozene, applied at 1·2 and 2·5 kg ha^{-1} of soil, did not increase the frequency, but thiram at the same rates and zineb or captan at 2·5 and 10 kg ha^{-1} did. *Theilaviopsis basicola* was not re-isolated from untreated soil seeded with this pathogen but recovery frequency increased significantly in soil treated with quintozene or thiram (1·2 kg ha^{-1}), thiram (2·5 kg ha^{-1}) or maneb (1·2 and 2·5 kg ha^{-1}) (Popov and Zdrozhevskaya, 1972). Agrosan GN dust increased the actinomycete antagonists of the bacterial pathogens *Xanthomonas malvacearum* and *Pseudomonas solanacearum* (Pugashetty and Rangaswami, 1969).

Of the ten test fungi used by Helling *et al.* (1974) for soil TLC assay of 33 fungicides, *Trichoderma viride* was the most sensitive to the greatest number of fungicides, applied at a rate of 100 µg. Since *T. viride* is an important antagonist of a number of pathogenic fungi, the results indicate the real possibility of an ecological imbalance arising as a result of fungicidal applications, which could be of great significance in the occurrence of epiphytotics in commercial crops.

Agnihotri (1973) found that with increasing rates of fenaminosulf (1·1 to 35 kg ha^{-1}) in soil, numbers of *Pythium* propagules decreased, whilst Pinckard (1970) found that sodium hexachlorophene (0·1 to 1·6 kg ha^{-1}) altered the soil bacterial composition to such an extent that one bacterial species predominated and this was antagonistic to *Rhizoctonia, Pythium, Phytophthora, Fusarium* and other fungi in soil. In this latter case the effect of the fungicide in controlling disease fungi is by alteration of the ecological balance in favour of an antagonistic bacterium.

MacNeill and Nicholson (1974) noted differences in growth habit and tolerance of different *Fusarium* spp. towards benomyl. The concentration of benomyl and exposure time were critical for adaptation to tolerance. Whilst the method is of limited value as a taxonomic tool, it demonstrates the variable response of pathogens in soil to applications of fungicides. Peeples (1974) found that *Trichoderma* sp. was frequently present in soils treated with benomyl at rates of 2·2–89·6 kg ha^{-1}.

If the findings of Helling *et al.* (1974) are excluded on the basis that their method cannot be equated realistically to soil conditions, the limited amount of information on fungicide effects on antagonists and phytopathogens gives a ratio of positive to negative effects of 4·0, indicating that both pathogens and antagonists have a high probability of not being adversely affected by fungicides. The relatively limited data available on effects of fungicides on phytopathogens in soil compared with that available for non-phytopathogens may be responsible for this somewhat alarming suggestion. It is acknowledged that many soil-borne plant pathogens are readily controlled by particular fungicides, but these fungicides are usually specifically recommended for control of the pathogen concerned and these effects are widely documented. What is not known to the same extent is whether fungicides, appropriate for the control of a particular pathogen, can stimulate other pathogens in soil. Possible reasons why only limited information on this aspect of soil microbiology is available are (i) that during the course of fungal counts few investigators undertake to identify every fungal colony appearing, (ii) phytopathogens which are stimulated by pesticide-treatment of soil may not be capable of growth on the particular medium employed or may be relatively slow-growing and hence rapidly overgrown by other fungi, or (iii) increases in the numbers of other pathogens in the treated soil may not be noticed until a suitable host crop is planted.

3. *Effects on soil algae*

Audus (1970b) reported that nabam at the very low concentration of 1 ppm adversely affects *Chlorella* and *Anabaena*. Venkataraman and Rajyalakshmi (1972), investigating the effects of fungicides on N_2-fixing blue–green algae, found that most of the strains of *Anabaena* could tolerate 100 ppm Ceresan, but one strain was sensitive to 0·01 ppm. Zineb was lethal to some strains of *Anabaena* and *Nostoc* even at the lowest concentrations used, but some strains grew well at 50 ppm, whilst *Tolypothrix tenuis* and *Aulosira fertilissima*, frequently introduced into rice paddies, tolerated high concentrations of Ceresan and zineb. *Chlorella sorokiniana* was sensitive to most of the 33 fungicides tested on soil TLC plates by Helling *et al.* (1974).

4. *Effects on combined activities of the soil microbial biomass*

(a) *Cellulolytic activity and organic matter degradation.* The effect ratio of 0·5 for fungicides on fungal and actinomycete populations suggests that fungicides may have a profound effect on reactions in which these two major groups play an important role. On the other hand, the effect ratio for bacteria of 3·5 may compensate for decreases in fungal and actinomycete involvement in cellulolysis and organic matter degradation, provided that bacteria are capable of the same enzymatic roles. A further point which may influence the degradation of cellulose and organic matter is that in spite of the obvious suppression of fungal populations by fungicides, the surviving population may be capable of continuing the degradation of organic matter, especially if species shifts are in favour of organisms with this capability. For example, in an arable soil, TCMTB at 5–30 kg ha^{-1} caused a species shift resulting in the displacement of cellulolytic fungi such as *Fusarium, Penicillium* and *Aspergillus* by cellulolytic *Streptomyces* (Voets and Vandamme, 1970), while at rates of 3–30 kg ha^{-1} in humus sand, benomyl had no effect on the decomposition of organic matter by surviving (resistant) species of fungi and other microorganisms (Hofer *et al.*, 1971).

Domsch (1960) suggests that some of the fungi which are inhibited by fungicides can be characterized by enzymic potentialities of wide and narrow substrate utilization and that for each sensitive species other frequently occurring fungi can take-over a "lost" function. This applied to the high fungicide tolerance of *Chaetomium* and *Penicillium* spp. and their rich complement of enzymes. Domsch (1970) also reports that captan seriously delayed the formation of cellulases by soil organisms.

Benomyl (2 kg ha^{-1}) had no effect on the cellulolytic activity of organisms in sandy soils (van Faassen, 1974). Baroux and Sechet (1974), however, report that low rates of copper sulphate in a sand inhibited cellulolysis, but the effect could be overcome by the addition of lime and colloidal humic acids.

(b) *Respiration.* Concentrations of 100–500 ppm of dodine and captan, or 400 and 500 ppm ethirimol, decreased the O_2-uptake by *Rhizobium trifolii* whilst lower concentrations of ethirimol (100–300 ppm) were stimulatory and dimethirimol, triforine, carboxin and oxycarboxin had little effect (Fisher, 1976). In liquid media a 50% decrease in respiration rates of *Fusarium oxysporum* f.sp. *melonis* and *Saccharomyces cerevisiae* was caused by benomyl ($3 \cdot 5 \times 10^{-6}$ M) and benomyl, thiabendazole, carbendazim, thiophanate and thiophanate-methyl at 250×10^{-6} M all inhibited mitochondrial respiration and oxidative phosphorylation (Decallone *et al.*, 1975). The inhibitory effects of copper oxychloride on *Azotobacter chroococcum* depended on its concentration and time of contact with the cell suspensions (Gil Diaz-Ordoñez *et al.*, 1970).

In autoclaved soil seeded with *Achromobacter* sp. and treated with the commercial formulation of benomyl (Benlate), O_2-uptake was greater than in soil with the active ingredient only, whilst another *Achromobacter* sp. isolated from an orchard soil, had a higher O_2-uptake in soil plus benomyl, inorganic salts and glucose than in soil containing Benlate (Weeks and Hedrick, 1975); indicating the potential differences of isolates and also the effects of the soil's nutrient status and the presence of formulation chemicals on activities of microorganisms. That benomyl does not have much inhibitory impact on soil respiration was also shown by Hofer *et al.* (1971), Peeples (1974) and van Faassen (1974).

Fenaminosulf (low and high rates) retarded respiration in soil amended with glucose and paddy straw for a period of 60 days (Karanth and Vasantharajan, 1973). Quintozene added at 10 kg ha^{-1} to soil amended with glucose delayed the time of peak O_2-uptake by *c.* 24 h compared with amended soil without the fungicide and in soil amended with chitin, respiration was strongly suppressed by quintozene at 5 kg ha^{-1} (Lockwood, 1970). A fine sandy loam treated with 1·0 and 10 kg ha^{-1} Lanstan exhibited a depressed respiratory activity, but quintozene and TCNB only affected O_2-uptake at concentrations of 100 kg ha^{-1} (Caseley and Broadbent, 1968).

In examining more than 50 experimental fungicides at rates up to 20 kg ha^{-1} in a loam soil, Domsch (1970) recorded four types of response:

   (i) Complete inhibition of O_2-uptake.

   (ii) Short-term inhibition followed by recovery with no significant difference to the total O_2-uptake.

   (iii) Initial inhibition followed by a manifold increase in O_2-uptake, indicating fungicide decomposition.

   (iv) No influence on O_2-uptake, indicating the lack of side effects.

The same worker also found that inhibitions of respiratory activity by, for example, captan depended on the nature of the substrate material upon which soil microorganisms grew. In soils with or without glucose, cellulose or chitin, captan inhibited respiration and the more resistant the substrate, the longer was the period of inhibited respiration (Domsch, 1970).

Based upon the limited information reported here, fungicides may be considered to be potentially inhibitory towards respiratory activity in soil since the effect ratio obtained is 0·4.

(c) *Other soil-enzymic activity.* Benomyl at 3·0 to 30 kg ha^{-1} in humus sand did not affect dehydrogenase, amylase or catalase activity (Hofer *et al.*, 1971) and confirmation of the lack of effect of benomyl (2·0 kg ha^{-1}) on amylase activity in sandy soils was given by van Faassen (1974). Fenaminosulf decreased dehydrogenase activity and retarded glucose oxidation in soil (Karanth and Vasantharajan, 1973) while TCMTB (5 to 30 kg ha^{-1}) in an arable soil inhibited saccharase, urease, phosphatase and β-glucosidase activity in proportion to the concentration of the compound (Voets and Vandamme, 1970). Degradation of cutin, pectin and glucose was severely delayed in soil treated with captan (Domsch, 1970).

The summation of effects of only five fungicides on a total of 13 individually-assessed enzymic activities gives an effect ratio of 0·44, indicating that fungicides in general may be harmful to soil enzymic activity.

(d) *Ammonification.* In an acid lateritic clay (pH 5·0) and an alluvial loam (pH 7·7), maneb and anilazine at up to 24 kg ha^{-1} had no persistent effect on ammonification, but at 96 kg ha^{-1} inhibition occurred and there was no synergistic effect when the chemicals were applied together (Dubey and Rodriguez, 1970). With increasing concentrations of Verdasan (0·1–5 kg ha^{-1}), thiram (0·5–25 kg ha^{-1}) and captan (0·5–25 kg ha^{-1}) applied to soil (pH 6·7; organic carbon, 6·9%), ammonification increased (Wainwright and Pugh, 1973). In contrast, TCMB inhibited ammonification in a humus soil (pH 5·7) with increasing concentration up to 30 kg ha^{-1} (Voets and Vandamme, 1970). Ammonification was also increased in soil (pH 5·3) treated with benomyl, captan, quintozene or thiram (Wainwright and Pugh, 1974), but was inhibited by low rates of copper sulphate in a sand inoculated with a soil suspension (Baroux and Sechet, 1974).

The ratio of effects for the foregoing fungicides is 1·3, suggesting that ammonification is generally favoured by fungicide treatment of soil. This may be due to decreased fungal growth and competition, or increased bacterial ammonification of the protein of killed fungi.

5. *Miscellaneous effects*

Pesticides may alter the root exudations of plants. Such alterations could affect the rhizosphere populations. Benomyl and thiram at 5 or 25 kg ha^{-1} and Verdasan at 1 or 5 kg ha^{-1} in a grass soil caused qualitative and quantitative changes in the free amino acid content of the soil (Wainwright and Pugh, 1975b). High rates of fungicides generally reduced the amount of amino acid N, but low rates caused increases.

Wainwright and Pugh (1974), who used laboratory-incubated and field soils treated with benomyl, captan, thiram and quintozene, noted increases in the amounts of exchangeable elements such as Mn, Zn, Cu, K and Na in some cases but no effect or decreases (e.g. with Zn) in other cases.

Elemental sulphur caused acidification of soil which impaired biological activities except for sulphur-oxidizing and sulphate-reducing bacteria and some fungi (Krol et al., 1972).

C. INSECTICIDES

Past reviewers of the effects of insecticides on soil microbial populations and activities have stressed the fact that insecticides do not in general have much effect, except at concentrations greatly exceeding normal recommended field rates. While this is undoubtedly still true, there is increasing evidence to suggest that even at normal field rates insecticides may have some impact on soil microorganisms.

1. *Effects on soil bacteria and their activities*

(a) *Bacterial numbers.* At very high rates (10 000 ppm), aldrin, chlordane, dieldrin, endrin and heptachlor have a selective effect on bacterial growth *in vitro* in that they completely inhibit a wide range of Gram-positive organisms without affecting Gram-negative types. This effect has been attributed (Audus, 1970b) to a blockage in the electron transport chain, but it is also manifest that there must be some factor(s) associated with the chemical composition of the cell wall and/or membrane that causes such selectivity.

Heptachlor (25 ppm) killed 63% of soil bacterial isolates while at 100 ppm, 89% of the isolates were prevented from developing (Shamiyeh and Johnson, 1973). Diazinon (*c.* 3000 ppm) inhibited the growth of bacterial isolates although one degradation product (2-isopropyl-4-methyl-6-hydroxypyrimidine) had no inhibitory effect whilst another (diethylphosphorothioate) slightly increased growth (Robson and Gunner, 1970).

In studies with sandy loam, calcareous clay and clay soils, Gawaad *et al.* (1972b) applied chlordecone, endrin and lindane at *c.* $1 \cdot 0$ kg ha^{-1}, and pirimiphos-ethyl and fonofos (Dyfonate) at *c.* $0 \cdot 5$ kg ha^{-1}. Numbers of bacteria generally decreased during the first week after insecticide applications, rapidly increased in the second week and then gradually reverted to normal.

In a loam soil of pH $7 \cdot 2$, high levels (*c.* 6000 kg ha^{-1}) of commercially formulated lindane or 600–6000 kg ha^{-1} of formulated toxaphene caused a slight stimulation of bacteria for a short period, whilst in drier conditions, anaerobic and spore-forming bacterial populations decreased (Naumann, 1971b). In greenhouse-incubated loam, numbers of Gram-positive bacteria increased at the expense of Gram-negative types when the soil was treated with parathion-methyl (288 kg a.i. ha^{-1}) and the autochthonous bacteria of low nutrient requirement were greatly decreased five weeks after treatment; total bacterial population numbers increased during the first five days after treatment, decreased over the next ten days and then increased greatly above the control level for the next 20–25 days (Naumann, 1971a). Using a wider concentration range of parathion-methyl (0·15 to 150 kg a.i. ha^{-1}) in the same soil-type, incubated under greenhouse conditions, Naumann (1970a) found that the lowest rates gave brief increases in bacterial population numbers and the highest rate resulted in greater increases. Under field conditions, the loam soil showed an increased population with the high rate that lasted for 26 weeks but in dry years the increase was delayed (Naumann, 1970a). Anaerobic bacteria were initially suppressed with rates of 0·15–300 kg a.i. ha^{-1} but then increased (Naumann, 1970b). A comparison of the effects of parathion-methyl at 1·5 kg a.i. ha^{-1} added to fallow loam and loam under vegetative cover revealed slight decreases and increases in bacterial numbers for a period of 20 weeks in the former but a significantly decreased population for the first five weeks followed by a recovery over the next 15 weeks in the latter (Naumann, 1972).

Dieldrin (c. 200 kg ha^{-1}) had no effect on bacterial population numbers in soil and a similar effect was obtained with aldrin, diazinon and thionazin at 10 kg ha^{-1} (Harris, 1969). Further effects of insecticides on soil bacterial populations are shown in Table 15. From the data presented in the text and Table 15, the overall effect ratio for insecticides on bacterial populations in soil is 1·3, indicating predominant stimulation.

(b) *Nitrification.* Growth of *Nitrobacter agilis* in aerated cultures was inhibited by chlordane, TDE, heptachlor or lindane at 10 ppm but not by aldrin, whilst lindane, TDE, heptachlor and chlordane also partially inhibited NO_2^--oxidation (Winely and San Clemente, 1970). On the other hand, Mishra *et al.* (1972) found that neither lindane nor phorate, at normal and ten times normal rates, had any overall effect on nitrification although *Nitrosomonas* sp. was slightly affected. Earlier, Garretson and San Clemente (1968) also showed *Nitrosomonas* to be more sensitive than *Nitrobacter* to high rates of lindane (as well as malathion). Kuseske *et al.* (1974) found that at rates of up to 500 ppm, propoxur and aldicarb were toxic to both *Nitrosomonas* sp. and *Nitro-bacter* sp.

Shin-Chsiang Lin *et al.* (1972), using loam soil, tested the effects of eight insecticides on nitrification. Carbofuran and carbophenothion had no effect, trichlorfon inhibited at 5 and 50 kg ha^{-1} for a short period, fonofos, trichloronate and chlorpyrifos showed persistent inhibition at 50 kg ha^{-1} and aldicarb and propoxur at 50 kg ha^{-1} showed marked inhibition of nitrification. Using small plots of a sandy clay loam and a clay soil, Gawaad *et al.* (1972b, 1973a) applied chlordecone or endrin at 22 kg ha^{-1} and fonofos or pirimiphos-ethyl at

11 kg ha^{-1} and found that nitrification decreased during the first two weeks, increased in the third and fourth weeks and then returned to normal after 7–8 weeks. Nitrite accumulated in the soil during the first two weeks, indicating a probable inhibitory effect on *Nitrobacter* spp. Additional data on the effects of insecticides on soil nitrification are given in Table 16. The effect ratio derived from the table and text is *c.* 0·82, indicating a predominantly inhibitory influence of insecticides on nitrification. Whilst this may be agronomically acceptable, it is an undesirable soil microbial ecological effect.

(c) *Denitrification.* In sandy loam, clay loam and clay soils, Gawaad *et al.* (1972b, 1973b) found that chlordecone or endrin at 22 kg ha^{-1} and fonofos or pirimiphos-ethyl at 11 kg ha^{-1} increased denitrification. Audus (1970b) reports that in soils previously incubated anaerobically to increase numbers of denitrifiers, lindane had little effect on active denitrification of added nitrate and that parathion had no effect on denitrifiers at normal to 100 times normal field rates. Numbers of denitrifying bacteria in a loam soil were not affected by parathion-methyl applied at up to 300 kg ha^{-1} (Naumann, 1970b).

(d) *Rhizobia and legume nodulation.* Several strains of rhizobia have been found to be resistant to insecticides, some are stimulated while others are sensitive. The implications are that strains should be chosen which are resistant to the insecticide to be used (Audus, 1970b). Results of sensitivity testing of rhizobia against insecticides are often at variance with nodulation data obtained in soil experiments. DDT, lindane and toxaphene have been implicated in nodulation inhibition.

Kapusta and Rouwenhorst (1973) tested the interaction of ten insecticides and *Rhizobium japonicum* strains in pure culture and found that only disulfoton, at 7·5 ppm, inhibited a mixture of strains whilst the other insecticides had no effect. When screening individual strains, these workers observed different sensitivities of six strains to disulfoton and carbaryl. Disulfoton reduced growth of all strains and only two strains successfully grew at the highest rate (24 ppm) whilst with carbaryl, two strains were completely inhibited by the highest concentration (40 ppm) and one strain was resistant (Kapusta and Rouwenhorst, 1973). Lindane and fonofos at 10 and 50 ppm, respectively, inhibited glucose utilization of an effective strain of *Rhizobium trifolii* and increased it in an ineffective strain; conversely, diazinon (50 ppm) increased glucose consumption in the effective strain (Salem, 1971).

A group of insecticides were tested for inhibition of rhizobia on agar medium and the most sensitive and similar in all cases were *R. trifolii* and R. leguminosarum (Shin-Chsiang Lin et al., 1972). Increasing concentrations of disulfoton, carbofuran and endrin inhibited Rhizobium sp. (Oblisami et al., 1973). Dimethoate (1·89 ppm) allowed normal growth, whereas 3·78 ppm completely inhibited two strains of R. meliloti (Staphorst and Strijdom, 1974). The inclusion of trichlorfon in media at rates up to 1285 ppm suppressed growth of R. leguminosarum and R. trifolii and shifted their optimum pH and temperatures for growth (Salama et al., 1973). Five insecticides were tested against four strains of rhizobia by Daitloff (1970a) using insecticide emulsions absorbed on to beads which were applied, wet or dry, to agar plates seeded with the rhizobia. The order of toxicity of wet beads was dimethoate > lindane > isobenzan > endrin > dieldrin and wet beads produced larger zones of inhibition than dry beads, especially in the case of dimethoate.

Treatment of Medicago sativa seed with dimethoate at rates of 37·8 to 226·8 ml a.i. 50 kg^{-1} seed five days before inoculation with two strains of R. meliloti decreased plant yield in quartz sand culture but did not affect nodulation of the plants, except at the higher rate (Staphorst and Strijdom, 1974). When seed was treated with 113·4 ml a.i. 50 kg^{-1} seed and then inoculated two to nine days later, the yield of plants was better than those inoculated immediately after treatment. Dimethoate, at 0·3 ppm a.i. as a foliar spray applied to inoculated and non-inoculated lucerne, had no significant effect on nodulation but sprayed plants were smaller than those not sprayed (Staphorst and Strijdom, 1974).

Carbofuran (16 kg ha^{-1}) had no effect whereas phorate (8 kg ha^{-1}), fensulfothion (20 kg ha^{-1}) and heptachlor (18 kg ha^{-1}) each increased the size but decreased the numbers of nodules on plants in a loam soil. None of these compounds affected the plant weight, yield, leghaemoglobin or nitrogen content (Kulkarni et al., 1974).

Under laboratory conditions, lindane, DDT, heptachlor and aldicarb significantly reduced, whilst phorate increased, numbers of nodules per plant on bean and clover, whereas under field conditions (and higher application rates) phorate, aldicarb and lindane did not affect nodulation of clover roots (Gawaad et al., 1972a).

Further effects of insecticides on rhizobia and legumes are given in Table 17. From the data in the text and Table 17 an effect ratio of 0·78 is derived, indicating a general inhibition of rhizobia and nodulation by insecticides.

(e) Free-living, nitrogen-fixation. Lindane (10 ppm), fonofos (50 ppm) and diazinon (50 ppm) initially inhibited glucose utilization and N_2-fixation in Azotobacter chroococcum and A. agile but later these functions increased (Salem and Gulyás, 1971). Lindane and DDT, at the equivalent of field rate and 50 times field rate, had no significant effect on N_2-fixation in *A. vinelandii* (Mackenzie and MacRae, 1972).

In clay soil treated with 0·22 and 4·4 kg ha^{-1} of lindane, or dieldrin at 2·5 and 50 kg ha^{-1}, population numbers of azotobacters increased, but lindane decreased and dieldrin increased the numbers of anaerobic N_2-fixers (Mahmoud *et al.*, 1970). Parathion-methyl (0·15 kg a.i. ha^{-1} briefly increased populations of N_2-fixing azotobacters in a loam soil (pH 7·2), and higher rates (15, 150 and 300 kg a.i. ha^{-1}) also increased population numbers, but only after a transient decrease (Naumann, 1970b). Chlorpyrifos, applied at field rates to a clay loam (pH 5·3), inhibited the aerobic N_2-fixing population (Sivasithamparam, 1970) whilst disulfoton at field rates inhibited the anaerobic, but increased the aerobic, N_2-fixing populations in the rhizosphere of cotton plants (Mahmoud *et al.*, 1972). In soil inoculated with *A. macrocytogenes*, phoxim (1, 10 and 100 kg ha^{-1}) decreased the N_2-fixing capacity of this organism (Eisenhardt, 1975). Schradan, at field rates, increased azotobacter numbers, whilst lindane or heptachlor at rates exceeding normal field rates, promoted aerobic N_2-fixing activity (Jaiswal, 1967).

On the information available, the effect ratio for insecticides on free-living, N_2-fixing bacteria is 1·75, suggesting that insecticides in general may not be inhibitory to this class of microorganism in soil.

2. *Effects on soil fungi and actinomycetes*

In summarizing the results of other workers, Audus (1970b) states that insecticides at normal field application rates have little effect on population numbers of fungi or actinomycetes, but that actinomycetes are more sensitive in culture than in soil and that at high rates of application, insecticides may sometimes increase population numbers of fungi.

(a) *Growth.* Fungi, isolated from a prairie soil, were grown on agar medium in the presence of 40 ppm of either lindane, parathion, phorate, carbaryl or aldrin (Cowley and Lichtenstein, 1970). Aldrin was the least toxic since only 7 out of 17 fungi were inhibited, whilst lindane affected 15, parathion 16, phorate 13 and carbaryl 15 out of 17 fungi. Threshold concentrations varied according to the insecticide and species of fungus. *Aspergillus fumigatus*, for example, was the least affected by the five insecticides but was susceptible to 2 and 5 ppm rates of aldrin. Phorate at 1 and 2 ppm stimulated *A. fumigatus*, which showed a threshold concentration for phorate of 10 ppm. Lindane was very toxic towards *Fusarium oxysporum*, even at 1 ppm. The addition of yeast extract, asparagine and some ammonium salts to insecticide-containing media reduced the susceptibility of fungi to individual insecticides (Cowley and Lichtenstein, 1970).

Mucor alternans, isolated from DDT-contaminated soil, could degrade the insecticide but the presence of other fungi or of lindane, parathion or fonofos restricted the DDT-degrading activity without affecting vegetative growth of the fungus (Anderson and Lichtenstein, 1972). The ability of 597 fungal and actinomycete isolates from a silt loam soil to grow on media supplemented with 10 or 25 ppm of heptachlor was tested by Shamiyeh and Johnson (1973) who found that of 287 actinomycete isolates, 16% were inhibited by the lower concentration and 38% by the higher concentration. Total inhibition occurred in 5% of actinomycetes exposed to 10 ppm and 16% exposed to 25 ppm. Of 310 fungal isolates, 26% and 35% were inhibited by 10 and 25 ppm heptachlor, respectively, while none and 4% were totally inhibited at the two respective concentrations.

Dieldrin (5 ppm) inhibited growth and macromolecular synthesis of *Dictyostelium discoideum*. Shrinkage of the mycelium and eventual lysis occurred within 5 to 10 h exposure to dieldrin but washing the mycelium after exposure annulled the effect (Bushway and Hanks, 1976).

Naumann (1970a, b, 1971a) found that parathion-methyl at 0·15, 1·5, 15, 150 or 300 kg a.i. ha^{-1} significantly increased numbers of actinomycetes in a loam soil. The high rates (150 and 300 kg a.i. ha^{-1}) decreased numbers of fungal propagules and altered the species spectrum. In greenhouse-incubated loam (10–14°C), parathion-methyl (300 kg a.i. ha^{-1}) initially depressed numbers of actinomycetes for c. 15 days, after which numbers increased to nearly twice as many as in untreated soil. Further investigations (Naumann, 1972) revealed that at 12–15°C, parathion-methyl (300 kg a.i. ha^{-1}) depressed populations of actinomycetes for longer periods (15 weeks) than at 20°C (10–12 weeks). Fungi, however, appeared to survive better at 12–15° than at 20°C over a five week period. In fallow loam soil in the field, 1·5 kg a.i. parathion-methyl ha^{-1} suppressed actinomycetes, but in loam with vegetative cover the same rate greatly increased populations of actinomycetes. Lindane and toxaphene formulated powders, at rates of 600 to 6000 kg ha^{-1}, did not affect fungal population numbers (Naumann, 1971b).

Treatment of sandy loam, calcareous clay and clay soils with c. 22 kg a.i. ha^{-1} lindane, endrin or chlordecone, or with c. 11 kg a.i. ha^{-1} pirimicarb or fonofos, decreased fungal and actinomycete populations during the first week, followed by a rapid increase two weeks after treatment and finally a return to pre-treatment population numbers

(Gawaad *et al.*, 1972b, 1973a). Carbaryl and DDT, at field rates, completely inhibited a number of saprophytic fungi in soil, but in some cases fungi which were susceptible to DDT were found to be stimulated by carbaryl and both insecticides increased the incidence of *Aspergillus* spp., *Paecilomyces* spp. and *Acrophialophora* sp. (Varshney and Gaur, 1972).

Further examples of the effects of insecticides on soil fungi and actinomycetes are summarized in Table 18. The effect ratio derived from the data presented in the text and Table 18 is 1·43, indicating that in general insecticides are more stimulatory than inhibitory.

A very important aspect of the effects of pesticides on non-target organisms was investigated by Ko and Lockwood (1970), who showed that dieldrin was transferred from lysing fungal mycelia to an actinomycete and also from plant tissue to a fungus growing on these tissues. The total amount of dieldrin transferred from the donor fungus to other microorganisms was low. However, since ^{32}P was also transferred in the same way as dieldrin and since ^{32}P was found to be concentrated within soil invertebrates, presumably through predation of the carrier microorganisms, it is not inconceivable that dieldrin (and other pesticides) may also accumulate in soil invertebrates.

(b) *Pathogenic fungi and their antagonists.* Both DDT and carbaryl at field rates completely inhibited *Alternaria, Helminthosporium, Sepe-donium* and *Trichoderma* spp., whilst DDT also inhibited *Rhizoctonia solani, Botrytis* sp., *Phoma humicola* and *Macrophoma* sp., all of which were stimulated by carbaryl. The incidence of *Curvularia lunata* and *Fusarium chlamydosporum* was decreased by both insecticides (Varshney and Gaur, 1972). Parathion-methyl (300 kg a.i. ha^{-1}) was toxic to *Fusarium* spp. (Naumann, 1970a).

3. *Effects on soil algae*

Sethunathan and MacRae (1969) flooded a clay soil (pH 6·6) containing 2 or 20 kg a.i. ha^{-1} of diazinon and found an increased population of algae in the free-standing water over the soil's surface. In another clay soil (pH 5·3) treated with chlorpyrifos, Sivasithamparam (1970) also found an increased algal population. In loam soil (pH 7·2), parathion-methyl at rates up to 300 kg a.i. ha^{-1} had no effect on soil algae (Naumann, 1970b).

Further information regarding effects of insecticides on algae, including some soil algae, is included in Chapter 8.

4. *Effects on combined activities of the microbial biomass*

(a) *Cellulolytic activity and organic matter degradation.* Mayaudon (1974) investigated the mineralization of ^{14}C-labelled DL-glutamic acid in soil treated with heptachlor, aldrin and prophos. Prophos and heptachlor at maximum rates inhibited mineralization by 15% and 25% respectively, while at minimum rates each insecticide decreased $^{14}CO_2$ evolution by not more than 10%. With lindane (0·22 and 4·4 kg ha^{-1}) and dieldrin (2·5 and 50 kg ha^{-1}) on clay soil supplemented with 2% compost, Mahmoud *et al.* (1970) found that numbers of cellulose decomposers were stimulated by lindane but decreased by dieldrin, although organic matter content of the soil was not affected by either insecticide. In moist loam (pH 7·2), formulated lindane at 6000 kg ha^{-1} or toxaphene powder at 600 and 6000 kg ha^{-1} increased the population of cellulose-decomposing microorganisms, but in drier soils population numbers decreased (Naumann, 1971b). With parathion-methyl at 0·15 to 300 kg a.i. ha^{-1} in loam soil, the population of cellulose decomposers increased after a transient decrease, but under drier soil conditions the increase was delayed (Naumann, 1970b). At 300 kg a.i. ha^{-1} parathion-methyl in the same loam soil, the numbers of cellulolytic microorganisms were again decreased (Naumann, 1972).

In a sandy silt-clay (pH 8·3 and 0·67% O.M.), glucose utilization was stimulated and then reduced by propoxur at 2·5 or 12·5 kg ha^{-1} and completely inhibited by 125 kg ha^{-1} (Gupta *et al.*, 1975). The presence of formulated propoxur at a rate of 50 kg ha^{-1} in a sandy loam (pH 7·4) did not prevent the decomposition of the insecticide carrier, corn cob grits (Kuseske *et al.*, 1974). At rates more than three times that normally recommended for seed treatment, endrin stimulated organic matter decomposition (Bollen and Tu, 1971) whilst chlorpyrifos at field rates in a clay loam increased the population numbers and activity of cellulose decomposers (Sivasithamparam, 1970). In three surface soils the application of Counter (0·1 to 100 kg ha^{-1}) had no effect on cellulose decomposition for six weeks (Laveglia and Dahm, 1974).

On balance, the effect of insecticides on cellulose and/or organic matter decomposition, does not appear to be harmful since the ratio of positive to negative effects was 1·1. Audus (1970b) suggests that cellulolysis is generally only adversely affected by very high rates of insecticides (e.g. DDT at 5 times, parathion > 70 times, chlordane > 20 times and dieldrin > 200 times the normally recommended field rates).

(b) *Respiration*. Although O_2-uptake by diazinon-tolerant soil bacteria was unaffected by 30 ppm diazinon, with some diazinon-sensitive species O_2-uptake on some substrates was decreased by at least 40% and degradation products of diazinon stimulated respiration of a diazinon-sensitive species (Robson and Gunner, 1970). Initial suppression, followed by increased O_2-uptake, occurred in soil treated with parathion-methyl (up to 300 kg ha^{-1}) (Naumann, 1970c).

Salonius (1972) found no effect of either DDT or fenitrothion, at the rate-equivalent of 112 kg ha^{-1} on respiration of three soil types. However, Tate (1974) found that 50 kg a.i. ha^{-1} DDT, fenitrothion or fensulfothion inhibited soil respiration, although at 0·75 kg a.i. ha^{-1} neither these insecticides nor carbofuran were inhibitory. Further data on effects of insecticides on soil respiration are included in Table 19, an effect ratio of 2·0 indicating general stimulation, or lack of inhibition. Since none of the major groups of soil microorganisms appear to be adversely affected by insecticides, this conclusion is more likely to be due to stimulated metabolic activity than to, *inter alia*, the uncoupling of oxidative phosphorylation.

(c) *Other soil-enzymic activities*. Dieldrin (5 ppm) inhibited RNA, DNA and protein synthesis in *Dictyostelium discoideum* cultures exposed to the insecticide for one hour (Bushway and Hanks, 1976). Tsirkov (1970) reported that hexachlordane and heptachlor decreased urease and catalase activities in meadow soils whilst lindane and dieldrin increased the activity of these enzymes and lindane also increased proteolytic activity. Similarly, lindane, aldrin and heptachlor did not prevent enzymic hydrolysis of urea in arable or meadow soils (Verstraeten and Vlassak, 1973). At rate-equivalents of 15 kg a.i. ha^{-1}, or lower, parathion-methyl increased dehydrogenase activity in soil but higher rates (150 and 300 kg a.i. ha^{-1}) completely inhibited activity (Naumann, 1970c), more so at 12–15°C than at 20°C (Naumann, 1972). At rates of 50, 200 and 1000 ppm, accothion, malathion and thimet inhibited soil urease activity, but ureolytic microbes developed tolerance to these organophosphates (Lethbridge and Burns, 1976). Based on the limited data, the effect ratio for insecticides on soil enzymes is 2·0.

(d) *Ammonification*. In clay soil supplemented with 2% compost, lindane at 0·22 and 4·4 kg ha^{-1} or dieldrin at 2·5 and 50 kg ha^{-1} initially suppressed and subsequently increased ammonification (Mahmoud *et al.*, 1970). Lindane (500 kg ha^{-1}) or camphechlor (50–500 kg ha^{-1}) in loam soil (pH 7·2) stimulated ammonification in moist soils but suppressed population numbers in drier soils (Naumann,

1971b). The ammonification of peptone added to a sandy silt-clay (pH 8·3; 0·67% O.M.) was stimulated by propoxur at 2·5, 12·5 and 125 kg ha^{-1}, especially by the highest rate (Gupta *et al.*, 1975). Using rates three-times the maximum recommended for seed dressing, Bollen and Tu (1971) found that endrin had no appreciable effect on ammonification in a forest soil. Endrin and chlordecone, each at 22 kg ha^{-1}, and fonofos or pirimiphos-ethyl at 11 kg ha^{-1} in sandy clay loam or a clay soil caused ammonification to fluctuate in the first four weeks after treatments were applied but activity returned to normal after seven to eight weeks (Gawaad *et al.*, 1972b, 1973b).

Diazinon, chlorpyrifos, thionazin and trichloronate, each applied at 1·0 or 10 kg ha^{-1} to a sandy loam, increased ammonification in the soil (Tu, 1969). Chlorpyrifos in a clay soil of pH 5·3 was also found by Sivasithamparam (1970) to increase population numbers and activity of ammonifiers. Parathion-methyl (0·15, 1·5, 15, 150 and 300 kg a.i. ha^{-1}), applied to a loam soil, increased numbers of ammonifying microorganisms, but the three highest rates first transiently decreased them (Naumann, 1970b). The decreases and subsequent increases in ammonifying populations depended on soil moisture level, being greatest at 60% M.H.C. (Naumann, 1972).

Counter, at rates of 0·1 to 10 kg ha^{-1} in three soil-types, had no effect on ammonification for six weeks (Laveglia and Dahm, 1974). However, according to Audus (1970b) the mixed phosphorodithioate, Demeton, is probably one of the most toxic of insecticides tested on ammonification since it inhibits the process at sub-normal field rates.

The overall effect of insecticides on ammonification, judging from the ratio of 1·84 derived from the data available, is stimulatory rather than inhibitory. Part of this effect may arise from the killing of soil insects which would thus provide a greater amount of substrate material for these organisms.

5. *Miscellaneous effects*

Mobilization of phosphate in a sandy silt-clay was not affected by applications of 2·5, 12·5 or 125 kg ha^{-1} of propoxur (Gupta *et al.*, 1975). In forest soil receiving endrin at three times the normally recommended rates for seed dressing, the oxidation of sulphur was not affected (Bollen and Tu, 1971). Whilst dieldrin at 5 ppm in culture solution adversely affected growth and macromolecule syntheses in *Dictyostelium discoideum*, the uptake of uracil and amino acids was not affected (Bushway and Hanks, 1976).

Diazinon, chlorpyrifos, thionazin and trichloronate at rates of 1·0 or 10 kg ha^{-1} had no effect on phosphate mobilization or on sulphur oxidation in a sandy loam (Tu, 1969). However, chlorpyrifos in a clay soil stimulated sulphate-reducers and heterotrophic iron-precipitating organisms over a period of 3 months, and phosphate-mobilizing bacteria were temporarily inhibited (Sivasithamparam, 1970). Oxidation of sulphur in three soil types treated with Counter (0·1–10 kg ha^{-1}) was not affected over a period of six weeks (Laveglia and Dahm, 1974). Flooding a clay soil with diazinon (40 ppm) caused an increase in the number of actinomycetes, especially those producing brown pigments (Sethunathan and MacRae, 1969).

D. SOIL FUMIGANTS AND OTHER PESTICIDAL CHEMICALS

Some fumigant preparations are used for the microbiological "sterilization" of soil for greenhouse or plant nursery purposes and, by inference, belong in the fungicide category. Other fumigants are used for weed elimination and belong in the herbicide category. Still others are designed for eliminating nematodes and soil insects and some serve two or more functions. In mitigation, therefore, fumigants are classed here as a separate category of chemicals with varied usages.

In most cases, fumigants, by virtue of their volatility, may be expected to have shorter residence time in soil than fungicides, herbicides or insecticides. This does not imply, however, that their effects on non-target microorganisms will be less severe but it does suggest that recovery of the affected ecosystem by re-growth or re-invasion may be more rapid than with particulate residues of other pesticides. Where effects on a non-target organism do appear to persist for several months as a result of fumigant treatment, this usually can be attributed to the non-sporing nature of the organism or at least to its dependence upon nutritional or synergistic requirements which have been affected by the fumigant. Conversions of rates of application into kg or litres ha^{-1} in this case have taken into account the depth of soil to which fumigants were applied, unless otherwise stated.

1. *Effects on soil bacteria and their activities*

(a) *Bacterial numbers.* Metham-sodium, at 2000 ppm in agar media, totally inhibited the growth of five *Bacillus* strains (Langkramer, 1970). The concentration used in this instance was *c.* 150 times greater than that normally recommended, on the basis of a.i. ha^{-1}.

Naumann (1970d, 1971a, 1972) showed that the effects of fumigants on bacterial populations in soil depended on rates of application, soil temperature and moisture content and whether the fumigants were applied under greenhouse or field conditions. For example, under greenhouse conditions, metham-sodium (72 kg a.i. ha^{-1}) and dazomet (60 kg a.i. ha^{-1}) briefly decreased and then increased bacterial population numbers in a loam soil. At 558 kg a.i. ha^{-1}, metham-sodium decreased numbers for slightly longer (c. 20–25 days) before increases were obtained, whilst dazomet at 2550 kg a.i. ha^{-1} severely decreased the population numbers for the entire 40 day observation period. Allyl alcohol (c. 240 litres ha^{-1}) in the same soil in the greenhouse caused a temporary decrease (c. 20–25 days) which was even shorter under field conditions. Also under field conditions, metham-sodium (558 kg a.i. ha^{-1}), dazomet (255 kg a.i. ha^{-1}) and formalin (720 litres a.i. ha^{-1}) showed shorter periods of decreased populations than under greenhouse conditions. At soil temperatures of 12–15°C decreases in population numbers incurred by the fumigants were longer than at 20°C, whilst with soil moisture contents of 60% MHC the bacterial counts were higher overall than at 40% and 80% MHC in the presence of fumigants. An interesting effect of metham-sodium and allyl alcohol treatments was the increase in Gram-positive and decrease in Gram-negative bacterial populations (Naumann, 1971a).

A typical agricultural loamy sand (pH 6·1; 1·3% O.M.), treated with fensulfothion or carbofuran, each at 0·1 and 0·5 kg ha^{-1}, D-D at 12 and 60 kg ha^{-1}, or with Vorlex at 3 and 18 kg ha^{-1}, was incubated at 5°C and 28°C. Only Vorlex at 3 kg ha^{-1} and D-D at 60 kg ha^{-1} significantly decreased the bacterial population numbers two days after incubation at 5°C, whilst D-D (60 kg ha^{-1}) increased population numbers after two weeks, and all other treatments showed no significant differences (Tu, 1973b). In a separate study using a sandy loam (pH 6·1 and 1·7% O.M.), Tu (1973a) found that ethoprophos (0·5 and 3·0 kg a.i. ha^{-1}), N-serve (2·0 and 4·0 kg a.i. ha^{-1}), 1,3-dichloropropene (as Telone) at 2·5 and 15 kg ha^{-1} and Vorlex (3 and 18 kg ha^{-1}) had no significant effect on bacterial population numbers at either 5°C or 28°C. The different effects of Vorlex (3 kg ha^{-1}) in the loamy sand and sandy loam soils, both incubated at 5°C, can only be attributed to differences in the type of resident bacterial populations in the two soils.

Dibromochloropropane, diazinon, carbathion and thiazon, all at recommended rates of application to a chernozem, decreased the numbers of butyrate-oxidizing organisms but increased the total numbers of viable bacteria (Abdel'Yussif et al., 1976). However, in acid clay and alkaline alluvial loam, dibromochloropropane (and D-D) severely depressed the bacterial population (Dubey et al., 1975). At normally recommended field rates (c. 5–20 kg a.i. ha^{-1}), the nematicide fenamiphos temporarily reduced the total bacterial population in a sandy soil (Hugé, 1974).

Fumigated nursery soil showed a rapid increase in numbers of bacteria and rhizosphere bacteria were also more prevalent than before treatment with 170, 340 or 680 kg ha^{-1} of methyl bromide (Danielson and Davey, 1970). In a terra rossa soil (pH 5·7) and a sandy loam (pH 5·3), methyl bromide at 2440 kg ha^{-1} caused 50 and 90% decreases in numbers of bacterial spores in the two soils, respectively, and also caused sharp reductions in the numbers of aerobic soil bacteria in the terra rossa soil (Ridge and Theodorou, 1972). Fluorescent pseudomonads also decreased in this soil but after 35 days both the fluorescent pseudomonads and total aerobic bacterial counts increased above those of untreated soil. Dazomet (85 kg ha^{-1}), in the terra rossa soil only, exerted a more persistent effect, with greatly decreased fluorescent pseudomonad counts for up to 68 days, probably due to a residue of fumigant granules. After 107 days both the total aerobic bacterial and fluorescent pseudomonad counts increased to above that of the untreated control soil. Rhizosphere bacterial numbers were higher in treated than in the untreated terra rossa soil but in the sandy loam the untreated soil had the higher counts (Ridge and Theodorou, 1972).

The effect ratio of the fumigants cited above is 1·0, indicating a probability of 0·5 of bacterial population numbers being adversely affected.

(b) *Nitrification*. Under field conditions, metham-sodium (558 kg a.i. ha^{-1}) and dazomet (255 kg a.i. ha^{-1}) increased numbers of nitrifying bacteria in a loam soil and, whilst allyl alcohol (*c*. 240 litres ha^{-1}) and formalin (720 litres a.i. ha^{-1}) inhibited nitrifiers, under less dry soil conditions increased numbers were recorded (Naumann, 1970d). Naumann (1971a) later showed that nitrifiers in the loam soil under greenhouse conditions were inhibited for 38 days after treatment with dazomet (408 kg a.i. ha^{-1}) and for 13 days with allyl alcohol (*c*. 240 litres ha^{-1}) but metham-sodium (558 kg a.i. ha^{-1}) greatly increased the nitrifying population 38 days after treatment. Formalin (720 litres ha^{-1}) decreased the population in the first week after treatment to a much greater extent at 12–15°C than at 20°C, whilst in the fifth week the nitrifiers had increased more at 20°C than at 12–15°C. At the lower temperatures, metham-sodium (558 kg a.i. ha^{-1}) had an inhibitory effect that was not apparent at 20°C, whilst dazomet (408 kg a.i. ha^{-1}) inhibited to approximately the same extent at 12–15°C as at 20°C (Naumann, 1972). In acid and alkaline clay-loam soils, dazomet (10 kg ha^{-1}) transiently inhibited the oxidation of NH_4^+ ions (Smith and Weeraratna, 1975).

Metham-sodium and D-D inhibited nitrification in sandy soils, the extent of inhibition depending on the time of application; metham-sodium was less toxic to nitrifiers than D-D and recovery from the inhibitory effects was more rapid in marine sediments than in sandy soil (Lebbink and Kolenbrander, 1974). Both dazomet and D-D inhibited nitrifiers and their re-establishment in the fumigated soil was very slow (Singh and Prasad, 1974). The fact that the autotrophic nitrifiers do not form spores is important in considering the rate of their re-establishment following pesticide treatment. Another factor that could influence re-establishment of nitrification is soil tempera-ture. Thus, whilst D-D inhibits nitrifiers the possibility of their re-invasion of the fumigated soil is greater with soil temperatures that permit the bacteria to multiply more rapidly (Kampfe, 1973). Lower temperatures may also mean a longer residence time (and hence inhibitory effect) of the fumigant. Zayed et al. (1968) found that in loam soil, D-D decreased both numbers and activity of nitrifiers for periods up to 45 days after which the toxic effects gradually dissipated. Dubey et al. (1975) found that both D-D and dibromochloropropane at c. 6 kg ha^{-1} inhibited nitrification in an acid clay and alkaline alluvial loam by virtue of their inhibition of Nitrosomonas spp. and that D-D was the more toxic of the two nematicides. On the other hand, whilst D-D (224·6 kg ha^{-1}) inhibited nitrification in a loamy sand the popu-lation numbers of Nitrosomonas spp. and Nitrobacter spp. were found (Elliot et al., 1974) to vary considerably and inconsistently. The same authors found that ethoprophos (as Mocap) at 11·2 kg a.i. ha^{-1} had no effect on nitrification.

The nitrification of native soil nitrogen in a sandy loam was stimu-lated by ethoprophos (0·5 kg a.i. ha^{-1}) and 1,3-dichloropropene (2·5 kg ha^{-1}) at a soil temperature of 5°C two weeks after application, whilst in the same soil supplemented with c. 20 kg ammonium sulphate ha^{-1} and maintained at 28°C, ethoprophos (0·5 and 3·0 kg a.i. ha^{-1}), N-serve (2 and 4 kg a.i. ha^{-1}), 1,3-dichloropropene (2·5 and 15 kg ha^{-1}) and Vorlex (3 and 18 kg ha^{-1}) inhibited the nitrification process four weeks after application (Tu, 1973a). In further investigations using a loamy sand at the same two incubation temperatures, Tu (1973b) found that whilst fensulfothion (0·1 and 0·5 kg ha^{-1}), carbofuran (0·1 and 0·5 kg ha^{-1}), D-D (12 and 60 kg ha^{-1}) and Vorlex (3 and 18 kg ha^{-1}) gave no indication of inhibition, nitrification of native soil nitrogen at either temperature was slow. In soil fortified with c. 20 kg of ammonium sulphate ha^{-1}, a slow nitrification was obtained in treated soils at 5°C but at 28°C nitrification occurred after a lag of 14–28 days. After eight weeks, oxidation of the added NH_4^+-nitrogen was very marked in soil treated with fensulfothion (0.5 kg ha^{-1}), carbofuran (both rates) and Vorlex (both rates) (Tu, 1973b).

In comparing the effects of two fumigants and N-serve on nitrification in three soils at temperatures of $5°$, $25°$ and $40°C$ Thiagalingam and Kanehiro (1971) found methyl bromide at 490 kg ha^{-1} to be a more effective inhibitor in all three soils regardless of temperature. Ethylene dibromide (78 litres ha^{-1}) and N-serve ($c.$ 1.0 kg ha^{-1}) were also inhibitory. Methyl bromide has been cited as a nitrification inhibitor (Maw and Kempton, 1973), giving rise to a lack of nitrate-nitrogen in soil which adversely affects certain vegetable crops.

The effect ratio derived from the above data is 0.32, indicating a strongly inhibitory effect by fumigants on soil nitrification.

(c) *Denitrification.* Numbers of denitrifying bacteria in a loam soil (pH 7.2) were decreased after treatment with metham-sodium, dazomet, formalin and allyl alcohol, but in most cases the inhibitory effect was dissipated after five weeks (Naumann, 1970d, 1971a). Indirect evidence that denitrification in a loamy sand may not have been affected by fensulfothion, carbofuran, D-D or Vorlex is given by Tu (1973b). Audus (1970b) states that numbers of denitrifying bacteria were increased in soil treated with chloropicrin, ethylene dibromide (EDB) and D-D at field rates of application.

(d) *Rhizobia and legume nodulation.* Audus (1970b) states that EDB and D-D have been reported to improve nodulation in soya bean but it is not known whether the effect is on the microorganisms or the host plant. Kulkarni *et al.* (1975) found that EDB, ethylene dichloride/carbon tetrachloride mixture (EDCT), chloropicrin and phosphene adversely affected the nodulation and yield of groundnuts.

(e) *Free-living, nitrogen-fixing bacteria.* Normal field rates of fenamiphos applied to sandy soil caused a temporary reduction in numbers of aerobic N_2-fixing bacteria three days after application but after 38–40 days the nematicide had no effect (Hugé, 1974). In loam soil, metham-sodium (568 kg a.i. ha^{-1}) and dazomet (255 kg a.i. ha^{-1}) severely decreased the population of azotobacters (particularly *Azotobacter chroococcum*) and whilst allyl alcohol ($c.$ 240 litres ha^{-1}) and formalin (720 litres a.i. ha^{-1}) increased numbers of N_2-fixing bacteria overall, *A. chroococcum* was adversely affected; under greenhouse conditions metham-sodium, dazomet and allyl alcohol, at the aforementioned rates, also severely inhibited total numbers of N_2-fixing bacteria and at $12–15°C$ the inhibitory effects were more pronounced than at 20°C (Naumann, 1970d, 1971a, 1972). Normal field rates of allyl alcohol (200 litres ha^{-1}) markedly suppressed numbers of *Azotobacter* spp. in soil or in culture (Domsch, 1960), whilst Langkramer (1970) demonstrated that metham-sodium at 1000 and 2000 ppm totally inhibited *A. chroococcum* in culture.

2. Effects on soil fungi and actinomycetes

(a) *Growth*. Whilst metham-sodium (10 000 ppm) inhibited growth of the mycorrhizal fungus *Suillus variegatus* on agar, Fungicidin (50 ppm) stimulated it (Sobotka, 1970).

Metham-sodium, dazomet, allyl alcohol and formalin at rates of 568 kg a.i. ha^{-1}, 255 kg a.i. ha^{-1}, 240 litres ha^{-1} and 720 litres ha^{-1}, respectively, severely decreased numbers of fungal propagules and also affected the distribution of types of fungi in a loam soil under field conditions, although actinomycetes were not adversely affected (Naumann, 1970d). Under greenhouse conditions, however, metham-sodium, dazomet and allyl alcohol decreased the incidence of actinomycetes (Naumann, 1971a). At 12–15°C, fungal and actinomycete counts were decreased more than at 20°C and at soil moisture contents of 40% MHC the actinomycetes were more severely affected by metham-sodium, dazomet, allyl alcohol and formalin than at 60% and 80% MHC. Fungi were less numerous in formalin, dazomet and metham-sodium-treated soils at 12–15°C than at 20°C (Naumann, 1972).

1,3-Dichloropropene (as Telone) at 15 kg ha^{-1} and Vorlex at 18 kg ha^{-1} reduced fungal population numbers significantly in sandy loam incubated at 5°C for 14 days (Tu, 1973a). However, in loamy sand, Tu (1973b) found that fensulfothion (0·1 and 0·5 kg ha^{-1}), carbofuran (0·1 and 0·5 kg ha^{-1}), D-D (12 and 60 kg ha^{-1}) and Vorlex (3 and 18 kg ha^{-1}) had comparatively little effect on soil fungal propagules.

Fungi were severely depressed in an acidic clay and an alkaline alluvial loam treated with dibromochloropropane or D-D applied at rates consistent with normal field rates (Dubey *et al.*, 1975). Dibromochloropropane, diazinon, carbathion and thiazon also depressed numbers of fungal propagules in a clay loam receiving normal rates of application (Abdel'Yussif *et al.*, 1976). A reduction in the fungal population of soil was observed ten days after applications of D-D, ethylene dibromide, dibromochloropropane, fensulfothion and thionazin but 25 days after treatments the fungal count increased two to five times higher than the pre-application counts (Midha and Nandwana, 1974).

Danielson and Davey (1970) found that methyl bromide (170, 340 and 680 kg ha^{-1}) almost completely inhibited the fungi in a forest nursery soil and after seven months the population was less than 10% of untreated soil with the two highest rates of application. Actinomycetes showed a slight decrease with each treatment rate but recovered quickly to numbers above those of untreated soil. In the rhizosphere the fungal population was similar to that in non-rhizosphere soil but mycorrhizal formation was retarded for up to three weeks after treatment with the 680 kg ha^{-1} rate (Danielson and Davey, 1970).

In a terra rossa soil, methyl bromide (2440 kg ha^{-1}) practically eliminated all fungi in the first two days after fumigation and dazomet also severely and persistently decreased the number of propagules (Ridge and Theodorou, 1972). In a sandy loam, methyl bromide initially severely reduced fungal numbers but recolonization was more rapid in this soil. In both soils, some genera present in the untreated soil plots were eliminated by fumigation and did not re-appear, while other genera not detected in untreated soils were found in the fumigated plots. Infection of roots by a seed-inoculated mycorrhizal fungus, *Rhizopogon luteolus*, was increased by methyl bromide fumigation of the terra rossa soil but decreased in the sandy loam soil (Ridge and Theodorou, 1972). In a sandy soil (pH 5·4) and silt loam (pH 5·9), methyl bromide (2275 kg ha^{-1}) had no effect on numbers of actinomycetes, but increased fungal propagules (Menge and French, 1976).

Foliar sprays of subamycin (10 ppm) increased the numbers of fungi per gram of soil but 20 ppm decreased them and also altered the species composition of the fungal microflora (Gupta, 1974).

The effect ratio obtained from the above data is 0·55, indicating that fumigants in general have an adverse effect on soil fungi.

(b) *Pathogenic fungi and their antagonists*. Methyl bromide (284–340 kg ha^{-1}) and Vorlex (*c*. 200 litres ha^{-1}) gave excellent weed control as well as control of *Stromatinia gladioli* and *Fusarium oxysporum* f. *gladioli* diseases of gladiolus (Milholland, 1969). In assessing the ability of soil fumigants to kill fungi within plant material, Ebbels (1970) found that soil treated with 38% formaldehyde (formalin), diluted 1:26 and watered on at 110 000 litres ha^{-1}, gave better control of *Cercosporella herpotrichoides* in surface litter than either D-D (896 kg ha^{-1}) or dazomet (448 kg ha^{-1}). In loam soil treated with D-D and chloropicrin at *c*. 30 kg ha^{-1} applied 150 mm below the surface, pathogenic fungi such as *Ophiobolus graminis*, *C. herpotrichoides* and *Fusarium roseum* were controlled by the fumigants and control was found to be better below the point of injection of fumigants than above it (Ebbels, 1970). Fensulfothion at rates equivalent to 10, 20 and 40 kg a.i. ha^{-1} applied to soil containing *Sclerotium rolfsii* and *Trichoderma* spp. did not affect growth of *S. rolfsii* but reduced development of sclerotial initials and eliminated formation of mature sclerotia; colonization of the pathogen by *Trichoderma* spp. was not affected (Rodriguez-Kaban *et al.*, 1976).

3. *Effects on soil algae*

Algal population numbers in loam soil were severely reduced by metham-sodium, dazomet, allyl alcohol and formalin (Naumann, 1970d, 1971a, 1972).

4. *Effects on combined activities of the microbial biomass*

(a) *Cellulolytic activity and organic matter decomposition.* Allyl alcohol, formalin, metham-sodium and dazomet had a temporary initial retarding effect on cellulolytic bacteria in a field-treated loam (Naumann, 1970d). In greenhouse-incubated loam, however, the same fumigants decreased the numbers of cellulolytic microorganisms for a period of 38 days and cellulolytic activity was generally decreased more at 12–15°C than at 20°C (Naumann, 1971a) and at moisture 80% MHC than at 60% or 40% MHC (Naumann, 1972).

The data of Jenkinson and Powlson (1970), who used a silt loam (pH 6·4) and a loamy sand (pH 7·1) repeatedly fumigated with formalin (1140 kg a.i. ha^{-1}) and a silt loam (pH 8·0) treated once with methyl bromide (980 kg ha^{-1}), suggest that fumigation (especially with formalin) eliminates a fraction of the soil organic matter, so that a second exposure to formalin causes a smaller flush of decomposition. Restoration of the formalin-eliminated organic matter was not complete even after several years of cropping of the soils (Jenkinson and Powlson, 1970). Singhal *et al.* (1975) noted that soil treated with low and high rates of D-D underwent increased organic matter decomposition with time, whilst Tu (1973b) reported that fensulfothion, carbofuran, D-D and Vorlex did not adversely affect O_2-consumption resulting from decomposition of native organic matter in soil. The organic matter content of a saline soil was also decreased following application of dibromochloropropane at different rates (Singhal and Singh, 1973).

An effect ratio of 0·62 indicates that fumigants are mainly inhibitory towards organic matter decomposition in soil.

(b) *Respiration*. Tu (1973a) investigated the effects of ethoprophos (0·5 and 3·0 kg a.i. ha^{-1}), N-serve (2 and 4 kg a.i. ha^{-1}), 1,3-dichloro-propene (2·5 and 15 kg ha^{-1}) and Vorlex (3 and 18 kg ha^{-1}) on a sandy loam (pH 6·1; 1·7% O.M.), with and without added glucose, at 15° and 30°C. With the exception of N-serve, the addition of glucose increased O_2-uptake in treated soils at both temperatures. More O_2 was consumed in the soil treated with increasing ethoprophos concentrations and this was attributed to possible microbial degradation of the chemical. N-serve depressed respiration at 30°C for 50 h, whilst at 15°C there was no effect. Vorlex and 1,3-dichloropropene had fewer effects on respiration in glucose-free soil at the lower temperature in the early stages of incubation but appreciably more O_2 was consumed at 30°C than at 15°C after 144 h. Vorlex and 1,3-dichloropropene significantly inhibited respiration in glucose-amended soil for 70 and 50 h, respectively, at 15°C, and for 20 h at 30°C with both treatments (Tu, 1973a). A marked increase in respiration of soil without glucose was evident at 30°C with high rates (0·5 kg ha^{-1}) of fensulfothion and carbofuran; Vorlex and D-D depressed respiration in the early stages of incubation of the glucose-amended soil, the inhibition being longer at 15°C than at 30°C (Tu, 1973b).

In a loam soil of pH 7·2, formalin (1440 litres a.i. ha^{-1}) and dazomet (428 kg a.i. ha^{-1}) treatments resulted in a very large increase in O_2-consumption but metham-sodium (558 kg a.i. ha^{-1}) caused only a slight increase (Naumann, 1972). A silt loam (pH 6·4) and a loamy sand (pH 7·4), both repeatedly fumigated with formalin (1140 kg a.i. ha^{-1}), and a silt loam (pH 8·0) treated once with methyl bromide (980 kg ha^{-1}), respired at similar rates to untreated soils (Jenkinson and Powlson, 1970). By contrast, after exposure to chloro-form or γ-irradiation, the previously fumigated soils respired less rapidly than the control soils; an effect attributed to the elimination of a section of the soil biomass by formalin and methyl bromide fumigation. Recovery from the effect was not complete even after several years (Jenkinson and Powlson, 1970). In acid (pH 5·1) and alkaline (pH 7·5) clay loam soils, dazomet at 10 kg ha^{-1} transiently reduced the extent of CO_2 evolution and then increased it (Smith and Weeraratna, 1975).

The effect ratio of 1·4 in this case indicates that fumigants are mainly non-inhibitory towards soil respiratory activity but this effect could be due to uncoupling of oxidative phosphorylation since in most of the foregoing cases fumigants have had a largely negative effect on the various populations and activities of soil microorganisms and especially O.M. degradation.

(c) *Other soil-enzymic activities.* Using field-treated loam (pH 7·2), Naumann (1970d) found that metham-sodium (568 kg a.i. ha^{-1}) increased, but that dazomet (255 kg a.i. ha^{-1}) decreased, dehydrogenase activity. Allyl alcohol (*c.* 240 litre ha^{-1}) at first greatly increased dehydrogenase activity but after the initial two weeks, activity decreased for four weeks and then fluctuated for the remaining ten weeks. Formalin (720 litres a.i. ha^{-1}) severely decreased soil dehydrogenase activity for most of the 16 week study. Under greenhouse conditions, Naumann (1972) established that dehydrogenase activity, in the presence of metham-sodium and dazomet, was lower overall (irrespective of stimulatory or inhibitory effects) at 12–15°C than at 20°C, especially in the case of dazomet-treated loam, whilst formalin-treated soil showed less temperature difference in its dehydrogenase activity. Markert (1974) found that urease activity in soils was inhibited by 50 litres or 20 kg ha^{-1} of metham-sodium, dazomet and D-D, whereas dibromochloropropane, diazinon, carbathion and thiazon increased invertase and catalase activity in soil, urease and phosphatase activity were decreased (Abdel'Yussif *et al.*, 1976).

On the limited amount of information given here, fumigants have a general inhibitory effect on soil enzymatic activity since the effect ratio obtained is 0·66.

(d) *Ammonification.* Field-treated loam (pH 7·2) receiving metham-sodium (558 kg a.i. ha^{-1}) or dazomet (255 kg a.i. ha^{-1}) showed initial decreases in populations of ammonifying microorganisms but after two weeks both treatments showed numbers which were equal to or greater than the untreated soil. However, allyl alcohol (*c.* 240 litres ha^{-1}) and formalin (720 litres a.i. ha^{-1}) inhibited ammonifying microorganisms for a period of eight weeks (Naumann, 1970d). In shorter-term experiments on the same loam soil under greenhouse conditions, Naumann (1971a) found that metham-sodium and dazomet showed the same pattern of an initial decrease in numbers of ammonifiers and allyl alcohol again decreased the population numbers. Inhibitory effects were prolonged at 12–15°C compared with the effects at 20°C. Soil moisture content of 60% MHC was more favourable to the growth of ammonifiers in the presence of metham-sodium than values of 40% or 80%. With dazomet, 40% and 60% were better for growth than 80% MHC while formalin and allyl alcohol-treated soil allowed better growth at 60% MHC than at 40% or 80% MHC (Naumann, 1972).

Treatment of a sandy loam (pH 6·1; 1·7% O.M.) incubated at 5° or 28°C with ethoprophos, N-serve, 1,3-dichloropropene or Vorlex had only minor effects on ammonification. In soil supplemented with peptone, however, a significant stimulatory effect was observed with N-serve (2 and 4 kg a.i. ha^{-1}) and 1,3-dichloropropene (2·5 and 15 kg ha^{-1}), but inhibition of ammonification was obtained with Vorlex (3 and 18 kg ha^{-1}) after the first week of incubation and with the low rate of Vorlex, only, during the second week; all of these effects occurred at 5°C (Tu, 1973a). Using a loamy sand (pH 6·1; 1·3% O.M.) treated with fensulfothion (0·1 and 0·5 kg ha^{-1}), carbofuran (0·1 and 0·5 kg ha^{-1}), D-D (12 and 60 kg ha^{-1}) or Vorlex (3 and 18 kg ha^{-1}) the ammonification of indigenous organic nitrogen was not found to be inhibited at incubation temperatures of 5°C or 28°C, but D-D at 60 kg ha^{-1} and Vorlex (both rates) markedly stimulated ammonification after four weeks at the higher temperature (Tu, 1973b). Ammonification of peptone added to the soil was adversely affected by all four fumigants at both rates of application in soil incubated at 5°C but this effect was of short duration and no adverse effects were noted either in subsequent weeks or at 28°C (Tu, 1973b). The treatment of a loam soil, supplemented with peptone, by D-D (no rates given) had no effect on ammonification activity or the numbers of ammonifying microorganisms (Zayed et al., 1968) whereas in a cotton soil low and high rates of D-D were found to decrease the soil O.M. content and to increase the availability of nitrogen (Singhal et al., 1975). A similar effect was found in a sodic saline soil treated with 10–50 litres ha^{-1} of dibromochloropropane (Singhal and Singh, 1973).

The effect ratio of 1·2 obtained from the above data suggests that the ammonifying population in soil may benefit from the killing effect of fumigants on other soil organisms, a factor in common with herbicides, fungicides and insecticides.

5. Miscellaneous effects

Oxidation of sulphur in a sandy loam was inhibited by ethoprophos (3 kg a.i. ha^{-1}), N-serve (2 and 4 kg a.i. ha^{-1}), Vorlex (3 and 18 kg ha^{-1}) and 1,3-dichloropropene (2·5 and 15 kg ha^{-1}) (Tu, 1973a). However, in a loamy sand sulphur oxidation was not significantly affected by fensulfothion (0·1 and 5 kg ha^{-1}), carbofuran (0·1 and 0·5 kg ha^{-1}), D-D (12 and 60 kg ha^{-1}) or Vorlex (3 and 18 kg ha^{-1}) (Tu, 1973b). The conflicting results with Vorlex may reflect differences in the strains of sulphur-oxidizing bacteria in the two soils.

Phosphate availability was increased in soil treated with low rates of D-D, but high rates inhibited mobilization of phosphate (Singhal et al., 1975). A similar effect on phosphate mobilization, and decrease in pH, was found in an acid clay and alkaline alluvial soil treated with increasing concentrations of D-D, whilst dibromochloropropane increased phosphate mobilization and increased the pH of both of the soils (Dubey *et al.*, 1975).

Pseudomonas fluorescens, Rhizobium leguminosarum and *Bacillus subtilis* strains released greater amounts of amino acids in the presence of D-D than in untreated media. This effect, also obtained in soil seeded with the bacterial isolates, is suggested as being a possible explanation of stimulated plant growth in soils fumigated with D-D and other fumigants (Altman, 1969). Applications of dazomet (10 kg ha^{-1}) to acid (pH 5·1) and alkaline (pH 7·5) clay loam soils increased the availability of manganese through increased microbial activity (Smith and Weeraratna, 1975).

IV. Summary of General Effects of Pesticides

The effect ratios derived under each soil microbial parameter, whenever sufficient data permitted, are listed in Table 20. These ratios can only be used for making broad comparisons of the data reviewed in this chapter and cannot be used for making predictions or definitive analyses of the true effects of all pesticidal compounds on all non-target microorganisms. From the ratios obtained it is possible to state that, on the data reviewed here, bacterial numbers in soil are not generally adversely affected by any of the four groups of pesticides, whereas nitrification is more sensitive to fungicides, insecticides and other pesticides than it is to herbicides. Rhizobia and legume nodulation appear more vulnerable to insecticides than to fungicides or herbicides, although both of the latter two groups may tend to be slightly more inhibitory than either stimulatory or without effect.

Soil fungi and actinomycetes are not as susceptible to herbicides or insecticides as they are to fungicides and others, many of which are fumigant fungicides. This generalized result, like the effect ratio for herbicides on algae, is not unexpected. What is unexpected is the very high positive (non-inhibitory) effect of fungicides in general on pathogenic microorganisms and their antagonists. It is necessary to emphasize that these findings are not concerned with the effects of fungicides on pathogens known to be controlled by particular fungicides, but on other pathogenic fungi (and their antagonists) which are not the intended target for the particular fungicide. Herbicides, on the other hand, appear to be effective in controlling the growth of many pathogens and antagonists of pathogens.

Cellulolytic activity and organic matter (O.M.) degradation are not adversely affected by the majority of herbicides and insecticides reviewed here. Other pesticides, however, many of which are fumigant fungicides, appear to be predominantly inhibitory to these organisms. Since many fungi and actinomycetes are very actively involved in O.M. degradation, it is interesting to note that the effect ratios for herbicides and insecticides on these organisms follow the same positive (non-inhibitory) trend as the effect ratios for cellulolytic activity and O.M. degradation. Soil respiratory activity, on the other hand, which can be said to be indicative of O.M. degradation, does not correlate well with effect ratios for the latter, especially in the case of other pesticides. Other soil enzymic activities appear more susceptible to fungicides and other pesticides than to either herbicides or insecticides in the data reviewed, while ammonification seems to benefit from the application of all groups of pesticides to soil.

V. Conclusions

In previous years, some reviewers of the effects of pesticides on non-target organisms have claimed, justifiably, that adverse effects were mainly the result of high rates of application and values of 100 kg a.i. ha^{-1} were frequently cited. It is now apparent that effects of one kind or another are being obtained with present day "normal field rates" which are often well below 10 kg a.i. ha^{-1}, and occasionally effects are obtained with sub-normal field rates. This is undoubtedly due to the considerable advances that have been made in pesticide technology and, in particular, to the higher chemical reactivity of modern pesticidal chemicals which are increasingly deliberately aimed at known biochemical pathways.

As a general observation accruing from the data reviewed in this chapter, the trend of effects of pesticides on soil microorganisms indicates that many chemicals, even at normal field rates, have some form of effect since even stimulations can be regarded as being undesirable in certain cases. In most instances the effects noted are transient, lasting a few days or a few weeks, but in other cases extended periods of effects have been recorded. Without doubt the more transient the particular effect the better, but based upon even the most theoretical average for microbial propagation time in soil, it is difficult to judge just how transient an effect should be to qualify as unimportant. The inhibition of nitrification for one month may be regarded as unimportant in agronomic terms but the inhibition of a natural antagonist of a plant pathogen for one week may be sufficient for the pathogen to gain a hold on a crop.

Some effects of pesticides, applied at normal field rates, are disturbing in their practical implication. The increase, for example, in plant pathogenic infestations of crops following pesticide treatments, the enormous impact of some pesticides on plant rhizosphere populations or the profound changes in species composition in pesticide-treated soils, indicate a very real necessity for microbial ecological monitoring programmes. Kreutzer (1970), in discussing the re-infestation of pesticide-treated soil points out that whilst the goal is to destroy organisms that are harmful to higher plants, most pesticides are not selective enough to differentiate between harmful and useful microorganisms. Kreutzer (1970) concludes that "we must learn the identity of the majority of our friends and foes" and "we must discover the substrate preferences of these organisms and learn more about the conditions in nature which favour or restrict their growth and development". In essence, are soil microbiologists delving deep enough? There are perhaps only 11 or 12 aspects which are regularly and routinely dealt with by soil microbiologists, whereas the actual number of microbiological parameters that could be investigated is very much greater. When the possible interrelationship of each parameter with all others is considered, the number of lines of investigation becomes enormous and this fact alone may be the reason for the apparent reluctance of soil microbiologists to branch out into a larger area of monitoring the effects of pesticides.

Attempts to widen the sphere of studies could be regarded as defeating the purpose of investigations by either diluting the thoroughness of each category of investigation or prolonging the investigation beyond the useful life of either the chemical or the investigator. Nonetheless, certain areas of additional effort are indicated. Very little has been done to monitor or record effects of pesticides on, *inter alia*, bacterio- and actino-phage, mycoparasitism, nematode-trapping fungi, useful phylloplane microorganisms, or teratogenic effects. In comparison, witness the effort applied to monitoring effects of pesticides on nitrification. Could it be the latter is far more easy to monitor than the former?

Finally, with regard to those parameters which are most frequently studied in relation to pesticide effects, it is in some small measure both gratifying and reassuring to know that, in general, saprophytic soil microorganisms are very resilient and that re-colonization of pesticide-decimated soil will occur. There is, however, also the danger of complacency in accepting this and our present approaches to soil microbial ecology.

TABLE 1

Effects of some herbicides on soil bacteria in culture

Herbicide				
Group	Concentration (ppm)	Media conditions	Effects	Reference
Anilines	benefin (200)*	Solid + liquid	Slightly inhibited *Bdellovibrio bacteriovorus*.	Wehr and Klein (1971)
	p-chloroformanilide (6·6; 16·6; 33)	(−)	None.	Khikmatov *et al.* (1969) Kosinkiewicz (1973)
Carbamates and carbanilates	chlorpropham (10–50)	Liquid	*Arthrobacter globiformis* growth retarded, but increased glutamic acid production.	
	chlorpropham (10–100)	Liquid	*Bacillus mesentericus* temporarily inhibited. *Pseudomonas* sp. unaffected.	Balicka and Krezel (1969)
	chlorpropham (0·2–1000)	Liquid	Inhibition, depending on concentration and presence of surfactants.	Balicka *et al.* (1969)
	EPTC (200)* IPC (200)*	Solid and liquid	Moderate inhibition of *Bd. bacteriovorus*.	Wehr and Klein (1971)
	pebulate (1000)	Liquid	Morphological changes in *Pseudomonas fluorescens*, *Arthrobacter* sp. and *Corynebacterium*. sp.	Sobieszczanski (1969a)
Diazines	1. pyrazon (50 000–100 000)	Solid	1. Slight inhibition of *Pseudomonas* spp.	Zharasov (1969)
	2. pyrazon (25 000)	Solid	2. Toxic to *Bacillus* spp.	
Halogenated aliphatics	TCA (500–50 000)	Solid	*Pseudomonas* spp. unaffected.	Zharasov (1969)
Phenols	DNBP (200)*	Solid and liquid	*Bd. bacteriovorus* inhibited.	Wehr and Klein (1971)

Phenoxy-acids	2,4-D (2·2–3·3)	Liquid	*Escherichia coli* unaffected until stationary phase, then viable count decreased. *Aerobacter* sp. and *Clostridium* sp. similarly affected. Cell wall formation possibly affected.	Lamartinere *et al.* (1969)
	2,4-D (0·2–1·1)	Aerobic and anaerobic, liquid	*Bacillus subtilis*, *B. megaterium*, *Corynebacterium fasciens* sensitive. Anaerobic clostridia less sensitive. Facultative Gram+ve and −ve aerobes less affected aerobically than anaerobically.	Hart and Larson (1966)
	2,4-D (200)* 2,4,5-T (200)* MCPA (200)*	Solid and liquid	Moderate inhibition of *Bd. bacteriovorus*.	Wehr and Klein (1971)
Pyridines	picloram (200)*	Solid and liquid	*Bd. bacteriovorus* slightly inhibited	Wehr and Klein (1971)
Quaternaries	1. paraquat (1–2)	Liquid	1. *E. coli* growth depressed; recovered if medium diluted.	Davison and Papirmeister (1971)
	2. paraquat (10–12)		2. *E. coli* almost completely inhibited; recovered if medium diluted.	
	3. paraquat (1800–2000)		3. 90% of cells killed after 60 min exposure.	
	1. paraquat (5–10)	Solid	1. Count increased.	Kokke (1970)
	2. paraquat (1–10)	Solid	2. Fewer yellow and fluorescent colonies.	
Triazines	atrazine (1–1000)	Liquid	Amino acid secretion increased or decreased, depending on species and concentration.	Balicka and Bilodub-Pantera (1964)
	atrazine (1000)	Liquid	Morphological changes in *Ps. fluorescens*, *Arthrobacter* sp. and *Corynebacterium* sp.	Sobieszczanski (1969a)

* "Recommended field rate".
Refer to *Notes on tabulated data and abbreviations* (p. 319).

TABLE 1 (continued)

Group	Herbicide Concentration (ppm)	Media conditions	Effects	Reference
Triazines (continued)	atrazine (5–10 000) simazine (5–10 000)	Liquid	Spore-forming bacteria more sensitive than non-sporing types.	Kozlova et al. (1964); Nepomiluev et al. (1966b)
	atrazine (100) simazine (100) prometryne (100)	Liquid	Soil isolates strongly inhibited.	Krezel and Musial (1969)
	atrazine (>10 000) simazine (>10 000) prometryne (1000)	Liquid	Toxic to rhizosphere bacteria, especially if surfactants present.	Balicka et al. (1969)
	prometryne (10–100)	Liquid	Inhibited Bacillus spp. and their antibiotic formation. Pseudomonas spp. unaffected.	Balicka and Krezel (1967, 1969)
	prometryne (5–100)	Liquid	Bacillus spp. sensitive at acid pH values. Pseudomonas sp. resistant at all pH values.	Balicka (1969)
Ureas	linuron (1·5–4·0)	Liquid	A. globiformis growth retarded but increased glutamic acid production. No effect if herbicide addition delayed.	Kosinkiewicz (1973)
	linuron (10)	Liquid	Bacillus brevis-antibiotic secretion inhibited, growth unaffected. B. oligonitrophilus-increased antibiotic secretion. B. polymyxa-growth decreased, antibiotic secretion unaffected.	Kosinkiewicz (1970)

linuron (5–50)	Liquid	Bacillus spp. sensitive at acid pH values. Pseudomonas sp. resistant at all pH values.	Balicka (1969)
linuron (10–100)	Liquid	Inhibited Bacillus spp. and their antibiotic formation. Pseudomonas spp. not affected.	Balicka and Krezel (1967, 1969)
linuron (1000)	Liquid	Toxic to Arthrobacter sp., Corynebacterium sp., Bacillus spp.	Sobieszczanski (1969a)
1. linuron (0·2–1000) monolinuron (0·2–1000) chloroxuron (0·2–1000)	Liquid	1. Inhibition, according to concentration and presence of surfactants.	Balicka et al. (1969)
2. linuron (10) 3. linuron (0·2–100) linuron (10) monolinuron (100) chloroxuron (100)	Liquid Solid Liquid	2. Marked inhibition of isolates. 3. B. mycoides inhibited. Soil isolates strongly inhibited.	Krezel and Musial (1969)
1. linuron (12–16) 2. diuron (200)*	1. Liquid 2. Solid	1. Inhibited Bd. bacteriovorus. 2. Moderately inhibited Bd. bacteriovorus.	Wehr and Klein (1971)
metobromuron (0·003–30) chlorbromuron (0·005; 0·5; 5) chlortoluron (10)	Liquid	B. cereus growth decreased initially; recovery rapid and complete.	Macek and Milharcic (1972)
Uracils phenobenzuron (100) bromacil (200)*	Liquid Solid	No effect on soil isolates. Slightly inhibited Bd. bacteriovorus.	Catroux and Fournier (1971) Wehr and Klein (1971)

* "Recommended field rate".

Refer to Notes on tabulated data and abbreviations (p. 319).

TABLE 2

Effects of some herbicides on soil bacteria *in situ*

Group	Herbicide Concentration (Kg or litre ha^{-1})	Soil type	Effects	Reference
Acylamides	diphenamid (6; 8) alachlor (7) propachlor (7)	Red and alluvial Heavy clay degraded chernozem	Non-sporing populations depressed. Annual applications (7 yr) decreased populations, especially in dry years.	Mickovski (1972) Husarova (1972a)
Phthlates	endothal (2; 10; 20)	Six soils	Populations either unaffected or increased.	Au (1968)
Anilines	trifluralin (1·1)	Sandy loam	Annual applications (5 yr) decreased populations.	Breazeale and Camper (1970)
	p-chloroformanilide (20–100)	Grey serozem, 65% MHC	Increased populations with rates up to 50 kg.	Khikmatov et al. (1969)
	dinitramine (0·1; 0·2; 0·4)	Silty loam	Increased populations.	Belles (1972)
Benzoics and related	chlorfenac (1·5)	Loam, 3·5% o.m.	Retarded populations for c. 20 days.	Geshtovt et al. (1970)
	dicamba di-methylamine (0·12)	Light loam derno-podzol	Increased *Clostridium acetobutylicum* but decreased *C. pasteurianum* and *C. butylicum*.	Kozyrev and Laptev (1972)
	dichlobenil (7·5)	Chernozem-smolnitsa	Inhibited populations for c. 90 days.	Nikolova and Bakalivanov (1972)
Carbamates and carbanilates	EPTC (–)	(–)	Increased populations.	Zhukova and Konoshevich (1967)
	EPTC (4·5)	Pale chestnut light loam, pH 8; 2% o.m.	Stimulated growth.	Chulakov and Zharasov (1967)

EPTC (4–6)	Loamy chernozem 5–7% o.m.	High rate decreased population during first 3–4 weeks.	Balkova and Semikhatova (1969)
EPTC (3–4·5)	Pale and dark chestnut	Retarded spore-forming population.	Chulakov and Zharasov (1971)
EPTC (3; 4·5)	Dark chestnut heavy loam, 4% o.m.	Low rate increased, high rate inhibited populations.	Zharasov (1972)
EPTC (5)	Leached medium loam chernozem	Increased populations.	Matsneva and Semikhatova (1973)
chlorpropham (285)	Volcanic ash	Increased populations.	Matsuguchi and Ishizawa (1969)
pebulate (4)	(–)	Increased populations but numbers of C. pasteurianum decreased.	Barona (1974)
barban (20) carbyne (140·8) solvents of carbyne (140·8)	Sandy loam, pH 5; 7·1% o.m.	Increased populations, especially with Carbyne and solvents.	Quilt et al. (in press)
Diazines 1. phenazone (3–6) 2. phenazone (8–18)	Leached derno-carbonate or derno-podzol	1. No effects 2. Inhibited populations.	Mezharaupe (1967)
pyrazon (3; 4·5)	Dark chestnut heavy loam, 4% o.m.	Low rate increased, high rate decreased populations.	Zharasov (1972)
pyrazon (3)	Light loam derno-podzol	Annual applications for 5 years increased populations.	Chunderova et al. (1971)
pyrazon (1·5–3·5; 15–35)	(–)	Lowest rates, no effect to slightly increased populations. High rates decreased spore-formers.	Simon (1971)
Halogenated aliphatics dalapon (10)	(–)	Increased populations.	Krykhanov and Pashkovskaya (1967)

Refer to *Notes on tabulated data and abbreviations* (p. 319).

TABLE 2 (continued)

Group	Herbicide Concentration (Kg or litre ha⁻¹)	Soil type	Effects	Reference
Halogenated aliphatics (continued)	dalapon (8)	Meadow-type medium loam, 1·5% o.m.	Temporarily decreased populations.	Tulabaev (1970)
	dalapon (20)	Irrigated brown soil, pH 7–8	Initially decreased populations. Recovered in c. 30 days.	Akopyan et al. (1974)
	dalapon (0·15–0·75)	Red clay	Decreased populations.	Sharma and Saxena (1972)
	dalapon (0·5–360) TCA (30–360)	Light brown soil	Decreased spore-formers.	Zhakharyan (1970)
	dalapon (1·0; 10) MCA (1·0; 10) DCA (1·0; 10) TCA (1·0; 10)	Clay soil (anaerobic)	Inhibited methane-producing bacteria. Adding CaCO₃ decreased period of inhibition.	Laskowski and Broadbent (1970)
	TCA (15–30; 150–300)*	(–)	Lower rates no effect to slightly increased populations. High rates decreased spore-formers.	Simon (1971)
	TCA (5)	Chernozem, 3% o.m.	Increased populations but decreased numbers of C. pasteurianum.	Barona (1974)
	TCA (18)	Pale chestnut light loam, pH 8; 2% o.m.	Increased populations.	Chulakov and Zharasov (1967)
Phenols	PCP (4) PCP (1·5)	Six soils Leached chernozem, low o.m.	Decreased populations. Increased populations.	Au (1968) Tyagny-Ryadno (1967).

	Compound (rate)	Soil type	Effect	Reference
	PCP (12–120)	Alluvial clay loam, pH 5·7; 2·7% carbon	Increased populations. *Pseudomonas* spp. predominated at high rates. Anaerobes increased.	Ishizawa and Matsuguchi (1966)
	dinoseb acetate (1·2–2)	Sandy loam	Decreased populations slightly. Selective effect.	Hoffman-Kakol and Kwarta (1972a)
	dinoseb acetate (1·5)* dinoseb triethanol-amine (1·5)*	Peaty bog, pH 5·5; 47% o.m.	Slightly increased populations.	Nepomiluev and Kuzyakina (1967)
	DNOC (5–10)*	Loam	Annual applications (1/yr) had no permanent effects.	Resz (1968)
Phenoxy-acids	2,4-D (22·5)	Sandy loam	Decreased populations following 5 applications of 4·5 kg ha^{-1} per annum (4 yr)	Breazeale and Camper (1970)
	2,4-D (1–2)*	Loam	Annual applications (1/yr) had no permanent effects.	Resz (1968)
	2,4-D (1–2)	Peaty-bog, pH 5·5; 47% o.m.	Decreased spore-formers and bacterial activity.	Nepomiluev and Kuzyakina (1967)
	2,4-D (1·5)	Leached chernozem, low o.m.	No effect.	Tyagny-Ryadno (1967)
	2,4-D amine (0·9) 2,4-D (0·5–0·75)	Loam, 3·5% o.m. Light loam derno-podzol	Decreased populations for c. 20 days. Annual applications (5. yr) slightly increased populations.	Geshtovt et al. (1970) Chunderova et al. (1971)
	2,4-D (1·1: 30; 40)	(—)	Lowest rate, no effect. Both high rates decreased populations.	Singh (1971)
	MCPA (1·0–1·5)	Sandy loam	No effect.	Hoffman-Kakol and Kwarta (1972b)
	MCPA (1·0)	Light loam derno-podzol	Increased numbers of *C. pasteurianum*.	Kozyrev and Laptev (1972)
Phthalaics	chlorthal-dimethyl (11·4)	Silt loam and silty clay loam	No effects.	Fields et al. (1967)

* "Recommended field rate".
Refer to *Notes on tabulated data and abbreviations* (p. 319).

TABLE 2 (continued)

Herbicide		Soil type	Effects	Reference
Group	Concentration (Kg or litre ha^{-1})			
Phthalaics (continued)	chlorthal-dimethyl (4·5; 13·5) naptalam (5)	(—) Medium loam, 1·5% o.m.	No effects. No effects.	Bone and Kuntz (1968) Tulabaev (1970)
Quaternaries	paraquat (1·14; 11·4) Gramoxone (5·7; 57) Additives of Gramoxone (5·7; 57) paraquat (0·57)	Calcareous loam, pH 8; 7·5% o.m. Field and laboratory treated Silt loam, pH 6·3; 0·11% o.m.	Slight decreases immediately after treatment, fluctuations, no severe effects. Decreased populations.	Anderson and Drew (1976a, c) Tu and Bollen (1968a)
Triazines	atrazine (0–20)	Chernozem (calcareous)	Decreased populations.	Pantera (1972) Bakalivanov (1971) Prozorova (1967)
	atrazine (10; 20)	Weak to medium podzol, 7% o.m.	Increased populations, especially spore-forming bacilli.	
	atrazine	Heavy clay degraded chernozem, >2% o.m.	Applied annually (7 yr), decreased populations.	Husarova (1972a)
	atrazine (2; 6) simazine (2; 6)	Medium clay loam, pH 5·3; 2% humus	No effect.	Tsvetkova (1966)
	atrazine (0·5) simazine (0·5)	Red sandy loam	Decreased populations.	Spirama Raju and Rangaswami (1971)
	atrazine (3–4) simazine (3–4)	Calcareous heavy loam, pH 6·8–7·2	Some species increased but total population decreased.	Bobyshev and Lapchenkov (1966)
	atrazine (5·2) simazine (5·2)	Loam (with/without nitrogen fertilizer)	Increased populations, only where nitrogen added.	Fink et al. (1968)

Treatment	Soil	Effect	Reference
atrazine (1·0; 3; 5) simazine (1·0; 3; 5)	Forest soil	Initial decrease then return to normal population.	Milkowska and Gorzelak (1966)
atrazine (8)	Alluvial and terra rossa soils	Spore-formers increased but total population decreased.	Goguadze (1966, 1969)
simazine (4–10) atrazine (2–100)* simazine (5–250)* prometryne (0·5–50)*	(—)	Low rates, no effect to slightly increased populations. High rates decreased populations, especially spore-formers.	Simon (1971)
atrazine (4·5; 10) simazine (4·5; 10) chlorazine (4·5; 10)	Peaty-bog, pH 5·5; 47% o.m.	Dry years – decreased populations. Wet years – increased then decreased populations. 2nd application sometimes increased populations.	Nepomiluev and Kuzyakina (1967)
simazine (2; 50)	1. Chernozem medium clay, pH 7·2; 1·5% o.m. 2. Sandy soil, pH 5·8; 0·9% o.m.	1. Low rate, no effect. High rate decreased then increased population. 2. Low rate, no effect. High rate decreased then increased population, but decrease lasted longer.	Kulinska (1967a, b)
simazine (1·5)	Leached chernozem, low o.m.	Increased populations, especially C. pasteurianum	Tyagny-Ryadno (1967)
simazine (8)	Irrigated brown soil.	Decreased populations but spore-formers increased.	Akopyan and Agaronyan (1968)
simazine (1·1; 30; 40)	(—)	Low rate, no effect. Two high rates decreased populations.	Singh (1971)
simazine (1·0; 10)	Red sandy loam	Decreased populations, especially at high rate.	Balasubramanian and Siddaramappa (1974)
simazine (2–100)	Derno-podzol	Low rates (2–6 kg) decreased aerobic population. Spore-formers decreased at 8 kg. Anaerobes decreased by 10 kg.	Kuzyakina (1971)

* "Recommended field rate".
Refer to Notes on tabulated data and abbreviations (p. 319).

TABLE 2 (continued)

Group	Herbicide Concentration (Kg or litre ha^{-1})	Soil type	Effects	Reference
Triazines (continued)	simazine (1·5–2·5)	Sod-podzol	Decreased spore-formers; non-spore formers resistant.	Nepomiluev et al. (1966b)
	1. simazine (0·4)	Light loam derno-podzol	1. Increased butyric acid bacteria – especially C. pasteurianum.	Kozyrev and Laptev (1972)
	2. prometryne (2)		2. Increased then decreased butyric acid bacteria. Increased C. acetobuty-licum.	
	prometryne (2)	Armenian vineyard	Increased populations.	Krykhanov and Pashkovskaya (1967)
	prometryne (1·0)	Loam, 3·5% o.m.	Decreased populations for c. 20 days.	Geshtovt et al. (1970)
	prometryne (2)	Leached medium loam chernozem	Increased populations.	Matsneva and Semikhatova (1973)
	1. prometryne (2)	1. Derno-podzolic loam (field)	1. No effect.	Samokhvalov (1973)
	2. prometryne (0·004–10 000)	2. Derno-podzolic loam (laboratory)	2. Populations increased.	
	1. prometryne (2·5; 25)	1. Leached cher-nozem 60% MHC	1. Increased populations.	Vasil'ev and Enkina (1967)
	2. prometryne (2·0–25)	2. Normal cher-nozem.	2. No effect.	
	prometryne (2–3) trietazine (5) atraton (3)	Leached medium loam chernozem, pH 6·0; 4% o.m.	Decreased populations.	Mikhailova (1968)
	terbutryne (2·5) methoprotryne (2·5)	Alluvial	Increased aerobic spore-formers.	Micev and Bubalov (1969a)

Ureas			
monuron (—)	(—)	Decreased populations.	Krykhanov and Pashkovskaya (1967)
monuron (8)	Alluvial	Increased spore-formers but total population decreased.	Goguadze (1969)
1. monuron (8)	Irrigated brown soil	1. Decreased populations 50%	Akopyan and Agaronyan (1968)
2. diuron (8)	Medium loam red soil	2. Decreased population 4-fold	
monuron (12–20) diuron (12–20)	Light clay, 3·1–3·5% o.m.	Decreased some species.	Spiridinov et al. (1970)
monuron (12–16) diuron (8)	1. Meadow bog	No effect.	Spiridinov et al. (1972)
1. monuron (8–16) diuron (8–16)		1. Monthly applications decreased populations.	Spiridinov et al. (1973)
2. monuron (12–20) diuron (12–20)	2. Red krasnozem	2. Monthly applications decreased populations.	
diuron (0·5)	Red sandy loam	Decreased populations.	Spirama Raju and Rangaswami (1971)
diuron (0·15–0·75)	(—)	Lowest rates increased populations.	Sharma and Saxena (1971)
DCU (18)	Pale chestnut light loam, pH 8; 2% o.m.	Increased populations.	Chulakov and Zharasov (1967)
monolinuron (0·5–1000)	(—)	Increased populations.	Beck (1969)
phenobenzuron (10)	Sandy loam, 1·4% o.m. Limestone soil, 9·0% o.m.	No effect.	Catroux and Fournier (1971)
Uracils			
lenacil (1·2–1·7)	Dark chestnut heavy loam. Pale chestnut light loam	No effect.	Chulakov and Zharasov (1973)
lenacil (1·0)	Chernozem, 3·0% o.m.	Increased populations but some species decreased.	Barona (1974)

Refer to *Notes on tabulated data and abbreviations* (p. 319).

TABLE 2 (continued)

Group	Herbicide Concentration (Kg or litre ha⁻¹)	Soil type	Effects	Reference
Mixtures	Alipur (–)	Peaty-bog, pH 5·5; 47% o.m.	Decreased activity for c. 30 days.	Nepomiluev and Kuzyakina (1967)
	simazine+amino-triazole (10)	Apple orchard soil	Decreased population.	Teuteberg (1968)
	Phytar 560 (c. 100; 300)	Sandy loam	Low rate, no effect. High rate increased populations.	Breazeale and Camper (1970)
	Printazol (0·25–2·0)	Rich loam, pH 7; 3·3% o.m.	Decreased, then increased populations.	Bertrand and de Wolf (1972)
	1. EPTC (5)+prometryne (2) 2. Alipur (1·2)	Leached medium loam	Increased populations, especially Bacillus spp.	Matsneva and Semikhatova (1973)
	PCP (4)+endothal (4)	Six soils	Decreased populations.	Au (1968)
	VCS-438 (1·0)+dicamba (0·5)	Rhizosphere soil of maize	Inhibited spore-formers.	Micev and Bubalov (1969c)
	Lumeton (3)	Rhizosphere soil of maize	Increased aerobic spore-formers.	Micev (1970)
	prometryne (1·0) and 2,4-D (0·9)	Loam, 3·5% o.m.	Prometryne followed by 2,4-D slightly increased populations.	Geshtovt et al. (1970)
	1. Alipur (2·4) 2. Alipur (2·4)+dalapon (6·0)	Pale and dark chestnut loams	No effect.	Chulakov and Zharasov (1971)
	2,4-D (0·8)+dicamba dimethylamine (0·75)	Light loam derno-podzol	Anaerobes unaffected. Other species decreased.	Kozyrev and Laptev (1972)
	Agelon (3)	Alluvial	No effect.	Micev and Bubalov (1972)

	atrazine (1·66)+ prometryne (0·8)	Heavy clay degraded chernozem	Applied annually (7 yr) decreased populations.	Husarova (1972a)
	1. lenacil (1·0)+ DCU (5) 2. lenacil (1·0)+ pebulate (÷) 3. lenacil (1·0)+ TCA (5)	Chernozem	Increased populations.	Barona (1974)
Other	sodium cacodylate (10·2; 30·7)	Alluvial loam	Temporarily increased populations.	Malone (1971)

* Recommended field rate.
Refer to *Notes on tabulated data and abbreviations* (p. 319).

TABLE 3

Effects of herbicides on nitrifiers and nitrification in soil

Herbicide				
Group	Concentration (Kg or litre ha^{-1})	Soil type	Effects	Reference
Acylamides	diphenamid (2·4; 4·8)	Alluvial soil, 2% o.m.	Increased activity (field plots). No effect in laboratory.	Horowitz et al. (1974)
	diphenamid (0·6)	Light alluvial meadow, pH 7·0; low o.m.	No effect.	Rankov (1968)
	diphenamid (6; 8) diphenamid (0·1–4)*	Red and alluvial Clay	Decreased activity for 60 days. Initially inhibited then increased activity.	Mickovski (1972) Hulin and Horowitz (1973)
	allidochlor (7)	Medium podzol loam, 1·9–2·5% o.m.	Temporarily decreased then increased population and activity.	Isaeva (1967)
	alachlor (0·6–1·2)*	Silt loam, pH 6; 2% o.m.	No significant effect.	Thorneburg and Tweedy (1973)
	propanil (3·14; 5·3)	West Bengal paddy soil	Decreased population.	De and Mukhopadhyay (1971)
Phthalates	endothal (8)	Medium podzol loam, 1·9–2·5% o.m.	Temporarily decreased then increased population and activity.	Isaeva (1967)
Anilines	endothal (2; 20)	Six soils	No effect.	Au (1968)
	trifluralin (0·67; 1·34)	Alluvial soil, 2% o.m.	Increased activity (field plots). No effect in laboratory.	Horowitz et al. (1974)
	trifluralin (1·0–6)	Podzolic loamy chernozem	Toxic to nitrifiers	Hamdi and Tewfik (1969)

	trifluralin (0·45–1·8)	1. Alluvial meadow, 2. Alluvial meadow, leached meadow cinnamon, humus carbonate, leached smonitza	1. Increased activity. 2. Decreased activity but less effect in heavy textured soils.	1. Rankov and Elenkov (1970); Rankov (1971a) 2. Rankov (1972; 1973)
	trifluralin (3)	Bog soil	Decreased activity.	Tyunyaeva et al. (1974)
	trifluralin (0·6; 1·2)	Silt loam, pH 6; 2% o.m.	No significant effect.	Thorneburg and Tweedy (1973)
	trifluralin (≤1 300)	Calcareous loam	Disposal of herbicide on soil severely decreased activity.	Naidu (1972)
Benzoics and related	dinitramine (0·1)	Sandy loam	No effect.	Belles (1972)
	ioxynil (1·32; 13·2, 132 a.i.)	Sandy loam	Inhibited activity.	van Schreven et al. (1970)
	ioxynil (1·5)	Podzolic loam, pH 7	No effect.	Hauke-Pacewiczowa and Krolowa (1968)
	dicamba (0·12)	Derno-podzolic medium loam	Inhibited activity.	Bezuglov et al. (1973)
	chlorthiamid (1·0; 10; 100) dichlobenil (10; 100) 2,3,6-TBA	Soil, pH 6·7	Decreased activity.	Helweg (1972a)
Carbamate and carbanilate	1. barban (20) Carbyne (140·8) 2. Solvents of Carbyne (140·8)	Brown soil	Decreased activity.	Chandra (1964)
	EPTC (1·5)	Sandy loam, pH 5; 7·1% o.m.	1. Completely inhibited for c. 18 weeks. NH_4^+ ions accumulated. 2. Temporary decrease in activity. Nitrobacter spp. transiently inhibited.	Quilt et al., pers. comm.
	EPTC (1·5)	(–)	Increased activity.	Zhukova and Konoshevich (1967)

* "Recommended field rate".
Refer to *Notes on tabulated data and abbreviations* (p. 319).

TABLE 3 (continued)

Group	Herbicide Concentration (Kg or litre ha^{-1})	Soil type	Effects	Reference
Carbamate and carbanilate (continued)	EPTC (4)	Derno-podzol light or medium loams, pH 6; 2–3% o.m.	Increased activity.	Chunderova et al. (1968)
	EPTC (4·5)	Pale chestnut light loam, pH 8; 2% o.m.	Increased activity.	Chulakov and Zharasov (1967)
	EPTC (8)	Meadow-type medium loam, 1·5% o.m.	No effect.	Tulabaev (1970)
	EPTC (3; 4·5)	1. Dark chestnut heavy loam, 4% o.m.	1. Increased activity.	Zharasov (1971, 1972)
		2. Pale chestnut medium loam	2. No effect.	
	EPTC (0·75–1·5) phenmedipham (0·75–1·5)	(—)	No effect.	Zhukova and Botin'eva (1974)
	EPTC (4·5) sulfallate (6)	Medium podzolic loam, 1·9–2·5% o.m.	Increased activity.	Isaeva (1967)
	chlorpropham (100)*	Black alluvial and sandy soils	Decreased activity.	Hauke-Pacewiczowa (1971)
	chlorpropham (10; 100)	Volcanic clay loam	Inhibited activity.	Matsuguchi and Ishizawa (1968, 1969)
	chlorpropham (10)	Sandy clay loam	Increased activity.	Taha et al. (1972)

chlorprophæm (1·2)	Light alluvial meadow, pH 7; low o.m. and heavy leached meadow-cinnamon, pH 7; 2% o.m. (—)	No effect.	Rankov (1968)
Diazines			
pebulate (4–6)	Alluvial, 2% o.m. (—)	Increased activity.	Balicka and Sobieszczanski (1969b)
pyrazon (1·6; 3·2)	Alluvial, 2% o.m.	Increased activity, (field plots). No effect in laboratory.	Horowitz et al. (1974)
pyrazon (0·75–1·5)	(—)	No effect.	Zhukova and Botin'eva (1974)
pyrazon (4)	Medium podzolic loam, 1·9–2·5% o.m.	Increased activity.	Isaeva (1967)
pyrazon (4)	Derno-podzol light or medium loams, pH 6; 2–3% o.m.	Increased activity.	Chunderova et al. (1968)
pyrazon (3; 4·5; 12–36)	Dark chestnut heavy loam, 4% o.m. and pale chestnut medium loam	Increased activity (field plots) with 3 and 4·5 kg rates. Decreased activity in laboratory with 12–36 kg rates.	Zharasov (1971, 1972)
pyrazon (3)	Light loam derno-podzol	Annual applications (5 yr) had no adverse effect.	Chunderova et al. (1971)
pyrazon (30–40)*	Black alluvial or sandy soils	Decreased populations and activity.	Hauke-Pacewiczowa (1971)
phenazone (3–18)	Leached derno-carbonate and derno-podzol	Inhibited activity.	Mezharaupe (1967)

* "Recommended field rate".
Refer to *Notes on tabulated data and abbreviations* (p. 319).

TABLE 3 (continued)

Group	Herbicide Concentration (Kg or litre ha^{-1})	Soil type	Effects	Reference
Halogenated aliphatics	dalapon (127·5; 1275)	Sandy loam, 65% M.H.C.	Low rate slightly, and high rate strongly decreased activity.	van Schreven et al. (1970)
	dalapon (10)	Medium podzol loam, 1·9–2·5% o.m.	Slightly increased activity.	Isaeva (1967)
	dalapon (8)	Meadow-type medium loam, 1·5% o.m.	Decreased activity.	Tulabaev (1970)
	dalapon (0·8–20)*	Sandy loam	No effect.	Leiderman et al. (1971)
	TCA (15–30)*			
	TCA (5; 15)	Leached chernozem, low o.m.	Low rate, no effect. High rate decreased then increased populations.	Lobanov and Poddubnaya (1967)
	TCA (18)	Pale chestnut light loam, pH 8; 2% o.m.	Increased activity.	Chulakov and Zharasov (1967)
	TCA (18; 198)	Pale chestnut medium loam	Low rate, no effect. High rate inhibited activity.	Zharasov (1971)
Phenols	dinoseb 20% (2–4)	Light loamy medium podzol, pH 5·6–6.	No effect.	Nepomiluev et al. (1966a)
	dinoseb acetate (15–25)*	Black alluvial and soils	Decreased activity and population.	Hauke-Pacewiczowa (1971)
	dinoseb acetate (1·2–2)	Sandy loam	Decreased activity.	Hoffman-Kakol and Kwarta (1972a)

Compound (rate)	Soil	Effect	Reference
dinoseb acetate (1.5–2.5)*	(—)	Inhibited activity.	Szember et al. (1973)
dinoseb acetate (1.5–2.5)	Four soils	Inhibited activity. Less effect in heavier textured soils.	Rankov (1972, 1973)
PCP (2; 20)	Alluvial clay loam	High rate decreased activity.	Ishizawa and Matsuguchi (1966)
PCP (100)	Volcanic clay loam	Inhibited activity.	Matsuguchi and Ishizawa (1968, 1969)
DNOC (10)	Light alluvial meadow pH 7; heavy leached meadow-cinnamon, pH 7	Decreased population.	Rankov (1968)
DNOC (2–10)*	Light loam derno-podzol	Annual applications (5 yr) increased activity.	Chunderova et al. (1971)
Phenoxy-acids			
2,4-D (1.5)	Normal chernozem	Increased activity.	Fisyunov (1969a)
2,4-D (2)	Silty medium loam, pH 5.3; 4% o.m.	Increased activity.	Boiko et al. (1969)
2,4-D (1.0–2)	Sandy loam	No effect.	Leiderman et al. (1971)
2,4-D (—)	Derno-podzol	Decreased activity.	Abueva and Bagaev (1975)
2,4-D (1.5)	Leached chernozem	Increased activity.	Tyagny-Ryadno (1967)
2,4-D (1.5)	Heavy loam	Increased activity.	Fisyunov (1969b)
2,4-D (0.75)	Silty medium loam, pH 5.3, 4% o.m.	Increased activity.	Boiko et al. (1969)
2,4-D (1.5–15)	Derno-podzol	Increased population.	Abueva (1970)
2,4-D (1.1; 30; 40)	(—)	Lowest rate, no effect. High rates decreased activity.	Singh (1971)

* "Recommended field rate".
Refer to *Notes on tabulated data and abbreviations* (p. 319).

TABLE 3 (continued)

Group	Herbicide Concentration (Kg or litre ha^{-1})	Soil type	Effects	Reference
Phenoxy-acids (continued)	2,4-D (0·2)	Medium loam chernozem	No effect.	Kudzin et al. (1973)
	2,4-DB (1·5–3)*	Brown soil	Decreased activity.	Chandra (1964)
	2,4–D ester (1·0–2·5)*	(–)	No effect.	Deshmukh and Shrikhande (1974)
	2,4-DES (5–10)	Heavy loams	Low rates increased, high rates decreased activity.	Zavarzin (1966)
	2,4-DEP (6; 30)			
	2,4-D (1·5–15)*	Sandy and clay soils	No effect.	Lozano-Calle (1970)
	2,4,5-T (2–20)*			
	2,4,5-T (11 300)	Calcareous loam	Disposal of herbicide on soil decreased activity.	Naidu (1972)
	2,4,5-T (4; 40)	Sandy clay and silty clay	High rates decreased activity.	Torstensson (1974)
	MCPA (4; 40)			
	2,4-D (0·5–0·75)	Light loam derno-podzol	Annual individual applications (5 yr) had no effect, but mixtures decreased activity.	Chunderova et al. (1971)
	2,4-DB (0·5–0·75)			
	MCPA (2·5)			
	MCPB (2·0)			
	MCPA (2·0; 3·8)	West Bengal paddy soil	Decreased population.	De and Mukhopadhyay (1971)
	MCPA (5)	Podzolic loam, pH 7	No effect.	Hauke-Pacewiczowa and Krolowa (1968)
	MCPA (20–25)*	Black alluvial and sandy soils	Decreased populations and activity.	Hauke-Pacewiczowa (1971)
	MCPA (1·0–1·5)	Sandy loam	Decreased activity.	Hoffman-Kakol and Kwarta (1972b)

	dichlorprop (2·6; 255) mecoprop (10; 100)	Sandy loam 65% MHC	Low rates decreased activity slightly. High rates, strongly.	van Schreven et al. (1970)
Phthalics	chlorthal-dimethyl (7·5; 15)	Alluvial soil, 2% o.m.	Increased activity.	Horowitz et al. (1974)
	chlorthal-dimethyl (1·5–3)	Serozem-meadow	Highest rate inhibited activity.	Tulabaev (1972)
	naptalam (5)	Meadow-type medium loam, 1·5% o.m.	No effect.	Tulabaev (1970)
Pyridines	picloram (0·2–10)	Acid volcanic ash, silt loam, neutral alluvium, and marine sand	Inhibited activity.	Dubey (1969)
	picloram (0·48; 4·8; 48)	Sandy loam, 65% MHC	No effect.	van Schreven et al. (1970)
	picloram (0·5) picloram (0·07)	Heavy clay Derno-podzol medium loam	No effect. Decreased activity, especially in dry conditions.	Grover (1972) Bezuglov et al. (1973)
	picloram (2) picloram (0·6; 1·2)	Alluvial soil Silt loam, pH 6; 2% o.m.	Increased population. No significant effect.	Goguadze et al. (1972) Thorneburg and Tweedy (1973)
Quaternaries	paraquat (0·6; 6)	Silt loam, pH 6·3; 0·11% o.m.	No significant effect.	Tu (1966), Tu and Bollen (1968a)
	paraquat (1·14; 11·4) Gramoxone (5·6; 56) Additives of Gramoxone (5·6; 56)	Calcareous loam, pH 8; 7·5% o.m.	No significant effect.	Anderson and Drew (1976b)
	paraquat (0·6; 1·2)	Silt loam, pH 6; 2% o.m.	No significant effect.	Thorneburg and Tweedy (1973)

* "Recommended field rate".
Refer to *Notes on tabulated data and abbreviations* (p. 319).

TABLE 3 (continued)

Group	Herbicide Concentration (Kg or litre ha^{-1})	Soil type	Effects	Reference
Triazines	atrazine (1–40)*	Leached cher- nozem	Low rates, no effect. High rates inhibited activity.	Peshakov et al. (1969)
	atrazine (1·0; 10)	Black clay loam	Decreased activity.	Setty et al. (1970)
	atrazine (4)	Loamy sand, pH 5·3; 2·4% humus	Annual applications (13 yr) temporarily decreased population.	Voets et al. (1974)
	atrazine (10; 20)	Weak to medium podzol, 7% o.m.	Increased activity.	Prozorova (1967)
	atrazine (3)	Fine medium loam and fine sand	No effect in loam, increased activity in sand.	Nedorost (1967)
	atrazine (2; 4·5)	Loamy clay	No effect.	Budoi and Budoi (1967)
	atrazine (10)	Leached sandy clay, 1·0–1·6% o.m.	No effect.	Peshakov et al. (1969)
	atrazine (4; 10)	Heavy loam cher- nozem, 4–5% o.m.	No effect.	Pilipets and Litvinov (1972)
	simazine (1·0–4)	Podzol, 2% humus	Increased activity.	Tsvetkova (1966)
	atrazine (3) simazine (3)	Medium loam meadow 1·2–2·2% o.m.	No effect (field plots); decreased activity in laboratory.	Tulabaev and Azimbegov (1967)
	atrazine (3) simazine (3)	Calcareous cher- nozem	Increased activity.	Tarlapan and Zakhariya (1966)
	atrazine (1·0; 2; 4) simazine (1·0; 2; 4)	Medium loam meadow bog	Two lowest rates increased, high rate decreased activity.	Spiridinov and Spiridinova (1971)

Treatment (kg ha⁻¹)	Soil	Effect	Reference
atrazine (0·5–2·0); simazine (0·5–2·0)	Alluvial sandy loam, pH 7·9	Increased activity.	Joshi *et al.* (1976)
atrazine (1–4)*; simazine (1–4)*	Sandy loam	No significant effects.	Leiderman *et al.* (1971)
atrazine (8–12); simazine (4–12)	Heavy loam, pH 6·3	Increased activity.	Zavarzin and Belyaeva (1966)
atrazine (4·5; 10); simazine (4·5; 10)	Peaty bog, pH 5·5; 47% o.m.	Little effect.	Nepomiluev and Kuzyakina (1967)
atrazine (2; 3); simazine (2–3)	Heavy loam chernozem, low o.m.	No effect, except atrazine (3 kg ha^{-1}) decreased activity.	Fisyunov (1969b)
atrazine (30–40)*; simazine (30–40)*	Black alluvial and sandy soils	Decreased population and activity.	Hauke-Pacewiczowa (1971)
atrazine (0·2–3)*; simazine (0·2–3)*	Medium loam chernozem	In presence of γ-BHC and/or N.P.K. fertilizer, activity increased.	Kudzin *et al.* (1973)
atrazine (1·2; 5); simazine (1·2; 5)	Silt loam, pH 6; 2% o.m.	No significant effect.	Thorneburg and Tweedy (1973)
atrazine (1–4); simazine (1–4); chlorazine (2–6)	Normal chernozem	No effect.	Fisyunov (1969a)
atrazine (15); simazine (15); prometryne (15)	Leach black calcareous clay	Increased population and activity.	Andreeva-Fetvadzhieva and Kolcheva (1969)
simazine (2)	Silty medium loam, pH 5·3, 4% o.m.	Increased population and activity.	Boiko *et al.* (1969)
simazine (0·25; 1·0)	Sandy loam	Low rate, no effect. High rate inhibited activity.	Nayyar *et al.* (1970)
simazine (6; 8)	Moderate loam carbonate chernozem	Decreased activity.	Ivanov (1974)

* "Recommended field rate".
Refer to *Notes on tabulated data and abbreviations* (p. 319).

TABLE 3 (continued)

Group	Herbicide Concentration (Kg or litre ha^{-1})	Soil type	Effects	Reference
Triazines (continued)	simazine (1·0)	Loam, pH 4·9	Increased heterotrophic nitrification.	Smith and Weeraratna (1974)
	simazine (1·0; 10)	Red sandy loam	Low rate increased activity.	Balasubramanian and Siddaramappa (1974)
	simazine (1·5–5)	Sod-podzol	No effect.	Nepomiluev et al. (1966b)
	simazine (1·0; 3; 6·9)*	Loam	Two highest rates inhibited activity.	Farmer et al. (1965)
	simazine (8–12)	Heavy loam	Increased activity.	Zavarzin (1966)
	simazine (2; 50)	Medium clay chernozem, pH 7·2; 1·5% o.m.	Low rate increased, high rate inhibited activity.	Kulinska (1967b)
	simazine (8)	Irrigated brown soil	No significant effect.	Akopyan and Agaronyan (1968)
	simazine (1·1; 30; 40)	(—)	Low rate, no effect. High rates decreased activity.	Singh (1971)
	simazine (2; 6)	Derno-podzolic light loam	Low rate increased, high rate temporarily decreased activity.	Bliev and Kozlova (1972)
	simazine (1·0; 10)	Red sandy loam, pH 5·3	Increased activity.	Balasubramanian et al. (1973)
	simazine (2; 20)	Sandy and silty clay loams	No effect.	Torstensson (1974)
	simazine (2–100)	Derno-podzolic loam	Only highest rate decreased activity.	Kuzyakina (1971)
	1. simazine (2) 2. prometryne (4)	Light alluvial and heavy leached meadow cinnamon, pH 7; 2% o.m.	1. No effect. 2. Decreased population.	Rankov (1968)
	prometryne (1·0–2·5)	Heavy cis-Caucasian cher-	Increased population and activity.	Yaraslovskaya et al. (1969)

Herbicide (rate)	Soil type	Effect	Reference
prometryne (2·5; 25)	Leached chernozem and chernozem	Increased population and activity. High rate delayed increase.	Vasil'ev and Enkina (1967)
prometryne (1·0–2·5)	Leached chernozem	Inhibited activity.	Voinova et al. (1970)
prometryne (0·1; 1·0; 5)	Clay soil	Highest rates transiently decreased activity.	Hulin and Horowitz (1973)
prometryne (10) ametryne (10)	Acid volcanic ash, silt loam, neutral alluvium, and marine sand	Inhibited activity.	Dubey (1969)
prometryne (2–3) atraton (3) trietazine (5)	Leached medium loam chernozem, pH 6; 4% o.m.	Increased activity.	Mikhailova (1968)
prometryne (2·5) propazine (2) ametryne (2–4) chlorazine (4–6)	Light loam derno-podzol Sandy loam Light loamy medium podzol, pH 5·6–6	Annual applications (5 yr) increased activity No significant effect. No effect.	Chunderova et al. (1971) Leiderman et al. (1971) Nepomiluev et al. (1966a)
methoprotryne 25% (2·5)	Alluvial soil	Inhibited activity.	Micev and Bubalov (1969a)
terbutryne (2·25–6·75)	Alluvial or bog-alluvial	No effect.	Mickovski (1967)
Ureas			
linuron (0·5–2)* linuron (0·5–2)*	Clay loam, pH 5·35 Sandy loam	Decreased activity. Initially decreased, then increased activity.	Sivasithamparam (1970) Beckmann and Pestemer (1975)
linuron (5–20)*	Black alluvial and sandy soil	Decreased population and activity.	Hauke-Pacewiczowa (1971)
linuron (2; 20)	Sandy and silty clay loams	High rate decreased activity.	Torstensson (1974)

* "Recommended field rate".
Refer to *Notes on tabulated data and abbreviations* (p. 319).

TABLE 3 (continued)

Group	Herbicide Concentration (Kg or litre ha^{-1})	Soil type	Effects	Reference
Ureas (continued)	linuron (1·5) monolinuron (1·5)	Light loam podzol	No effect.	Kozaczenko and Sobieraj (1973)
	linuron (2) monolinuron (10)	Sandy clay chernozem, 1·0-1·6% o.m.	Retarded activity.	Peshakov et al. (1969)
	1. linuron (0·2)	Light alluvial meadow, pH 7, heavy leached	1. Decreased populations.	Rankov (1968)
	2. monolinuron (0·2)	meadow-cinnamon, pH 7; 2% o.m.	2. No effect.	
	linuron (0·5-2) monolinuron (1·0-3)	(—)	No effect.	Kozaczenko and Sobieraj (1973)
	linuron (0·5-2)* monolinuron (1·0-3)*	(—)	Increased activity.	Balicka and Sobieszczanski (1969b)
	linuron (0·5-2)* monolinuron (1·0-3)* fluometuron (1·0-2)	Leached chernozem	Inhibited activity.	Voinova et al. (1970)
	1. linuron (10-15) diuron (10-15) 3,4-dichloro-phenylurea	Loam soil, pH 7-7·4	1. Inhibited activity.	Corke and Thompson (1970)
	2. Degradation product of 3,4-dichloro-phenylurea		2. Nitrosomonas sp inhibited.	

Herbicide (rate)	Soil	Effect	Reference
linuron (50) diuron (50) monuron (50) fluometuron (50) chloroxuron (50) metoxuron (50)	Sandy loam soils, pH 6·1–7·4; 1·3–1·6% o.m.	Inhibited activity.	Grossbard and Marsh (1974)
1. monuron (1·5) diuron (1·5) 2. fenuron (1·5)	Meadow-type chernozem	1. Little effect. 2. Decreased populations.	Tulabaev and Tamikaev (1968)
1. monuron (6; 12) neburon (4; 8) 2. neburon (8) 3. monuron (6)	Heavy loam soils	1. Low rate increased activity. High rate decreased activity. 2. Second application increased activity. 3. Second application decreased activity.	Zavarzin (1966)
monuron (4) diuron (6–12)	Brown heavy loam, pH 6·3	Increased activity.	Zavarzin and Belyaeva (1966)
monuron (8) diuron (8)	Irrigated brown soil	Decreased populations (field plots); no effect in lab.	Akopyan and Agaronyan (1968)
monuron (2·5–5)* neburon (2–3)	Light loam derno-podzol	Annually applied (5 yr) increased activity.	Chunderova et al. (1971)
diuron (2)	Acid volcanic ash, silt loam, neutral alluvium and marine sand	Inhibited activity.	Dubey (1969)
diuron (0·1–0·4) diuron (0·8; 1·6) neburon (0·9; 1·8) fluometuron (1·6; 3·2)	Clay loam Alluvial, 2% o.m.	No significant effect. Increased activity.	Kuramato et al. (1970) Horowitz et al. (1974)

* "Recommended field rate".
Refer to *Notes on tabulated data and abbreviations* (p. 319).

TABLE 3 (continued)

Group	Herbicide Concentration (Kg or litre ha^{-1})	Soil type	Effects	Reference
Ureas (continued)	fluometuron (0·1; 1·0; 10; 100)	Sandy loam	Two highest rates decreased activity.	Bozarth et al. (1969)
	1. fluometuron (1·5; 3)	Serozem-meadow soil	1. High rate decreased activity.	Tulabaev (1972)
	2. noruron (1·5; 3)		2. Increased activity.	
	DCU (5; 15)	Leached chernozem, low o.m.	Low rate, no effect. High rate decreased then increased activity.	Lobanov and Poddubnaya (1967)
	DCU (18)	Pale chestnut light loam, pH 8; 2% o.m.	Increased activity.	Chulakov and Zharasov (1967)
	fenuron (1·5)	Medium loam meadow 1·2–2·2% o.m.	Persistent decrease in activity.	Tulabaev and Azimbegov (1967)
	phenobenzuron (10; 100)	Sandy loam, 1·4% o.m.; Limestone soil	No effect.	Catroux and Fournier (1971)
	metoxuron (100–200)*	Calcareous soil	Slightly decreased activity.	Simon et al. (1973)
Uracils	bromacil (4; 8)	Alluvial, 2% o.m.	Increased activity.	Horowitz et al. (1974)
	bromacil (10)	Sandy	Decreased activity.	Pancholy and Lynd (1969)
	bromacil (0·1; 10; 100)	Clay	Two highest rates inhibited activity.	Hulin and Horowitz (1973)
Mixtures	Alipur (4)	Light loam podzol	No effect.	Kozaczenko and Sobieraj (1973)

Alipur (2–4;	Light loamy medium podzol, pH 5·6-6	No effect.	Nepomiluev et al. (1966a)
Alipur (0·6)	Medium podzol loam 1·4–2·5% o.m.	Initially decreased then increased population and activity.	Isaeva (1967)
1. Alipur (0·6; 1·8) 2. Murbetol (7·6; 9·1) Alipur (2·4)	Sod-podzolic loam	1. Transiently decreased then increased activity. 2. Decreased activity.	Isaeva and Novogrudskaya (1966)
Murbetol (22·2)	Pale chestnut and medium loams	Slightly decreased activity in wet conditions.	Zharasov (1971)
Murbetol (40; 80)	Leached chernozem, low o.m.	Low rate, no effect. High rate decreased activity.	Lobanov and Poddubnaya (1967)
Amitrole-T (8·75; 87·5; 875)	Sandy loam, 65% MHC	Low rate transiently decreased activity. High rates inhibited activity.	van Schreven et al. (1970)
aminotriazole 38% +simazine 20% +MCPA 10% (10)	Sandy clay chernozem, 1-1·6% o.m.	Decreased activity.	Peshakov et al. (1969)
prometryne 40% + simazine 15% (2)	Light alluvial meadow, pH 7; heavy leached meadow cinnamon, pH 7	No effect.	Rankov (1968)
Gesaran (2·25; 4·5; 6·75)	Alluvial or bog-alluvial	Lowest rates, no effect. High rates transiently decreased activity.	Mickowski (1967)
Lumetron (3; 6; 9) PCP (4) + endothal (2)	Six soils	Decreased activity.	Au (1968)

* "Recommended field rate".
Refer to Notes on tabulated data and abbreviations (p. 319).

TABLE 3 (continued)

Group	Herbicide		Effects	Reference
	Concentration (Kg or litre ha⁻¹)	Soil type		
Mixtures (continued)	MCPA + dichlorprop (23·5; 235)	Sandy loam, 65% MHC	Low rate slightly decreased, high rate strongly inhibited activity.	van Schreven *et al.* (1970)
	dalapon (20) + 2,4-D(2)	Derno-podzolic light loam	Increased activity.	Bliev and Kozlova (1972)
Others	sodium chlorate (6·75–150)	Various soils (laboratory treated)	Inhibited *Nitrobacter* sp.	Karki *et al.* (1972)
	sodium chlorate (150)	Sandy loam (field study)	No effect.	Karki *et al.* (1973)

Refer to *Notes on tabulated data and abbreviations* (p. 319).

TABLE 4

Effects of herbicides on denitrification in soil

Group	Herbicide	Concentration (Kg or litre ha^{-1})	Soil type	Effects	Reference
Acylamides	propanil (10·5)		Paddy soil	Large decrease in population and activity.	De and Mukhapadhyay (1971)
Diazines	maleic hydrazide (2–5)*		(—)	Not greatly affected.	Sethunathan (1970a,b)
	pyrazon (4)		Medium loam derno-podzol, 1·5% o.m.	Increased population and activity.	Karasavich and Isaeva (1969)
Halogenated aliphatics	TCA (5; 15)		Leached cher-nozem	Low rate, no effect; high rate inhibited activity.	Lobanov and Poddubnaya (1967)
Phenols	PCP (1·5)		Leached cher-nozem	Increased population.	Tyagny-Ryadno (1967)
	PCP (0·8; 1.6; 3)		Clay loam, 1·3% C; alluvial, 1·1% C; volcanic light clay, 7·4% C; alluvial clay loam, 2·7% C	Increased population.	Ishizawa and Matsuguchi (1966)
Phenoxy-acids	2,4-D(1–3)*		(—)	Not greatly affected.	Sethunathan (1970a,b)
	2,4-D (1·5)		Leached cher-nozem	Increased population.	Tyagny-Ryadno (1967)
	2,4,5-T (4; 40) MCPA (4; 40)		Sandy clay and silty clay	No effect.	Torstensson (1974)

* "Recommended field rate".
Refer to *Notes on tabulated data and abbreviations* (*p.* 319).
C (carbon)

TABLE 4 (continued)

Group	Herbicide Concentration (Kg or litre ha^{-1})	Soil type	Effects	Reference
Phenoxy-acids (continued)	MCPA (2·25)	Paddy soil	Large decrease in populations and activity.	De and Mukhopadhyay (1971)
Phthalaics	chlorthal-dimethyl (1·5–3)	Serozem-meadow soil	Slight decrease in population but rapidly recovering.	Tulabaev (1972)
Quaternaries	diquat (1–5)*	Lake sediment	No effect.	Atkinson (1973)
Triazines	atrazine (3)	Medium loam meadow soil, 1·2–2·2% o.m.	Increased population and activity.	Tulabaev and Azimbegov (1967)
	simazine (3)			
	atrazine (10)	Medium to heavy sandy clay leached chernozem, 1–1·6% o.m.	No effect.	Peshakov et al. (1969)
	simazine (10)	Orchard soil	Annual applications (3 yr) decreased population	Kuryndina (1965)
	simazine (1·5)	Leached cher-nozem	Increased population.	Tyagny-Ryadno (1967)
	simazine (2; 20)	Sandy clay and silty clay	Increased population.	Torstensson (1974)
	simazine (2–100)	Derno-podzol	No effect except at highest rate which decreased population and activity.	Kuzyakina (1971)
Ureas	prometryne (0·5–1·5)* linuron (2; 20)	(–) Sandy clay and silty clay	Decreased population and activity. No effect.	Zhabyuk (1973) Torstensson (1974)
	linuron (0·5–2) monolinuron (0·5–2)	Medium to heavy sandy clay leached chernozem 1·0–1·6% o.m.	Decreased population and activity.	Peshakov et al. (1969)

fenuron (1·5)	Medium loam meadow soil, 1·2–2·2% o.m.	Increased population and activity.	Tulabaev and Azimbegov (1967)
fluometuron (1·5–3)	Serozem-meadow soil	Decreased population and activity.	Tulabaev (1972)
metoxuron (100–400)*	Calcareous soil	No significant effect.	Simon et al. (1973)
DCU (5; 15)	Leached chernozem	Low rate, no effect; High rate temporarily decreased populations.	Lobanov and Poddubnaya (1967)
Mixtures lenacil (1) + pebulate (4)	Chernozem, 3% o.m.	Increased activity.	Barona (1974)
prometryne (1)* + TCA (8)	(–)	Increased population and activity.	Zhabyuk (1973)
Murbetol (7·6; 9·1)	Sod-podzol loam	High rate decreased population and activity.	Isaeva and Novogrudskaya (1966)
Murbetol (40; 80)	Leached chernozem	No effect.	Lobanov and Poddubnaya (1967)
Others sodium chlorate (6·75; 20·3; 67·5)	Acid sandy soil, low o.m.	Small and large decreases in population according to rate applied.	Karki et al. (1972)
sodium chlorate (150)	Fallow soil or turfed soil	No effect.	Karki et al. (1973)
sodium chlorate (6·75; 20·3; 67·5)	Acidic sandy soil, low o.m.	Retarded or inhibited activity.	Karki and Kaiser (1974)
Methurin (1·5–3)	Serozem-meadow soil	Decreased populations and activity.	Tulabaev (1972)

* "Recommended field rate".
Refer to Notes on tabulated data and abbreviations (p. 319).

TABLE 5

Some effects of herbicides in soil on *Rhizobium* spp. and nodulation of legumes

Group	Herbicide Concentration (Kg or litre ha⁻¹)	Soil type	Effects	Reference
Anilines	trifluralin (0·36; 0·74; 1·1)	Sand and silt loam mixtures	Decreased nodulation.	Kust and Struckmeyer (1971)
	trifluralin (1·0)	Loamy sand	Nodulation reduced but not population.	Brock (1972)
	trifluralin (0·5–1·0)*	Red latosol	No effect.	Lopes et al. (1971)
	trifluralin (1–6)	Loam	Applied before planting, increased nodulation. Applied at planting, inhibited nodulation.	Hamdi and Tewfik (1969)
	trifluralin (5)* nitralin (5)*	Silt loam	Affected nodulation.	Dunigan et al. (1970)
	nitralin (1·1, 5·6, 11·2)	Silt loam, pH 5·7; 1·3% o.m.	No effect on population but nodule weight decreased.	Kapusta and Rouwenhorst (1973)
Benzoics and related	chloramben (4)	Dark chestnut light loam, pH 7–8	Slightly toxic in some cases, increased nodulation sometimes	Avrov et al. (1968)
	chloramben (10–20)*	Silt loam	Affected nodulation.	Dunigan et al. (1970)
	chloramben (5)	Irrigated light chestnut, 2% o.m.	Decreased nodulation.	Shkola (1970)
Carbamate and carbanilate	carbetamide (2)	Loamy sand	Nodulation reduced, population not.	Brock (1972)
	chlorpropham (3·4; 16·8; 33·6)	Silt loam, pH 5·7; 1·3% o.m.	No effect except at highest rate which decreased nodule weight.	Kapusta and Rouwenhorst (1973)
	chlorpropham (–)	Dark chestnut light loam, pH 7–8	Inhibited nodulation.	Avrov et al. (1968)
	EPTC (3–6)*	Red latosol	No effect.	Lopes et al. (1971)
	vernolate (15)*	Silt loam	Affected nodulation.	Dunigan et al. (1970)

Diazines	phenazone (3–6; 8–18)	Leached derno-carbonate and a derno-podzol	Inhibited population.	Mezharaupe (1967)
Halogenated aliphatics	dalapon (0·5–360) TCA (30–360)	Light brown	Inhibited population.	Zhakharyan (1970)
Phenols	dinoseb (2)*	Peaty	No effect on nodulation.	Hauke-Pacewiczowa (1969)
	dinoseb acetate (1·5–2·5)	Leached meadow-cinnamon and alluvial meadow	Increased nodulation.	Rankov et al. (1966)
	dinoseb amine (2·4) dinoseb 20% (2–4)	Light loamy medium podzol, pH 5-6–6	Increased nodulation.	Nepomiluev et al. (1966a)
	dinoseb acetate (1·5–2·5)	Leached meadow-cinnamon and alluvial meadow soil	Increased nodulation.	Rankov et al. (1966)
	dinoseb (15–50)* DNOC (150–200)*	Sandy loam	No significant decrease in population.	Jensen (1969)
	PCP (0·5)	Dark chestnut light loam	No effect.	Avrov et al. (1968)
Phenoxy-acids	2,4-DB (1·4)	Sandy loam	Decreased nodulation and efficiency of nitrogen-fixation.	Garcia and Jordan (1969)
	2,4-D (1; 5; 10) MCPA (1·0)	(—)	Decreased nodulation.	Mickovski (1966)
	1. 2,4-DB (1·5–3)* 2. MCPA (0·2–2·3)* 3. MCPB (1·0–3·0)*	Peaty soil	1. No effect. 2. No effect. 3. Decreased nodulation.	Hauke-Pacewiczowa (1969)
	2,4-D (3–6)* MCPA (2–4;*	Sandy loam	No effect on populations.	Jensen (1969)

* "Recommended field rate".

Refer to *Notes on tabulated data and abbreviations* (p. 319).

TABLE 5 (continued)

Group	Herbicide	Concentration (Kg or litre ha^{-1})	Soil type	Effects	Reference
Quaternaries		Gramoxone (5·6; 56) chlormequat (1; 40) Gramoxone (5; 20)	Sand culture (—)	Inhibited population. Slightly decreased nodule weight.	Manninger et al. (1972) Mouchova and Apltauer (1971)
Triazines	1. atrazine (1–4)* 2. simazine (0·5–4)*		Podzol	1. Decreased nodulation. 2. Residual activity increased plant weight but 3 successive years application decreased nodulation in the 4th year.	Avrov (1966)
	simazine (0·05–0·1)		Podzolic loam	No effect at low rate but higher rate decreased nodulation, protein and amino acid content.	Hauke-Pacewiczowa (1969)
	simazine (0·01; 0·05)		Podzolic loam (pot trials)	Low rate decreased N_2-fixation. Higher rate inhibited plant development.	Hauke-Pacewiczowa (1970)
	simazine (0·1–1·0)		(—)	Morphological changes in nodules. Higher rate inhibited nodulation.	Thomas and Hammond (1971)
	1. prometryne (2–3) 2. atraton (3) trietazine (5)		Leached medium loam chernozem; pH 6; 4% o.m.	1. No effect on population. 2. Decreased population.	Mikhailova (1968)
	prometryne (5)* prometryne (2·5)		Silt loam Irrigated light chestnut, 2% o.m.	Decreased nodulation. Nodulation initially decreased, then increased.	Dunigan et al. (1970) Shkola (1970)
	prometryne (2·0–3·5)		Acid brown forest soil	No effect.	Borbély and Kecskés (1972)

prometryne (2–3)*	Leached meadow-cinnamon and alluvial meadow soils	Increased nodulation.	Rankov et al. (1966)
chlorazine (4–6)	Light loamy medium podzol, pH 5·6–6	Increased nodulation.	Nepomiluev et al. (1966a)
Ureas monolinuron (1–2)*	Leached meadow-cinnamon and alluvial meadow soils	Increased nodulation.	Rankov et al. (1966)
linuron (5–7)*	Silt loam	Decreased nodulation.	Dunigan et al. (1970)
monolinuron (2–3·5) + neburon (4·5–5·9)	Acidic brown forest soil	Little to no effect on populations or nodulation.	Borbély and Kecskés (1972)
Mixtures 1. simazine (2·5) + 2,3,6-TBA (1·0) 2. simazine (2) + DCU (5)	Light chestnut soils	1. Decreased nodulation and plant weight. 2. Increased nodulation and plant growth.	Malanina and Nikitin (1971)
2,4-DB (1·4) + dalapon (4·5)	Sandy loams	Decreased nodulation and efficiency of fixation.	Garcia and Jordan (1969)
Alipur (2–4)	Light loamy medium podzol, pH 5·6–6	Increased nodulation.	Nepomiluev et al. (1966a)
prometryne 40% + simazine 15% (2–3·5)	Acidic brown forest soil	No effect on populations or nodulation.	Borbély and Kecskés (1972)

* "Recommended field rate".
Refer to *Notes on tabulated data and abbreviations* (p. 319).

TABLE 6

Effects of herbicides on free-living, nitrogen-fixing bacteria in soil

Herbicide		Concentration (Kg or litre ha^{-1})	Soil type	Effects	Reference
Group					
Acylamides	diphenamid (4–6)*		Leached meadow-cinnamon and alluvial soils	No effect on populations.	Rankov et al. (1966)
Anilines	diphenamid (6; 8)		Red and alluvial	No effect.	Mickovski (1972)
	trifluralin (1–6)		Loamy soil	No effect.	Hamdi and Tewfik (1969)
Benzoics and related	ioxynil (0·5; 50)*		Polder soil, 65% MHC	Decreased azotobacter populations.	van Schreven et al. (1970)
	ioxynil (1·5)		Podzolic loam, pH 7	No effect on anaerobic forms but azotobacters increased.	Hauke-Pacewiczowa and Krolowa (1968)
	chloramben (5)		Irrigated light chestnut, 2% o.m.	Slightly decreased azotobacters.	Shkola (1970)
	dichlobenil (7·5)		Chernozem-smolnitsa	No effect.	Nikolova and Bakalivinov (1972)
Carbamates and carbanilates	chlorpropham (−)		Sandy clay loam	Decreased populations.	Taha et al. (1972)
	chlorpropham (1–2)*		Leached meadow-cinnamon and alluvial soils	No effect on populations.	Rankov et al. (1966)
	eptam (4·5)		Pale chestnut light loam, pH 8; 2% o.m.	No effect.	Chulakov and Zharasov (1967)
	eptam (8)		Meadow-type medium loam, 1·5% o.m.	No effect.	Tulabaev (1970)

	Herbicide (rate)	Soil	Effect	Reference
	eptam (3; 4·5)	Dark chestnut heavy loam, 4% o.m.	No effect.	Zharasov (1972)
Diazines	1. phenazone (3–6)	1. Leached derno-carbonate	1. No effect.	Mezharaupe (1967)
	2. phenazone (8–18)	2. Leached derno-carbonate	2. Inhibited azotobacters.	
	3. phenazone (12) pyrazon (3; 4·5)	3. Derno-podzol Dark chestnut heavy loam, 4% o.m.	3. No effect. No effect.	Zharasov (1972)
Halogenated aliphatics	TCA (75)	(−)	Annual applications (5 yr) decreased azotobacter populations.	Pantera (1972)
	TCA (6)	Chernozem	No effect.	Sosnovskaya and Pashchenko (1965)
	TCA (5; 15)	Leached chernozem	Low rate had no effect. High rate decreased population of azotobacters.	Lobanov and Poddubnaya (1967)
	TCA (18)	Pale chestnut light loam, pH 8; 2% o.m.	No effect.	Chulakov and Zharasov (1967)
	dalapon (0·15; 0·3; 0·45; 0·6; 0·75)	Red clay	Two lowest rates had no effect. Higher rates decreased azotobacter populations.	Sharma and Saxena (1972)
	dalapon (0·9; 90)*	Polder soil, 65% MHC	Decreased azotobacter populations.	van Schreven et al. (1970)
	dalapon (8;	Meadow-type medium loam, 1·5% o.m.	Decreased azotobacter populations.	Tulabaev (1970)

* "Recommended field rate".
Refer to *Notes on tabulated data and abbreviations* (p. 319).

TABLE 6 (continued)

Group	Herbicide Concentration (Kg or litre ha^{-1})	Soil type	Effects	Reference
Phenols	dinoseb acetate (1·5–2·5)*	Leached meadow-cinnamon and alluvial soils	No effect on numbers of *Azotobacter chroococcum* or *Clostridium pasteurianum*	Rankov *et al.* (1966)
	DNOC 50% (15)	Leached meadow-cinnamon and alluvial soils	Decreased azotobacter populations.	Rankov *et al.* (1966)
	dinoseb acetate (1·2–2)	Sandy loam	Temporarily decreased populations.	Hoffman-Kakol and Kwarta (1972a)
Phenoxy-acids	MCPA (5)	Podzolic loam, pH 7	Increased population of azotobacters.	Hauke-Pacewiczowa and Krolowa (1968)
	MCPA (30–225)* mecoprop (2; 200)* dichlorprop (2·5; 250)*	Polder soil, 65% MHC	Decreased azotobacter populations.	van Schreven *et al.* (1970)
	MCPA (0·3–2·25)*	Leached meadow-cinnamon and alluvial soils	No effect on *A. chroococcum* or *C. pasteurianum* numbers.	Rankov *et al.* (1966)
	2,4-D (2; 100)	Chernozem	Low rate increased population of azotobacter. High rate decreased population.	Yurkevich and Tolkachev (1972a)
	2,4-D (1·5)	Chernozem	No effect.	Sosnovskaya and Pashchenko (1965)
	2,4-DEP (0–8)	Light loam pale chestnut, 1·7% o.m.	No effect.	Lazareva and Blokhin (1969)

Group	Herbicide (rate)	Recommended field rate	Soil type	Effect	Reference
	2,4-D (1·1; 30; 40)	(—)		Lowest rate had no effect. Higher rates decreased azotobacter population.	Singh (1971)
	2,4-D (0·4)	(—)		Numbers of *A. chroococcum* increased. Higher rates than 0·4 were toxic.	Sharma and Saxena (1974)
	2,4-D (0·7)		Calcareous heavy loam, pH 6·8–7·2	Increased azotobacter population.	Bobyshev and Lapchenkov (1966)
Phthalaics	naptalam (—)		Meadow-type medium loam, 1·5% o.m.	No effect.	Tulabaev (1970)
	chlorthal-dimethyl (1·5; 3)		Serozem-meadow	Low rate temporarily decreased, high rate inhibited azotobacters.	Tulabaev (1972)
Pyridines	picloram (2)		Alluvial	Increased populations.	Goguadze *et al.* (1972)
	picloram (100–300)*		Polder soil, 65% MHC	Decreased azotobacter population.	van Schreven *et al.* (1970)
Quaternaries	diquat (1·1–2)*		Lake sediment	N_2-fixation inhibited.	Atkinson (1973)
	1. paraquat dichloride (1·14; 11·4) 2. Gramoxone (5·6; 56)		Calcareous loam, pH 8; 7·5% o.m.	No effect for first 3 weeks. Populations varied thereafter.	Anderson and Drew (1976c)
Triazines	atrazine (1–4)* simazine (1–4)* prometryne (0·5–1·5)*		Leached meadow-cinnamon and alluvial soils	No effect on numbers of *A. chroococcum* or *C. pasteurianum*.	Rankov *et al.* (1966)
	atrazine (3) simazine (3)		Medium loam meadow soil	Decreased azotobacter population.	Tulabaev and Azimbegov (1967)
	atrazine (20) prometryne (20)		Light loam	Annual applications (5 yr) decreased azotobacter population.	Pantera (1972)

* "Recommended field rate".
Refer to *Notes on tabulated data and abbreviations* (. p. 319).

TABLE 6 (continued)

Herbicide		Soil type	Effects	Reference
Group	Concentration (Kg or litre ha^{-1})			
Trazines (continued)	atrazine (4)	Loamy sand, pH 5·3	Annual applications (12 yr) increased azotobacter population.	Voets et al. (1974)
	atrazine (3) simazine (3)	Chernozem	No effect on populations.	Sosnovskaya and Pashchenko (1965)
	atrazine (3–4) simazine (3–4)	Calcareous heavy loam, pH 6·8–7·2	Increased azotobacter population.	Bobyshev and Lapchenko (1966)
	1. atrazine (1–4) simazine (1–4) 2. prometryne (2)	(—)	1. Temporarily decreased azotobacter population. 2. Increased population.	Krykhanov and Pashkovskaya (1967)
	atrazine (10)	Medium to heavy sandy clay, 1–1·6% o.m.	Decreased azotobacter population.	Peshakov et al. (1969)
	atrazine (8) simazine (10)	Alluvial	Increased populations.	Goguadze (1969)
	1. atrazine (0·5) 2. simazine (0·5)	Red sandy loam	1. Increased azotobacters. 2. Inhibited azotobacters.	Spirama Raju and Rangaswami (1971)
	atrazine (4)	Heavy clay degraded chernozem, >2% o.m.	Annual applications (7 yr) increased azotobacter population.	Husarova (1972a)
	atraton (3) prometryne (2–3) trietazine (5)	Leached medium loam chernozem, pH 6; 4% o.m.	Increased azotobacter population.	Mikhailova (1968)
	simazine (12–14)	1. Alluvial krasnozem, pH 7 2. Acid soil, pH 5	1. Increased N$_2$-fixing capacity of azotobacters and Beijerinckia spp. 2. Inhibited azotobacters.	Goguadze (1968)

Herbicide (rate)	Soil	Effect	Reference
simazine (3·0)	Alluvial Smonitsa	Decreased azotobacter population.	Micev and Bubalov (1969a)
simazine (4·0)		Increased numbers and N$_2$-fixing capacity of azotobacters.	Bakalivanov and Nikolova (1972)
simazine (4; 200)	Chernozem	Low rate increased azotobacter population. High rate decreased population.	Yurkevich and Tolkachev (1972a)
simazine (1–4)* prometryne (0·5–1·5)*	Leached meadow-cinnamon and alluvial soils	No effect on numbers of *A. chroococcum* and *Cl. pasteurianum*.	Rankov et al. (1966)
simazine (8)	Irrigated brown soil	Decreased azotobacter population.	Akopyan and Agaronyan (1968)
simazine (3)	Alluvial	Decreased populations in soil but increased populations in rhizosphere of crop.	Micev and Bubalov (1969b)
simazine (1·1; 30; 40)	(–)	Lowest rate had no effect. High rates decreased azotobacter population.	Singh (1971)
simazine (4·5; 450)	(–)	No effect.	Misra and Gaur (1971)
simazine (4)	Chernozem	Inhibited azotobacters.	Yurkevich and Tolkachev (1972b)
prometryne (2·5)	Irrigated light chestnut, 2% o.m.	Decreased azotobacter population.	Shkola (1970)
prometryne (2–4)	Light loam pale chestnut, 1·7% o.m.	Decreased azotobacter population.	Lazareva and Blokhin (1969)
prometryne (5–7)	Cinnamon, alluvial and skeletal soils	Decreased azotobacter population, but no loss of N$_2$-fixation.	Goguadze and Gogiya (1971)
prometryne (2)	Leached medium loam chernozem	Increased azotobacter population.	Matsneva and Semikhatova (1973)

* "Recommended field rate".
Refer to *Notes on tabulated data and abbreviations* (p. 319).

TABLE 6 (continued)

Herbicide		Soil type	Effects	Reference
Group	Concentration (Kg or litre ha^{-1})			
Triazines (continued)	terbutryne (2·25; 4·5; 6·75)	Alluvial or bog-alluvial soils	Two lowest rates had no effect on azotobacters. High rate sometimes inhibited.	Mickovski (1967)
Ureas	linuron (0·5–2·5)*	Leached meadow-cinnamon and alluvial soils	No effect on numbers of A. chroococcum and Cl. pasteurianum.	Rankov et al. (1966)
	linuron (0·5–2) monolinuron (0·5–2)	Medium to heavy sandy clay chernozem, 1–1·6% o.m.	Increased azotobacter population.	Peshakov et al. (1969)
	linuron (1·5) monolinuron (1·5) monolinuron (100)	Light loam podzol (–)	Decreased numbers of A. chroococcum but later increased them. Prolonged exposure decreased azotobacter population.	Kozaczenko and Sobieraj (1973) Beck (1969)
	1. monuron (1·5) diuron (1·5) 2. fenuron (1·5)	Meadow soil	1. No effect. 2. Decreased azotobacter population.	Tulabaev and Tamikaev (1968)
	monuron (8–10) diuron (6–8)	1. Alluvial kras-nozem, pH 7 2. Acid soil, pH 5	1. No effects. 2. Inhibition of azotobacters.	Goguadze (1968)
	monuron (15)	(–)	Annual applications (5 yr) decreased azotobacter population.	Pantera (1972)
	monuron (5–30)*	(–)	Decreased azotobacter population but recovered within 3 months.	Krykhanov and Pashkovskaya (1967)

monuron (8), diuron (8)	Irrigated brown soil	Decreased azotobacter population.	Akopyan and Agaronyan (1968)
monuron (8)	Alluvial	Increased populations.	Goguadze (1969)
diuron (0·5)	Red sandy loam	Increased azotobacter population.	Spirama Raju and Rangaswami (1971)
diuron (0·15–0·75)	(–)	Decreased *A. chroococcum* population.	Sharma and Saxena (1971)
DCU (5; 15)	Leached chernozem	Low rate had no effect. High rate decreased azotobacter population.	Lobanov and Poddubnaya (1967)
DCU (18)	Pale chestnut light loam, pH 8; 2% o.m.	No effect.	Chulakov and Zharasov (1967)
fluometuron (1·5–3), noruron (1·5–3)	Serozem-meadow	Low rate temporarily decreased azotobacter population, but recovered later. High rate inhibited azotobacters.	Tulabaev (1972)
siduron (11·25)	Turfed soil	Annual applications (3 yr) decreased azotobacter population.	Fields and Hemphill (1968)
Mixtures Alipur (–)	Leached meadow-cinnamon and alluvial soils	At recommended field rates, no effect on *A. chroococcum* or *Cl. pasteurianum*.	Rankov *et al.* (1966)
Alipur (4)	Light loam podzol	Initially decreased numbers of *A. chroococcum*, then increased them.	Kozaczenko and Sobieraj (1973)
Alipur (1·2–2)	Light loam pale chestnut, 1·7% o.m.	Decreased azotobacter population.	Lazareva and Blokhin (1969)
Alipur (1·2), eptam (5) + prometryne (2)	Leached medium loam chernozem	Increased azotobacter populations.	Matsneva and Semikhatova (1973)

* "Recommended field rate".
Refer to *Notes on tabulated data and abbreviations* (p. 319).

TABLE 6 (continued)

Group	Herbicide Concentration (Kg or litre ha^{-1})	Soil type	Effects	Reference
Mixtures (continued)	Agelon (2·5)	Heavy clay degraded chernozem, >2% o.m.	Annual applications (7 yr) had little effect.	Husarova (1972a)
	simazine (3) + 2,3,6–TBA (1)	Pale chestnut soils	Increased azotobacter populations.	Malanina and Nikitin (1971)
	simazine (1·5) + terbutryne (1·5)	Alluvial	Slight inhibition of azotobacters in soil but increased population in the rhizosphere of the crop.	Micev and Bubalov (1969b)
	aminotriazole 38% + simazine 20% + MCPA 16% (10)	Medium to heavy sandy clay leached chernozem, 1–1·6% o.m.	No effect.	Peshakov et al. (1969)
	Gesaran (2·25; 4·5; 6·75) Lumeton (3; 6; 9)	Alluvial or bog-alluvial soil	Two lowest rates had no effect.	Mickovski (1967)
	Gesaprim (3) VCS 438 (1) + dicamba (0·5) dalapon (20) + 2,4-DA (2)	Alluvial	Little to no effect.	Micev and Bubalov (1969c)
	dalapon (60) + 2,4-DA (6)	Podzolic light loam; moderate loam; podzolic compact; thick heavy loam	Increased N$_2$-fixation during first 10 days.	Bliev (1973)

Others	Murbetol (40; 80)	Leached chernozem	Increased azotobacter population and numbers of *Cl. pasteurianum*	Lobanov and Poddubnaya (1967)
	methurin (1·5–3)	Serozem-meadow	Low rate temporarily decreased azotobacter population, then increased it. High rate inhibited azotobacters.	Tulabaev (1972)
	KCNO	Leached meadow-cinnamon and alluvial soils	No effect on numbers of *A. chroococcum* or *Cl. pasteurianum*	Rankov *et al.* (1966)

* "Recommended field rate".
Refer to *Notes on tabulated data and abbreviations* (p. 319).

TABLE 7

Effects of herbicides on fungal and actinomycete populations in soil

Group	Herbicide Concentration (Kg or litre ha^{-1})	Soil type	Effects	Reference
Acylamide	diphenamid (6; 8) propanil (2; 3; 4)	Red and alluvial Paddy	Decreased fungi. Populations of *Aspergillus* and *Fusarium* spp. decreased; later, increased.	Mickovski (1972) Chowdhury *et al.* (1972)
	propachlor (7)	Heavy clay degraded chernozem, >2% o.m.	Applied annually (7 yr) decreased fungi.	Husarova (1972a)
Phthlates	1. endothal (2; 20) 2. endothal di-NN-di-methylcocoamine (10)	1. Six soil types 2. Sandy	1. No effect. 2. Increased moulds.	Au (1968)
Anilines	trifluralin (0·9–419)	Alluvial meadow	Variable effect on numbers of fungi and actinomycetes.	Rankov (1971b)
	trifluralin (1·1)	Sandy loam	Fungi and actinomycetes increased following annual applications (5 yr).	Breazeale and Camper (1970)
	trifluralin (3; 9; 30)	Bog and peaty-gley	Lowest rate had no effect. Higher rates decreased fungi and actinomycetes.	Tyunyaeva *et al.* (1974)
	p-chloroformanilide (20–100)	Grey serozem, 65% MHC	Increased actinomycetes. Up to 50 kg increased fungi.	Khikmatov *et al.* (1969)
	dinitramine (0·1; 0·2; 0·4)	Silty loam	Increased actinomycetes.	Belles (1972)
Benzoics and related	dichlobenil (7·5)	Chernozem-smolnitza	Inhibited fungi and actinomycetes for up to 90 days.	Nikolova and Bakalivanov (1972)

dicamba (0·12)	Derno-podzolic medium loam	Increased or decreased fungi, according to moisture content of soil.	Bezuglov et al. (1973)
nitrofen (1·5)	Sandy	Increased fungi.	Gowda and Patil (1972)
nitrofen (2; 4; 6)	Paddy	Populations of Aspergillus and Fusarium spp. decreased. Later, increased.	Chowdhury et al. (1972)
Carbamates and carbanilates			
EPTC (3–6)*	(—)	Decreased actinomycetes.	Rakhimov and Rybina (1963)
EPTC (4)	Derno-podzol light or medium loams, pH 6; 2–3% o.m.	Decreased fungi and sometimes actinomycetes.	Chunderova et al. (1968)
EPTC (4–6)	Loamy chernozem, 5–7% o.m.	Decreased fungi.	Balkova and Semikhatova (1969)
EPTC (4–5)	Pale chestnut soil	Penicillium spp. and Aspergillus niger predominated after treatment.	Zharasov et al. (1972)
EPTC (4–5)	Leached medium loam chernozem	No consistent effect on fungi.	Matsneva and Semikhatova (1973)
EPTC (3; 4·5; 18)	Dark chestnut heavy loam, 4% o.m.	Decreased fungi. Actinomycetes not affected except at highest rate.	Zharasov (1972)
chlorpropham (0·1–1)*	Sandy clay loam	Inhibited actinomycetes.	Taha et al. (1972)
chlorpropham (0·46; 45·8)	Volcanic clay loam	Decreased fungi and actinomycetes (mostly at high rate).	Matsuguchi and Ishizawa (1969)
molinate (0·75–1·5)	Pale sandy loam chernozem	No effect; but at 8–10 times field rate, fungi and actinomycetes were sensitive.	Chulakov and Zharasov (1973)
1. Carbyne (140·8)	Sandy loam, pH 5; 7·1% o.m.	1. Numbers decreased, spectrum of species narrowed. Penicillia predominated.	Quilt et al. (in press)

* "Recommended field rate".
Refer to Notes on tabulated data and abbreviations (p. 319).

TABLE 7 (continued)

Group	Herbicide Concentration (Kg or litre ha^{-1})	Soil type	Effects	Reference
Carbamates (continued)	2. Solvents of Carbyne (140·8), barban (20)	Sandy loam, pH 5; 7·1% o.m.	2. Smaller decrease in numbers.	Quilt et al. (in press)
Diazines	1. phenazone (3–6)	1. Leached derno-carbonate and derno-podzol	1. No effect on actinomycetes.	Mezharaupe (1967)
	2. phenazone (4–6)	2. Leached derno-carbonate	2. Inhibited fungi.	
	3. phenazone (8–18)	3. Leached derno-carbonate	3. Inhibited fungi and actinomycetes.	
	pyrazon (4)	Derno-podzol light or medium loam, pH 6; 2–3% o.m.	Fungi were sensitive.	Chunderova et al. (1968)
	pyrazon (1–4)*	Chernozem medium clay pH 7·2; 1·5% o.m.	No effect.	Kulinska (1967a)
	pyrazon (4)	Medium loam derno-podzol, 1·5% o.m.	Increased actinomycetes. Fungi decreased.	Karasavich and Isaeva (1969)
	pyrazon (3; 4·5)	Dark chestnut heavy loam, 4% o.m.	Decreased fungi. Actinomycetes were decreased by 4–8 times the rates shown.	Zharasov (1972)
	pyrazon (4·8)	Podzolic light loam	Decreased fungi and actinomycetes.	Pranaitene (1970)
	pyrazon (3)	Light loam derno-podzol	Annual applications (5 yr) had no effect.	Chunderova et al. (1971)
	pyrazon (4·5)	Pale chestnut soil	Penicillium spp. and Aspergillus niger	Zharasov et al. (1972)

Halogenated aliphatics	TCA (8)	(—)	Increased fungi and actinomycetes.	Zhabyuk (1973)
	TCA (30–360) dalapon (0·5–360)	Light brown soil	Toxic to actinomycetes.	Zhakharyan (1970)
	TCA (12–18)	Pale chestnut soil	*Penicillium* spp. and *A. niger* predominated after treatment.	Zharasov *et al.* (1972)
	dalapon (0·3–1·5)* dalapon (8)	Meadow-type medium loam, 1·5% o.m.	Decreased actinomycetes. Decreased fungi and actinomycetes.	Rakhimov and Rybina (1963) Tulabaev (1970)
	dalapon (0·15–0·75)	Red clay	No effect on fungi; actinomycetes increased.	Sharma and Saxena (1972)
	dalapon (20)	Irrigated brown soil, pH 7–8	Decreased fungi but increased actinomycetes.	Akopyan *et al.* (1974)
Phenols	dinoseb acetate (1·5–2·5)*	(—)	Increased fungi and actinomycetes.	Kaszubiak (1970)
	dinoseb acetate (0·8–400)	Alluvial meadow	Variable effect on numbers of fungi and actinomycetes.	Rankov (1971b)
	dinoseb acetate (1·2–2)	Sandy loam	Decreased actinomycetes; selective effect on fungi.	Hoffman-Kakol and Kwarta (1972a)
	PCP (12–120; 2–20)	Alluvial clay loam, pH 5·7; volcanic clay loam	Low rates increased actinomycetes. High rates decreased fungi and actinomycetes. Sometimes low rates decreased both groups.	Ishizawa and Matsuguchi (1966)
	PCP (4)	Six soil types	Decreased fungi and actinomycetes.	Au (1968)
	PCP (1·5)	Leached chernozem	Increased fungi but decreased actinomycetes.	Tyagny-Ryadno (1967)
	DNOC (1–10)*	Loam	Annual applications had no effect.	Rintelen (1968)
	DNOC (1–10)*	Loam	Temporarily decreased populations but soon recovered.	Resz (1968)

* "Recommended field rate".

Refer to *Notes on tabulated data and abbreviations* (p. 319).

TABLE 7 (continued)

Group	Herbicide Concentration (Kg or litre ha^{-1})	Soil type	Effects	Reference
Phenols (continued)	DNOC (2–3)	Light loam derno-podzol	Annual applications (5 yr) had no effect on fungi or actinomycetes.	Chunderova et al. (1971)
	DNPP (field rates)	Chernozem medium clay, pH 7.2; 1.5% o.m.	No effect on actinomycetes.	Kulinska (1967a)
Phenoxy-acids	2,4-D (22.5)	Sandy loam	No effect on fungi or actinomycetes following 5 applications of 4.5 kg ha^{-1} yr^{-1} for 4 years.	Breazeale and Camper (1970)
	2,4-D (2)	Chernozem	Initially decreased and then increased fungi and actinomycetes.	Yurkevitch and Tolkachev (1972a)
	2,4-D (0.3–2.3)* MCPA (0.3–2.3)*	Loam	Annual applications had no effect on fungi or actinomycetes.	Rintelen (1968)
	2,4-D (5–20) MCPA (5–20) MCPB (5–20)	Alluvial	Lower rates adversely affected numbers of Penicillium and Fusarium spp. but recovered in 60 days. Effects with higher rate lasted longer.	Mickovski et al. (1968)
	2,4-D (0–0.45; 0.45–4.5)	Red clay	Increased fungi but decreased actinomycetes. Rates higher than 0.45 kg were toxic to fungi.	Sharma and Saxena (1974)
	2,4-D (0.3–2.3)*	Peaty-bog, pH 5.5; 47% o.m.	Reduced activity of actinomycetes for 1 month.	Nepomiluev and Kuzyakina (1967)
	2,4-D (0.3–2.3)* MCPA (0.3–2.3)*	Loam	Temporarily decreased populations but soon recovered.	Resz (1968)
	2,4-D (0.3–2.3)* 2,4,5-T (0.5–3)* MCPA (0.3–2.3)*	Chernozem medium clay, pH 7.2; 1.5% o.m.	No effect on actinomycetes.	Kulinska (1967a)

	Compound (rate*)	Soil	Effect	Reference
	2,4-D (1·5)	Leached chernozem	No effect on fungi or actinomycetes.	Tyagny-Ryadno (1967)
	2,4-DEP (0-8)	Light loam pale chestnut soil, 1·7% o.m.	Decreased fungi and actinomycetes.	Lazareva and Blokhin (1969)
	2,4-D amine (0·9)	Dark chestnut loam, 3·5% o.m.	Decreased fungi and actinomycetes.	Geshtovt et al. (1970)
	2,4-D (0·5–0·75) MCPA (1·0–2·5) MCPB (2·0)	Light loam derno-podzol	Annual applications (5 yr) had no effects on fungi or actinomycetes.	Chunderova et al. (1971)
	2,4-D (1·1; 30; 40)	(—)	Lowest rate increased, higher rates decreased fungal population.	Singh (1971)
	2,4-D (1–2) MCPA (1–2;	Alluvial	Increased fungi and actinomycetes.	Bakalivinov and Nikolova (1971)
	2,4-D (0·4) MCPA (1–1·5)	Pale sandy loam Sandy loam	No effect on fungi or actinomycetes. No effect on fungi or actinomycetes.	Chulakov and Zharasov (1973) Hoffman-Kakol and Kwarta (1972b)
Phthalaics	chlorthal dimethyl (11)	Silt loam and silty clay loam	No effect on fungi or actinomycetes.	Fields et al. (1967)
	chlorthal dimethyl (21·6)	New York and Colorado soils	Annual applications (5 yr) increased actinomycetes	Tweedy et al. (1968)
	chlorthal dimethyl (1·5–3)	Serozem-meadow soil	Low rates decreased then increased fungi and actinomycetes. High rate increased fungi.	Tulabaev (1972)
	chlorthal dimethyl (4·5; 13·5)	(—)	Decreased fungi but not actinomycetes.	Bone and Kuntz (1968)
Pyridines	picloram (2)	Alluvial	Decreased fungi.	Goguadze et al. (1972)
	picloram (0·07)	Derno-podzol medium loam	Increased or decreased populations depending on soil moisture content.	Bezuglov et al. (1973)

* "Recommended field rate".
Refer to *Notes on tabulated data and abbreviations* (p. 319).

TABLE 7 (continued)

Group	Herbicide Concentration (Kg or litre ha^{-1})	Soil type	Effects	Reference
Quaternaries	paraquat (0·6)	Silt loam, pH 6·3; 0·11% o.m.	Decreased mould population but increased *Streptomyces* spp.	Tu and Bollen (1968b)
	paraquat (5·6; 112)	Fine sand	Low rate increased fungi but high rate decreased.	Camper *et al.* (1973)
	paraquat (1·14; 11·4) Gramoxone (5·6; 56)	Calcareous loam, pH 8; 7·5% o.m.	Fungal and actinomycete populations fluctuated slightly.	Anderson and Drew (1976a,c)
Triazines	atrazine (1; 3; 5) simazine (1; 3; 5)	Forest soil	Delayed increase in populations of fungi and actinomycetes.	Milkowska and Gorzelak (1966)
	atrazine (4)	Loamy sand, pH 5·3; 2·44% humus	Annual applications (13 yr) had no effect on fungi.	Voets *et al.* (1974)
	1. atrazine (3) 2. simazine (3)	Medium loam, meadow soil 1·2–2·2% o.m.	1. Decreased fungal population. 2. Increased fungal population.	Tulabaev and Azimbegov (1967)
	atrazine (4·5; 10) simazine (4·5; 10)	Peaty-bog, pH 5·5, 47% o.m.	Actinomycetes increased in wet seasons but decreased in dry years. Fungi equally prevalent in wet and dry years.	Nepomiluev and Kuzyakina (1967)
	atrazine (5·2) simazine (5·2)	(—)	No effect on numbers of fungi but species variation occurred.	Fink *et al.* (1968)
	atrazine (3; 6) simazine (3; 6)	Medium clay loam pH 5·3; 2% humus	No effect.	Tsvetkova (1966)
	atrazine (1–4)* simazine (1–4)*	Chernozem medium clay, pH 7·2; 1·5% o.m.	No effect on actinomycetes.	Kulinska (1967a)

Herbicide (kg/ha)	Soil type	Effect	Reference
atrazine (10)	...odzol, 1·6 o.m. Medium to heavy sandy clay chernozem, 1-1·6% o.m.	and affected pigments. No effect on fungi but increased actinomycetes	Peshakov et al. (1969)
atrazine (8; simazine (10)	Alluvial	Decreased actinomycetes; increased fungi.	Goguadze (1969)
atrazine (1-5) simazine (1-5) propazine (1-5) prometryne (1-5)	Derno-podzolic soils	Decreased fungi, except in loam soil. Propazine eliminated *Fusarium* sp. and simazine reduced it in light loams. Heavier soils, less effect.	Gorlenko et al. (1969)
atrazine (0·5) simazine (0·5) atrazine (4)	Red sandy loam Heavy clay degraded chernozem	Decreased actinomycetes. Only atrazine increased fungi. Annually applied for 7 years, decreased populations of fungi and actinomycetes.	Spirama Raju and Rangaswami (1971) Husarova (1972a)
atrazine (4; 10)	Heavy loam chernozem	Decreased fungi.	Pilipets and Litvinov (1972)
simazine (3) simazine (4)	Alluvial Smonitza	No effect on *Stachybotrys* spp. Decreased fungi and actinomycetes.	Micev and Bubalov (1969a) Bakalivinov and Nikolova (1972)
simazine (4)	Chernozem	Initially decreased, then increased fungi and actinomycetes.	Yurkevich and Tolkachev (1972a)
simazine (1-50)*	Red sandy loam	Increased actinomycetes. No clear effect on fungi.	Balasubramanian and Siddaramappa (1974)
simazine (1·5-2·5) simazine (2; 50)	Sod-podzol Chernozem medium clay, pH 7·2; 1·5% o.m.; sandy soil, pH 5·8; 0·9% o.m.	Decreased fungi and actinomycetes. No effect on fungal or actinomycete populations or their antibiotic properties.	Nepomiluev et al. (1966b) Kulinska (1967b)

* "Recommended field rate".
Refer to *Notes on tabulated data and abbreviations* (p. 319).

TABLE 7 (continued)

Group	Herbicide Concentration (Kg or litre ha^{-1})	Soil type	Effects	Reference
Triazines (continued)	simazine (1·5)	Leached chernozem	Increased growth of fungi but suppressed actinomycetes.	Tyagny-Ryadno (1967)
	simazine (8)	Irrigated brown soil	Decreased fungi.	Akopyan and Agaronyan (1968)
	simazine (0·75)	Podzolized light loam	Decreased fungi.	Pranaitene (1970)
	simazine (2–4)	1. Pale grey forest soil	1. Retarded.	Khrushcheva (1971)
		2. Degraded chernozem	2. Increased growth of mycorrhizal fungi on roots.	
	simazine (1·1; 30; 40)	(–)	Lowest rate increased fungal population. Higher rates decreased it.	Singh (1971)
	simazine (0·5–10)	Chernozem smolnitza	Decreased actinomycete population with increasing concentration.	Vlagov et al. (1972b)
	simazine (1–4)*	Sandy soil	Decreased, then increased actinomycetes.	Gowda and Patil (1972)
	simazine (4·5)	Several soil types	Increased numbers of mycorrhizal fungi on roots.	Smith (1973)
	simazine (2–100)	Derno-podzol	No effect on fungi. Actinomycetes decreased by rates of 8 kg or more.	Kuzyakina (1971)
	simazine (4–10)	Acid alluvial and terra rossa soils	Actinomycetes and moulds were resistant but Fusaria were susceptible.	Goguadze (1966)
	prometryne (2·5; 25)	1. cis-Caucasian leached chernozem	1. Increased fungi.	Vasil'ev and Enkina (1967)
		2. Siberian chernozem	2. Increased fungi but decreased actinomycetes.	

Treatment	Soil	Effect	Reference
atraton (3) trietazine (5) prometryne	loam chernozem, pH 6; 4% o.m.	...effect on fungi.	Mikhailova (1968)
prometryne (0·5-1·5)*	Leached chernozem	Altered the ratio of actinomycetes to other soil organisms.	Voinova et al. (1970)
	Rhizosphere of Serradella sp.	Increased fungi and actinomycetes.	Kaszubiak (1970)
prometryne (0·5-1·5)*	(–)	Increased fungi and actinomycetes.	Zhabyuk (1973)
prometryne (2-4)	Light loam pale chestnut soil, 1·7% o.m.	Decreased fungi and actinomycetes.	Lazareva and Blokhin (1969)
prometryne (1·0)	Loam, 3·5% o.m.	Decreased fungal and actinomycete populations.	Geshtovt et al. (1970)
prometryne (1-2)* propazine (2)	Light loam derno-podzol	Annual applications (5 yr) had no effect on fungi or actinomycetes.	Chunderova et al. (1971)
prometryne (5-7)	Cinnamon, alluvial and skeletal	Actinomycetes, Penicillium spp. not affected.	Goguadze and Gogiya (1971)
prometryne (0·25-1·0)	Paddy soil	Populations of Aspergillus, Fusarium spp. decreased but later recovered.	Chowdhury et al. (1972)
prometryne (2)	Leached medium loam chernozem	No consistent effects on fungi.	Matsneva and Semikhatova (1973)
prometryne (2)	Derno-podzolic loam	In field plots, no effect. In laboratory studies, fungi and actinomycetes were sensitive.	Samokhvalov (1973)
1. propazine (3) 2. propazine (6; 10) 3. propazine (6; 10)	1. Calcareous and alkaline chernozem 2. Calcareous chernozem 3. Saline soils	1. No effect on fungi or actinomycetes. 2. Decreased both populations. 3. Only actinomycetes decreased.	Putintsev and Putintseva (1966)
1. terbutryne 50% (2·5) 2. methoprotryne 25% (2·5)	Alluvial	1. Actinomycetes and fungi inhibited in soil but not in rhizosphere. 2. Fungi inhibited.	Micev and Bubalov (1969a)

* "Recommended field rate".
Refer to Notes on tabulated data and abbreviations (p. 319).

TABLE 7 (continued)

Group	Herbicide Concentration (Kg or litre ha^{-1})	Soil type	Effects	Reference
Ureas	linuron (0·5–2)* monolinuron (0.5–2)*	Leached chernozem	Little effect on fungi or actinomycetes.	Peshakov et al. (1969)
	linuron (0·5–2)* monolinuron (0·5–2)* fluometuron (1–2)	Leached chernozem	Altered the ratio of actinomycetes to other soil organisms.	Voinova et al. (1970)
	linuron (2) monolinuron (1·5–2) chloroxuron (5)	Alluvial meadow soil	Increased actinomycetes	Rankov (1972)
	monolinuron (100)	(–)	Decreased fungal population after prolonged exposure.	Beck (1969)
	monuron (8)	Alluvial	Decreased actinomycete and increased fungal populations.	Goguadze (1969)
	monuron (1–10)*	Chernozem medium clay, pH 7·2; 1·5% o.m.	No effect on actinomycetes.	Kulinska (1967a)
	monuron (8) diuron (8)	Irrigated brown soils	Decreased fungal population.	Akopyan and Agaronyan (1968)
	monuron (12–20) diuron (12–20)	Medium loam red soil	Little effect on fungi.	Spiridinov et al. (1970)
	monuron (8; 12–16) diuron (8; 12–16)	Light clay, 3·1–3·5% o.m.	Decreased fungi.	Spiridinov et al. (1972)
	monuron (8–20) diuron (8–20)	Meadow bog soil and red krasnozem	Decreased fungal populations and activity.	Spiridinov et al. (1973)

439

Herbicide (rate)	Effect	Soil	Reference
fenuron (2–30)*			
1. monuror (60)	1. Decreased fungi and actinomycetes. 2. less effect on both groups.	Silt and water of drains	Gel'tser et al. (1972)
2. diuron (60) monuron (1–10)* neburon (2–3)*	Annual applications (5 yr) did not affect fungi or actinomycetes.	Light loam derno-podzol (—)	Chunderova et al. (1971)
diuron (0·15–0·75)	Fungi and actinomycetes increased by rates up to 0·3. Actinomycetes decreased at higher rates.		Sharma and Saxena (1971)
diuron (0·5)	Fungal population increased but actinomycetes decreased.	Red sandy loam	Spirama Raju and Rangaswami (1971)
fluometuron (0·5)	Fungi increased; actinomycetes decreased.	Sandy loam	Trull (1969)
fluometuron (8–10)	Actinomycetes and Penicillium spp. not affected.	Cinnamon, alluvial and skeletal	Goguadze and Gogiya (1971)
1. fluometuron (1·5–3)	1. Decreased fungal population.	Serozem-meadow soil	Tulabaev (1972)
2. noruron (1·5–3; 1·3–)	2. Affected species composition. Decreased fungal populations.	Sandy loam soils, pH 6·1–7·4; 1·3–1·6% o.m.	Grossbard and Marsh (1974)
metoxuron (5; 50)			
siduron (11·3)	No effects on fungal or actinomycete populations.	Turfed soil	Fields and Hemphill (1968)
Uracils			
isocil (3–5)	Actinomycetes and Penicillium spp. not affected.	Cinnamon, alluvial and skeletal	Goguadze and Gogiya (1971)
lenacil (1·2–1·7)	No effect on fungi and actinomycetes.	Dark chestnut heavy loam and pale chestnut light loam	Chulakov and Zharasov (1973)

* "Recommended field rate".
Refer to *Notes on tabulated data and abbreviations* (p. 319).

TABLE 7 (continued)

Group	Herbicide Concentration (Kg or litre ha^{-1})	Soil type	Effects	Reference
Mixtures	simazine + aminotriazole (10) Agelon (3)	Orchard soil	Decreased actinomycete populations after 2 yr.	Teuteberg (1968)
		Alluvial	Fungi increased but actinomycetes decreased.	Micev and Bubalov (1972)
	simazine (1·5) + terbutryne (1·5) Alipur (1·2–2)	Alluvial	No effect on *Stachybotrys* spp.	Micev and Bubalov (1969b)
		Light loam pale chestnut soil, 1·7% o.m.	Fungal and actinomycete populations decreased.	Lazareva and Blokhin (1969)
	1. Alipur (0·6; 1·8)	Sod-podzol loam	1. Fungal population retarded by high rate.	Isaeva and Novogrudskaya (1966)
	2. Murbetol (7·6; 9·1)		2. Fungal population retarded; actinomycetes decreased by high rate only.	
	propazine (10) + thiram (4) + γ-BHC (4)	Calcareous saline chernozem	Decreased fungal and actinomycete populations.	Putintsev and Putintseva (1966)
	VCS-438 (1·0) + dicamba (0·5)	Alluvial	Actinomycete populations initially decreased; subsequently increased.	Micev and Bubalov (1969c)
	Lumeton (3)	Alluvial	Actinomycete populations initially decreased; subsequently increased.	Micev (1970)
	Lumeton (4)	Derno-podzolic medium loam	Decreased fungi.	Bezuglov et al. (1973)

		Leached chernozem	No consistent effects on fungi.	Matsneva and Semikhatova (1973)
	Eptam (5) + prometryne (2) Alipur (1:2)			
	atrazine (1·6) + prometryne (0·8)	Heavy clay degraded chernozem, >2% o.m.	Annual applications (7 yr) decreased fungi and actinomycete populations.	Husarova (1972a)
	Ustinex GL (10)	Loess loam	No effect on fungi.	Teuteberg (1970)
	CaCN$_2$ (200–400)*	Loam	Annual applications had no effect on fungi.	Rintelen (1968)
Others	CaCN$_2$ (200–400)*	Loam	Annual applications (11 yr) had no effect on fungi or actinomycetes.	Resz (1968)
	sodium cacodylate (10·2; 30·7)	Alluvial loam	Decreased fungal population.	Malone (1971)
	methurin (1·5–3)	Serozem-meadow	Decreased fungal population and inhibited activity of actinomycetes.	Tulabaev (1972)

* "Recommended field rate".
Refer to *Notes on tabulated data and abbreviations* (p. 319).

TABLE 8

Effects of herbicides on phytopathogenic fungi and their antagonists

Group	Herbicide	Concentration[a]	Soil type, or media conditions and organisms	Effects	Reference
Acylamides	propachlor	1. 100–300 ppm 2. 400–500 ppm	Liquid culture (Sclerotium rolfsii)	1. Sclerotia germinated well. 2. Sclerotia germination suppressed.	Clerk and Bimpong (1969)
Anilines		trifluralin (0·06–4·0 kg)	Sandy loam and clay soils, (Fusarium oxysporum f. sp. vasinfectum)	Increased chlamydospore production and germination.	Curl et al. (1969); Tang et al. (1970)
		1. trifluralin (0·01–1·0 ppm) 2. trifluralin (0·625–10 kg)	1. Liquid culture (S. rolfsii) 2. Sandy loam (S. rolfsii)	1. Mycelial dry-weight and glucose, phosphate and nitrate utilization decreased. 2. CO_2 evolution increased by lowest rates.	Rodriguez-Kabana et al. (1969)
		trifluralin (6·25 ppm)	Sandy loam (S. rolfsii)	Decreased respiration.	Curl et al. (1967)
		trifluralin (0–100 ppm)	Liquid culture (F. oxysporum f. sp. vasinfectum)	Little effect on mycelium production, glucose or nitrate utilization.	Rodriguez-Kabana and Curl (1972)
		1. trifluralin (0·05–0·35 kg) 2. trifluralin (0·5–5·0 ppm)	1. Sandy loam and sand (Rhizoctonia solani) 2. Liquid culture (Rhizoctonia solani)	1. Increased incidence of disease on cotton seedlings. 2. Inhibited growth with increasing rate.	Neubauer and Avizohar-Hershenson (1973)

Group	Herbicide (concentration)	Soil (pathogen)	Effect	Reference
	trifluralin (0·84 kg)	Sandy and clay soils (*F. oxysporum* f. sp. *pisi* and *R. solani*)	Susceptibility of peas to both pathogens increased.	Grau (1974)
	trifluralin (0·1 kg) benefin (0·1 kg) butralin (0·1 kg) dinitramine (0·1 kg) isopropalin (0·1 kg) nitralin (0·1 kg)	Loamy sand (*R. solani, F. oxysporum*, f. sp. *lycopersici*; *Verticillium dahliae*)	Generally increased resistance of plants to pathogens. Benefin and isopropalin had no effect.	Grinstein *et al.* (1976)
Benzoics and related	DMPA (10; 17 kg)	Sterile and non-sterile soil (*Pythium debaryanum*)	Decreased incidence of damping-off in peas. 8 annual applications of 17 kg did not increase virulence of pathogen in turf.	Fields and Hemphill (1967)
Carbamate and carbanilate	1. EPTC (10–100 ppm)	1. Liquid culture (*S. rolfsii*)	1. Decreased mycelium formation and glucose, nitrate and phosphate utilization.	Rodriguez-Kabana *et al.* (1970)
	2. EPTC (0·1–1·0 kg)	2. Sandy loam (*S. rolfsii*)	2. Growth increased by 0·5 and 1·0 kg rates.	
	1. EPTC (1–40 ppm)	1. Liquid culture (*S. rolfsii, Trichoderma viride*)	1. *S. rolfsii* mycelium formation increased by 1–10 ppm; decreased by 40 ppm. *T. viride* not affected.	Peeples and Curl (1970)
	2. EPTC (0·1–4·0 kg)	2. Sandy loam (*S. rolfsii*)	2. Growth increased.	
	EPTC (0·25–2·0 kg)	Sandy loam (*S. rolfsii, Sclerotina trifoliorum*)	Inhibited mycelium formation with increasing concentration.	Curl and Rodriguez-Kabana (1971)

Refer to *Notes on tabulated data and abbreviations* (p. 319).

[a] Concentrations: kg ha^{-1} or ppm.

TABLE 8 (continued)

Group	Herbicide Concentration[a]	Soil type, or media conditions and organisms	Effects	Reference
Carbamate and carbanilate (continued)	EPTC (0–100 ppm)	Liquid culture (*F. oxysporum* f. sp. *vasinfectum*)	Increased mycelium formation up to 25 ppm EPTC but decreased at higher rates. Utilization of glucose and nitrate decreased at higher rates.	Rodriguez-Kabana and Curl (1972)
	asulam (1·5 kg)	Mushroom casing (*Verticillium malthousei*)	Decreased growth.	Poppe (1972)
Halogenated aliphatics	dalapon (7·5 kg)	Mushroom casing (*Verticillium malthousei*)	Decreased growth.	Poppe (1972)
Phenoxy-acids	2,4-D (1–1000 ppm) 2,4,5-T (1–1000 ppm)	Culture (*Curvularia lunata*)	No increase in spore germination.	Saxena and Gupta (1969)
	2,4-D (0·005–1·0 ppm) TPA (0·005–1·0 ppm)	Agar and liquid media (*Ceratocystis ulmi*)	Inhibited linear growth of pathogen. Increased or decreased some pectolytic enzymes depending on rate.	Sekhon *et al.* (1974)
Quaternaries	paraquat (1-25–100 kg)	Sandy loam (*S. rolfsii*.	CO_2-evolution decreased by all rates. Nitrate assimilation decreased at high rates.	Curl *et al.* (1967)
	paraquat (0–100 ppm)	Liquid culture (*F. oxysporum* f. sp. *vasinfectum*)	Increased mycelium formation. Least increase at 75 ppm. Highest rates decreased glucose and nitrate utilization.	Rodriguez-Kabana and Curl (1972)

	paraquat (0·25–2·0 kg)	Sandy loam (S. rolfsii, S. trifoliorum)	Little effect on sclerotial formation.	Curl and Rodriguez-Kabana (1971)
	paraquat (1·0)	Mushroom casing (V. malthousei, Mycogone perniciosa)	Decreased growth.	Poppe (1972)
	paraquat (10–1000 ppm) Gramoxone (10–1000 ppm a.i.)	Agar culture (Septoria nodorum, S. tritici)	Suppressed mycelial growth, spore production and germination.	Jones and Williams (1971)
	atrazine (100–1000 ppm) prometryne (100–1000 ppm)	Culture (S. rolfsii)	At 100–300 ppm, sclerotia germinated well. Above 400 ppm, germination suppressed.	Clerk and Bimpong (1969)
Triazines	atrazine (0·3–8·0 kg)	Sandy loam, pH 6·4 (S. rolfsii, T. viride)	Maximum CO_2 evolved from S. rolfsii at 0·8 kg rate. Higher rates decreased CO_2. T. viride grew well on all rates. CO_2 evolved increased with all rates above 0·8 kg.	Rodriguez-Kabana et al. (1968)
	atrazine (0·2–2·0 kg)	Sandy loam (S. rolfsii)	Addition of atrazine to soil plus glucose decreased growth. In absence of added glucose effects of atrazine were slight.	Pitts et al. (1970)
	1. atrazine (8–80 ppm)	1. Liquid culture (F. oxysporum f. sp. vasinfectum)	1. Ratio of CO_2 released: NO_3^- utilized was highest for 20–80 ppm rates.	Rodriguez-Kabana and Curl (1970)
	2. atrazine (0·8–8·0 kg)	2. Sandy loam (F. oxysporum f. sp. vasinfectum)	2. $CO_2:NO_3^-$ ratio highest for 2–8 kg rates.	

Refer to *Notes on tabulated data and abbreviations* (p. 319).
[a] Concentrations: kg ha^{-1} or ppm.

TABLE 8 (continued)

Group	Herbicide Concentration[a]	Soil type, or media conditions and organisms	Effects	Reference
Triazines (continued)	simazine (0·5–10 kg)	Chernozem smolnitza (various actinomycetes)	Actinomycetes attacking Fusaria were most tolerant; those attacking *Bacillus tumefaciens* were least tolerant.	Vlagov *et al.* (1972b)
	simazine (0·5–10 kg)	Chernozem smolnitza (various fungi)	Inhibition of "useful" fungi increased with rate but no effect on others.	Vlagov *et al.* (1972a)
	simazine (4·5 kg)	Mushroom casing (*V. malthousei*)	Controlled growth of pathogen.	Poppe (1972)
	prometryne (0·1–8·0 kg)	Sandy loam (*F. oxysporum* f. sp. *vasinfectum*)	At 2 kg, increased respiration and nitrate utilization. At 2–8 kg, increased chlamydospore formation. Less effect at lower rates.	Chopra *et al.* (1970)
	1. prometryne (0·15–8·0 kg)	1. Sandy loam (*F. oxysporum* f. sp. *vasinfectum*)	1. Low rates had no effect but 2·0 and 8·0 kg rates increased spore density.	Chopra and Curl (1968)
	2. prometryne (0·2–2·0 kg)	2. Clay loam (*F. oxysporum* f. sp. *vasinfectum*)	2. Fewer chlamydospores germinated.	
	prometryne (0·25–2·0 kg)	Sandy loam (*S. rolfsii, S. trifoliorum*)	Little effect on sclerotia production.	Curl and Rodriguez-Kabana (1971)
Ureas	diuron (2–4 kg)	Mushroom casing (*V. malthousei, M. perniciosa*)	Decreased growth.	Poppe (1972)

diuron (0.25–2 kg) fluometuron (0.25–2 kg)	Sandy loam (*S. rolfsii, S. trifoliorum*)	Inhibited mycelial formation.	Curl and Rodriguez-Kabana (1971)
fluometuron (1–20 ppm)	Liquid culture (*Aspergillus flavus*)	Dry weight of mycelium increased with rate. CO_2 evolution increased with time.	Chopra *et al.* (1973)
1. fluometuron (50 ppm) metobromuron (50 ppm)	1. Potato dextrose agar (*S. rolfsii*)	1. Inhibited production of sclerotia. Higher rates increased size of sclerotia.	Bozarth and Tweedy (1970)
2. fluometuron (25 ppm) metobromuron (25 ppm)	2. Liquid culture (*S. rolfsii*)	2. Inhibited formation of mycelium, increased acid production and altered polygalacturonase activity.	

Refer to *Notes on tabulated data and abbreviations* (p. 319).
[a] Concentrations: kg ha^{-1} or ppm.

TABLE 9

Effects of herbicides on soil algae *in situ*

Group	Herbicide			
	Concentration (Kg or litre ha^{-1})	Soil type	Effects	Reference
Acylamide	propanil (0·1; 0·25)	Sandy and sandy loam	Inhibited *Chlorococcum aplanosporum* and decreased numbers.	Sharabi (1969)
Carbamate and carbanilate	propanil (1–4)*	Paddy soil	No adverse effects on N$_2$-fixation.	Ibrahim (1972)
	1. EPTC (3–6)*	Paddy soil	1. Adversely affected N$_2$-fixation.	Ibrahim (1972)
	2. molinate (2–4)*		2. No adverse effect.	
	1. EPTC (5)	Gleyed heavy loam, >8% o.m.; peaty-podzol, 35% o.m.; light sandy loam, >3% o.m.	1. Decreased populations after 6 days. Later, blue–green algae were stimulated.	Mikhailova and Kruglov (1973)
	2. metham-Na (500)		2. Suppressed all growth in 2 soils, but *Chlorococcum* sp. and *Phormidium* sp. were tolerant in sandy soil.	
	chlorpropham (100–120)*	Alluvial black or sandy	Decreased populations and activity.	Hauke-Pacewiczowa (1971)
Diazines	pyrazon (15–30)*	Alluvial black or sandy	Decreased populations and activity.	Hauke-Pacewiczowa (1971)
Halogenated aliphatics	TCA (15–30)*	(–)	No decrease in population in moist soil; decreases in drier soil. In pots with high soil moisture, increases were found.	Balezina (1967)
Phenols	TCA (50; 75)	Light clay	Decreased populations.	Pantera (1970)
	DNOC (1–10)*	(–)	In pots with high soil moisture content, populations increased.	Balezina (1967)
	dinoseb acetate (15–25)*	Alluvial black or sandy	Decreased populations and activity.	Hauke-Pacewiczowa (1971)
	PCP (0·8–2·0)	Alluvial clay loam,	Decreased populations of diatoms, blue–green and filamentous forms.	Ishizawa and Matsuguchi (1966)

Group	Herbicide (rate)	Soil	Effect	Reference
			rate decreased populations.	Platonova (190/b)
	MCPA (3–23)*	loam chernozem Alluvial black or sandy (–)	Decreased populations and activity.	Hauke-Pacewiczowa (1971)
	MCPA 80% (0-3–2-3)*	(–)	In pots with high soil moisture content, populations increased.	Balezina (1967)
Quaternaries	paraquat (1·14; 11·4) Gramoxone (5·6; 56)	Calcareous loam, pH 8; 7·5% o.m.	Decrease in numbers during first 24 hr, then recovery.	Anderson and Drew (1976a)
Triazines	atrazine (0–0·05)	Sandy soils, low in o.m.	Chlorella sp. sensitive.	Atkins and Tchan (1967)
	atrazine (17·4)	Light loam	Decreased populations but recovered within 3 months.	Kiss (1966)
	atrazine (3–50) simazine (3–50) prometryne (3–50)	Sandy clay	Decreased populations for up to 5 months.	Pantera (1970)
	atrazine (5) simazine (5)	Gleyed heavy loam, >8% o.m.; peaty-podzol, 35% o.m.; light sandy loam, >3% o.m.	Decreased populations, especially blue-green algae. Some diatoms and Chlorococcum sp. in sandy soil were tolerant.	Mikhailova and Kruglov (1973)
	atrazine (10–40)* simazine (10–40)* simazine (2–12)	Alluvial black or sandy (–)	Decreased populations and activity.	Hauke-Pacewiczowa (1971)
	simazine (2–12)	(–)	In pots, with high soil moisture content, decreased populations especially of blue–green algae.	Balezina (1967)
Ureas	simazine (4) prometryne (2)	Leached medium loam chernozem loam chernozem	No effect.	Platonova (1967b)
	prometryne (2–3)	loam chernozem	No effect.	Platonova (1967a)
	linuron (0·5–2·5)*	Clay loam, pH 5·4	Decreased populations.	Sivasithamparam (1970)
	linuron (5–25)*	Alluvial black or sandy	Decreased populations and activity.	Hauke-Pacewiczowa (1971)

* "Recommended field rate".
Refer to *Notes on tabulated data and abbreviations* (p. 319).

TABLE 9 (continued)

Herbicide		Soil type	Effects	Reference
Group	Concentration (Kg or litre ha^{-1})			
Ureas (continued)	monolinuron (0·05)	(−)	Decreased populations.	Beck (1969)
	monuron (5–10)*	Gleyed heavy loam, >8% o.m.; peaty-podzol, 35% o.m.; light sandy loam, >3% o.m.	Decreased populations except diatom *Navicula atomus*. Very persistent effect.	Mikhailova and Kruglov (1973)
	monuron (5·2) diuron (5·2)	Light loam	Decreased populations, but recovered within 3 months.	Kiss (1966)
	monuron (10; 20) linuron (3–50)	Light and sandy clay	Decreased populations. Monuron effect lasted 4 years.	Pantera (1970)
	siduron (11·3)	Turfed soil	Annual applications (3 yr) had no effect.	Fields and Hemphill (1968)
Mixtures	atrazine 30% + mecoprop 11·5% + 2,4,5-T 5·2% (13·9) simazine 18% + aminotriazole 34% (13·9)	Light loam	Decreased populations, but recovered within 3 months.	Kiss (1966)
	picloram (0·28; 0·56; 1·12) + 2,4-D (1·12; 2·24; 4·48)	Clay loam, pH 8·2; 2·5% o.m.	No effect.	Arvik *et al.* (1971)

| simazine 33·3%+ prometryne 16·7% | Sandy clay loam and sandy loam | Increased populations at lower doses after 3 months. | Pantera (1970) |

* "Recommended field rate".
Refer to *Notes on tabulated data and abbreviations* (p. 319).

TABLE 10

Effects of herbicides on organic matter and cellulose-decomposing microorganisms in soils

Group	Herbicide Concentration (Kg or litres ha^{-1})	Soil type	Effects	Reference
Acylamides	1. diphenamid (4–6)* 2. propachlor (3·5–5)*	Alluvial meadow, leached meadow cinnamon, meadow bog, humic calcareous	1. No significant effects. 2. Increased activity.	Rankov (1970)
Anilines	diphenamid (6–8) propanil (10·2; 17·1) trifluralin (1–4) trifluralin (3; 15)	Red and alluvial Paddy soil Alluvial Leached heavy loam chernozem	Decreased activity. Decreased populations. Increased activity. Increased population of anaerobic cellulose decomposers.	Mickovski (1972) De and Mukhopadhyay (1971) Rankov and Elenkov (1970) Gruzdev et al. (1973)
	trifluralin (0·5–1)*	Alluvial meadow, leached meadow cinnamon, meadow bog, humic calcareous	Increased activity.	Rankov (1970)
Benzoics and related	ioxynil (1·5)	Podzolized loam, pH 7	Some cellulolytic organisms inhibited.	Hauke-Pacewiczowa and Krolowa (1968)
	nitrofen (2–3)*	Alluvial meadow, leached meadow cinnamon, meadow bog, humic calcareous	No significant effect.	Rankov (1970)
Carbamates and carbanilates	EPTC (3–6)*	(—)	Increased populations.	Zhukova and Konoshevich (1967)

	EPTC (4·5)	Pale chestnut light loam, pH 8; 2% o.m.	No effect.	Chulakov and Zharasov (1967)
	EPTC (8)	Medium meadow loam, 1·5% o.m.	No effect.	Tulabaev (1970)
	EPTC (3; 4·5)	Dark chestnut heavy loam, 4% o.m.	No effect.	Zharasov (1972)
	EPTC (3–50)*	Grey podzolized soil	No effect.	Nikolenko and Geller (1969)
	1. EPTC (3–6)* 2. chlorpropham (–) pebulate (4–6)*	(–)	1. Inhibited activity 2. No effect.	Sobieszczanski (1969b)
Diazines	pyrazon (50)	Grey podzolized soil	Decreased activity.	Nikolenko and Geller (1969)
	pyrazon (4·5)	Podzolized light loam	No effect.	Pranaitene (1970)
	pyrazon (3; 4·5)	Dark chestnut heavy loam, 4% o.m.	No effect.	Zharasov (1972)
	pyrazon (1·5–3; 75–150)*	(–)	Little effect at low rates. High rate decreased activity.	Simon (1971)
Halogenated aliphatics	TCA (6)	Chernozem	No effect.	Sosnovskaya and Pashchenko (1965)
	TCA (5; 15)	Leached chernozem	No effect.	Lobanov and Poddubnaya (1967)
	TCA (18)	Pale chestnut light loam, pH 8; 2% o.m.	No effect.	Chulakov and Zharasov (1967)
	TCA (8–30; 80–300)*	(–)	Low rates had little effect. High rates decreased activity.	Simon (1971)

* "Recommended field rate".
Refer to *Notes on tabulated data and abbreviations* (p. 319).

TABLE 10 (continued)

Group	Herbicide Concentration (Kg or litre ha^{-1})	Soil type	Effects	Reference
Halogenated aliphatics (continued)	TCA (8–30; 26·5) dalapon (0·6; 6; 15)	Sandy or sandy loam soils augmented with cellulose, o.m. and (NH$_4$)$_2$SO$_4$	Effects depended upon presence or absence of o.m., cellulose, nitrogen and moisture. Low rates generally no effect; high rates decreased activity.	Eglite (1969)
	dalapon (8)	Medium meadow loam, 1·5% o.m.	No effect.	Tulabaev (1970)
Phenols	dinoseb amine (2–4) dinoseb 20% (2–4)	Light loamy medium podzol, pH 5·6–6	No effect.	Nepomiluev et al. (1966a)
	dinoseb triethanol-amine (3)	Peaty-bog, 47% o.m., pH 5·5	No effect.	Nepomiluev and Kuzyakina (1967)
	dinoseb acetate (1·5–2·5)*	Alluvial meadow, leached meadow cinnamon, meadow bog, humic calcareous	No effect.	Rankov (1970)
Phenoxy-acids	PCP (1·5)	Leached chernozem	Increased populations.	Tyagny-Ryadno (1967)
	2,4-D (1·5)	Chernozem	No effect.	Sosnovskaya and Pashchenko (1965)
	2,4-D (1·5)	Leached chernozem	Inhibited activity.	Tyagny-Ryadno (1967)
	2,4-D (0·75)	Grey forest silty medium loam, pH 5·3; 4% o.m.	In field, o.m. decomposition increased. In pots, o.m. decomposition inhibited.	Boiko et al. (1969)

Herbicide (rate)	Soil	Effect	Reference
	soil, pH 6·6–7; 0·6% o.m.	1. ...ibited activity.	Szegi (1972)
	2. Chernozem, pH 7·3–7·5; 3·5% o.m.	2. No effect.	
2,4-DEP (0–8)	Light loam pale chestnut soil, 1·7% o.m.	No effect.	Lazareva and Blokhin (1969)
2,4,5-T (4; 40) MCPA (4; 40)	Sandy clay and silty clay	High rates decreased activity.	Torstensson (1974)
MCPA (5)	Podzolic loam, pH 7	Some cellulolytic organisms inhibited	Hauke-Pacewiczowa and Krolowa (1968)
MCPA (8; 15)	Paddy soil	Decreased populations at both rates; recovery took longer with high rates.	De and Mukhopadhyay (1971)
Phthalaics			
chlorthal dimethyl (1·5–3)	Serozem-meadow	No effect.	Tulabaev (1972)
naptalam (5)	Meadow medium loam, 1·5% o.m.	No effect.	Tulabaev (1970)
Pyridines			
picloram (2)	Alluvial	Decreased populations.	Goguadze et al. (1972)
Quaternaries			
paraquat (0·3–1·7)* diquat (0·3–1·7)*	Alluvial meadow, leached meadow cinnamon, meadow bog, humic calcareous	No significant effect.	Rankov (1970)
paraquat (250; 500)	1. Alluvial sand, pH 6·6–7; 0·6% o.m.	1. Lower rate inhibited activity.	Szegi (1972)
	2. Chernozem, pH 7·3–7·5; 3·5% o.m.	2. Higher rate inhibited activity.	

* "Recommended field rate".
Refer to *Notes on tabulated data and abbreviations* (p. 319).

TABLE 10 (continued)

Group	Herbicide Concentration (Kg or litre ha^{-1})	Soil type	Effects	Reference
Quaternaries (continued)	paraquat (1·14; 11·4) Gramoxone additives (5·6; 56·0)	Calcareous loam, pH 8; 7·5% o.m.	^{14}C-labelled plant material degradation increased. Slightly decreased degradation of water-insoluble o.m.	Anderson and Drew (1976b)
	Gramoxone (80)	1. Sandy soil 2. Chernozem	Inhibited activity.	Szegi and Gulyás (1971)
	Gramoxone (320)			
Triazines	atrazine (4)	Loamy sand, pH 5·3	Annual applications (c. 13 yr) decreased populations and activity.	Voets et al. (1974)
	atrazine (10; 20)	Weak to medium podzol, 7% o.m.	Increased populations.	Prozorova (1967)
	atrazine (4)	Heavy clay degraded chernozem, >2% o.m.	Annual applications (7 yr) decreased populations.	Husarova (1972a)
	atrazine (5; 10; 20) simazine (5; 10; 20)	Medium loam meadow-bog	Annually applied (3 yr), the two lower rates of each increased populations.	Spiridinov and Spiridinova (1971)
	atrazine (3) simazine (3)	Medium loam meadow, 1·2–2·2% o.m.	Decreased populations.	Tulabaev and Azimbegov (1967)
	atrazine (10–20) simazine (5–10)	Medium loam meadow, pH 4·9; 3% o.m.	Increased activity.	Spiridinov and Yakovlev (1968)
	atrazine (8) simazine (10)	Alluvial	Decreased populations.	Goguadze (1969)

Herbicide (rate)	Soil	Effect	Reference
atrazine (3) simazine (3) propazine (3)	Pale chestnut light loam, 1·6% o.m.	Decreased activity.	Stepanova (1967)
1. atrazine (4·5; 10) 2. simazine (2) chlorazine (8)	Peaty-bog, pH 5·5; 47% o.m.	1. Anaerobic populations and activity not affected; aerobic populations increased. 2. Anaerobic populations not affected; aerobic populations decreased.	Nepomiluev and Kuzyakina (1967)
atrazine (1–200)* simazine (1–200)* prometryne (0·5–60)*	(–)	Low rates had little effect or stimulated activity. Very high rates decreased activity.	Simon (1971)
1. atrazine (1–4)* 2. simazine (1–4)* prometryne (0·5–1·5)*	(–)	1. Decreased activity slightly. 2. No effect.	Sobieszczanski (1969b)
simazine (1·5–2·5) simazine (3)	Sod-podzol Chernozem	No effect. Decreased activity.	Nepomiluev et al. (1966b) Sosnovskaya and Pashchenko (1965)
simazine (1·5)	Leached chernozem	Decreased populations.	Tyagny-Ryadno (1967)
simazine (8)	Irrigated brown soil	Increased populations.	Akopyan and Agaronyan (1968)
simazine (2)	Grey forest silty medium loam, pH 5·3; 4% o.m.	Organic matter decomposition increased when fertilizers also present. Cellulolytic organisms not affected or slightly increased. In pots, o.m. decomposition inhibited.	Boiko et al. (1969)
simazine (3)	Alluvial	Inhibited cellulolytic bacteria and fungi.	Micev and Bubalov (1969b)

* "Recommended field rate".
Refer to *Notes on tabulated data and abbreviations* (p. 319).

TABLE 10 (continued)

Group	Herbicide Concentration (Kg or litre ha^{-1})	Soil type	Effects	Reference
Triazines (continued)	simazine (0·75)	Podzolized light loam	Decreased population.	Pranaitene (1970)
	simazine (0–500)	Alluvial sand, pH 6·6–7·0; 0·6% o.m. Chernozem, pH 7·3–7·5; 3–5% o.m.	No effect.	Szegi (1972)
	simazine (2; 20)	Sandy clay and silty clay	Increased population.	Torstensson (1974)
	simazine (0·1–1·35) propazine (1·35)	Sandy or sandy loam forest soils	Increased or decreased activity depending on presence of o.m., cellulose, (NH$_4$)$_2$SO$_4$ and moisture content.	Eglite (1969)
	prometryne (2·0; 25)	Leached chernozem	Increased population.	Nasil'ev and Enkina (1967)
	prometryne (2–4)	Light loam pale chestnut soil, 1·7% o.m.	No effect.	Lazareva and Blokhin (1969)
	prometryne (5–7)	Cinnamon, alluvial and skeletal soils	No effect.	Goguadze and Gogiya (1971)
	prometryne (15–60)	Leached chernozem	Decreased activity.	Voinova et al. (1970)
	chlorazine (4–6)	Light loamy medium podzol, pH 5·6–6	No effect.	Nepomiluev et al. (1966a)

	Compound (rate)	Soil type	Effect	Reference
	desmetryne (0·5–2)*	Alluvial meadow, leached meadow cinnamon, meadow bog and humic calcareous	Increased activity.	Rankov (1970)
	terbutryne (2·25–6·75)	Alluvial or bog-alluvial	Decreased populations and activity.	Mickovski (1967)
	1. terbutryne 50% (2·5) 2. methoprotryne 25% (2·5)	Alluvial	1. No effect. 2. Decreased population and activity.	Micev and Bubalov (1969a)
Ureas	linuron (0·5–2·5)* linuron (2; 20)	Clay loam, pH 5·4 1. Sandy clay 2. Silty clay	Increased population and activity. 1. Higher rate decreased population. 2. No effect.	Sivasithamparam (1970) Torstensson (1974)
	1. linuron (0·6–100) 2. monolinuron (0·5–3)* 3. chloroxuron (1–5·6)*	(—)	1. Inhibition of activity. 2. Slight inhibition of activity in field. 3. No effect.	Sobieszczanski (1969b)
	linuron (15–60) monolinuron (15–60) fluometuron (15–60)	Leached chernozem	Decreased activity.	Voinova et al. (1970)
	linuron (1·5) monolinuron (1·5)	Light loam podzol	Decreased population.	Kozaczenko and Sobieraj (1973)
	linuron (2–2·5) monolinuror (2–2·5) chloroxuron (5)	Straw-coloured, chernozem, brown, light brown and peaty soils	Inhibited population for 6 months.	Kozaczenko (1974)

* "Recommended field rate".
Refer to *Notes on tabulated data and abbreviations* (p. 319).

TABLE 10 (continued)

Group	Herbicide Concentration (Kg or litre ha^{-1})	Soil type	Effects	Reference
Ureas (continued)	linuron (0·5–3)* monolinuron (0·5–3)*	Alluvial meadow, leached meadow cinnamon, meadow bog and humic calcareous	Increased activity.	Rankov (1970)
	monuron (8)	Alluvial	Decreased population.	Goguadze (1969)
	monuron (8)	Irrigated brown soil	Increased populations.	Akopyan and Agaronyan (1968)
	diuron (8)			
	monuron (12–20) diuron (12–20)	Medium loam red.	Little toxic effect.	Spiridinov et al. (1970)
	monuron (8; 12–16) diuron (8; 12–16)	Light clay, 3·1–3·5% o.m.	Increased activity.	Spiridinov et al. (1972)
	1. monuron (8–16) diuron (8–16)	1. Meadow bog	1. Increased population.	Spiridinov et al. (1973)
	2. monuron (12–20) diuron (12–20)	2. Red Krasnozem	2. Decreased activity.	
	fenuron (1·5)	Medium loam meadow, 1·2–2·2% o.m.	Decreased population.	Tulabaev and Azimbegov (1967)
	1. fluometuron (1·5–3)	Serozem-meadow	1. Decreased activity.	Tulabaev (1972)
	2. noruron (1·5–3) fluometuron (8–10)	Cinnamon, alluvial and skeletal soils.	2. Altered species composition. No effect.	Goguadze and Gogiya (1971)

	DCU (18)	Pale chestnut light loam, pH 8; 2% o.m.	No effect.	Chulakov and Zharasov (1967)
Uracils	DCU (5; 15)	Leached chernozem	No effect.	Lobanov and Poddubnaya (1967)
	isocil (3–5)	Cinnamon, alluvial and skeletal soils	No effect.	Goguadze and Gogiya (1971)
	lenacil (1·0)	Chernozem, 3% o.m.	No effect.	Barona (1974)
Mixtures	Alipur (2–4)	Light loamy medium podzol, pH 5·6–6	No effect.	Nepomiluev et al. (1966a)
	Alipur (100+)	Grey podzolized soil	Inhibited activity.	Nikolenko and Geller (1969)
	Alipur (1·2–2)	Light loam pale chestnut, 1·7% o.m.	No effect.	Lazareva and Blokhin (1969)
	Alipur (4)	Light loam podzol	Decreased population.	Kozaczenko and Sobieraj (1973)
	Alipur (4)	Straw coloured, chernozem, brown, light brown and peaty soils	Inhibited population for 6 months.	Kozaczenko (1974)
	Alipur (2·4)–dalapon (6·0) Murbetol (40; 80)	Pale and dark chestnut soils	Decreased activity.	Chulakov and Zharasov (1971)
	Gesaran (2·25; 4·5; 6·75) Lumeton (3; 6; 9)	Leached chernozem	No effect.	Lobanov and Poddubnaya (1967)
		Alluvial or bog-alluvial	Decreased population.	Mickovski (1967)

* "Recommended field rate".

Refer to *Notes on tabulated data and abbreviations* (p. 319).

TABLE 10 (continued)

Herbicide				
Group	Concentration (Kg or litre ha^{-1})	Soil type	Effects	Reference
Mixtures (continued)	Gesaprim (3) VCS-438 (1·0)+ dicamba (0·5)	Alluvial	No effect.	Micev and Bubalov (1969c)
	atrazine (1·6)+ prometryne (0·8)	Heavy clay degraded chernozem	Annual applications (7 yr) decreased populations.	Husarova (1972a)
	simazine (1·5)+ terbutryne (1·5)	Alluvial soil	Slightly inhibited cellulolytic organisms.	Micev and Bubalov (1969b)
	simazine (3)+ 2,3,6-TBA dimethyl-amine (1·0)	Pale chestnut soil	Decreased populations.	Malanina and Nikitin (1971)
	propazine (10)+ thiram and γ-BHC (4)	Calcareous and saline chernozem soils	Decreased bacterial population.	Putintsev and Putintseva (1966)
	lenacil (1·0)+ TCA (5)	Chernozem, 3% o.m.	Decreased activity.	Barona (1974)
	dalapon (20; 60)+ 2,4-D (2; 6)	Podzolic light loam, moderate loam, podzolic compact, heavy loam	Increased activity in the heavy loam soil.	Bliev (1973)

Others				
Ca(CN)$_2$ (—)	(—)		Increased population.	Zhukova and Konoshevich (1967)
sodium chlorate (150)		Fallow or turfed soil	No effect.	Karki et al. (1973)
sodium cacodylate (10·2; 30·6)		Alluvial loam	Temporarily increased activity.	Malone (1971)
methurin (1·5–3)		Serozem-meadow	No effect.	Tulabaev (1972)
methurin (4)		Medium loam	Decreased population.	Kruglov et al. (1973)

* "Recommended field rate".
Refer to *Notes on tabulated data and abbreviations* (p. 319).

TABLE 11

Effects of herbicides on soil respiratory activity

Group	Herbicide Concentration (Kg or litre ha^{-1})	Soil type	Effects	Reference
Phthalates	endothal (2; 20; 200)	Sandy	Bimodal inhibition of O_2-uptake.	Au (1968)
Anilines	trifluralin (11 300)	Calcareous loam	Inhibition.	Naidu (1972)
	trifluralin (0·6+)	Sandy loam	Respiration of *Sclerotium rolfsii* decreased.	Curl et al. (1967)
	trifluralin (15)	Fine sandy loam	Decreased O_2-uptake.	Whitworth et al. (1965)
	trifluralin (1·0)	Clay,	Decreased O_2-uptake.	Liu and Cibes-Viade (1972)
Benzoics and related	chlorthiamid (10–20)*	Limestone, 3·2% o.m.	Decreased activity.	Berthold (1970)
	dichlobenil (2·8–13)*	Clay shale, 4·4% o.m.		
	chlorthiamid (10–20)*	Calcareous and shale soils	Decreased CO_2-evolution.	Walter (1970, 1971)
	dichlobenil (2·8–13)*			
	nitrofen (1–3)*	Sandy	No effect.	Gowda and Patil (1972)
	bensulide (4·8; 24; 96)	Fine sandy loam	Low rate increased activity. High rates delayed increase (O_2-uptake).	Whitworth et al. (1965)
	ioxynil (1·32; 13·2; 132 a.i.)	Sandy loam	Decreased CO_2-evolution, especially at highest rate.	van Schreven et al. (1970)
Carbamates and carbanilates	EPTC (6)	(−)	No effect.	Balicka and Sobieszczanski (1969a)
	chlorpropham (6)			
	pebulate (4)			
	chlorpropham (12)	Silty medium loam derno-podzol, pH 5·9	Decreased activity.	Zinchenko and Osinskaya (1969)

Diazines	1. barban (20) 2. carbyne (140·8) 3. carbyne formulation additives (140·8)	Sandy loam, pH 5; 7·1% o.m.	1, 2 and 3. Initial increase in CO_2-evolution. 2. Depressed CO_2 evolution in later stages.	Quilt et al. (in press)
	pyrazon (1·5–3; 15–30)*	(–)	Low rates had no effect; high rates decreased activity.	Simon (1971)
	pyrazon (1·5–3)	Chernozem medium clay, pH 7·2; 1·5% o.m.	No effect.	Kulinska (1967a)
	pyrazon (4)	Silty medium loam derno-podzol, pH 5·9	Decreased activity.	Zinchenko and Osinskaya (1969)
	bentazon (4)	Loam and loamy sand	No effect.	Jung (1972)
Halogenated aliphatics	TCA (7–30; 70–300)*	(–)	No effect at low rates; high rates decreased activity.	Simon (1971)
	TCA (1·0; 10) DCA (1·0; 10) MCA (1·0; 10)	Clay (anaerobic)	CO_2- and CH_4-evolution inhibited, especially by high rate.	Laskowski and Broadbent (1970)
	dalapon (1·0; 10) dalapon (0·6; 6)	Sandy or sandy loam	Increased or decreased CO_2-evolution depending on presence/absence of o.m., cellulose, $(NH_4)_2SO_4$ and H_2O.	Eglite (1969)
	dalapon (20)	Silty medium loam derno-podzol, pH 5·9	Decreased activity.	Zinchenko and Osinskaya (1969)
	dalapon (12·8; 128; 1280 a.i.)	Sandy loam	Decreased CO_2-evolution, especially at 2 highest rates.	van Schreven et al. (1970)

* "Recommended field rate".
Refer to *Notes on tabulated data and abbreviations* (p. 319).

TABLE 11 (continued)

Group	Herbicide Concentration (Kg or litres ha^{-1})	Soil type	Effects	Reference
Phenols	PCP (4) PCP (1–3)*	Six soil types Clay loam	Decreased activity. Increased activity.	Au (1968) Matsuguchi and Ishizawa (1968)
	dinoseb (5)	Silty medium loam derno-podzol, pH 5·9	Decreased activity.	Zinchenko and Osinskaya (1969)
Phenoxy-acids	2,4-D (1·0) 2,4-D (2)	Clay Silty, medium loam derno-podzol, pH 5·9	Decreased O_2-uptake. Decreased activity.	Liu and Cibes-Viade (1972) Zinchenko and Osinskaya (1969)
	2,4-D (1–2) MCPA (1–2)	Alluvial	Increased activity.	Bakalivinov and Nikolova (1971)
	2,4,5-T (11 300) MCPA (6·7)	Calcareous loam Sandy loam	Inhibited activity. Stimulatory and inhibitory effects on CO_2-evolution over a period of years. No effect on O_2-uptake in laboratory-treated soil.	Naidu (1972) Grossbard (1971)
	dichlorprop (2·5; 25; 250) mecoprop (1·0; 10; 100)	Sandy loam	All rates decreased CO_2-evolution at various stages.	van Schreven et al. (1970)
Phthalaics	chlorthal dimethyl (4·5; 8·2; 50)	Sandy clay loam	No effect.	Bone and Kuntz (1968)
Pyridines	picloram (24; 48) picloram (0·03–0·12)	Fine sandy loam Heavy clay	Increased O_2-uptake. CO_2-evolution, O_2-uptake not affected.	Whitworth et al. (1965) Grover (1972)
		Sandy loam polder	Increased CO_2-evolution at	van Schreven et al. (1970)

Quaternaries	paraquat (1·25–100) paraquat (10; 20; 40) paraquat (1·14; 11·4) Gramoxone (5·6; 56) Gramoxone formulation additives (5·6; 56)	Sandy loam (−) Calcareous loam, pH 8; 7·5% o.m. with and without grass cuttings	Decreased CO_2-evolution by *S. rolfsii*. Increased O_2-uptake. Minor and irregular changes in CO_2-evolution	Curl *et al.* (1967) Giardina *et al.* (1970) Anderson and Drew (1976b)
Triazines	atrazine (1·2)	Sandy loam (autoclaved)	Decreased CO_2-evolution from *S. rolfsii*. No effect on *Trichoderma viride*. Effects depended on C:N ratio.	Curl *et al.* (1966)
	atrazine (88) atrazine (4)	Fine sandy loam Heavy clay degraded chernozem	Decreased O_2-uptake. Decreased CO_2-evolution.	Whitworth *et al.* (1965) Husarova (1972a)
	atrazine (1–4)	Calcareous chernozem	Increased activity.	Bakalivanov (1971)
	atrazine (3) simazine (3) prometryne (3)	(−)	Increased O_2-uptake.	Balicka and Sobieszczanski (1969a)
	1. atrazine (2; 3) simazine (2; 3) chlorazine (2; 3) 2. propazine (2; 3)	Heavy and light loam chernozems	1. Low rates increased activity. 2. No effect.	Fisyunov (1969b)
	atrazine (1·0; 10; 25) ametryne (1·0; 10; 25)	Sandy clay loam and 2 clay soils	Small to moderate increase in activity.	Liu and Cibes-Viade (1972)
	atrazine (3–150)* simazine (3–150)* prometryne (3–150)*	(−)	Decreased CO_2-evolution.	Simon (1971)

* "Recommended field rate".
Refer to *Notes on tabulated data and abbreviations* (p. 319).

TABLE 11 (continued)

Herbicide		Soil type	Effects	Reference
Group	Concentration (Kg or litres ha^{-1})			
Triazines (continued)	atrazine (2; 50) simazine (2; 50)	Chernozem medium clay, pH 7·2; 1·5% o.m. Sand, pH 5·8; 0·9% o.m.	Low rates had no effect. High rates inhibited CO_2-evolution. Simazine did not affect O_2-uptake.	Kulinska (1967a,b)
	simazine (5)	Silty medium loam derno-podzol, pH 5·9	Decreased activity.	Zinchenko and Osinskaya (1969)
	simazine (0·1–1·4)	Sandy or sandy loam	Increased or decreased CO_2-evolution depending on presence/absence of o.m., cellulose, $(NH_4)_2SO_4$ and H_2O.	Eglite (1969)
	simazine (1–4)* simazine (1·0; 10)	Sandy Red sandy loam	Transiently decreased activity. Decreased activity.	Gowda and Patil (1972) Balasubramanian and Siddaramappa (1974)
	simazine (133) simazine (0·5–10)	Sandy loams Nine soils, variable o.m.	Decreased CO_2-evolution. Inhibited CO_2-evolution. High o.m. decreased inhibition.	Marsh et al. (1974) Chandra et al. (1960)
	simazine (7·5)	Sandy loam	Applied annually (7 yr), decreased CO_2-evolution.	Grossbard (1970)
	prometryne (15–60)	Leached chernozem	Increased CO_2-evolution.	Voinova et al. (1970)
Ureas	linuron (130) linuron (5)	Sandy loams Sandy loam	Decreased CO_2-evolution. Applied annually, increased and then decreased CO_2-evolution.	Marsh et al. (1974) Grossbard (1970, 1971)
	linuron (0·3; 3) monolinuron (0·3; 3)	1. Sandy 2. Loess	1. High rate decreased activity. 2. No effect.	Manninger and Szava (1972)

Herbicide (rate)	Soil type	Effect	Reference
monolinuron (3), chloroxuron (7), linuron (15–60), monolinuron (15–60), fluometuron (15–60)	(—)	No effect on O_2-uptake.	Balicka and Sobieszczanski (1969a)
1. linuron (1·0; 10; 25)	Leached cher-nozem	Increased CO_2-evolution.	Voinova et al. (1970)
2. fluometuron (1·0; 10; 25), diuron (1·0; 10; 25)	Sandy clay loam and 2 clay soils	1. High concentrations increased activity. 2. Low concentrations increased activity.	Liu and Cibes-Viade (1972)
1. linuron (5; 50), 2. metoxuron (0·5; 5; 50), 3. fluometuron (50), chloroxuron (50), monuron (50), diuron (50)	Sandy loam, pH 7·4; 1·5% o.m. Coarse sandy loam, pH 6·1; 1·6% o.m.	1. Decreased CO_2-evolution. 2. Low rates had no effect but highest rate decreased CO_2-evolution. 3. Initially increased, then decreased CO_2-evolution.	Grossbard and Marsh (1974)
monolinuron (0·05–100)	(—)	Decreased CO_2-evolution.	Beck (1969)
monuron (5–100), diuron (5–100)	Nine soil types, variable o.m.	Decreased activity for 28 days then recovered, high o.m. decreased inhibition.	Chandra et al. (1960)
monuron (10)*, diuron (10)*, fenuron (10)*	Silt loam	Decreased CO_2-evolution.	Doxtader (1968)
diuron (88)	Fine sandy loam	Decreased O_2-uptake.	Whitworth et al. (1965)
diuron (5)	Silty medium loam derno-podzol, pH 5·9	Sometimes increased activity.	Zinchenko and Osinskaya (1969)

* "Recommended field rate".
Refer to *Notes on tabulated data and abbreviations* (p. 319).

TABLE 11 (continued)

Group	Herbicide Concentration (Kg or litres ha^{-1})	Soil type	Effects	Reference
Ureas (continued)	fluometuron (0·1–100) chloroxuron (0·1–10)	Sandy loam Three clay soils, pH 5·2; pH 6·2; pH 6·5	Variable effects. No effect, or slight increase in CO_2-evolution.	Bozarth et al. (1968, 1969) Odu and Horsfall (1971)
Uracils	terbacil (1·0; 10; 25)	Sandy clay loam and 2 clay soils	Small to moderate increase in activity.	Liu and Cibes-Viade (1972)
Mixtures	dalapon (20; 60) + 2,4-D (2; 6)	Podzolic light loam, moderate loam, podzolic compact and a heavy loam	Low rate decreased but high rate increased activity in some soils.	Bliev (1973)
	atrazine (1·6) + prometryne (0·8)	Heavy clay degraded chernozem, >2% o.m.	Decreased CO_2-evolution.	Husarova (1972a)
	simazine + aminotriazole (10) simazine + MCPA + aminotriazole (15)	Orchard soils	Decreased CO_2-evolution.	Teuteberg (1968)
	propham 40% + diuron 5% (field rate)	Chernozem medium clay, pH 7·2; 1·5% o.m.	No effect.	Kulinska (1967a)
	PCP (4) + endothal (2)	Six soil types	Decreased activity.	Au (1968)

4-CPA isopropyl ester 40% + atrazine 30% (0·3; 3) after linuron	1. Loess soil 2. Sandy soil	1. No effect. 2. High rate decreased activity.	Manninger and Szava (1972)
amitrole-T (35; 350; 3500)	Sandy loam polder	All rates decreased CO_2-evolution.	van Schreven et al. (1970)
paraquat + diquat (—) atrazine + mecoprop (—) simazine + amino-triazole (—)	Calcareous and shale soils	Decreased CO_2-evolution in first year after treatment, increased CO_2-evolution in second year.	Walter (1970, 1971)
Others sodium chlorate (150)	Fallow or turfed acid sandy soil	Decreased CO_2-evolution.	Karki et al. (1973)
sodium chlorate (67·5)	Acid sandy soil + lucerne powder	Increased CO_2-evolution.	Karki and Kaiser (1974)
1. cacodylic acid (10; 100) 2. MSMA (10; 100)	Forest soil	1. Decreased CO_2-evolution from forest litter. 2. Decreased CO_2-evolution from litter and soil.	Bollen et al. (1974)
methurin (3)	Silty medium loam derno-podzol, pH 5·9	Decreased activity.	Zinchenko and Osinskaya (1969)

Refer to *Notes on tabulated data and abbreviations* (p. 319).

TABLE 12

Effects of herbicides on soil enzymic activity

Herbicide		Soil type	Effects	Reference
Group	Concentration (Kg or litres ha^{-1})			
Acylamides Anilines	propanil (40) trifluralin (11 300) trifluralin (3) trifluralin (3; 15)	Loam Calcareous loam Bog soil Leached heavy loam chernozem	Increased peroxidase. Decreased urease. Decreased urease. No effect on catalase, urease, phosphatase or invertase.	Lay and Ilnicki (1974) Naidu (1972) Tyunyaeva et al. (1974) Gruzdev et al. (1973)
Benzoics and related	chlorthiamid (15–30)* dichlobenil (2·8–13)*	Limestone, 3·2% o.m. Shale, 4·4% o.m.	Decreased dehydrogenase and increased urease.	Berthold (1970)
	chlorthiamid (15–30)* dichlobenil (2·8–13)*	Calcareous and shale soils	Decreased dehydrogenase, phosphatase, urease, β-glucosidase, saccharase.	Walter (1970, 1971)
Carbamates and carbanilates	1. barban (20)	Sandy loam, pH 5; 7·1% o.m.	1, 2. Slightly inhibited then increased phosphatase. Glucose- and cellobiose-utilization inhibited. Amylase increased.	Quilt et al. (in press)
	2. carbyne (140·8) 3. Solvents of carbyne (140·8)		2. Urease increased. 3. Slightly inhibited phosphatase. Urease increased.	
	chlorpropham (12)	Silty medium loam derno-podzol, pH 5·9	Decreased urease, transiently inhibited catalase.	Zinchenko and Osinskaya (1969)
	chlorpropham (3; 7; 10; 100)	Sandy and loamy soils	Decreased invertase, urease, asparaginase. Catalase and gelatinase not affected.	Krezel and Musial (1969)

	molinate (0·75–1·5)	Pale sandy loam chernozem	No effect on soil enzymes.	Chulakov and Zharasov (1973)
Diazines	pyrazon (4)	Silty medium loam derno-podzol, pH 5·9	Decreased then increased catalase. Urease, invertase not affected.	Zinchenko and Osinskaya (1969)
	pyrazon (4–8)	Silty medium loam derno-podzol, pH 5·9; 1·3% o.m.	Decreased urease and invertase but not catalase (cf: previous notation above).	Zinchenko et al. (1969)
	pyrazon (3)	Light loam derno-podzol	5th application decreased soil enzyme activities.	Chunderova et al. (1971)
Halogenated aliphatics	dalapon (20)	Silty medium loam derno-podzol, pH 5·9	Decreased urease, catalase; increased invertase.	Zinchenko and Osinskaya (1969)
	dalapon (10)	Grassland oxisol, pH 5·9	Increased urease, decreased proteinase.	Namdeo and Dube (1973a)
Phenols	dinoseb (5)	Silty medium loam derno-podzol, pH 5·9	Decreased urease and catalase.	Zinchenko and Osinskaya (1969)
	dinoseb (5)	Silty medium loam derno-podzol, pH 5·9; 1·3% o.m.	Increased urease, decreased catalase. Invertase not affected (cf: Previous notation above).	Zinchenko et al. (1969)
	DNOC	Light loam derno-podzol	5th application decreased soil enzyme activities.	Chunderova et al. (1971)
	1. PCP (2–120)	1. Alluvial clay loam pH 5·7; 2·7% carbon	1. Increasing rate increased oxidation of iron.	Ishizawa and Matsuguchi (1966)
	2. PCP (0·8–3)	2. Clay loam, alluvial, volcanic light clay, alluvial clay loam.	2. Fluctuations in sulphate reduction.	

* "Recommended field rate".
Refer to *Notes on tabulated data and abbreviations* (p. 319).

TABLE 12 (continued)

Group	Herbicide Concentration (Kg or litres ha^{-1})	Soil type	Effects	Reference
Phenoxy-acids	2,4-D (1–2)	1. Fallow chernozem	1. Catalase not affected.	Mereshko (1969)
		2. Cropped chernozem	2. Increased catalase.	
	2,4-D (2)	Silty medium loam derno-podzol, pH 5·9; 1·3% o.m.	Initially, no effect. Later experiments showed increased urease, catalase, invertase.	Zinchenko et al. (1969)
	2,4-D (0·4)	Pale sandy loam chernozem	No effect.	Chulakov and Zharasov (1973)
	2,4-DB (1·26)	Arguidol soil	Urease unaffected.	Leiderman et al. (1971)
	2,4-D (2; 100)	Chernozem	Urease unaffected; phosphatase inhibited.	Yurkevitch and Tolkachev (1972a)
	2,4-D (0·75)	Grey forest silty medium loam, pH 5·3; 4% o.m.	Increased phosphatase-producing bacteria.	Boiko et al. (1969)
	1. 2,4-D (2).	Silty medium loam derno-podzol, pH 5·9	1. Decreased urease and catalase. Increased invertase.	Zinchenko and Osinskaya (1969)
	2. 2,3,6-TBA (4)		2. Invertase, urease unaffected. Decreased catalase (cf: earlier notation on 2,4-D).	
	2,4-D (0·5–0·75) MCPA (1·0)	Light loam dernopodzol	5th application decreased soil enzyme activities.	Chunderova et al. (1971)
	2,4,5-T (11 300)	Calcareous loam	Decreased urease.	Naidu (1972)
	2,4,5-T (4; 40) MCPA (4; 40)	Sandy clay and silty clay	Proteinases unaffected.	Torstensson (1974)

	Herbicide (rate)	Soil	Effect	Reference
	MCPA (1–1·5)	Sandy loam	Temporarily increased proteinases.	Hoffman-Kakol and Kwarta (1972b)
Pyridines	picloram (0·07)	Derno-podzolic medium loam	Decreased phosphatase.	Bezuglov et al. (1973)
Quaternaries	paraquat (10; 20; 40)	(—)	Decreased proteinases, cellulase; urease unaffected.	Giardina et al. (1970)
	1. paraquat (3·75) 2. paraquat (3·75) + urea fertilizer	Grassland oxisol, pH 5·9	1. Increased urease. 2. Increased urease and proteinase.	Namdeo and Dube (1973a)
	paraquat (0·6 a.i.)	Silt loam, pH 6·3; 0·11% o.m.	Decreased sulphur oxidation.	Tu and Bollen (1968a)
	paraquat (1·14; 11·4) Gramoxone (5·6; 56) Additives of Gramoxone (5·6; 56)	Calcareous loam, pH 8; 7·5% o.m.	Variable increases in urease, amylase, proteinases and sucrose oxidation. Phosphatases, peroxidase decreased. Lipoxidase unaffected.	Anderson and Drew (1976b, c)
Triazines	atrazine (1–4)*	Calcareous and shale	Increased urease.	Walter (1970, 1971)
	atrazine (4; 10)	Heavy loam chernozem, 4–5% o.m.	Phosphatase unaffected.	Pilipets and Litvinov (1972)
	atrazine (4)	Loamy sand, pH 5·3	Annual applications for c. 13 years decreased phosphatase, saccharase, β-glucosidase, urease.	Voets et al. (1974)
	1. atrazine (10; 100) 2. simazine (10; 100) prometryne (10; 100)	Sandy and loamy	1. Increased invertase, asparaginase. Inhibited urease. Catalase, gelatinase unaffected. 2. Increased invertase, asparaginase, dehydrogenase. Inhibited urease. Catalase, gelatinase unaffected.	Krezel and Musial (1969)

* "Recommended field rate".
Refer to Notes on tabulated data and abbreviations (p. 319).

TABLE 12 (continued)

Group	Herbicide Concentration (Kg or litres ha^{-1})	Soil type	Effects	Reference
Triazines (continued)	atrazine (1–4)* simazine (1–4)*	1. Fallow chernozem 2. Cropped chernozem	1. Catalase unaffected. 2. Increased catalase.	Mereshko (1969)
	1. atrazine (1·0)	Silty medium loam derno-podzol, pH 5·9; 1·3% o.m.	1. Decreased urease, invertase, catalase.	Zinchenko et al. (1969)
	2. simazine (1·0) prometryne (1·0)		2. Increased urease, catalase. Invertase unaffected.	
	atrazine (2; 10) simazine (2; 10) prometryne (2; 10)	Podzolized medium loam chernozem, or podzolized grey silty loam	Phosphatase slightly affected in first year but increased in second year. Low rates increased urease, high rates inhibited.	Manorik and Malickenko (1969)
	atrazine (5; 10; 20) simazine (5; 10; 20) prometryne (5; 10; 20)	Medium loam meadow-bog	Little effect on soil enzymes in laboratory soil. In field soil, simazine (5; 10) increased catalase, decreased dehydrogenase, proteinase.	Spiridinov and Spiridinova (1973)
	atrazine (1·6 a.i.) simazine (1·6 a.i.) ametryne (1·6 a.i.)	Arguidol soil	Urease unaffected.	Leiderman et al. (1971)
	simazine (4; 200)	Chernozem	Increased phosphatase with low rate. Inhibited by high rate. Urease unaffected.	Yurkevitch and Tolkachev (1972a)

simazine (6–12)	Leached loam chernozem	Increased phosphatase.	Arshidinov et al. (1974)
simazine (2)	Grey forest silty medium loam, pH 5·3; 4% o.m.	Increased phosphatase.	Boiko et al. (1969)
simazine (2; 50)	Chernozem medium clay, pH 7·2; 1·5% o.m. Sandy soil, pH 5·8; 0·9% o.m.	Low rate: urease, saccharase, catalase unaffected. High rate: inhibited all 3. Catalase unaffected in sandy soil.	Kulinska (1967b)
simazine (5)	Silty medium loam derno-podzol, pH 5·9	Increased urease, invertase. Catalase decreased then increased (cf: earlier notation under atrazine, simazine and prometryne).	Zinchenko and Osinskaya (1969)
simazine (2; 20)	Sandy clay and silty clay	Proteinases unaffected.	Torstensson (1974)
simazine (1·0; 10)	Red sandy loam	Irregular changes in amylase and invertase.	Balasubramanian and Siddaramappa (1974)
prometryne (2) propazine (2)	Light loam derno-podzol	No effect until 5th application then decreased soil enzymes.	Chunderova et al. (1971)
Ureas linuron (1–10)*	Clay loam, pH 5·4 (incubated in dark)	Increased sulphate reduction, iron precipitation.	Sivasithamparam (1970)
linuron (2; 20)	Sandy clay and silty clay.	Proteinases unaffected.	Torstensson (1974)
linuron (1–10)*	Sandy loam	Decreased dehydrogenase.	Beckmann and Pestemer (1975)
1. linuron (10–100) 2. monolinuron (10–100) chloroxuron (10–100)	Sandy and loamy	1. Inhibited urease, asparaginase. 2. Increased invertase, asparaginase. Inhibited urease. None affected catalase or gelatinase.	Krezel and Musial (1969)

* "Recommended field rate".
Refer to *Notes on tabulated data and abbreviations* (p. 319).

TABLE 12 (continued)

	Herbicide				
Group	Concentration (Kg or litres ha^{-1})	Soil type	Effects		Reference
Ureas (continued)	monolinuron (0·05–100)*	(–)	Decreased dehydrogenase.		Beck (1969)
	monolinuron (3) noruron (2)	Silty medium loam, derno-podzol, pH 5·9; 1·3% o.m.	Little to no effect.		Zinchenko et al. (1969)
	monuron neburon	Light loam derno-podzol	5th application decreased soil enzyme activities.		Chunderova et al. (1971)
	monuron (8–16) diuron (8–16)	Light clay, 3·1–3·5% o.m.	Increased catalase, peroxidase.		Spiridinov et al. (1972)
	1. monuron (8–16) diuron (8–16)	1. Meadow bog soil	1. Increased catalase, peroxidase, dehydrogenase. Monthly applications decreased them.		Spiridinov et al. (1973)
	2. monuron (12–20) diuron (12–20)	2. Red krasnozem soil	2. No effects. Monthly applications decreased catalase, peroxidase, dehydrogenase.		
	diuron (5)	Silty medium loam derno-podzol, pH 5·9	Increased catalase. Urease, invertase unaffected		Zinchenko and Osinskaya (1969)
	diuron (6)	Leached loam chernozem	Increased phosphatase.		Arshidinov et al. (1974)
	fluometuron (1·5–3)	Serozem meadow soil	Decreased phosphatase.		Tulabaev (1972)
	chloroxuron (1–10)	Three clay soils, pH 5·2; 6·2; 6·5	Dehydrogenase unaffected.		Odu and Horsfall (1971)
	metoxuron (100–400)*	Calcareous	Little effect on proteinases.		Simon et al. (1973)

	Herbicide (rate)	Soil	Effect	Reference
Uracils	lenacil (1·2–1·7)	Dark chestnut heavy loam and pale chestnut light loam	Soil enzymes unaffected.	Chulakov and Zharasov (1973)
	lenacil (1·0)	Chernozem, 3% o.m.	Decreased phosphatase.	Barona (1974)
Mixtures	lenacil (1·0) + DCU (5) lenacil (1·0) + TCA (5)	Chernozem, 3% o.m.	Decreased phosphatase.	Barona (1974)
	atrazine + mecoprop (–) simazine + aminotriazole + MCPA (–) paraquat + diquat (–)	Limestone soil, 3·2% o.m. and clay shale, 4·4% o.m.	Decreased dehydrogenase. Nitrogen-containing herbicides increased urease.	Berthold (1970)
	dalapon (20; 60) + 2,4-D (2; 6)	Podzolic light loam, podzolic compact, moderate loam and heavy loam	Amylase and invertase unaffected.	Bliev (1973)
	1. paraquat – diquat (–) 2. simazine + amino-triazine + MCPA (–) atrazine + mecoprop (–)	Calcareous and shale soils	1. Increased urease, β-glucosidase, saccharase. 2. Increased β-glucosidase, saccharase.	Walter (1970, 1971)
Others	sodium chlorate (150)	Turfed plots of acid sandy soil	Decreased dehydrogenase.	Karki et al. (1973)

* "Recommended field rate".
Refer to *Notes on tabulated data and abbreviations* (p. 319).

TABLE 12 (continued)

Herbicide				
Group	Concentration (Kg or litres ha^{-1})	Effects	Soil type	Reference
Others (continued)	sodium chlorate (6·8; 20·3; 67·5)	Highest rate decreased catalase.	Acid sandy soil	Karki and Kaiser (1974)
	methurin (3)	Decreased urease. Decreased then increased catalase. Increased invertase.	Silty medium loam derno-podzol, pH 5·9	Zinchenko and Osinskaya (1969)
	methurin (4)	Increased urease. Decreased invertase, proteinase.	Derno-podzolic sandy loam and medium loam	Kruglov et al. (1973)

Refer to *Notes on tabulated data and abbreviations* (p. 319).

TABLE 13

Effects of herbicides on soil microbial ammonification

Group	Herbicide, Concentration (Kg or litres ha⁻¹)	Soil type	Effects	Reference
Acylamides	propachlor (1·0)	Heavy clay degraded chernozem	Applied annually (7 yr) decreased numbers, especially in dry years.	Husarova (1972a, b)
Phthalates	endothal (2; 20)	Six soils	No effect.	Au (1968)
Anilines	trifluralin (2–1000)	1. Alluvial meadow 2. Heavy textured, high o.m.	1. Inhibited activity. 2. Less inhibition.	Rankov (1971b)
	trifluralin (1–4)	Alluvial meadow	Increased activity.	Rankov and Elenkov (1970)
	trifluralin (11 300)	Calcareous loam	No effect.	Naidu (1972)
	ioxynil (1·5)	Podzolic loam, pH 7·0	No effect.	Hauke-Pacewiczowa and Krolowa (1968)
Benzoics and related	chlorthiamid (2·8; 28, 280) dichlobenil (2·8; 28, 280)	Alluvial meadow	No effect.	Helweg (1972a)
Carbamates and carbanilates	EPTC (–)	(–)	Increased activity.	Zhukova and Konoshevich (1967)
	EPTC (4)	Light or medium loam	Increased activity.	Chunderova et al. (1968)
Diazines	phenazone (3–6; 8–18)	Leached derno-carbonate and derno-podzol	Low range, no effect. High range inhibited activity.	Mezharaupe (1967)

Refer to *Notes on tabulated data and abbreviations* (p. 319).

TABLE 13 (continued)

Group	Herbicide Concentration (Kg or litres ha^{-1})	Soil type	Effects	Reference
Diazines (continued)	pyrazon (4)	Light or medium loam	Increased activity.	Chunderova et al. (1968)
	pyrazon (4·8)	Podzolized light loam	Decreased population.	Pranaitene (1970)
Halogenated aliphatics	dalapon (10–15)	Brown heavy loam, pH 6·3	Increased activity.	Zavarzin and Belyaeva (1966)
	dalapon (10)	(−)	Increased activity.	Krykhanov and Pashkovskaya (1967)
	dalapon (8)	Meadow type medium loam, 1·5% o.m.	Decreased population.	Tulabaev (1970)
	TCA (25)	Podzolized light loam	Increased populations.	Pranaitene (1970)
Phenols	dinoseb acetate (2–1000)	1. Alluvial meadow 2. Heavier textured, high o.m.	1. Inhibited activity. 2. Less inhibition.	Rankov (1971b)
	dinoseb acetate (5; 50) PCP (12–120)	Loess Alluvial clay loam, pH 5·7; 2·7% carbon	No effect. Inhibited activity.	Uziak et al. (1971) Ishizawa and Matsuguchi (1966)
Phenoxy-acids	2,4-D (0·75)	Silty medium loam, pH 5·3; 4% o.m.	Increased activity.	Boiko et al. (1969)
	2,4-D (1·5)	Leached chernozem, low o.m.	No effect.	Tyagny-Ryadno (1967)
	2,4-D (1·5–15)	Derno podzol	Increased population.	Abueva (1970)

Group	Herbicide (rate)	Soil	Effect	Reference
	2,4-D (1·1; 30; 40)	(—)	Lowest rate increased activity.	Singh (1971)
	2,4-D (1–2)	Alluvial	Increased activity.	Bakalivinov and Nikolova (1971)
	MCPA (1–2); 2,4-DEP (0–8)	Light loam, 1·7% o.m.	No effect.	Lazareva and Blokhin (1969)
	2,4-DEP (6–30); 2,4-DES (5–10)	Heavy loam.	Increased activity after 10 months.	Zavarzin (1966)
	2,4,5-T (11 ±00)	Calcareous loam	No effect.	Naidu (1972)
Phthalaics	chlorthal-dimethyl (1·5–3)	Serozem-meadow	High rate inhibited activity.	Tulabaev (1972)
Pyridines	picloram (0·1–1·0)	Silty clay loam, and silt loam	Increased activity.	Tu and Bollen (1969)
	picloram (2)	Alluvial	Increased population.	Goguadze et al. (1972)
Quaternaries	paraquat (0·6)	Silt loam, pH 6·3; 0·11% o.m.	Decreased activity.	Tu and Bollen (1968a)
Triazines	atrazine (4)	Loamy sand, pH 5·3; 2·4% humus	Annual applications (c. 13 yr) increased population.	Voets et al. (1974)
	atrazine (2; 4·5)	Red–brown loamy clay	Little effect.	Budoi and Budoi (1967)
	atrazine (10)	Sandy clay leached chernozem 1–1·6% o.m.	Decreased activity.	Peshakov et al. (1969)
	atrazine (4)	Heavy clay degraded chernozem	Annual applications (7 yr) decreased population.	Husarova (1972a, b)
	atrazine (3); simazine (3)	Medium loam, 1·2–2·2% o.m.	Decreased activity.	Tulabaev and Azimbegov (1967)
	atrazine (4·5; 10); simazine (4·5; 10)	Peaty-bog, pH 5·5; 47% o.m.	Little effect.	Nepomiluev and Kuzyakina (1967)

Refer to *Notes on tabulated data and abbreviations* (p. 319).

TABLE 13 (continued)

Group	Herbicide Concentration (Kg or litres ha^{-1})	Soil type	Effects	Reference
Triazines (continued)	1. atrazine (1–4)* 2. prometryne (2)	(—)	1. Transiently decreased population. 2. Increased activity.	Krykhanov and Pashkovskaya (1967)
	simazine (2)	Silty medium loam, pH 5·3; 4% o.m.	Increased activity.	Boiko et al. (1969)
	simazine (1·5–2·5)	Sod-podzol	No effect.	Nepomiluev et al. (1966b)
	simazine (4–8; 8–12)	Heavy loam	Decreased activity after 10 months.	Zavarzin (1966)
	simazine (2)	Derno-podzol light or medium loams, pH 6; 2–3% o.m.	Increased activity.	Chunderova et al. (1968)
	simazine (1·1; 30; 40)	(—)	Lowest rate increased activity.	Singh (1971)
	simazine (2–100)	Derno-podzol	Temporarily decreased activity by 30 and 100 kg rates.	Kuzyakina (1971)
	simazine (12) trietazine (8)	Brown heavy loam, pH 6·3	Inhibited activity.	Zavarzin and Belyaeva (1966)
	prometryne (1–2·5)	Chernozem	Increased population 4–5 fold.	Yaroslavskaya et al. (1969)
	prometryne (0·5–1·5)*	Leached chernozem	Increased population and activity.	Voinova et al. (1970)
	prometryne (2·5; 25)	Chernozems	Increased activity.	Vasil'ev and Enkina (1967)
	prometryne (2–4)	Light loam pale chestnut, 1·7% o.m.	No effect.	Lazareva and Blokhin (1969)
	1. terbutryne 50% (2·5) 2. methoprotryne 25% (2·5)	Alluvial	1. Inhibited activity. 2. Increased activity.	Micev and Bubalov (1969a)

	Herbicide (rate)	Soil	Effect	Reference
Ureas	linuron (0·5–2·5)*	Clay loam, pH 5·4	Increased population and activity.	Sivasithamparam (1970)
	linuron (3; 30)	Sandy	Slightly decreased population.	Uziak et al. (1971)
	linuron (0·5–2) monolinuron (0·5–2)	Sandy clay leached chernozem 1–1·6% o.m.	Increased activity.	Peshakov et al. (1969)
	linuron (2) monolinuron (1–2) chloroxuron (5)	Alluvial meadow and leached cinnamon	Increased activity.	Rankov (1972)
	linuron (0·5–2·5)* monolinuron (0·5–3)* Cotoran (0·5–4)*	Leached chernozem	Increased population and activity.	Voinova et al. (1970)
	monuron (6–12)	(–)	Temporarily decreased population.	Krykhanov and Pashkovskaya (1967)
	1. monuron (60) 2. diuron (50)	Silt	1. Decreased activity. 2. No effect.	Gel'tser et al. (1972)
	monuron (6–12) neburon (4–8)	Heavy loam	Increased then decreased activity.	Zavarzin (1966)
	diuron (1·5) fenuron (1·5)	Medium loam meadow soil, 1·2–2·0% o.m.	Inhibited activity.	Tulabaev and Azimbegov (1967)
	fluometuron (1·5–3) noruron (1·5–3)	Serozem-meadow soil	Low rates decreased populations, high rates inhibited activity.	Tulabaev (1972)
Mixtures	Alipur (1·2–2)	Light loam pale chestnut, 1·7% o.m.	No effect.	Lazareva and Blokhin (1969)
	Alipur (4; 40) atrazine (1·6) + prometryne (0·8)	Loess Heavy clay degraded chernozem	No effect. Annual applications (7 yr) decreased population.	Uziak et al. (1971) Husarova (1972a)

* "Recommended field rate".
Refer to *Notes on tabulated data and abbreviations* (p. 319).

TABLE 13 (continued)

Herbicide		Soil type	Effects	Reference
Group	Concentration (Kg or litres ha^{-1})			
Mixtures (continued)	aminotriazole 38% + simazine 20% + MCPA 16% (10)	Sandy clay chernozem, 1–1·6% o.m.	Increased activity.	Peshakov *et al.* (1969)
	Murbetol (10; 17·5)	Derno-podzol light or medium loam, pH 6; 2–3% o.m.	Increased populations.	Chunderova *et al.* (1968)
	Printazol (0·25–2)	Rich loam, pH 7, 3·3% o.m.	Up to 1 litre, retarded then increased activity, but 2 litres inhibited.	Bertrand and de Wolf (1972)
	PCP (4) + endothal (2)	Six soils	Decreased populations.	Au (1968)
	VCS438 (1·0) + dicamba (0·5) Gesaprim (3)	Rhizosphere soil of maize	Decreased population initially. Increased population later.	Micev and Bubalov (1969c)
	lenacil (1·0) + pebulate (4) lenacil (1·0) + DCU (5) lenacil (1·0) + TCA (5)	Chernozem, 3% o.m.	Decreased population.	Barona (1974)
Others	sodium chlorate (150)	Grassed or fallow acid sandy soils.	No effect.	Karki *et al.* (1973)

Refer to *Notes or tabulated data and abbreviations* (p. 319).

TABLE 14

Effects of fungicides on fungi and actinomycetes in soil

Fungicide				
Group	Concentration (Kg or litres ha^{-1})	Soil type	Effects	Reference
Benzoics and related	fenaminosulf (1·1–4·4; 8·8–35)	Clay loam, pH 6·4; 3·5% o.m.	Low range – no effect. Higher range decreased fungi. Highest rate decreased actinomycetes and eliminated *Pythium* spp.	Agnihotri (1973)
	quintozene (1·0)	Loam, pH 6·7; 3·8% o.m. + alfalfa powder	Rate of colonization of alfalfa affected. Overall decrease of 20–80%. Qualitative and quantitative changes in fungi and actinomycetes.	Katan and Lockwood (1970)
	2-(thiocyanomethyl-thio) benzothiazole (5–30)	Arable soil, pH 5·7; humus 1·73%	Decrease in some fungal spp. and increase in *Streptomyces* spp.	Voets and Vandamme (1970)
Carbamate, carbonate, carbanilate	benomyl (3–30)	Humus sand	Caused fungistasis, but no effect on numbers of fungi or actinomycetes. Some spp. incorporated benomyl into mycelia.	Hofer *et al.* (1971)
	benomyl (1·0)	Humus-poor greenhouse soil	No effect.	Kaastra-Howeler and Gams (1973)
	benomyl (0·1–10)	Arable soil	No effect other than some fungistasis.	Raynal and Ferrari (1973)
	benomyl (2·2–89·6)	Silt loam, fine sand, sandy loam	No effect on numbers. *Aspergilli, Penicillia,* isolated from treated soils.	Peeples (1974)
	benomyl (2; 5; 10)	1. Sandy, pH 7·0; 8·9% o.m.	1. Increased numbers of actinomycetes.	van Faassen (1974)

Refer to *Notes on tabulated data and abbreviations* (p. 319).

TABLE 14 (continued)

Group	Fungicide Concentration (Kg or litres ha^{-1})	Soil type	Effects	Reference
Carbamate, carbonate, carbonilate (continued)	benomyl	2. Sandy, pH 7·4; 9·8% o.m.	2. Decreased actinomycetes 200 fold.	Van Faassen (1974)
		3. Sandy, pH 7·3; 9·7% o.m.	3. Decreased actinomycetes 1000 fold.	
	benomyl(−)	Mushroom casing soils	Inhibited mycelial growth of *Agaricus bisporus*. Effect decreased by increasing o.m.	Dutsch (1975)
	maneb, zineb (repeated applications annually)	(−)	Fungal population limited to *Penicillium citrinum*.	Dubey (1970)
	maneb (6; 24; 96)	Acid lateritic clay and alkaline loam	Decreased numbers of fungi and actinomycetes. Qualitative changes.	Dubey and Rodriguez (1974)
	thiram (0·1; 1·0)	Silt loam	Initially decreased fungal populations but recovered later.	Houseworth and Tweedy (1973)
Phthalimide	captan (0·1; 1·0)	Silt loam	Initially decreased fungal populations but recovered later.	Houseworth and Tweedy (1973)
	captan (−)	(−)	Decreased actinomycetes.	Mahmoud *et al.* (1972)
	captan (0·5–10)	Soil under grass pH 6·7; 2·9% organic carbon	Fungi decreased. Actinomycetes increased by 2·5 kg rate.	Wainwright and Pugh (1975a)

Pyrimidines and related	triarimol (2·0)	Soil under grass pH 6·7; 2·9% organic carbon	Fungi decreased.	Wainwright and Pugh (1975a)
Organophosphates	Kitazin (0·01–0·02 a.i.)	Paddy soil	Decreased numbers of fungi in rhizosphere and rhizoplane.	Sullia (1969)
Triazines	anilazine (6; 24; 96)	Acid lateritic clay and alkaline loam	Decreased numbers of fungi and actinomycetes.	Dubey and Rodriguez (1974)
Isoxazalone	drazoxolon (5)	Soil under grass, pH 6·7; 2·9% organic carbon	Fungi decreased.	Wainwright and Pugh (1975a)
Others	tribasic copper (repeated applications)	(—)	Decreased numbers of fungi. Population limited to *P. citrinum*.	Dubey (1970)

Refer to *Notes on tabulated data and abbreviations* (p. 319).

TABLE 15

Effects of some insecticides on soil bacteria *in situ*

Insecticide				
Group	Concentration (Kg or litres ha^{-1})	Soil type	Effects	Reference
Benzoics and related	lindane (–)	(–)	Low rates increased; high rates decreased numbers.	Lodha and Lal (1969)
	lindane (22)	Sandy loam, calcareous clay and clay	Numbers decreased, then increased and reverted to normal.	Gawaad *et al.* (1973a)
	lindane (0·45–2·25 a.i.)	Rice soils	No effect.	Nair *et al.* (1973)
	lindane (0·22; 4·4)	Clay soil + 2% compost	Increased numbers.	Mahmoud *et al.* (1970)
	lindane (30)	Chernozem	Decreased numbers in rhizosphere of maize.	Tsirkov (1971)
	DDT (massive doses)	Forest soil	No effect.	Salonius (1972)
Cyclodienes	dieldrin (2·5; 50)	Clay soil + 2% compost	Increased, then decreased numbers.	Mahmoud *et al.* (1970)
	dieldrin (150)	Chernozem	Decreased numbers in rhizosphere of maize.	Tsirkov (1971)
	endrin (–)	(–)	Low rates increased; high rates decreased numbers	Lodha and Lal (1969)
	endrin (0·45–2·25 a.i.)	Rice soils	No effect.	Nair *et al.* (1973)
	endrin (1·5)*	Sandy loam	No appreciable effect.	Bollen and Tu (1971)
	heptachlor (1·1; 7·9; 55·3)	Silt loam	Numbers increased.	Shamiyeh and Johnson (1973)
	heptachlor (150)	Chernozem	Decreased numbers in rhizosphere of maize.	Tsirkov (1971)

Dithioates and thioates	chlorpyrifos (–)	Clay loam, pH 5·3	Increased numbers.	Sivasithamparam (1970)
	chlorpyrifos (1·0; 10) thionazin (1·0; 10) trichloronate (1·0; 10) diazinon (1·0; 10)	Sandy loam	Numbers decreased initially then recovered.	Tu (1969)
	diazinon (3·4)	(–)	Selective effect on bacteria in rhizosphere.	Gunner (1970)
	disulfoton (50)	Volcanic ash soil	Decreased numbers.	Kobayashi and Katsura (1968)
	fonofos (11)	Sandy loam, calcareous clay and clay	Numbers decreased, then increased and reverted to normal.	Gawaad et al. (1973a)
	malathion (4·2; 5; 6)	(–)	Decreased numbers.	Swaminathan and Sullia (1969)
	malathion (0·2) parathion-methyl (0·3)	(–)	Increased numbers.	Yurovskaya and Zhulinskaya (1974)
Pentadiene derivatives	chlordecone (22 a.i.)	Sandy loam, calcareous loam and clay	Number decreased, then increased and reverted to normal.	Gawaad et al. (1973a)
	mirex (20–2000)	Sandy, (0·98% o.m.; forest sandy, 4·94% o.m.; sand, 0·63% o.m.	No effect.	Jones and Hodges (1974)
Phosphonates	trichlorphon (1·3)	(–)	Increased numbers.	Yurovskaya and Zhulinskaya (1974)
Others	Sevidol (0·45–2·25)	Rice soils	No effect.	Nair et al. (1973)
	Tendrin (22 a.i.)	Sandy loam, calcareous clay and clay	Numbers decreased then increased and reverted to normal.	Gawaad et al. (1973a)

* "Recommended field rate".
Refer to *Notes on tabulated data and abbreviations* (p. 319).

TABLE 16

Effects of insecticides on nitrifiers and nitrification in soil

Group	Insecticide Concentration (Kg or litres ha⁻¹)	Soil type	Effects	Reference
Benzoics and related	lindane (0·1; 1·0; 2·5; 10; 100)	Sandy loam	No effect (0·1, 1·0); Inhibition (>2·5).	Bardiya and Gaur (1970)
	lindane (0·22; 4·4)	Clay + 2% compost	Numbers of nitrifying bacteria severely depressed.	Mahmoud et al. (1970)
	lindane (500)	Loam, pH 7·2; 1·49% carbon	Slight stimulation in moist soil; numbers depressed in drier soil.	Naumann (1971b)
	lindane (1·0)	(–)	Inhibited activity.	Jaiswal et al. (1971)
	lindane (0·3; 1·5; 3 a.i.)	Arable loam, pH 5·8–6·6; 1·7% o.m. and meadow loam, pH 6·3–7; 3·1% o.m.	Only the two highest rates inhibited. Temperature affected degree of inhibition.	Verstraeten and Vlassak (1973)
	lindane (0·2; 2)*	(–)	No overall effect.	Mishra et al. (1972)
	DDT (>0·5; >10)	Alluvial sandy loam	No effect on rate of nitrification but numbers of nitrifiers decreased.	Gaur and Pareek (1971)
	DDT (0·75; 10; 50)	Silt loam	Low rate, no effect. High rates decreased activity.	Ross (1974)
	isobenzan (0·4; 0·75)	(–)	Inhibited activity.	Jaiswal et al. (1971)
Carbamates	carbofuran (0·75; 10; 50)	Silt loam	Indirect, slightly stimulatory effect.	Ross (1974)
	propoxur (0·5 a.i.) aldicarb (0·5 a.i.)	Sandy loam, pH 7·4	20%–30% inhibition at high rate.	Kuseske et al. (1974)

	125)	pH 8·3; 0·67% o.m.	Lowest two rates of each, no effect. All other rates inhibited (NO$_2^-$ accumulated).	Gupta et al. (1975)
Cyclodienes	aldrin (0·1; 1·0; 2·5; 10; 100) dieldrin (0·1; 1·0; 2·5; 10; 100)	Sandy loam	Lowest two rates of each, no effect. All other rates inhibited (NO$_2^-$ accumulated).	Bardiya and Gaur (1970)
	dieldrin (2·5; 50)	Clay + 2% compost	Numbers of nitrifying bacteria severely depressed.	Mahmoud et al. (1970)
Dithioates and thioates	endrin (1·8)*	Sandy loam	No effect.	Bollen and Tu (1971)
	chlorpyrifos (−)	Clay loam, pH 5·3	Decreased activity.	Sivasithamparam (1970)
	chlorpyrifos (1·0; 10) diazinon (1·0; 10) trichloronate (1·0; 10) thionazin (1·0; 10)	Sandy loam	No effect or slight inhibition.	Tu (1969)
	Counter (0·1–10)	Three soils, pH 5·7; 6·3; 7·0	No inhibition.	Laveglia and Dahm (1974)
	disulfoton (60)	Volcanic ash	Increased activity.	Kobayashi and Katsura (1968)
	disulfoton (−)	Sandy clay loam	Decreased numbers of nitrifiers.	Mahmoud et al. (1972)
	fenitrothion (0·75; 10; 50)	Silt loam	Lowest rate, no effect. High rates initially decreased, but subsequently increased activity.	Ross (1974)
	parathion-methyl (0·15; 1·5; 15; 150; 300 a.i.)	1. Loam, pH 7·2; 1·49% carbon	1. Lowest two rates decreased numbers; top three rates initially decreased then increased numbers.	1. Naumann (1970b)
		2. Loam, pH 7·2; 1·49% carbon	2. Highest rate decreased numbers less at 12°–15°C than at 20°C during first three weeks. In 5th week, numbers greatly increased at 20°C.	2. Naumann (1972)
Phenyls	chlorfenvinphos (1·0; 10; 100)	Alluvial meadow	Two lowest rates, no effect. High rate inhibited activity.	Helweg (1972b)
Mixtures	toxaphene (600–6000)	Loam, pH 7·2	Slight stimulation in moist soil but numbers depressed in drier soil.	Naumann (1971b)

* "Recommended field rate".
Refer to Notes on tabulated data and abbreviations (p. 319).

TABLE 17

Some effects of insecticides in soil on *Rhizobium* spp. and nodulation of legumes

Group	Insecticide	Concentration (Kg or litres ha^{-1})	Soil type	Effects	Reference
Benzoics and related	lindane	(5·3; 106)	Clay	Low rate, slight effect on nodulation. High rate, inhibited nodulation.	Selim *et al.* (1970)
	lindane isobenzan	(2·2 g kg^{-1} seed) (2·2 g kg^{-1} seed)	Skeletal podzolic clay	Reduced nodulation.	Daitloff (1970a)
	lindane	(100)	Sandy	Decreased efficiency of nodules on lucerne. No effect on red clover.	Salem *et al.* (1971)
	lindane	(5; 50; 500)*	Alluvial	Decreased nodulation and efficiency of nodules.	Mishra and Gaur (1975)
	DDT	(0·1; 0·5; 1·0; 4·0; 10; 100)	Alluvial sandy loam, pH 7·8; 0·53% o.m. 1% manure added to some treatments	Lowest 2 rates increased nodulation. Middle 2 rates decreased nodulation. Highest 2 rates inhibited nodulation. Leghaemoglobin increased by lower rates. Manure alleviated effects.	Gaur and Pareek (1969); Pareek and Gaur (1970)
	DDT	(0·5–100)	Alluvial sandy loam, pH 7·8; 0·53% o.m. 1% manure added to some treatments	Rates of 0·5–4·0 kg increased nodulation and efficiency of N$_2$-fixation. Higher rates decreased nodulation. Manure alleviated effects.	Pareek and Gaur (1969)
Carbamates	carbaryl	(2·8; 14; 28)	Oven-dried soil.	No effect on numbers of *R. japonicum* or nodulation.	Kapusta and Rouwenhorst (1973)

Group	Chemical (rate)	Soil	Effect	Reference
Cyclodienes	dieldrin (5·9; 118)	Clay	No effect on nodulation.	Selim et al. (1970)
	dieldrin (2·2 g kg^{-1} seed)	Skeletal podzolic clay	Reduced nodulation.	Daitloff (1970a)
	endrin (2·2 g kg^{-1} seed)			
	endrin (1·1 a.i.)	(−)	No effect on nodulation or activity of rhizobia.	Swamiappan and Chandy (1975)
Dithioates and thioates	1. diazinon (80); 120)	1. Sandy chernozem	1. Decreased nodulation.	Salem et al. (1971)
	2. fonofos (1–100)* dimethoate (2·2 g kg^{-1} seed)	2. Chernozem Skeletal podzolic clay	2. Decreased nodulation. Decreased nodulation.	Daitloff (1970a)
	disulfoton (1·7; 8·4; 16·8)	Oven dry soil.	No effect on numbers of R. japonicum or nodulation.	Kapusta and Rouwenhorst (1973)
	phorate (1·1 a.i.)	(−)	Increased nodulation.	Swamiappan and Chandy (1975)
Phenyls	chlorfenvinphos (1·1 a.i.)	(−)	No effect on nodulation or activity of rhizobia.	Swamiappan and Chandy (1975)
Phosphonates	trichlorfon (1 5)*	Soil + manure	No effect on nodulation.	Salama et al. (1974)

* "Recommended field rate".
Refer to *Notes on tabulated data and abbreviations* (p. 319).

TABLE 18

Effect of insecticides on fungi and actinomycetes in soil

Insecticide				
Group	Concentration (Kg or litres ha^{-1})	Soil type	Effects	Reference
Benzoics and related	lindane (0·22; 4·4) lindane (30) BHC (30) BHC (1·0; 10; 100 a.i.)	Clay + 2% compost Rhizosphere soil (maize) 1. Loamy sand, pH 6·9 2. Clay loam, pH 7·7 3. Sandy loam, pH 7·5	Increased numbers of actinomycetes. Highly toxic. 1. No significant effect. 2. No significant effect. 3. Decreased fungi at highest rate.	Mahmoud et al. (1970) Tsirkov (1971) Tu (1975)
	DDT (112)	Sombric humo-ferric, mini humo-ferric and orthic ferro-humic pod-zols	No qualitative changes. No effects.	Salonius (1972)
Carbamates	aldicarb (0·5; 50 a.i.) propoxur (0·5; 50 a.i.)	Sandy loam, pH 7·4	Low rate, no effect. High rate increased numbers of fungi and actinomycetes.	Kuseske et al. (1974)
Cyclodienes	dieldrin (2·5; 50) dieldrin (150) heptachlor (150) heptachlor (1·1; 7·9; 55·3)	Clay + 2% compost Rhizosphere soil (maize) Silt loam	Decreased numbers of actinomycetes. Highly toxic Fungi decreased. Numbers of tolerant species increased.	Mahmoud et al. (1970) Tsirkov (1971) Shamiyeh and Johnson (1973)

Dithioates and thioates	diazinon (2; 20 a.i.)	Clay, pH 6·6; 2% o.m.	Large increase in actinomycete numbers.	Sethunathan and MacRae (1969)
	diazinon (1·0; 10) chlorpyrifos (1·0; 10) trichloronate (1·0; 10) thionazin (1·0; 10)	Sandy loam	Decreased fungi in first 7 days, then recovered.	Tu (1969)
	diazinon (3·4)	(−)	Large increase in numbers of actinomycetes. No effect on fungal population.	Gunner et al. (1966), Gunner (1970)
	malathion (4·2; 5; 6)	(−)	No effect on fungi. Increased actinomycete populations (especially anti-bacterial types)	Swaminathan and Sullia (1969)
	malathion (0·18) parathion-methyl (0·27)	(−)	No effects.	Yurovskaya and Zhulinskaya (1974)
	disulfoton (−)	Rhizosphere soil (cotton)	Decreased actinomycetes.	Mahmoud et al. (1972)
Phosphonates	chlorpyrifos (−) trichlorphon (1·3)	Clay loam, pH 5·3 (−)	Increased actinomycetes. No effects.	Sivasithamparam (1970) Yurovskaya and Zhulinskaya (1974)
Pentadiene derivatives	mirex (20; 200; 1000; 2000)	1. Sandy (pasture), 0·63% o.m. 2. Sandy (pasture), 0·98% o.m. 3. Sandy (forest), 4·94% o.m.	1. No effects. 2. No effects. 3. Large increase in Gliocladium catenulatum at high rate; two highest rates decreased, but lowest rate increased, actinomycetes.	Jones and Hodges (1974)

Refer to *Notes on tabulated data and abbreviations* (p. 319).

TABLE 19

Effects of insecticides on soil respiratory activity

Insecticide				
Group	Concentration (Kg or litres ha^{-1})	Soil type	Effects	Reference
Benzoics and related	lindane (0·1–100)	Alluvial	Low rates, no effect. Higher rates decreased CO_2 evolution.	Bardiya and Gaur (1968)
	lindane (6000)	Loam, pH 7·2; 1·49% carbon	Slightly stimulated activity.	Naumann (1971b)
	lindane (22 a.i.)	Sandy loam, calcareous clay and clay	CO_2-evolution decreased, then increased and returned to normal.	Gawaad et al. (1973a)
	lindane (1·0–100 a.i.)	Loamy sand, pH 6·9; 0·64% carbon. Sandy loam, pH 7·5; 3·42% carbon. Clay loam, pH 7·7; 2·3% carbon	Increased O_2-uptake.	Tu (1975)
Carbamates	1. aldicarb (0·5; 50)	Sandy loam, pH 7·4	1. Little effect on CO_2-evolution.	Kuseske et al. (1974)
	2. propoxur (0·5; 50)		2. Increased O_2-uptake due to presence of organic carrier. Insecticide had little effect.	
	propoxur (2·5; 12·5; 125)	Sandy silt-clay, pH 8·3; 0·67% o.m.	Two lowest rates decreased CO_2-evolution, but recovered. High rate decreased CO_2-evolution, after initial increase.	Gupta et al. (1975)
	Zectran (0·17)	Forest soil with or without litter	No effect on CO_2-evolution.	Bollen et al. (1970)

Group	Compound (rate)	Soil	Effect	Reference
Cyclodienes	aldrin (0·1–100) dieldrin (0·1–100)	Alluvial	Low rates, no effect. Higher rates decreased CO_2-evolution.	Bardiya and Gaur (1968)
Dithioates and thioates	diazinon (1·0; 10) chlorpyrifos (1·0; 10) thionazin (1·0; 10) trichloronate (1·0; 10) fonofos (11 a.i.)	Sandy loam	Increased respiration.	Tu (1969)
		Sandy loam, calcareous clay and clay	CO_2-evolution decreased, then increased, and returned to normal.	Gawaad et al. (1973a)
	Counter (0·1–10)	3 soils, pH 5·7; 6·3; 7·0	No effect on CO_2-evolution.	Laveglia and Dahm (1974)
Pentadiene derivatives	chlordecone (22 a.i.)	Sandy loam, calcareous clay and clay	CO_2-evolution decreased, then increased and returned to normal.	Gawaad et al. (1973a)
Pyrimidine	pirimiphos-ethyl (11 a.i.)	Sandy loam, calcareous clay and clay	CO_2-evolution decreased, then increased and returned to normal.	Gawaad et al. (1973a)
Mixture	camphechlor (600; 6000)	Loam, pH 7·2; 1·49% carbon	Slightly stimulated activity.	Naumann (1971b)

Refer to *Notes on tabulated data and abbreviations* (p. 319).

TABLE 20

Summary of pesticide effect ratios on microbial processes in soil

Parameter	Herbicides	Fungicides	Insecticides	Other Pesticides
Bacterial numbers	1·20	3·50	1·30	1·00
Nitrification	1·40	0·54	0·82	0·32
Denitrification	1·82	i.d.a.	i.d.a.	i.d.a.
Rhizobia and legume nodulation	0·94	1·00	0·78	i.d.a.
Free-living, N_2-fixation	1·65	i.d.a.	1·75	i.d.a.
Fungi and actinomycetes	1·09	0·5	1·43	0·55
Pathogens and their antagonists	0·81	4·00	i.d.a.	i.d.a.
Algae	0·45	i.d.a.	i.d.a.	i.d.a.
Cellulolytic activity and O.M. degradation	1·31	i.d.a.	1·10	0·62
Respiratory activity	0·91	0·40	2·00	1·40
Other enzymic activity	1·70	0·44	2·00	0·66
Ammonification	1·74	1·30	1·84	1·20

i.d.a. = insufficient data available.

Refer to *Notes on tabulated data and abbreviations* (p. 319).

Notes on ratios of effects of pesticides (p. 320).

References

ABDEL'YUSSIF, R. M., ZINCHENKO, V. A. and GRUZDEV, G. S. (1976). The effect of nematicides on the biological activity of soil. *Izv. timiryazev. sel.'-khoz. Akad.* No. 1, 206–214. (*Soils and Fert.* **39**, 3982).

ABUEVA, A. A. (1970). The effect of 2,4-D on the growth of micro-organisms and on nitrogen metabolism in a derno-podzolic soil. *Khimiya sel'Khoz.* **8**, 601–604. (*Weed Abstr.* **21**, 1122).

ABUEVA, A. A. and BAGAEV, V. B. (1975). The transformation of fertilizer nitrogen in a derno-podzolic soil under the influence of 2,4-D. *Izv. timiryazev. sel.'-khoz. Akad.* No. 2, 127–130. (*Soils and Fert.* **38**, 4167).

AFIFI, N. M., MOHARRAM, A. A., HAMDI, Y. A. and ABDEL-MALEK, Y. (1969). Sensitivity of *Rhizobium* species to certain fungicides. *Arch. Mikrobiol.* **66**, 121–128.

AGNIHOTRI, V. P. (1971). Persistence of Captan and its effects on microflora, respiration and nitrification of a forest nursery soil. *Can. J. Microbiol.* **17**, 377–383.

AGNIHOTRI, V. P. (1973). Effect of Dexon on soil microflora and their ammonification and nitrification activities. *Indian J. exptl. Biol.* **11**, 213–216.

AKOPYAN, E. A. and AGARONYAN, A. G. (1968). The effect of monuron, diuron and simazine on the microflora of vineyard soils under Armenian conditions. *Khimiya sel'Khoz.* **6**, 50–51. (*Soils and Fert.* **33**, 3133).

AKOPYAN, E. A., AVAKYAN, B. P. and KALANTAROV, A. A. (1974). The effect of dalapon on soil microflora of vineyards on the Ararat plain. *Khimiya sel'Khoz.* **12**, 538–539. (*Weed Abstr.* **24**, 1364).

ALTMAN, J. (1969). Effect of chlorinated C_3 hydrocarbons on amino acid production by indigenous soil bacteria. *Phytopathology* **59**, 762–766.

ANDERSON, J. P. E. and LICHTENSTEIN, E. P. (1972). Effects of various soil fungi and insecticides on the capacity of *Mucor alternans* to degrade DDT. *Can. J. Microbiol.* **18**, 553–560.

ANDERSON, J. R. and DREW, E. A. (1976a). Effects of pure paraquat dichloride, 'Gramoxone W' and formulation additives on soil microbiological activities. I. Estimation of soil biomass in laboratory treated soil. *Zbl. Bakt.* Abt. II, **131**, 125–135.

ANDERSON, J. R. and DREW, E. A. (1976b). Effects of pure paraquat dichloride, 'Gramoxone W' and formulation additives on soil microbiological activities. II. Effects on respiration, mineralisation and nitrification in laboratory treated soil. *Zbl. Bakt.* Abt. II, **131**, 136–147.

ANDERSON, J. R. and DREW, E. A. (1976c). Effects of pure paraquat dichloride, 'Gramoxone W' and formulation additives on soil microbiological activities. III. Estimations of soil microflora and enzyme activity in field treated soil. *Zbl. Bakt.* Abt. II, **131**, 247–258.

ANDREEVA-FETVADZHIEVA, N. and KOLCHEVA, B. (1969). Influence of some triazine herbicides on the nitrate nitrogen content of a chernozem smolnitsa. *Nauchni Trud. vissh selkostop. Inst. Georgi Dimitrov.* **20**, 49–69. (*Weed Abstr.* **20**, 1870).

ANDRUSZEWSKA, A. (1973). Investigations of the influence of herbicides, used in forestry, on soil fungi especially on mycorrhizal fungi. *Zesz. nauk. Akad. Rolnicz. Szczecin*. No. 39, 3–21. (*Weed Abstr.* **24**, 1185).

ARSHIDINOV, A. A., ISIN, M. M. and ZHARASOV, Sh. U. (1974). The effect of herbicides on microbiological activity and the nutrient status of soil under a young orchard on the slopes of the Zailiiskii Alatau. *Khimiya sel'Khoz.* **12**, 132–134. (*Weed Abstr.* **23**, 2040).

ARVIK, J. H., WILLSON, D. L. and DARLINGTON, L. C. (1971). Response of soil algae to picloram-2,4-D mixtures. *Weed Sci.* **19**, 276–278.

ATKINS, C. A. and TCHAN, Y. T. (1967). Study of soil algae. 6. Bioassay of atrazine and the prediction of its toxicity in soils using an algal growth method. *Pl. Soil* **27**, 432–442.

ATKINSON, G. (1973). Effects of diquat on the microbiological organisms in the soil of lakes. Proc. Polln. Res. Conf., June 1973, Wairakei, N.Z. DSIR Information Series No. 97, 529–539.

AU, F. H. F. (1968). Effect of endothal and certain other selective herbicides on microbial activity in six different soils. *Ph.D. Thesis, Oregon State University.* (*Weed Abstr.* **18**, 966).

AUDUS, L. J. (1970a). The action of herbicides and pesticides on the microflora. *Meded. Fac. Landb. Rijksuniv. Gent* **35**, 465–492.

AUDUS, L. J. (1970b). The action of herbicides on the microflora of the soil. *Proc. 10th Br. Weed Control Conf.* 1036–1051.

AVROV, O. E. (1966). Effect of herbicides on nodule bacteria and nodule formation in legumes. *Dokl. vses. Akad. sel'.-khoz. Nauk* No. 3, 16–19. (*Weed Abstr.* **16**, 673).

AVROV, O. E., BELOUS, A. G., ZHURBINA, N. S. and ZAVERYUKHIN, V. I. (1968). The effect of various herbicides on nodule bacteria of soybeans. *Khimiya sel'Khoz.* **6**, 767–768. (*Weed Abstr.* **19**, 1358).

BABAK, N. M. (1968). The sensitivity of *Azotobacter* to some antibiotics and herbicides. *Microbiology (Mikrobiologiya)* **37**, 283–288.

BACKMAN, P. A., RODRIGUEZ-KABANA, R. and WILLIAMS, J. C. (1975). The effect of peanut leafspot fungicides on the non-target pathogen *Sclerotium rolfsii*. *Phytopathology* **65**, 773–776.

BAKALIVINOV, D. (1971). Effect of the herbicide atrazine on the microflora and its detoxication in calcareous chernozem. In "Vtori Kongres po Mikrobiologiya" (I. Pashev, Ed.), IV, Chast, Sofia, 1969, *Bulg. Akad. Nauk* (1971), 259–262. (*Soils and Fert.* **36**, 607).

BAKALIVINOV, D. (1972). Biological activity of certain herbicides on microscopic soil fungi. *Symp. Biol. Hung.* **11**, 373–377. (*Weed Abstr.* **22**, 1920).

BAKALIVINOV, D. and NIKOLOVA, G. (1971). The effect of the herbicides 2,4-D and Dicotex on the microflora of alluvial soil. In "Vtori Kongres po Mikrobiologiya" (I. Pashev, Ed.), IV, Chast, Sofia, 1969, *Bulg. Akad. Nauk* (1971), 263–266. (*Weed Abstr.* **22**, 2451).

BAKALIVINOV, D. and NIKOLOVA, G. (1972). Effect of simazine on the microflora of a smonitza. *Prvi Nationalen Kongres po Pochvoznanie* Bulgaria 1969 (1972), 221–226. (*Soils and Fert.* **36**, 3617).

BALASUBRAMANIAN, A. and SIDDARAMAPPA, R. (1974). The effect of simazine application on certain microbiological and chemical properties of a red sandy loam soil. *Mysore J. agric. Sci.* **8**, 214–219.

BALASUBRAMANIAN, A., SIDDARAMAPPA, R. and OBLISAMI, G. (1973). Effect of simazine and Dithane M-45 on nitrification in soil. *Pesticides* **7**, 14.

BALEZINA, L. S. (1967). Effect of some herbicides on development of soil algae. *Mikrobiologiya* **36**, 163–167.

BALICKA, N. (1969). The effect of herbicides on soil microflora. 6. The effect of pH on susceptibility of some bacteria to herbicides. *Acta microbiol. pol.* (Ser. B). 1 **(18)**, 43–45. (*Weed Abstr.* **19**, 1346).

BALICKA, N. and BILODUB-PANTERA, H. (1964). The influence of atrazine on some soil bacteria. *Acta microbiol. pol.* **13**, 149–152.

BALICKA, N. and KREZEL, Z. (1967). The influence of herbicides on antagonism between various bacteria. *Abstr. 6th Int. Congr. Pl. Protection*, Vienna (1967), 386–387.

BALICKA, N. and KREZEL, Z. (1969). The influence of herbicides upon the antagonism between *Bacillus* sp. and *Pseudomonas phaseoli*. *Weed Res.* **9**, 37–42.

BALICKA, N. and SOBIESZCZANSKI, J. (1969a). The effect of herbicides on soil microflora. 1. The effect on the number of soil micro-organisms in a field experiment. *Acta microbiol. pol.* 1 **(18)**, 3–6. (*Soils and Fert.* **32**, 3682).

BALICKA, N. and SOBIESZCZANSKI, J. (1969b). The effect of herbicides on soil microflora. 3. The effect of herbicides on ammonification and nitrification in soil. *Acta microbiol. pol.* 1 **(18)**, 7–10. (*Soils and Fert.* **32**, 3683).

BALICKA, N., SOBIESZCZANSKI, J. and NIEWIADOMA, T. (1969). The effect of herbicides on soil microflora. 4. The action of herbicides on soil microorganisms. *Acta microbiol. pol.* 1 **(18)**, 11–14.

BALKOVA, E. N. and SEMIKHATOVA, O. A. (1969). Soil and rhizosphere microflora under the influence of EPTC in sugarbeet plantings. *Khimiya sel'Khoz.* **7**, 846–849. (*Weed Abstr.* **20**, 2400).

BARDIYA, M. C. and GAUR, A. C. (1968). Influence of insecticides on carbon dioxide evolution from soil. *Indian J. Microbiol.* **8**, 233–238.

BARDIYA, M. C. and GAUR, A. C. (1970). Effect of some chlorinated hydrocarbon insecticides on nitrification in soil. *Zbl. Bakt.* Abt. II, **124**, 552–555.

BARNES, R. D., BULL, A. T. and POLLER, R. C. (1973). Studies on the persistence of the organotin fungicide fentin acetate (triphenyltin acetate) in the soil and on surfaces exposed to light. *Pestic. Sci.* **4**, 305–315.

BARONA, V. P. (1974). Mixtures of selective herbicides and their effects on soil nutrients and microflora. *Agrokhimiya* **11**, 131–134. (*Weed Abstr.* **23**, 2723).

BAROUX, J. and SECHET, J. (1974). The toxicity of copper towards the microflora of vineyard soils. *Annali di Microbiol. ed Enzimol.* **24**, 125–136.

BECK, T. (1969). Model trials on the behaviour of monolinuron. 3. Influence on the biological activity of soil in comparison with other soil herbicides. *Mitt. biol. Bund Anst. Ld-U. Forstiv.* **132**, 71–72.

BECKMANN, E. O. and PESTEMER, W. (1975). The effect of various organic manures on herbicide decomposition and biological activity in soil. *Landw. Forsch.* **28**, 24–33.

BELLES, W. S. (1972). Effect of dinitramine on soil microorganisms. *Proc. N. cent. Weed Control Conf.* **27**, 50–51.

BERTAGNOLLI, B. L. and NADAKAVUKAREN, M. J. (1970). Effect of 2,4-dichloro-phenoxyacetic acid on the fine structure of *Chlorella pyrenoidosa* Chick. *J. Phycol.* **6**, 98–100.

BERTHOLD, W. (1970). Investigations on the enzyme activities of herbicide-treated vineyard soils. *7th Int. Congr. Pl. Protection, Paris* 1970, 295–297. (*Weed Abstr.* **20**, 2391).

BERTRAND, D. and de WOLF, A. (1972). The influence of certain herbicides on the soil microflora and agronomic effect. *C.r. hebd. Séanc. Acad. Agric. Fr.* **58**, 1469–1473.

BEZUGLOV, V. G., MINENKO, A. K. and SHELESTOV, E. P. (1973). The effect of dicamba, Tordon 22K and Lumeton on weeds and soil microflora. *Khimiya sel'Khoz.* **11**, 854–856. (*Weed Abstr.* **23**, 2917).

BLIEV, O. K. (1973). Effect of herbicides on biological activity of soils. *Pochvovedenie* **7**, 61–68.

BLIEV, Yu. K. and KOZLOVA, L. M. (1972). The effect of different rates of herbicides on soil fertility. *Khimiya sel'Khoz.* **10**, 850–853. (*Weed Abstr.* **22**, 1210).

BLYTHE, T. O. and FRANS, R. E. (1972). The effect of fluometuron on the growth of *Chlorella pyrenoidosa*. *Proc. Sth. Weed Sci. Soc.* (1972), 420.

BOBYSHEV, V. G. and LAPCHENKOV, G. Ya. (1966). The effect of herbicides on soil microflora. *Khimiya sel'Khoz.* **4**, 30–31. (*Weed Abstr.* **17**, 1937).

BOIKO, V. S., DEGTYAREVA, M. G. and KAZAKOVA, I. P. (1969). The effect of 2,4-D and simazine on the microflora and nutrient status of grey forest soils. *Khimiya sel'Khoz.* **7**, 203–207. (*Weed Abstr.* **19**, 899).

BOLLAG, J. M. and NASH, C. L. (1974). Effect of chemical structure of phenylureas and anilines on the denitrification process. *Bull. Environ. Contam. Toxicol.* **12**, 241–248.

BOLLEN, W. B. and TU, C. M. (1971). Influence of endrin on soil microbial populations and their activity. *U.S. Dept. Agr. Forest Serv. Res. Paper* PNW-114, 4 p.

BOLLEN, W. B., LU, K. C. and TERRANT, R. F. (1970). Effect of Zectran on microbial activity in a forest soil. *U.S. Dept. Agr. Forest Serv. Res. Note* PNW-124, 10 p.

BOLLEN, W. B., NORRIS, L. A. and STOWERS, K. L. (1974). Effect of cacodylic acid and MSMA on microbes in forest floor and soil. *Weed Sci.* **22**, 557–562.

BONE, H. T. and KUNTZ, J. E. (1968). The breakdown of Dacthal and its effects on certain processes of soil microbes. *Proc. N. cent. Weed Control Conf.* **23**, 34.

BORBÉLY, I. and KECSKÉS, M. (1972). The influence of ureas and s-triazines on rhizobia and grain yield of *Lupinus luteus* L. *Symp. Biol. Hung.* **11** 437–444. (*Weed Abstr.* **22**, 1917).

BOZARTH, G. A. and TWEEDY, B. G. (1970). Interaction of some pesticides with *Sclerotium rolfsii*. *Phytopathology* **60**, 1285.

BOZARTH, G. A., FUNDERBURK, H. H. and CURL, E. A. (1968). The effect of fluometuron on soil respiration and degradation of ^{14}C-fluometuron in sandy loam soil. *Proc. Sth. Weed Conf.* **21**, 340.

BOZARTH, G. A., FUNDERBURK, H. H. and CURL, E. A. (1969). Interaction of fluometuron and soil microorganisms. *Abstr. Meet. Weed Sci. Soc. Am. 1969*, 236.

BREAZEALE, F. W. and CAMPER, N. D. (1970). Bacterial, fungal and actinomycete populations in soils receiving applications of 2,4-dichlorophenoxyacetic acid and trifluralin. *Appl. Microbiol.* **19**, 379–380.

BROCK, J. L. (1972). Effects of the herbicides trifluralin and carbetamide on nodulation and growth of legume seedlings. *Weed Res.* **12**, 150–154.

BUDOI, G. and BUDOI, I. (1967). The effect of atrazine on some physical and biochemical properties of the soil and on the stock of weed seeds in the soil. *Lucr. stiint. Inst. Agron. Nicolae Balcescu* (Ser. A), **10**, 369–379. (*Weed Abstr.* **19**, 1809).

BUSHWAY, R. J. and HANKS, A. R. (1976). Pesticide inhibition of growth and macromolecular synthesis in the cellular slime mold *Dictyostelium discoideum*. *Pestic. Biochem. Physiol.* **6**, 254–260.

CAMPER, N. D., MOHEREK, E. A. and HUFFMAN, J. (1973). Changes in microbial populations in paraquat-treated soil. *Weed Res.* **13**, 231–233.

CASELEY, J. C. and BROADBENT, F. E. (1968). The effect of five fungicides on soil respiration and some nitrogen transformations in yolo fine sandy loam. *Bull. Environ. Contam. Toxicol.* **3**, 58–64.

CATROUX, G. and FOURNIER, J-C. (1971). Effect of N-benzoyl-N-(3,4-dichloro-phenyl)-N'N'-dimethylurea on the microflora of soils. *C.r. 6ᵉ Conf. Commte Francais de Lutte contre les Mauvaises Herbes (Columa) 1971*, 69–78. (*Weed Abstr.* **21**, 2666).

CHANDRA, P. (1964). Herbicide effects on certain soil microbial activities in some brown soils of Saskatchewan. *Weed Res.* **4**, 54–63.

CHANDRA, P. and GROVER, R. (1967). Residual effect of numerous herbicides on the soil microorganisms and their activities in a prairie soil of Canada. *Abstr. 6th Int. Congr. Pl. Protection, Vienna 1967*, 604–605.

CHANDRA, P., FURTICK, W. R. and BOLLEN, W. B. (1960). The effects of four herbicides on microorganisms in nine Oregon soils. *Weeds* **8**, 589–598.

CHINN, S. H. F. (1973). Effect of eight fungicides on microbial activities in soil as measured by a bioassay method. *Can. J. Microbiol.* **19**, 771–777.

CHOPRA, B. K. and CURL, E. A. (1968). Effect of prometryne on sporulation and fungistasis of *Fusarium oxysporum* f. sp. *vasinfectum* in soil. *Phytopathology* **58**, 1047.

CHOPRA, B. K., CURL, E. A. and RODRIGUEZ-KABANA, R. (1968). Effect of prometryne on growth of *Fusarium oxysporum* f. sp. *vasinfectum* in soil culture. *Phytopathology* **58**, 726–727.

CHOPRA, B. K., CURL, E. A. and RODRIGUEZ-KABANA, R. (1970). Influence of prometryne in soil on growth-related activities of *Fusarium oxysporum* f. sp. *vasinfectum*. *Phytopathology* **60**, 717–722.

CHOPRA, B. K., COLLINS, L. C. and OSUJI, F. (1973). Effects of Cotoran (fluometuron) on the growth of *Aspergillus flavus* in soil and liquid culture. *Acta Phytopath. Acad. Scient. Hung.* **8**, 43–47.

CHOWDHURY, A. K., MUKHOPADHYAY, A. K. and MUKHOPADHYAY, S. (1972). Influence of certain herbicides on fungal population in soil. *Indian Phytopath.* **25**, 188–194.

CHULAKOV, Sh. A. and ZHARASOV, Sh. U. (1967). Effects of herbicides on the microflora population of light chestnut soils of the Alma-Ata region. *Vest. sel'Khoz. Nauki, Alma-Ata* **10**, 122–124. (*Weed Abstr.* **19**, 2630).

CHULAKOV, Sh. A. and ZHARASOV, Sh. U. (1971). Root absorbed herbicides and soil microorganisms. *Vest. Akad. Nauk kazakh. SSR* **27**, 35–40. (*Weed Abstr.* **21**, 2655).

CHULAKOV, Sh. A. and ZHARASOV, Sh. U. (1973). The biological activity of southern soils of Kazakhstan with the use of herbicides. *Izv. Akad. Nauk. kazakh. SSR, Ser. Biol.* **11**, 7–13. (*Weed Abstr.* **22**, 3038).

CHUNDEROVA, A. I., SOFINSKII, A. M. and ZUBETS, T. P. (1968). The toxicity of herbicides to soil microflora in sugarbeet crops. *Khimiya sel'Khoz.* **6**, 280–282. (*Weed Abstr.* **18**, 465).

CHUNDEROVA, A. I., ZUBETS, T. P. and SOFINSKII, A. M. (1971). The effect of herbicides on the soil microflora with their systemic use in the crop rotation. *Khimiya sel'Khoz.* **9**, 527–530. (*Weed Abstr.* **21**, 3267).

CLERK, G. C. and BIMPONG, C. E. (1969). Influence of three herbicides on germination of sclerotium of *Sclerotium rolfsii. Ghana J. Sci.* **9**, 115–118.

CORKE, C. T. and THOMPSON, F. R. (1970). Effect of some phenylamide herbicides and their degradation products on soil nitrification. *Can. J. Microbiol.* **16**, 567–571.

COWLEY, G. T. and LICHTENSTEIN, E. P. (1970). Growth inhibition of soil fungi by insecticides and annulment of inhibition by yeast extract or nitrogenous nutrients. *J. gen. Microbiol.* **62**, 27–34.

CURL, E. A. and RODRIGUEZ-KABANA, R. (1971). Influence of herbicides on sclerotial production by *Sclerotium rolfsii* and *Sclerotina trifoliorum* in soil. *Fitopatologia* **5**, 10–14. (*Weed Abstr.* **23**, 693).

CURL, E. A., RODRIGUEZ-KABANA, R. and FUNDERBURK, H. H. (1966). Effect of carbon and nitrogen amendments on growth of *Sclerotium rolfsii* and *Trichoderma viride* in atrazine-treated soil. *Phytopathology* **56**, 874.

CURL, E. A., RODRIGUEZ-KABANA, R. and FUNDERBURK, H. H. (1967). Effect of dipyridyl and toluidine herbicides on growth response of *Sclerotium rolfsii* in soil. *Phytopathology* **57**, 7.

CURL, E. A., TANG, A. and RODRIGUEZ-KABANA, R. (1969). Effect of trifluralin on production and germination of chlamydospores of *Fusarium oxysporum* f. sp. *vasin-fectum* in soil. *Phytopathology* **59**, 1023.

DAITLOFF, A. (1970a). The effects of some pesticides on root nodule bacteria and subsequent nodulation. *Aust. J. exp. Agric. Anim. Husb.* **10**, 562–567.

DAITLOFF, A. (1970b). Overcoming fungicide toxicity to *Rhizobia* by insulating with a polyvinyl acetate layer. *J. Aust. Inst. agric. Sci.* **36**, 293–294.

DANIELSON, R. M. and DAVEY, C. B. (1970). Microbial recolonization of a fumigated nursery soil. *Forest Sci.* **15**, 368–380.

DAVISON, C. L. and PAPIRMEISTER, B. (1971). Bacteriostasis of *Escherichia coli* by the herbicide paraquat. *Proc. Soc. exp. Biol. Med.* **136**, 359–364.

DE, B. K. and MUKHOPADHYAY, S. (1971). Effect of MCPA and Stam F-34 on the occurrence of some nutritional groups of bacteria in the rice fields of West Bengal. *Int. Rice Commn. Newsl.* **20**, 35–39.

DEBONA, A. C. and AUDUS, L. J. (1970). Studies on the effects of herbicides on soil nitrification. *Weed Res.* **10**, 250–263.

DECALLONE, J. R., GENOT, M. and MEYER, J. A. (1975). Effects of benomyl, carbendazim and thiophanates on respiration and oxidative phosphorylation of *Fusarium oxysporum* and *Saccharomyces cerevisiae*. *Pestic. Sci.* **6**, 113–120.

DESHMUKH, V. A. and SHRIKHANDE, J. G. (1974). Effect of pre- and post-emergence treatment of herbicides on soil microflora and two microbial processes. *J. Indian Soc. Soil Sci.* **22**, 36–42.

DICKINSON, C. H. (1973). Interactions of fungicides and leaf saprophytes. *Pestic. Sci.* **4**, 563–574.

DOBEREINER, J. (1968). Non-symbiotic nitrogen-fixation in tropical soils. *Pesq. Agrapec. Bras.* **3**, 1–6.

DOMSCH, K. H. (1960). Die Wirkung von Bodenfungiciden. V. Empfindlichkeit von Bodenorganismen *in vitro*. *Z. PflKrankh. PflPath. PflSchutz.* **67**, 213–216.

DOMSCH, K. H. (1970). Effects of fungicides on microbial populations in soil. *In* "Pesticides in the Soil: Ecology, Degradation and Movement" (G. E. Guyer, Ed.), pp. 42–46. Michigan State University, East Lansing.

DOMSCH, K. H. and PAUL, W. (1974). Simulation and experimental analysis of the influence of herbicides on soil nitrification. *Arch. Mikrobiol.* **97**, 283–301.

DOXTADER, K. G. (1968). Inhibition of microbial growth by phenylurea herbicides. *Bact. Proc.* **4**, A21.

DOXTADER, K. G. and ALEXANDER, M. (1966). Nitrification by heterotrophic soil organisms. *Proc. Soil Sci. Soc. Am.* **30**, 351–355.

DUBEY, H. D. (1969). Effect of picloram, diuron, ametryne and prometryne on nitrification in some tropical soils. *Proc. Soil Sci. Soc. Am.* **33**, 893–896.

DUBEY, H. D. (1970). A nitrogen deficiency disease of sugarcane probably caused by repeated pesticide application. *Phytopathology* **60**, 485–487.

DUBEY, H. D. and RODRIGUEZ, R. L. (1970). Effect of Dyrene and maneb on nitrification and ammonification and their degradation in tropical soils. *Proc. Soil Sci. Soc. Am.* **34**, 435–439.

DUBEY, H. D. and RODRIGUEZ, R. L. (1974). Changes in soil microflora following application of fungicides Dyrene and maneb to tropical soils. *J. Agric. Univ. P. Rico* **58**, 78–86.

DUBEY, H. D., RIERA, A. and RODRIGUEZ, R. L. (1975). Effect of the nematicides Nemagon and DD on mineralization, nitrification, soil microbial population and soil fertility status of two tropical soils. *J. Agric. Univ. P. Rico* **59**, 43–50.

DUNIGAN, E. P., ALLEN, L. D. and FREY, J. P. (1970). Effects of selected herbicides on the nodulation of soyabeans. *La Agric.* **13**, 6–7.

DUTSCH, G. A. (1975). Decreasing the biological activity of benomyl by the use of various types of casing soils in mushroom cultivation. *Phytopath. Z.* **83**, 14–22.

EBBELS, D. L. (1970). Effect of fumigants on fungi in buried wheat straw. *Trans. Br. mycol. Soc.* **54**, 227–232.

EGLITE, A. K. (1969). The effect of herbicides on the microbiological activity and mineralization of nitrogen in nursery and forest soils. *Riga. Zinante* 1969, 157–168. (*Weed Abstr.* **21**, 2658).

EISENHARDT, A. R. (1975). Influence of the insecticide phoxim on symbiotic and non-symbiotic nitrogen-fixation determined by the acetylene reduction method. *Tidsskr. PlAvl.* **79**, 254–258.

EL-KHADEM, M., NAGIB, M. M. and TEWFIK, M. S. (1973). The effect of certain nitro-herbicides on the growth of some soil inhabiting fungi. *Zbl. Bakt.* Abt. II, **128**, 780–786.

ELLIOT, J. M., MARKS, C. F. and TU, C. M. (1974). Effect of the nematicides D-D and Mocap on soil nitrogen, soil microflora, population of *Pratylenchus penetrans* and flue-cured tobacco. *Can. J. Pl. Sci.* **54**, 801–809.

EYLAR, J. R. and SCHMIDT, E. L. (1959). A survey of heterotrophic micro-organisms from soil for ability to form nitrite and nitrate. *J. gen. Microbiol.* **20**, 473–481.

FARLEY, J. D. and LOCKWOOD, J. L. (1968). The suppression of actinomycetes by PCNB in culture media used for enumerating soil bacteria. *Phytopathology* **58**, 714–715.

FARLEY, J. D. and LOCKWOOD, J. L. (1969). Reduced nutrient competition by soil microorganisms as a possible mechanism for pentachloronitrobenzene-induced disease accentuation. *Phytopathology* **59**, 718–724.

FARMER, F. H., BENOIT, R. E. and CHAPPELL, W. E. (1965). Simazine, its effects on nitrification and decomposition by soil micro-organisms. *Proc. N. E. Weed Control Conf.* **19**, 350–354.

FIELDS, M. L. and HEMPHILL, D. D. (1967). The influence of DMPA on damping-off of peas by *Pythium debaryanum* ATCC 9998. *Weeds* **15**, 281–282.

FIELDS, M. L. and HEMPHILL, D. D. (1968). Influence of siduron and its degradation products on soil microflora. *Weed Sci.* **16**, 417–420.

FIELDS, M. L., DER, R. and HEMPHILL, D. D. (1967). Influence of DCPA on selected soil microorganisms. *Weeds* **15**, 195–197.

FINK, R. J., FLETCHALL, O. H. and CALVERT, O. H. (1968). Relation of triazine residues to fungal and bacterial colonies. *Weed Sci.* **16**, 104–105.

FISHER, D. J. (1976). Effects of some fungicides on *Rhizobium trifolii* and its symbiotic relationship with white clover. *Pestic. Sci.* **7**, 10–18.

FISHER, D. J. and CLIFTON, G. (1976). Effect of fungicides on *Rhizobium. Long Ashton Research Station Report,* **1975** (publ. 1976), 126.

FISYUNOV, A. V. (1969a). Effect of herbicides on nitrate content and nitrification in ordinary chernozem. *Agrokhimiya* No. 2, 122–126.

FISYUNOV, A. V. (1969b). Effect of triazine derivatives on soil respiration. *Agrok-himiya* No. 3, 112–116.

FOLDESY, R. G., MINEO, L. and MAJUMDAR, S. K. (1972). The effect of 2,4,5-trichlorophenoxyacetic acid on two nitrogen-fixing bacteria. *Proc. Pa Acad. Sci.* **46**, 23–24.

GAPONENKOV, T. K. and NOGUMANOVA, S. T. (1969). The activity of pectolytic enzymes under the influence of herbicides. *Sel'khoz. Biol.* **4**, 298–299. (*Weed Abstr.* **19**, 2632).

GARCIA, M. M. and JORDAN, D. C. (1969). Action of 2,4-DB and dalapon on the symbiotic properties of *Lotus corniculatus* (birdsfoot trefoil). *Pl. Soil* **30**, 317–333.

GARRETSON, A. L. and SAN CLEMENTE, C. L. (1968). Inhibition of nitrifying chemolithotrophic bacteria by several insecticides. *J. econ. Ent.* **61**, 285–288.

GAUR, A. C. and MISRA, K. C. (1972). Effect of 2,4-dichlorophenoxyacetic acid on the growth of *Rhizobium* spp. *in vitro. Indian J. Microbiol.* **12**, 45–46.

GAUR, A. C. and PAREEK, R. P. (1969). Effect of dichlorodiphenyl-trichloroethane (DDT) on leghaemoglobin content of root nodules of *Phaseolus aureus* (green gram). *Experientia* **25**, 777.

GAUR, A. C. and PAREEK, R. P. (1971). Tolerance of nitrification to DDT. *Indian J. Ent.* **33**, 368–370.

GAWAAD, A. A. A., EL-MINSHAWY, A. M. and ZEID, M. (1972a). Studies on soil insecticides. VIII. Effect of some soil insecticides on broad beans and Egyptian clover nodule forming bacteria. *Zbl. Bakt.* Abt. II, **127**, 290–295.

GAWAAD, A. A. A., HAMMAD, M. H. and EL-GAYAR, F. H. (1972b). Studies on soil insecticides. X. Effects of some soil insecticides on soil micro-organisms. XI. Effect of some insecticides on the nitrogen transformations in treated soils. *Zbl. Bakt.* Abt. II, **127**, 290–300.

GAWAAD, A. A. A., HAMMAD, M. H. and EL-GAYAR, F. H. (1973a). Effect of some insecticides on soil microorganisms. *Agrokém. Talajt.* **22**, 161–168.

GAWAAD, A. A. A., HAMMAD, M. H. and EL-GAYAR, F. H. (1973b). Effect of insecticides on nitrogen transformations in soil. *Agrokém. Talajt.* **22**, 169–174.

GEL'TSER, Yu. G., GEPTNER, V. A. and STONOV, L. D. (1972). The effect of herbicides on the microorganisms of the silt and water in collector drains of the chardzhou Oasis, Turkmenian SSR. *Agrokhimiya* **9**, 119 123. (*Weed Abstr.* **22**, 249).

GERGELY, Z. (1973). Effect of herbicides on some *Nocardia* strains. *Agrartudomanyi Egyetem Kozlemenyei* (1973), 165–172. (*Soils and Fert.* **38**, 2295).

GESHTOVT, Yu. N., ZHARASOV, Sh. U. and ADEKENOV, N. (1970). The effect of herbicides on the microflora of dark chestnut soils under maize crops in the Karagandinskaya Province. *Vest. sel'Khoz. Nauki, Kazakhskoi SSR.* **13**, 35–38. (*Weed Abstr.* **21**, 2106).

GIARDINA, M. C., TOMATI, U. and PIETROSANTI, W. (1970). Hydrolytic activities of the soil treated with paraquat. *Meded. Fac. Landb. Rijksuniv. Gent* **35**, 615–626.

510   J. R. ANDERSON

GIL DIAZ-ORDOÑEZ, I., MORALES, J. and MARTIN GONZALES, A. (1970). Effect of copper oxychloride on the respiration of *Azotobacter. Meded. Fac. Landb. Rijksuniv. Gent* **35**, 493–496.

GILLBERG, B. O. (1971). On the effects of some pesticides on *Rhizobium* and isolation of pesticide-resistant mutants. *Arch. Mikrobiol.* **75**, 203–208.

GIZBULLIN, N. G. and SOLOV'EVA, E. P. (1968). Effect of the herbicides Alipur, Murbetol and Dalapon on soil microflora. *Microbiol. Zh. Kiev* **30**, 354–357. (*Soils and Fert.* **32**, 1207).

GOGUADZE, V. D. (1966). The effect of herbicides on soil microflora. *Khimiya sel'Khoz.* **4**, 41–45. (*Weed Abstr.* **17**, 451).

GOGUADZE, V. D. (1968). Effects of herbicides on the development of azotobacters in some soils of western Georgia. *Agrokhimiya* **5**, 99–103. (*Soils and Fert.* **31**, 2566).

GOGUADZE, V. D. (1969). Changes in the composition of soil microflora with the use of herbicides at different stages of development of grape vines on alluvial soil. *Subtrop. Kul'tury* No. 5, 139–143.

GOGUADZE, V. D. and GOGIYA, L. N. (1971). The effect of herbicides on the development of soil microflora in cinnamon – alluvial – skeletal soils of Abkhaziya. *Subtrop. Kul'tury* No. 4, 161–167.

GOGUADZE, V. D., GOBRONIDZE, E. A. and GOGIYA, L. G. (1972). The residual effect of the herbicide Tordon 22K on the microflora and nutrient status of the soil. *Subtrop. Kul'tury* No. 3, 157–159.

GORING, C. A. I., GRIFFITH, J. D., O'MELIA, F. C., SCOTT, H. H. and YOUNGSON, C. R. (1967). The effect of Tordon on microorganisms and soil biological processes. *Down to Earth* **22**, 14–17.

GORLENKO, M. V., LEBEDEVA, G. F. and MANTUROVSKAYA, N. V. (1969). Triazine derivatives and the soil mycoflora. *Agrokhimiya* **6**, 122–128. (*Soils and Fert.* **33**, 1250).

GOWDA, T. K. S. and PATIL, R. B. (1972). Effect of pesticides applied to the soil on the biological activity of the soil. *Biochem. J.* **128**, 56P–57P.

GRAU, C. R. (1974). The effect of Treflan on root diseases of pea. *Proc. Am. Phytopath. Soc.* **1**, 29.

GRINSTEIN, A., KATAN, J. and ESHEL, Y. (1976). Effect of dinitroaniline herbicides on plant resistance to soil-borne pathogens. *Phytopathology* **66**, 517–522.

GROSSBARD, E. (1970). Effect of herbicides on the symbiotic relationship between *Rhizobium trifolii* and white clover. *Occasional Symp. Br. Grassland Soc. 6: White Clover Research*, 47–59.

GROSSBARD, E. (1971). The effect of repeated field applications of four herbicides on the evolution of carbon dioxide and mineralization of nitrogen in soil. *Weed Res.* **11**, 263–275.

GROSSBARD, E. and MARSH, J. A. P. (1974). The effect of several substituted urea herbicides on the soil microflora. *Pestic. Sci.* **5**, 609–623.

GROVER, R. (1972). Effect of picloram on some soil microbial activities. *Weed Res.* **12**, 112–114.

GRUZDEV, G. S., MOZGOVOI, A. F. and KUZ'MINA, I. V. (1973). Effect of treflan on biological activity of soil under sunflower. *Izv. timiryazev. sel.'-khoz. Akad.* No. 4, 136–141. (*Soils and Fert.* **37**, 250).

GUNNER, H. B. (1963). Nitrification by *Arthrobacter globiformis*. *Nature, Lond.* **197**, 1127–1128.

GUNNER, H. B. (1970). Microbial ecosystem stress induced by an organophosphate insecticide. *Meded. Fac. Landb. Rijksuniv. Gent* **35**, 581–597.

GUNNER, H. B., ZUCKERMAN, B. M., WALKER, R. W., MILLER, C. W., DEUBERT, K. H. and LONGLEY, R. E. (1966). The distribution and persistence of diazinon applied to plant and soil and its influence on rhizosphere and soil microflora. *Pl. Soil* **25**, 249–264.

GUPTA, V. K. (1974). Effect of foliar applications of subamycin on rhizosphere and rhizoplane mycoflora. *Indian Phytopath.* **27**, 267.

GUPTA, K. G., SUD, R. K., AGGARWAL, P. K. and AGGARWAL, J. C. (1975). Effect of Baygon (2-isopropoxyphenyl-N-methyl carbamate) on some soil biological processes and its degradation by a *Pseudomonas* sp. *Pl. Soil* **42**, 317–325.

GUSTEROV, G., BRANKOVA, R. and VLAGOV, S. S. (1972). The influence of some herbicides on the development and the antibiotic activity of actinomycete-antagonists with anti-fungal activity spectra. *Symp. Biol. Hung.* **11**, 359–363. (*Weed Abstr.* **22**, 1922).

HAMDI, Y. A. and TEWFIK, M. S. (1969). Effect of the herbicide trifluralin on nitrogen fixation in *Rhizobium* and *Azotobacter* and on nitrification. *Acta microbiol. pol.* Ser. B. **1** (18), 53–57.

HARRIS, C. R. (1969). Insecticide pollution and soil organisms. *Proc. ent. Soc. Ont.* **100**, 14–28.

HART, L. T. and LARSON, A. D. (1966). Effect of 2,4-dichlorophenoxyacetic acid on different metabolic types of bacteria. *Bull. Environ. Contam. Toxicol.* **1**, 108–120.

HAUKE-PACEWICZOWA, T. (1969). Influence of herbicide treatments on the symbiosis of leguminous plants with *Rhizobium*. *Pam. Pulawski* **37**, 241–259.

HAUKE-PACEWICZOWA, T. (1970). Effect of simazine residues in the soil on the symbiotic fixation of nitrogen by legumes. *Meded. Fac. Landb. Rijksuniv. Gent* **35**, 497–503.

HAUKE-PACEWICZOWA, T. (1971). The effect of herbicides on the activity of soil microflora. *Pam. Pulawski* **46**, 5–48. (*Weed Abstr.* **22**, 1214).

HAUKE-PACEWICZOWA, T. and KROLOWA, M. (1968). The influence of MCPA and ioxynil on the soil microflora of the oat rhizosphere. *Pam. Pulawski* **31**, 17–28.

HELWEG, A. (1972a). Influence of chlorthiamid and dichlobenil on CO_2 liberation, ammonification and nitrification in soil. *Tidsskr. PlAvl* **76**, 145–155.

HELWEG, A. (1972b). Chlorfenvinphos; persistence and influence on nitrogen metabolism in soil. *Tidsskr. PlAvl* **76**, 519–527.

HELLING, C. S., DENNISON, D. G. and KAUFMAN, D. D. (1974). Fungicide movement in soils. *Phytopathology* **64**, 1091–1100.

HEMPHILL, D. D. and FIELDS, M. L. (1967). Effects of prolonged use of certain herbicides on soil microorganisms. *Abstr. Meet. Weed Soc. Am.* 36.

HOFER, I., BECK, T. and WALLNÖFER, P. (1971). Effects of the fungicide benomyl on the microflora of soil. *Z. PflKrankh. PflPath. PflShutz.* **78**, 399–405.

HOFFMAN-KAKOL, I. and KWARTA, H. (1972a). The effect of Aretit on blue lupin, weeds and soil microflora. *Zesz. nauk. Akad. Rolnicz. Szczecin.* No. 38, 91–103. (*Weed Abstr.* **23**, 2046).

HOFFMAN-KAKOL, I. and KWARTA, H. (1972b). Investigating the effect of MCPA on weeds, the yield of spring barley and the soil microflora. *Zesz. nauk. Akad. Rolnicz. Szczecin* No. 38, 105–116.

HOROWITZ, M., BLUMENFIELD, T., HERZLINGER, G. and HULIN, N. (1974). Effects of repeated applications of ten soil-active herbicides on weed population, residue accumulation and nitrification. *Weed Res.* **14**, 97–109.

HOUSEWORTH, L. D. and TWEEDY, B. G. (1973). Effect of atrazine in combination with captan or thiram upon fungal and bacterial populations in the soil. *Pl. Soil* **38**, 493–500.

HUBBELING, N. and BASU CHAUDHARY, K. C. (1970). Mutagenic effect of a herbicide on *Verticillium dahliae. Meded. Fac. Landb. Rijksuniv. Gent* **35**, 627–635.

HUGÉ, P. L. (1970). Contribution to the study of the influence of methabenzthiazuron on soil microorganisms. *Meded. Fac. Landb. Rijksuniv. Gent* **35**, 811–827.

HUGÉ, P. L. (1973). Contribution to the study of the influence of metribuzin on several important soil chemical, physical and microbiological properties. Herbicides and soil. *Proc. Eur. Weed Res. Counc. Symp.* Versailles, 1973, 31–40.

HUGÉ, P. L. (1974). Effect of phenamiphos on some of the main chemical, physical and microbiological characteristics of soils. *Meded. Fac. Landb. Rijksuniv. Gent* **39**, 1247.

HULIN, N. and HOROWITZ, M. (1973). Effect of herbicides on nitrification. *Phytoparasitica* **1**, 64.

HUSAROVA, M. (1972a). Microflora in the rhizosphere of maize: Perennial effect of herbicides. *Ved. Pr. vyzk. Ust. Kukur. Trnave*, No. 6, 45–69. (*Weed Abstr.* **22**, 3040).

HUSAROVA, M. (1972b). The use of herbicides over a number of years and changes of nitrogen ($NH_4 + NO_3$) in a maize monoculture. *Rostlinná výroba* **18**, 959–970. (*Weed Abstr.* **22**, 3041).

IBRAHIM, A. N. (1972). Effect of certain herbicides on growth of nitrogen-fixing algae and rice plants. *Symp. Biol. Hung.* **11**, 445–448.

ISAEVA, L. I. (1967). The effect of herbicide application on root nutrient uptake in sod-podzolic soils. *Agrokhimiya* **4**, 122–125. (*Weed Abstr.* **17**, 1933).

ISAEVA, L. I. and NOVOGRUDSKAYA, E. D. (1966). The effect of various rates of herbicides on soil microflora. *Vest. sel.'-khoz. Nauki, Mosk.* **11**, 57–59. (*Weed Abstr.* **16**, 1901).

ISHAQUE, M. and CORNFIELD, A. H. (1976). Evidence for heterotrophic nitrification in an acid Bangladesh soil lacking autotrophic nitrifying organisms. *Trop. Agric.* **53**, 157–160.

ISHAQUE, M., CORNFIELD, A. H. and CAWSE, P. A. (1971). Effect of gamma-irradiation of an acid tea soil from East Pakistan (Bangladesh) on nitrogen mineralization and nitrification during subsequent incubation. *Pl. Soil* **35**, 201–204.

ISHIZAWA, S. and MATSUGUCHI, T. (1966). Effects of pesticides and herbicides upon microorganisms in soil and water under waterlogged conditions. *Bull. natn. Inst. agric. Sci., Tokyo* (B) **16**, 1–90.

IVANOV, A. I. (1974). The effect of simazine on the nutrient regime of soil. *Agrokhimiya* **11**, 113–115. (*Soils and Fert.* **38**, 288).

JAISWAL, S. P. (1967). Effect of some pesticides on soil microflora and their activity. *J. Res. Punjab Agric. Univ.* **4**, 223–226.

JAISWAL, S. P., VERMA, A. K. and BABU, C. N. (1971). Studies on the effect of some agro-chemicals on nitrogen transformations in soil. 2. Ammonium nitrogen carrier. *J. Res. Punjab Agric. Univ.* **8**, 447–450.

JENKINSON, D. S. and POWLSON, D. S. (1970). Residual effects of soil fumigation on soil respiration and mineralization. *Soil Biol. Biochem.* **2**, 99–108.

JENSEN, H. L. (1969). The effect of various herbicides on root nodule bacteria. *Tidsskr. PlAvl* **73**, 309–317.

JENSEN, V. (1968). The plate count technique. *In* "The Ecology of Soil Bacteria" (T. R. G. Gray and D. Parkinson, Eds), pp. 158–170. Liverpool University Press, Liverpool.

JONES, A. S. and HODGES, C. S. (1974). Persistence of mirex and its effects on soil microorganisms. *J. agric. Fd Chem.* **22**, 435–439.

JONES, D. G. and WILLIAMS, J. R. (1971). Effect of paraquat on growth and sporulation of *Septoria nodorum* and *Septoria tritici*. *Trans. Br. mycol. Soc.* **57**, 351–357.

JORDAN, D. C. and GARCIA, M. M. (1969). Interactions between 2,4-DB and the root nodule bacteria of *Lotus corniculatis*. *Pl. Soil* **30**, 360–372.

JOSHI, O. P., SACHDEV, M. S., SAHRAWAT, K. L. and KOHLI, B. N. (1976). Effect of simazine and atrazine on the mineralization of fertilizer and manure nitrogen. *Pl. Soil* **44**, 367–375.

JUNG, J. (1972). The effect of the herbicide bentazon on soil respiration and nitrification in two soils. *Z. PflErnähr. Dung. Bodenk.* **133**, 18–23.

KAASTRA-HOWELER, L. H. and GAMS, W. (1973). Preliminary study on the effect of benomyl on the fungal flora in a greenhouse soil. *Neth. J. Pl. Path.* **79**, 156–158.

KAISER, P. and REBER, H. (1966). Unpublished data quoted by Kaiser, P., Pochon, J. J. and Cassini, R. (1970). *Residue Rev.* **32**, 211–233.

KAISER, P. and REBER, H. (1970). Interactions between simazine and the rhizosphere microorganisms of maize. *Meded. Fac. Landb. Rijksuniv. Gent* **35**, 689–705.

KAMPFE, K. (1973). Results of the combined application of anhydrous ammonia and a mixture of dichloropropane and dichloropropylene (D-D) as nitrification inhibitor in autumn. *Archiv. Acker-und Pflanzen. Bodenk.* **17**, 827–835.

KAPUSTA, G. and ROUWENHORST, D. L. (1973). Interaction of selected pesticides and *Rhizobium japonicum* in pure culture and under field conditions. *Agron. J.* **65**, 112–115.

KARANTH, N. G. K. and VASANTHARAJAN, V. N. (1973). Persistence and effect of Dexon on soil respiration. *Soil Biol. Biochem.* **5**, 679–684.

KARASAVICH, E. K. and ISAEVA, L. I. (1969). The effect of Pyramin on the microflora of a derno-podzolic soil. *Khimiya sel'Khoz.* **7**, 452–453. (*Weed Abstr.* **21**, 501).

KARKI, A. B. and KAISER, P. (1974). Effect of sodium chlorate on soil microorganisms, their respiration and enzyme activity. II. Laboratory incubation. *Rev. Ecol. Biol. Sol* **11**, 477–498.

KARKI, A. B., KAISER, P. and POCHON, J. J. (1972). Effect of sodium chlorate on nitrifying and denitrifying microorganisms in soil. Ecological study in the laboratory. *C.r. hebd. Séanc. Acad. Sci., Paris, Sci. Nat.* 274D, 2809–2814.

KARKI, A. B., COUPIN, L., KAISER, P. and MOUSSIN, M. (1973). Effects of sodium chlorate on soil microorganisms, their respiration and enzymatic activity. I. Ecological study in the field. *Rev. Ecol. Biol. Sol* **10**, 3–11.

KARPIAK, S. and IWANOWSKI, H. (1969). The effect of herbicides on soil microflora. VII. Respiration of bacteria isolated from maize rhizosphere. *Acta microbiol. pol.* Ser. B. **1** (**18**) 47–52.

KASZUBIAK, H. (1966). The effect of herbicides on *Rhizobium*. 1. Susceptibility of *Rhizobium* to herbicides. *Acta microbiol. pol.* **15**, 357–363.

KASZUBIAK, H. (1968a). The effect of herbicides on *Rhizobium*. 2. Adaptation of *Rhizobium* to Afalon, Aretit and Liro-Betarex. *Acta microbiol. pol.* **17**, 41–49.

KASZUBIAK, H. (1968b). The effect of herbicides on *Rhizobium*. 3. Influence of herbicides on mutation. *Acta microbiol. pol.* **17**, 51–57.

KASZUBIAK, H. (1970). Effect of herbicides on microorganisms in the rhizosphere of *Serradella*. *Meded. Fac. Landb. Rijksuniv. Gent* **35**, 543–550.

KATAN, J. and LOCKWOOD, J. L. (1970). Effect of pentachloronitrobenzene on colonization of alfalfa residues by fungi and streptomycetes in soil. *Phytopathology* **60**, 1578–1582.

KECSKÉS, M. (1970). Comparative investigations of the action of fungicides on *Rhizobium leguminosarum* Frank and its symbiosis with *Vicia sativa* L. *Meded. Fac. Landb. Rijksuniv. Gent* **35**, 505–514.

KECSKÉS, M. and VINCENT, M. J. (1969a). The effect of some fungicides on *Rhizobium leguminosarum* species. 1. Laboratory investigations. *Agrokém. Talajt.* **18**, 57–70.

KECSKÉS, M. and VINCENT, M. J. (1969b). The effect of some fungicides on *Rhizobium leguminosarum* species. 2. Investigations in the light chamber and glasshouse. *Agrokém. Talajt.* **18**, 461–472.

KHIKMATOV, A., RAKHIMOV, A. and MUSLIMOV, Z. (1969). Effect of *n*-chlorformanilide on soil microflora. *Agrokhimiya* **6** No. 9, 111–112.

KHRUSHCHEVA, E. P. (1971). The effect of simazine on the growth of maize mycorrhiza. *Khimiya sel'Khoz.* **9**, 692–693. (*Weed Abstr.* **21**, 3272).

KISS, Á. (1966). Herbicides and soil algae. *Novenyvedelem* **2**, 217–224. (*Weed Abstr.* **18**, 976).

KISS, Á. (1967). Microbiological examination of the action of herbicides used in vineyards in soils developed on loess. *Agrokém. Talajt.* **16**, 11. (*Soils and Fert.* **30**, 2975).

KISSNÉ, —. and KISS, M. (1966). Investigating the genotypic effects of herbicides on individual species of *Streptomyces. Agrartud. Egyet. Közl. Gödöllö*, 1966, 100–114. (*Weed Abstr.* **18**, 1952).

KO, W-H. and FARLEY, J. D. (1969). Conversion of pentachloronitrobenzene to pentachloroaniline in soil and the effect of these compounds on soil microorganisms. *Phytopathology* **59**, 64–67.

KO, W-H. and LOCKWOOD, J. L. (1968). Accumulation and concentration of chlorinated hydrocarbon pesticides by microorganisms in soil. *Can. J. Microbiol.* **14**, 1075–1078.

KO, W-H. and LOCKWOOD, J. L. (1970). Transfer of ^{32}P and dieldrin among selected microorganisms. *Rev. Ecol. Biol. Sol* **7**, 465–470.

KOBAYASHI, T. and KATSURA, S. (1968). The soil application of insecticides. 4. Effects of systemic organophosphates on soil nitrification and on the growth and yield of potatoes. *Jap. J. Appl. Ent. Zool.* **12**, 53–63.

KOKKE, R. (1970). Pesticide and herbicide interaction with microbial ecosystems. *Antonie van Leeuwenhoek* **36**, 580–581.

KOSINKIEWICZ, B. (1970). The influence of Aphalon on the production of antibiotic substances by soil bacteria of the genus *Bacillus. Meded Fac. Landb. Rijksuniv. Gent* **35**, 673–680.

KOSINKIEWICZ, B. (1973). The effect of linuron and chlorpropham on glutamic acid production by *Arthrobacter globiformis. Acta microbiol. pol.* Ser. B, **5** (**22**), 145–150.

KOZACZENKO, H. (1974). Studies on the biological activity of urea herbicides. *Zesz. nauk. Akad. roln. techn. Olsztyn. Rolnic.* No. 8, 58. (*Weed Abstr.* **24**, 1621).

KOZACZENKO, H. and SOBIERAJ, W. (1973). The influence of soil moisture content on the phytotoxicity of Afalon, Aresin and Alipur and their effect on the soil microflora. *Biul. Warzywn.* **14**, 161–178. (*Weed Abstr.* **23**, 2039).

KOZLOVA, E. I., BELOUSOVA, A. A. and VANDAR'EVA, V. S. (1964). Effect of simazine and atrazine on the development of soil microorganisms. *Agrobiologiya* **2**, 271–277. (*Biol. Abstr.* **45**, 88189).

KOZYREV, M. A. and LAPTEV, A. A. (1972). The effect of herbicides used in the crop rotation on the microflora of a derno-podzolic soil. *Khimiya sel'Khoz.* **10**, 614–617. (*Weed Abstr.* **22**, 1916).

KRASIL'NIKOV, N. A. (1967). Microbes and chemicals against pests. *Sel'Khoz. Biologiya* **2**, 857–865. (*Weed Abstr.* **17**, 1929).

KRATKA, J. and UJEVIC, I. (1970). Seed treatment of winter vetch with fungicides and their effect on formation of root nodules. *Pol'nohospodarstvo* **16**, 203–213. (*Soils and Fert.* **34**, 444).

KRETSCHMAR, G., MACH, F. and GUNTHER, G. (1970). Effect of 3-amino-1,2,4-triazol (amitrole) on the ultrastructure of *Azotobacter. Z. Allg. Mikrobiol.* **10**, 1–15.

KREUTZER, W. A. (1970). The re-infestation of treated soil. *In* "Ecology of Soil-Borne Plant Pathogens: Prelude to Biological Control" (K. F. Baker and W. C. Snyder, Eds), pp. 495–507. University of California Press, Los Angeles.

KREZEL, Z. and KOSINKIEWICZ, B. (1972). The effect of herbicides on the morphology of colonies and antibiotic activity of some streptomycetes. *Acta microbiol. pol.* Ser. B, **4 (21)**, 3–8.

KREZEL, Z. and LESZCZYNSKA, D. (1970). The effect of herbicides on the antibiotic activity of *Streptomyces griseus*. *Meded. Fac. Landb. Rijksuniv. Gent* **35**, 655–661.

KREZEL, Z. and MUSIAL, M. (1969). The effect of herbicides on soil microflora. II. Effect of herbicides on enzymatic activity of the soil. *Acta microbiol. pol.* Ser. B **1 (18)**, 93–97.

KROL, M., MALISZEWSKA, W. and SIUTA, J. (1972). Biological activity of soils strongly polluted with sulphur. *Polish J. Soil Sci.* **5**, 25–33.

KRUGLOV, Yu. V., GERSH, N. B. and BEI-BIENKO, N. V. (1973). The effect of Methurin on the biological activity of soil. *Khimiya sel'Khoz.* **11**, 294–296. (*Weed Abstr.* **23**, 540).

KRYKHANOV, L. I. and PASHKOVSKAYA, Z. V. (1967). The effect of various herbicides on the microflora and nitrogen regime of the soil. *Trudy ural. nauchno-issled. Inst. sel'Khoz.* **7**, 66–74. (*Weed Abstr.* **18** 2455).

KUDZIN, Yu. K., FISYUNOV, A. V., CHERNYAVSKAYA, N. A. and MAKAROVA, A. Ya. (1973). The change in the nitrate content of a typical chernozem soil under the influence of mineral fertilizers and pesticides. *Dokl. Vses. Akad. Sel'Khoz. Nauk. imeni V.I. Lenina* No. 9, 13–15. (*Weed Abstr.* **23**, 1691).

KULINSKA, D. (1967a). The effect of herbicides on oxygen uptake by soil. *Roczn. Nauk. roln.* (A) **93**, 125–130. (*Weed Abstr.* **18**, 960).

KULINSKA, D. (1967b). The effect of simazine on soil microorganisms. *Roczn. Nauk. roln.* (A) **93**, 229–262. (*Weed Abstr.* **18**, 980).

KULINSKA, D. and ROMANOV, I. (1970). Investigations on the effect of two herbicides, monolinuron and simazine, on some selected bacterial strains. *Meded. Fac. Landb. Rijksuniv. Gent* **35**, 551–558.

KULKARNI, J. H., SARDESHPANDE, J. S. and BAGYARAJ, D. J. (1974). Effect of four soil-applied insecticides on the symbiosis of *Rhizobium* sp. with *Arachis hypogaea* Linn. *Pl. Soil* **40**, 169–172.

KULKARNI, J. H., SARDESHPANDE, J. S. and BAGYARAJ, D. J. (1975). Effect of seed fumigation on the symbiosis of *Rhizobium* sp. with *Arachis hypogaea* Linn. *Zbl. Bakt.* Abt. II **130**, 41–44.

KURAMATO, M., HAAG, H. P. and SARRUGE, J. R. (1970). Effect of diuron on the nitrifying power of two Sao Paulo soils. *An. Esc. Sup. Agric. Luiz de Queiroz* **27**, 117–123. (*Soils and Fert.* **35**, 5006).

KURYNDINA, T. I. (1965). The effect of herbicides on the microflora of orchard soil. *Sb. nauch. Rab. Inst. Sadov I.V. Michurina* **II**, 113–116. (*Weed Abstr.* **17**, 1938).

KUSESKE, D. W., FUNKE, B. R. and SCHULZ, J. T. (1974). Effects and persistence of Baygon (propoxur) and Temik (aldicarb) insecticides in soil. *Pl. Soil* **41**, 255–269.

KUST, C. A. and STRUCKMEYER, B. E. (1971). Effects of trifluralin on growth, nodulation and anatomy of soyabeans. *Weed Sci.* **19**, 147–152.

KUZYAKINA, T. I. (1971). Changes in the microflora populations of a derno-podzolic soil with the use of different rates of simazine. Dokl. TSKhA, No. 172, 142–145. (*Weed Abstr.* **24**, 820).

LAMARTINERE, C. A., HART, L. T. and LARSON, A. D. (1969). Delayed lethal effect of 2,4-dichlorophenoxyacetic acid on bacteria. *Bull. Environ. Contam. Toxicol.* **4**, 113–119.

LANGKRAMER, O. (1970). Determination of the effect of pesticides on soil micro-organisms in pure culture by means of a laboratory technique. *Zbl. Bakt.* Abt. II **125**, 713–722.

LASKOWSKI, D. A. and BROADBENT, F. E. (1970). Effect of chlorinated aliphatic acids on gas and volatile fatty acid production in anaerobic soil. *Proc. Soil Sci. Soc. Am.* **34**, 72–77.

LAVEGLIA, J. and DAHM, P. A. (1974). Influence of AC92100 (Counter) on microbial activities in three Iowa surface soils. *Environ. Ent.* **33**, 528–533.

LAY, M. M. and ILNICKI, R. D. (1974). Peroxidase activity and propanil degradation in soil. *Weed Res.* **14**, 111–113.

LAZAREVA, N. V. and BLOKHIN, V. G. (1969). The effect of prometryne, Falone and alipur on the soil microflora. *Khimiya sel'Khoz.* **7**, 609–611. (*Weed Abstr.* **21**, 497).

LEBBINK, G. and KOLENBRANDER, G. J. (1974). Quantitative effect of fumigation with 1,3-dichloropropene mixtures and with metham sodium on the soil nitrogen status. *Agric. Environ.* **1**, 283–292.

LEELAVATHY, K. M. (1969). Effects of growth regulating substances on fungi. *Can. J. Microbiol.* **15**, 713–721.

LEIDERMAN, J., HINOJO, J. M. and FOGLIATA, F. A. (1971). Effect of different herbicides applied to sugarcane on urea nitrification. Part I. *Revta ind. agric. Tucumán* **48**, 7–14.

LEMBECK, W. J. and COLMER, A. R. (1967). Effect of herbicides on cellulose decomposition of *Sporocytophaga myxococcoides. Appl. Microbiol.* **15**, 300–303.

LESZCZYNSKA, D. (1970). The effect of some herbicides on yeast from the rhizosphere of cultivated plants. *Meded. Fac. Landb. Rijksuniv. Gent* **35**, 637–645.

LETHBRIDGE, G. and BURNS, R. G. (1976). Inhibition of soil urease by organophosphorus insecticides. *Soil Biol. Biochem.* **8**, 99–102.

LIU, L. C. and CIBES VIADE, H. R. (1972). Effect of various herbicides on the respiration of soil microorganisms. *J. Agric. Univ. P. Rico* **56**, 417–425.

LOBANOV, V. E. and PODDUBNAYA, L. P. (1967). The effect of herbicides on the microflora and nutrient regime of soil under sugar beet. *Khimiya sel'Khoz.* **5**, 32–34. (*Weed Abstr.* **17**, 924).

LOCKWOOD, J. L. (1970). Ecological effects of PCNB. *In* "Pesticides in the Soil: Ecology, Degradation and Movement" (G. E. Guyer, Ed.), pp. 47–50. Michigan State University, East Lansing.

LODHA, B. K. and LAL, P. B. (1969). Effect of pesticide treatments of seeds on the rhizosphere microflora of some field crops. *J. Indian Soc. Soil Sci.* **17**, 75–78.

LOPES, E. S., DEUBER, R., FORSTER, R., GARGANTINI, H. and BULISANI, E. A. (1971). Effect of the herbicides EPTC and trifluralin and of *Rhizobium* inoculation on the nodulation and yield of *Phaseolus vulgaris*. *Bragantia* **30**, 109–116.

LOZANO-CALLE, J. M. (1970). Effects of 2,4-D and 2,4,5-T ethyl esters on the microflora of a sandy soil and a clay soil. *Meded. Fac. Landb. Rijksuniv. Gent* **35**, 599–614.

MACEK, J. and MILHARCIC, L. (1972). The effect of some urea herbicides on the growth of *Bacillus cereus* Frankl. et Frankl. *Fragm. herb. Croatica* No. 16, 1–6. (*Weed Abstr.* **22**, 2704).

MACKENZIE, K. A. and MacRAE, I. C. (1972). Tolerance of the nitrogen-fixing system of *Azotobacter vinelandii* for four commonly used pesticides. *Antonie van Leeuwenhoek* **38**, 529–535.

MACNEILL, B. H. and NICHOLSON, D. (1974). The selective effects of benomyl on species of Fusarium. *Proc. Am. Phytopath. Soc.* **1**, 139.

MAHMOUD, S. A. Z., SELIM, K. G. and EL-MOKADEM, T. (1970). Effect of dieldrin and lindane on soil microorganisms. *Zbl. Bakt.* Abt. II **125**, 134–149.

MAHMOUD, S. A. Z., TAHA, S. M., ABDEL-HAFEZ, A. M. and HAMED, A. S. (1972). Effect of some pesticides on rhizosphere microflora of cotton plants. 1. Insecticides and fungicides. *Egypt. J. Microbiol.* **7**, 39–52.

MAKAWI, A. A. M. and ABDEL-GHAFFAR, A. S. (1970). The effect of some pesticides on the growth of root-nodule bacteria. *Un. Arab Rep. J. Microbiol.* **5**, 109–117.

MALANINA, E. I. and NIKITIN, P. D. (1971). Effect of mixtures of soil herbicides on the microflora of light chestnut soils. *Vest. sel.'-khoz. Nauki* **10**, 122–124.

MALONE, C. R. (1971). Response of soil microorganisms to non-selective vegetation control in a fescue meadow. *Soil Biol. Biochem.* **3**, 127–131.

MANORIK, A. V. and MALICKENKO, S. M. (1969). The effect of symmetrical triazines on phosphatase and urease activity in the soil. *Fiz. Biokhim. kul't. Rast.* **1**, 173–178. (*Weed Abstr.* **21**, 1114).

MANNINGER, E. and SZAVA, J. (1972). Behaviour of some herbicides used for weed control in vineyards on different soils. *Symp. Biol. Hung.* **11**, 369–372.

MANNINGER, E., BAKONDI, E. and TAKATS, T. (1972). The effect of Gramoxone on N-fixing microorganisms. *Symp. Biol. Hung.* **11**, 401–404.

MARKERT, S. (1974). On the possibility of regulating the process of transformation of urea applied to the soil by commercial pesticides. *Trans. 10th Int. Congr. Soil Sci.* **IX**, 133–140.

MARSH, J. A. P., DAVIES, H. A. and GROSSBARD, E. (1974). The effect of a single field application of very high rates of linuron and simazine on carbon dioxide evolution and transformation of nitrogen within soil. *Proc. 12th Br. Weed Control Conf.* 53–58.

MARTIN, H. and WORTHING, C. R. (1974). "Pesticide Manual," 4th edition, British Crop Protection Council.

MATSNEVA, N. G. and SEMIKHATOVA, O. A. (1973). Effect of herbicides on the microflora of a solonetzic chernozem. *Khimiya sel'Khoz.* **11**, 222–224. (*Soils and Fert.* **37**, 2529).

MATSUGUCHI, T. and ISHIZAWA, S. (1968). Effect of agrochemicals on micro-organisms and their activities in soil (Part 1). Effect of pentachlorophenol (PCP). *J. Sci. Soil Manure, Tokyo* **39**, 241–246.

MATSUGUCHI, T. and ISHIZAWA, S. (1969). Effect of agrochemicals on micro-organisms and their activities in soil (Part 2). Effect of isopropyl-*N*-(3-chloro-phenyl) carbamate (CIPC). *J. Sci. Soil Manure, Tokyo* **40**, 20–25.

MAW, G. A. and KEMPTON, R. J. (1973). Methyl bromide as a soil fumigant. *Soils and Fert.* **36**, 41–47.

MAYAUDON, J. (1974). Respiration determinations with labelled C: the specific effects of pesticides on the biological activity of soils under beet. *Annali Microbiol. Enzimol.* **24**, 57–62.

MENGE, J. A. and FRENCH, D. W. (1976). Determining inoculum potentials of *Cylindrocladium floridanum* in cropped and chemically-treated soils by quantitative assay. *Phytopathology* **66**, 862–867.

MERESHKO, M. Ya. (1969). The effect of herbicides on the biological activity of the microflora of chernozem soils. *Mykrobiol. Zh.* **31**, 525–529. (*Weed Abstr.* **20**, 419).

MEZHARAUPE, V. A. (1967). Effect of phenazone and other herbicides on develop-ment of soil microorganisms. *Mikroorg. Rast. Trudy Ìnst. Mikrobiol. Akad. Nauk. SSR* 65–83. (*Weed Abstr.* **17**, 2937).

MICEV, N. (1970). The effect of the herbicide Lumeton on the soil and rhizosphere microflora of wheat. *Savr. Poljopr.* **18**, 245–249.

MICEV, N. and BUBALOV, M. (1969a). The effect of some herbicides on the soil and rhizosphere microflora of wheat c.v. Mara. *Godišen Zb. zemjod.-šum. Fak. Univ. Skopje* **21**, 13–20.

MICEV, N. and BUBALOV, M. (1969b). The effect of some herbicides on aerobic cellulolytic microflora and *Azotobacter* in the soil and in the maize rhizosphere. *Agrokém. Talajt.* **18**, 71–75.

MICEV, N. and BUBALOV, M. (1969c). The effects of the herbicides Gesaprim 1798 and VCS − 438 + dicamba on the soil and rhizosphere microflora of maize. *Godišen Zb. zemjod.-šum. Fak. Univ. Skopje* **22**, 19–24.

MICEV, N. and BUBALOV, M. (1972). Interaction between soil microflora and the herbicide Agelon. *Symp. Biol. Hung.* **11**, 379–384.

MICKOVSKI, M. D. (1966). The influence of hormone herbicides upon the growth of some species of rhizobia as well as upon the germination and nodulation of *Medicago sativa* and *Trifolium repens*. *Godišen Zb. zemjod.-šum. Fak. Univ. Skopje* **19**, 5–25.

MICKOVSKI, M. D. (1967). Effects of the herbicides Gesaran 2079, Lumeton 2412 and Igran 1866 on the microflora of soil. *Ann. Fac. Agric. Univ. Skopje* **20**, 43–54.

MICKOVSKI, M. D. (1972). Relationship between herbicide Dymid and soil micro-organisms. *Symp. Biol. Hung.* **11**, 397–400.

MICKOVSKI, M. D., SEAMAN, A. and WOODBINE, M. (1968). Effect of the hormone weedkillers 2,4-D, MCPA and MCPB on soil fungi. *Ann. Fac. Agric. Univ. Skopje* **21**, 60–75.

MIDHA, S. K. and NANDWANA, R. P. (1974). Effect of certain nematicides on the population of soil microorganisms. 1. Quantitative and qualitative changes in the fungal flora. *Indian Phytopath.* **27**, 312–315.

MIKHAILOVA, M. F. (1968). Nutrient regime and microbiological activity of soil under pea crops treated with herbicides. *Khimiya sel'Khoz.* **6**, 204–206. (*Weed Abstr.* **18**, 476).

MIKHAILOVA, E. I. and KRUGLOV, Yu. V. (1973). Effect of some herbicides on the algal flora of soil. *Pochvovedenie* No. 8, 81–85.

MILHOLLAND, R. D. (1969). Effect of soil fumigation on disease control and yield of gladiolus in south-eastern North Carolina. *Pl. Dis. Reptr* **53**, 132–136.

MILKOWSKA, A. and GORZELAK, A. (1966). Effect of atrazine and simazine on the soil microflora in weed control in forest nurseries. *Sylwan* **110**, 13–22.

MISHRA, K. C. and GAUR, A. C. (1975). Influence of treflan, lindane and ceresan on different parameters of symbiotic nitrogen fixation and yield in *Cicer arietinum*. *Zbl. Bakt.* Abt. II **130**, 598–602.

MISHRA, M. M., NEELAKANTAN, S. and KHANDELWAL, K. C. (1972). Effect of lindane and thimet on nitrification. *Haryana Agric. Univ. J. Res.* **2**, 283–285. (*Soils and Fert.* **38**, 176).

MISRA, K. C. and GAUR, A. C. (1971). Tolerance of *Azotobacter* to some herbicides. *Indian J. Weed Sci.* **3**, 99–103.

MOUCHOVA, H. and APLTAUER, J. (1971). Factors influencing the nodulation and yield of *Vicia faba* L. *Abstr. Commun. 9th Ann. Mtg. Czech. Soc. Microbiol.* (*Weed Abstr.* **22**, 248).

NAIDU, S. M. (1972). Pesticide residues and their effects on microbial activities following massive disposal of pesticides in the soil. *Diss. Abstr. Int.* B. **33**, 2894.

NAIR, K. S., RAMAKRISHNAN, C. and SITHANANTHAM, S. (1973). Effect of soil application of insecticides on the nutrient status, bacterial population and yield in rice soils. *Madras agric. J.* **60**, 441–443.

NAMDEO, K. N. and DUBE, J. N. (1973a). Residual effect of urea and herbicides on hexosamine content and urease and proteinase activities in a grassland soil. *Soil Biol. Biochem.* **5**, 855–859.

NAMDEO, K. N. and DUBE, J. N. (1973b). Herbicidal influence on growth sensitivity and mutagenic transformation in Rhizobia. *Indian J. exptl. Biol.* **11**, 114–116.

NAUMANN, K. (1970a). Zur Dynamik der Bodenmikroflora nach Anwendung von Pflanzenschutzmitteln. I. Freilandversuche über die Wirkung von Parathion-methyl auf die Bakterien-und Strahlenpilzpopulation des Boden. *Zbl. Bakt.* Abt. II **124**, 743–754.

NAUMANN, K. (1970b). Zur Dynamik der Bodenmikroflora nach Anwendung von Pflanzenschutzmitteln. II. Die Reaktion verschiedener physiologischer Gruppen von Bodenbakterien auf den Einsatz von Parathion-methyl im Freiland. *Zbl. Bakt.* Abt. II **124**, 755–765.

NAUMANN, K. (1970c). Zur Dynamik der Bodenmikroflora nach Anwendung von Pflanzenschutzmitteln. IV. Untersuchungen über die Wirkung von Parathion-methyl auf die Atmung und die Dehydrogenase-aktivität des Bodens. *Zbl. Bakt.* Abt. II **125**, 119–133.

NAUMANN, K. (1970d). Zur Dynamik der Bodenmikroflora nach Anwendung von Pflanzenschutzmitteln. VII. Die Wirkung einiger Entseuchungsmittel auf die Bodenmikroorganismen. *Zbl. Bakt.* Abt. II **125**, 478–491.

NAUMANN, K. (1971a). Veranderungen in der zusammensetzung der Bodenbakterien-flora nach Einbringung von Pflanzenschutzmitteln in den Boden. *Zbl. Bakt.* Abt. II **126**, 530–544.

NAUMANN, K. (1971b). Dynamics of the soil microflora following the application of pesticides. VI. Trials with the insecticides gamma-BHC and toxaphene. *Pedobiologia* **11**, 286–295.

NAUMANN, K. (1971c). Schädigen Pflanzenschutzmittel die Mikroorganismen des Bodens? *Wissenschaft. Fortschritt* **21**, 318–321.

NAUMANN, K. (1972). Die Wirkung einiger Umweltfaktoren auf die Reaktion der Bodenmikroflora gengenüber Pflanzenschutzmitteln. *Zbl. Bakt.* Abt. II **127**, 379–396.

NAYYAR, V. K., RANDHAWA, N. S. and CHOPRA, S. L. (1970). Effect of simazine on nitrification and microbial populations in a sandy-loam soil. *Indian J. agric. Sci.* **40**, 445–451.

NEDEROST, J. (1967). The effect of atrazine and of some cultivation methods of maize on the nitrate content of the soil. *Sb. vys. Sk. zemed. Ceskych. Budejovicich* **13**, 1–19. (*Weed Abstr.* **18**, 2923).

NEPOMILUEV, V. F. and KUZYAKINA, T. I. (1967). The effect of herbicides on the microflora of a peaty-bog soil. *Izv. timiryazev. sel.'-khoz. Akad.* No. 4, 84–91.

NEPOMILUEV, V. F. and KUZYAKINA, T. I. (1972). The effect of atrazine and simazine on soil fungi and herbicide decomposition in the soil. *Biol. Nauki* **15**, 127–131. (*Weed Abstr.* **23**, 2044).

NEPOMILUEV, V. F., BEBIN, S. I. and KUZYAKINA, T. I. (1966a). The effect of herbicides on the microflora of a sod-podzol soil under field beans. *Izv. timiryazev. sel.'-khoz. Akad.* No. 4, 88–94.

NEPOMILUEV, V. F., GRECHIN, I. P. and KUZYAKINA, T. I. (1966b). The effect of simazine on microflora and microbiological activity of a sod–podzol soil with different methods of treatment. *Izv. timiryazev. sel.'-khoz. Akad.* No. 2, 128–136.

NEUBAUER, R., AVIZOHAR-HERSHENSON, Z. (1973). Effect of the herbicide trifluralin on *Rhizoctonia* disease in cotton. *Phytopathology* **63**, 651–652.

NIKOLENKO, Zh. I. and GELLER, I. A. (1969). Minimal toxic concentrations of herbicides for certain groups of soil microorganisms. *Khimiya sel'Khoz.* **7**, 213–214. (*Weed Abstr.* **19**, 895).

NIKOLOVA, G. and BAKALIVANOV, D. (1972). Microbiological activity and breakdown of Casoran herbicide in the soil. *Symp. Biol. Hung.* **11**, 385–389.

NOVAKOVA, J. (1974). Effect of kaolinite on the mineralisation of peptone. *Rostlinna výroba* **20**, 813–816. (*Soils and Fert.* **38**, 3461).

OBLISAMI, G., BALARAMAN, K., VENKATARAMANAN, C. V. and RANGASWAMI, G. (1973). Effect of three granular insecticides on the growth of *Rhizobium* from redgram. *Madras agric. J.* **60**, 462–464.

ODU, C. T. I. and HORSFALL, M. A. (1971). Effect of chloroxuron on some microbial activities in soil. *Pestic. Sci.* **2**, 122–125.

PANCHOLY, S. K. and LYND, J. Q. (1969). Bromacil interactions in plant bioassay, fungi cultures and nitrification. *Weed Sci.* **17**, 460–463.

PANTERA, H. (1966). The effect of atrazine on soil bacteria in relation to the substrate. *Zesz. Probl. Postep. Nauk Roln.* **60**, 169–176. (*Weed Abstr.* **17**, 2927).

PANTERA, H. (1970). The effect of herbicides on algae in the soil. *Meded. Fac. Landb. Rijksuniv. Gent* **35**, 847–854.

PANTERA, H. (1972). Influence of high rates of herbicide on some groups of soil microorganisms. *Pam. Pulawski* **51**, 69–76.

PAREEK, R. P. and GAUR, A. C. (1969). Effect of dichloro-diphenyl-trichloroethane (DDT) on the nodulation, growth, yield and nitrogen uptake of *Pisum sativum* inoculated with *Rhizobium leguminosarum. Indian J. Microbiol.* **9**, 93–99.

PAREEK, R. P. and GAUR, A. C. (1970). Effect of dichloro-diphenyltrichloro-ethane (DDT) on symbiosis of *Rhizobium* sp. with *Phaseolus aureus* (green gram). *Pl. Soil* **33**, 297–304.

PEEPLES, J. L. (1974). Microbial activity in benomyl-treated soils. *Phytopathology* **64**, 857–860.

PEEPLES, J. L. and CURL, E. A. (1970). Effect of the herbicide EPTC on growth and enzymatic activity of *Sclerotium rolfsii* and *Trichoderma viride. Phytopathology* **60**, 586.

PESHAKOV, G., RAIKOV, E. and TSVETANOV, D. (1969). Investigating the effect of various herbicides on the soil microflora and on ammonification and nitrification in the soil. *Pochv. Agrokhim.* **4**, 89–94. (*Weed Abstr.* **20**, 1856).

PILIPETS, S. G. and LITVINOV, I. A. (1972). The direct and residual effects of atrazine on the soil microflora of maize inter-rows. *Trudy khar'kov. sel'Khoz. Inst.* No. 12, 142–146. (*Weed Abstr.* **22**, 3046).

PINCKARD, J. A. (1970). Microbial antagonism encouraged by the monosodium salt of hexachlorophene. *Phytopathology* **60**, 1308.

PITTS, G., CURL, E. A. and RODRIGUEZ-KABANA, R. (1970). Effect of glucose on growth response of *Sclerotium rolfsii* in atrazine-treated soil culture. *Phytopathology* **60**, 587.

PLAMADEALA, B. (1972). Effects of certain herbicides on the fungi *Rhizoctonia solani* Kühn, *Alternaria* sp. and *Fusarium* sp. *An. Inst. Cercet. pent. Cult. Cart. Sfec. zahar, Brasov* **3**, 365–372. (*Weed Abstr.* **24**, 817).

PLATONOVA, V. P. (1967a). The effect of prometryne on soil algae. *Zap. Voronezh. sel'Khoz. Inst.* **34**, 41–44. (*Weed Abstr.* **18**, 1470).

PLATONOVA, V. P. (1967b). The effect of 2,4-D, simazine and prometryne on soil algae. *Trudy kirov. sel'Khoz. Inst.* **20**, 215–221. (*Weed Abstr.* **18**, 1953).

POPOV, V. I. and ZDROZHEVSKAYA, S. D. (1972). Characteristics of the action of fungicides in soil. *Trudy vses. Inst. Zashch. Rast.* No. 35, 270–276. (*Soils and Fert.* **37**, 2190).

POPPE, J. A. (1972). Fungicidal action of some herbicides in mushroom culture. *Meded. Fac. Landb. Rijksuniv. Gent* **37**, 705–712.

PRANAITENE, E. V. (1970). The effect of different cultivation treatments and herbicides on the soil microflora. *Mat. Pribalt. Sov. Zashch. Rast. Elgava* 8–12. (*Weed Abstr.* **21**, 2653).

PRASAD, I. (1970). Mutagenic effects of the herbicide 3,4-dichloropropionanilide and its degradation products. *Can. J. Microbiol.* **16**, 369–372.

PROZOROVA, M. I. (1967). The effect of atrazine on the biological activity and microflora of a forest soil. *Izv. sib. Otdel. Akad. Nauk SSSR* **5**, 32–35. (*Weed Abstr.* **18**, 979).

PUGASHETTY, B. K. and RANGASWAMI, G. (1969). Antagonistic actinomycetes in the rhizosphere of cotton seedlings as influenced by seed treatments before sowing. *Mysore J. agric. Sci.* **3**, 241–244.

PUTINTSEV, N. I. and PUTINTSEVA, L. A. (1966). The effect of propazine and TMTD (thiram) with the γ-isomer of HCCH (BHC). *Khimiya sel'Khoz.* **4**, 49–52. (*Weed Abstr.* **17**, 942).

QUILT, P., GROSSBARD, E. and WRIGHT, S. J. L. (submitted). Effects of the herbicide barban and its commercial formulation Carbyne on soil microorganisms. *J. appl. Bact.*

RAKHIMOV, A. A. and RYBINA, V. F. (1963). Effect of some herbicides on soil microflora. *Uzbek. biol. Zh.* 74–76. (*Biol. Abstr.* **45**, 83599).

RANKOV, V. (1968). The effect of some herbicides on nitrification in soils. *Pochv. Agrokhim.* **3**, 81–90. (*Weed Abstr.* **18**, 1955).

RANKOV, V. (1970). The effect of certain herbicides on soil cellulose-decomposing activity. *Pochv. Agrokhim.* **5**, 73–80. (*Weed Abstr.* **21**, 1113).

RANKOV, V. (1971a). Effect of Treflan on certain soil microorganisms *in vitro*. *Pochv. Agrokhim.* **6**, 113–120. (*Soils and Fert.* **35**, 350).

RANKOV, V. (1971b). Effect of Aretit and Treflan herbicides on the microflora of different soil types. *B"lgarskata Akad. Nauk* (1971), 253–257. (*Soils and Fert.* **36**, 605).

RANKOV, V. (1972). Effect of some urea-derived herbicides on the microflora of soil. "P"rvi Nat. Kongr. Pochv. Bulg. 1969 (1972), 227–232. (*Soils and Fert.* **36**, 4067).

RANKOV, V. (1973). Nitrification activity of certain soils treated with herbicides and fertilizers. *Pochv. Agrokhim.* **8**, 125–134. (*Soils and Fert.* **37**, 2518).

RANKOV, V. and ELENKOV, E. (1970). The effect of Treflan on soil microflora. *Pochv. Agrokhim.* **5**, 127–136. (*Soils and Fert.* **34**, 405).

RANKOV, V. and GENCHEV, S. (1969). An investigation into the effect of the Dymid preparation on the growth and certain physiological features of *Azotobacter chroococcum*. *C.r. Acad. Sci. agric. Bulg.* **2**, 247–252.

RANKOV, V., ELENKOV, E., SURLEKOV, P. and VELEV, B. (1966). Effect of some herbicides on development of nitrogen-fixing bacteria. *Agrokhimiya* No. 4, 115–120.

RAYNAL, G. and FERRARI, F. (1973). Persistence of soil-incorporated benomyl and its effects on soil fungi. *Phyt. Phytopharm.* **22**, 259–272.

RESZ, A. (1968). Studies on long-term herbicide trials. 2. Quantitative changes in soil microflora through the use of certain herbicides for eleven years. *Sonderh. Z. PflKrankh. PflPath. PflSchutz* No. 4, 147–150.

RICHARDSON, L. T. (1970). Effects of atrazine on growth response of soil fungi. *Can. J. Pl. Sci.* **50**, 594–596.

RIDGE, E. H. and THEODOROU, C. (1972). The effect of soil fumigation on microbial recolonization and mycorrhizal infection. *Soil Biol. Biochem.* **4**, 295–305.

RIECK, C. E. and McCALLA, T. M. (1969). Stimulatory effects of 4-amino-3,5,6-trichloropicolinic acid on the growth of *Aspergillus. Bact. Proc.* p. 4.

RINTELEN, J. (1968). Studies on long-term herbicide trials. 3. Does the repeated use of annual applications of the same herbicide damage the species spectrum of the soil microflora? *Sonderh. Z. PflKrankh. PflPath. PflSchutz* No. 4, 151–161.

ROBSON, H. and GUNNER, H. B. (1970). Differential response of soil microflora to diazinon. *Pl. Soil* **33**, 613–621.

RODRIGUEZ-KABANA, R. and CURL, E. A. (1970). Effect of atrazine on growth of *Fusarium oxysporum* f. sp. *vasinfectum. Phytopathology* **60**, 65–69.

RODRIGUEZ-KABANA, R. and CURL, E. A. (1972). Effect of paraquat, trifluralin and EPTC on the growth of *Fusarium oxysporum* f. sp. *vasinfectum* in liquid culture. *Fitopatologia* **6**, 7–17: (*Weed Abstr.* **23**, 694).

RODRIGUEZ-KABANA, R., CURL, E. A. and FUNDERBURK, H. H. (1966a). Effect of atrazine, paraquat and EPTC (ethyl-*N,N*-dipropylthiolcarbamate) on growth of *Sclerotium rolfsii. Phytopathology* **56**, 897.

RODRIGUEZ-KABANA, R., CURL, E. A. and FUNDERBURK, H. H. (1966b). Effect of four herbicides onngrowth of *Rhizoctonia solani. Phytopathology* **56**, 1332–1333.

RODRIGUEZ-KABANA, R., CURL, E. A. and FUNDERBURK, H. H. (1967a). Effect of atrazine on growth response of *Sclerotium rolfsii* and *Trichoderma viride. Can. J. Microbiol.* **13**, 1343–1349.

RODRIGUEZ-KABANA, R., CURL, E. A. and FUNDERBURK, H. H. (1967b). Effect of paraquat on growth of *Sclerotium rolfsii* in liquid culture and soil. *Phytopathology* **57**, 911–915.

RODRIGUEZ-KABANA, R., CURL, E. A. and FUNDERBURK, H. H. (1968). Effect of atrazine on growth activity of *Sclerotium rolfsii* and *Trichoderma viride* in soil. *Can. J. Microbiol.* **14**, 1283–1288.

RODRIGUEZ-KABANA, R., CURL, E. A. and FUNDERBURK, H. H. (1969). Effect of trifluralin on growth of *Sclerotium rolfsii* in liquid culture and soil. *Phytopathology* **59**, 228–232.

RODRIGUEZ-KABANA, R., CURL, E. A. and PEEPLES, J. L. (1970). Growth response of *Sclerotium rolfsii* to the herbicide EPTC in liquid culture and soil. *Phytopathology* **60**, 431–436.

RODRIGUEZ-KABANA, R., BACKMAN, P. A., KARR, G. W. and KING, P. S. (1976). Effects of the nematicide fensulfothion on soilborne pathogens. *Pl. Dis. Reptr.* **60**, 521–524.

ROSS, D. J. (1974). Influence of four pesticide formulations on microbial processes in a New Zealand pasture soil. II. Nitrogen mineralization. *N.Z. Jl agric. Res.* **17**, 9–17.

SALAMA, A. M., MOSTAFA, I. Y. and EL-ZAWAHRY, Y. A. (1973). Insecticides and soil microorganisms. I. Effect of Dipterex on the growth of *Rhizobium leguminosarum* and *Rhizobium trifolii* as influenced by temperature, pH and type of nitrogen. *Acta biol. hung.* **24**, 25–30.

SALAMA, A. M., MOSTAFA, I. Y. and EL-ZAWAHRY, Y. A. (1974). Insecticides and soil microorganisms. II. Effect of Dipterex on nodule formation in broad bean and clover plants under different manurial treatments. *Acta biol. hung.* **25**, 239–246.

SALEM, S. H. (1971). Effects of some insecticides on the physiological activity of effective and ineffective strains of *Rhizobium trifolii*. *Agrokém. Talajt.* **20**, 368–376.

SALEM, S. H. and GULYÁS, F. (1971). Effect of insecticides on the physiological properties of *Azotobacter*. *Agrokém. Talajt.* **20**, 377–388.

SALEM, S. H., SZEGI, J. and GULYÁS, F. (1971). The influence of some insecticides on the symbiosis of rhizobia and legume plants. *Agrokém. Talajt.* **20**, 581–582.

SALONIUS, P. O. (1972). Effect of DDT and fenitrothion on forest soil microflora. *J. econ. Ent.* **65**, 1089–1090.

SAMOKHVALOV, A. N. (1973). The effect of prometryne on the microflora of a derno-podzolic soil. *Vest. Mosk. Univ.* **6**, 87–92. (*Weed Abstr.* **24**, 229).

SANSING, N. G. and CHO, Y. P. (1970). Lack of effect of *s*-triazine herbicides on nucleic acid metabolism in bacteria. *Proc. sth. Weed Control Conf.* 320.

SAXENA, R. B. L. and GUPTA, U. S. (1969). Influence of some plant hormones and amino acids on germination and sporulation of a few cellulolytic fungi. *Labdev J. Sci. Tech.* (Part B) **7**, 266–271. (*Weed Abstr.* **19**, 2341).

SCHRÖDER, I., MEYER, M. and MÜCKE, D. (1970). Effects of the herbicides 2,4-D, aminotriazole, atrazine, chlorpropham and chlorflurenol on nucleic acid biosynthesis in the ascomycete *Neurospora crassa*. *Weed Res.* **10**, 172–177.

SEKHON, H. S., AHLUWALIA, B. and ABDOU, A. (1974). *In vitro* effects of herbicides on fungal pectolytic enzymes. *Proc. Am. Phytopath. Soc.* **1**, 53.

SELIM, K. G., MAHMOUD, S. A. Z. and EL-MOKADEM, M. T. (1970). Effect of dieldrin and lindane on the growth and nodulation of *Vicia faba*. *Pl. Soil* **33**, 325–329.

SETHUNATHAN, N. (1970a). Foliar sprays of growth regulators and rhizosphere effect in *Cajanus cajan* Millsp. I. Quantitative changes. *Pl. Soil* **33**, 62–70.

SETHUNATHAN, N. (1970b). Foliar sprays of growth regulators and rhizosphere effect in *Cajanus cajan* Millsp. II. Qualitative changes in the rhizosphere and certain metabolic changes in the plant. *Pl. Soil* **33**, 71–80.

SETHUNATHAN, N. and MacRAE, I. C. (1969). Some effects of diazinon on the microflora of submerged soils. *Pl. Soil* **30**, 109–112.

SETTY, R. A., BALIGAR, V. C. and PATIL, S. V. (1970). Effect of atrazine on the rate of nitrification in black clay loam soil. *Mysore J. agric. Sci.* **4**, 111–113.

SHAMIYEH, N. B. and JOHNSON, L. F. (1973). Effect of heptachlor on numbers of bacteria, actinomycetes and fungi in soil. *Soil Biol. Biochem.* **5**, 309–314.

SHARABI, N. EL-DIN (1969). Influence of propanil (3,4-dichloropropionanilide) and related compounds on growth of the alga *Chlorococcum aplanosporum* in soil and in solution culture. Ph.D. Thesis, Rutgers Univ., New Brunswick, N.J., U.S.A.

SHARABI, N. EL-DIN and PRAMER, D. (1973). A spectrophotofluorimetric method for studying algae in soil. *Bull. Ecol. Res. Comm.* (Stockholm) **17**, 77–84.

SHARMA, L. N. and SAXENA, S. N. (1971). Effect of diuron (3,4-dichlorophenyl-1,1-dimethyl urea) on soil microflora. *Andhra agric. J.* **17**, 108–111.

SHARMA, L. N. and SAXENA, S. N. (1972). Influence of dalapon (2,2-dichloropropionic acid) on biotic potential of soil. *Andhra agric. J.* **18**, 74–78.

SHARMA, L. N. and SAXENA, S. N. (1974). Influence of 2,4-D on soil microorganisms with special reference to *Azotobacter*. *J. Indian Soc. Soil Sci.* **22**, 168–171.

SHIN-CHSIANG LIN, FUNKE, B. R. and SCHULZ, J. T. (1972). Effects of some organophosphate and carbamate insecticides on nitrification and legume growth. *Pl. Soil* **37**, 489–496.

SHKOLA, A. I. (1970). The effect of prometryne and amiben (chlorambin) on the development of nitrogen fixing bacteria. *Vest. sel.'-khoz. Nauki, Alma-Ata* **13**, 34–37. (*Weed Abstr.* **20**, 1859).

SIMON, J. C., JAMET, P., LEMAIRE, J. M. and JOUAN, B. (1973). The effect of introducing gelatin and a substituted urea on the microbial life of a soil. *Sci. Agron. Rennes, École Nat. Sup. Agron.* 231–244. (*Soils and Fert.* **38**, 1456).

SIMON, L. (1971). Effects of herbicides on microorganisms. *Folia Microbiol.* **16**, 516.

SINGH, I. and PRASAD, S. K. (1974). Effect of some nematicides on nematodes and soil microorganisms. *Indian J. Nematol.* **3**, 109–133.

SINGH, K. (1971). The effect of 2,4-D and simazine on total bacteria, fungi, *Azotobacter*, ammonification and nitrification under field conditions. *Pestic. India* (6), 14–17. (*Weed Abstr.* **21**, 3275).

SINGHAL, J. P. and SINGH, C. P. (1973). Effect of nemagon on some soil properties. *Indian J. agric. Sci.* **43**, 280–284.

SINGHAL, J. P., KHAN, S. and GUPTA, G. K. (1975). Effect of D-D mixture on availability of some plant nutrients in a black cotton soil. *J. Indian Soc. Soil Sci.* **23**, 109–112.

SIVASITHAMPARAM, K. (1970). Some effects of an insecticide (Dursban) and a weedkiller (Linuron) on the microflora of a submerged soil. *Riso* **19**, 339–346.

SMITH, J. E. and SHENNAN, J. L. (1966). The effect of substituted phenoxyacetic and phenoxybutyric acids on the growth and respiration of *Aspergillus niger*. *J. gen. Microbiol.* **42**, 293–300.

SMITH, J. R. (1973). The effect of herbicides on the ectotrophic mycorrhizal fungi of pine. Ph.D. Thesis Univ. London, U.K.

SMITH, M. S. and WEERARATNA, C. S. (1974). A study of the effect of simazine on soil microbial activity and available nitrogen. *Trans. 10th Int. Congr. Soil Sci.* III. 173–178.

SMITH, M. S. and WEERARATNA, C. S. (1975). Influence of some biologically active compounds on microbial activity and on the availability of plant nutrients in soils. II. Nitrapyrin, dazomet, 2-chlorobenzamide and tributyl-3-chlorobenzyl-ammonium bromide. *Pestic. Sci.* **6**, 605–615.

SOBIESZCZANSKI, J. (1968). The influence of different herbicides upon the growth and development of cellulolytic microorganisms. *Annls Inst. Pasteur, Paris* **115**, 613–616.

SOBIESZCZANSKI, J. (1969a). The effects of herbicides on soil microflora. VIII. The effect of herbicides on growth and morphology of some species of bacteria. *Acta microbiol. pol.* (Ser. B), 1 (**18**), 99–104.

SOBIESZCZANSKI, J. (1969b). The effect of herbicides on soil microflora. V. Growth and activity of cellulolytic microorganisms. *Acta microbiol. pol.* (Ser. B), 1 (**18**), 39–42.

SOBOTKA, A. (1970). Testing the effect of pesticides on mycorrhizal fungi in forest soils. *Zbl. Bakt.* Abt. II **125**, 723–730.

SOMMER, K. (1970). Effect of different pesticides on nitrification and nitrogen metabolism in soils. *Sonderh. Landw. Forsch.* **25**, 22–30.

SOSNOVSKAYA, E. A. and PASHCHENKO, P. D. (1965). The effect of herbicides on microflora of the soil under maize. *Mat. tez. VI. Konf. Khim. sel'Khoz. Orenburg* 179–182. (*Weed Abstr.* **16**, 1066).

SPIRAMA RAJU, K. and RANGASWAMI, G. (1971). Studies on the effect of herbicides on soil microflora. *Indian J. Microbiol.* **11**, 25–29.

SPIRIDINOV, Y. Y. and SPIRIDINOVA, G. S. (1971). The effect of the systematic use of atrazine and simazine on some groups of micro-organisms in soil. *Trudy vses. nauchoissled. Inst. Udobr. Agrotekh. Agropochv.* No. 51, 264–270. (*Soils and Fert.* **37**, 3353).

SPIRIDINOV, Y. Y. and SPIRIDINOVA, G. S. (1973). Effect of repeated application of symmetrical triazines on biological activity of soil. *Agrokhimiya* No. 3, 122–131.

SPIRIDINOV, Y. Y. and YAKOVLEV, A. I. (1968). The effect of symmetrical triazines on cellulolytic soil microorganisms. *Mikrobiologiya* **37**, 137–141.

SPIRIDINOV, Y. Y., SKHILADZE, V. Sh. and SPIRIDINOVA, G. S. (1970). The effect of diuron and monuron in a red soil of the moist subtropics of Adzhariya. *Subtrop. Kul'tury* (3), 152–160. (*Weed Abstr.* **21**, 2642).

SPIRIDINOV, Y. Y., SKHILADZE, V. Sh. and SPIRIDINOVA, G. S. (1972). The activity of diuron and monuron in a meadow-bog soil of the moist sub-tropics of Adzhariya. *Subtrop. Kul'tury* (1), 150–155. (*Weed Abstr.* **22**, 2703).

SPIRIDINOV, Y. Y., SKHILADZE, V. Sh., SPIRIDINOVA, G. S. and MEPARISHVILI, U. Kh. (1973). The effect of diuron and monuron on the biological activity of sub-tropical soils. *In* "The Behaviour, Conversion and Analysis of Pesticides and their Metabolites in Soil." *Proc. 1st All-Union Conf. Pushchino-na-Oke, 1973.* Akad. Nauk. 116–118. (*Weed Abstr.* **24**, 819).

STAPHORST, J. L. and STRIJDOM, B. W. (1974). Effect of treatment with a dimethoate insecticide on nodulation and growth of *Medicago sativa* L. *Phytophylactica* **6**, 205–208.

STAPHORST, J. L. and STRIJDOM, B. W. (1976). Effects on rhizobia of fungicides applied to legume seed. *Phytophylactica* **8**, 47–54.

STEPANOVA, Z. A. (1967). The effect of triazines on the microflora of pale-chestnut soils. *Vest. sel'Khoz. Nauki. Mosk.* **12**, 41–45. (*Weed Abstr.* **16**, 1912).

STOTZKY, G. (1966). Influence of clay minerals on microorganisms. III. Effect of particle size, cation exchange capacity and surface area on bacteria. *Can. J. Microbiol.* **12**, 1235–1242.

SUD, R. K. and GUPTA, K. G. (1972). On the sensitivity of isolates of *Rhizobium* spp. and *Azotobacter chroococcum* to TMTD and its degradation product NaDDC. *Arch. Mikrobiol.* **85**, 19–22.

SUD, R. K., KUMAR, P. and GUPTA, K. G. (1973). On the effect of two herbicides on micro- and macro-symbionts. *Zbl. Bakt.* Abt. II **128**, 423–425.

SULLIA, S. B. (1969). The fungicide Kitazin and the mycoflora of rice. *Proc. Indian Acad. Sci.* **69**, 295–305.

SWAMIAPPAN, M. and CHANDY, K. C. (1975). Effect of certain granular insecticides on the nodulation by nitrogen-fixing bacteria in cowpea (*Vigna sinensis* L.). *Curr. Sci.* **44**, 558.

SWAMINATHAN, R. and SULLIA, S. B. (1969). Influence of the pesticide malathion on groundnut (*Arachis hypogaea* L.) microflora. *Curr. Sci.* **38**, 282–284.

SZABO, I. (1964). The effect of two herbicides on the soil microflora and root nodule formation in pea. *Mosonmagy. Agrártud. Föisk. Közl.* 7. *Kem. Talajt.* 26–33. (*Soils and Fert.* **28**, 1709).

SZEGI, J. (1970). Effect of some herbicides on the growth of cellulose decomposing microscopic fungi. *Meded. Fac. Landb. Rijksuniv. Gent* **35**, 559–561.

SZEGI, J. (1972). Effect of a few herbicides on the decomposition of cellulose. *Symp. Biol. Hung.* **11**, 349–354.

SZEGI, J. and GULYÁS, F. (1971). Influence of Gramoxone on cellulose decomposition in soils. *Agrokém. Talajt.* **20**, 590–598.

SZEGI, J., GULYÁS, F., MANNINGER, E. and ZAMORY, E. (1974). The influence of Gramoxone on nitrogen-fixing microorganisms. *Trans. 10th Int. Congr. Soil Sci.* III, 179–184.

SZEMBER, A., GOSTKOWSKA, K. and FURCZAK, J. (1973). The intensity of ammonification and nitrification in the soil under legumes in monoculture and in crop rotation with constant application of herbicides, Part 1. *Pol. J. Soil Sci.* **6**, 141–147.

TAHA, S. M., MAHMOUD, S. A. Z., ABDEL-HAFEZ, A. M. and HAMED, A. S. (1972a). Effect of some pesticides on rhizosphere microflora of cotton plants. 2. Herbicides. *Egypt. J. Microbiol.* **7**, 53–61.

TAHA, S. M., MAHMOUD, S. A. Z. and SALEM, S. H. (1972b). Effect of pesticides on rhizobium inoculation, nodulation and symbiotic N-fixation of some leguminous plants. *Symp. Biol. Hung.* **11**, 423–429.

TANG, A. CURL, E. A. and RODRIGUEZ-KABANA, R. (1970). Effect of trifluralin on inoculum density and spore germination of *Fusarium oxysporum* f. sp. *vasinfectum* in soil. *Phytopathology* **60**, 1082–1086.

TARLAPAN, M. I. and ZAKHARIYA, V. P. (1966). Einfluss von Triazinen auf den NPK Gehalt im Boden. *Agrokhimiya* **11**, 115.

TATE, K. R. (1974). Influence of four pesticide formulations on microbial processes in a New Zealand pasture soil. 1. Respiratory activity. *N.Z. Jl agric. Res.* **17**; 1–7.

TEUTEBERG, A. (1968). The effect of herbicide applications on antibiotically active microorganisms in the soil. *Z. PflKrankh. PflPath. PflSchutz.* **75**, 72–86.

TEUTEBERG, A. (1970). A further report on the effect of herbicides on the soil microflora in orchards. *Erwerbsobstbau* **12**, 69–71.

THIAGALINGAM, K. and KANEHIRO, Y. (1971). Effect of two fumigating chemicals and 2-chloro-6-trichloro-methylpyridine and temperature on nitrification of added ammonia in Hawaiian soils. *Trop. Agric.* **48**, 357–364.

THOMAS, M. A. and HAMMOND, H. D. (1971). The effect of simazine, kinetin and *Rhizobium phaseoli* on legume nodulation and morphogenesis in *Phaseolus vulgaris* L., c.v. "Red Kidney". *Ohio J. Sci.* **71**, 21–29.

THORNEBURG, R. P. and TWEEDY, J. A. (1973). A rapid procedure to evaluate the effect of pesticides on nitrification. *Weed Sci.* **21**, 397–399.

TORSTENSSON, L. (1974). Effects of MCPA, 2,4,5-T, linuron and simazine on some functional groups of soil microorganisms. *Swedish J. agric. Res.* **4**, 151–160.

TRULL, J. H. (1969). The responses of some soil and water microorganisms to the herbicide Cotoran. Ph.D. Thesis, Univ. Mississippi.

TSIRKOV, Y. I. (1970). Effect of the organic chlorine insecticides hexachlorane, heptachlor, lindane and dieldrin on activity of some soil enzymes. *Pochv. Agrokhimiya* **4**, 85–88. (*Soils and Fert.* **33**, 4620).

TSIRKOV, Y. I. (1971). Effect of the insecticides BHC, heptachlor, lindane and dieldrin on rhizosphere microflora in maize. *Bulg: Akad. Nauk.* (1971), 267–272. (*Soils and Fert.* **36**, 606).

TSVETKOVA, S. D. (1966). The effect of simazine and atrazine on the microflora of weakly podzolic soil. *Les. Khoz.* **19**, 25–26. (*Weed Abstr.* **18**, 473).

TU, C-M. (1966). Interaction between dipyridilium herbicides and microbes in the soil. Ph.D. Thesis, Oregon State Univ., Corvalis.

TU, C-M. (1969). Effects of four organophosphorus insecticides on the activity of microorganisms in soil. *Bact. Proc.* p. 3.

TU, C-M. (1973a). Effects of Mocap, N-Serve, Telone and Vorlex at two temperatures on populations and activities of microorganisms in soil. *Can. J. Pl. Sci.* **53**, 401–405.

TU, C-M. (1973b). The temperature dependent effect of residual nematicides on the activities of soil microorganisms. *Can. J. Microbiol.* **19**, 855–859.

TU, C-M. (1975). Interaction between lindane and microbes in soils. *Arch. Mikrobiol.* **105**, 131–134.

TU, C-M. and BOLLEN, W. B. (1968a). Effect of paraquat on microbial activities in soil. *Weed Res.* **8**, 28–37.

TU, C-M. and BOLLEN, W. B. (1968b). Interactions between paraquat and microbes in the soil. *Weed Res.* **8**, 38–45.

TU, C-M. and BOLLEN, W. B. (1969). Effect of Tordon herbicides on microbial activities in three Willamette Valley soils. *Down to Earth* **25**, 15–17.

TULABAEV, B. D. (1970). The effect of herbicides on soil microflora in cotton crops. *Uzbek. biol. Zh.* **14**, 11–14.

TULABAEV, B. D. (1972). The effect of different rates of herbicides on the microflora of a serozem-meadow soil under cotton crops. *Khimiya sel'Khoz.* **10**, 776–780. (*Weed Abstr.* **22**, 1911).

TULABAEV, B. D. and AZIMBEGOV, N. (1967). The effect of triazine and urea derivatives on soil microflora. *Khimiya sel'Khoz.* **5**, 43–44. (*Weed Abstr.* **16**, 2388).

TULABAEV, B. D. and TAMIKAEV, S. (1968). Effect of herbicides on meadow soil microflora. *Uzbek. biol. Zh.* No. 2, 14–17. (*Soils and Fert.* **32**, 2018).

TWEEDY, B. G. and LOEPPKY, C. A. (1967). The effect of Cotoran, simazine and atrazine upon the respiratory pathways of soil organisms. *Abstr. 6th Int. Congr. Pl. Prot., Vienna*, 409–410.

TWEEDY, B. G. and LOEPPKY, C. A. (1968). The use of ^{14}C-labelled glucose, glucuronate and acetate to study the effect of atrazine, simazine and fluometuron on glucose catabolism in selected plant pathogenic fungi. *Phytopathology* **58**, 1522–1531.

TWEEDY, B. G., TURNER, N. and ACHITUV, M. (1968). The interactions of soil-borne microorganisms with DCPA. *Weed Sci.* **16**, 470–473.

TYAGNY-RYADNO, M. G. (1967). Effect of herbicides on the microflora and agrochemical properties of the soil. *Trudy Kamenets-podolsk. sel'Khoz. Inst.* **9**, 43–48. (*Weed Abstr.* **18**, 1440).

TYUNYAEVA, G. N., MINENKO, A. K. and PEN'KOV, L. A. (1974). The effect of treflan on the biological properties of soil. *Agrokhimiya* No. 6, 110–114.

UHLIG, S. K. (1966). Studies on the interaction between chlorbisethylamine-s-triazine (simazine) and mycorrhiza-forming fungi. *Wiss. Z. Tech. Univ. Dresden* **15**, 639–641. (*Weed Abstr.* **16**, 1063).

UZIAK, S., LEONIAK, K. and GOSTKOWSKA, K. (1971). The influence of some herbicides on the selected groups of microorganisms in loess and sandy soils. *Ann. Univ. Marie Curie–Sklodowska, C*, **26**, 57–72. (*Weed Abstr.* **21**, 3266).

VALASKOVA, E. (1967). The sensitivity of soil fungi to the effects of herbicides. *Abstr. 6th Int. Congr. Pl. Prot., Vienna, 1967*, 606–607.

VALASKOVA, E. (1968). Susceptibility of soil fungi to herbicides. *Pflanzenschutzberichte* **38**, 135–146.

VALASKOVA, E. (1969). The effect of herbicides on soil fungi. *Ochr. Rost.* **5**, 55–64. (*Weed Abstr.* **20**, 1410).

van FAASSEN, H. G. (1974). Effect of the fungicide benomyl on some metabolic processes and on numbers of bacteria and actinomycetes in the soil. *Soil Biol. Biochem.* **6**, 131–133.

van SCHREVEN, D. A., LINDENBERGH, D. J. and KORIDON, A. (1970). Effect of several herbicides on bacterial populations and activity and the persistence of these herbicides in soil. *Pl. Soil* **33**, 513–532.

VARSHNEY, T. N. and GAUR, A. C. (1972). Effect of DDT and Sevin on soil fungi. *Acta Microbiol. Acad. Sci. Hung.* **19**, 97–102.

VASIL'EV, D. S. and ENKINA, O. V. (1967). The effect of prometryne on the microbiological activity of soil under sunflower. *Dokl. Vses. Akad. sel'Khoz. Nauk.* (12), 10–12. (*Weed Abstr.* **17**, 1939).

VENKATARAMAN, G. S. and RAJYALAKSHMI, B. (1971). Interactions between pesticides and soil microorganisms. *Indian J. exptl. Biol.* **9**, 521–522.

VENKATARAMAN, G. S. and RAJYALAKSHMI, B. (1972). Relative tolerance of nitrogen-fixing blue–green algae to pesticides. *Indian J. agric. Sci.* **42**, 119–121.

VERSTRAETEN, L. M. J. and VLASSAK, K. (1973). The influence of some chlorinated hydrocarbon insecticides on the mineralisation of nitrogen fertilizers and plant growth. *Pl. Soil* **39**, 15–28.

VLAGOV, S. S., DAMYANOVA, L., GUSTEROV, G. and KAMENOVA, L. (1972a). Action of simazine on the antibiotic activity of microscopic soil fungi. *Symp. Biol. Hung.* **11**, 365–368.

VLAGOV, S. S., KAMENOVA, L., GUSTEROV, G. and DAMYANOVA, L. (1972b). Effect of simazine on the antibiotic activity of actinomycetes. *Symp. Biol. Hung.* **11**, 355–358.

VODERBERG, V. K. (1961). Abhangigkeit der Herbizid-Wirkung auf Bodenmik-roorganismen vom Nährsubstrat. *Nachr. dt. PflSchutzdienst.* **15**, 21–40.

VOETS, J. P. and VANDAMME, E. (1970). Effect of 2-(thiocyanomethylthio) benzo-thiazole on the microflora and enzymes of the soil. *Meded. Fac. Landb. Rijksuniv. Gent* **35**, 563–580.

VOETS, J. P., MEERSCHMAN, M. and VERSTRAETE, W. (1974). Soil microbiological and biochemical effects of long-term atrazine applications. *Soil Biol. Biochem.* **6**, 149–152.

VOINOVA, Zh., VLAKHOVA, S. and PETKOVA, P. (1970). Microbiological processes in soil treated with various herbicides. *Pochv. Agrokhim.* **4**, 103–110. (*Soils and Fert.* **33**, 4575).

WAINWRIGHT, M. and PUGH, G. J. F. (1973). The effects of three fungicides on nitrification and ammonification in soil. *Soil Biol. Biochem.* **5**, 577–584.

WAINWRIGHT, M. and PUGH, G. J. F. (1974). The effects of fungicides on certain chemical and microbial properties of soils. *Soil Biol. Biochem.* **6**, 263–267.

WAINWRIGHT, M. and PUGH, G. J. F. (1975a). Effect of fungicides on the numbers of microorganisms and frequency of cellulolytic fungi in soils. *Pl. Soil* **43**, 561–572.

WAINWRIGHT, M. and PUGH, G. J. F. (1975b). Changes in the free amino acid content of soil following treatment with fungicides. *Soil Biol. Biochem.* **7**, 1–4.

WALTER, B. (1970). Effects of various herbicides on structure and microbiology of the soil. *Sonderh. Z. PflKrankh. PflPath. PflSchutz* **5**, 29–31.

WALTER, B. (1971). Investigations on CO_2 production and the enzymatic activity of a vineyard soil after treatment with herbicides. *C.r. Conf. Cté f. Lutte contr. Mauv. Herbes*, Columa, 79–82. (*Weed Abstr.* **21**, 3268).

WEEKS, R. E. and HEDRICK, H. G. (1975). Influence of a systemic fungicide on oxygen uptake by soil microorganisms. *Soil Sci.* **119**, 280–284.

WEGRZYN, T. (1971). The effect of some herbicides on *Azotobacter chroococcum*. *Acta microbiol. pol.* **20**, 131–134.

WEHR, N. B. and KLEIN, D. A. (1971). Herbicide effects on *Bdellovibrio bacteriovorus* parasitism of a soil pseudomonad. *Soil Biol. Biochem.* **3**, 143–149.

WHITWORTH, J. W., GARNER, W. and WILLIAMS, B. C. (1965). The influence of herbicides on soil metabolism. *Res. Prog. Rept. West. Weed Contr. Conf.* **1965**, 117–118.

WILKINSON, V. (1969). Ecological effects of diquat. *Nature, Lond.* **224**, 618–619.

WILKINSON, V. and LUCAS, R. L. (1969a). 'Gramoxone W': its effects on spores and mycelia of *Rhizopus stolonifer*. *Trans. Br. mycol. Soc.* **53**, 297–299.

WILKINSON, V. and LUCAS, R. L. (1969b). Effects of herbicides on the growth of soil fungi. *New Phytol.* **68**, 709–719.

WILKINSON, V. and LUCAS, R. L. (1969c). Effects of constituents of 'Gramoxone W' on rates of respiration of soil fungi. *Weed Res.* **9**, 288–295.

WINELY, C. L. and SAN CLEMENTE, C. L. (1968). Inhibition by certain pesticides of the nitrite oxidation of *Nitrobacter agilis*. *Bact. Proc.* A 6311.

WINELY, C. L. and SAN CLEMENTE, C. L. (1969). The effect of pesticides on nitrite oxidation by *Nitrobacter agilis*. *Bact. Proc.* 4.

WINELY, C. L. and SAN CLEMENTE, C. L. (1970). Effects of pesticides on nitrite oxidation by *Nitrobacter agilis*. *Appl. Microbiol.* **19**, 214–219.

YANG, J-S., FUNDERBURK, H. H. and CURL, E. A. (1968). Interaction of dipyridylium herbicides and soil microorganisms. *Abstr. Meet. Weed Sci. Soc. Am.* 38.

YAROSLAVSKAYA, P. M., VASIL'EV, D. S. and AGARKOVA, N. T. (1969). Effect of prometryne on nutrient regime of leached chernozem. *Agrokhimiya* No. 1, 97–101.

YURKEVICH, I. V. and TOLKACHEV, N. Z. (1972a). Effect of different doses of 2,4-D and simazine on the microflora of a typical chernozem. *Khimiya sel'Khoz.* **10**, 696–698. (*Soils and Fert.* **37**, 1219).

YURKEVICH, I. V. and TOLKACHEV, N. Z. (1972b). The effect of separate and combined application of simazine and nitrofoska on the yield of maize and the soil microflora in the Don basin. *Trudy khar'kov. sel'Khoz. Inst.* No. 172, 173–176. (*Weed Abstr.* **22**, 3047).

YUROVSKAYA, Y. M. and ZHULINSKAYA, V. A. (1974). The behaviour of organophosphorus insecticides in soil. *Khimiya sel'Khoz.* **12**, 38–41. (*Soils and Fert.* **38**, 5026).

ZAVARZIN, V. I. (1966). The effect of herbicides on some agro-chemical properties of soil. *Agrokhimiya* No. 4, 121–124.

ZAVARZIN, V. I. and BELYAEVA, T. V. (1966). The effect of herbicides on the soils content of mineral plant nutrients. *Khimiya sel'Khoz.* **4**, 34–36. (*Weed Abstr.* **17**, 439).

ZAYED, M. N., ABD-EL NASSER, M., ABD-EL MALEK, Y. and MONIB, M. (1968). Effect of D-D on nitrogen transformations in soil. *Zbl. Bakt.* Abt. II **122**, 527–532.

ZHABYUK, F. V. (1973). The effect of prometryne, sodium trichloroacetate and their combination on soil microflora and potato yield. *Mikrobiol. Zh.* **35**, 394. (*Soils and Fert.* **38**, 4207).

ZHAKHARYAN, S. V. (1970). Relationships between dalapon and trichloroacetic acid herbicides and the microflora of light brown soils. *Symp. Biol. Hung.* **11**, 391–395.

ZHARASOV, Sh. U. (1969). The effect of herbicides on the growth of microorganisms. *Vest. sel'-Khoz. Nauki, Alma-Ata* **12**, 94–98. (*Weed Abstr.* **20**, 1854).

ZHARASOV, Sh. U. (1971). The effect of herbicides on the intensity of nitrification. *Vest. sel'Khoz. Nauki, Kazakhskoi* **14**, 41–43. (*Weed Abstr.* **21**, 2663).

ZHARASOV, Sh. U. (1972). The effect of herbicides on the microflora of dark chestnut soils under sugarbeet crops in Alma-Ata Province. *Vest. sel'Khoz. Nauki, Kazakhskoi* **15**, 25–29. (*Weed Abstr.* **21**, 2657).

ZHARASOV, Sh. U., TSUKERMAN, G. M. and CHULAKOV, Sh. A. (1972). The effect of herbicides used in sugarbeet crops on soil fungi. *Khimiya sel'Khoz.* **10**, 617–619. (*Weed Abstr.* **22**, 1919).

ZHUKOVA, P. S. and BOTIN'EVA, A. M. (1974). The effectiveness of eptam, pyramine and betanal in red beet fields and their residues in soil and plants. *Khimizat. sel'Khoz.* **12**, 53–56. (*Soils and Fert.* **38**, 6104).

ZHUKOVA, P. A. and KONOSHEVICH, Z. I. (1967). The effect of herbicides on the microflora and agricultural and chemical properties of the soil. *Puti Povysh. Urozh. ovoshchn., Minsk* 1967, 138–150. (*Weed Abstr.* **17**, 2923).

ZINCHENKO, V. A. and OSINSKAYA, T. V. (1969). Alteration of biological activity of soil during incubation with herbicides. *Agrokhimiya* No. 9, 94–101.

ZINCHENKO, V. A., OSINSKAYA, T. V. and PROKUDINA, N. A., (1969). The effect of herbicides on the biological activity of the soil. *Khimiya sel'Khoz.* **7**, 850–853. (*Weed Abstr.* **20**, 2390).

Chapter 8

Interactions of Pesticides with Micro-Algae

S. JOHN L. WRIGHT

School of Biological Sciences, University of Bath, Bath, England

I. Introduction

Micro-algae are microscopic, photosynthetic organisms. They range from unicellular to colonial and filamentous types and embrace the aquatic phytoplankton populations important in global photosynthesis, oxygen evolution and primary production. Eukaryotic micro-algae include the Euglenophyta (flagellates), Chlorophyta (green

algae), Pyrrophyta (dinoflagellates), Chrysophyta and the Bacillario-
phyta (diatoms), whilst the prokaryotic types are the Cyanophyta
(blue–green algae or cyanobacteria).

Micro-algae typically inhabit aquatic environments, the soil surface
and other exposed locations. Some micro-algae live in associations
with other organisms, for example with invertebrates, fungi (forming
lichens) and plants (including ferns and cycads). Whilst most algae are
ecologically beneficial, some are troublesome to man.

Diatoms, green algae, dinoflagellates and chrysophytes are
significant organisms in the large aquatic environments, while the
blue–green algae may assume particular importance in the soil and·
associated environments such as irrigation and drainage systems and
in lakes and ponds. Blue–green algae can colonize exposed or barren
surfaces and help to maintain soil fertility through carbon fixation,
synthesis of substances supporting plant growth and the binding of
soil particles (Shields and Durrell, 1964). Probably the most noted
activity is the ability of some species to fix atmospheric nitrogen. This
contributes to the level of combined nitrogen in tropical and
temperate soils and is particularly significant in the productivity of
rice paddies which receive no added fertilizers. Inoculation of N_2-
fixing blue–green algae into paddy irrigation water is known to
increase rice yields (Watanabe, 1962). Of the many texts on the
micro-algae, those of Venkataraman (1969), Fogg (1965), Round
(1973) and Jackson (1964) are of general interest, whilst Carr and
Whitton (1973) and Fogg et al. (1973) provide a wealth of information
on blue–green algae.

As inhabitants of environments which are directly or indirectly
entered by pesticides, there are distinct opportunities for algae to
encounter pesticides. An awareness of the potential for biological
reactivity of pesticides must consider these important microorganisms. An attempt is made here to review information on the
various aspects which are involved.

In this chapter the values for pesticide concentrations are usually
given in standardized units (ppm or ppb) to facilitate comparison of
the data of different workers. Where necessary, therefore, conversion
of the concentrations quoted by the original authors are based on the
following:

$$\text{Part per billion } (10^9), \text{ppb} = \mu\text{g litre}^{-1}$$
$$\text{Part per million, ppm} = \mu\text{g ml}^{-1} \text{ or mg litre}^{-1}$$
$$1 \text{ ppm} = 1000 \text{ ppb}; \, 0\cdot001 \text{ ppb} = 1 \text{ ng litre}^{-1}$$

The chemical names of pesticides, which are generally referred to
by common names in the text, are given at the end of the book.

II. Pesticide Uptake, Accumulation and Metabolism by Micro-algae

Since micro-algae have fundamental affinities with the cells of higher plants, it is to be expected that they will be vulnerable to some herbicides, especially those which affect photosynthesis. There is also the possibility that these organisms, like higher plants, might detoxify or metabolize herbicides to some extent.

There have been several studies (see Section IIIB) on the action of herbicides on micro-algae. However, by comparison with insecticides for instance, herbicide uptake and metabolism by algae has not been widely studied and we know little of the contribution made by such organisms to herbicide decomposition in soil and water environments.

Widely located at the base of food chains and in general having a high surface area:volume cell characteristic, algal populations have a significant potential for sorption of, and reaction with, pesticides. This has been especially recognized with respect to the persistent, lipid-partitioning organochlorine insecticides and the phytoplankton which are so important in pelagic food chains; reflecting the general concern for biological magnification of organochlorine pesticides (Rice and Sikka, 1973a, b). Indeed, algae were considered to be useful for studying pesticide accumulation (Vance and Drummond, 1969) and both eukaryotic and prokaryotic algal types are known to metabolize some pesticides.

A. ALGAL DEGRADATION OF AROMATIC COMPOUNDS

The ability to degrade aromatic rings has been described for numerous heterotrophic microorganisms, especially bacteria. Although there have been comparatively few reports of the fission of aromatic and heterocyclic substrates by micro-algae, sufficient data have been obtained to suggest that this property is possessed or acquired by different types of algae. It is therefore possible that micro-algae may be more important in degradation of pesticides, at least the aromatic compounds, than has generally been supposed.

Craigie et al. (1965) gave evidence for the destruction of the aromatic ring of (U-^{14}C) phloroglucinol in axenic cultures of algae, according to the release of $^{14}CO_2$. An adaptive mechanism was suggested by the fact that a higher order of phloroglucinol-degrading activity was obtained after pre-conditioning the algae to the substrate.

Several types of unicellular marine algae in axenic cultures were tested by Vose *et al.* (1971) for the ability to degrade the aromatic nucleus of phenylalanine, an amino acid known to occur in planktonic algae. Of the 22 algal species, nine degraded (*ring*-^{14}C) phenylalanine in the light and four species did so in the dark. In addition to $^{14}CO_2$, all active organisms also produced a volatile radio-labelled metabolite which was probably non-aromatic. Knutsen (1972) has reported degradation of the pyrimidine uracil to CO_2 and β-alanine in synchronous cultures of *Chlorella fusca*.

B. INSECTICIDES

1. *Organochlorines*

(a) *DDT*. Some algal species show a marked capacity to concentrate DDT from the surrounding medium, although the degree of accumulation varies with DDT concentration and the algal species (Table 1).

Hill and McCarty (1967) reported that DDT sorption by a mixed algal culture was approximately twice that adsorbed from water by bentonite clay. Since algae are numerous in lake and stream water, it was considered that pesticide-containing algal cells would contribute to the eventual anaerobic decay of the pesticide in bottom sediments, by settling to the bottom after death.

An indication that algae could collect DDT, or its metabolites, from water was obtained by Ware *et al.* (1968) who examined DDT levels in plants and algae in irrigation canals. The algae, especially the filamentous *Cladophora*, accumulated higher residues of the DDT metabolite DDE than did plants, prompting the suggestion that *Cladophora* could serve as an indicator of DDT contamination of water. Similarly, blue–green algae removed from streams were found to contain DDT, this being consistent with use of the pesticide on the stream watersheds (Worthen, 1973).

Södergren (1968) studied the mechanism of DDT uptake by a *Chlorella* sp. and found that (^{14}C) DDT at a concentration of 0·6 ppb was rapidly taken up (within 15 s) by *Chlorella* cells. The primary uptake was attributed to a passive, physical process of absorption rather than active assimilation. This observation was endorsed by the fact that dead cells took up DDT at a similar rate to live cells, and was later substantiated by Rice and Sikka (1973a). Södergren (1968) concluded that the rate of penetration of DDT into algal cells was probably equal to its rate of diffusion in water, and noted that DDT accumulation in the algae in continuous culture induced morphological changes and cell clumping. A further observation of Södergren (1968) was that at concentrations higher than 0·6 ppb DDT in algal cultures, filtration or centrifugation could not be used to efficiently separate algae from DDT in the medium. Bowes (1972) also recorded the unsuitability of centrifugation for this purpose. Since most workers subsequent to Södergren (1968) have relied on such techniques and used higher levels of DDT, Södergren's observation may be of significance. However, more recently Adamich *et al.* (1974) have reported that extracellular DDT particles can be separated from *Chlorella* cells by centrifugation in a sucrose-based (Ficoll) density gradient. Since approximately 99% of the extracellular DDT particles could be removed, the technique of Adamich *et al.* should allow more accurate and complete kinetic studies of the interactions of algae with DDT at concentrations above its water solubility (1·2 ppb).

Without any apparent adverse effect, axenic cultures of *Anacystis nidulans* and *Scenedesmus obliquus* concentrated DDT 849 and 626 times, respectively, from a growth medium initially containing 1 ppm of the insecticide (Gregory *et al.*, 1969). These values were very much higher than that obtained by the same workers for the flagellate *Euglena gracilis*, although not as high as that for a ciliate protozoan, *Paramaecium multimicronucleatum*. In all cases the concentration factors were higher than those obtained with the organophosphorus insecticide parathion; however, no DDT metabolites were detected. Keil and Priester (1969) reported that the marine diatom *Cylindrotheca closterium* concentrated DDT 190-fold from a medium containing 0·1 ppm of the insecticide. Small amounts of the less toxic metabolite DDE were formed, by dehydrodechlorination, this being the only DDT metabolite of the diatom detected. In view of their tendency for intracellular storage of oil, it was considered possible that diatoms might serve as "pick-up" organisms for oil-soluble pesticides which might then be detoxified (Keil and Priester, 1969). Later studies by Miyazaki and Thorsteinson (1972), using (^{14}C) pp'-DDT, established that DDT was transformed to the non-toxic DDE by several pure cultures of freshwater diatoms isolated from a mosquito breeding site. Furthermore, although DDE was detected in small quantities, it was again the only metabolite formed. Like DDT, its metabolite pp'-DDE was accumulated to a high degree by *Chlorella* cells, only a small part of the added DDE remaining in the aqueous medium (Södergren, 1971).

Vance and Drummond (1969) incorporated higher pp'-DDT concentrations (up to 20 ppm) in cultures of green and blue–green algae. The algae were generally quite resistant to the toxic effects, even though DDT was concentrated at least 100-fold from the medium;

TABLE 1

Algal accumulation and metabolism of DDT

Algal type	Species	DDT in medium (ppb)	DDT accumulation factor[a] or % removed from medium	Metabolites detected	Reference
Green algae (Chlorophyta)	Scenedesmus obliquus	1000	626	None	Gregory et al. (1969)
	Scenedesmus quadricauda	1000	134	N.R.	Vance and Drummond (1969)
	Ankistrodesmus amalloides	0·72	616 (max.)	DDE and TDE	Neudorf and Khan (1975)
	Dunaliella tertiolecta	80	up to 97·8% gross	DDE	Bowes (1972)
	Dunaliella sp.			DDOH, DDE, DDNS, TDE	Patil et al. (1972)
Blue–green algae (Cyanophyta)	Oedogonium sp.	1000	270	N.R.	Vance and Drummond (1969)
	Synechococcus elongatus	99	84%	DDE	Worthen (1973)
	Anacystis nidulans	1000	849	None	Gregory et al. (1969)
	Microcystis aeruginosa	1000	230	N.R.	Vance and Drummond (1969)
	Anabaena cylindrica	1000	268	N.R.	Vance and Drummond (1969)
	Skeletonema costatum	0·7	up to 38 400	DDE	Rice and Sikka (1973a)
Diatoms (Bacillariophyta)	Cyclotella nana	0·7	up to 58 100	DDE	Rice and Sikka (1973a)
	Cylindrotheca closterium	100	190	DDE	Keil and Priester (1969)
	Nitzchia sp.	710	N.R.	DDE	Miyazaki and Thorsteinson (1972)

	S. costatum	80	up to 95·7% gross	DDE	Bowes (1972)
	C. nana	80	up to 96·6% gross	DDE	Bowes (1972)
	Thalassiosira fluviatilis	80	up to 94·4% gross	DDE	Bowes (1972)
Dinoflagellates (Pyrrophyta)	*Amphidinium carteri*	Ambient	80 000	N.R.	Cox (1970)
	Amphidinium carteri	0·7	up to 9600	DDE	Rice and Sikka (1973a)
	Amphidinium carteri	80	up to 93·2% gross	None	Bowes (1972)
Euglenophyta	*Euglena gracilis*	1000	99	None	Gregory *et al.* (1969)
	Tetraselmis chuii	0·7	up to 6300	DDE	Rice and Sikka (1973a)
Coccolithophorid	*Syracosphaera* sp.	Ambient	25 000	N.R.	Cox (1970)
Other	Plankton samples			TDE	Patil *et al.* (1972)

[a] Accumulation or magnification factor, usually expresses ratio of cellular DDT to extracellular level.
N.R. Not reported.

they also degraded the pesticide to a slight extent. These authors considered that whereas algae are generally more resistant than higher members of the food chain to the effects of chlorinated pesticides, they are very efficient potentiators of pesticide residues within the food web.

Cox (1970) interpreted Södergren's (1968) results of rapid, passive (^{14}C) DDT-uptake by *Chlorella* in terms of a partition mechanism and considered that the much higher concentrations of DDT used by some workers might affect the partition coefficients of organisms for DDT residues in water. Cox, therefore, determined (^{14}C) DDT uptake by pure cultures of marine phytoplankton at concentrations equivalent to natural low ambient levels of DDT. The concentration factors obtained, 25 000-fold and 80 000-fold for *Syracosphaera* sp. and *Amphidinium carteri*, respectively, far exceeded that reported for the diatom *Cylindrotheca closterium* by Keil and Priester (1969).

Further evidence for a high degree of pp'-DDT uptake and its metabolism to some extent to DDE by axenic algae, especially marine diatoms, was obtained by Bowes (1972) using a concentration of 80 ppb DDT in a seawater medium.

Worthen (1973), studying the interaction of blue–green algae with pp'-DDT, obtained results essentially similar to those of other workers who had used green algae and diatoms. Thus *Synechococcus elongatus* in axenic culture quickly removed a high percentage of the 99 ppb DDT added to the growth medium. The accumulated level of DDT was subsequently maintained by the algae, there being no effect on growth or cell morphology, and a small proportion of the DDT was converted to DDE.

A comprehensive study by Rice and Sikka (1973a) on the uptake and accumulation of DDT by different types of axenic marine algae revealed significant differences between species in the capacity to accumulate DDT. Although all organisms tested rapidly took up DDT from a medium containing 1 ppb, the diatoms *Skeletonema costatum* and *Cyclotella nana* accumulated the highest levels. The amount of DDT accumulated by algae was linearly related to its concentration in the medium, with variation in the ranges for different species. Similarly, increasing cell density caused an increased, but disproportionate, overall removal of DDT from the medium. Conversely, as cell density increased, the degree to which individual cells concentrated DDT decreased. Furthermore, the ability of algal species to concentrate DDT was inversely correlated with cell size. The DDT concentration factors in the diatoms especially (e.g. 58 100-fold for *C. nana*) agreed with values obtained by Cox (1970) for marine phytoplankton (coccolithophorids and dinoflagellates) and far exceeded the diatom DDT concentration values obtained by Keil and Priester (1969) and Gregory *et al.* (1969). Rice and Sikka (1973a) considered that the concentration factor difference could be due to the use of different algal cell densities and DDT concentrations. The latter seems particularly likely since the solubility limit of DDT in water is 1·2 ppb. The detection of DDE as the only metabolite of DDT in all algal species examined by Rice and Sikka (1973a) confirmed the findings of other workers and suggested the widespread occurrence of this mechanism in algae.

In a study of the uptake and metabolism of DDT by the freshwater green alga, *Ankistrodesmus amalloides*, Neudorf and Khan (1975) also noticed that whilst the total amount of (^{14}C) DDT absorbed from medium containing 0·72 ppb was higher in denser cell suspensions, the adsorption/absorption efficiency of individual cells decreased with increasing cell density. Likewise, the DDT magnification or concentration ratios declined with increasing cell density. Neudorf and Khan (1975) found that in addition to DDE, the usual algal metabolite of DDT, very small amounts of DDD (TDE) were also formed. TDE was also detected in marine plankton samples, whilst TDE, DDNS, DDOH and DDE were the DDT metabolites in pure cultures of *Dunaliella* sp. (Patil *et al.*, 1972).

Uptake of methoxychlor (methoxy-DDT) by actively growing green algae and diatoms was reported by Butler *et al.* (1975).

(b) *Aldrin, dieldrin and endrin.* Vance and Drummond (1969) reported the uptake and concentration of aldrin, dieldrin and endrin by species of green and blue–green algae growing in media containing the pesticides at concentrations up to 20 ppm. Wheeler (1970) added (^{14}C) dieldrin (0·3 ppm) to cultures of *Chlorella pyrenoidosa* and noticed that although it was readily taken up by the algae, no metabolites were formed. The rate of dieldrin uptake was not as rapid as had been reported by other workers for DDT and this was attributed to the difference in water solubility and lipid affinity of the two compounds. Wheeler's data indicated that dieldrin became bound to subcellular organelles. However, his suggestion that algae might be utilized as pesticide scavengers does not appear to take the possible food-chain consequences into consideration.

Six species of axenic marine algae, representing four taxonomic divisions, rapidly took up (^{14}C) dieldrin but formed no detectable metabolites of the insecticide (Rice and Sikka, 1973b). The dieldrin concentration factors in algae were lower than those obtained by the same authors for algal concentration of DDT (Rice and Sikka, 1973a). Uptake of dieldrin (Rice and Sikka, 1973b) by some algae was linearly related to the dieldrin concentration supplied, even at levels

(1000 ppb) far exceeding its maximum water solubility (100 ppb). This indicated, as had been suggested by Menzel *et al.* (1970), that the algae might either incorporate dieldrin as small particles, or that saturation was maintained while the algae concentrated the pesticide from solution. Total dieldrin uptake increased with increasing cell concentration (Rice and Sikka, 1973b) but uptake was not correlated with number of cells per unit mass. Although Rice and Sikka did not detect any algal metabolites of dieldrin, photodieldrin has been identified as a product of (^{14}C) dieldrin in some microbial cultures isolated from dieldrin-contaminated lake water (Matsumura *et al.*, 1970), and in marine sediment and algal samples collected by Patil *et al.* (1972). Photodieldrin, a "terminal residue", is more toxic to many biological systems than is dieldrin (Matsumura *et al.*, 1970; Neudorf and Khan, 1975). Patil *et al.* (1972) reported also that aldrin was converted to dieldrin and trans-aldrindiol, and endrin to ketoendrin, in the marine samples containing algae, plankton and other micro-organisms, but not in plain seawater.

Neudorf and Khan (1975) have confirmed that dieldrin and photodieldrin are less readily taken up and concentrated by algae than is DDT, which has lower water-solubility, and the alga *Ankistrodesmus amalloides* had very low affinity for photodieldrin, the least lipophilic compound tested.

(c) *Lindane.* This insecticide, which contains at least 99% gamma-BHC, was less readily accumulated by *Chlorella pyrenoidosa* from an aqueous medium containing 0·18 ppb lindane (Södergren, 1971) than was DDT, which again has much lower affinity for water (Södergren, 1968). However, Sweeney (1969) reported lindane uptake and metabolism by the green algae *Chlorella vulgaris* and *Chlamydomonas reinhardtii* in media containing up to 4·5 ppm insecticide. A non-toxic metabolite, 1,3,4,5,6-pentachlorocyclohex-1-ene, was formed. It was considered that algal metabolism might account for the low levels of lindane detected in the Great Lakes as compared with other organo-chlorines applied at similar rates within the watershed (Sweeney, 1969).

(d) *Mirex.* Marine algae accumulated mirex 10 000-fold from water containing 0·2 ppb (Hollister *et al.*, 1975).

2. *Organophosphorus compounds*

The metabolism of some organophosphorus insecticides by algae (Chlorella) was first reported by Ahmed and Casida (1958). Other workers have since examined the uptake and breakdown of the commonly used organophosphates.

(a) *Malathion*. Uptake of (^{14}C) malathion (100 ppm) by axenic cultures of *Chlorella pyrenoidosa*, with some conversion to malaoxon, a known metabolite of malathion in lettuce and wheat, was reported by Christie (1969).

(b) *Parathion*. Sato and Kubo (1964) found that parathion in rice paddies was degraded within a few days and that the presence of algae greatly accelerated the degradation rate *in vitro*.

Parathion, at 1 ppm in the medium, was concentrated to about the same extent by different types of algae without adverse effect (Gregory *et al.*, 1969). As with DDT, the concentration factor in the ciliate *Paramaecium multimicronucleatum* was higher than that in the algae. With the exception of *Euglena gracilis*, the concentration factors for parathion were very much lower than for DDT in all organisms.

In a different type of study, Mackiewicz *et al.* (1969) showed that parathion was transformed in the presence of the roots of gnotobiotically-grown beans (*Phaseolus vulgaris*) only when cells of the alga *Chlorella pyrenoidosa* were added. Metabolites formed included aminoparathion and an unidentified sulphur-containing compound, the latter even being translocated to aerial parts of the bean plants. It was thus demonstrated that pesticide metabolites found in plants could be formed by other organisms. However, the significance of such an alga in the rhizosphere of natural beans seems doubtful. Subsequent studies (Zuckerman *et al.*, 1970), in which (^{35}S) parathion was added to cultures of the same *C. pyrenoidosa* strain, axenically grown, confirmed that aminoparathion was the principal metabolite of parathion. Aminoparathion, which was detected at high levels in the medium and was not formed chemically, was non-insecticidal at levels tested. Subsidiary metabolites, unidentified, included two sulphur-containing products and one containing the intact phenyl ring but no sulphur (Zuckerman *et al.*, 1970).

(c) *Diazinon*. Uptake of this, another phosphorothionate insecticide and acaricide, by several algae has been reported by Butler *et al.* (1975).

3. *Methylcarbamate*

C. pyrenoidosa cultures took up (^{14}C) Sevin (carbaryl) from the medium, but the compound was not altered appreciably by the alga in acidic medium and above 0·1 ppm inhibited growth (Christie, 1969).

C. HERBICIDES

1. Phenoxy-acids

(a) *2,4-D.* Wedding and Erickson (1957) established that the extent of uptake of 2,4-D by *C. pyrenoidosa* cells was greatest at low pH values. Subsequently, Swets and Wedding (1964) showed that 2,4-D uptake by *Chlorella* was increased under conditions which tended to increase the rate of glycolysis or reduce TCA cycle activity and it was suggested that after entering the cell 2,4-D was activated by a reaction involving acetyl CoA. Valentine (1973) and Valentine and Bingham (1974) confirmed that algal uptake of 2,4-D occurred more readily at low pH, recording a pH of 4·7 for uptake by *Scenedesmus quadricauda*. Uptake of 2,4-D from water by this organism was hindered at lowered temperature, but was favoured in darkness. At pH 4·7, 3-hydroxy-2,4-D was detected as a major metabolite of (*ring*-^{14}C) 2,4-D, mostly inside cells, whilst 5-hydroxy-2,4-D was found extracellularly. The observation that the percentage removal of 2,4-D by *Scenedesmus* cells increased at concentrations below 1 ppm prompted the suggestion that such an organism would be effective in removing low residual levels of 2,4-D in certain natural surface waters, provided the pH was low enough (Valentine and Bingham, 1974). Although Voight and Lynch (1974) reported low uptake of 2,4-D by eukaryotic and prokaryotic algae, again the highest value (with the eukaryote, *Coelastrum microporum*) occurred at low pH. Butler *et al.* (1975), however, found that cultures of several algae removed significant amounts of 2,4-D from a medium containing 0·01 ppm 2,4-D.

(b) *MCPA and MCPB.* Unicellular green algae absorbed MCPB to a greater extent than MCPA, which has lower lipid solubility (Kirkwood and Fletcher, 1970). Greatest uptake occurred in *Chlamydomonas globosa*, an alga of relatively large cell size and thin cell wall. As was the case with 2,4-D (see above), absorption of MCPA and MCPB generally increased with decreasing pH, and Kirkwood and Fletcher considered that uptake of the latter herbicides was optimal at pH values favouring their movement as undissociated molecules. Further evidence for the importance of pH level as a factor determining algal absorption, and hence activity, of MCPB and MCPA was given by Fletcher *et al.* (1970).

2. Triazine compounds

Having noticed that simazine was detoxified in soil to which a culture of the alga *Chlorosarcina* had been added, Kruglov and Paromenskaya

(1970) examined uptake and metabolism of the herbicide in pure cultures of algae. Both the *Chlorosarcina* sp. (simazine-resistant) and *Ankistrodesmus braunii* (simazine-sensitive) took up (^{14}C) simazine from a liquid medium containing 5·16 ppm simazine, bound the herbicide to cellular protein, and also metabolized it.

Uptake of atrazine by *Scenedesmus quadricauda*, in contrast to 2,4-D, was favoured by slightly alkaline pH and unaffected by darkness (Valentine, 1973). Algal degradation of atrazine was not detected by Valentine (1973) or Butler *et al.* (1975). However, under optimal light conditions amitrole uptake by *Scenedesmus* occurred and the herbicide was metabolized.

3. *Other herbicides*

Chlorella vulgaris degraded ^{14}C-labelled fluorodifen (Preforan) and fluometuron to a small extent, forming metabolites similar to those in bacteria and fungi, whilst metobromuron was not appreciably affected (Tweedy *et al.*, 1969). Uptake of (^{14}C) propanil by *Chlorococcum aplanosporum* occurred rapidly, by simple diffusion (Sharabi, 1969).

In my laboratory evidence was obtained for degradation of asulam in cultures of a *Chlorella* sp. The organism had been isolated from soil columns showing profuse algal growth whilst percolated with an asulam solution (200 ppm). The *Chlorella* grew vigorously in cultures containing 50 ppm asulam, completely eliminating supernatant toxicity within 7 days as determined by an algal bioassay (Wright, 1972). However, there was no decline in u.v. absorbance of the culture supernatants at 257 nm (λ max asulam), nor loss of supernatant toxicity when dense suspensions of dead cells were incubated with the herbicide. It is possible that the results could be explained by the conversion of asulam to sulphanilamide, which has similar u.v. absorbance characteristics but only half the toxicity of asulam according to the algal bioassay. The foregoing data on asulam have yet to be substantiated by chemical and chromatographic analysis using axenic algal cultures.

A non-axenic culture of *Chlorella pyrenoidosa* removed propanil (5 ppm) from the growth medium and formed the less toxic 3,4-dichloroaniline (DCA), an established microbial metabolite of propanil. More propanil degradation was attributable to *C. pyrenoidosa* than to the associated bacteria (Wright and Walker, unpublished data). DCA was also detected when dense suspensions of some species of blue–green algae (especially non-axenic cultures) were incubated with propanil (Wright *et al.*, 1977). Recent studies in our laboratory have established that axenic cultures of some blue–green algae are also able to degrade propanil in this way, and that some species can hydrolyse the phenylcarbamates propham and chlorpropham to the corresponding aniline compounds (Wright and Maule, unpublished data).

Algal absorption of the hydroxybenzonitrile herbicides ioxynil and bromoxynil was increased at higher concentrations and lower pH levels (Fletcher *et al.*, 1970). Whilst absorption of ioxynil was also enhanced at higher temperature (35°C), there was a lower level of uptake in the dark than in light conditions.

D. FUNGICIDES

The only study of fungicide metabolism by algae of which I am aware was that by Teal (1974), who reported the degradation of the systemic fungicide ethirimol in cultures of three species of green algae. Teal raised the question of the possible use of algae such as the common *Chlorella* as a tool for elucidation of toxicant metabolism pathways in plants.

E. ALGICIDES

Whilst algistatic and algicidal chemicals should persist long enough to be effective when used for algal control, it is desirable that such compounds should be biodegradable in order to safeguard the quality of the environment (usually aquatic) in which they are used. Fitzgerald (1975) gave evidence for the participation of algae in the degradation of such chemicals when applied at minimal effective concentrations.

III. Effects of Pesticides on Micro-algae

This section concerns pesticides other than those used as algicides. For some pesticides reasonably consistent data have been obtained by different workers regarding their effects on micro-algae. However, in very few cases is there complete information concerning the extent to which such effects are influenced by other factors, the variation in response of different algal types, the effects on natural algal populations in soil or water and the mechanisms by which pesticides affect algae. It is a step in the right direction that for a few pesticides the studies have included both pure cultures and natural populations. In recognition of the differences in chemical and toxicological properties of different pesticides, and the variation in experimental procedures of different workers (e.g. axenic and non-axenic cultures), caution is advised in comparing some of the data.

Other authors who have included data on the reactions of pesticides with algae and phytoplankton of aquatic environments are Mullison (1970a, b); Ware and Roan (1970); Luard (1973); Paris and Lewis (1973); see also *Note added in proof*, p. 602.

see also *Note added in proof*, p. 602.

A. INSECTICIDES

1. *Organochlorines*

There is evidence of considerable diversity in sensitivity among marine and freshwater phytoplankton to the effects of chlorinated hydrocarbons (Mosser *et al.*, 1972b; Luard, 1973) (see Table 2).

(a) *DDT*. Wurster (1968), who emphasized the important contribution of phytoplankton to global photosynthesis and the consequences of interference with this process, found that concentrations of DDT below 10 ppb inhibited photosynthesis in marine planktonic algae. At concentrations between 3·6 and 36 ppb *pp'*-DDT also inhibited photosynthetic CO_2 fixation in the green alga *Selanastrum capricormutum* (Lee *et al.*, 1976). Lee *et al.* (1976) presented evidence that *pp'*-DDT stimulated photorespiration and suppressed the incorporation of ^{14}C from $^{14}CO_2$ into the C_4-dicarboxylic acid pathway. The DDT-induced shift in metabolism from an efficient to a non-efficient pathway was interpreted to be through disruption of cyclic photophosphorylation. The toxicity of DDT to the diatom *Skeletonema costatum* increased with decreasing cell concentration and it was concluded that low levels of DDT in natural waters might have deleterious effects on the phytoplankton and hence organisms higher in the food chain (Wurster, 1968). A similar conclusion was reached by Södergren (1968), who reported that growth of a *Chlorella* sp. was affected by less than 0·3 ppb DDT.

The results of other studies of the effects of DDT on micro-algae seem somewhat at variance with these observations. It should be noted, however, that in many of the studies much higher levels of DDT have been used, far exceeding the water solubility (1·2 ppb).

DDT (0·6 ppm) was not lethal to marine phytoplankton cultures (Ukeles, 1962), neither did it affect algal growth or respiration at concentrations of 20 ppm and 1 ppm, respectively (Vance and

Drummond, 1969). Similarly, photosynthetic ^{14}C-fixation by *Scenedesmus quadricauda* was not significantly inhibited by DDT at concentrations up to 1 ppm, although the metabolite, DDE, was inhibitory (Luard, 1973). Bowes (1972) has also shown that DDE can be toxic to algae. Christie (1969) found that DDT did not affect growth of an axenic *Chlorella pyrenoidosa* culture at 10 ppm and even at 100 ppm was not significantly toxic. The lack of DDT influence on *Chlorella* was attributed to low solubility of the compound and degradative activity of the algae (Christie, 1969). Axenic cultures of green and blue–green algae were not adversely affected through accumulation of high levels (e.g. 630 ppm, 850 ppm) of DDT from the medium (Gregory *et al.*, 1969). Algal population densities were not significantly reduced by 100 ppm DDT, although ATP levels were lowered (Clegg and Koevenig, 1974). Growth of the green alga *Chlamydomonas reinhardii* was not appreciably affected by DDT at levels up to 20 ppm and there was no significant effect on ^{14}C uptake (Morgan, 1972). In contrast, both growth and ^{14}C-assimilation in low-density populations of *S. quadricauda* were inhibited by 0·1 ppm DDT (Stadnyk *et al.*, 1971).

Algal species isolated from various marine environments responded differently to DDT (Menzel *et al.*, 1970). Whereas the estuarine naked green flagellate *Dunaliella tertiolecta* was insensitive to 1000 ppb DDT, photosynthetic ^{14}C-uptake by the marine diatom *Skeletonema costatum* and the coccolithophorid *Coccolithus huxleyi* was significantly reduced above 10 ppb DDT. Menzel *et al.* (1970) further showed that cell division in the diatom, but not in *Coccolithus*, was prevented by 100 ppb DDT. The most DDT-sensitive organism tested by Menzel *et al.* (1970) was also a marine diatom, *Cyclotella nana*, in which ^{14}C-uptake and growth were affected at DDT levels of 1 ppb and 100 ppb, respectively. Consistent with these observations, Bowes (1972) reported that with the exception of *Skeletonema costatum*, whose growth was slowed, 80 ppb DDT was without significant effect on the growth or morphology of seven species of axenic marine phytoplankton. Similarly, neither growth nor cell morphology of the blue–green alga *Synechococcus elongatus* were adversely affected when 99 ppb DDT was added to axenic cultures in exponential growth (Worthen, 1973). This contrasts with the effects of lower levels of DDT ($<0·3$ ppb) on *Chlorella* cells in continuous culture, which caused cells to clump and exhibit irregular cell walls and chloroplasts (Södergren, 1968).

Mosser *et al.* (1972a) confirmed that *Dunaliella tertiolecta* was unaffected by 1000 ppb DDT, and that other organisms, including *Euglena gracilis* and *Chlamydomonas reinhardii* were also relatively resistant. Electron transport in isolated chloroplasts of the "DDT-resistant" alga *D. tertiolecta* was inhibited by DDE and DDT to the same extent (Bowes, 1972), suggesting that for this system the dehydrochlorination of DDT to DDE was not a detoxication process.

Mosser *et al.* (1972b) investigated the effects of DDT on mixed cultures of marine algae containing a "sensitive" diatom (*Thalassiosira pseudonana*) and a "resistant" green alga (*D. tertiolecta*) in equal proportions. It was found that *D. tertiolecta* was not inhibited at any DDT level tested. However, *T. pseudonana*, which grew faster and soon outnumbered *D. tertiolecta* in control cultures, was affected by 10 ppb DDT to the extent that its competitive success was significantly diminished, even though *T. pseudonana* was unaffected by 10 ppb DDT in pure culture. Mosser *et al.* (1972b) pointed out that DDT can occur in natural waters at levels equivalent to those which caused a marked change in species ratio in their experiments and considered the ecological implications of such a pesticide-induced alteration in phytoplankton populations, via effects on the selectively grazing zooplankton.

The relative resistance of *D. tertiolecta* to DDT is interesting in view of the lack of cell wall, which leaves its cell membrane exposed (Luard, 1973). Moreover, by inference, it seems unlikely that the resistance of *Scenedesmus quadricauda* to DDT (Vance and Drummond, 1969; Luard, 1973) can be accounted for by its more complex cell surface with a cellulose cell wall and extracellular pectic material (Luard, 1973).

Recognizing that in nature organisms may be exposed to several toxicants simultaneously, Mosser *et al.* (1974) investigated the effects of mixtures of organochlorines on the growth of the marine diatom *T. pseudonana* (formerly *Cyclotella nana*). Because of their ubiquity, chlorinated hydrocarbons were considered to be likely environmental co-contaminants. Results indicated that interactions between the organochlorines may occur, affecting their toxicity to phytoplankton. DDE, the universal pollutant derived from DDT, and polychlorinated biphenyls (PCB's) were far more inhibitory to *T. pseudonana* in combination than they were individually. In complete contrast, the addition of DDT (500 ppb) to cultures whose growth had been prevented by PCB's (50 ppb), substantially restored growth (Mosser *et al.*, 1974).

(b) *Aldrin, dieldrin, endrin.* Vance and Drummond (1969) reported that at concentrations up to at least 1 ppm, these insecticides had no significant effect on respiration of the green and blue–green algae tested. Furthermore, although growth of the blue–green alga *Microcystis aeruginosa* was prevented by less than 5 ppm dieldrin, aldrin and endrin, other algae resisted this and higher levels. At high concentration (100 ppm), aldrin and dieldrin lowered ATP levels, but not population densities, of algae (Clegg and Koevenig, 1974).

The four algal species tested by Menzel *et al.* (1970) differed markedly in their response to dieldrin and endrin. Thus, whereas the estuarine green flagellate *Dunaliella tertiolecta* was insensitive to the insecticides at concentrations up to 1000 ppb, the growth rate of *Skeletonema costatum* and *Coccolithus huxleyi* was reduced by 100 ppb. As was the case with DDT, *Cyclotella nana* was the most sensitive organism, its growth being completely inhibited by 100 ppb dieldrin and endrin, whilst photosynthetic ^{14}C-uptake was prevented at levels above 1 ppb.

Studies of the effects of dieldrin, endrin, aldrin, and some of their metabolites, on the growth of axenic cultures of blue–green algae were reported by Batterton *et al.* (1971). Of these insecticides, aldrin was the least inhibitory to the algae, and in general the marine isolate *Agmenellum quadruplicatum* was more tolerant than the freshwater *Anacystis nidulans*. At higher concentrations (up to 950 ppb), dieldrin and its metabolites including photodieldrin, depressed algal growth rates. The organisms were, however, less affected by ketoendrin than endrin. The observations of Batterton *et al.* (1971) that some insecticide metabolites or "terminal residues" were relatively toxic to algae prompted the laudable suggestion that studies on pesticide toxicity to primary producers in the ecosystem should include the metabolites and residues, and not be restricted to the parent compounds. Singh (1973) reported that endrin was less toxic than some other insecticides to nitrogen-fixing blue–green algae.

(c) *gamma-BHC*. At 5 ppm, γ-BHC (as lindane) did not prevent the growth of marine phytoplankton cultures and at 1 ppm even stimulated growth of a *Protococcus* sp. (Ukeles, 1962).

Surface application of γ-BHC at low (normal) rates to waterlogged soil under conditions simulating those of rice paddies appeared to affect only the diatom population, whilst at higher (10 × normal) application rates the nitrogen-fixing blue–green algae were also affected (Ishizawa and Matsuguchi, 1966). Following observations that normal applications (5 kg ha^{-1}) of γ-BHC for insect control in rice fields promoted more algal growth than in untreated plots, it was established in pot experiments that the insecticide selectively stimulated the indigenous soil blue–green algae (Raghu and MacRae, 1967). The stimulation of blue–greens, attributed to elimination by the γ-BHC of small predatory crustaceans, was considered as a possible added benefit of the use of γ-BHC in rice cultivation.

A somewhat different picture emerges from the studies of Singh (1973) on the effects of insecticides on pure cultures of blue–green algae, where, of the insecticides tested, a commercial preparation of BHC was generally the most inhibitory.

(d) *Other organochlorines.* At 0·2 ppb, mirex had no effect on algae (Hollister *et al.*, 1975) and at 100 ppb both mirex and methoxychlor only slightly inhibited algal growth (Kricher *et al.*, 1975).

(e) *Polychlorinated biphenyls (PCB's).* PCB's, which are similar in structure, persistence and biological effects to some organochlorine pesticides, have been widely detected as environmental pollutants in the Western World. Although they are not designed as pesticides, the PCB's find numerous industrial uses and since they may be analytically confused with DDT (Keil *et al.*, 1971) it is important to investigate their reactions with the biota. Several workers have reported that PCB's are toxic to algae and some evidence points to their being more toxic than DDT.

The marine diatom *Cylindrotheca closterium* took up and concentrated the PCB mixture Aroclor 1242, which at 100 ppb reduced growth, RNA and chlorophyll levels in this organism (Keil *et al.*, 1971). Similarly, at 200 ppb, Aroclor 1242 slowed the rate of growth and ^{14}C-uptake in *Chlamydomonas reinhardii*, whilst higher concentrations were lethal and more inhibitory than DDT (Morgan, 1972). Growth rates of the marine diatoms *Thalassiosira pseudonana* and *Skeletonema costatum* were reduced in the presence of 25 ppb PCB's and severely inhibited by 100 ppb, at which level DDT was only slightly inhibitory (Mosser *et al.*, 1972a). By comparison with the diatoms, other algae were less sensitive to PCB's and DDT, suggesting that selective inhibition of sensitive phytoplankton species by organochlorines might alter the species composition in natural algal communities. This view was endorsed by Mosser *et al.* (1972b) who found that at 1 ppb PCB's affected the species ratio in a mixture of *T. pseudonana* (sensitive) and *D. tertiolecta* (relatively resistant). This level of PCB, which was ten times lower than the DDT concentration causing a similar effect, represented that which can occur in some natural waters. Although *D. tertiolecta* was less sensitive to PCB's than was the green alga *Scenedesmus quadricauda*, both organisms were more sensitive to PCB's than to DDT and DDE (Luard, 1973).

2. Organophosphorus compounds

Of the wide range of pesticides screened, organophosphorus compounds were found to be relatively non-toxic to marine

TABLE 2

Effects of organochlorine insecticides on micro-algae

Insecticide	Effect	Algal type[a]	Insecticide conc. (ppb)	Reference
DDT	Growth inhibited	G	0·3	Södergren (1968)
		G	100	Stadnyk et al. (1971)
		D	100	Menzel et al. (1970)
		D	80	Bowes (1972)
		D	100	Mosser et al. (1972b)
		G	100 000	Christie (1969)
		G	100 000	Clegg and Koevenig (1974)
		G	1000	Mosser et al. (1972a)
		G	20 000	Morgan (1972)
		C	100	Menzel et al. (1970)
	Growth not appreciably affected	G	20 000	Vance and Drummond (1969)
		G	1000	Ukeles (1962)
		D	25	Mosser et al. (1972b)
		B–G	99	Worthen (1973)
		G	80	Bowes (1972)
		G	3600 to 36 000	Lee et al. (1976)
		G	100	Stadnyk et al. (1971)
	Photosynthesis inhibited	G	<10	Wurster (1968)
		G	<10	Wurster (1968)
		D	>10	Menzel et al. (1970)
		D	>1	Menzel et al. (1970)
		D	>10	Menzel et al. (1970)
		C	1000	Luard (1973)
	Photosynthesis not inhibited	G	1000	Menzel et al. (1970)
		G	20 000	Morgan (1972)
	Respiration not affected	G	1000	Vance and Drummond (1969)

Compound / effect	Organism[a]	Concentration	Reference
No morphological effects	D	80	Bowes (1972)
	B–G	99	Worthen (1973)
Cell clumping and cytological abnormalities	G	0.3	Södergren (1968)
CYCLODIENES (aldrin, dieldrin, endrin)		μM	
Chloroplast e⁻ transport inhibited; ATP levels reduced	G	20	Bowes (1972)
	G	100 000	Clegg and Koevenig (1974)
Growth inhibited	B–G	950	Batterton et al. (1971)
	B–G	50 000 (endrin)	Singh (1973)
Growth completely inhibited	B–G	<5000	Vance and Drummond (1969)
Cultures not killed	B–G	>15 000	Vance and Drummond (1969)
	G	>20 000	Vance and Drummond (1969)
Growth inhibited, or prevented	D	100 (dieldrin, endrin)	Menzel et al. (1970)
Population density not significantly altered; ATP levels reduced	G	100 000 (aldrin, dieldrin)	Clegg and Koevenig (1974)
Growth not affected	G	1000 (dieldrin, endrin)	Menzel et al. (1970)
Respiration not affected	G, B–G	1000	Vance and Drummond (1969)
Photosynthesis prevented	D	1 (dieldrin, endrin)	Menzel et al. (1970)
other compounds			
gamma-BHC (lindane)			
Growth inhibited	B–G	100 000	Singh (1973)
Growth reduced in some spp., but not in others	G	7500	Ukeles (1962)
mirex			
Growth stimulated (1 sp.)	G	1000	Ukeles (1962)
Growth and O_2-production not affected	G	0.2	Hollister et al. (1975)
methoxychlor			
Slight inhibition of growth	G	100	Kricher et al. (1975)
Slight inhibition of growth	G	100	Kricher et al. (1975)

[a] G – Green algae, B–G – Blue–green algae, D – Diatoms, C – Coccolithophorid.

phytoplankton (Ukeles, 1962). Christie (1969) observed that 100 ppm malathion had little significant effect on the green alga *Chlorella pyrenoidosa*, whilst cultures of blue–green and green algae were not adversely affected by accumulated levels (50 to 72 ppm) of parathion (Gregory *et al.*, 1969). Moore (1970) found that malathion and parathion were relatively non-inhibitory to the flagellate *Euglena gracilis*. Initial inhibition of N_2-fixation in blue–green algae caused by 100 ppm malathion was followed by recovery and stimulation (DaSilva *et al.*, 1975).

The nitrogen-fixing blue–green algae, *Cylindrospermum* sp. and *Aulosira fertilissima*, grew well in the presence of 300 and 400 ppm diazinon, respectively (Singh, 1973), and growth of other algal species was not affected by 100 ppm diazinon (Clegg and Koevenig, 1974). This compound, used for insect control in rice fields, had been shown by Sethunathan and MacRae (1969) to stimulate an increase in algal populations of the standing water of flooded soil in a way analogous to the effect reported for γ-BHC by Raghu and MacRae (1967). Diazinon (1 ppm) had no effect on cell numbers or photosynthesis in *S. quadricauda* (Stadnyk *et al.*, 1971) but at 100 ppm it reduced ATP levels in algae (Clegg and Koevenig, 1974). It would seem from the foregoing data that organophosphorus insecticides are relatively non-toxic to micro-algae. However, the finding that some of the "newer" organophosphorus insecticides, at less than 1 ppm, significantly lowered photosynthetic O_2-production by cultures of four species of marine phytoplankton (Derby and Ruber, 1971), signified the need to obtain further data on the anti-algal activity of this group of insecticides. Indeed, the larvacide Dursban, applied at the normal rate (1·2 ppb) to a pond, exerted a persistent reduction in growth of most of the phytoplankton (Brown *et al.*, 1976). At 2·4 ppb, however, Dursban had been found to stimulate growth of blue–green algae in artificial ponds (Hulbert *et al.*, 1972).

3. Carbamates

(a) *Carbaryl.* Ishizawa and Matsuguchi (1966) reported only slight suppression of the algae in flooded soil when naphthyl methylcarbamate (carbaryl) was applied at high application rates. However, Christie (1969) found that at 100 ppm the insecticide reduced the population in axenic cultures of *C. pyrenoidosa* by 30% and was inhibitory at concentrations as low as 0·1 ppm. Marine phytoplankton were also susceptible to carbaryl, which was lethal to two species at 1 ppm and to all five species tested at 10 ppm (Ukeles, 1962). At a lower level (0·1 ppm), carbaryl stimulated growth and ^{14}C-assimilation in low density populations of the freshwater alga *Scenedesmus quadricauda* (Stadnyk *et al.*, 1971).

(b) *Mexacarbate*. The growth rates of four different algae were not appreciably affected by Zectran, a mexacarbate formulation, at concentrations up to 1 ppm, although at 10 ppm growth was prevented (Snyder and Sheridan, 1974). Similar observations were made by Sheridan and Simms (1975) in both "field" and laboratory tests on freshwater algae, photosynthetic rate in the "field" algal samples being unaffected by up to 1 ppm Zectran. Higher levels of Zectran (12·5 ppm) were required to inhibit photosynthesis and respiration in blue–green algae, whilst the green alga *S. quadricauda* and the diatom *Navicula pelliculosa* were even less sensitive (Snyder and Sheridan, 1974). An interesting observation by Snyder and Sheridan was that motility of the blue–green alga *Oscillatoria terebriformis* was arrested by >0·5 ppm Zectran. However, it was considered that normal spray application of Zectran to lakes would pose little threat to the aquatic algae.

B. HERBICIDES

Evidence of food chain biological magnification problems with herbicides is lacking and in general these compounds probably exert only a temporary harmful effect on natural populations in aquatic environments (Mullison, 1970a). There are, however, many reports of the direct effects which herbicides can exert upon micro-algae.

1. *Phenoxy compounds*

(a) *2,4-D and 2,4,5-T*. Available data suggest that only at high concentrations does 2,4-D have adverse effects on algal populations, natural or cultured. An early indication of this was the fact that *Cladophora* was unaffected by the presence of 100 ppm 2,4-D for one month (Gerking, 1948), whilst Fitzgerald *et al.* (1952) found that 250 ppm 2,4-D was ineffective in controlling the growth of blue–green algae such as *Microcystis aeruginosa*. Similarly, concentrations of 2,4-D and 2,4,5-T up to 200 ppm were non-toxic to pure cultures of the green algae *Chlamydomonas*, *Chlorella* and *Scenedesmus* spp. (Vance and Smith, 1969). At 400 ppm, 2,4-D had no effect on the growth of a filamentous blue-green alga (*Cylindrospermum licheniforme*) or the unicellular green algae, *Chlorella vulgaris* and *Chlorococcum* sp. (Arvick *et al.*, 1971). However, Stadnyk *et al.* (1971)

reported a small reduction in biomass in cultures of *Scenedesmus quadricauda* by low concentrations of 2,4-D (0·1 and 1·0 ppm).

Relatively high concentrations (up to 100 ppm) of 2,4-D were required to reduce growth in agar cultures of *Scenedesmus*, *Chlorella*, *Chlamydomonas* and *Euglena* spp., with 25 ppm 2,4-D causing similar reductions in liquid media (Valentine and Bingham, 1974). *S. quadricauda* was the most sensitive and the flagellate *Euglena gracilis* the most resistant. Thomas *et al.* (1973), who used an agar diffusion technique to study the effects of several herbicides on the growth of *C. pyrenoidosa*, found 2,4-D to be non-inhibitory by comparison with most other compounds. Conversely, Poorman (1973) reported that *Euglena gracilis* was more susceptible to 2,4-D and 2,4,5-T than to other pesticides.

Several nitrogen-fixing species of blue–green algae tolerated levels (100 to 500 ppm) of 2,4-D much higher than are recommended for the treatment of rice field irrigation water (Venkataraman and Rajyalakshmi, 1971, 1972). These results suggested that the beneficial algae of rice paddies would not be endangered by the application of this pesticide. However, at levels equivalent to field application rates ($1·1 \times 10^{-2}$ M) 2,4-D prevented N_2-fixation by the blue–green algae *Nostoc punctiforme*, *N. muscorum* and *Cylindrospermum* sp. (Lundqvist, 1970). Cultures of the non-nitrogen-fixing blue–green alga *Anacystis nidulans* grew abundantly in the presence of 50 ppm 2,4-D, but were inhibited above 100 ppm (Voight and Lynch, 1974).

Consistent with the impression from pure-culture studies that 2,4-D was comparatively non-toxic to algae, Wojtalik *et al.* (1971) observed that 2,4-D applied at 20–40 lb acid equivalent per acre to reservoirs for aquatic weed control had no harmful effect on the natural phytoplankton populations.

Voight and Lynch (1974) reported a similar order of sensitivity to 2,4-D for eukaryotic and prokaryotic algae, although the herbicide carrier DMSO (dimethyl sulphoxide) was notably more inhibitory to the former group. In contrast, stimulation of algal growth and photosynthesis, observed with 10 ppm 2,4-D in formulation and not with pure 2,4-D, was attributed to a soluble component of the attaclay carrier (Walsh *et al.*, 1970).

Arvick *et al.* (1971) comprehensively examined the response of important soil algae to a commercial mixture (approx. 1:4) of picloram and 2,4-D under field and laboratory conditions. Their data from field experiments indicated that the herbicide mixture caused no significant change in the composition of the soil algal flora. In laboratory experiments, 50 ppm picloram alone was more inhibitory than 400 ppm 2,4-D to the growth of algae isolated from soil, whilst the picloram-2,4-D mixture (reported to act synergistically in plants) was less inhibitory than picloram alone. At lower levels, approximating to field application rates, there was no adverse effect on algal cultures or soil algal flora (Arvick *et al.*, 1971).

(b) *MCPA and MCPB.* At field-rate concentrations $(1.9 \times 10^{-2}\,\text{M})$ MCPA prevented blue–green algal N_2-fixation, although at very low concentration $(1 \times 10^{-5}\,\text{M})$ it was stimulated (Lundqvist, 1970). Initial inhibition of N_2-fixation in several blue–green algae by MCPA (20 ppm) was followed by partial recovery (DaSilva *et al.*, 1975).

The growth of two species of green algae isolated from soil was inhibited to a greater extent by MCPB than MCPA at high concentrations (50 to 1000 ppm), although at lower levels (5 ppm) the algae were hardly affected by the herbicides (Shennan and Fletcher, 1965). Similarly, growth, respiration and phosphate uptake by unicellular algae was inhibited more strongly by MCPB than MCPA (Kirkwood and Fletcher, 1970).

(c) *Fenoprop (silvex).* This herbicide, as Kuron, was the subject of several annual studies by Pierce into its ecological effects when sprayed onto a lake for control of aquatic weeds. Data from two of Pierce's reports are cited here. Following applications of Kuron in 1957 (which effectively killed the target weeds), there were no apparent adverse effects on planktonic algal populations, both numbers and species variety of which maintained normal patterns (Pierce, 1958). An application of Kuron in 1959, however, caused an initial temporary decline in plankton and filamentous algae (Pierce, 1960). Silvex, as a potassium salt formulation Kurosal SL, applied at the recommended rate (2 ppm) for controlling aquatic weeds in fish ponds, did not adversely affect the phytoplankton or zooplankton, although the filamentous green algae were controlled quite successfully (Cowell, 1965).

Although few quantitative data were given, Butler (1965) reported that silvex and 2,4,5-T were inhibitory to the "productivity" of mixed natural phytoplankton populations.

2. Triazines

Zweig *et al.* (1963) found that atrazine and simazine inhibited photosynthetic O_2-evolution in *C. pyrenoidosa*, and further evidence for atrazine's inhibitory action on algal photosynthesis has come from Walsh (1972), Hollister and Walsh (1973) and Valentine (1973). Whereas the photoautotrophic growth of *Chlamydomonas reinhardii* was inhibited by 0·5 ppm atrazine, its heterotrophic growth was not affected by 10 ppm (Loeppky and Tweedy, 1969). The same authors also reported marked differences in the sensitivity of *Chlorella* species to atrazine; for whilst the growth of *C. pyrenoidosa* was almost prevented by 5 ppm atrazine, growth of *C. vulgaris* was only partially affected. Atrazine (70 ppm) prevented growth and chlorophyll formation in synchronously grown *C. vulgaris* (Ashton *et al.*, 1966), but the inhibition was alleviated by the addition of glucose. Although starch, a normal storage material in *Chlorella*, did not accumulate in atrazine-treated cells, there were no anatomical abnormalities induced in the cell organelles, contrary to the effect on plant chloroplasts (Ashton *et al.*, 1966). Atrazine (LD_{50}, 2·5 ppm) was more inhibitory to growth of a *Chlorella* sp. than simazine (LD_{50}, 3·3 ppm) (Cho *et al.*, 1972).

A marked difference between algal species in their susceptibility to atrazine and simazine is evidenced by the fact that they were determined to be algicidal to some algae (Virmani, 1973) and algistatic to others (Shilo, 1965; Virmani, 1973). Moreover, concentrations of simazine up to 200 ppm were not toxic to three species of green algae and even stimulated one alga, *Chlamydomonas eugametos* (Vance and Smith, 1969).

Ametryne, which inhibited photosynthetic O_2-evolution in several marine unicellular algae (Walsh, 1972), was more toxic to such organisms than atrazine (Hollister and Walsh, 1973). Simetryne, also, inhibited algal growth (Adachi and Hamada, 1971).

Arvick *et al.* (1973) measured the response of algae to the presence of metribuzin both in soil and in liquid cultures. Whereas *in vitro* the growth of all species was reduced by metribuzin at 0·05 ppm and completely inhibited at 1·0 ppm, higher concentrations were required to damage the algal population in soil, where 10 ppm caused a significant reduction in numbers. From the latter observation it was considered that soil algal populations were unlikely to be affected by metribuzin at normal application rates, even though the compound is a photosynthesis inhibitor (Arvick *et al.*, 1973). The reaction of soil algae to herbicides such as the triazines is clearly an area warranting further research, for Mikhailova and Kruglov (1973), have reported that atrazine did cause detrimental changes in the soil algae.

3. *Phenylureas*

Of the large number of chemicals screened by Palmer and Maloney (1955) for algicidal activity, monuron was among the most toxic. Subsequent tests indicated that although green algae in pure cultures were generally more resistant to monuron than were blue–green algae and diatoms, the herbicide controlled natural populations of filamentous green algae in ponds (Maloney, 1958). Monuron also inhibited the development of some soil micro-algae (Mikhailova and Kruglov, 1973) and was more inhibitory than fenuron to the growth of *Chlorella pyrenoidosa* (Wright, 1975b). Addition of monuron (4 and 20 ppm) to exponentially growing cultures of a *Chlorella* sp. caused a decline in growth rate, followed by recovery several days later (Cho *et al.*, 1972), whereas a slower-growing monuron-resistant mutant of *Chlorella* was unaffected by similar additions.

Recognizing the importance of evaluating the tolerance of marine phytoplankton to pesticides used on commercial shellfish beds, Ukeles (1962) tested a wide range of toxicants, including phenylurea herbicides, for effects on the growth of marine algae in an enriched seawater medium. Diuron, lethal to all but one species at 0·004 ppm, was the most toxic of the phenylureas, and the relative order, if not the absolute values, of toxicity, diuron > monuron > (neburon) > fenuron (Ukeles, 1962), agreed with that reported by Geoghegan (1957) for autotrophic growth of *Chlorella vulgaris*, and by Cho *et al.* (1972) for *Chlorella* strain NMI and its resistant mutant M20 (Table 3). Similar observations on the relatively high toxicity of diuron to marine unicellular algae were made by Walsh and Grow (1971) and by Walsh (1972) who also showed that neburon was particularly inhibitory to such algae in axenic culture. A high order of toxicity of diuron was reported for *Scenedesmus quadricauda* populations, 0·1 ppm causing a drastic decline in cell numbers (Stadnyk *et al.*, 1971). *Chlorella ellipsoidea* was also sensitive to diuron (Adachi and Hamada, 1971).

TABLE 3

Toxicity of some urea herbicides to *Chlorella* species

Alga	Effect on growth	Herbicide, ppm		Reference
Chlorella vulgaris	Growth prevented (liquid medium)	diuron, monuron, fenuron,	0·1 0·5 5·0	Geoghegan (1957)
Chlorella "NMI"	50% inhibition (liquid medium)	diuron, monuron, neburon,	0·15 (1·9)[a] 0·75 (10·5)[a] 0·35 (2·0)[a]	Cho *et al.* (1972)
C. pyrenoidosa	Inhibition detectable (agar plate assay)	monuron, fenuron,	0·4 μg 1·0 μg	Wright (1975a)
C. pyrenoidosa	50% inhibition (liquid medium)	fenuron,	0·53	Wright (1972)

[a] Values for a herbicide-resistant mutant of *Chlorella* "NMI".

Using an agar-diffusion technique, Shilo (1965) demonstrated that monuron and diuron were algistatic towards unicellular and filamentous blue–green algae. Diuron also suppressed growth of the nitrogen-fixing blue–green algae *Tolypothrix tenuis* and *Aulosira fertilissima* (Venkataraman and Rajyalakshmi, 1971) and at 1 ppm inhibited growth of green and blue–green algae (Virmani, 1973).

Sikka and Pramer (1968), who showed that fluometuron inhibited growth of a *Chlorella* sp. and to a lesser extent *Euglena gracilis*, found that the addition of a carbon-source (e.g. glucose) considerably alleviated the inhibition. Such a glucose-mediated reduction in phenylurea inhibition of algae had also been observed by Geoghegan (1957).

Intra-generic differences in algal sensitivity to a phenylurea herbicide were clearly shown by Loeppky and Tweedy (1969) in studies of the effect of metobromuron on soil algae. Metobromuron was less inhibitory to autotrophic growth of the obligate photoautotroph *Chlamydomonas eugametos* than it was to *Chlamydomonas reinhardii*, which could also grow heterotrophically. Furthermore, whereas 10 ppm metobromuron inhibited autotrophic growth of *Chlorella pyrenoidosa*, the same concentration stimulated *C. vulgaris* (Loeppky and Tweedy, 1969).

Apart from the growth-inhibitory effects, several workers have reported on other physiological aspects of the action of phenylurea herbicides on micro-algae (Table 4). Inhibition of algal photosynthesis has been established with monuron (Zweig *et al.*, 1963), diuron (Zweig *et al.*, 1963; Zweig *et al.*, 1968; Walsh, 1972; Hollister and Walsh, 1973; Virmani, 1973), fluometuron (Sikka and Pramer, 1968) and neburon (Walsh, 1972; Hollister and Walsh, 1973). Inhibition of O_2-evolution by monuron in illuminated *C. pyrenoidosa* corresponded with an increase in fluorescence intensity (Zweig *et al.*, 1963). The red fluorescence of chlorophyll illuminated by blue light is well established and there is a good correlation between inhibition of O_2-evolution and fluorescence increase (Zweig, 1969). Zweig and Greenberg (1964) utilized the fluorescence phenomenon to study the rate of diffusion of photosynthesis inhibitors into *Chlorella* cells, finding that chlorinated urea herbicides entered rapidly. According to effects on O_2-evolution, Hollister and Walsh (1973) determined that of several different types of marine phytoplankton, the diatoms were generally less sensitive than other algal types to phenylurea (and triazine) herbicides.

Fluometuron caused a decrease in chlorophyll and protein levels in autotrophically-grown *Chlorella* and *Euglena*, although the effect was overcome when the organisms were grown heterotrophically, whilst respiration was not adversely affected (Sikka and Pramer, 1968). In

TABLE 4
Effects of phenylurea herbicides on micro-algae

Effect	Herbicide	Reference
Growth inhibited	monuron	Palmer and Maloney (1955); Geoghegan (1957); Maloney (1958); Ukeles (1962); Shilo (1965); Walsh and Grow (1971); Cho et al. (1972); Mikhailova and Kruglov (1973); Wright (1975b).
	diuron	Geoghegan (1957); Ukeles (1962); Shilo (1965); Adachi and Hamada (1971); Stadnyk et al. (1971); St. John (1971); Walsh and Grow (1971); Venkataraman and Rajyalakshmi (1971); Cho et al. (1972); Walsh (1972).
	neburon	Ukeles (1962); Walsh and Grow (1971); Cho et al. (1972); Walsh (1972).
	fenuron	Geoghegan (1957); Ukeles (1962); Walsh and Grow (1971); Wright (1972, 1975b).
	fluometuron	Sikka and Pramer (1968).
	metobromuron	Loeppky and Tweedy (1969).
Photosynthesis inhibited	monuron	Zweig et al. (1963).
	diuron	Zweig et al. (1963, 1968); Walsh (1972); Hollister and Walsh (1973); Virmani (1973).
	neburon	Walsh (1972); Hollister and Walsh (1973).
	fluometuron	Sikka and Pramer (1968).
	fluometuron	Sikka and Pramer (1968).
Chlorophyll level decreased	fluometuron	Sikka and Pramer (1968).
Protein content decreased	monuron, diuron, neburon, fenuron	Walsh and Grow (1971).
Carbohydrate levels decreased	monuron	Geoghegan (1957).
Stimulation of endogenous respiration and substrate utilization		
Decrease in photochemically produced ATP	diuron	St. John (1971).
N_2-fixation inhibited	monuron, linuron	DaSilva et al. (1975).
N_2-fixation stimulated after initial inhibition	linuron	DaSilva et al. (1975).

contrast, urea herbicides did not affect protein, chlorophyll or carotenoid content of cultures of marine algae, although they did reduce
the carbohydrate content of the algae, particularly at high salinity
(Walsh and Grow, 1971). It is interesting that *Dunaliella tertiolecta* was
the most resistant alga tested to urea herbicides (Walsh and Grow,
1971; Walsh, 1972). The relative resistance of this organism to DDT
was noted earlier in this chapter (Section III A.1 a.).

The endogenous respiration and utilization of low substrate levels
by *Chlorella vulgaris* were stimulated by monuron which, it was
considered, acted by uncoupling substrate oxidation and assimilation
(Geoghegan, 1957). Diuron reduced growth and the level of photochemically produced ATP in *Chlorella sorokiniana* equally, but did
not inhibit production of ATP by oxidative phosphorylation (St.
John, 1971).

Initial inhibition of N_2-fixation in blue–green algae caused by
monuron and linuron at 10 ppm was followed by recovery and even
stimulation in the case of linuron (DaSilva *et al.*, 1975).

4. Phenylcarbamates

These herbicides have received comparatively little attention regarding their effects on micro-algae. This may be due to the fact that they
are generally non-persistent and do not act primarily upon the photosynthetic process. They have, however, been the subject of some
investigations in our laboratory during studies on the interactions of
microbes with this group of herbicides.

Three phenylcarbamates tested for effects on the growth of
C. pyrenoidosa in liquid medium were inhibitory in the order
barban > chlorpropham > propham, with 50% reduction in growth
caused by 0·3, 2·7 and 13·7 ppm, respectively (Wright, 1972); but the
Chlorella sp. used by Cho *et al.* (1972) was apparently far less sensitive
than our strain to barban (although the barban levels used by these
workers, up to 18 ppm, exceed the water-solubility limit). Aniline
metabolites of these herbicides were non-toxic to *Chlorella* over the
same (Wright, 1972) or higher (Wright, 1975b) concentration ranges.
The relatively strong inhibitory action of barban on unicellular green
algae has also been demonstrated using an agar diffusion method
(Wright, 1975a, b). Virmani (1973) also noted that chlorpropham was
more inhibitory than propham to algal growth, whilst St. John (1971)
reported that chlorpropham inhibited growth and ATP production
(both oxidative and photochemical) in *Chlorella sorokiniana*.

In recent work (Wright, Whyard and Dawe, unpublished data) we
have observed that although blue–green algae were less sensitive than

TABLE 5

Inhibition of growth of green and blue–green algae by phenylcarbamate herbicides

	Concentration (ppm) causing 50% reduction in growth[a]				
Herbicide	Chlorella[b] pyrenoidosa	Anacystis[c] nidulans	Gloeocapsa[c] alpicola	Anabaena[d] cylindrica	Tolypothrix[d] tenuis
propham	13·7	52	41	70 (5)	64 (5)
chlorpropham	2·7	5·5 (1·0)	9·0	12·5 (8·2)*	13·2 (1·5)
barban	0·3	N.D.	N.D.	2·8 (2·0)* (1·0)	1·8

[a] data from Wright (1972) and Wright, Whyard and Dawe (unpublished data).
[b] unicellular green alga (eukaryotic).
[c] unicellular blue–green algae (prokaryotic).
[d] filamentous blue–green algae (prokaryotic).
N.D. Not determined.
( ) concentrations causing stimulation (5–10%) of growth.
( )* concentration causing 50% growth inhibition under N_2-fixing conditions.

the green alga *Chlorella* to phenylcarbamates, the order of potency was again barban > chlorpropham > propham. Unicellular blue–green algae (*Anacystis nidulans* and *Gloeocapsa alpicola*) were more sensitive to propham and chlorpropham than were the filamentous types (Table 5). *Anabaena cylindrica* was more sensitive to phenylcarbamates when grown without an added nitrogen source (nitrogen-fixing conditions), and at low concentrations, propham and barban stimulated growth of both *Anabaena cylindrica* and *Tolypothrix tenuis* (Wright *et al.*, unpublished data). The relative toxicity of chlorpropham and propham to micro-algae is in accordance with the relative toxicity of these compounds to barley, a sensitive plant (Clark and Wright, 1970). A striking effect of chlorpropham and propham on *T. tenuis* was a change in pigmentation, from dark brown–green to blue–green. A concentration of between 3 and 4 ppm chlorpropham was required to cause the effect, which occurred at whatever stage during active growth the chlorpropham was added and was readily reversed by removing the herbicide. Spectrophotometric examination of pigment extracts of *T. tenuis* indicated a decrease in the level of the red phycobiliprotein, *c*-phycoerythrin, and chlorophyll in the presence of chlorpropham, with a relative increase in the blue *c*-phycocyanin. A chlorpropham-induced pigmentation change (from blue–green to blue) has also been observed in the unicellular cyanophyte *Anacystis nidulans*.

Whilst our results have shown that at levels corresponding to those used in the field, chlorpropham disrupts algal physiology and growth in cultures, it would be interesting to evaluate effects under field conditions, especially rice-paddy conditions, in view of the use of this herbicide in rice cultivation (Ray, 1973).

5. Acylanilides

Dicryl inhibited photosynthetic O_2-evolution in *Chlorella* (Zweig *et al.*, 1963) and entered *Chlorella* cells more rapidly than other herbicides tested (Zweig and Greenberg, 1964). Dicryl has also been shown to suppress growth of blue–green algae (Shilo, 1965).

Propanil has also been found to be toxic to both green and blue–green algae (Table 6). Growth of *Chlorococcum aplanosporum* in liquid culture was reduced by 0·2 ppm propanil and prevented by 5 ppm, whilst in soil an estimated 50% reduction in growth of *C. aplanosporum* was caused by 2·5 ppm propanil (Sharabi, 1969). At 0·18 ppm, propanil caused 50% reduction in the autotrophic growth of *Chlorella pyrenoidosa* in liquid culture (Wright, 1972). Propanil inhibition of chlorophyll synthesis in *C. aplanosporum* generally paralleled and preceded growth inhibition, whilst cultures inhibited by concentrations between 0·5 and 1 ppm propanil could recover (Sharabi, 1969).

Concentrations of propanil causing 50% reduction in the growth of blue–green algae did not exceed 0·2 ppm (Table 6), whilst at levels below 0·03 ppm some algae were slightly stimulated (Wright *et al.*, 1977). A similar effect was reported by Ibrahim (1972), with 0·01 ppm propanil (as Stam) slightly stimulating growth and N_2-fixation in *T. tenuis* and *Calothrix brevissima*, whereas 0·1 ppm propanil was inhibitory. Propanil inhibition of chlorophyll synthesis (Hamdi *et al.*, 1970) and photosynthesis (Wright *et al.*, 1977) in blue–green algae has also been reported. A decline, followed by recovery, was observed when propanil was added to growing cultures of *A. cylindrica* (Wright *et al.*, 1977) and a similar phenomenon was noted earlier with the green alga, *Chlorella* (Wright and Walker, unpublished data).

The action of propanil on blue–green algae may be relevant to its use in rice fields and investigations under natural conditions are required. It is interesting to note, however, that propanil was less inhibitory to algae in the presence of soil than in culture media (Sharabi, 1969; Ibrahim, 1972; Wright *et al.*, 1977) and that *A. cylindrica* was less sensitive when grown *in vitro* under N_2-fixing conditions than it was when supplied with nitrate (Wright *et al.*, 1977).

The primary degradation product of propanil, 3,4-dichloroaniline, was less toxic to algae than the parent compound (Sharabi, 1969; Wright *et al.*, 1977). A secondary product, 3,3',4,4'-tetrachloroazobenzene did not inhibit algal growth or photosynthesis at levels tested (Sharabi, 1969).

6. *Bipyridyls*

In view of their non-selective action in killing plants, it is to be anticipated that bipyridyl herbicides will also be lethal to micro-algae. Shilo (1965) demonstrated the algicidal effect of diquat and paraquat on blue–green algae, which was only manifest in the light and which was accompanied by a reduction in chlorophyll, carotenoid and phycocyanin pigment levels. Similarly, a *Chlorella* strain capable of heterotrophic growth was far more sensitive to paraquat in the light than when grown in the dark (Shilo, 1965). The agar diffusion technique used by Thomas *et al.* (1973) indicated that diquat and paraquat were more inhibitory than other compounds to the growth of *C. pyrenoidosa*. Tsay *et al.* (1970) reported paraquat (2 ppm)-induced chlorosis of *Chlorella* cells and found that the herbicide's toxicity was reduced by Cu^{2+} ions (0·5 ppm), suggesting that copper-containing cytochrome oxidase may be directly involved with paraquat toxicity.

Cell filaments of blue–green algae were soon destroyed and pigmentation lost when paraquat was added to liquid cultures during growth, or to established lawn growth of the algae on agar medium (Balasooriya and Wright, unpublished data). Paraquat and diquat (25 ppm) were lethal to blue–green algae and, at 20 ppm, persistently inhibited N_2-fixation (DaSilva *et al.*, 1975). Paraquat has shown promise for use in stubble clearance in rice before sowing or transplanting (Anon, 1970; Patel, 1972). It seems unlikely, however, that practical applications of the herbicide would disrupt the blue–green algae to the extent observed *in vitro* owing to its rapid adsorption to vegetation and soil.

Detailed investigations by the group of Stokes and Turner into the action of diquat on the physiology and metabolism of *Chlorella vulgaris* established that in the light diquat irreversibly inhibited real photosynthesis and respiration, bleached the cells, and caused considerable damage to cell membranes and plastids (Turner *et al.*, 1970; Stokes *et al.*, 1970). Diquat rapidly stimulated dark respiration in *C. vulgaris*, and tracer studies showed that under dark conditions diquat accelerated starch breakdown and release of ^{14}C from carbohydrate metabolites (Stokes and Turner, 1970).

TABLE 6

Effects of propanil on micro-algae

Effect	Alga	Propanil concentration (ppm)	Reference
Growth stimulated	*Tolypothrix tenuis*[a]	0·01	Ibrahim (1972)
	Calothrix brevissima[a]	0·01	Ibrahim (1972)
	Anabaena cylindrica[a]	<0·03	Wright *et al.* (1977)
	Nostoc entophytum[a]	<0·03	Wright *et al.* (1977)
Growth inhibited	*Chlorococcum aplanosporum*[b]	0·2–1·0 (in culture)	Sharabi (1969)
		2·5 (in soil)	Sharabi (1969)
	Chlorella pyrenoidosa[b]	0·18 (50% inhibition)	Wright (1972)
	T. tenuis	0·18	Hamdi *et al.* (1970)
	T. tenuis	0·1	Ibrahim (1972)
	C. brevissima	0·1	Ibrahim (1972)
	T. tenuis	0·2 (50% inhibition)	Wright *et al.* (1977)
	A. cylindrica	0·09 (50% inhibition)	Wright *et al.* (1977)
	N. entophytum	0·17 (50% inhibition)	Wright *et al.* (1977)
	Gloeocapsa alpicola[a]	0·005 (50% inhibition)	Wright *et al.* (1977)
Growth prevented	*Ch. aplanosporum*	5	Sharabi (1969)
	C. pyrenoidosa	>2	Wright (1972)
	Anabaena variabilis[a]	} 5	} Wright *et al.* (1977)
	A. cylindrica		
	Nostoc muscorum[a]		
	N. entophytum		
	T. tenuis		
	G. alpicola		
	T. tenuis	} 10	} Ibrahim (1972)
	C. brevissima		

Effect	Species	Value	Reference
Photosynthesis prevented	T. tenuis	18	Hamdi et al. (1970)
	A. cylindrica T. tenuis N. entophytum	8	Wright et al. (1977)
Chlorophyll synthesis inhibited	T. tenuis Ch. aplanosporum	1·8 0·2	Hamdi et al. (1970) Sharabi (1969)
Lysis, followed by recovery in growth	A. cylindrica	up to 0·2	Wright et al. (1977)
Inhibition of growth followed by full recovery	C. pyrenoidosa Ch. aplanosporum	0·5 to 2·0 0·5 to 1·0	Wright and Walker (unpublished) Sharabi (1969)
N_2-fixation stimulated	T. tenuis C. brevissima	0·01 0·01	Ibrahim (1972) Ibrahim (1972)
N_2-fixation inhibited	T. tenuis C. brevissima	0·1	Ibrahim (1972)

[a] Blue–green algae   [b] Green algae

Apart from their widespread and important use in agriculture, the bipyridyls are also used for aquatic weed control. It is therefore desirable that these herbicides should be assessed for any deleterious effects on the ecology and quality of treated waters. In such a study, Silvo (1968) found that partial destruction of the phytoplankton, which occurred within a few days of an application of paraquat for weed control, could lead to an alteration in the phytoplankton composition upon recovery. Silvo also noted a high consumption of free oxygen in the water consequent upon paraquat-destruction of the algae. More recently, Brooker and Edwards (1973a) in a study of the ecological effects of two applications of paraquat (1 ppm and 0·6 ppm) for weed control in a fishing reservoir, reported significant toxic effects on the algal populations. However, these were temporary since certain species either recovered or others put in a natural seasonal appearance, within a matter of weeks. Gross oxygen production of the reservoir declined with the herbicide-induced death of plants, but was restored upon subsequent growth of filamentous and macrophytic algae (Brooker and Edwards, 1973b). Diquat (1 ppm) has also been shown to control freshwater algae temporarily (Robson et al., 1976).

7. Other herbicides

(a) *Aminotriazole (amitrole)*. The observation of Wolf (1962) that inhibition of growth of *C. pyrenoidosa* caused by aminotriazole (5 ppm) could be reversed by the addition of purine bases, supported the view that the herbicide inhibited purine synthesis. Aminotriazole, albeit at a much higher concentration (150 ppm), also inhibited the growth of other species of green algae (Vance and Smith, 1969), whilst at the intermediate level of 50 ppm it inhibited growth and reduced pigment levels (reversibly) in the blue–green alga *Anacystis nidulans* (Kumar, 1963). Initial inhibition of blue–green algal N_2-fixation by 20 ppm amitrole was followed by recovery and stimulation (DaSilva et al., 1975).

(b) *Diphenamid*. Autotrophic growth of the soil algae *Chlamydomonas reinhardii*, *Chlorella vulgaris* and *C. pyrenoidosa* was stimulated by less than 10 ppm diphenamid, although heterotrophic growth of *Ch. reinhardii* was inhibited (Loeppky and Tweedy, 1969). Thomas et al. (1973) also reported the relative lack of toxicity of diphenamid to *C. pyrenoidosa*.

(c) *Picloram*. The growth of *Cylindrospermum licheniforme* on a soil medium, and *Chlorella* and *Chlorococcum* spp. in liquid cultures, was reduced in the presence of a high concentration of picloram (50 ppm), which was more inhibitory than a mixture of picloram and 2,4-D (Arvick *et al.*, 1971) (see Section B.1 a.). Hardy (1966) found that the development of mixed algal populations inoculated into aquaria was not retarded by an application of 1 ppm picloram (as Tordon 22 K).

(d) *Glyphosate*. Photosynthesis in *Scenedesmus* was inhibited by glyphosate, higher temperature and light intensity increasing the effect (van Rensen, 1974).

(e) *Trichloracetates*. Cell division, and uptake of glucose, phosphate and nitrogen in *Chlorella vulgaris* were retarded by these herbicides (Hassall, 1961).

8. *Plant growth retardants*

Maleic hydrazide (1,2-dihydropyridazine-3,6-dione) was lethal to axenic cultures of the blue–green alga *Fischerella muscicola* at 4 mM, whereas at a sub-lethal level (2 mM) it reduced the carotenoid: chlorophyll ratio and caused cytological and morphological abnormalities (Singh and Subbaramaiah, 1970). Czerpak (1970) found that CCC (chlorocholine chloride) inhibited growth and lowered chlorophyll levels in all algae tested, although marked differences in sensitivity were apparent among different types.

C. FUNGICIDES

1. *Quinones*

In a survey of chemicals which might be suitable for short-term control of bloom-forming blue–green algae, Fitzgerald *et al.* (1952) gave particular attention to the quinones, which were generally found to be toxic to the algae at lower concentrations than other chemicals. Phenanthraquinone and especially 2,3-dichloronaphthaquinone (dichlone) were particularly potent inhibitors and appeared to be selectively toxic to bloom-forming species, all of which were killed by 5 ppb of 2,3-dichloronaphthaquinone (Fitzgerald *et al.*, 1952). In contrast, with few exceptions, the non-blooming blue–greens, diatoms and green algae were comparatively resistant to 15 ppb of 2,3-dichloronaphthaquinone.

Limited field trials with 2,3-dichloronaphthaquinone indicated that the compound was effective at low concentrations (10–100 ppb) in selectively killing bloom-forming blue–green algae (Fitzgerald *et al.*, 1952) and this was confirmed by larger-scale trials in a lake, where the green algae were unharmed (Fitzgerald and Skoog, 1954). Palmer and Maloney (1955) and Maloney and Palmer (1956) also found that 2,3-dichloronaphthaquinone was more toxic to blue–green algae (and diatoms) than to green algae, whilst Bisiach (1970) reported that 10 ppb of this compound prevented the growth of an *Anabaena* sp.

Whitton and MacArthur (1967) chose the non-blooming unicellular blue–green alga *Anacystis nidulans* for investigation of the action of two quinones and found the minimum lethal level of 2,3-dichloronaphthaquinone to be 140 ppb. This value would appear to substantiate the findings of earlier workers regarding the higher sensitivity of the blooming algae. An interesting observation by Whitton and MacArthur was that quinone toxicity to *A. nidulans* was markedly reduced on addition of culture supernatants from various blue–green algae.

Other workers, particularly the group of Zweig and Sikka, have investigated the action of quinone pesticides on eukaryotic algae. *Chlorella pyrenoidosa* Chick was used in these studies, which were concentrated on 2,3-dichloronaphthaquinone (now referred to as dichlone). Zweig *et al.* (1968) reported that most quinones were lethal to *C. pyrenoidosa* and that photosynthetic O_2-evolution was rapidly inhibited, especially by dichlone. Similarly, most quinones inhibited photosynthetic $^{14}CO_2$ fixation and affected the incorporation of ^{14}C into cellular constituents (Zweig *et al.*, 1972). However, the finding that quinones, including dichlone, reduced acetate photometabolism and dark fixation of $^{14}CO_2$ in *C. pyrenoidosa* (Sikka *et al.*, 1971) indicated that several cell processes were vulnerable, and that the inhibitory action of quinones was not restricted to photosynthetic reactions.

Zweig *et al.* (1968) noticed that quinones caused irreversible damage to *Chlorella* cells, whilst Sikka *et al.* (1973), having observed rapid efflux of ^{14}C-labelled intracellular material from quinone-treated *Chlorella*, concluded that these chemicals altered cell membrane permeability. The latter authors considered that the cell membrane was a primary site of action of dichlone, which might alter the configuration thereof by reaction with sulphydryl (–SH) groups. The dichlone-induced shrinkage and invagination of the *Chlorella* cell membrane and loss of normal structured cytoplasmic appearance (Schwelitz *et al.*, 1974) were considered consistent with an alteration of protein elements in the cells and compatible with the physiological changes reported previously.

2. Organic mercury compounds

Ethyl mercury phosphate was lethal to all marine phytoplankton species tested when incorporated at a level of 60 ppb in the culture media (Ukeles, 1962). Harriss *et al.* (1970) found that three organo-mercury fungicides, at less than 1 ppb, reduced growth and photosynthesis in the marine diatom *Nitzschia delicatissima* and also in a natural population of freshwater phytoplankton. These results, indicating that at least some marine and freshwater phytoplankton species were sensitive to organo-mercury compounds at levels below those proposed for water quality standards, suggested that entry of such compounds into natural waters should be prevented.

3. Other fungicides

(a) *Thiocarbamates.* Maneb was more inhibitory than zineb to the growth of *Euglena gracilis* (Moore, 1970) whilst nabam was toxic at all levels tested. Growth in enrichment cultures of freshwater algae was prevented initially by 1 ppm nabam (and captan) although growth subsequently occurred due to selection of resistant forms (Lazaroff, 1967).

(b) *Benzimidazoles.* Canton (1976) reported that MBC (methyl benzimidazol-2-yl carbamate) was more toxic than benomyl or thiophanate-methyl towards *C. pyrenoidosa*, with EC_{50} values of $0·34$, $1·4$ and $8·5$ ppm, respectively. This is an interesting observation since MBC, itself fungicidal, is a known metabolite of the other two fungicides.

(c) *Dichlorophen.* Gupta and Saxena (1974) tested Panacide, a formulation of dichlorophen, for its effects on cultures of green and blue–green algal species. The blue–green *Nostoc* sp. was found to be the most sensitive, even though it survived up to 10 ppm Panacide.

(d) *Ethirimol. C. pyrenoidosa* tolerated high concentrations of this systemic fungicide (Teal, 1974).

D. ANTIBIOTICS

Antibiotics have for some time been used for obtaining bacteria-free cultures of algae and protozoa (Provasoli *et al.*, 1951; see also Foter *et al.*, 1953; Vance, 1966), although with varying success since the algae may "protect" the bacteria from the antibiotics (Tchan and Gould,

1961). It has been found, however, that the algae may also be killed by the antibiotics (Bershova *et al.*, 1968). In view of their close affinity with the bacteria, the blue–green algae, or "cyanobacteria", are likely to be particularly sensitive among algae to antibiotics. Indeed, Foter *et al.* (1953) reported that in general the blue–green species tested were more susceptible to antibiotics than the green algae and diatoms; streptomycin and neomycin having the highest algicidal activity. Vance (1966) also found that these and several other antibiotics such as penicillin, aureomycin, tetracycline, erythromycin, chloramphenicol and bacitracin inhibited five test species of blue–green algae. Palmer and Maloney (1955) had also demonstrated substantial antibiotic activity against certain algal groups.

The culture filtrates of fungal and actinomycete isolates from various environments were assayed for anti-algal activity by Safferman and Morris (1962), who found that a high proportion of the actinomycetes in particular exerted specific effects against blue–green algae. Some of the actinomycetes were algicidal, but the suggestion (Safferman and Morris, 1962) that large-scale screening of isolates for an economically significant selective algicide does not appear to have been successfully followed up.

Kumar (1964, 1968) studied the antibiotic sensitivity of blue–green algae and reported that streptomycin, penicillin and mitomycin c were very inhibitory but neomycin less so. Streptomycin, an inhibitor of chlorophyll formation in plants, reduced the levels of chlorophyll a, phycocyanin and carotenoids in *Anacystis nidulans* and *Anabaena variabilis* (Kumar, 1964). Upon successive sub-culture in media containing mitomycin c, *A. nidulans* developed tolerance to the antibiotic, probably by adaptation (Kumar, 1968). In contrast, a mutational origin was suggested for the streptomycin- and penicillin-resistant strains of *A. nidulans* obtained through sub-culture in media containing streptomycin and penicillin, respectively (Kumar, 1964). Both mitomycin c and neomycin induced u.v.-resistant mutants but not streptomycin-resistant mutants in *A. nidulans* (Kumar, 1968). Indirect evidence for the possibility of episomal control of streptomycin-resistance in the same alga was obtained by Kumar *et al.* (1967), who showed that proflavine (a known "curer" of microbial episomes) caused a streptomycin-resistant strain to lose this property.

The toxicity of polymyxin B to *Anabaena cylindrica* and *A. nidulans* was decreased in the presence of non-dialysable extracellular polypeptide material of *An. cylindrica* (Whitton, 1965); an interesting phenomenon in view of the known association of *Bacillus polymyxa* with blue–green algae in nature. Action of polymyxin B on the cell membrane of *A. nidulans* was suggested by the fact that it caused a release of u.v. (260 nm) – absorbing materials from the cells (Whitton, 1967).

Cannon *et al.* (1971), who presented the first evidence for lysogeny in blue–green algae, also found that treatment of the lysogenic strain of *Plectonema boryanum* with mitomycin c induced a rapid increase in the titre of infectious virus. It was subsequently determined that lysogeny of *P. boryanum* could be established by treatment with the antibiotic chloramphenicol (Cannon and Shane, 1972). One may speculate on the possible ecological significance of the latter observation in view of the common occurrence of antibiotic-producing micro-organisms, although there are physico-chemical restrictions on the operation of antibiotic substances in the environment.

IV. The Use of Micro-algae for Pesticide Bioassay

Even with the increasing availability of sensitive analytical procedures, biological assays are essential for evaluating the toxicity of chemicals and are an important tool in detecting and quantifying environmental pollutants such as persistent pesticide residues. In some cases bioassays may be preferred for studies of pesticide persistence and movement, especially in the soil, where chemical analyses usually require elaborate extraction procedures.

Microorganisms have not been extensively used for pesticide assay (Addison and Bardsley, 1968). However, microbiological methods of analysis of other biologically active substances, e.g. vitamins, amino acids and antibiotics, are universally used. The many aspects of the subject of microbiological assay are covered by Kavanagh (1963, 1972a). Microbiological assays are relative, not absolute, methods in which responses of an organism to a chemical are compared with responses to standard preparations of known composition and concentration (Kavanagh, 1972c). The potential usefulness of micro-algae as assay organisms was indicated by Tchan (1959), who related the growth of mixed soil algae in liquid medium to the availability of individual mineral nutrients in soil.

In view of the fundamental physiological relations of micro-algae to the cells of higher plants, and the many aspects of algal cell integrity which are subject to the disruptive effects of pesticides (see previous section), it is to be expected that at least some of the diverse algae will be suitable for the bioassay of pesticides, especially herbicides. Such bioassays have been proposed and although many may not be as sensitive or specific as assays utilizing higher plants, the algal bioassays are generally more rapid and less demanding of facilities, space and the time of personnel. Important factors governing the usefulness of algal bioassays include the sensitivity and stability of the strains, the ease and rapidity with which the organisms are cultured and by which growth, or other indices of physiological state, can be measured. Any changes in sensitivity, for example by mutation, will clearly jeopardize the suitability of algae as assay organisms. Such a change was observed with a *Chlorella* strain which developed resistance to monuron and related photosynthetic inhibitors (Cho *et al.*, 1972).

This section outlines some techniques for algal bioassay of pesticides, especially herbicides. The methods described are divided according to the type of medium in or upon which the algae are cultured in the assay.

A. SOIL AND SOLID SUBSTRATES

In the method reported by Jansen *et al.* (1958) the algicidal activity of chemicals was detected using mixed algal cultures grown under greenhouse conditions on the surface of a smooth nutrient-moistened inert perlite substrate contained in cups. By this method, spot-inoculation of algae on the surface was followed by spray-application of the chemicals. Their inhibitory action was determined according to the reduced diameter of the subsequent film of growth relative to that attained in untreated controls in up to 28 days. Although this technique could be incorporated into conventional (plant) screening programmes in greenhouses, by comparison with most algal bioassays it is slow.

In the technique used for detecting phenylurea herbicides in soil (Pillay and Tchan, 1969, 1972), absorbent paper discs (12·7 mm diam.) were inoculated with a point source of green algal cells and placed on the levelled surface of herbicide-treated soil in Petri dishes. The algal growth was assessed visually after 24–30 h incubation at 26° under illumination. Although this method lacked sensitivity, advantages of speed, simplicity and low cost were claimed for rapid screening of phytotoxic residues in soil. A more quantitative approach was adopted by Kruglov and Paromenskaya (1970) who used a similar paper disc technique to determine simazine levels in soil and measured algal growth on the discs according to the absorbance of ethanolic extracts of pigments. This technique was later adopted by Kruglov and Kwiatkowskaja (1975) to estimate levels of phenylurea herbicides in soil, using *Chlorella vulgaris* – impregnated Seitz filter pads placed on the soil for three days.

Helling *et al.* (1971) described an algal method for detecting the leaching movement of pesticides on soil thin-layer chromatograms. By this method *Chlorella* cells suspended in an agar medium were aspirated onto the pesticide soil chromatograms and after one to two days incubation clear zones in the algal growth marked the position of pesticides. Apart from a range of herbicides, some insecticides and fungicides were also detectable on soil t.l.c. plates (Helling *et al.*, 1971).

B. ASSAYS IN LIQUID MEDIA

The use of liquid media permits quantification of the algal response to pesticides according to effects on a number of parameters including cell numbers and mass, pigment levels and rates of photosynthesis, respiration and growth.

The bioassay described by Price and Estrada (1964) could, it was claimed, supplement or even replace conventional screening for potential herbicides using plants. Based on inhibition of chlorophyll formation in cell suspensions of the flagellate *Euglena gracilis*, this rapid method (4–8 h) detected inhibitors which specifically affected chlorophyll synthesis, distinguishing between such compounds (e.g. ethionine) and those which inhibited chloroplast formation (e.g. amitrole).

Atkins and Tchan (1967) placed atrazine-treated soil samples inside cellophane dialysis sacks which were then submerged in cultures of a *Chlorella* sp. The atrazine level in the soil was determined according to the extent of inhibition of subsequent algal growth. Using *Chlorella vulgaris* Beijerinck, which responded to low herbicide concentrations, Addison and Bardsley (1968) were able to detect diuron in aqueous extracts prepared from soil which had been treated with the herbicide. The assay was faster than an oat bioassay and more specific than a chemical method.

Several other authors have described algal liquid culture bioassays, most commonly using *Chlorella* strains, (Ikawa *et al.*, 1969; Adachi and Hamada, 1971; Kratky and Warren, 1971a, b; Pillay and Tchan, 1970, 1972; Wright, 1972, 1975a). According to the kinetic analysis of inhibition of growth of *C. pyrenoidosa* by herbicides, this organism was considered to be well suited for bioassays (Gramlich and Frans, 1964). The assay methods of Adachi and Hamada (1971), Kratky and Warren (1971a, b) and Wright (1972, 1975a) were more rapid than many others. This was probably attributable to a higher growth rate promoted by aeration of cultures with air containing either 3% or 5% CO_2. Culture vessels ranged from Erlenmeyer flasks (Kratky and Warren, 1971a) to large test tubes with a conical base (Wright, 1972, 1975a). Pillay and Tchan (1970, 1972) found that the extent of inhibition of growth of the assay alga decreased with increasing inoculum cell density and suggested that the sensitivity of liquid culture assays might be improved using small inocula. Our own findings (unpublished) support this observation.

The *Chlorella* bioassay described by Kratky and Warren (1971a), which was at least as sensitive as most plant assays towards atrazine, was used to monitor field residues of atrazine and to study leaching of atrazine and terbacil in soil. When tested with 42 herbicides, the *Chlorella* assay was found to be more rapid than plant bioassays and was generally more sensitive to inhibitors of photosynthesis and respiration than were the plants (Kratky and Warren, 1971b). The *Chlorella* assay used in the author's laboratory (Wright, 1975a) has been used to monitor degradation of the herbicide propham in bacterial cultures and also to detect asulam in aqueous extracts of soil.

The insecticide parathion specifically inhibited motility of the flagellate *Euglena gracilis* at levels (10 ppb) lower than those preventing growth. This was used as a basis for the assay of the compound (Lazaroff, 1967).

Tchan *et al.* (1975) drew attention to an urgent need for a really rapid bio-analysis method for monitoring herbicide levels in irrigation waters so that spot-checks can be made and possible damage to irrigated crops avoided. The authors gave details of a novel and simple bioassay for herbicides in soil or water, which appears to be the most rapid technique yet reported. The assay, combining two biological systems, is based on the effect of herbicides on photosynthetic evolution of oxygen by algae and its measurement by the bioluminescence (light emission) system of photobacteria highly sensitive to low levels of oxygen. The method was specific for direct photosynthesis inhibitors and detected very low concentrations of urea herbicides in soil or water in one hour.

Several factors, including the incubation temperature, illumination, aeration, pH and adsorption to glassware or algal cells, may affect the interactants in an algal bioassay. It is therefore important to establish the stability of the assay system regarding both the alga and chemicals under test. As with all microbiological assays, it is important to use rigidly standardized procedures and scrupulously clean glassware, the latter being particularly relevant when studying herbicides which are known to be adsorbed on glass. The sensitivity of the assay alga should be checked periodically with herbicide standards.

C. AGAR PLATE ASSAYS

The use of agar media to demonstrate the antimicrobial action of chemicals has been widely used, proving especially useful and acceptable for the assay of antibiotics (see Kavanagh, 1963, 1972a). Such assays depend upon the diffusion of the test chemical into the agar from a reservoir, such as a well cut in the agar, a cup pressed into the agar surface, or a paper disc on the surface. The inhibitory action of the chemical on the microorganisms with which the agar is seeded is manifest by the appearance of zones around the diffusion source, in which no growth occurs. The size of the zone is usually found to relate directly to the concentration of antimicrobial agent, over a certain range. The theory of diffusion zones and the principles of diffusion assays, with particular reference to antibiotics, have been considered by Cooper (1963, 1972) and the factors determining the accuracy of diffusion assay by Kavanagh (1972b). It is reasonable to presume that similar factors will govern the diffusion of pesticides in agar and hence the formation of inhibition zones in algal growth on, or in, the agar.

Ikawa *et al.* (1969) studied the effects of toxic fungal and algal metabolites on *Chlorella pyrenoidosa* using an agar plate technique. By this method, suspensions of *C. pyrenoidosa* cells were incorporated into the agar medium and filter-sterilized test compounds were applied in dimethyl sulphoxide (DMSO) to sterile paper discs (1·3 cm diam.) which were placed on the agar surface. Zones in which growth was inhibited were formed around the discs within three days incubation under constant illumination. Using 6 mm diameter paper discs, Thomas *et al.* (1973) applied the technique of Ikawa *et al.* (1969) to a study of the inhibitory action of herbicides on the growth of *C. pyrenoidosa* (and *Bacillus subtilis*). Thomas *et al.* found that despite relative insensitivity to some herbicides, the agar plate algal method had some advantages for rapid testing. Diquat and paraquat were the most inhibitory of the compounds examined and in many cases the relative order of toxicity of herbicides towards *Chlorella* agreed with those established in liquid cultures by other workers.

In the paper disc algal bioassay method of Wright (1975a, b) the agar was evenly inoculated by spraying an algal cell suspension over the surface. Paper discs (6 mm diam.) were impregnated with herbicide in a volatile solvent (e.g. ethanol) and when dry these were placed on the agar surface when the latter had absorbed the inoculum. With unicellular homogeneously suspended algae such as *C. pyrenoidosa* and also a *Pleurococcus* (*Protococcus*) sp., which formed regular clusters of cells, spray inoculation was quicker and more effective in giving reproducible, even algal "lawns" than the use of spreading techniques. The agar plates were incubated at room temperature under constant illumination and results were obtained in two or three days. This assay was sensitive to herbicides of different modes of action and detected differences in anti-algal potency of the phenylcarbamate herbicides propham, chlorpropham and barban which paralleled the relative order of toxicity as determined by liquid culture assay (Wright, 1972, 1975a, b). The technique, which was also suitable for comparing the sensitivity of different algae to an individual herbicide (Fig. 1), has advantages of convenience, simplicity, speed and low cost. It has potential application for detection of toxic pesticide residues in soil or water and in the primary screening of chemicals for herbicidal and algicidal properties.

Wider screening of algae may be fruitful in indicating whether certain species or strains are exceptionally sensitive to particular groups, or even individual herbicides. For example, in a recent survey we have found *Chlorella vulgaris* Beijerinck to be more sensitive, by the agar plate assay, to barban than three other *Chlorella* spp., including the *C. pyrenoidosa* strain used previously.

In a study of the action of herbicides on the growth of blue–green algae, Shilo (1965) placed the herbicides in assay cups on agar cultures of the algae. Of the range of herbicides tested, all showed algistatic activity but only paraquat and diquat were algicidal; the latter effect being determined according to the formation of cleared zones in established algal growth. We have studied the effect of paraquat on nitrogen-fixing blue–green algae, using paraquat-impregnated paper discs applied to established dense "lawns" of *Anabaena cylindrica* and *Tolypothrix tenuis* (Balasooriya and Wright, unpublished data). The destructive effect of paraquat on the algae was apparent within two days illumination, with the appearance of light-coloured zones around the discs, which upon further incubation increased in size and became completely clear. Zone diameter was directly related to paraquat concentration applied to discs, the low limit of detection being $30 \mu l$ of a 1×10^{-5} M solution.

Cullimore (1975) has described a paper disc assay with green algae which can be used with agar or liquid media, and which involves inoculation of the paper disc, rather than the underlying growth medium, in a manner analogous to the earlier soil assay of Pillay and Tchan (1969, 1972). In Cullimore's method, the inoculated paper discs are placed on the medium in individual compartments of sub-divided dishes and the eventual growth of algae is measured according to absorbance at 600 nm using a spectrophotometer with reflectance attachment. This method, which can be used to assess a range of

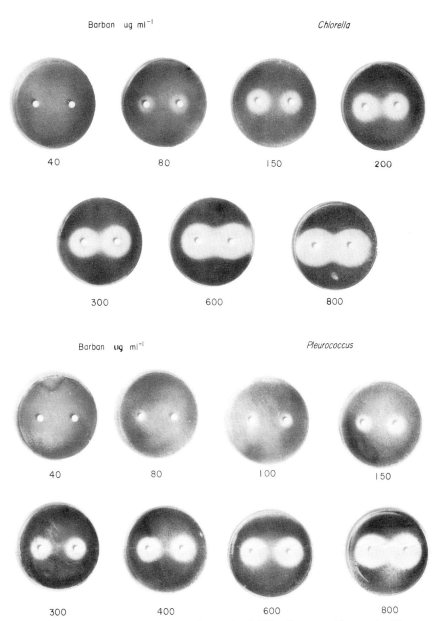

Fig. 1. Comparison of the inhibition of growth of *Chlorella pyrenoidosa* and a *Pleuro-coccus* (*Protococcus*) sp. by the herbicide bàrban. Paper discs (6 mm diam.) were impregnated (see Wright, 1975b) by immersion in ethanolic solutions of barban at the concentrations indicated. Plates were photographed after 3 days incubation (25°C) under constant illumination. Part of this figure is reproduced with the permission of Springer-Verlag, New York.

compounds for activity against different algal species under identical, easily replicated conditions, also has distinct possibilities for use in routine screening for potential herbicides and algicides.

.... "An assay can be no better than the sample preparation and the assay procedure .... the quality of an assay is related to the understanding by the analyst of the pitfalls of the assay" (Kavanagh, 1972c).

V. Troublesome Algae and Their Control

A. ALGAL PROBLEMS

The ecological and environmental importance of most algae are clouded somewhat by problems caused by a relatively small number of species. These problems are usually, but not always, associated with massed rapid growth of the algae under favourable conditions. In aquatic situations extensive growth may appear as a "bloom". The formation of successive blooms involving different types of algae complicates the adoption of control measures (Hawkins, 1972).

Eutrophication of aquatic environments, accelerated in recent years by abnormal nutrient loading from domestic and industrial wastes and drainage from agricultural land, is one cause of the increasing number of algal fouling problems, especially those associated with blue–green algae (Fitzgerald *et al.*, 1952; Echlin, 1966; Hawkins, 1972; Starmach, 1974).

Probably the earliest substantial record of a blue–green algal bloom in lake water and the characteristic limited existence of the bloom was enchantingly related by Thompson (1840). Other types of algal blooms, for example toxic blooms of marine dinoflagellates or "red tides", have also long been recognized. The Bible (Exodus 7: 14–25) records the poisoning of the river Nile in Egypt when "the fish died and the river stank and the Egyptians could not drink water from the Nile". Multiplication of algae is not always sufficient to account for large, dense blooms, so that passive concentration by wind and water currents is also implicated.

Problems caused by dense algal growths, especially blue–green algae, are usually associated with warm climatic regions; but the problems are not restricted to such organisms or geographical situations. Steel (1971) noted that sand filters in water treatment units in England were liable to be blocked by diatoms, including *Asterionella*, *Nitzchia*, and *Melosira* spp. In some temperate areas various species of the relatively large filamentous eukaroytic algae such as *Cladophora*,

Rhizoclonium and *Vaucheria* spp., which tend to form floating surface scums, are considered to be among the worst aquatic "weeds" in ponds and lakes (Robson, 1973).

The following examples indicate the socio-economic problems caused by blooms of planktonic algae, especially blue–greens (Crance, 1963; Shilo, 1971a; Hawkins, 1972):

 (i) Production of objectionable taints and odours in drinking water.
 (ii) Increased cost of water treatment due to filter clogging and tainting.
 (iii) Rendering of lake waters unsuitable for recreational purposes.
 (iv) Adverse effects on commercial fishing caused by toxic algal water-blooms and by deoxygenation of water upon decay of algal masses.
 (v) Direct toxic hazards to farm livestock and marine life.
 (vi) Detrimental effects on rice crops (Bisiach, 1970; Hawkins, 1972; Yamagishi and Hashizume, 1974).

Other algal problems, more usually associated with benthic types, including some filamentous green algae, are:

 (i) Formation of tangled mats which block channels and retard water flow over weirs and through gauges (Jordan *et al.*, 1962).
 (ii) Colonization of ships' hulls, causing increased frictional drag and higher operating costs.
 (iii) Growths on swimming pool walls and commercial glasshouses.
 (iv) Blockage of industrial cooling water systems (Fielden, 1969).
 (v) Growth on stonework leading to architectural disfiguration (Keen, 1971) and deterioration of antiquities (Fogg *et al.*, 1973).

Toxic algal blooms may appear in oceans, freshwater bodies and artificial fishponds (Shilo, 1967). Some of the algal poisons are highly potent and the poisoning of animals through consumption of water containing blooms of blue–green algae such as *Microcystis aeruginosa* and *Anabaena flos-aquae* has long been recognized in many parts of the world (Steyn, 1945; Shelubsky, 1951; Hughes *et al.*, 1958; Gorham, 1960; Gorham *et al.*, 1964). There was also suspected cases of human poisoning by such algae (Steyn, 1945). Marine dinoflagellates (e.g. *Gymnodium* spp.) and chrysophyte algae (e.g. *Prymnesium* and *Ochromonas* spp.) can kill fish (Shilo, 1967). Toxigenic algae may also become public health hazards through ingestion and concentration by shellfish consumed by humans.

Authoritative accounts of toxin formation by dinoflagellates, blue–green, green and chrysophyte algae are given by Schantz (1971), Gentile (1971) and Shilo (1971b), respectively.

B. CONTROL MEASURES

1. Physical methods

Chancellor (1958) and Hawkins (1972) noted several aspects of the design of water storage reservoirs, which aim at minimizing algal problems. Another approach is to create movement or flow and prevent the thermal stratification which would normally favour phytoplankton proliferation near the surface. The growth of plank- tonic blue–green algae is hindered by rapid water movement.

Shapiro (1973) found that the addition of CO_2, or lowering of the pH, stimulated a spectacular shift in a lake algal population from predominantly blue–green to green algae, with lower scum-forming tendency. The shift in population balance was considered similar to the change which occurs when stratified lakes are circulated mechanically or by compressed air. It was suggested (Shapiro, 1973) that injection of air, with or without added CO_2, should be investi- gated as a means of controlling blue–green algal growths in lakes. Good aeration of the water was considered to give the best protection against mass development of blue–green algae (Starmach, 1974).

The buoyancy of planktonic algae is fundamental to the formation of surface scums and it has been suggested that the inhibition or destruction of algal gas vacuoles might be used as a form of control (Fogg et al., 1973). Whilst ultrasonic and explosive devices, as sugges- ted, might have drastic effects on the integrity of the algae, the lack of selectivity for algae would seem to mitigate against their adoption in practice. An electrolytic method for killing algae was described by Paul et al. (1975).

2. Chemical methods

Chemicals can either be used to prevent development of algal growth (algistatic chemicals) or to kill existing populations (algicides). Fitzgerald (1975) discussed the evaluation of such chemicals and described a rapid test of algicide effectiveness, based on measurement of phosphate uptake (Fitzgerald, 1974).

Copper sulphate pentahydrate ($CuSO_4 . 5H_2O$), first advocated as an algicide in 1904, is widely used for controlling planktonic algae (Crance, 1963) and is even used in irrigation canals (Frank, 1970). Algae, however, vary widely in susceptibility to copper sulphate and resistant types can become established on removal of sensitive dominants (Chancellor, 1958; Hawkins, 1972). Although generally effective in controlling algal blooms, copper sulphate treatment is regarded as a temporary measure (Fogg *et al.*, 1973) and use is limited by its toxicity to fish. Recommended algicidal rates quoted by Chancellor (1958) ranged from 0·05 to 12 ppm, according to algal density, water quality and presence or absence of fish; usually the rate is below 1 ppm. For example, Crance (1963) found that copper sulphate levels as low as 0·05 to 0·08 ppm controlled growth of bloom-forming *Microcystis* in ponds, whilst 0·2 ppm gave "100% kill" of *Microcystis aeruginosa* cultures (Fitzgerald *et al.*, 1952). Toth and Riemer (1968) confirmed that recommended dosages of copper sulphate were adequate for algal control in farm ponds, but indicated that in practice the required concentrations might not be attained due to adsorption by bottom muds. The algicidal effectiveness of copper sulphate is also influenced by water temperature and hardness (Chancellor, 1958). In warm, soft water 0·25 ppm controls the algae, whereas in cool, hard water 2 ppm may be required, but at such levels sensitive fish are likely to be harmed (Keen, 1971). Palmer (1956) suggested that copper sulphate was less effective in hard water due to precipitation as copper carbonate.

Reporting on the different chemicals which showed promise as algicides, Palmer (1956) included quaternary ammonium compounds, rosin amines, quinones, activated silver, CMU (monuron), chlorophenate, antibiotics, organic zinc, copper and chlorine compounds.

The fungicide dichlone was selectively toxic towards noxious bloom-forming blue–green algae, which were killed by less than 5 ppb (Fitzgerald *et al.*, 1952), and was comparatively safe ecologically (Fitzgerald and Skoog, 1954). The selective high toxicity of dichlone to some blue–green algae was confirmed by Maloney and Palmer (1956) and Bisiach (1970). Although dichlone is clearly established as a potent inhibitor of blue–green algae, its toxicity may be reduced by the presence of algal extracellular products, as is also the case with copper sulphate (Whitton and MacArthur, 1967).

Maloney and Palmer (1956) evaluated several compounds and concluded that dodecylacetamido dimethyl benzyl ammonium chloride and rosin amine D acetate had the best potential as general algicides. On the other hand, zinc dimethyl dithiocarbamate, although good for algal control, was also highly toxic to fish (Palmer, 1956; Maloney and Palmer, 1956).

Herbicides used for aquatic weed and algal control in drainage and irrigation waters include 2,4-D, dalapon, dichlobenil, diquat, maleic hydrazide, acrolein, TCA, amitrole, some substituted ureas and triazines (Hawkins, 1972). Maloney (1958) used CMU for laboratory and field tests against a wide range of algae, following earlier findings (Palmer and Maloney, 1955) that it had good algicidal properties.

Results for the destruction and prevention of filamentous algal growth were good, although there was the danger that trees and other plants might be destroyed by seepage of monuron into surrounding soil. The related herbicide diuron has been used for algal control in water (Frank, 1970), although its use may be restricted by long persistence and possible accumulation in fish (Hawkins, 1972). Jordan *et al.* (1962) tested several herbicides for control of green algae growing in ponds and found diuron to be the most effective. Other herbicides providing effective control of the green algae were acrolein, copper sulphate and 2-amino-3-chloro-1,4-naphthoquinone, whilst at the concentrations used the triazines (simazine, atrazine, atraton and prometon) were comparatively ineffective. More recently, however, Yamagishi and Hashizume (1974) found that simetryne and prometryne (and pentachlorophenol) effectively controlled green algae in rice paddy fields.

Robson *et al.* (1976) reported that of 12 herbicides tested against troublesome fresh water algae, the methylthio-1,3,5-triazines, terbutryne and cyanatryn were the most active. These compounds killed algae at levels of 0·05 ppm.

Some blue–green algae are very sensitive to antibiotics (Foter *et al.*, 1953; Kumar, 1964, 1968). However, the use of such compounds for algal control would normally be uneconomic.

Other chemicals used for algal control include tin-containing paints on ships' hulls; fentin acetate and dichlone in rice paddies (Bisiach, 1970; Hawkins, 1972); sodium hypochlorite, formalin or dichlorophen, on exposed concrete surfaces, and copper sulphate in the concrete mix of underwater structures (Keen, 1971). There are numerous commercial algicides for use in swimming pools, many of which contain quaternary ammonium compounds or organic halides as active ingredients. Quaternary compounds, amines, phenates and chlorine are used in the treatment of industrial cooling waters (Fielden, 1969).

Fitzgerald (1975) has considered that most algicides, if used at minimal effective concentrations, do not pose a threat to aquatic environments due to their biodegradability at such levels. In contrast, the recalcitrant methyl mercuric chloride is unsuitable for use in aquatic systems.

3. Biological methods

Phytoplankton-eating fish, such as the silver carp, show promise for algal control in some situations and have been used in Eastern Europe (Robson, 1973).

Although protozoa are reported to greatly reduce the numbers of certain green algae in lakes (Canter and Lund, 1968), the significance of grazing protozoa in checking natural blue–green algal populations seems doubtful (Fogg *et al.*, 1973). Chytrid fungi pathogenic to blue–green algae are widely distributed in aquatic environments (see Canter, 1972). The potential for lysis of blue–green algae by moulds has received little attention until their recent isolation, from lake water and soil, in the author's laboratory (Redhead and Wright, 1978); where amoebal predators of blue–green algae have also been isolated.

A virus (cyanophage) which lysed blue–green algae was first isolated (from a sewage waste-stabilization pond) in the USA by Safferman and Morris (1963a). This virus, designated LPP-1, was highly specific for members of the non-blooming algal genera *Lyngbya*, *Plectonema* and *Phormidium* (Safferman and Morris, 1964a, b), possibly controlling them in nature (Safferman and Morris, 1967). The widespread occurrence of this and other cyanophages has since been established by their isolation from aquatic and soil environments in many countries (see Stewart and Daft, 1976).

Bacteria capable of lysing blue–green algae have been widely found in freshwater habitats, sewage treatment plants, waste-stabilization ponds, fishponds and the soil (see Stewart and Daft, 1976). Many of these bacteria were non-fruiting myxobacteria, but other species of lytic bacteria and also actinomycetes have been reported (e.g. Safferman and Morris, 1962, 1963b; Reim *et al.*, 1974). The data of Gunnison and Alexander (1975) suggested that whereas green algae were susceptible to microbial cellulases, the cell wall of the prokaryotic blue–green algae was attacked by lysozyme.

Safferman and Morris (1964a) suggested the possibility of using cyanophages for selective control of algae. Although viral algicides might be economically feasible there is the potential problem of the development of virus-resistant algal mutants (Shilo, 1971a). The number of cyanophages so far investigated in detail is limited and mostly restricted to those which attack non-blooming algae. It is reasonable to anticipate, however, that virulent cyanophages of the bloom-forming algae may be discovered in the future.

Although lytic bacteria have lower overall potential merit as algicides (Stewart and Daft, 1976), they may in fact be a better practical proposition than the viruses. Their chief virtues in this respect are that they attack a wider host spectrum, cause rapid lysis, and may be cultured in bulk free from host cells. There are not many reports of the evaluation of bacteria for algal control, but Daft *et al.* (1975) observed extensive destruction of a *Microcystis* bloom in an impounded section of a reservoir when lytic myxobacteria were added.

Stewart and Daft (1976) have drawn attention to the potential practical problems of applying and maintaining high density populations
of the lytic bacteria, and also to the possible deleterious consequences
of the rapid lysis of algal masses resulting in deoxygenation and release
of pigments, odours and toxins.

The practical solution to the problems of algal blooms may lie in
preventing their formation by nutrient removal (Fitzgerald *et al.*,
1952), especially phosphate removal (Shapiro, 1973). Such a pro-
phylactic approach is only likely to succeed where the nutrients or
pollutants entering the water are known.

VI. Conclusions

Micro-algae, by virtue of their habitation of most soil and aquatic
situations, may encounter a wide range of pesticides aimed at other
biotic targets. An appreciation of the reactions of these important
microorganisms to synthetic chemical toxicants is at least desirable
and ethically obligatory.

By comparison with heterotrophic microorganisms such as bacteria
and fungi, the micro-algae have received little attention regarding
their ability to metabolize and detoxify foreign chemicals such as
pesticides. Although it is difficult to determine the contribution of
algae to the degradation of pesticides in the ecosystem, even in aquatic
locations, there is evidence that these organisms may absorb pesticides
and in some cases transform them. Indeed, since representatives of
both the eukaryotic and prokaryotic types are known to metabolize at
least some pesticides, there is reason to believe that such activity may
be far more common than has apparently been generally considered by
investigators in the field of microbial degradation of pesticides. The
potential for algal degradation of pesticides, especially soil-applied
herbicides, should therefore be given more attention.

Whilst the full significance of interactions between pesticides and
non-target micro-algae may not yet be fully appreciated, there is
clearly an awareness of the possible hazard of pesticide concentration
in these organisms at the primary trophic level and thence other
organisms in the aquatic food chain ascendency. Wherever there are
indications of damage to natural algal populations through normal use
of pesticides, thorough investigations of the nature, extent, and
ecological and practical consequences of such effects should be made.
It is also important that efforts are made to identify any effects on algal
populations which might be caused by the pesticide formulation additives, as distinct from the active ingredient *per se*.

With the exception of unicellular green algae of the *Chlorella* type and, to an extent, some marine diatoms, data are lacking regarding the vulnerability of individual algal groups to the effects of pesticides, although generic and specific differences in sensitivity are apparent in some cases. Whilst at very low concentrations some pesticides stimulate algal activity, in the majority of cases at higher concentrations the pesticides are deleterious. Such concentrations are often below the normal practical application rates. Among the established anti-algal effects of different pesticides are: changes in species composition of mixed algal populations in soil and aquatic environments; inhibition of population growth, photosynthesis, pigment formation, motility, nitrogen-fixation and protein synthesis; effects on cell membrane permeability; cytological changes; lysis; mutation and lysogeny induction; alteration of carbohydrate metabolism. Quantification of some of these parameters is used in the various algal bioassays for pesticides.

Detailed studies of the mode of action of pesticides on algal cells have been made in relatively few cases. However, micro-algae would seem to be very convenient experimental material for investigating the reactivity of potential herbicides with typical features of plant physiology such as chlorophyll synthesis, cell replication, photosynthesis and cell membrane composition and function.

In order to obtain a fuller picture of the interactions of micro-algae with pesticides some emphasis could be given to certain aspects which have hitherto received insufficient attention, in some cases for obvious reasons of practical difficulty. These include: investigations of the relative pesticide sensitivities of different algal types, especially with regard to the typical algal flora in situations in which particular pesticides are normally used or to which they may be naturally transported; the effects of pesticides as used in practice, with particular reference to the properties of commercial formulation ingredients; the effects of pesticides, their metabolites and residues, on the activities and balances of natural algal populations; the possibility of development of algal resistance to pesticide action; and determination of the extent to which pesticide–algae interactions are influenced by extrinsic factors. Specific data on algal reactions with pesticides can only be obtained using axenic cultures under controlled experimental conditions. However, in relating such findings to what may occur under natural conditions it is important to consider the possible intervention of other organisms, especially bacteria which typically co-habit ecological niches with the algae.

There is little evidence to suggest that micro-algae in general are widely endangered through man's current activities with pesticides. Nonetheless it is clearly established that these organisms may react with many pesticides at levels at or below normal application rates to the detriment of either the pesticides, the algae, or organisms which depend upon them. Whereas biodegradability is a desirable characteristic of a pesticide, its indiscriminate disruption of primary production, oxygen evolution and nitrogen-fixation is totally unacceptable and its potential for such behaviour must be determined.

References

ADACHI, M. and HAMADA, K. (1971). *Chlorella* assay of the activity of herbicides. *Zasso Kenkyu* **11**, 54–58.

ADAMICH, M., TOWLE, A. and LUNAN, K. D. (1974). Ficoll density gradient separation of extracellular DDT from *Chlorella. Bull. Environ. Contam. Toxicol.* **12**, 562–566.

ADDISON, D. A. and BARDSLEY, C. E. (1968). *Chlorella vulgaris* assay of the activity of soil herbicides. *Weed Sci.* **16**, 427–429.

AHMED, M. K. and CASIDA, J. E. (1958). Metabolism of some organophosphorus insecticides by microorganisms. *J. econ. Ent.* **51**, 59–63.

ANON (1970). Weed control in transplanted rice. *Rep. int. Rice Res. Inst.* (1968). 181–195.

ARVICK, J. H., WILLSON, D. L. and DARLINGTON, L. C. (1971). Response of soil algae to picloram-2,4-D mixtures. *Weed Sci.* **19**, 276–278.

ARVICK, J. H., HYZAK, D. L. and ZIMDAHL, R. L. (1973). Effect of metribuzin and two analogs on five species of algae. *Weed Sci.* **21**, 173–175.

ASHTON, F. M., BISALPUTRA, T. and RISLEY, E. B. (1966). Effect of atrazine on *Chlorella vulgaris. Am. J. Bot.* **53**, 217–219.

ATKINS, C. A. and TCHAN, Y. T. (1967). Studies of soil algae. VI. Bioassay of atrazine and the prediction of its toxicity in soils using an algal growth method. *Pl. Soil* **27**, 432–442.

BATTERTON, J. C., BOUSH, G. M. and MATSUMURA, F. (1971). Growth response of blue–green algae to aldrin, dieldrin, endrin and their metabolites. *Bull. Environ. Contam. Toxicol.* **6**, 589–594.

BERSHOVA, O. I., KOPTEVA, Z. P. and TANTSYURENKO, E. V. (1968). The interrelations between blue–green algae, the causative agents of "water-bloom", and bacteria. *In* "Tsvetenie Vody" (A. V. Topachevsky, Ed.), pp. 100–117. (English trans.), Dumka Naukova, Kiev.

BISIACH, di M. (1970). Primi dati sulle infestazioni algali da Cianoficee nelle risaie italiane. *Riso* **19**, 129–134.

BOWES, G. W. (1972). Uptake and metabolism of 2,2-bis (*p*-chlorophenyl)-1,1,1-trichloroethane (DDT) by marine phytoplankton and its effect on growth and chloroplast electron transport. *Pl. Physiol.* **49**, 172–176.

BROOKER, M. P. and EDWARDS, R. W. (1973a). Effects of the herbicide paraquat on the ecology of a reservoir. I. Botanical and chemical aspects. *Freshwat. Biol.* **3**, 157–175.

BROOKER, M. P. and EDWARDS, R. W. (1973b). Effects of the herbicide paraquat on the ecology of a reservoir. III. Community metabolism. *Freshwat. Biol.* **3**, 383–389.

BROWN, J. R., CHOW, L. Y. and DENG, C. B. (1976). The effect of Dursban upon fresh water phytoplankton. *Bull. Environ. Contam. Toxicol.* **15**, 437–441.

BUTLER, G. L., DEASON, T. R. and O'KELLEY, J. C. (1975). Loss of five pesticides from cultures of twenty one planktonic algae. *Bull. Environ. Contam. Toxicol.* **13**, 149–152.

BUTLER, P. A. (1965). Commercial fishery investigations. Circular No. 226, U.S. Dept. Int., Fish and Wildlife Service, pp. 65–77.

CANNON, R. E. and SHANE, M. S. (1972). The effect of antibiotic stress on protein synthesis in the establishment of lysogeny of *Plectonema boryanum*. *Virology* **49**, 130–133.

CANNON, R. E., SHANE, M. S. and BUSH, V. N. (1971). Lysogeny of a blue–green alga, *Plectonema boryanum*. *Virology* **45**, 149–153.

CANTER, H. M. (1972). A guide to the fungi occurring on planktonic blue–green algae. *In* "Taxonomy and Biology of Blue–Green Algae" (T. V. Desikachary, Ed.), pp. 145–159. University of Madras, Madras.

CANTER, H. M. and LUND, J. W. G. (1968). The importance of protozoa in controlling the abundance of planktonic algae in lakes. *Proc. Linn. Soc. Lond.* **179**, 203–219.

CANTON, J. H. (1976). The toxicity of benomyl, thiophanate-methyl and BCM to four freshwater organisms. *Bull. Environ. Contam. Toxicol.* **16**, 214–218.

CARR, N. G. and WHITTON, B. A. (Eds) (1973). "The Biology of Blue–Green Algae". Blackwell Scientific Publications, Oxford.

CHANCELLOR, A. P. (1958). "The control of aquatic weeds and algae". H.M.S.O., London.

CHO, K. Y., TCHAN, Y. T. and LO, E. H. M. (1972). Resistance of *Chlorella* to monuron, a herbicide inhibiting photosynthesis. *Soil Biol.* **16**, 18–21.

CHRISTIE, A. E. (1969). Effects of insecticides on algae. *Wat. Sewage Wks* **116**, 172–176.

CLARK, C. G. and WRIGHT, S. J. L. (1970). Degradation of the herbicide isopropyl *N*-phenylcarbamate by *Arthrobacter* and *Achromobacter* spp. from soil. *Soil Biol. Biochem.* **2**, 217–226.

CLEGG, T. J. and KOEVENIG, J. L. (1974). The effect of four chlorinated hydrocarbon pesticides and one organophosphate pesticide on ATP levels in three species of photosynthetic freshwater algae. *Bot. Gaz.* **135**, 368–372.

COOPER, K. E. (1963). The theory of antibiotic inhibition zones. *In* "Analytical Microbiology" (F. Kavanagh, Ed.), pp. 1–86. Academic Press, New York and London.

COOPER, K. E. (1972). The theory of antibiotic diffusion zones. *In* "Analytical Microbiology" (F. Kavanagh, Ed.), vol. 2, pp. 13–30. Academic Press, New York and London.

COWELL, B. C. (1965). The effects of sodium arsenite and silvex on the plankton populations in farm ponds. *Trans. Am. Fish. Soc.* **98**, 371–377.

COX, J. L. (1970). Low ambient level uptake of ^{14}C-DDT by three species of marine phytoplankton. *Bull. Environ. Contam. Toxicol.* **5**, 218–221.

CRAIGIE, J. S., McLACHLAN, J. and TOWERS, G. H. N. (1965). A note on the fission of an aromatic ring by algae. *Can. J. Bot.* **43**, 1589–1590.

CRANCE, J. H. (1963). The effects of copper sulphate on *Microcystis* and zooplankton in ponds. *Progve Fish Cult.* **25**, 198–202.

CULLIMORE, D. R. (1975). The *in vitro* sensitivity of some species of Chlorophyceae to a selected range of herbicides. *Weed Res.* **15**, 401–406.

CZERPAK, R. (1970). The effect of CCC and diethylamine hydrochloride on certain species of algae belonging to Cyanophyceae, Chlorophyceae and Diatomeae. *Acta Hydrobiol.* **12**, 143–151.

DAFT, M. J., McCORD, S. and STEWART, W. D. P. (1975). Ecological studies on algal-lysing bacteria in fresh waters. *Freshwat. Biol.* **5**, 577–596.

DaSILVA, E. J., HENRIKSSON, L. E. and HENRIKSSON, E. (1975). Effect of pesticides on blue–green algae and nitrogen-fixation. *Arch. Environ. Contam. Toxicol.* **3**, 193–204.

DERBY, S. B. and RUBER, E. (1971). Primary production: depression of oxygen evolution in algal cultures by organophosphorus insecticides. *Bull. Environ. Contam. Toxicol.* **5**, 553–558.

ECHLIN, P. (1966). The blue–green algae. *Sci. Am.* **214**, 74–83.

FIELDEN, T. B. (1969). Biocides in water conservation. *Process Biochem.* **4**, 35.

FITZGERALD, G. P. (1974). Shortcut methods test algicides. *Wat. Sewage Wks* **121**, 85–87.

FITZGERALD, G. P. (1975). Are chemicals used in algae control biodegradable? *Wat. Sewage Wks* **122**, 82–85.

FITZGERALD, G. P. and SKOOG, F. (1954). Control of blue–green algae blooms with 2,3-dichloronaphthoquinone. *Sewage ind. Wastes* **26**, 1136–1140.

FITZGERALD, G. P., GERLOFF, G. C. and SKOOG, F. (1952). Studies on chemicals with selective toxicity to blue–green algae. *Sewage ind. Wastes* **24**, 888–896.

FLETCHER, W. W., KIRKWOOD, R. C. and SMITH, D. (1970). Investigations on the effect of certain herbicides on the growth of selected species of micro-algae. *Meded. Fac. Landb. Rijksuniv. Gent* **35**, 855–867.

FOGG, G. E. (1965). "Algal Cultures and Phytoplankton Ecology". Athlone Press, London.

FOGG, G. E., STEWART, W. D. P., FAY, P. and WALSBY, A. E. (1973). "The Blue–Green Algae". Academic Press, London.

FOTER, M. J., PALMER, C. M. and MALONEY, T. E. (1953). Antialgal properties of various antibiotics. *Antibiotics Chemother.* **3**, 505–508.

FRANK, P. A. (1970). Degradation and effects of herbicides in water. *In* "FAO International Conference on Weed Control". Weed Sci. Soc. Am.

GENTILE, J. H. (1971). Blue–green and green algal toxins. *In* "Microbial Toxins, Vol. VII. Algal and Fungal Toxins" (S. Kadis, A. Ciegler and S. J. Ajl, Eds), pp. 27–66. Academic Press, New York and London.

GEOGHEGAN, M. J. (1957). The effect of some substituted methylureas on the respiration of *Chlorella vulgaris* var. *viridis*. *New Phytol.* **56**, 71–80.

GERKING, S. D. (1948). Destruction of submerged aquatic plants by 2,4-D. *J. Wildlife Mgemnt.* **12**, 221–227.

GORHAM, P. R. (1960). Toxic waterblooms of blue–green algae. *Can. vet. J.* **1**, 235–245.

GORHAM, P. R., McLACHLAN, J., HAMMER, U. T. and KIM, W. K. (1964). Isolation and culture of toxic strains of *Anabaena flos-aquae* (Lyngb.) de Breb. *Verh. int. Verein. Limnol.* **15**, 796–804.

GRAMLICH, J. V. and FRANS, R. E. (1964). Kinetics of *Chlorella* inhibition by herbicides. *Weeds* **12**, 184–189.

GREGORY, W. W., REED, J. K. and PRIESTER, L. E. (1969). Accumulation of parathion and DDT by some algae and protozoa. *J. Protozool.* **16**, 69–71.

GUNNISON, D. and ALEXANDER, M. (1975). Basis for the susceptibility of several algae to microbial decomposition. *Can. J. Microbiol.* **21**, 619–628.

GUPTA, G. S. and SAXENA, P. N. (1974). Effect of Panacide on some green and blue–green algae. *Curr. Sci.* **43**, 492–493.

HAMDI, Y. A., EL-NAWAWY, A. S. and TEWFIK, M. S. (1970). Effect of herbicides on growth and nitrogen-fixation of the alga, *Tolypothrix tenuis*. *Acta Microbiol. Pol.* ser. B, **2**, 53–56.

HARDY, J. L. (1966). Effect of Tordon herbicide on aquatic chain organisms. *Down to Earth* **22**, 11–13.

HARRISS, R. C., WHITE, D. B. and MACFARLANE, R. B. (1970). Mercury compounds reduce photosynthesis by plankton. *Science, N.Y.* **170**, 736–737.

HASSALL, K. A. (1961). Toxicity of trichloracetates to *Chlorella vulgaris*. *Physiol. Plant.* **14**, 140–149.

HAWKINS, A. F. (1972). Control of algae. *Outlook on Agriculture* **7**, 21–26.

HELLING, C. S., KAUFMAN, D. D. and DIETER, C. T. (1971). Algal bioassay detection of pesticide mobility in soils. *Weed Sci.* **19**, 685–690.

HILL, D. W. and McCARTY, P. L. (1967). Anaerobic degradation of selected chlorinated hydrocarbon pesticides. *J. Wat. Pollut. Control Fed.* **39**, 1259–1277.

HOLLISTER, T. A. and WALSH, G. E. (1973). Differential responses of marine phytoplankton to herbicides: oxygen evolution. *Bull. Environ. Contam. Toxicol.* **9**, 291–295.

HOLLISTER, T. A., WALSH, G. E. and FORESTER, J. (1975). Mirex and marine unicellular algae: accumulation, population growth and oxygen evolution. *Bull. Environ. Contam. Toxicol.* **14**, 753–759.

HUGHES, E. O., GORHAM, P. R. and ZEHNDER, A. (1958). Toxicity of a unialgal culture of *Microcystis aeruginosa*. *Can. J. Microbiol.* **4**, 225–236.

HULBERT, S. H., MULLA, M. S. and WILLSON, H. (1972). Effects of an organophosphorus insecticide on the phytoplankton, zooplankton and insect populations of fresh-water ponds. *Ecol. Monogr.* **42**, 269–299.

IBRAHIM, A. N. (1972). Effect of certain herbicides on growth of nitrogen-fixing algae and rice plants. *Symp. Biol. Hung.* **11**, 445–448.

IKAWA, M., MA, D. S., MEEKER, G. B. and DAVIS, R. P. (1969). Use of *Chlorella* in mycotoxin and phycotoxin research. *J. agric. Fd Chem.* **17**, 425–429.

ISHIZAWA, S. and MATSUGUCHI, T. (1966). Effects of pesticides and herbicides upon microorganisms in soil and water under waterlogged condition. *Bull. Nat. Inst. Agric. Sci.*, Tokyo, Ser. B. No. 16, 1–90.

JACKSON, D. F. (Ed.) (1964). "Algae and Man". Plenum Press, New York.

JANSEN, L. L., GENTNER, W. A. and HILTON, J. L. (1958). A new method for evaluating potential algicides and determination of the algicidal properties of several substituted-urea and *s*-triazine compounds. *Weeds* **6**, 390–398.

JORDAN, L. S., DAY, B. E. and HENDRIXSON, R. T. (1962). Chemical control of filamentous green algae. *Hilgardia* **32**, 433-441.

KAVANAGH, F. (Ed.) (1963). "Analytical Microbiology". Academic Press, New York.

KAVANAGH, F. (Ed.) (1972a). "Analytical Microbiology", vol. 2. Academic Press, New York.

KAVANAGH, F. (1972b). An approach to accurate diffusion assays. *In* "Analytical Microbiology" (F. Kavanagh, Ed.), vol. 2, pp. 31–42. Academic Press, New York.

KAVANAGH, F. (1972c). Introduction. *In* "Analytical Microbiology" (F. Kavanagh, Ed.), vol. 2, pp. 1–12. Academic Press, New York.

KEEN, R. (1971). Controlling algae and other growths on concrete. Advisory Note No. 45·020, Cement and Concrete Assocn (London).

KEIL, J. E. and PRIESTER, L. E. (1969). DDT uptake and metabolism by a marine diatom. *Bull. Environ. Contam. Toxicol.* **4**, 169–173.

KEIL, J. E., PRIESTER, L. E. and SANDIFER, S. H. (1971). Polychlorinated biphenyl (Aroclor 1242): Effects of uptake on growth, nucleic acids, and chlorophyll of a marine diatom. *Bull. Environ. Contam. Toxicol.* **6**, 156–159.

KIRKWOOD, R. C. and FLETCHER, W. W. (1970). Factors influencing the herbicidal efficiency of MCPA and MCPB in three species of micro-algae. *Weed Res.* **10**, 3–10.

KNUTSEN, G. (1972). Degradation of uracil by synchronous cultures of *Chlorella fusca*. *Biochim. biophys. Acta* **269**, 333–343.

KRATKY, B. A. and WARREN, G. F. (1971a). A rapid bioassay for photosynthetic and respiratory inhibitors. *Weed Sci.* **19**, 658–661.

KRATKY, B. A. and WARREN, G. F. (1971b). The use of three simple, rapid bioassays on forty-two herbicides. *Weed Res.* **11**, 257–262.

KRICHER, J. C., UREY, J. C. and HAWES, M. L. (1975). The effects of mirex and methoxychlor on the growth and productivity of *Chlorella pyrenoidosa*. *Bull. Environ. Contam. Toxicol.* **14**, 617–620.

KRUGLOV, J. W. and KWIATKOWSKAJA, L. B. (1975). The algae as indicatory culture of herbicide contamination in soil. *Roczniki glebozn.* **26**, 145–148.

KRUGLOV, Yu. V. and PAROMENSKAYA, L. N. (1970). Detoxication of simazine by microscopic algae. *Microbiology (Mikrobiologiya)* **39**, 139–142.

KUMAR, H. D. (1963). Inhibition of growth and pigment production of a blue–green alga by 3-amino-1,2,4-triazole. *Ind. J. Pl. Physiol.* **6**, 150–155.

KUMAR, H. D. (1964). Streptomycin- and penicillin-induced inhibition of growth and pigment production in blue–green algae and production of strains of *Anacystis nidulans* resistant to these antibiotics. *J. exptl. Bot.* **15**, 232–250.

KUMAR, H. D. (1968). Inhibitory action of the antibiotics mitomycin c and neomycin on the blue–green alga *Anacystis nidulans. Flora, Jena* **159**, 437–444.

KUMAR, H. D., SINGH, H. N. and PRAKASH, G. (1967). The effect of proflavine on different strains of the blue–green alga *Anacystis nodulans. Pl. Cell Physiol., Tokyo* **8**, 171–179.

LAZAROFF, N. (1967). Algal response to pesticide pollutants. *Bact. Proc.* G.149, p. 48.

LEE, S. S., FANG, S. C. and FREED, V. H. (1976). Effect of DDT on photosynthesis of *Selanastrum capricormutum. Pestic. Biochem. Physiol.* **6**, 46–51.

LOEPPKY, C. and TWEEDY, B. G. (1969). Effects of selected herbicides upon growth of soil algae. *Weed Sci.* **17**, 110–113.

LUARD, E. J. (1973). Sensitivity of *Dunaliella* and *Scenedesmus* (Chlorophyceae) to chlorinated hydrocarbons. *Phycologia* **12**, 29–33.

LUNDQVIST, I. (1970). Effect of two herbicides on nitrogen fixation by blue–green algae. *Svensk bot. Tidskr.* **64**, 460–461.

MACKIEWICZ, M., DEUBERT, K. H., GUNNER, H. B. and ZUCKERMAN, B. M. (1969). Study of parathion biodegradation using gnotobiotic techniques. *J. agric. Fd Chem.* **17**, 129–130.

MALONEY, T. E. (1958). Control of algae with chlorophenyl dimethylurea. *J. Am. Wat. Wks Ass.* **50**, 417–422.

MALONEY, T. E. and PALMER, C. M. (1956). Toxicity of six chemical compounds to thirty cultures of algae. *Wat. Sewage Wks* **103**, 509–513.

MATSUMURA, F., PATIL, K. C. and BOUSH, G. M. (1970). Formation of "Photodieldrin" by microorganisms. *Science, N.Y.* **170**, 1206–1207.

MENZEL, D. W., ANDERSON, J. and RANDTKE, A. (1970). Marine phytoplankton vary in their response to chlorinated hydrocarbons. *Science, N.Y.* **167**, 1724–1726.

MIKHAILOVA, E. I. and KRUGLOV, Y. V. (1973). Effect of some herbicides on soil algoflora. *Pochvovedenie* **8**, 81–85.

MIYAZAKI, S. and THORSTEINSON, A. J. (1972). Metabolism of DDT by fresh water diatoms. *Bull. Environ. Contam. Toxicol.* **8**, 81–83.

MOORE, R. B. (1970). Effects of pesticides on growth and survival of *Euglena gracilis* Z. *Bull. Environ. Contam. Toxicol.* **5**, 226–230.

MORGAN, J. R. (1972). Effects of Aroclor 1242[R] (a polychlorinated biphenyl) and DDT on cultures of an alga, protozoan, daphnid, ostracod, and guppy. *Bull. Environ. Contam. Toxicol.* **8**, 129–137.

MOSSER, J. L., FISHER, N. S., TENG, T. C. and WURSTER, C. F. (1972a). Polychlorinated biphenyls: toxicity to certain phytoplankters. *Science, N.Y.* **175**, 191–192.

MOSSER, J. L., FISHER, N. S. and WURSTER, C. F. (1972b). Polychlorinated biphenyls and DDT alter species composition in mixed cultures of algae. *Science, N.Y.* **176**, 533–535.

MOSSER, J. L., TENG, T. C., WALTHER, W. G. and WURSTER, C. F. (1974). Interactions of PCB's, DDT and DDE in a marine diatom. *Bull. Environ. Contam. Toxicol.* **12**, 665–668.

MULLISON, W. R. (1970a). Effects of herbicides on water and its inhabitants. *Weed Sci.* **18**, 738–750.

MULLISON, W. R. (1970b). The significance of herbicides to non-target organisms. *Proc. Suppl. 24th Ann. Meeting N.E. Weed Control Conf.* 111–147.

NEUDORF, S. and KHAN, M. A. Q. (1975). Pick-up and metabolism of DDT, dieldrin and photodieldrin by a fresh water alga (*Ankistrodesmus amalloides*) and a micro-crustacean (*Daphnia pulex*). *Bull. Environ. Contam. Toxicol.* **13**, 443–450.

PALMER, C. M. (1956). Evaluation of new algicides for water supply purposes. *J. Am. Wat. Wks Ass.* **48**, 1133–1137.

PALMER, C. M. and MALONEY, T. E. (1955). Preliminary screening for potential algicides. *Ohio J. Sci.* **55**, 1–8.

PARIS, D. F. and 'LEWIS, D. L. (1973). Chemical and microbial degradation of ten selected pesticides in aquatic systems. *Residue Rev.* **45**, 95–124.

PATEL, N. P. (1972). Rice research in Fiji 1960–1970. Part III. Weed Control. *Fiji agric. J.* **34**, 27–34.

PATIL, K. C., MATSUMURA, F. and BOUSH, G. M. (1972). Metabolic transformations of DDT, dieldrin, aldrin, and endrin by marine microorganisms. *Environ. Sci. Technol.* **6**, 629–632.

PAUL, S. K., CHARI, K. V. and BHATTACHARYYA, B. (1975). Electrolytic control of algae. *J. Am. Wat. Wks Ass.* **67**, 140–141.

PIERCE, M. E. (1958). The effect of the weedicide Kuron upon the flora and fauna of two experimental areas of Long Pond, Duchess County, N.Y. *Proc. NEast Weed Control Conf.* **12**, 338–343.

PIERCE, M. E. (1960). Progress report of the effects of Kuron upon the biota of Long Pond, Duchess County, N.Y. *Proc. NEast Weed Control Conf.* **14**, 472–475.

PILLAY, A. R. and TCHAN, Y. T. (1969). Paper disc method for rapid bioassay of s-phenylurea herbicides. *Soil Biol.* **10**, 23–25.

PILLAY, A. R. and TCHAN, Y. T. (1970). The use of a soil alga for the bioassay and study of triallate (herbicide). *Soil Biol.* **12**, 20–22.

PILLAY, A. R. and TCHAN, Y. T. (1972). Study of soil algae. VII. Adsorption of herbicides in soil and prediction of their rate of application by algal methods. *Pl. Soil* **36**, 571–594.

POORMAN, A. E. (1973). Effects of pesticides on *Euglena gracilis*. I. Growth studies. *Bull. Environ. Contam. Toxicol.* **10**, 25–28.

PRICE, C. A. and ESTRADA, M-T. G. (1964). Chlorophyll formation in *Euglena* as a test for herbicides. *Weeds* **12**, 234–235.

PROVASOLI, L., PINTNER, I. and PACKER, L. (1951). Use of antibiotics in obtaining pure cultures of algae and protozoa. *Proc. Soc. Protozool.* **2**, 6.

RAGHU, K. and MacRAE, I. C. (1967). The effect of the gamma-isomer of benzene hexachloride upon the microflora of submerged rice soils. I. Effect upon algae. *Can. J. Microbiol.* **13**, 173–180.

RAY, B. R. (1973). Weed control in rice – a review. *Ind. J. Weed Sci.* **5**, 60–72.

REDHEAD, K. and WRIGHT, S. J. L. (1978). Isolation and properties of fungi that lyse blue–green algae. *Appl. environ. Microbiol.* **35**, 962–969.

REIM, R. L., SHANE, M. S. and CANNON, R. E. (1974). The characterization of a *Bacillus* capable of blue–green bactericidal activity. *Can. J. Microbiol.* **20**, 981–986.

RICE, C. P. and SIKKA, H. C. (1973a). Uptake and metabolism of DDT by six species of marine algae. *J. agric. Fd Chem.* **21**, 148–152.

RICE, C. P. and SIKKA, H. C. (1973b). Fate of dieldrin in selected species of marine algae. *Bull. Environ. Contam. Toxicol.* **9**, 116–123.

ROBSON, T. O. (1973). Recent trends in weed control in freshwater. *OEPP/EPPO Bull.* **3**, 5–17.

ROBSON, T. O., FOWLER, M. C. and BARRETT, P. R. F. (1976). Effect of some herbicides on freshwater algae. *Pestic. Sci.* **7**, 391–402.

ROUND, F. E. (1973). "The Biology of the Algae", 2nd edition. Edward Arnold, London.

SAFFERMAN, R. S. and MORRIS, M. E. (1962). Evaluation of natural products for algicidal properties. *Appl. Microbiol.* **10**, 289–292.

SAFFERMAN, R. S. and MORRIS, M. E. (1963a). Algal virus: Isolation. *Science, N.Y.* **140**, 679–680.

SAFFERMAN, R. S. and MORRIS, M. E. (1963b). The antagonistic effect of Actinomycetes on algae found in waste stabilization ponds. *Bact. Proc.* 14.

SAFFERMAN, R. S. and MORRIS, M. E. (1964a). Control of algae with virus. *J. Am. Wat. Wks Ass.* **56**, 1217–1224.

SAFFERMAN, R. S. and MORRIS, M. E. (1964b). Growth characteristics of the blue–green algal virus LPP-1. *J. Bact.* **88**, 771–775.

SAFFERMAN, R. S. and MORRIS, M. E. (1967). Observations on the occurrence, distribution, and seasonal incidence of blue–green algal viruses. *Appl. Microbiol.* **15**, 1219–1222.

SATO, R. and KUBO, H. (1964). The water pollution caused by organophosphorus insecticides in Japan. *Adv. Wat. Pollut. Res.* **1**, 95–99.

SCHANTZ, E. J. (1971). The dinoflagellate poisons, *In* "Microbial Toxins, vol. 7, Algal and Fungal Toxins" (S. Kadis, A. Ciegler and S. J. Ajl, Eds), pp. 3–26. Academic Press, New York.

SCHWELITZ, F. D., SIKKA, H. C., SAXENA, J. and ZWEIG, G. (1974). Ultrastructural changes in isolated spinach chloroplasts and in *Chlorella pyrenoidosa* Chick (Emerson strain) treated with dichlone. *Pestic. Biochem. Physiol.* **4**, 379–385.

SETHUNATHAN, N. and MacRAE, I. C. (1969). Some effects of diazinon on the microflora of submerged soils. *Pl. Soil* **30**, 109–112.

SHAPIRO, J. (1973). Blue–green algae: Why they become dominant. *Science, N.Y.* **179**, 382–384.

SHARABI, N. E-D. (1969). Influence of propanil (3′,4′-dichloropropionanilide) and related compounds on growth of the alga *Chlorococcum aplanosporum* in soil and in solution culture. Ph.D. Thesis. Rutgers – The State University, U.S.A. (*Weed Abstr.* **20**, 434).

SHELUBSKY, M. (1951). Observations on the properties of a toxin produced by *Microcystis*. *Verh. int. Verein. theor. angew. Limnol.* **11**, 362–366.

SHENNAN, J. L. and FLETCHER, W. W. (1965). The growth *in vitro* of micro-organisms in the presence of substituted phenoxyacetic and phenoxybutyric acids. *Weed Res.* **5**, 266–274.

SHERIDAN, R. P. and SIMMS, M. A. (1975). Effect of the insecticide Zectran (mexacarbate) on several algae. *Bull. Environ. Contam. Toxicol.* **13**, 565–569.

SHIELDS, L. M. and DURRELL, L. W. (1964). Algae in relation to soil fertility. *Bot. Rev.* **30**, 92–128.

SHILO, M. (1965). Study on the isolation and control of blue–green algae from fish ponds. *Bamidgeh* **17**, 83–93.

SHILO, M. (1967). Formation and mode of action of algal toxins. *Bact. Rev.* **31**, 180–193.

SHILO, M. (1971a). Biological agents which cause lysis of blue–green algae. *Mitt. int. Verein. theor. angew. Limnol.* **19**, 206–213.

SHILO, M. (1971b). Toxins of Chrysophyceae. *In* "Microbial Toxins, vol. 7, Algal and Fungal Toxins" (S. Kadis, A. Ceigler and S. J. Ajl, Eds), pp. 67–103. Academic Press, New York.

SIKKA, H. C. and PRAMER, D. (1968). Physiological effects of fluometuron on some unicellular algae. *Weed Sci.* **16**, 296–299.

SIKKA, H. C., CARROLL, J. and ZWEIG, G. (1971). Effect of certain quinone pesticides on acetate photometabolism and dark CO_2 fixation in *Chlorella*. *Pestic. Biochem. Physiol.* **1**, 381–388.

SIKKA, H. C., SAXENA, J. and ZWEIG, G. (1973). Alteration in cell permeability as a mechanism of action of certain quinone pesticides. *Pl. Physiol.* **51**, 363–367.

SILVO, O. E. J. (1968). The influence of paraquat, a herbicide used in aquatic weed control, upon fish, phytoplankton and oxygen content of water, and its disappearance from the water. *Suom. Kalatal.* **34**, 20. (*Weed Abstr.* **18**, 462).

SINGH, P. K. (1973). Effect of pesticides on blue–green algae. *Arch. Mikrobiol.* **89**, 317–320.

SINGH, R. N. and SUBBARAMAIAH, K. (1970). Effects of chemicals on *Fischerella muscicola* (Thuret) Gom. *Can. J. Microbiol.* **16**, 193–199.

SNYDER, C. E. and SHERIDAN, R. P. (1974). Toxicity of the pesticide Zectran on photosynthesis, respiration, and growth in four algae. *J. Phycol.* **10**, 137–139.

SÖDERGREN, A. (1968). Uptake and accumulation of C^{14}-DDT by *Chlorella* sp. (Chlorophyceae). *Oikos* **19**, 126–138.

SÖDERGREN, A. (1971). Accumulation and distribution of chlorinated hydrocarbons in cultures of *Chlorella pyrenoidosa* (Chlorophyceae). *Oikos* **22**, 215–220.

ST. JOHN, J. B. (1971). Comparative effects of diuron and chlorpropham on ATP levels in *Chlorella*. *Weed Sci.* **19**, 274–276.

STADNYK, L., CAMPBELL, R. S. and JOHNSON, B. T. (1971). Pesticide effect on growth and ^{14}C assimilation in a freshwater alga. *Bull. Environ. Contam. Toxicol.* **6**, 1–8.

STARMACH, K. (1974). Bioecological characteristics of blue–green algae and conditions of their development. A review. *Pol. Arch. Hydrobiol.* **21**, 29–39.

STEEL, J. A. (1971). Factors affecting algal blooms. *In* "Microbial Aspects of Pollution" (G. Sykes and F. A. Skinner, Eds), pp. 201–213. Academic Press, London.

STEWART, W. D. P. and DAFT, M. J. (1976). Algal lysing agents of freshwater habitats. *In* "Microbiology in Agriculture, Fisheries and Food" (F. A. Skinner and J. G. Carr, Eds), pp. 63–90. Academic Press, London.

STEYN, D. G. (1945). Poisoning of animals and human beings by algae. *S. Afr. J. Sci.* **41**, 243–244.

STOKES, D. M. and TURNER, J. S. (1970). The effects of the dipyridyl diquat on the metabolism of *Chlorella vulgaris*. III. Dark metabolism: effects on respiration rate and path of carbon. *Aust. J. biol. Sci.* **24**, 433–447.

STOKES, D. M., TURNER, J. S. and MARKUS, K. (1970). The effects of the dipyridyl diquat on the metabolism of *Chlorella vulgaris*. II. Effects of diquat in the light on chlorophyll bleaching and plastid structure. *Aust. J. biol. Sci.* **23**, 265–274.

SWEENEY, R. A. (1969). Metabolism of lindane by unicellular algae. *Proc. 12th Conf. Great Lakes Res.* (1969), 98–102.

SWETS, W. A. and WEDDING, R. T. (1964). The effect of dinitrophenol and other inhibitors on the uptake and metabolism of 2,4-dichlorophenoxyacetic acid by *Chlorella*. *New Phytol.* **63**, 55–72.

TCHAN, Y. T. (1959). Study of soil algae. III. Bioassay of soil fertility by algae. *Pl. Soil* **10**, 220–232.

TCHAN, Y. T. and GOULD, J. (1961). Use of antibiotics to purify blue–green algae. *Nature, Lond.* **192**, 1276.

TCHAN, Y. T., ROSEBY, J. E. and FUNNELL, G. R. (1975). A new rapid specific bioassay method for photosynthesis inhibiting herbicides. *Soil Biol. Biochem.* **7**, 39–44.

TEAL, G. (1974). The metabolism of the systemic fungicide ethirimol by plants. Ph.D. Thesis, University of Reading.

THOMAS, V. M., BUCKLEY, L. J., SULLIVAN, J. D. and IKAWA, M. (1973). Effect of herbicides on the growth of *Chlorella* and *Bacillus* using the paper disc method. *Weed Sci.* **21**, 449–451.

THOMPSON, W. (1840). On a minute alga which colours Ballydrain Lake, in the county of Antrim. *Ann. Nat. Hist.* **5**, 75–83.

TOTH, S. J. and RIEMER, D. N. (1968). Precise chemical control of algae in ponds. *J. Am. Wat. Wks Ass.* **60**, 367–371.

TSAY, S. F., LEE, J. M. and LYND, J. Q. (1970). The interactions of Cu^{++} and CN^- with paraquat phytotoxicity to a *Chlorella*. *Weed Sci.* **18**, 596–598.

TURNER, J. S., STOKES, D. M. and GILMORE, L. B. (1970). The effects of the dipyridyl diquat on the metabolism of *Chlorella vulgaris*. I. Gas exchange in the light. *Aust. J. biol. Sci.* **23**, 43–61.

TWEEDY, B. G., LOEPPKY, C. and ROSS, J. A. (1969). Degradation of preforan, fluometuron, and metobromuron by selected soil microorganisms. *Phytopathology* **59**, 1054.

UKELES, R. (1962). Growth of pure cultures of marine phytoplankton in the presence of toxicants. *Appl. Microbiol.* **10**, 532–537.

VALENTINE, J. P. (1973). The influence of four algae on herbicide residues in water. *Diss. Abstr. Int. B.* **34**, 1251–1252.

VALENTINE, J. P. and BINGHAM, S. W. (1974). Influence of several algae on 2,4-D residues in water. *Weed Sci.* **22**, 358–363.

van RENSEN, J. J. S. (1974). Effects of N-(phosphonomethyl) glycine on photosynthetic reactions in *Scenedesmus* and in isolated spinach chloroplasts. *In* "Proceedings of the Third International Congress on Photosynthesis" (M. Avron, Ed.), pp. 683–687. Elsevier, Amsterdam.

VANCE, B. D. (1966). Sensitivity of *Microcystis aeruginosa* and other blue–green algae and associated bacteria to selected antibiotics. *J. Phycol.* **2**, 125–128.

VANCE, B. D. and DRUMMOND, W. (1969). Biological concentration of pesticides by algae. *J. Am. Wat. Wks Ass.* **61**, 360–362.

VANCE, B. D. and SMITH, D. L. (1969). Effects of five herbicides on three green algae. *Texas J. Sci.* **20**, 329–337.

VENKATARAMAN, G. S. (1969). "The Cultivation of Algae". Ind. Counc. Agric. Res., New Delhi.

VENKATARAMAN, G. S. and RAJYALAKSHMI, B. (1971). Tolerance of blue–green algae to pesticides. *Curr. Sci.* **40**, 143–144.

VENKATARAMAN, G. S. and RAJYALAKSHMI, B. (1972). Relative tolerance of nitrogen-fixing blue–green algae to pesticides. *Ind. J. agric. Sci.* **42**, 119–121.

VIRMANI, M. (1973). Effects of six herbicides on three common fresh water algae. Ph.D thesis, Utah State Univ. (*Diss. Abstr. Int. B.* **33**, 5699).

VOIGHT, R. A. and LYNCH, D. L. (1974). Effects of 2,4-D and DMSO on prokaryotic and eukaryotic cells. *Bull. Environ. Contam. Toxicol.* **12**, 400–405.

VOSE, J. R., CHENG, J. Y., ANTIA, N. J. and TOWERS, G. H. N. (1971). The catabolic fission of the aromatic ring of phenylalanine by marine planktonic algae. *Can. J. Bot.* **49**, 259–261.

WALSH, G. E. (1972). Effects of herbicides on photosynthesis and growth of marine unicellular algae. *Hyacinth Contr. J.* **10**, 45–48.

WALSH, G. E. and GROW, T. E. (1971). Depression of carbohydrate in marine algae by urea herbicides. *Weed Sci.* **19**, 568–570.

WALSH, G. E., KELTNER, J. M. and MATTHEWS, E. (1970). Effects of herbicides on marine algae. U.S. Fish and Wildlife Circular No. 335, pp. 10–12. Dept. of the Interior, Washington.

WARE, G. W. and ROAN, C. C. (1970). Interaction of pesticides with aquatic micro-organisms and plankton. *Residue Rev.* **33**, 15–45.

WARE, G. W., DEE, M. K. and CAHILL, W. P. (1968). Water florae as indicators of irrigation water contamination by DDT. *Bull. Environ. Contam. Toxicol.* **3**, 333–338.

WATANABE, A. (1962). Effect of nitrogen-fixing blue–green alga *Tolypothrix tenuis* on the nitrogenous fertility of paddy soils and on the crop yield of rice plant. *J. Gen. appl. Microbiol.* **8**, 85–91.

WEDDING, R. T. and ERICKSON, L. C. (1957). The role of pH in the permeability of *Chlorella* to 2,4-D. *Pl. Physiol.* **32**, 503–512.

WHEELER, W. B. (1970). Experimental absorption of dieldrin by *Chlorella*. *J. agric. Fd Chem.* **18**, 416–419.

WHITTON, B. A. (1965). Extracellular products of blue–green algae. *J. gen. Microbiol.* **40**, 1–11.

WHITTON, B. A. (1967). Studies on the toxicity of polymyxin B to blue–green algae. *Can. J. Microbiol.* **13**, 987–993.

WHITTON, B. A. and MacARTHUR, K. (1967). The action of two toxic quinones on *Anacystis nidulans*. *Archiv. Mikrobiol.* **57**, 147–154.

WOJTALIK, T. A., HALL, T. F. and HILL, L. O. (1971). Monitoring ecological conditions associated with wide-scale applications of DMA 2,4-D to aquatic environments. *Pestic. Monit. J.* **4**, 184–203.

WOLF, F. T. (1962). Growth inhibition of *Chlorella* induced by 3-amino-1,2,4-triazole, and its reversal by purines. *Nature, Lond.* **193**, 901–902.

WORTHEN, L. R. (1973). Interception and degradation of pesticides by aquatic algae. Rpt. PB-223/506, prepd. for Office of Water Resources Research. Distrib. N.T.I.S. U.S. Dept. Commerce.

WRIGHT, S. J. L. (1972). The effect of some herbicides on the growth of *Chlorella pyrenoidosa*. *Chemosphere* **1**, 11–14.

WRIGHT, S. J. L. (1975a). Use of micro-algae for the assay of herbicides. *In* "Some Methods for Microbiological Assay" (R. G. Board and D. W. Lovelock, Eds), pp. 257–269. Academic Press, London.

WRIGHT, S. J. L. (1975b). A simple agar plate method, using micro-algae, for herbicide bio-assay or detection. *Bull. Environ. Contam. Toxicol.* **14**, 65–70.

WRIGHT, S. J. L., STAINTHORPE, A. F. and DOWNS, J. D. (1977). Interactions of the herbicide propanil and a metabolite, 3,4-dichloroaniline, with blue–green algae. *Acta phytopath. Hung.* **12**, 51–60.

WURSTER, C. F. (1968). DDT reduces photosynthesis by marine phytoplankton. *Science, N.Y.* **159**, 1474–1475.

YAMAGISHI, A. and HASHIZUME, A. (1974). Ecology of green algae in paddy fields and their control with chemicals. *Zasso Kenkyu* **18**, 39–43.

ZWEIG, G. (1969). Mode of action of photosynthesis inhibitor herbicides. *Residue Rev.* **25**, 69–79.

ZWEIG, G. and GREENBERG, E. (1964). Diffusion studies with photosynthetic inhibitors *Biochim. biophys. Acta* **79**, 226–233.

ZWEIG, G., TAMAS, I. and GREENBERG, E. (1963). The effect of photosynthesis inhibitors on oxygen evolution and fluorescence of illuminated *Chlorella*. *Biochim. biophys. Acta* **66**, 196–205.

ZWEIG, G., HITT, J. E. and McMAHON, R. (1968). Effect of certain quinones, diquat, and diuron on *Chlorella pyrenoidosa* Chick (Emerson strain). *Weed Sci.* **16**, 69–73.

ZWEIG, G., CARROLL, J., TAMAS, I. and SIKKA, H. C. (1972). Studies on effects of certain quinones. II. Photosynthetic incorporation of $^{14}CO_2$ by *Chlorella*. *Pl. Physiol.* **49**, 385–387.

ZUCKERMAN, B. M., DEUBERT, K., MACKIEWICZ, M. and GUNNER, H. (1970). Studies on the biodegradation of parathion. *Pl. Soil* **33**, 273–281.

Note Added in Proof

The following papers are among those which have come to the author's attention since the manuscript of this chapter was prepared:
Butler *et al.* (1975), describing the effects of atrazine, 2,4-D, methoxychlor, carbaryl and diazinon on the growth of 36 isolates of planktonic algae; O'Kelley and Deason (1976) describing the toxicity, sorption and degradation of twelve pesticides (including insecticides, fungicides and herbicides) in cultures of fresh water algae, and Cullimore and McCann (1977) reporting on the influence of the herbicides 2,4-D, trifluralin, MCPA and TCA on the algal populations of a grassland soil. A review of interactions of pesticides and algae has also been compiled by Butler (1977).

REFERENCES

BUTLER, G. L. (1977). Algae and pesticides. *Residue Rev.* **66**, 19–62.

BUTLER, G. L., DEASON, T. R. and O'KELLEY, J. C. (1975). The effect of atrazine, 2,4-D, methoxychlor, carbaryl and diazinon on the growth of planktonic algae. *Br. phycol. J.* **10**, 371–376.

CULLIMORE, D. R. and McCANN, A. E. (1977). Influence of four herbicides on the algal flora of a prairie soil. *Pl. Soil* **46**, 499–510.

O'KELLEY, J. C. and DEASON, T. R. (1976). Degradation of pesticides by algae. E.P.A.-600/3-76-022. Natnl. Tech. Inform. Service, Springfield, Virginia, U.S.A.

Chapter 9

Pesticides and the Micro-Fauna of Soil and Water

CLIVE A. EDWARDS

Rothamsted Experimental Station, Harpenden, Hertfordshire, England

I. Introduction

Although there is a considerable literature on the influence of pesticides on soil micro-fauna (Edwards and Thompson, 1973) we have much less information on interactions of pesticides with the micro-fauna in the aquatic environment (Ware and Roan, 1970; Thompson and Edwards, 1974).

There are several reasons for including a discussion of the relationships between pesticides and the micro-fauna under the general heading of pesticide microbiology. Firstly, the activities of microbes and the micro-fauna in soil and water are so intimately interlinked that any factor that influences any particular groups of organisms must indirectly affect many other organisms that interact with them. To consider the effects of pesticides on microbes in isolation, would be to ignore these complex interactions. Furthermore, the main distinction between microbes and the micro-fauna tends to be artificial and is usually based on size, with organisms less than 20 μm in length or diameter considered to be microbes and animals between 20 μm and 2·0 mm long classified as micro-fauna. However, although the protozoa range in size from 10–160 μm they are usually included with the microbes.

In soil, other organisms which are considered to be true micro-fauna include Tardigrada, Rotifera and Nematoda, which range in size from 160 to 2000 μm. Many of the smaller species of Acarina and Collembola are also included in the micro-fauna and will be considered as such in this chapter.

Most of the smaller soil animals live in the films of moisture that surround the soil particles and have the same environmental requirements as aquatic organisms, many species being common to both soil and water. In water, these groups include Protozoa, Rotifera and Nematoda as well as the Hydrozoa, Polyzoa and Crustacea. In both soil and water the immature stages of many organisms can also be considered as micro-fauna, but will not be discussed in this chapter. In terms of numbers, the smaller the organisms, the larger their populations tend to be, so that some of the smallest organisms such as bacteria make a major contribution to the total biomass (Wallwork, 1970).

Our knowledge of the roles of the micro-fauna in terrestrial and aquatic systems is still very poor, so that functional interpretation of the effects of pesticides on these organisms is extremely difficult; furthermore, we still have little data on the overall effects of pesticides on many of these little-known groups. All that can be attempted in this chapter is to review the data available and produce some generalizations and speculation on the overall effects of pesticides on terrestrial and aquatic biota and the ecosystems in which they live.

II. Methods of Studying the Effects of Pesticides on the Soil and Water Micro-fauna

A. FIELD STUDIES

To study the effects of pesticides on the micro-fauna in the field it is essential (i) to have adequate methods of assessing populations of these organisms, (ii) to have suitable control populations that are not treated with pesticides, for comparison, and (iii) to have some understanding of the dynamics of the pesticide in the system under study. These requirements have not been even partially fulfilled for more than a few experiments with different pesticides or ecosystems. Unfortunately, many of the studies reported in the literature are based on the application of a pesticide to soil or water with subsequent studies of population changes, but no reference to a control situation, so that seasonal or local changes cannot be distinguished from those caused by the pesticide.

Ideally, for field studies, areas of land that do not vary greatly in soil type, or ecologically similar bodies of water, are necessary, so that the effects of treatment with a pesticide can be compared with a control system or systems; usually some degree of replication is necessary. However, such sites are hard to find. Field experiments that study the effects of pesticides on the soil micro-fauna are easier to organize than those on the aquatic fauna. Soil is a much more static system than water and there is much less movement of pesticides or biota from one area to another. Moreover, methods of sampling populations of terrestrial micro-fauna are better developed than those aimed at assessing numbers of the micro-fauna of aquatic systems. In some of our pesticide experiments, we have attempted to create superficially identical aquatic systems by digging artificial ponds and stocking them as uniformly as possible with a range of organisms, but even this has proved to be very difficult because unknown factors tend to encourage the multiplication of some organisms at the expense of others.

Even when suitable sites are available, the problem of adequate estimation of populations of the micro-fauna remains. In soil, the problems of extracting arthropods from soil have been reviewed by Edwards and Fletcher (1970) who recommended high controlled-gradient heat extraction from numerous soil cores as being most efficient for mites and springtails. The various methods of extraction of nematodes from soil have been described and evaluated by Southey (1959) and Oostenbrink (1971) and of protozoa by Heal (1971). None of these methods assess absolute population numbers and they differ very much in efficiency with the particular group of organisms concerned. The assessment of populations of aquatic microorganisms is even more difficult due to their mobility and the continuous nature of aquatic systems.

B. LABORATORY ASSAYS

It is much easier to test the toxicity of pesticides to the micro-fauna in laboratory assays than in the field. The techniques used in such assays are extremely variable and many have been summarized by Busvine (1971). Some of the workers who have tested pesticides against aquatic micro-fauna in this way include Wurster (1968), Ukeles (1962), Butler (1965) and a summary of many of the results has been made by Lawrence (1962). Fewer workers have tested pesticides against soil micro-fauna in laboratory assays (Scopes and Lichtenstein, 1967); see also *Note added in proof*, p. 622.

Some of the aquatic micro-fauna are extremely sensitive to pesticides and have been used as test organisms in pesticide residue assays. In particular, the cladoceran water flea *Daphnia* has been extensively used for this purpose. However, not all micro-crustaceans are equally sensitive, ostracods and copepods being killed much less readily than cladocerans. For instance a copepod, *Cyclops*, can tolerate fenthion concentrations of up to 1 ppm before 100% are killed whereas as little as $0 \cdot 00009$ ppm·killed 100% of a *Daphnia* population (Ruber, 1965). Cultures of ciliate protozoa accumulated parathion and DDT without adverse effects (Gregory *et al.*, 1969). Isobenzan, fensulfothion and diazinon also had no effect on protozoa (Moeed, 1975). Among the soil-inhabiting microarthropods, only springtails such as *Folsomia* have been used in laboratory pesticide assays and these are not particularly satisfactory because the Collembola as a group tend to be rather insensitive to many pesticides.

It is often difficult to relate results obtained from laboratory assays to those from the field, but they do provide some evidence of relative sensitivity of organisms to pesticides and indicate those that may be potential ecological hazards.

C. INDIRECT METHODS

The difficulties of assessing the effects of pesticides on individual organisms may be overcome, to some extent, by techniques that assess the functional effects of pesticides on key processes in soil or water systems, particularly those processes associated with primary and secondary productivity and nutrient cycling.

1. *Respiration*

The measurement of consumption of oxygen and production of carbon dioxide can be used as an index of total bioactivity in both soil and water and a number of workers have investigated the influence of pesticides on such gaseous exchange. Unfortunately, the relative contributions of different organisms to overall gaseous turnover is extremely difficult to differentiate. Furthermore, although a pesticide may decrease oxygen consumption by killing some organisms, it may also increase oxygen consumption, at the same time, by providing additional substrates for microbial activity, either in the form of decomposing bodies or the pesticide molecules. Hence, the overall effect of a pesticide may be to increase or decrease soil respiration. Bartha *et al.* (1967) found that most pesticides decreased soil respiration but we have found the converse in most of our experiments. There is also some evidence that pesticides can influence the oxygenation of aquatic systems. Such studies have shown that the micro-fauna make a significant contribution to organic matter breakdown because degradation is influenced proportionately to the effect of particular pesticides on these organisms.

2. *Organic matter breakdown*

Many microorganisms and also animals belonging to the micro-fauna use dead and decaying organic matter as a food substrate, and play an important role in breaking down this material and making the basic elements available once more for the growth of other organisms. Several workers have studied the influence of pesticides on organic matter breakdown in soil by burying leaf material in nylon mesh bags (Kurcheva, 1960; Crossley and Hoglund, 1962; Edwards and Heath, 1963; Edwards, 1965; Edwards *et al.*, 1969), and studying its rate of degradation.

There has been much less work of this kind in aquatic systems although it has been suggested that the rate of breakdown of submerged woody materials could be a good indicator of pesticide effects.

III. Direct Effects of Pesticides on Micro-fauna Populations

A pesticide that reaches soil or water can have a number of direct effects upon organisms that live in these environments. Not only may they be killed, but alternatively there may be changes in their growth rate and metabolic processes, changes in reproductive patterns and changes in behaviour. We have little information on the direct non-lethal effects of pesticides on the micro-fauna, so of necessity most of the discussion will be confined to toxic effects.

A. SOIL

In soil, most of our knowledge of the effects of pesticides on the micro-fauna relates to arthropod species. We know little of the effects of pesticides on protozoa except that DDT, at rates up to 200 ppm, had no effect (Smith and Wenzell, 1947; Jones and Moyle, 1963). However, MacRae and Vinckx (1973) reported that DDT (5 and 50 ppm) inhibited soil protozoa for three months and lindane for one month. We know nothing of the effects of pesticides on tardigrades.

Information on the effects of pesticides (other than nematicides) on nematode populations is sparse (Clayton and Ellis, 1947; Grigoreva, 1952; Stockli, 1952; French et al., 1959; Edwards, 1965; Corbett and Webb, 1968; Way and Scopes, 1968; Wicke, 1968; Whitehead and Storey, 1970). Most of these workers concluded that pesticides had little direct effect on nematode populations.

However, many workers have investigated the influence of pesticides on micro-arthropods. Some of the more important studies being those of Grigoreva (1952), Richter (1953), Dobson and Lofty (1956), Sheals (1956), Baring (1957), Karg (1961, 1963, 1964), Edwards (1970), Edwards and Dennis (1960), Edwards and Lofty (1971), Edwards et al. (1967a, 1967b, 1968), Bund (1965), Griffiths et al. (1967), Voronova (1968), Way and Scopes (1968). Such studies are easier than on the groups of larger arthropods because it is much easier to assess field populations of soil mites and springtails. Many of the results are conflicting and were discussed fully by Edwards and Thompson (1973); several tendencies are now clear.

(i) Organochlorine and carbamate insecticides have a greater influence on the soil micro-fauna than organophosphates.

(ii) All insecticides except aldrin are very toxic to predatory mites.

(iii) Most insecticides decrease the species diversity of micro-arthropods.

(iv) Fumigants and other nematicides tend to be very toxic to soil micro-arthropods.

(v) Fungicides, herbicides and molluscicides tend to have little direct effect on the soil micro-fauna.

(vi) Pesticides often result in increases in numbers of some micro-arthropods when their principal predators are killed (Table 1, Figs 1, 2).

B. WATER

Much less information is available on the effects of pesticides on the aquatic micro-fauna than on the soil micro-fauna. This is probably

TABLE 1

Some investigations into effects of pesticides on soil micro-arthropod populations

Pesticide	Change in numbers			Reference
	Acarina			
	Predators	Other spp.	Collembola	
aldrin	0	−	−	Edwards (1965)
BHC	−	−	−	Sheals (1956)
	−	+	+	Richter (1953)
	−	−	−	Karg (1964)
	−	−	−	Bund (1965)
DDT	−	−	+`	Wallace (1954)
	−	−	+	Sheals (1956)
	−	−	+	Edwards et al. (1967a)
	−	−	+	Bund (1965)
chlordane	−	−	−	Edwards (1965)
heptachlor		−	−	Fox et al. (1964)
	−	−	−	Edwards (1965)
chlorfenvinphos	−	−	+	Edwards et al. (1968)
diazinon	−	+	+	Edwards et al. (1967b)
disulfoton	−	−	−	Edwards et al. (1967b)
	−	−		Abdellatif and Reynolds (1967)
fenitrothion	−	+		Raw et al. (1965)
malathion		+	0	Voronova (1968)
menazon	0	0	0	Way and Scopes (1968)
parathion	−	−	+	Edwards et al. (1967b)
	−	−	+	Wallace (1959)
	−	−	−	Richter (1953)
phorate		−	−	Way and Scopes (1968)
	−	−	−	Edwards et al. (1967b)
thionazin	−	+	+	Griffiths et al. (1967)

− decrease in numbers,   + increase in numbers,   0 no effect.

due to a much more extensive use of pesticides in or on soil than in aquatic systems. Perhaps, more investigations should be made in aquatic environments because there can be run-off or leaching from soil into aquatic systems. Moreover, aquatic systems are sometimes seriously contaminated by pesticides from surplus pesticide containers, effluent from factories and other sources.

Most studies into the effects of pesticides on aquatic organisms have concentrated either on the bottom-living fauna or on insect larvae. Relatively few workers have investigated their effects on the micro-fauna.

There are few reports of the effects of pesticides on populations of aquatic protozoa, rotifers, nematodes, Hydrozoa or Polyzoa. Most of the sparse data from field experiments available relate to the micro-crustacea. There are considerable laboratory data on the toxicity of pesticides to cladocerans and ostracods (Frear and Boyd, 1967). Some aquatic herbicides were toxic to *Daphnia magna* at concentrations below those used in practice (Crosby and Tucker, 1966). The toxicity of several pesticides to *Daphnia pulex* is summarized in Table 2.

TABLE 2

Toxicity of some pesticides to *Daphnia pulex*

Pesticide	Toxicity (48h LC_{50}, in ppm)
azinphosmethyl	3
carbaryl	6
2,4-D	3200
DDT	0·4
diazinon	0·9
dichlobenil	3700
dieldrin	250
diuron	1400
endrin	20
fenac	4500
heptachlor	42
lindane	460
malathion	2
pyrethrins	25
fenoprop	2400
sodium arsenite	1800
toxaphene	15
trifluralin	240

Adapted from Cope (1966).

A very slight decrease in micro-crustacean populations after treatment with recommended doses of DDT occurred in six ponds (Jones and Moyle, 1963). Grzenda *et al.* (1964) found that toxaphene and BHC had little effect on zooplankton. Nicholson *et al.* (1962) reported that parathion had little effect on cladocerans, copepods or ostracods in a farm pond.

In an investigation into the effects of TDE on aquatic organisms in Clear Lake, California, numbers of only a few species of zooplankton were reported to be decreased (Cook and Conners, 1963). Fenthion had little effect on numbers of copepods and ostracods in a pond

(Patterson and Windeguth, 1964). However, cladocerans have been reported to be killed by chlorpyrifos, parathion and methyl parathion (Mulla and Khasawinah, 1969). Other workers also reported that chlorpyrifos killed cladocerans and copepods but not rotifers (Hurlbert *et al.*, 1970). Two molluscicides, sodium pentachlorophenate and niclosamide decreased numbers of zooplankton in ponds (Shiff and Garnet, 1961), whilst rotenone and toxaphene, used as piscicides, reduced zooplankton populations in reservoirs (Hoffman and Olive, 1961).

A few herbicides such as simazine and diquat have been shown to be toxic to plankton (Tatum and Blackburn, 1962; Walker, 1964), but others such as fenoprop and sodium arsenite were not (Pierce, 1960; Cowell, 1965; Cope, 1966).

In the few studies where the effects of pesticides on different species of the aquatic micro-fauna were studied, most workers reported that crustacea were much more susceptible to pesticides than rotifers or protozoa (Kasahara, 1962).

IV. Indirect Effects of Pesticides on Micro-fauna Populations

Few pesticides are very specific in their effects, although they tend to have the greatest direct effects on those groups of organisms they are designed to kill. However, many nematicides which are usually broad spectrum biocides and the carbamate compounds can also have insecticidal, acaricidal, nematicidal and fungicidal effects.

No organisms live in soil or water in isolation; they interact with many other organisms. If certain groups of organisms are partially or completely eliminated by pesticides, the balance changes and other organisms may multiply to fill the ecological niche that has been created.

A. SOIL

We still know remarkably little about interactions between different soil organisms; indeed, some of our knowledge of such interactions is derived from studies of the effects of pesticides. For example, predatory mites are extremely sensitive to most insecticides (except aldrin) and prey on many other species of mites and also on springtails. When predatory mites are killed by pesticides, particularly organophosphate

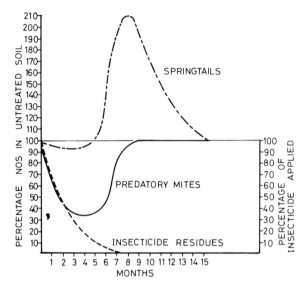

Fig. 1. Changes in numbers of onychiurid springtails and predatory mites after treatment of soil with an organophosphorus insecticide, chlorfenvinphos.

insecticides, there are commonly increases in numbers of many other species of mites and springtails (Sheals, 1956; Edwards and Dennis, 1960; Bund, 1965; Edwards *et al.*, 1967b) (Fig. 1). Often an interaction can be even more closely linked between species or groups, for instance, increased numbers of a particular genus of onychiurid springtails were reported to occur as a result of treatment of soil with an organophosphate insecticide chlorfenvinphos which killed the rhodacarid mites that are their principal predators (Edwards and Thompson, 1973) (Fig. 2).

Other indirect effects can also occur. Thus, numbers of micro-arthropods are often changed drastically after the use of herbicides because their available food supply is altered (Edwards, 1970). Similarly, it seems probable that if the soil micro-flora should be changed drastically as a result of the use of herbicides, it is likely that numbers of the soil micro-fauna will be influenced.

B. WATER

There is less evidence of such indirect effects of pesticides in water. However, some increases in numbers of the micro-fauna have been reported after the use of pesticides. For instance, numbers of zooplankton, particularly copepods and cladocerans, increased

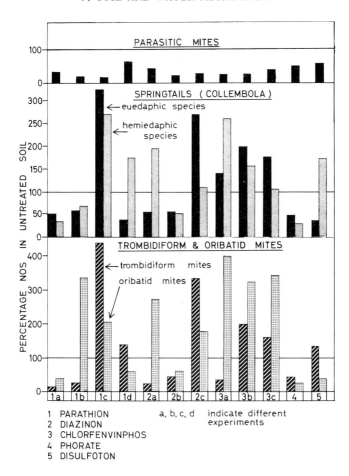

Fig. 2. Increases and decreases in mite and springtail numbers after treatment with organophosphorus insecticides.

markedly a few days after application of copper sulphate (Crance, 1963). Blue–green algae increased in numbers after application of lindane (50 kg ha^{-1}) to submerged Philippine rice soils (Raghu and MacRae, 1967) and this was attributed to the elimination of small algal-feeding animals. Other workers obtained similar results; diazinon treatments (2 and 20 kg ha^{-1}) increased actinomycete and algal populations in submerged rice soils (Sethunathan and MacRae, 1969). There is little direct evidence, but it seems likely that aquatic herbicides must affect some of the micro-fauna by removing some of their food supply. Many other such indirect effects are possible.

V. Uptake of Pesticides into the Micro-fauna

Many pesticides, particularly the organochlorine insecticides, are lipophilic and tend to concentrate into the fatty tissues of animals. This uptake can be from the food or the medium in which the animals live. There has been considerable discussion as to whether there is progressive concentration through trophic levels in a food chain (Carson, 1963; Robinson *et al.*, 1967; Woodwell *et al.*, 1967; Khan and Khan, 1974) but Moriarty (1973) pointed out that there are other possible explanations for the observations that animals in the higher trophic levels tend to have larger residues of pesticides in their tissues, the most simple explanation being based on the balance between pesticide intake and excretion rates.

A. SOIL

There is considerable evidence that the larger animals that live in soil, such as earthworms, slugs and beetles, can concentrate insecticides into their tissues (Edwards and Thompson, 1973) but very little evidence that animals belonging to the micro-fauna can do this. Most of the data available on the micro-fauna come from one group of workers who have shown that both springtails (Butcher *et al.*, 1969) and mites (Aucamp and Butcher, 1971) can take up DDT into their tissues and actually degrade some of it to DDE. It seems likely that such a phenomenon is common to other animals and pesticides, but the small size of the animals has discouraged studies into the uptake of other pesticides.

B. WATER

Even in soil, there is reason to believe that most pesticides are taken up into the tissues or organisms from the aqueous phase between the soil particles, possibly by simple partitioning. Most organic pesticides have very low water solubilities. Examples of water solubilities of organochlorine insecticides, in ppm, at 20–25°C are: DDT 0·0012; aldrin <0·05; dieldrin <0·1; endrin <0·1; toxaphene 3·0, and lindane 10·0. Living organisms tend to accumulate many pesticides from the water in which they live (Cope, 1966; Ware and Roan, 1970) (Fig. 3), with concentration factors ranging from 5 (Hannon *et al.*, 1970) through 16 000 (Keith, 1964) to 141 000 (Johnson *et al.*, 1971) (see also

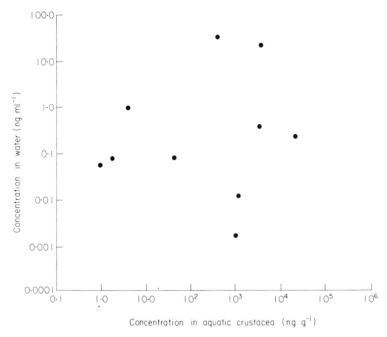

Fig. 3. Concentration of insecticide residues from water into crustacea (data from various sources).

Table 3). Although most reports are of organochlorine pesticides accumulating in the micro-fauna, one group of workers reported considerable accumulations of parathion in algae and protozoa (Gregory *et al.*, 1969). It has now been shown clearly that aquatic organisms acquire most of the pesticide residues in their tissues from the water in which they live rather than from the food they eat. For instance, when a fish (*Poecilia reticulata*), a crustacean (*Daphnia magna*) and an alga were exposed to dieldrin in water separately, the concentration factors were 49 000 for the fish, 14 000 for the crustacean and 1300 for the alga. When the fish were kept in water containing no dieldrin and fed *Daphnia* containing dieldrin residues, they accumulated less than one tenth as much (Reinert, 1967, 1972). These results were confirmed for other species (Chadwick and Brocksen, 1969).

We have little evidence that aquatic micro-fauna can degrade pesticides but it seems extremely probable that they can. Johnson *et al.* (1971) reported that DDT and aldrin could be degraded by freshwater invertebrates.

TABLE 3

Residues of pesticides in aquatic micro-fauna

Organism	Pesticide	Mean Residue ($\mu g\ kg^{-1}$)	Reference
plankton	DDT	5000	Keith (1964)
copepods	DDT	250	Cox (1971)
zooplankton	DDT	40	Woodwell et al. (1967)
plankton	DDT	2–107	Williams and Holden (1973)
micro-zooplankton	DDT	30	Robinson et al. (1967)
plankton	DDT	0·2–34·0	Giam et al. (1973)
plankton	DDT	5	Hannon et al. (1970)
crustacea	DDT	4	Kerr and Vass (1973)
zooplankton	DDT	3	Kerr and Vass (1973)
plankton	BHC	10	Bailey and Hannum (1967)
plankton	BHC	2	Hannon et al. (1970)
plankton	dieldrin	1–230	Williams and Holden (1973)
micro-plankton	dieldrin	20	Robinson et al. (1967)
plankton	dieldrin	7	Hannon et al. (1970)
plankton	dieldrin	5	Bailey and Hannum (1967)
plankton	heptachlor	1·1	Hannon et al. (1970)
plankton	toxaphene	5·0	Hannon et al. (1970)
Paramecium	parathion	964	Gregory et al. (1969)

VI. Conclusions

1. The effects of pesticides on the micro-fauna of soil and water have been little studied.

2. Our lack of knowledge of the effects of pesticides on the micro-fauna reflects a paucity of information on the fundamental processes in soil and water systems.

3. The indirect effects of pesticides on the micro-fauna of soil and water such as by changing food supplies can be at least as drastic as the direct toxic effects.

4. Pesticides applied to soil and water tend to upset the balance between different organisms so that populations of some species decrease but others increase.

5. Pesticides are readily taken up into the tissues of the micro-fauna in both soil and water.

References

ABDELLATIF, M. A. and REYNOLDS, H. T. (1967). Toxic effects of granulated disulfoton on soil arthropods. *J. econ. Ent.* **60**, 281–283.

AUCAMP, J. L. and BUTCHER, J. W. (1971). Conversion of DDT to DDE by two mite species. *Rev. Ecol. Biol. Sol.* **8**, 635–638.

BAILEY, T. E. and HANNUM, J. R. (1967). Distribution of pesticides in California. *J. sanit. Engng Div. Am. Soc. civ. Engrs* **93**, 27–43.

BARING, H. H. (1957). Die Milben fauna eines Ackerbodens und ihre Beeinflussung durch Pflanzenschutz-mittel. II. Der Einfluss von Pflanzenschutzmitteln. *Z. angew. Ent.* **41**, 17–51.

BARTHA, R., LANZILOTTA, R. P. and PRAMER, D. (1967). Stability and effects of some pesticides in soil. *Appl. Microbiol.* **15**, 67–75.

BUND, C. F. van de (1965). Changes in the soil fauna caused by the application of insecticides. *Boll. Zool. agr. Bachic.* **7**, 185–212.

BUSVINE, J. R. (1971). "A Critical Review of the Techniques for Testing Insecticides". Commonwealth Agricultural Bureaux, London.

BUTCHER, J. W., KIRKNEL, E. and ZABIK, M. (1969). Conversion of DDT to DDE by *Folsomia candida* (Willem). *Rev. Ecol. Biol. Sol.* **6**, 291–298.

BUTLER, P. A. (1965). Effects of herbicides on estuarine fauna. *Proc. 5th Weed Control Conf.* **18**, 576–580.

CARSON, R. (1963). "Silent Spring". Hamish Hamilton, London.

CHADWICK, G. G. and BROCKSEN, R. W. (1969). Accumulation of dieldrin by fish and selected fish-food organisms. *J. Wildl. Mgmt* **33**, 693–700.

CLAYTON, C. V. and ELLIS, D. E. (1947). Benzene hexachloride fails to control the root knot nematode. *Pl. Dis. Reptr* **31**, 487–489.

COOK, S. F. and CONNERS, J. D. (1963). The short-term side-effects of the insecticide treatment of Clear Lake, Lake County, California in 1962. *Ann. ent. Soc. Am.* **56**, 819–824.

COPE, O. B. (1966). Contamination of the freshwater ecosystem by pesticides. *J. appl. Ecol.* **3** (Suppl.), 33–34.

CORBETT, D. C. M. and WEBB, R. M. (1968). Effect of herbicides in minimum tillage on nematode population. *Rep. Rothamsted Exp. Stn.* 158–159.

COWELL, B. C. (1965). The effects of sodium arsenite and silvex on the plankton populations in farm ponds. *Trans. Am. Fish Soc.* **98**, 371–377.

COX, J. L. (1971). Uptake assimilation and loss of DDT residues in *Euphausia pacifica* an euphausid shrimp. *Fish Bull. Calif.* **69**, 627–633.

CRANCE, J. H. (1963). The effects of copper sulfate on *Microcystis* and zooplankton in ponds. *Progve Fish Cult.* **25**, 198–202.

CROSBY, D. G. and TUCKER, R. K. (1966). Toxicity of aquatic herbicides to *Daphnia magna*. *Science, N.Y.* **154**, 289–290.

CROSSLEY, D. A. Jr. and HOGLUND, M. (1962). A litter bag method for the study of microarthropods inhabiting peat litter. *Ecology* **43**, 571–574.

DOBSON, R. M. and LOFTY, J. R. (1956). Observations on the effects of BHC on the soil fauna of arable land. *VIth Int. Congr. Sci. Soil., Paris* 203–205.

EDWARDS, C. A. (1965). Effects of soil insecticides on soil invertebrates and plants. *In* "Ecology and the Industrial Society" (G. Goodman, Ed.), pp. 239–261. Blackwell Scientific Publications, Oxford.

EDWARDS, C. A. (1970). Effects of herbicides on the soil fauna. *Proc. 10th Br. Weed Control Conf.* **3**, 1052–1062.

EDWARDS, C. A. and DENNIS, E. B. (1960). Some effects of aldrin and dieldrin on the soil fauna of arable land. *Nature, Lond.* **188**, 767.

EDWARDS, C. A. and FLETCHER, K. E. (1970). Assessment of terrestrial invertebrate populations. *In* "Methods of Study in Soil Ecology" (J. Phillipson, Ed.), pp. 57–66. UNESCO, Paris.

EDWARDS, C. A. and HEATH, G. W. (1963). The role of soil animals in breakdown of leaf material. *In* "Soil Organisms" (J. Doeksen and J. van der Drift, Eds), pp. 76–84. North-Holland Publ. Co., Amsterdam.

EDWARDS, C. A. and LOFTY, J. R. (1971). Nematicides and the soil fauna. *Proc. VIth Br. Insecticide, Fungicide Conf.* 158–166.

EDWARDS, C. A. and THOMPSON, A. R. (1973). Pesticides and the soil fauna. *Residue Rev.* **45**, 1–79.

EDWARDS, C. A., DENNIS, E. B. and EMPSON, D. W. (1967a). Pesticides and the soil fauna. Effects of aldrin and DDT in an arable field. *Ann. appl. Biol.* **60**, 11–22.

EDWARDS, C. A., THOMPSON, A. R. and LOFTY, J. R. (1967b). Changes in soil invertebrate populations caused by some organophosphate insecticides. *Proc. IVth Br. Insecticide, Fungicide Conf.* **1**, 48–55.

EDWARDS, C. A., THOMPSON, A. R. and BEYNON, K. I. (1968). Some effects of chlorfenvinphos, an organophosphorus insecticide on populations of soil animals. *Rev. Ecol. Biol. Sol.* **5**, 199–224.

EDWARDS, C. A., REICHLE, D. E. and CROSSLEY, D. A. (1969). Experimental manipulation of soil invertebrate populations for trophic studies. *Ecology* **50**, 495–498.

FOX, C. J. S., CHISHOLM, D. and STEWART, D. K. R. (1964). Effect of consecutive treatments of aldrin and heptachlor on residues in rutabagas and carrots and on certain soil arthropods and yield. *Can. J. Pl. Sci.* **44**, 149–156.

FREAR, D. E. H. and BOYD, J. E. (1967). Use of *Daphnia magna* for the microbioassay of pesticides. 1. Development of standardised technique for rearing *Daphnia* and the preparation of dosage-mortality curves for pesticides. *J. econ. Ent.* **60**, 1228–1236.

FRENCH, N., LICHTENSTEIN, E. P. and THORNE, G. (1959). Effects of some chlorinated hydrocarbon insecticides on nematode populations in soils. *J. econ. Ent.* **52**, 861–865.

GIAM, C. S., WONG, M. K., HANKS, A. R., SACKETT, W. M. and RICHARDSON, R. L. (1973). Chlorinated hydrocarbons in plankton from the Gulf of Mexico and Northern Caribbean. *Bull. Environ. Contam. Toxicol.* **9**, 376–382.

GREGORY, W. W., REED, J. K. and PRIESTER, L. E. (1969). Accumulation of parathion and DDT by some algae and protozoa. *J. Protozool.* **16**, 69–71.

GRIFFITHS, D. C., RAW, F. and LOFTY, J. R. (1967). The effects on soil fauna of insecticides tested against wireworms (*Agriotes* sp.) in wheat. *Ann. appl. Biol.* **60**, 479–490.

GRIGOR'EVA, T. G. (1952). The action of hexachlorane introduced into the soil on soil fauna. *Rev. appl. Ent.* (A), **41**, 336.

GRZENDA, A. R., LAUER, C. J. and NICHOLSON, H. P. (1964). Water pollution by insecticides in an agricultural river basin. II. The zooplankton, bottom fauna and fish. *Limnol. Oceanogr.* **9**, 318–323.

HANNON, M. R., GREICHUS, Y. A., APPLEGATE, R. L. and FOX, A. C. (1970). Ecological distribution of pesticides in Lake Poinsett, S. Dakota. *Trans. Am. Fish. Soc.* **99**, 496–500.

HEAL, O. W. (1971). Protozoa. *In* "Methods of Study in Quantitative Soil Ecology" (J. Phillipson, Ed.), pp. 51–57, I.B.P. Handbook No. 18. Blackwell, London.

HOFFMAN, D. A. and OLIVE, J. R. (1961). The effects of rotenone and toxaphene upon plankton of two Colorado reservoirs. *Limnol. Oceanogr.* **6**, 219–222.

HURLBERT, S. H., MULLA, M. S., KEITH, J. O., WESTLAKE, W. E. and DUSCH, M. E. (1970). Biological effects and persistence of Dursban in freshwater ponds. *J. econ. Ent.* **63**, 43–52.

JOHNSON, B. T., SAUNDERS, C. R., SANDERS, H. O. and CAMPBELL, R. S. (1971). Biological magnification and degradation of DDT and aldrin by fresh water invertebrates. *J. Fish. Res. Bd Can.* **28**, 705–709.

JONES, B. R. and MOYLE, J. B. (1963). Populations of plankton animals and residual chlorinated hydrocarbons in soils of six Minnesota ponds treated for control of mosquito larvae. *Trans. Am. Fish. Soc.* **92**, 211–215.

KARG, W. (1961). Über die wirkung von hexachlor-cyclohexane auf die boden-biozonose unter besonderer berucksichtigung der Acarina. *NachrBl. dt. PflSchutzdienst, Berl.* **15**, 23–33.

KARG, W. (1963). Bodenbiologische Untersuchung von Kohlfeldern nach berergnungen mit HCH oder trichlorphon praparaten. *NachrBl. dt. PflSchutzdienst, Berl.* **17**, 157–162.

KARG, W. (1964). Untersuchungen über die wirkung von dinitroorthokresol (DNOC) auf die mikroarthropoden des bodens unter berucksichtigung zwischen mikroflora und mesofauna. *Pedobiologia* **4**, 138–157.

KASAHARA, S. (1962). Studies on the biology of the parasitic copepod *Lernaea cyrinacea* Linn. and the methods of controlling this parasite in fish culture. *Contr. Fish. Lab. Facult. Agr. Univ. Tokyo* **3**, 103.

KEITH, J. O. (1964). Relation of pesticides to water-fowl mortality at Tule Lake Refuge. *U.S. Fish Wildlife Serv., Denver Wildlife Res. Centre Ann. Progr. Report* 14 pp.

KERR, S. R. and VASS, W. P. (1973). Pesticide residues in aquatic invertebrates. *In* "Environmental Pollution by Pesticides" (C. A. Edwards, Ed.), pp. 134–180. Plenum Press, London.

KHAN, H. M. and KHAN, M. A. Q. (1974). Biological magnification of photodieldrin by food chain organisms. *Arch. Env. Contam. Toxicol.* **2**, 289–301.

KURCHEVA, G. F. (1960). The role of invertebrates in decomposition of oak litter. *Pochvovedenie* **4**, 16–23.

LAWRENCE, J. M. (1962). Aquatic herbicide data. *Agr. Handbook* No. 231.

MacRAE, I. C. and VINCKX, E. (1973). Effect of lindane and DDT on populations of protozoa in a garden soil. *Soil Biol. Biochem.* **5**, 245–247.

MOEED, A. (1975). Effects of isobenzan, fensulfothion and diazinon on invertebrates and microorganisms. *N.Z. Jl exp. Agric.* **3**, 181–185.

MORIARTY, F. (1973). Pesticides: Significance and Implications of Biological Accumulation. *AGPP: MISC/10.* F.A.O., Rome.

MULLA, M. S. and KHASAWINAH, A. M. (1969). Laboratory and field evaluation of larvicides against chironomid midges. *J. econ. Ent.* **62**, 37–41.

NICHOLSON, H. P., WEBB, H. J., LAUER, G. J., O'BRIEN, R. E., GRZENDA, A. R. and SHANKLIN, D. W. (1962). Insecticide contamination in a farm pond. Part I, Origin and duration. Part II, Biological effects. *Trans. Am. Fish. Soc.* **91**, 213–222.

OOSTENBRINK, M. (1971). Nematodes. *In* "Methods of Study in Quantitative Soil Ecology" (J. Phillipson, Ed.), pp. 72–82, I.B.P. Handbook No. 18. Blackwell, London.

PATTERSON, R. S. and VON WINDEGUTH, D. L. (1964). The effects of Baytex on some aquatic organisms. *Mosquito News* **24**, 46–49.

PIERCE, M. E. (1960). Progress report of the effect of Kuron upon the biota of Long Pond, Duchess County, N.Y. *Proc. NEast Weed Control Conf.* **14**, 472–475.

RAGHU, K. and MacRAE, I. C. (1967). The effect of the gamma-isomer of benzene hexachloride upon the microflora of submerged rice soils. 1. Effect upon algae. *Can. J. Microbiol.* **13**, 173–180.

RAW, F., LOFTY, J. R. and GRIFFITHS, D. C. (1965). Soil Fauna: Wireworms. *Rep. Rothamsted Exp. Stn.* 188.

REINERT, R. E. (1967). The accumulation of dieldrin in an alga (*Scenedesmus obliquus*), Daphnia (*Daphnia magna*), guppy (*Lebistes reticulatus*) food chain. Ph.D. Dissertation. Univ. Michigan, Ann. Arbor.

REINERT, R. E. (1972). Accumulation of dieldrin in an alga (*Scenedesmus obliquus*), *Daphnia magna* and the guppy (*Poecilia reticulata*). *J. Fish. Res. Bd Can.* **29**, 1413–1418.

RICHTER, G. (1953). Die Auswirkung von Insekticiden auf die terricole Makrofauna. *NachrBl. dt. PflSchutzdienst, Berl.* **7**, 61–72.

ROBINSON, J., RICHARDSON, A., CRABTREE, A. N., COULSON, J. C. and POTTS, G. R. (1967). Organochlorine residues in marine organisms. *Nature, Lond.* **214**, 1307–1311.

RUBER, E. (1965). The effects of certain mosquito larvicides on cultures of micro-crustacea. *Proc. New Jers. Mosq. Exterm. Ass.* 207–210.

SCOPES, N. E. A. and LICHTENSTEIN, E. P. (1967). The use of *Folsomia fimetaria* and *Drosophila melanogaster* as test insects for the detection of insecticide residues. *J. econ. Ent.* **60**, 1539–1541.

SETHUNATHAN, N. and MacRAE, I. C. (1969). Some effects of diazinon on the microflora of submerged soils. *Pl. Soil* **30**, 109–112.

SHEALS, J. G. (1956). Soil population studies. 1. The effects of cultivation and treatment with insecticides. *Bull. ent. Res.* **4**, 803–822.

SHIFF, C. J. and GARNET, B. (1961). The short-term effects of three molluscicides on the microflora and microfauna of small biologically stable ponds in Southern Rhodesia. *Bull. W.H.O.* **25**, 543–547.

SMITH, N. R. and WENZELL, M. E. (1947). Soil microorganisms are affected by some of the new insecticides. *Proc. Soil Sci. Soc. Am.* **12**, 227–233.

SOUTHEY, J. F. (1959). Plant Nematology. *M.A.F.F. Tech. Bull.* No. 7.

STOCKLI, A. (1952). Studies on soil nematodes with special reference to the nematode content of forest grassland and arable soils. *Z. PflErnähr. Düng. Bodenk.* **59**, 97–139.

TATUM, W. M. and BLACKBURN, R. D. (1962). Preliminary study of the effects of diquat on the natural bottom fauna and plankton in two sub-tropical ponds. *Proc. SEast. Ass. Game Fish Commnrs* **16**, 301–307.

THOMPSON, A. R. and EDWARDS, C. A. (1974). Effects of pesticides on nontarget invertebrates in freshwater and soil. *In* "Pesticides in Soil and Water" (W. D. Guenzi, Ed.), pp. 341–386. Soil Sci. Soc. Am., Madison.

UKELES, R. (1962). Growth of pure cultures of marine phytoplankton in the presence of toxicants. *Appl. Microbiol.* **10**, 532–537.

VORONOVA, L. D. (1968). The effect of some pesticides on the soil invertebrate fauna in the South Taiga zone in the Perm region (U.S.S.R.). *Pedobiologia* **8**, 507–525.

WALKER, C. R. (1964). Simazine and other *s*-triazine compounds as aquatic herbicides in fish habitats. *Weeds* **12**, 134–139.

WALLACE, M. M. H. (1954). The effect of DDT and BHC on the population of the lucerne flea (*Sminthurus viridis* L.) (Collembola) and its control by predatory mites, *Biscirus* spp. (Bdellidae). *Aust. J. agric. Res.* **5**, 148–155.

WALLACE, M. M. H. (1959). Insecticides for the control of the lucerne flea *Sminthurus viridis* (L.) and the red-legged earth mite, *Halotydeus destructor* (Tuck.) and their effects on population numbers. *Aust. J. agric. Res.* **10**, 160–170.

WALLWORK, J. A. (1970). "Ecology of Soil Animals". McGraw-Hill, London.

WARE, G. W. and ROAN, C. C. (1970). Interaction of pesticides with aquatic microorganisms and plankton. *Residue Rev.* **33**, 15–45.

WAY, M. J. and SCOPES, N. E. A. (1968). Studies on the persistence and effects on soil fauna of some soil-applied systemic insecticides. *Ann. appl. Biol.* **62**, 119–214.

WHITEHEAD, A. G. and STOREY, G. (1970). Control of potato cyst-nematode. *Rep. Rothamsted Exp. Stn.* 199–203.

WICKE, E. (1968). Weed control in fruit – Side effects on the soil fauna. *Z. Pflanzenkr. PflPath. PflSchutz* **4**, 163–167.

WILLIAMS, R. and HOLDEN, A. V. (1973). Organochlorine residues from plankton. *Mar. Polln. Bull.* **4**, 109–111.

WOODWELL, G. M., WURSTER, C. F. and ISAACSON, P. A. (1967). DDT residues in an east coast estuary. A case of biological concentration of a persistent insecticide. *Science, N.Y.* **156**, 821–824.

WURSTER, C. F. (1968). DDT reduces photosynthesis by marine phytoplankton. *Science, N.Y.* **159**, 1471–1475.

Note Added in Proof

Contributions to the information on pesticide interactions with protozoa have been made by Prescott and Olson (1972) and Prescott *et al.* (1977), who described the effects of several pesticides on the growth, *in vitro*, of the free-living amoeba *Acanthamoeba castellanii.*

REFERENCES

PRESCOTT, L. M. and OLSON, D. L. (1972). The effect of pesticides on the soil amoeba *Acanthamoeba castellanii. Proc. S.D. Acad. Sci.* **51**, 136–141.

PRESCOTT, L. M., KUBOVEC, M. K. and TRYGGESTAD, D. (1977). The effects of pesticides, polychlorinated biphenyls and metals on the growth and reproduction of *Acanthamoeba castellanii. Bull. Environ. Contam. Toxicol.* **18**, 29–34.

Chapter 10

Microbial Degradation of Insecticides

FUMIO MATSUMURA AND HERMAN J. BENEZET[*]

*Department of Entomology, University of Wisconsin,
Madison, Wisconsin, USA*

I. Introduction

Insecticidal chemicals which are introduced into the environment eventually face various natural "weathering" forces that alter their locations and chemical characteristics. The level of insecticidal residues at the site of application starts declining, first at relatively fast rates, and then usually settles down to somewhat slower rates of

[*] Present address: Department of Entomology, University of Missouri, Columbia, Missouri, USA.

disappearance. The major environmental factors affecting the rate of residue disappearance are: evaporation, wash-off and leaching, dust-carried air transport, bio-system mediated transfer, bio-concentration, alteration by sunlight, and metabolic changes by biological systems.

It must be noted here that all these forces, except those belonging to the last two categories, represent mere translocation phenomena. When one considers the environment as a whole such translocations of pesticide residues do not actually contribute to the overall decline of pesticide levels. Rather, such phenomena can create further troubles by causing accumulation of pesticidal residues in unexpected areas, such as food commodities and wildlife.

Among the biologically mediated changes in insecticides, microbial processes play by far the most significant role. There are several reasons why such a conclusion can be reached. The majority of insecticides eventually find their way to soil and aquatic environments, whether they are directly aimed at them or not. Even in the aquatic environments the bulk of insecticidal residues are tied up in sediments, making the soil-related systems the largest reservoir of pesticide residues in the environment. Soil is rich with microorganisms and microbial activities occur in many niches, including those inaccessible to sunlight and other weathering forces, or where no other biological forms are active. In addition, because of their ability to adapt and proliferate, and because of their total biomass and surface area, microorganisms must be considered to represent the prime force of environmental alteration of insecticidal residues.

Insecticides belong to a group of chemicals that may become environmental micro-pollutants. In other words, they are not present in large quantities to contaminate the environments physically. Rather, they are present in such minute quantities that, but for their biological activities, their mere existence would not have been the centre of attention. One may therefore wonder why microorganisms should bother to metabolize the compounds that are normally present in such small amounts and which, in addition to being insecticidal, have properties that generally make them non-nutritional and resistant to biological degradation.

Basically one can assume two different types of insecticidal pollution. Firstly, the low level environmental contamination as mentioned above and, second, the large scale spillings, such as in treatment pools, loading docks, accidental spillings, sewer or disposal outlets of insecticide plants, etc. In the latter cases growth of insecticide-tolerant or even insecticide-utilizing microorganisms might be expected. Since soil is generally regarded as being of a nutritionally carbon-deficient status, it is possible that microorganisms which can utilize insecticides may be selected, as long as the total input of these organic chemicals is relatively large. However, where insecticide residues are present in low amounts, it is not logical to assume that evolution of particular microbes or adaptive metabolic activities can take place easily. Thus the majority of microbial metabolism of insecticidal residues must be considered to be incidental. That is, these chemicals, unless they are very toxic to the microorganisms, are metabolized by the non-specific enzyme systems which happen to be present. The answer to the question as to why microorganisms pick up the low level residues rests upon the overall chemical characteristics of insecticidal compounds, as explained below.

II. Chemical Characteristics of Insecticidal Compounds

It has been known for some time that the rate of insecticide disappearance in the environment is generally specific to each insecticidal compound. Indeed, it is difficult to establish such processes for degradation of the more stable compounds, such as mirex. It may therefore be appreciated that the chemical characteristics of the insecticidal materials are very important in determining the pattern of microbial metabolism.

Despite the inclusion of many different classes of compounds under the name of insecticides, modern organic insecticides have several group-specific chemical characteristics. They are generally small organic molecules (molecular weight 200 to 400), biologically active and lipophilic. The most important parameter for a compound to be insecticidal is that it must penetrate through the insect cuticle and often through the nerve sheath, both of which serve as formidable barriers to the generally polar organic compounds. For this purpose, insecticidal chemicals are designed to be non-polar, making them lipophilic and sparsely water-soluble. The presence of a carboxylic acid, a hydroxyl or a quaternary ammonium moiety, for instance, drastically curtails the insecticidal activity of any given compound. The majority of insecticidal chemicals are hydrocarbons and include both aromatic and aliphatic esters, ethers and thio-ethers, primary and secondary amines, amides, phosphates, thiocyanates and ketones. Also, generally speaking, the majority of insecticides are designed to last for a long time in the environment so as to warrant their effectiveness in the field.

Several chemical groups have been adopted for insecticidal use, the major ones being organophosphorus esters, carbamate esters and chlorinated compounds. Because of the very nature of insecticidal chemicals, they are expected to have a high affinity for biological systems. All those cases of bio-accumulation of insecticidal residues in wildlife and agricultural products attest to this tendency. It is also easy to understand why the problem of bio-accumulation becomes more acute in aquatic environments. In soil the prime translocation force is water. The level of microbial activities is also affected by the water contents of the micro-environments, in addition to other factors such as available nutrients. In aqueous situations the insecticidal molecules can be expected to seek, or partition into biological materials. Such processes can occur even at very low insecticide concentrations. The microbial cell membranes can also be expected to be rather permeable to these lipophilic and small organic molecules.

III. Characteristics of Microbial Metabolism

It is understandable that many pesticide scientists assume that in general the patterns of microbial metabolism are similar to those already found in animals, particularly mammalian species, as the field of microbial metabolism of pesticides lags far behind that of comparable studies in mammalian species. However, a simple deduction of the existence of similar mechanisms and processes in these two different groups of organisms is very hazardous.

Firstly, the purpose of metabolism of all xenobiotics in higher animals is, eventually, to convert them into excretable forms. This is generally achieved by first converting the xenobiotic to more polar metabolic products and then, if they are still difficult to excrete, by conjugating them with polar, water-soluble ligands. Secondly, in higher animals the processes of primary metabolism of xenobiotics are centralized in a few specialized organs. In the case of the liver, its metabolic pattern is largely determined by the activity of an oxidative detoxification system involving mixed-function oxidases.

On the contrary, the predominant metabolic activities in the microbial world are meant for production of energy. Generally, it is not even possible to define xenobiotics in this case, since most of the organic materials can serve as a source of energy to at least some microorganisms. However, a few groups of chemicals are foreign to microorganisms. Among insecticidal compounds, the halogen-containing chemicals, particularly halogenated aromatics, must be regarded as foreign, or unusable material as such, for microorganisms.

Another characteristic of microbial metabolism is the adaptability of microorganisms through mutation and induction, particularly towards chemicals that are initially toxic to them. The case of penicillin-resistance in bacteria through induction of penicillinase (see e.g. Pollack, 1957) is well known. Such microbial adaption has also been observed in the pesticide field. For example, Raghu and MacRae (1966) noticed that the second addition of BHC (benzene hexachloride) to a submerged rice field resulted in a higher BHC-degrading metabolic activity, indicating the evolution of BHC-metabolizing microorganisms in the first incubation period.

The metabolic activities of microorganisms encompass many different types of biological processes (see e.g. Doelle, 1969), which are not found in any other organisms. They include fermentation, some types of anaerobic metabolism, chemolithotrophic metabolism, metabolism through exoenzymes, to name a few.

In general, microbial contributions to metabolic alteration of insecticides may be classified in several categories as shown in Table 1.

TABLE 1

General classification of microbial metabolism of insecticides

A. Enzymatic
    1. Incidental metabolism of insecticides which cannot themselves serve as energy sources
       (a) Metabolism by wide-spectrum, all-purpose enzymes (hydrolases, oxidases, etc.)
          (i) Insecticides as substrates
          (ii) Insecticides as electron acceptors or donors
       (b) Co-metabolism of compounds structurally very similar to the natural substrates
    2. Catabolism: insecticides serve as energy sources
    3. Detoxification metabolism: serving as a resistance mechanism

B. Non-enzymatic
    1. Participation in photochemical reactions
    2. Contribution through pH changes
    3. Through production of organic and inorganic reactants
    4. Through production of co-factors

A. ENZYMATIC PROCESSES

In classifying the enzymatic metabolism an important criterion has been adopted; that is, whether the microorganisms derive energy from the process or not. The consideration is significant from both an academic standpoint and from the practical view, for upon administration of high doses of insecticides proliferation of insecticide-degrading microbes is expected only in the cases involving catabolic processes (Class A2 metabolism, Table 1).

On the other hand, in both classes A1 and A3 metabolism (Table 1), the microorganisms depend for their survival upon other more accessible carbon sources, such as glucose, which should increase their general metabolic activities. Thus one could control the rate of microbial metabolism by changing either the amounts of insecticides or other carbon sources added, depending upon the types of microbial degradation activities. It may be generalized, therefore, that incidental metabolism is the more prevalent form of microbial metabolism when the amount of insecticide is low in comparison with other carbon sources. Catabolic metabolism could occur when the amount of insecticide is high, coupled with the favourable chemical structure of the insecticide that allows it to be microbially degradable and utilizable as a carbon source.

Another way to distinguish these classes of metabolic activities is to induce insecticide-metabolizing microbes by the use of insecticides themselves (in the presence and absence of other carbon sources). Strictly speaking, in the absence of other carbon sources only the microbes performing the class A2 metabolism should be inducible and able to flourish by using the insecticide as a sole carbon source. In the presence of other carbon sources, administration of high doses of insecticides can cause two different types of response. If the insecticide is toxic to the organisms, it may stimulate the growth of resistant strains which can either metabolically or otherwise detoxify it or withstand the toxic action (class A3 metabolism, Table 1). If the insecticide is not toxic to the microbes, the class A1 metabolism will continue, depending upon the availability of other carbon sources.

The above classification is not made on the basis of microbial species or the individual insecticides. On the contrary, the nature of metabolism can be influenced largely by environmental factors, which in turn affect the physiological conditions of the microbes. It is therefore possible that the same microorganisms might metabolize an insecticide differently according to the environmental conditions.

Benezet and Matsumura (1974), for instance, studied the change in metabolic patterns of mexacarbate (4-dimethylamino-3,5 xylylmethylcarbamate; Zectran) by using mainly two species of microorganisms, and found that the microorganisms generally metabolized mexacarbate by the class A1(a) metabolism, yielding a hydrolytic, or decarbamylation product in the presence of mannitol or glucose as the major carbon source. Upon removal of most of these easily available

carbon sources, the microorganisms changed the pattern of degradation to enhance *N*-dealkylation reactions (Fig. 1). Furthermore, the
overall speed of mexacarbate metabolism increased in the absence of
other available carbon sources, indicating a shift to a catabolic
metabolism.

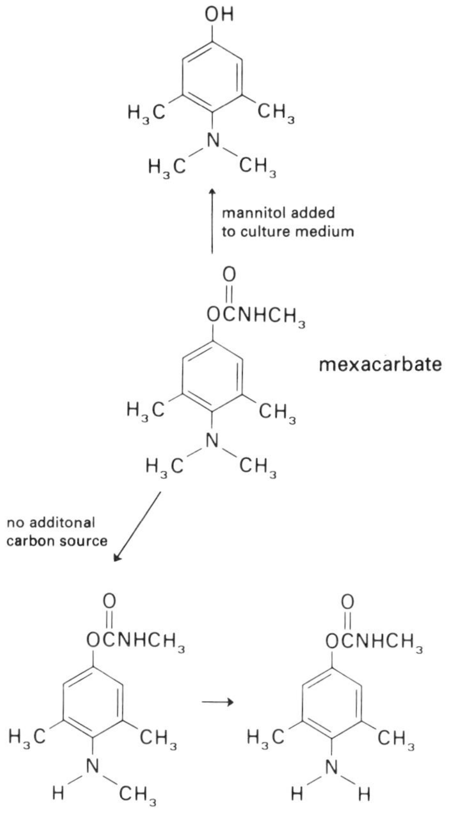

Fig. 1. Microbial metabolism of mexacarbate (Bezenet and Matsumura, 1974).

It was not tested whether mexacarbate could be toxic to some
microorganisms and so induce class A3 metabolism. However, it is
conceivable that the insecticide can become toxic at very high concentrations and that at such concentrations the survival of individuals (or
strains or species in a mixed colony) with detoxification capabilities is
favoured. Thus, it is at least theoretically possible that these three
different types of metabolism can take place in a given circumstance
within the same species.

The metabolic processes related to co-metabolism may require some explanation. It has been pointed out by Horvath and Alexander (1970) that in studies of chlorinated aromatic pesticides, it may be possible to select a microbial strain capable of degrading the pesticide by enriching the medium with a non-chlorinated analogue of the pesticide. By such an approach, even very stable and usually non-degradable pesticides might be made susceptible to microbial attack. In such cases, however, the metabolic ·pathways are not totally completed to mineralization (i.e., to yield CO_2, H_2O and HCl) and the processes themselves are seldom energy-yielding. The cause of such an incomplete metabolism can be that the products are either toxic or unacceptable as a substrate for the subsequent enzyme systems, despite the fact that at the initial step of metabolism, enzyme systems must have accepted these insecticides (unnatural substrates) as substitutes. A clear separation of co-metabolic reactions from other types of incidental metabolism is not always possible, since sometimes the corresponding natural substrates may be unknown. For all practical purposes, however, the term co-metabolism is used only for the cases where the natural substrates are well known, or have been artificially selected so that they are structurally close to the pesticidal molecules themselves. Generally, the substrate specificity in such cases appears to be rather narrow, for example, a *Hydrogenomonas* sp. selected for diphenylmethane can co-metabolize bis-(*p*-chlorophenyl) methane (DDM) but not DDT (1,1,1-trichloro-2,2-di-(4-chloro-phenyl)ethane) as such, thus facilitating the distinction of co-metabolic phenomena from others. More work, however, is needed to clarify the extent of co-metabolism of insecticidal materials.

B. NON-ENZYMATIC PROCESSES

The processes by which microbial activities contribute to the overall alteration of insecticidal molecules by non-enzymatic mechanisms are not well studied.

It is known that some pesticidal chemicals can be photochemically altered in the environment and there are two ways in which microbial products can promote photochemical reactions. Firstly, microbial products can act as photosensitizers by absorbing the energy from light and transmitting it to the insecticidal molecule. We have been able to show, for instance, that an aqueous extract from heat-sterilized blue–green algal cultures promoted photochemical degradation of DDT (Benezet and Matsumura, unpublished data). Another way that microbial products can facilitate such photochemical reactions is to serve as donors or acceptors, for example of hydrogen and OH⁻, which are often needed for photochemical reactions.

The effect of pH is often neglected in the field of pesticide metabolism despite numerous reports on the pH-dependent reactions of relatively labile molecules, both in soil and *in vitro*. Large pH changes are often associated with microbial activities together with the changes in nutritional sources, particularly in aqueous media. Initially degradation of proteins causes more alkaline pHs, then with carbohydrate metabolism the pH becomes acid. Martens (1972) observed, for instance, that at values above pH 7 hydrolysis of endosulfan was due mainly to chemical reaction. Ross and Tweedy (1973) noted that up to 50% of added chlordimeform was hydrolysed upon incubation for three weeks with a heat-sterilized soil originally containing a mixed population of microorganisms. While the actual occurrence of microbial pH effects in nature might be difficult to document, it is certainly easy to demonstrate the phenomenon *in vitro*, where during the decay period the pH of spent culture media often becomes very low. Incubation of labile insecticides such as TEPP (tetraethyl pyrophosphate) with such spent medium under sterile conditions would certainly cause breakdown of the insecticides.

Little attention has been paid so far to the importance of the microbial formation of organic products capable of reacting with pesticides. Such reactants of microbial origin can be postulated to include amino acids, peptides, alkylating agents such as acetyl CoA, methylcobalamine, S-adenosylmethionine and organic acids. Conceivably, organic amines and nucleic acids could also react with some insecticidal chemicals but no such incidence has been reported. Insecticidal chemicals are known to react with amino acids, particularly with an –SH moiety. As for alkylation, Cserjesi and Johnson (1972) reported finding methylated pentachlorophenol in a liquid culture of *Trichoderma virgatum*. Similarly, Tweedy *et al.* (1970) have demonstrated an acetylation product, *p*-bromoacetanilide, in the products of incubation of *Talaromyces wortmanii* and *Fusarium oxysporum* with metobromuron (3-(4-bromophenyl)-1-methoxy-1-methylurea). Presumably acetylation takes place on 4-bromoaniline, a proposed initial degradation product of metobromuron. Ross and Tweedy (1973) found 4'-chloro-2'-methylmalonanilic acid as the major microbial reaction product of chlordimeform, indicating that a conjugation reaction took place between malonic acid and 4-chloro-*o*-toluidine, the major hydrolysis product of chlordimeform. However, in none of the above cases was the proof for an actual non-enzymatic conjugation reaction offered. However, these types of compounds are the ones which have been shown to be capable of conjugating with reactive groups of chemicals. Production of such compounds by micro-organisms could conceivably facilitate the overall rate of microbial alteration of insecticidal chemicals, whether the subsequent conjugation mechanism is actually enzymatic or not.

Much less is known about the contribution of inorganic reactants to the alteration of insecticides or derivatives. Certainly various metallic and some non-metallic ions and gases, such as H_2S, O_2 and H_2, are known to react with organic chemicals, though some of the reactions require rather extreme conditions of, for example, light, heat and pressure. Walker and Stojanovic (1973) reported that most of the microbial metabolic products of malathion in an *Arthrobacter* sp. were in the form of potassium salts. Hydrogen sulphide is known to react with mercuric and other metallic ions which are also used as insecticides.

Finally, microbial production of co-factors which are used in both enzymatic and non-enzymatic reactions should not be overlooked. Co-factors are here defined as the organic molecules which promote the overall reactions involving an organic chemical without becoming a part of the reaction product derived from that chemical. For example, Miskus *et al.* (1965) found that dechlorination of DDT to form TDE (DDD) proceeded non-enzymatically in the presence of reduced porphyrins added to lake water. The presence of such porphyrin residues in natural lake water could certainly be traced to microbial activities, particularly to those of photosynthetic microorganisms. French and Hoopingarner (1970) studying DDT metabolism in *Escherichia coli* found that FADH acts as a co-factor for dechlorination of DDT. Other possible candidates are glutathione, NAD^+, $NADP^+$, NADH, NADPH and cytochromes.

IV. Common Processes of Insecticide Metabolism

A. HYDROLYTIC PROCESSES

As mentioned earlier, one of the insecticidal moieties most commonly present is the ester linkage. This includes all of the organophosphate and carbamate insecticides as well as pyrethroids. Hydrolysis of ester, halide, ether and amide bonds (Fig. 2) usually gives rise to non-toxic products.

There is a considerable body of evidence that hydrolytic activities are much more prevalent in the microbial world than in any other biological group. For instance, the normal microbial metabolic product of carbaryl (1-naphthyl methylcarbamate) is 1-naphthol

A. Ester hydrolysis

$$ROC(O)NHCH_3 \rightarrow ROH + HOC(O)NHCH_3$$
$$(R'O_2)P(O)OR'' \rightarrow (R'O)_2P(O)OH + R''OH$$
$$R'C(O)OR'' \rightarrow R'C(O)OH + R''OH$$

B. Ether hydrolysis

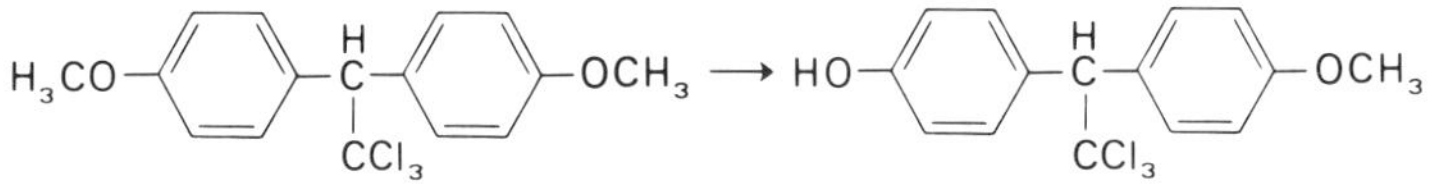

methoxychlor

C. Hydrolysis of labile halogens

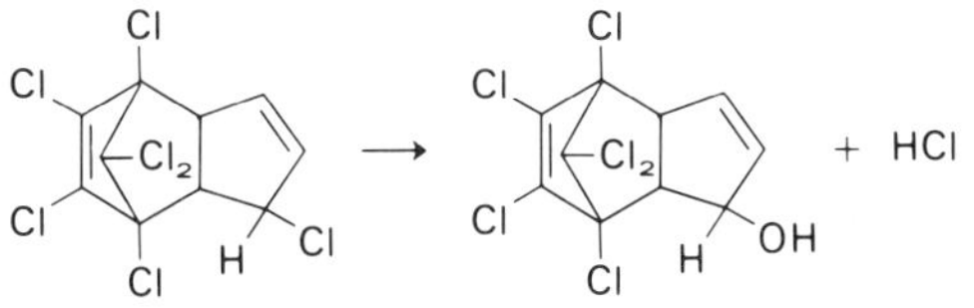

heptachlor

D. Amide hydrolysis

$$R'C(O)NR''R''' \rightarrow R'C(O)OH + HNR''R'''$$

Fig. 2. Examples of the most commonly observed hydrolytic processes of insecticides.

(Matsumura and Boush, 1968; Liu and Bollag, 1971a, b), in contrast to the mammalian metabolism of this insecticide (Dorough et al., 1963; Dorough and Casida, 1964; Leeling and Casida, 1966; Knaak and Sullivan, 1967), which mostly results in oxidation products. Similarly, in the case of mexacarbate, Benezet and Matsumura (1974) found that the major metabolic product in *Trichoderma viride* was the phenol of mexacarbate (i.e., the hydrolysis product) as compared with metabolism in animals which gave various oxidation products.

Even in the case of organophosphates, which are largely degraded via hydrolytic processes, this general trend is clearly observable. The major microbial degradation product of diazinon (diethyl 2-isopropyl-6-methyl-4-pyrimidinyl phosphorothionate) is 2-isopropyl-4-methyl-6-hydroxypyrimidine, a product of hydrolysis at the P–O bond (Laanio et al., 1972) and not the hydroxylation products or the glutathione-S-aryl transferase products which are prevalent in animal species. In fact, there have been almost no records of microbial activities to form P=O analogues from P=S compounds. Such a reaction is extremely common throughout the animal and the plant kingdom. Thus, in animals diazinon is expected to become diazoxon, which is actually the toxic principle of the insecticide owing to its high cholinesterase inhibitory property. The lack of such reports in the microbial world is indeed striking in view of the existence of a wide variety of biological systems which carry out such a reaction.

Perhaps the reason for such hydrolytic reactions being common in the microbial world is that many of the organisms excrete hydrolytic enzymes outside the cells (exoenzymes), particularly the fungi. Nearly all the exoenzymes liberated by microorganisms seem to be related to the metabolism of large molecules, so that such compounds may be reduced to smaller fragments to permit their passage through the cell membrane (Pollack, 1962). Various soils also contain exoenzymes which are hydrolytic in nature (Skujiņš, 1967). Thus by definition most of the hydrolytic reactions belong to the type A1(a) metabolism (Table 1), where incidental metabolism takes place by the action of broad spectrum enzymes.

Getzin and Rosefield (1968) found that hydrolysis of malathion in soil was hardly affected by a gamma radiation treatment (4 Mrad) that should destroy all viable microbial cells, whilst autoclaving, involving a high temperature that destroys both exoenzymes and micro-organisms, eliminated 90% of the malathion degradation activity of the soil. Molozhanova et al. (1973) found by using cultures of *Penicillium cyclopium*, *Aspergillus niger*, *Mucor mucedo*, *Trichoderma viride* and *Actinomyces griseus* that between one fourth and one half of the degradation of trichlorphon (dimethyl 2,2,2-trichloro-1-hydroxy-ethylphosphonate) was due to the culture supernatant, indicating the exoenzymatic origins of the degradation enzymes. The major microbial degradation process on trichlorphon is likely to be hydrolytic since Zayed et al. (1965) had already demonstrated the major metabolic products to be O-methyl-2,2,2-trichloro-1-hydroxyethyl phosphonic acid and 2,2,2-trichloro-1-hydroxyphosphonic acid by using cultures of *A. niger* and *Penicillium notatum* (Fig. 2).

B. REDUCTIVE SYSTEMS

The work which first showed that the major conversion product of parathion in microorganisms is aminoparathion (Cook, 1957) and not the products of oxidative reactions such as para-oxon diethyl-thiophosphoric acid, is now regarded as a classic demonstration of the basic difference between microbial and animal metabolism (Fig. 3). Similarly, both fenitrothion and EPN are converted to their corresponding amino-analogues by *Bacillus subtilis* (Miyamoto et al., 1966).

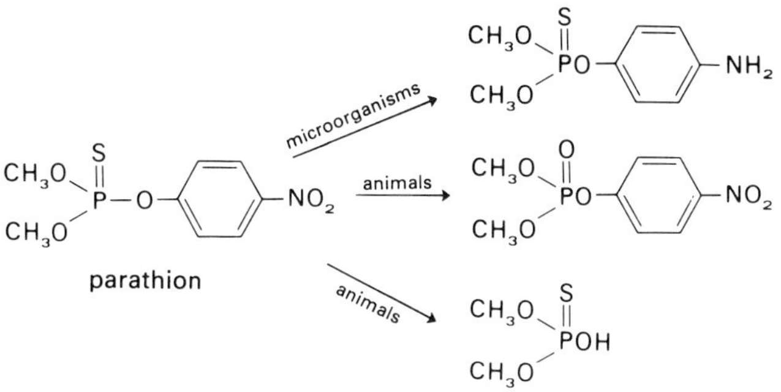

Fig. 3. Metabolism of parathion.

Another important microbial reaction on insecticidal chemicals is reductive dechlorination. The reaction proceeds by replacing a chlorine atom on a non-aromatic carbon with a hydrogen atom (Fig. 4), the best known case being the conversion of DDT to TDE (DDD).

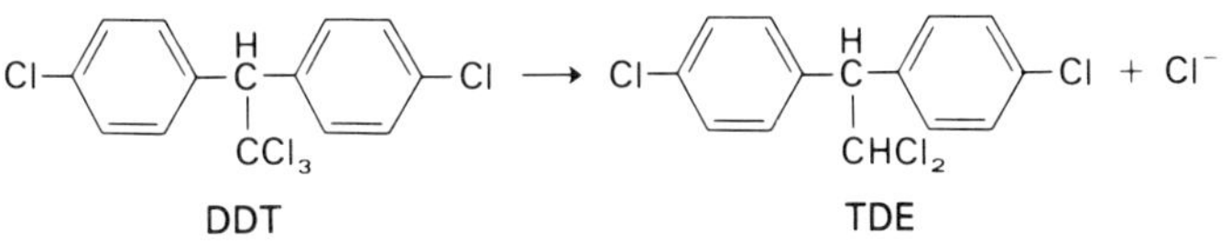

Fig. 4. Anaerobic metabolism of DDT by microorganisms.

French and Hoopingarner (1970) found that in *Escherichia coli* this reaction was stimulated by the presence of reduced FAD. The enzyme system was located in the cell membrane, but the presence of the cytoplasmic supernatant also increased the rate of dechlorination. The addition of TCA cycle intermediates, which shift the overall state of metabolism to more oxidative processes, resulted in the loss of dechlorination activities. Other insecticides which are known to go through such dechlorination reactions are gamma-BHC (Tsukano and Kobayashi, 1972; Benezet and Matsumura, 1973) and endrin (Matsumura et al., 1971).

Another type of reductive process was reported by Matsumura et al. (1968), who found a small amount of aldrin in the microbial incubation product of dieldrin, indicating a reversal of the common epoxidation reaction (Fig. 11).

C. OXIDATION

While the extent of oxidative metabolism in the microbial world is somewhat less than that found in other biological systems, there are several oxidative reactions that occur very widely. They are:

1. *Epoxidation of cyclodienes* such as aldrin and heptachlor to corresponding epoxides (see Section V.A.3.).

2. *Oxidation of thioethers* to sulphoxides and sulphones (see Section V.B.).

3. *Oxidative dealkylation of alkylamines* (see Section V.C.).

4. *Aromatic ring opening.*

One very important reaction that takes place only in the microbial world is the process of opening of the aromatic ring. The system is operated by a series of oxidative ring hydroxylation (including epoxidation) reactions. Ring hydroxylation can occur at a chlorine-attached aromatic carbon, as for example in 2,4-D (2,4-dichlorophenoxyacetic acid), in contrast to reductive dechlorination reactions on chlorinated hydrocarbons. The degree of such hydroxylation reactivity decreases drastically as the number of chlorine substitutions on an aromatic ring increases. Thus, 2,4,5-T (2,4,5-trichloroacetic acid) is almost always more persistent than 2,4-D, and polychlorinated biphenyls with fewer chlorines degrade faster than the highly chlorinated ones (Furukawa and Matsumura, unpublished data).

Focht and Alexander (1970a, b, c) selected a *Hydrogenomonas* sp. by using diphenyl methane (true metabolism) and found that the more chlorinated analogues such as bis-(*p*-chlorophenyl) methane (DDM) were less vigorously metabolized. The final product of DDM degradation was *p*-chlorophenylacetic acid, indicating the opening of one of the chlorinated aromatic rings. Similarly, Ahmed and Focht (1972) demonstrated a ring opening process in two species of *Achromobacter* which were originally selected by using nonchlorinated biphenyl, or *p*-chlorobiphenyl. The process of ring opening of *p*-chlorobiphenyl has been proposed by these workers (Fig. 5). *p*-Nitrophenol, a degradation product of parathion, can also be degraded through a ring opening reaction (Munnecke and Hsieh, 1974).

5. *Decarboxylation*

Decarboxylation reactions are common oxidative reactions. Miyazaki *et al.* (1969) for instance found that 4,4'-dichlorobenzilic acid (DBA), a hydrolysis product of both chlorobenzilate (ethyl 4,4'-dichloroben-

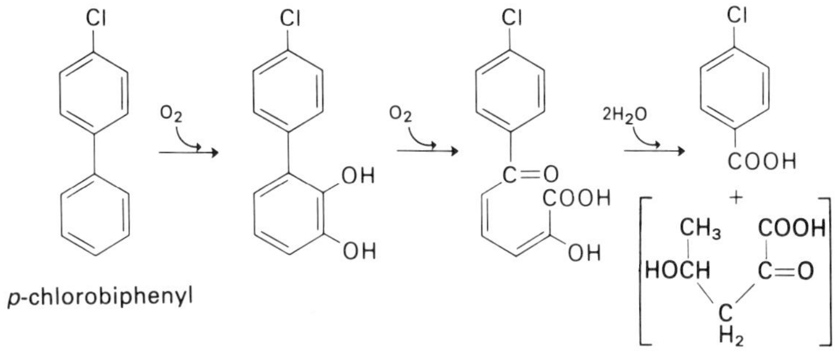

Fig. 5. Ring opening of p-chlorobiphenyl by two species of *Achromobacter* (Ahmed and Focht, 1972).

zilate) and chloropropylate (isopropyl 4,4'-dichlorobenzilate) gave rise to 4,4'-dichlorobenzophone (DBP) in a strain of the yeast *Rhodotorula gracilis* (Fig. 6). The process was stimulated when citric acid was added to the culture medium and inhibited by 2-ketoglutaric acid, indicating the necessity of promoting an oxidative activity in decarboxylating DBA.

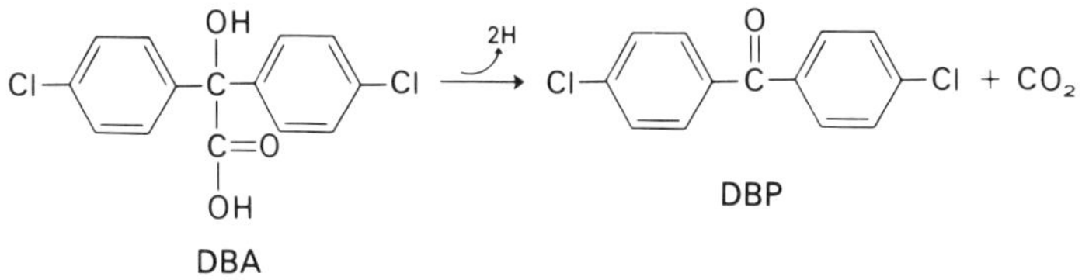

Fig. 6. Oxidative decarboxylation of 4,4'-dichlorobenzilic acid (DBA) to 4,4'-dichlorobenzophone (DBP) by *Rhodotorula gracilis* (Miyazaki *et al.*, 1969).

Other oxidative reactions include β-oxidation, conversion of alcohols and aldehydes to acids, and dehydrogenation, but they are much less frequently observed among metabolic activities on insecticides.

D. OTHER METABOLIC REACTIONS RELATED TO INSECTICIDE DEGRADATION

1. *Dehydrochlorination*

It is well known that DDT is dehydrochlorinated to give 1,1-dichloro-2,2-bis-(p-chlorophenyl) ethylene (DDE) (Fig. 7). The reaction involves an elimination of HCl, and is hence called dehydrochlorination. In higher animals the reaction is mediated by glutathione

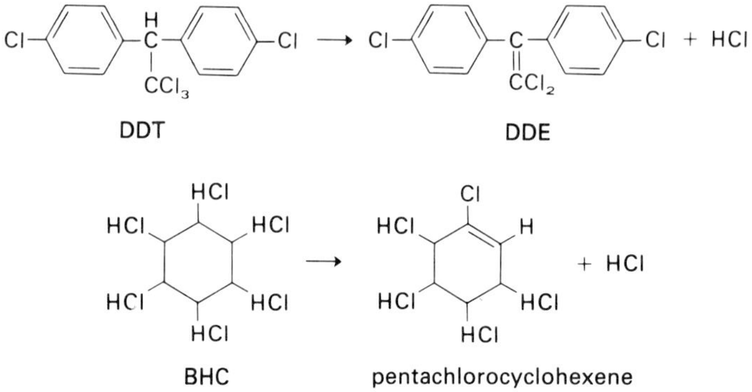

DDT → DDE + HCl

BHC → pentachlorocyclohexene + HCl

Fig. 7. Dehydrochlorination reactions.

(GSH), while the evidence is lacking in the microbial world as to whether GSH plays any significant role. Typically the reaction takes place between the saturated chlorinated carbon and the adjacent hydrogen on the neighbouring carbon, the net result being the appearance of olefinic compounds. Another example of such a reaction is that occurring on BHC to form pentachlorocyclohexene (Fig. 7).

2. *Isomerization*

It has been accidentally observed that various cyclodiene compounds go through isomerization reactions to yield more stable products. For example, Matsumura *et al.* (1970) observed that in several microbial species dieldrin is converted to photodieldrin by the formation of an

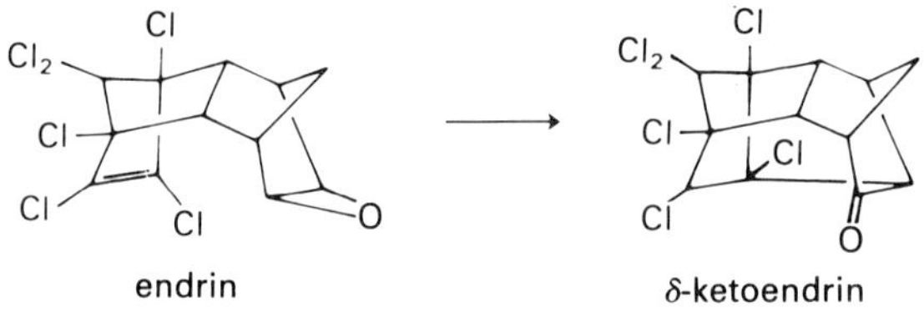

endrin → δ-ketoendrin

Fig. 8. Isomerization of endrin to delta-ketoendrin by several species of micro-organisms.

intramolecular bridge. Similarly δ-ketoendrin (Fig. 8) is formed from endrin (Matsumura *et al.*, 1971). It is not at all clear why micro-organisms go through such an isomerization reaction which does not yield any energy. However, from the viewpoint of environmental toxicology such reactions are very important since they give rise to other unsuspected environmental contaminants of which the toxicity and hazards are often unknown (Matsumura, 1974).

V. Metabolism of Individual Insecticides by Microorganisms

A. CHLORINATED HYDROCARBON INSECTICIDES

1. DDT (1,1,1-trichloro-2,2-di-(4-chlorophenyl) ethane)

DDT is one of the most extensively studied insecticides. The primary metabolic mechanism which has been studied is the reductive dechlorination of DDT (Fig. 9) with the formation of 1,1-dichloro-2,2-bis-(p-chlorophenyl)ethane (TDE or DDD). This mechanism was first described by Kallman and Andrews (1963) in yeast. Barker and Morrison (1964) detected TDE in the gut of mice fed DDT, which had been kept after death for a period of time. This degradation was later determined to be microbial and *Proteus vulgaris* was isolated (Barker and Morrison, 1965) which could degrade DDT mainly to TDE. Several minor metabolites including 1,1-bis-(p-chlorophenyl)-2-chloro-ethylene (DDMU) and bis-(p-chlorophenyl) methane (DDM), were also formed (Barker and Morrison, 1965; Barker et al., 1965). Numerous microorganisms have been isolated from animal and plant sources which could degrade DDT to TDE. These included *Escherichia coli* and *Aerobacter aerogenes* isolated from rat faeces (Mendel and Walton, 1966; Mendel et al., 1967), anaerobic bacteria from stable fly gut (Stenerson, 1965), rumen fluid microflora (Miskus et al., 1965), rat intestinal microflora (Braunberg and Beck, 1968), *E. coli* (Langlois, 1967) and plant pathogens (Johnson et al., 1967). TDE (DDD) was also found in wildlife samples and animal tissues (Finley and Pillmore, 1963).

The formation of TDE from DDT is also a common reaction among soil microorganisms (Guenzi and Beard, 1967). Chacko et al. (1966) isolated numerous actinomycetes (*Nocardia* sp., *Streptomyces aureofaciens*, *Streptomyces cinnamoneus*, *Streptomyces viridochromogenes*) from soil which readily degraded DDT to TDE. These organisms, however, required another carbon source to facilitate degradation. Soil fungi not only produce TDE, and small amounts of 2,2,2-trichloro-2,2-bis-(p-chlorophenyl) ethanol (dicofol), but some variants can produce 2,2-bis-(p-chlorophenyl) acetic acid (DDA), or 1,1-dichloro-2,2-bis-(p-chlorophenyl) ethylene (DDE) exclusively (Matsumura and Boush, 1968). A large number of microorganisms isolated from soil were able to degrade DDT (Patil et al., 1970). Thus, *Trichoderma viride*, nine *Pseudomonas* spp., a *Micrococcus* sp., three *Bacillus* spp., and an *Arthrobacter* sp., all converted DDT to TDE; some of these organisms also formed a dicofol-like metabolite and DDA (Fig. 9).

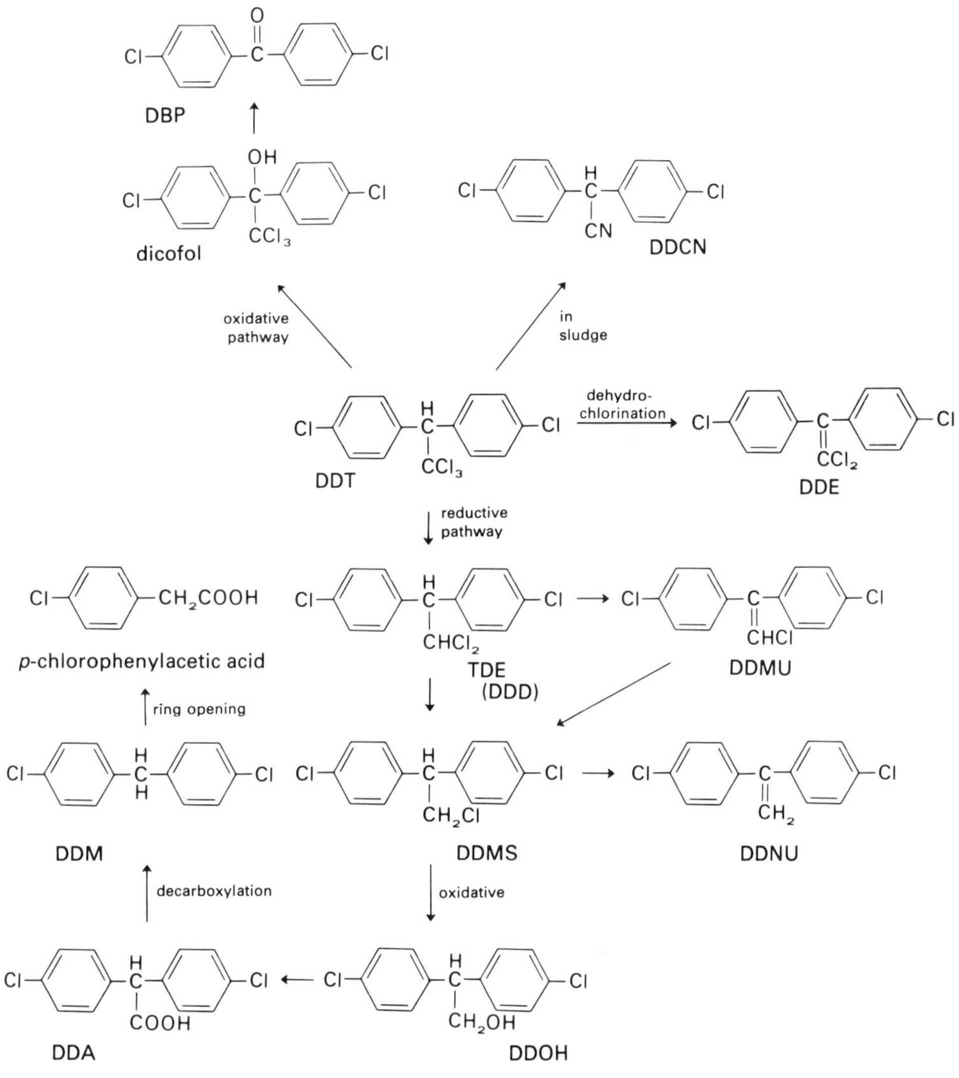

Fig. 9. Pathways of DDT metabolism in microorganisms.

Of environmental interest is the discovery by a number of researchers that the metabolism of DDT by microorganisms could be affected by nutrient additives. Johnsen *et al.* (1971) enhanced the bio-degradation ability of soil by adding cattle manure. DDT was readily degraded in amended soil to TDE, whereas in non-amended soil metabolism of the insecticide was slow. Glucose and diphenylmethane, a bio-degradable analogue of DDT, were both found to enhance metabolism of DDT to TDE, DDE and DBP (4,4'-dichloro-benzophenone) in sewage (Pfaender and Alexander, 1973) (Fig. 9).

Marine microorganisms also have the ability to degrade DDT. Patil *et al.* (1972) studied the degradation of DDT under oceanic conditions, and in pure cultures of micro-algae (*Dunaliella* sp. and *Agmenellum quadruplicatum*) and bacteria. The majority of the organisms showed an ability to degrade DDT mainly to TDE, with small amounts of 1-chloro-bis-(*p*-chlorophenyl) ethane (DDMS), 2,2-bis-(*p*-chlorophenyl) ethanol (DDOH) and DDA.

A strain of *Hydrogenomonas*, isolated by Focht and Alexander (1970a) has been extensively studied for its ability to degrade DDT. Although unable to utilize DDT as a sole carbon source, this microorganism was found to cleave one of the rings of bis-(*p*-chlorophenyl) methane (DDM), a product of DDT metabolism, to yield *p*-chloro-phenylacetate under aerobic conditions. It is of particular interest that a strain of the mould *Fusarium*, when it was grown with the *Hydro-genomonas*, showed the ability to degrade DDM to H_2O, CO_2 and HCl (Focht, 1972). Pfaender and Alexander (1972) studied the metabolism of DDT by cell extracts and whole cells of *Hydrogenomonas* sp. Under anaerobic conditions TDE, DDMS and DBP were formed and when whole cells and oxygen were subsequently added, *p*-chloro-phenylacetic acid was detected (Fig. 9). Later, Focht and Joseph (1974) isolated strains of *Pseudomonas, Acetomonas, Acinetobacter* and a *Cladosporium* sp. from a sewage culture by using 1,1-diphenylethylene as an enrichment agent. The compound was first hydrated to form 2,2-diphenylethanol and was then degraded through fission of one of the benzene rings.

Several metabolic studies of the microbial degradation of DDT have been undertaken. Wedemeyer (1966) was the first to implicate a particular enzyme system for the reductive dechlorination of DDT to TDE. In cell-free preparations of *Aerobacter aerogenes*, the use of selective metabolic inhibitors indicated reduced Fe(II) cytochrome oxidase as the enzyme system responsible for this degradation. French and Hoopingarner (1970) studied reductive dechlorination in the membranous fraction of *E. coli* cells and found that dechlorination was stimulated by components in the cytoplasm, but did not utilize electrons produced by oxidation of Krebs' cycle intermediates. The formation of TDE depends upon the reduction of FAD and occurs only under anaerobic conditions. Another important conversion product of DDT appears to be 1,1-dichloro-bis-(*p*-chlorophenyl) methyl cyanide (DDCN). Two schools, Albone *et al.* (1972) and Jensen *et al.* (1972), have simultaneously reported the discovery of *pp'*-DDCN in sewage sludges. The former workers incubated radiolabelled and non-labelled pp'-DDT at 37°C for 88 days in a vessel containing anaerobic sewage sludge from a water treatment plant. When the latter workers incubated DDT with sludge at 20°C for up to eight days under nitrogen atmosphere DDCN was second to TDE as a major metabolic product formed. In both cases involvement of microbial activity was implicated.

2. *BHC (Hexachlorocyclohexane, HCH)*

BHC disappears rapidly from the soil environment. At least part of this disappearance can be attributed to microbial activity (Bradbury, 1963). Raghu and MacRae (1966) found that γ-BHC disappeared faster after a second application, implying an "evolved" microbial degradation. It is generally agreed that the bio-degradation of BHC is greatest in submerged soil and under anaerobic conditions (MacRae *et al.*, 1967). Newland *et al.* (1969) found that the highest percentage of water-soluble metabolites of γ-BHC were formed in artificial lake impoundments under anaerobic conditions.

Several metabolites of γ-BHC have been detected under natural conditions. Generally, 1,3,4,5,6-pentachlorocyclohex-1-ene (γ-PCCH) is considered to be the major metabolic product (Yule *et al.*, 1967), although several other metabolites have also been identified. Tsukano and Kobayashi (1972) detected and identified 3,4,5,6-γ-tetrachloro-cyclohex-1-ene (γ-BTC) as a metabolite of γ-BHC in submerged rice field soil (Fig. 10). Benezet and Matsumura (1973) also identified γ-BTC, together with α-BHC, as metabolites of γ-BHC in semi-natural conditions in Pearl Harbour, Hawaii.

Direct evidence of microbial degradation of BHC was first obtained by Allen (1955) who found that *Clostridium sporogenes* and *Bacillus coli* both converted BHC to 1,2,3,5-tetrachlorobenzene. Several *Clostridium* spp. degraded γ-BHC under anaerobic conditions (MacRae *et al.*, 1969) and Sethunathan *et al.* (1969) found an unknown metabolite formed by a *Clostridium* sp. This metabolite now seems to be γ-BTC, which was later identified by other researchers (Tsukano and Kobayashi, 1972). The *Clostridium* sp. degraded only a minute amount of γ-BHC to CO_2 (Sethunathan and Yoshida, 1973a).

Very few metabolic studies are documented for the microbial degradation of γ-BHC. Data from our laboratory (Benezet and Matsumura, 1973) indicate that in *Pseudomonas putida*, NAD is required for the conversion of γ-BHC to γ-BTC and α-BHC. Additional unpublished data from our laboratory also indicate that in *Ps. putida* reduced FAD promotes complete degradation of the γ-BTC ring to CO_2 (Fig. 10).

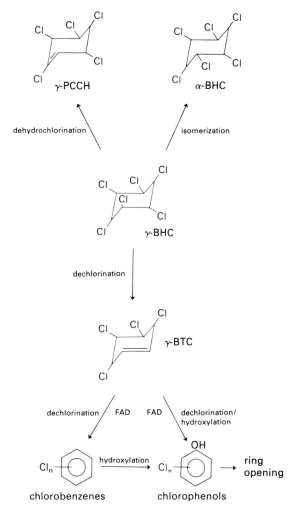

Fig. 10. Microbial degradation of γ-BHC.

Two species of algae have been studied for their ability to degrade γ-BHC. *Chlorella vulgaris* and *Chlamydomonas reinhardtii* both convert γ-BHC to γ-PCCH (Sweeney, 1969).

3. Chlorinated cyclodienes

(a) *Dieldrin and aldrin* (see Table 2). The oxidative metabolism in soil leading to the formation of an epoxy-ring from the unsaturated carbon–carbon bond of the unchlorinated ring of aldrin was first described by Kiigemagi *et al.* (1958). Soil microorganisms have been implicated in this epoxidation process (Lichtenstein and Schulz, 1960). The first direct evidence that microorganisms have the ability to convert aldrin to the epoxide, dieldrin, was given by Korte *et al.* (1962) who found that cultures of *Aspergillus niger, Aspergillus flavus, Penicillium notatum* and *Penicillium chrysogenum* converted aldrin to dieldrin and four other metabolites. Tu *et al.* (1968) isolated a number of soil microorganisms (*Trichoderma, Fusarium, Penicillium, Aspergillus, Nocardia, Storestomyces* and *Micromonospora*) which rapidly converted aldrin to dieldrin, although it was also found that aldrin disappeared from some of the cultures without the formation of dieldrin. Patil *et al.* (1970) isolated 13 microorganisms, including *Trichoderma viride, Pseudomonas* spp., a *Micrococcus* sp. and a *Bacillus* sp., which had the ability to degrade aldrin. One of the metabolites was identified as 6,7-*trans*-dihydroxydihydroaldrin (*trans*-aldrindiol) (Fig. 11). An alga, *Dunaliella* sp., also degraded aldrin, converting it to dieldrin and *trans*-aldrindiol (Patil *et al.*, 1972).

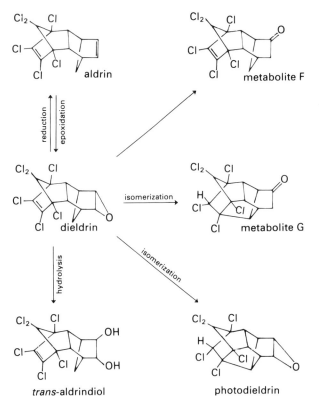

Fig. 11. Pathways of dieldrin and aldrin transformations by microorganisms.

Dieldrin is one of the more stable chlorinated hydrocarbons. Matsumura and Boush (1967) screened approximately 600 microbial isolates and found only ten strains capable of degrading dieldrin. A number of metabolites were detected, including one identified as an aldrindiol. Other researchers have found that aldrindiol was the only metabolite formed by *Aerobacter aerogenes* (Wedemeyer, 1968) and *Pseudomonas melophthora* (Boush and Matsumura, 1967). Several species of *Trichoderma, Fusarium* and *Aspergillus* were capable of degrading dieldrin (Tu *et al.*, 1968), whilst the mould *Mucor alternans* showed 26% breakdown of dieldrin in two days (Anderson *et al.*, 1970).

An extensive study of dieldrin metabolism by a *Pseudomonas* sp. (Shell 33) has been reported by Matsumura *et al.* (1968). This organism produced 6-keto derivatives as major metabolites, an alcohol, an aldehyde and an acid. Later studies (Matsumura *et al.*, 1970) also identified photodieldrin (Fig. 11) as a bacterial metabolite. An interesting aspect of dieldrin metabolism was reported by Bixby *et al.* (1971). A soil fungus, *Trichoderma koningii*, cleaved the rings of dieldrin, liberating some of the carbons as CO_2. More recently, Vockel and Korte (1974) studied microbial degradation of (^{14}C) dieldrin in *Pseudomonas marginalis, Ps. fluorescens, Ps. syringae, Ps. morsprunorum, Ps. glycinea* and *Neurospora* sp., all of which were found in a soil sample contaminated with aldrin and other organochlorine compounds. They found photodieldrin, dihydroaldrin-6,7-*trans*-diol (*trans*-aldrindiol) and a trace of 6-ketoaldrin (metabolite G, Fig. 11) among the possible dieldrin metabolic products of these microorganisms.

Several algal cultures, including a *Dunaliella* sp., had the ability to metabolize dieldrin (Patil *et al.*, 1972). The metabolites were photodieldrin, *trans*-aldrindiol, and several unidentified compounds, although in this particular instance the role of light in producing these products by photochemical reactions must be considered along with the true biological degradation ability of the algae.

(b) *Endrin, heptachlor and chlordane* (Table 2). Patil *et al.* (1970) isolated 20 microorganisms, including *Trichoderma viride, Pseudomonas* spp., a *Micrococcus* sp., an *Arthrobacter* sp. and *Bacillus* spp., which degraded endrin. Ketoendrin was the only metabolite identified. Matsumura *et al.* (1971) reported at least seven metabolites of endrin, which in addition to the ketoendrin included ketones and aldehydes with five and six chlorine atoms. An unidentified metabolite and ketoendrin were formed by algae (Patil *et al.*, 1972).

The first evidence that heptachlor is degraded by microorganisms was indicated when less of the epoxide was recovered from sterilized than from unsterilized soil treated with heptachlor (Lichtenstein and Schulz, 1960). The β-dihydroheptachlor was very slowly degraded by *Aspergillus niger* and *Penicillium urticae* (Poonawalla and Korte, 1968). Miles *et al.* (1969) studied the degradation of heptachlor by 92 microbial isolates from soil. Bacteria oxidized heptachlor to heptachlor epoxide, followed by hydrolysis in aqueous media to 1-*exo*-hydroxyheptachlor epoxide. Both fungi and bacteria oxidized heptachlor, chlordene and 1-*exo*-hydroxychlordene to the respective epoxides. The same researchers (Miles *et al.*, 1971) have studied the degradation of heptachlor and its epoxide by a mixed culture of soil microorganisms. Heptachlor and the epoxide were both readily degraded to chlordene and 1-*exo*-hydroxychlordene respectively. 1-*exo*-hydroxychlordene and its epoxide were both stable to microbial action. Bourquin *et al.* (1972) added technical heptachlor to mixed cultures of microorganisms isolated from a cutting fluid of the Texaco Co. via pesticide enrichment methods. They observed rapid formation of 1-hydroxychlordene. They also identified chlordene, 1-hydroxy-2,3-epoxychlordene, and heptachlor epoxide among the metabolic products. Iyengar and Rao (1973) studied the rate of disappearance of chlordane and heptachlor using cultures of *Aspergillus niger*. Neither compound could serve as a sole carbon source for this organism, although up to $12 \cdot 5 \ \mu g \ ml^{-1}$ of chlordane and heptachlor were degraded in 48 hours. Chlordane-adapted *A. niger* could utilize chlordane, heptachlor and aldrin, but not BHC or DDT.

Weisgerber *et al.* (1974), on the other hand, studied the metabolism of (^{14}C) heptachlor in soil. They found hydroxydihydroheptachlor, 1-methoxychlordene, 1-hydroxychlordene and heptachlor epoxide as the metabolic products in the soil.

4. *Miscellaneous chlorinated hydrocarbons* (see Table 2)

(a) *Mirex.* Very little has been reported on the microbial degradation of this highly persistent compound. However, Andrade and Wheeler (1974) have shown the degradation of mirex by sewage sludge microorganisms under anaerobic, but not aerobic, conditions.

(b) *Methoxychlor.* Anaerobic cultures of *Aerobacter aerogenes* produced 1,1-dichloro-2,2-bis-(*p*-methoxyphenyl) ethylene and 1,1-dichloro-2,2-bis-(*p*-methoxyphenyl) ethane from methoxychlor (Mendel *et al.*, 1967).

(c) *Endosulfan*. Martens (1972) studied endosulfan degradation by soil microorganisms and concluded that fungi predominantly produce endosulphate, while the major metabolic product in bacteria is endosulfan diol. At pH values above 7 chemical hydrolysis predominated, while at pH values below 7 microbial degradation became significant. Other metabolic products identified were endoether, endohydroxyether, chlorendic acid and endolactate. Actinomycetes, in particular, appear to produce several of these metabolites in about equal quantities. El-Zorgani and Omer (1974) incubated α- and β-endosulfan with a culture of *A. niger* and found the major metabolic product to be endosulfan diol, in contrast to the finding of Martens (1972). Approximately 30% of the endosulfan was converted to this metabolic product from both endosulfan isomers under the experimental conditions used by El-Zorgani and Omer (1974).

B. ORGANOPHOSPHORUS INSECTICIDES

1. *Parathion (diethyl 4-nitrophenyl phosphorothionate)*

This compound presents very little in the way of residue problems in the environment. It is readily degraded in soil, mainly by hydrolytic and oxidative means. Miller *et al.* (1967) suggested that the microflora of flooded cranberry bog was responsible for the breakdown of parathion to aminoparathion (Fig. 12) and two unidentified metabolites. Lichtenstein and Schulz (1964) found that degradation of parathion in

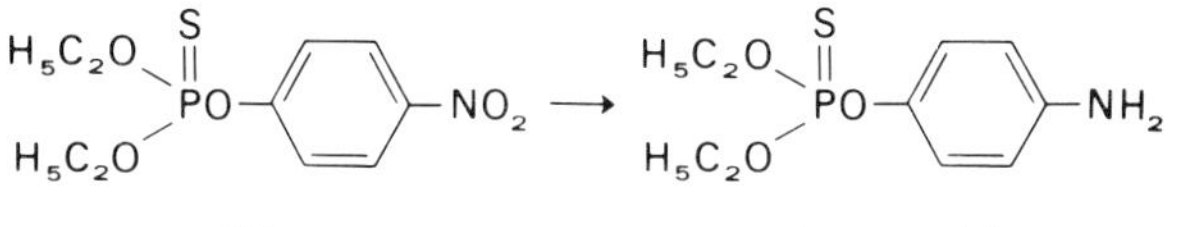

parathion   aminoparathion

Fig. 12. Reductive metabolism of parathion.

soil was by hydrolysis and reduction, the major end products being aminoparathion, *p*-nitrophenol and *p*-aminophenol. Parathion was relatively persistent in autoclaved soil, thereby implicating microbial degradation. The microbial population in natural lake sediment was implicated in parathion degradation by Graetz *et al.* (1970) who found that aminoparathion was the major metabolite in aerobic and anaerobic cultures. The hydrolysis of parathion in river water may also be biological in nature (Eichelberger and Lichtenberg, 1971). The persistence of parathion in submerged rice soil was studied by Sethunathan and Yoshida (1973a, c), where parathion was reduced to amino-parathion.

The metabolism of parathion by specific microorganisms is quite extensively recorded. Yasuno *et al.* (1965) reported that *Bacillus subtilis*, isolated from polluted water, reduced parathion to amino-parathion. Boush and Matsumura (1967) isolated *Pseudomonas melo-phthora*, a bacterial symbiote of the apple maggot, which had the ability to degrade a number of insecticides including parathion. The same researchers (Matsumura and Boush, 1968) reported that *T. viride* readily degraded parathion through its oxidative system. An extensive study by Hirakoso (1968) on the inactivating effects of microorganisms on parathion revealed several bacterial species which formed aminoparathion from parathion: *Bacillus subtilis*, *B. mega-therium*, *B. cereus*, *B. macerans*, *Pseudomonas aeruginosa*, *Ps. ovalis*, *Ps. aureofaciens*, *Alcaligenes viscolactis*, *Escherichia freundii*, *E. coli*, *Ser-ratia plymuthica* and *Achromobacter eurydice*. *Rhizobium japonicum* and *Rhizobium meliloti* metabolized parathion primarily to aminoparathion although 10% was hydrolysed to O,O-diethyl thiophosphoric acid (Mick and Dahm, 1970). Reduced glutathione enhanced the hydro-lysis reaction, whereas NADPH was without effect. Cell-free extracts of a *Flavobacterium* sp. hydrolysed parathion at the P—O—C nitro phenyl linkage to form *p*-nitrophenol, which was the end product in this study (Sethunathan and Yoshida, 1972). Parathion was utilized as the sole carbon source by this organism and it is probable that the parathion metabolism proceeds past *p*-nitrophenol via a ring opening in this species, as in the case reported by Munnecke and Hsieh (1974). Rao and Sethunathan (1974) used an enrichment culture technique to isolate *Penicillium waksmani* from a flooded acid sulphate soil and found that parathion was initially converted to aminoparathion. The culture containing parathion later showed signs of vigorous growth, with a concurrent increase in the amounts of polar metabolites pro-duced. Sethunathan and Yoshida (1973b) also obtained another microbial strain by a similar enrichment technique. This organism, a *Flavobacterium* sp., converted parathion directly to *p*-nitrophenol and the hydrolytic enzyme appeared to be constitutive.

Some algae also have the ability to degrade parathion. Sato and Kubo (1964) reported that algae accelerated degradation in the rice paddy fields. Parathion disappeared at the rate of 65% in eight days in cultures of *Chlorella pyrenoidosa* (Ahmed and Casida, 1958). An interesting result was reported by Mackiewicz *et al.* (1969). Bean plants grown under aseptic conditions did not metabolize parathion whereas added algae (*Chlorella* sp.) readily converted parathion to aminoparathion and an unidentified metabolite. An extensive study was made on the degradation of parathion by the alga *Chlorella pyrenoidosa* (Zuckerman *et al.*, 1970). Four metabolites were formed, one of which was aminoparathion. Two of the remaining metabolites were toxic, both compounds containing phenyl rings, but only one contained sulphur, whilst the remaining non-toxic metabolite lacked a phenyl ring.

2. *Diazinon (diethyl 2-isopropyl-6-methyl-4-pyrimidinyl phosphorothionate)*

The first indication that diazinon was degraded by microorganisms was given by Gunner *et al.* (1966), who found increases in the actinomycete population in soil treated with diazinon. Getzin (1967) showed indirectly that soil microorganisms degraded ring-labelled diazinon to $^{14}CO_2$, with 35% of the labelled ring carbon released as CO_2 in 20 weeks. No appreciable release of $^{14}CO_2$ occurred in soil fumigated with propylene oxide. Microorganisms also apparently readily degraded the product of hydrolysis of diazinon, 2-isopropyl-6-methyl-4-hydroxy pyrimidine (IMHP). The hydrolysis of diazinon has been shown to be greatly facilitated by the presence of microorganisms isolated from diazinon-treated soil (Trela *et al.*, 1968).

Sethunathan and co-workers have extensively studied the degradation of diazinon in rice paddy water and soil, and by microorganisms isolated from these locations. In submerged soil, microorganisms played an important role in the degradation of diazinon to the less toxic hydrolysis product IMHP (Sethunathan and Yoshida, 1969), whilst degradation of diazinon was slower in sterilized soil. Microorganisms were not important in the hydrolysis of diazinon in acidic soils (Sethunathan and MacRae, 1969). Repeated application of diazinon to soil rice paddies increased the ability of the soil to degrade this compound (Sethunathan and Pathak, 1972). Diazinon was rapidly hydrolysed to IMHP (75 h) followed by complete degradation to CO_2 in a further 25 h in this soil, apparently by microorganisms. *Streptomyces* sp. isolated from paddy soil could degrade diazinon aerobically in the presence of glucose (Sethunathan and MacRae, 1969). Two bacteria, *Arthrobacter* sp. (Sethunathan and Pathak, 1971) and *Flavobacterium* sp. (Sethunathan, 1972) also had the ability to degrade diazinon. The *Flavobacterium* sp. utilized diazinon as sole carbon source in a mineral medium, evolving CO_2 (Sethunathan and Yoshida, 1973b). The first step of the metabolism was found to be hydrolytic, yielding 2-isopropyl-6-methyl-4-hydroxy pyrimidine.

Several other workers have studied the degradation of diazinon by specific microorganisms. Boush and Matsumura (1967) found that

Pseudomonas melophthora produced 27% water- and 2% solvent-soluble metabolites from diazinon after 24 h; whilst in similar studies the soil fungus *Trichoderma viride* degraded diazinon (Matsumura and Boush, 1968). An *Arthrobacter* sp. and a *Streptomyces* sp., when incubated separately with diazinon, did not produce any metabolites (Gunner and Zuckerman, 1968). However, when both were incubated together diazinon was rapidly metabolized (84% in seven days).

3. *Malathion (S-(1,2-di(ethoxycarbonyl)ethyl) dimethyl phosphorothiolothionate)*

The degradation of malathion in an aqueous system proceeded slowly for the first seven days, and very rapidly thereafter (Konrad *et al.*, 1969). It was suggested that microbial action was responsible. Walker and Stojanovic (1973) studied the biological and chemical degradation of malathion in several types of soil. Degradation was more rapid under non-sterile than under sterile conditions, indicating microbial involvement, and was directly related to the amount of organic matter in the soil. These authors found malathion mono-carboxylic acid, dicarboxylic acid, potassium dimethylphosphorothioate, potassium dimethylphosphorodithioate and *O*-desmethyl malathion potassium salt among the incubation products. They concluded that all the products except the last mentioned compound were due to metabolic activities of an *Arthrobacter* sp.

Two microorganisms, *T. viride* and a *Pseudomonas* sp., having the ability to degrade malathion were isolated by Matsumura and Boush (1966). Carboxylic acid derivatives of malathion constituted the major metabolites. This would suggest the presence of carboxylesterases in microorganisms. They were indeed able to extract such an enzyme by an acetone-powdering technique and demonstrated its activity *in vitro*. Tiedje and Alexander (1967) reported that soil pseudomonads convert malathion to the monoacid. Chen *et al.* (1969) suggested that only the α-monoacid of malathion was produced in a biological system.

4. *Other organophosphorus insecticides*

Several other phosphate insecticides (Table 2) have had cursory examination in relation to microbial degradation. Probably the first indication that these insecticides were degraded by microorganisms came from Ahmed and Casida (1958) who found that *Pseudomonas fluorescens*, *Chlorella pyrenoidosa*, *Torulopsis utilis* and *Thiobacillus thiooxidans* were capable of degrading phorate and dialkyl phenyl phosphates. *T. thiooxidans*, a sulphur-oxidizing bacterium, had the least effect on the oxidation of the thioether bond of phorate, indicating that phorate is not accepted as an alternative for natural sulphides by this autotrophic organism.

The implication that microorganisms degrade carbophenothion (as Trithion) in loamy soils was made by Menn *et al.* (1960). Autoclaving and fumigating the soils with sodium *n*-methyl dithiocarbamate significantly increased the persistence of carbophenothion. Similar results were found for phosmet (as Imidan) with respect to stability in autoclaved soil (Menn *et al.*, 1965). Lichtenstein and Schulz (1964) reported that methyl parathion was degraded faster in non-autoclaved soil and in soils amended with glucose.

Boush and Matsumura (1967) found that both dichlorvos and DFP were degraded by *Pseudomonas melophthora*. Only 8% of dichlorvos and 33% of DFP remained after 24 h in the presence of this microorganism.

Hirakoso (1968) studied the effect of 32 bacterial isolates on chlorpyrifos (as Dursban), fenitrothion, dichlorvos and fenthion. Chlorpyrifos was stable in the media containing all the species of bacteria tested, but loss of activity of fenitrothion, dichlorvos and fenthion was caused by four, seven and one bacterial species, respectively.

Flashinski and Lichtenstein (1974) studied metabolism of (ethoxy-^{14}C) and (ring-^{14}C)-fonofos (as Dyfonate) by soil fungi. They found that *Mucor plumbeus* and *Rhizopus arrhizus* were most active in degrading this insecticide. The metabolic products identified were: dyfoxon, ethylethoxyphosphonothioic acid, ethylethoxyphosphonic acid, methyl phenyl sulphoxide and methyl phenyl sulphone. The hexane extracts from cultures of Dyfonate-degrading fungi were non-toxic to fruit flies, probably indicating further degradation of toxic intermediates, whereas extracts from fungi only capable of slowly metabolizing the insecticide were toxic to the same bioassay organism.

C. CARBAMATE INSECTICIDES

There have been relatively few studies on the microbial degradation of this type of compound. Boush and Matsumura (1967) studied the degradation of carbaryl (1-naphthyl methylcarbamate) by *Pseudomonas melophthora* and found that the major metabolite was 1-naphthol (46%), with a small amount of a polar product (6%). In a subsequent study Matsumura and Boush (1968) showed that two soil microorganisms, *T. viride* and a *Pseudomonas* sp., could also degrade carbaryl. Microbial isolates from soil were able to accelerate the hydrolysis of carbaryl to 1-naphthol (Bollag and Liu, 1971).

Liu and Bollag (1971a) isolated a fungus, *Gliocladium roseum*, from soil which had been treated with carbaryl for four weeks. This fungus was found to metabolize carbaryl to 1-naphthyl *N*-hydroxymethylcarbamate, 4- and 5-hydroxy-1-naphthyl methylcarbamate. Liu and Bollag (1971b) further demonstrated that *N*-desmethylcarbaryl was formed together with *N*-hydroxymethylcarbaryl in *Aspergillus terreus*, and, to a lesser extent, in *Aspergillus flavus*. The same workers (Bollag and Liu, 1972) also tested 18 fungal isolates for their ability to degrade carbaryl. They reported that 13 of the isolates produced the *N*-hydroxymethyl analogue and 11 of them also produced 4- and 5-hydroxy-1-naphthyl methylcarbamate as metabolic products. It must be mentioned that these workers administered a relatively high dose of carbaryl in an aerobic culture condition.

Bux insecticide, a mixture of *m*-(1-methylbutyl)-phenyl methylcarbamate and *m*-(1-ethylpropyl)-phenyl methylcarbamate, was readily degraded by soil organisms at the carbamic ester linkage (Tucker and Pack, 1972). There was a significant difference in the metabolism of the carbamate in sterile and non-sterile soil samples. Two metabolites were identified: CO_2 derived from the carbonyl group and *m*-(1-hydroxy-1-methylbutyl) phenylcarbamate from the one isomer. These studies were performed on silt soil from Iowa and silt loam soil from Nebraska.

Bacteria, moulds and fungi were screened by Benezet and Matsumura (1974) for their ability to degrade mexacarbate (4-dimethylamino 3,5-xylyl methylcarbamate). All the microorganisms tested degraded this carbamate into one or more of the following products: 4-dimethylamino 3,5-xylenol, 4-methylamino 3,5-xylyl-methylcarbamate, 4-formamido 3,4-xylyl-methylcarbamate, 4-amino 3,4-xylyl-methylcarbamate and several unidentified metabolites. Modification of the metabolism of the carbamate was possible by the addition of co-factors, inhibitors, and by varying the carbon source available to the microorganisms.

D. MISCELLANEOUS INSECTICIDES

A relatively new group of insecticides, used previously as acaricides, are the formamidines. Chlordimeform (*N*-(4-chloro-*o*-tolyl)-*N*,*N* dimethylformamidine) is readily degraded by several soil microorganisms (Johnson and Knowles, 1970). The major metabolites were *N*-formyl-4-chloro-*o*-toluidine and 4-chloro-*o*-toluidine 48 hours after addition of chlordimeform. Other workers (Ross and Tweedy, 1973) have identified a malonic acid conjugate 4'-chloro-2'-methyl-malonanilic acid, formed after three weeks dark incubation with soil added to a synthetic medium. This compound accounted for 22% of the total added radioactivity and appears to be of microbial origin, since only the hydrolysis product (50% of the total) was recovered from incubation mixtures with sterile soil. Such a novel conjugation mechanism has not been reported before, and its significance is of great interest.

Botanicals were probably among the first organic insecticides used by man. The microbial degradation of the nicotine alkaloids has been studied by a number of researchers. Initial decomposition of nicotine, 1-methyl-2-(3-pyridyl)pyrrolidine, by microorganisms involves the conversion to N-methyl-3(2-pyrrye)pyridinium hydroxide (Choman *et al.*, 1954). Wada and Yamasaki (1953, 1954) detected both pseudooxynicotine, and γ-keto-γ-(3-pyridyl) butyric acid as metabolites of nicotine by *Pseudomonas* sp. Nicotine incubated with *Pseudomonas nicotinophaga* gave rise to both 3-succinoyl pyridine and 3-succinoyl-6-hydroxypyridine (Frankenburg and Vaitekunas, 1955). This degradation initially proceeds through N-methylmyosmine with direct decomposition of the pyridine nucleus (Tabuchi, 1955a, b). The pseudomonads utilized this alkaloid as a carbon and nitrogen source converting it to aliphatic compounds, oxalic and succinic acids, and ammonia. γ-(3-Pyridyl)-methylamino butyric acid is believed to be an intermediate of nicotine metabolism by *Pseudomonas*. A metabolite of nicotine formed by the soil bacterium *Arthrobacter oxydans* was identified as 1,6-dihydroxynicotine (Hochstein and Rittenburg, 1959a, b). Later studies identified 6-hydroxypseudooxynicotine as another metabolite (Hochstein and Rittenburg, 1960). Wada (1957) studied the metabolism of two other alkaloids found in tobacco, nornicotine and anabasine. Nornicotine, 2-(3-pyridyl) pyrrolidine, was converted by soil isolates to 3-succinoyl, 6-hydroxy pyridine and 6-hydroxymyosmine after 3–4 days. Anabasine was converted to 3-glutaroyl-6-hydroxypyridine and 1′,6′-dihydro-6-hydroxyanabasine.

VI. Conclusions

Certain key reactions are commonly employed by microorganisms in metabolism of insecticides. These include reductive dechlorination, dehydrochlorination, aromatic nitro-reduction, hydrolysis of esters, amides and ethers, and various oxidation reactions such as hydroxylation, epoxidation and decarboxylation.

An interesting observation is that most of these reactions are not really energy-yielding processes and that microorganisms themselves seldom derive any immediate benefit from such metabolic activities. This fact, coupled with the generally low levels of insecticides present in the environment, led us to conclude that most of the microbial metabolism is executed in a fashion such that the reactions are incidental to the energy production of the corresponding microorganism. In this paper, a term "incidental metabolism" has been coined to describe these reactions. Naturally, in the laboratory one could utilize relatively high levels of labile insecticidal substrates and adjust incubation conditions to select microorganisms which can grow on insecticides as sole carbon sources. Such metabolic activities could conceivably occur in nature, where a large quantity of insecticidal substrates are available.

Among the group-specific reactions, those concerned with chlorinated insecticides appear to be most significant from the standpoint of pollution. It may not be fortuitous that the most stable insecticides against microbial attack are highly chlorinated hydrocarbons and aromatics, since such molecules are not found among the natural microbial nutrients. The processes of dechlorination must be considered to be very important rate-limiting steps. Basically there are four major types of metabolic reactions on the chlorinated insecticides; reductive dechlorination, dehydrochlorination, oxidative hydroxylation to replace chlorine atoms, and hydrolytic reaction on labile chlorines such as occurs on heptachlor. The problem is that none of the above reactions have been really carefully studied. For instance, we know that γ-BHC is microbiologically converted to γ-PCCH, and DDT to DDE, through dehydrochlorination activities, but no enzymic descriptions of such reactions have been made. Another example is the dechlorination reaction. The importance of this system is apparent. For instance, DDT does not serve as a substrate for a ring-opening system for the *Hydrogenomonas* sp. unless it is first reductively dechlorinated to form DDM under anaerobic conditions, showing clearly that the dechlorination process is the limiting step. While there have been some studies on the mechanisms of dechlorination of DDT (particularly the important work by French and Hoopingarner, 1970) and BHC, they are not sufficient to indicate the significance or substrate spectrum. For example, we do not even know whether such systems can dechlorinate aromatics. Without such a system one must assume that most of the highly chlorinated members of the polychlorinated biphenyls (PCBs) are going to remain in the environment (Furukawa and Matsumura, unpublished data).

There has been some confusion with regard to the term metabolism. Some people, notably microbiologists, tend to use it in a strict sense to mean the complete energy-yielding reactions, while others such as pesticide scientists interpret it as any reaction which changes the original molecule. The effort we have made in this chapter (Table 1) hopefully clarifies the situation. Further clarification is needed, however, in the area of true metabolic activities, where a part of the insecticidal molecule may be utilized as a carbon source. Such a process is conceivable, since it is not really necessary for many organisms to degrade the substrate all the way to CO_2 and water in order to gain energy; witness the microbial fermentation processes. Examples can also be found in metabolism of the phenoxyalkanoate herbicides, of which various esters or long chain acid moieties are known to serve as carbon sources for microorganisms. Such cases could well be demonstrated in the field of insecticide metabolism.

Finally, more efforts will be needed to clarify the role of conjugation activities in the overall process of insecticide degradation in microorganisms. Certainly, one must be careful in drawing comparison with the conjugation systems in animals, but such metabolic activities through synthetic routes, as shown for pentachlorophenol (Cserjesi and Johnson, 1972) and p-bromoacetanilide (Tweedy et al., 1970), must be studied further and their meaning explained.

In the final analysis one must approach the problem of microbial metabolism of insecticides with a very flexible attitude. It should not be assumed either that all insecticides can be eventually degraded by microorganisms (e.g. the microbial infallibility theory), or that complete energy-yielding metabolism on insecticidal molecules is not possible. The important thing is to identify one by one the key metabolic reactions and the stable end products in order to provide the necessary information to understand the processes, tendencies and roles of microorganisms in altering the characters of this important group of environmental pollutants.

TABLE 2

List of insecticidal chemicals cited

Common name	Structure	Chemical name
aldrin (contains 95% HHDN)		1,2,3,4,10,10-hexachloro-1,4,4a,5,8,8a-hexahydro-exo-1,4-endo-5,8-dimethanonaphthalene (HHDN)
gamma-BHC		1,2,3,4,5,6-hexachloro-cyclohexane (gamma isomer)

Table 2 – continued

Common name	Structure	Chemical name
carbaryl		1-naphthyl methylcarbamate
carbophenothion	C_2H_5O, S, PS—CH_2S—Cl, C_2H_5O	S-(4-chlorophenyl-thiomethyl) O,O-diethyl phosphorodithioate
chlordane		1,2,4,5,6,7,8,8-octa-chloro-3a,4,7,7a-tetrahydro-4,7-methanoindane (mixed isomers)
chlorobenzilate		ethyl 4,4′-dichlorobenzilate
chlorodimeform		N′-(4-chloro-o-tolyl)-N,N-dimethylformamidine
chloropropylate		isopropyl 4,4′-dichlorobenzilate
chlorpyrifos		O,O-diethyl O-3,5,6-trichloro-2-pyridyl phosphorothioate
pp′-DDT		1,1,1-trichloro-2,2-di-(4-chlorophenyl)ethane

Table 2 – continued

Common name	Structure	Chemical name
diazinon		O,O-diethyl O-(2-isopropyl-6-methyl-4-pyrimidinyl) phosphorothioate
dichlorvos		2,2-dichlorovinyl O,O-dimethyl phosphate
dieldrin (contains 85% HEOD)		1,2,3,4,10,10-hexachloro-6,7-epoxy-1,4,4a,5,6,7,8,8a-octahydro-exo-1,4-$endo$-5,8-dimethanonaphthalene (HEOD)
DFP		O,O-diisopropyl phosphorofluoridate
endosulfan		6,7,8,9,10,10-hexachloro-1,5,5a,6,9,9a-hexahydro-6,9-methano-2,4,3-benzodioxathiepin 3-oxide
endrin		1,2,3,4,10,10-hexachloro-6,7-epoxy-1,4,4a,5,6,7,8,8a-octahydro-exo-1,4-exo-5,8-dimethanonaphthalene
EPN		O-ethyl O-4-nitrophenyl phenylphosphonothioate
fenitrothion		O,O-dimethyl O-(nitro-3-tolyl) phosphorothioate

Table 2 – continued

Common name	Structure	Chemical name
fenthion		O,O-dimethyl O-(3-methyl-4-methylthiophenyl) phosphorothioate
fonofos (dyfonate)		ethyl S-phenyl ethylphosphonodithioate
heptachlor		1,4,5,6,7,8,8-heptachloro-3a,4,7,7a-tetrahydro-4,7-methanoindene
malathion		S-(1,2-di(ethoxycarbonyl)ethyl)dimethyl phosphorothiolothionate
methoxychlor		1,1,1-trichloro-2,2-di-(4-methoxyphenyl) ethane
metalkamate (bufencarb)		3-ethylpropylphenyl methylcarbamate
		3-methylbutylphenyl methylcarbamate
metobromuron		3-(4-bromophenyl)-1-methoxy-1-methylurea
mexacarbate		4-dimethylamino-3,5-xylyl methylcarbamate

Table 2 – continued

Common name	Structure	Chemical name
mirex		dodecachlorooctahydro-1,3,3-methano-2H-cyclobuta(cd) pentalene
nicotine		1-methyl-2-(3-pyridyl)pyrrolidine
parathion		O,O-diethyl O-(p-nitrophenyl) phosphorothioate
phorate		O,O-diethyl S-(ethyl thiomethyl)phosphorodithioate
phosmet		O,O-dimethyl phthalimidomethyl phosphorodithioate
TEPP		tetraethyl pyrophosphate
trichlorphon		dimethyl 2,2,2-trichloro-1-hydroxyethylphosphonate

References

AHMED. M. K. and CASIDA, J. E. (1958). Metabolism of some organophosphorus insecticides by microorganisms. *J. econ. Ent.* **51**, 59 63.

AHMED, M. K. and FOCHT, D. D. (1972). Degradation of polychlorinated biphenyls by two species of *Achromobacter*. *Can. J. Microbiol.* **19**, 47–52.

ALBONE, E. S., EGLINGTON, G., EVANS, N. C. and RHEAD, M. M. (1972). Formation of bis (*p*-chlorophenyl)-acetonitrile (*pp'*-DDCN) from *pp'*-DDT in anaerobic sewage sludge. *Nature, Lond.* **240**, 420–421.

ALLAN, J. (1955). Loss of biological efficiency of cattle-dipping wash containing benzene hexachloride. *Nature, Lond.* **175**, 1131–1132.

ANDERSON, J. P. E., LICHTENSTEIN, E. P. and WHITTINGHAM, W. F. (1970). Effect of *Mucor alternans* on the persistence of DDT and dieldrin in culture and in soil. *J. econ. Ent.* **63**, 1595–1599.

ANDRADE, P. S. L. and WHEELER, W. B. (1974). Biodegradation of mirex by sewage sludge organisms. *Bull. Environ. Contam. Toxicol.* **11**, 415–416.

BARKER, P. S. and MORRISON, F. O. (1964). Breakdown of DDT to DDD in mouse tissue. *Can. J. Zool.* **42**, 324–325.

BARKER. P. S. and MORRISON, F. O. (1965). The metabolism of TDE by *Proteus vulgaris. Can. J. Zool.* **43**, 652–654.

BARKER, P. S., MORRISON, F. O. and WHITAKER, R. S. (1965). Conversion of DDT to DD by *Proteus vulgaris*. A bacterium isolated from the intestinal flora of a mouse. *Nature, Lond.* **205**, 621–622.

BENEZET, H. J. and MATSUMURA, F. (1973). Isomerization of γ-BHC to α-BHC in the environment. *Nature, Lond.* **243**, 480–481.

BENEZET, H. J. and MATSUMURA, F. (1974). Factors influencing the metabolism of mexacarbate by microorganisms. *J. agric. Fd Chem.* **22**, 427–430.

BIXBY, M. W., BOUSH, G. M. and MATSUMURA, F. (1971). Degradation of dieldrin to carbon dioxide by a soil fungus *Trichoderma koningii. Bull. Environ. Contam. Toxicol.* **6**, 491–494.

BOLLAG, J. M. and LIU, S. Y. (1971). Degradation of Sevin by soil microorganisms. *Soil. Biol. Biochem.* **3**, 337–345.

BOLLAG, J. M. and LIU, S. Y. (1972). Hydroxylations of carbaryl by soil fungi. *Nature, Lond.* **236**, 177–178.

BOURQUIN, A. W., ALEXANDER, S. K., SPEIDEL, H. K., MANN, J. E. and FAIR, J. F. (1972). Microbial interactions with cyclodiene pesticides. *Develop. Ind. Microbiol.* **13**, 264–276.

BOUSH, G. M. and MATSUMURA, F. (1967). Insecticidal degradation by *Pseudomonas melophthora* the bacterial symbiote of the apple maggot. *J. econ. Ent.* **60**, 918–920.

BRADBURY, F. R. (1963). The systematic action of benzene hexachloride seed dressings. *Ann. appl. Biol.* **52**, 361–370.

BRAUNBERG, R. C. and BECK. V. (1968). Interaction of DDT and gastro-intestinal microflora of the rat. *J. agric. Fd Chem.* **16**, 451–453.

CHACKO, C. I., LOCKWOOD, J. L. and ZABIK, M. (1966). Chlorinated hydrocarbon pesticides: degradation by microbes. *Science, N.Y.* **154**, 893–895.

CHEN, P. R., TUCKER, W. P. and DAUTERMAN, W. C. (1969). Structure of biologically produced malathion monoacid. *J. agric. Fd Chem.* **17**, 86–90.

CHOMAN, B. R., ABDEL-GHAFFAR, A. S., REID, J. J. and CONE, J. F. (1954). Bacterial decomposition of nicotine. *Proc. Soc. Am. Bacteriol.* **54**, 21.

COOK, J. W. (1957). *In vitro* destruction of some organophosphate pesticides by bovine rumen fluid. *J. agric. Fd Chem.* **5**, 859–863.

CSERJESI, A. J. and JOHNSON, E. L. (1972). Methylation of pentachlorophenol by *Trichoderma virgatum*. *Can. J. Microbiol.* **18**, 45–49.

DOELLE, H. W. (1969). "Bacterial metabolism", pp. 353–399, Academic Press, New York.

DOROUGH, H. W. and CASIDA, J. E. (1964). Nature of certain carbamate metabolites of the insecticide Sevin. *J. agric. Fd Chem.* **12**, 294–304.

DOROUGH, H. W., LEELING, N. C., and CASIDA, J. E. (1963). Nonhydrolytic pathway in metabolism of *n*-methylcarbamate insecticides. *Science, N.Y.* **140**, 170–171.

EICHELBERGER, J. W. and LICHTENBERG, J. J. (1971). Persistence of pesticides in river water. *Environ. Sci. Technol.* **5**, 541–544.

EL-ZORGANI, G. A. and OMER, M. E. H. (1974). Metabolism of endosulfan isomers by *Aspergillus niger*. *Bull. Environ. Contam. Toxicol.* **12**, 182–185.

FINLEY, R. B. and PILLMORE, R. E. (1963). Conversion of DDT to DDD in animal tissue. *Am. Inst. biol. Sci. Bull.* **13**, 41–42.

FLASHINSKI, S. J. and LICHTENSTEIN, E. P. (1974). Metabolism of Dyfonate by soil fungi. *Can. J. Microbiol.* **20**, 399–411.

FOCHT, D. D. (1972). Microbial degradation of DDT metabolites to carbon dioxide, water and chloride. *Bull. Environ. Contam. Toxicol.* **7**, 52–56.

FOCHT, D. D. and ALEXANDER, M. (1970a). DDT metabolites and analogues: ring fission by *Hydrogenomonas*. *Science, N.Y.* **170**, 91–92.

FOCHT, D. D. and ALEXANDER, M. (1970b). Aerobic cometabolism of DDT analogues by *Hydrogenomonas* sp. *J. agric. Fd Chem.* **19**, 20–22.

FOCHT, D. D. and ALEXANDER, M. (1970c). Bacterial degradation of diphenylmethane, a DDT model substrate. *Appl. Microbiol.* **20**, 608–611.

FOCHT, D. D. and JOSEPH, H. (1974). Degradation of 1,1-diphenylethylene by mixed cultures. *Can. J. Microbiol.* **20**, 631–635.

FRANKENBERG, W. G. and VAITEKUNAS, A. A. (1955). Chemical studies on nicotine degradation by microorganisms derived from the surface of tobacco seeds. *Archs. Biochem. Biophys.* **58**, 509–512.

FRENCH, A. L. and HOOPINGARNER, R. A. (1970). Dechlorination of DDT by membranes isolated from *Escherichia coli*. *J. econ. Ent.* **63**, 756–759.

GETZIN, L. W. (1967). Metabolism of diazinon and zinophos in soils. *J. econ. Ent.* **60**, 505–508.

GETZIN, L. W. and ROSEFIELD, I. (1968). Organophosphorous insecticide degradation by heat-labile substances in soil. *J. agric. Fd Chem.* **16**, 598–601.

GRAETZ, D. A., CHESTERS, G., DANIEL, T. C., NEWLAND, L. W. and LEE, C. B. (1970). Parathion degradation in lake sediments. *J. Wat. Pollut. Control. Fed.* **42**, 76–94.

GUENZI, W. D. and BEARD, W. E. (1967). Anaerobic biodegradation of DDT to DDD in soil. *Science, N.Y.* **156**, 1116–1117.

GUNNER, H. B. and ZUCKERMAN, B. M. (1968). Degradation of diazinon by synergistic microbial action. *Nature, Lond.* **217**, 1183–1184.

GUNNER, H. B., ZUCKERMAN, B. M., WALKER, R. W., MILLER, C. W., DEUBERT, K. H. and LONGLEY, R. E. (1966). The distribution and persistence of diazinon applied to plant and soil and its influence on rhizosphere and soil microflora. *Pl. Soil.* **25**, 249–264.

HIRAKOSO, S. (1968). Inactivating effects of microorganisms on insecticidal activity of Dursban. *World Health Organisation* WHO/VBC/68–63, 1–4.

HOCHSTEIN, L. I. and RITTENBERG, S. C. (1959a). The bacterial oxidation of nicotine. *J. biol. Chem.* **234**, 151–155.

HOCHSTEIN, L. I. and RITTENBERG, S. C. (1959b). The bacterial oxidation of nicotine. II. The isolation of the first oxidative product and its identification as (1)-6-hydroxynicotine. *J. biol. Chem.* **234**, 156–160.

HOCHSTEIN, L. I. and RITTENBERG, S. C. (1960). The bacterial oxidation of nicotine. III. The isolation and identification of 6-hydroxypseudooxynicotine. *J. biol. Chem.* **235**, 795–799.

HORVATH, R. S. and ALEXANDER, M. (1970). Cometabolism of m-chlorobenzoate by an *Arthrobacter. Appl. Microbiol.* **20**, 254–258.

IYENGAR, L. and RAO, A. V. S. P. (1973). Metabolism of chlordane and heptachlor by *Aspergillus niger. J. gen. appl. Microbiol.* **19**, 321–324.

JENSEN, S., GOETHE. R. and KINDSTEDT, M. O. (1972). Bis-(p-chlorophenyl)-acetonitrile (DDCN), a new DDT derivative formed in anaerobic digested sewage sludge and lake sediment. *Nature, Lond.* **240**, 421–422.

JOHNSON, R. E., LIN, C. S. and COLLYARD, K. J. (1971). Influence of soil amendments on the metabolism of DDT in soil. *Proc. IUPAC Second Int. Congr. of Pest. Chem.* **6**, 139–156.

JOHNSON, B. T., GOODMAN, R. N. and GOLDBERG, H. S. (1967). Conversion of DDT to DDD by pathogenic and saprophytic bacteria associated with plants. *Science, N.Y.* **157**, 560–561.

JOHNSON, B. T. and KNOWLES, C. O. (1970). Microbial degradation of the acaricide N-(4-chloro-o-tolyl)-N,N-dimethylformamidine. *Bull. Environ. Contam. Toxicol.* **5**, 158–163.

KALLMAN, B. J. and ANDREWS, A. K. (1963). Reductive dechlorination of DDT to DDD by yeast. *Science, N.Y.* **141**, 1050–1051.

KIIGEMAGI, U., MORRISON, H. E., ROBERTS, J. E. and BOLLEN, W. B. (1958). Biological and chemical studies on the decline of soil insecticides. *J. econ. Ent.* **51**, 198–204.

KNAAK, J. B. and SULLIVAN, L. J. (1967). The metabolism of 2-methyl-2(methyl-thio)propionaldehyde O-(methylcarbamoyl)oxime in the rat. *J. agric. Fd Chem.* **14**, 573–578.

KONRAD, J. G., CHESTERS, G. and ARMSTRONG, D. E. (1969). Soil degradation of malathion, phosphorodithioate insecticide. *Soil Sci. Soc. Am. Proc.* **33**, 259–262.

KORTE, F., LUDWIG, G. and VOEGL, J. (1962). Umwandlung von aldrin (^{14}C) und dieldrin (^{14}C) durch mikroorganismen, leberhomogenate, und moskito-larven. *Ann. Chem.* **656**, 135–140.

LAANIO, T. L., DUPUIS, G. and ESSER, H. O. (1972). Fate of ^{14}C-labelled diazinon in rice, paddy soil and pea plants. *J. agric. Fd Chem.* **20**, 1213–1219.

LANGLOIS, B. E. (1967). Reductive dechlorination of DDT by *Escherichia coli*. *J. Dairy Sci., Wash.* **50**, 1168–1170.

LEELING, N. C. and CASIDA, J. E. (1966). Metabolites of carbaryl (1-naphthyl methylcarbamate) in mammals and enzymatic systems for their formation. *J. agric. Fd Chem.* **14**, 281–290.

LICHTENSTEIN, E. P. and SCHULZ, K. R. (1960). Epoxidation of aldrin and heptachlor in soils as influenced by autoclaving, moisture and soil types. *J. econ. Ent.* **53**, 192–197.

LICHTENSTEIN, E. P. and SCHULZ, K. R. (1964). The effects of moisture and microorganisms on the persistence and metabolism of some organophosphorus insecticides in soils, with special emphasis on parathion. *J. econ. Ent.* **57**, 618–627.

LIU, S. -Y. and BOLLAG, J. M. (1971a). Metabolism of carbaryl by a soil fungus. *J. agric. Fd Chem.* **19**, 487–490.

LIU, S. -Y. and BOLLAG, J. M. (1971b). Carbaryl decomposition to 1-naphthyl carbamate by *Aspergillus terreus*. *Pestic. Biochem. Physiol.* **1**, 366–372.

MACKIEWICZ, M., DEUBERT, K. H., GUNNER, H. B. and ZUCKERMAN, B. M. (1969). Study of parathion biodegradation using gnotobiotic techniques. *J. agric. Fd Chem.* **17**, 129–130.

MacRAE, I. C., RAGHU, K. and CASTRO, T. F. (1967). Persistence and biodegradation of four common isomers of benzene hexachloride in submerged soils. *J. agric. Fd Chem.* **15**, 911–914.

MacRAE, I. C., RAGHU, K. and BAUTISTA, E. M. (1969). Anaerobic degradation of the insecticide lindane by *Clostridium* sp. *Nature, Lond.* **221**, 859–860.

MARTENS, R. (1972). Der abbau von endosulfan durch mikroorganismen des bodens. *Schriftenr. Ver. Wasser-Boden-Lufthyg. Berlin-Dahlem* **37**, 167–173.

MATSUMURA, F. (1974). Microbial degradation of pesticides. *In* "Survival in Toxic Environments" (M. A. Q. Khan and J. P. Bederka, Eds), pp. 129–154, Academic Press, New York.

MATSUMURA, F. and BOUSH, G. M. (1966). Malathion degradation by *Trichoderma viride* and a *Pseudomonas species*. *Science, N.Y.* **153**, 1278–1280.

MATSUMURA, F. and BOUSH, G. M. (1967). Dieldrin: degradation by soil microorganisms. *Science, N.Y.* **156**, 959–961.

MATSUMURA, F. and BOUSH, G. M. (1968). Degradation of insecticides by a soil fungus, *Trichoderma viride*. *J. econ. Ent.* **61**, 610–612.

MATSUMURA, F., BOUSH, G. M. and TAI, A. (1968). Breakdown of dieldrin in the soil by a microorganism. *Nature, Lond.* **219**, 965–967.

MATSUMURA, F., PATIL, K. C. and BOUSH, G. M. (1970). Formation of "photodieldrin" by microorganisms. *Science, N.Y.* **170**, 1206–1207.

MATSUMURA, F., KHANVILKAR, V. G., PATIL, K. C. and BOUSH, G. M. (1971). Metabolism of endrin by certain soil microorganisms. *J. agric. Fd Chem.* **19**, 27–31.

MENDEL, J. L. and WALTON, M. S. (1966). Conversion of pp'-DDT to pp'-DDD by intestinal flora of the rat. *Science, N.Y.* **151**, 1527–1528.

MENDEL, J. L., KLEIN, A. K., CHEN, J. T. and WALTON, M. S. (1967). Metabolism of DDT and some other chlorinated organic compounds by *Aerobacter aerogenes*. *J. Ass. off. agric. Chem.* **50**, 897–903.

MENN, J. J., PATCHETT, G. G. and BATCHELDER, G. H. (1960). The persistence of trithion, an organophosphorus insecticide, in soil. *J. econ. Ent.* **53**, 1080–1082.

MENN, J. J., McBAIN, J. B., ADELSON, B. J. and PATCHETT, G. G. (1965). Degradation of *N*-(mercaptomethyl)phthalimide-*S*-(*O,O*-dimethyl-phosphorodithioate) (Imidan) in soils. *J. econ. Ent.* **58**, 875–878.

MICK, D. L. and DAHM, P. A. (1970). Metabolism of parathion by two species of *Rhizobium*. *J. econ. Ent.* **63**, 1155–1159.

MILES, J. R. W., TU, C. M. and HARRIS, C. R. (1969). Metabolism of heptachlor and its degradation products by soil microorganisms. *J. econ. Ent.* **62**, 1334–1337.

MILES, J. R. W., TU, C. M. and HARRIS, C. R. (1971). Degradation of heptachlor epoxide and heptachlor by a mixed culture of soil microorganisms. *J. econ. Ent.* **64**, 839–841.

MILLER, C. W., TOMLINSON, W. E. and NORGREN, R. L. (1967). Persistence and movement of parathion in irrigation waters. *Pestic. Monit. J.* **1**, 47–48.

MISKUS, R. P., BLAIR, D. P. and CASIDA, J. E. (1965). Conversion of DDT to DDD by bovine rumen fluid, lake water, and reduced porphyrins. *J. agric. Fd Chem.* **13**, 481–483.

MIYAMOTO, J., KITAGAWA, K. and SATO, Y. (1966). Metabolism of organophosphorus insecticides by *Bacillus subtilis* with special emphasis on sumithion. *Jap. J. exp. Med.* **36**, 211–225.

MIYAZAKI, S., BOUSH, G. M. and MATSUMURA, F. (1969). Metabolism. of ^{14}C-chlorobenzilate and ^{14}C-chloropropylate by *Rhodotorula gracilis*. *Appl. Microbiol.* **18**, 972–976.

MOLOZHANOVA, YE.G., REMIZOVA, L. B. and BRANTSEVICH, L. G. (1973). Detoksikatsiya kylorofosa pochvennymi mikroorganizmami. *Khim. Sel. Khoz.* **11**, 761–762. Indirectly cited from Pesticide Abstracts (1974) vol 7, p. 795. Abstract No. 74-2898.

MUNNECKE, D. M. and HSIEH, D. P. H. (1974). Microbial decontamination of parathion and *p*-nitrophenol in aqueous media. *Appl. Microbiol.* **28**, 212–217.

NEWLAND, L. W., CHESTERS, G. and LEE, G. B. (1969). Degradation of γ-BHC in simulated lake impoundments as affected by aeration. *J. Wat. Pollut. Control Fed.* **41**, 174–188.

PATIL, K. C., MATSUMURA, F. and BOUSH, G. M. (1970). Degradation of endrin, aldrin, and DDT by soil microorganisms. *Appl. Microbiol.* **19**, 879–881.

PATIL, K. C., MATSUMURA, F. and BOUSH, G. M. (1972). Metabolic transformation of DDT, dieldrin, aldrin, and endrin by marine microorganisms. *Environ. Sci. Technol.* **6**, 629–632.

PFAENDER, F. K. and ALEXANDER, M. (1972). Extensive microbial degradation of DDT *in vitro* and DDT metabolism by natural communities. *J. agric. Fd Chem.* **20**, 842–846.

PFAENDER, F. K. and ALEXANDER, M. (1973). Effect of nutrient additions on the apparent cometabolism of DDT. *J. agric. Fd Chem.* **21**, 397–399.

POLLACK, M. R. (1957). Penicillin-induced resistance to penicillin in cultures of *Bacillus cereus*. *In* "Drug Resistance in Microorganisms" (G. E. W. Wolstenholme and C. M. O'Connor, Eds), pp. 78–95. Little, Brown and Co., Boston, Mass.

POLLACK, M. R. (1962). Exoenzymes. *In* "The Bacteria" (I. C. Gunsalas and R. Y. Stanier, Eds), vol. 4, p. 121. Academic Press, New York.

POONAWALLA, N. H. and KORTE, F. (1968). Metabolism of β-dihydro heptachlor ^{14}C in soil and by microorganisms. *J. agric. Fd Chem.* **16**, 15–16.

RAGHU, K. and MacRAE, I. C. (1966). Biodegradation of the gamma isomer of benzene hexachloride in submerged soils. *Science, N.Y.* **154**, 263–264.

RAO, A. V. and SETHUNATHAN, N. (1974). Degradation of parathion by *Penicillium waksmani* Zaleski isolated from flooded acid sulphate soil. *Arch. Mikrobiol.* **97**, 203–208.

ROSS, J. A. and TWEEDY, B. G. (1973). Malonic acid conjugation by soil micro-organisms of a pesticide-derived aniline moiety. *Bull. Environ. Contam. Toxicol.* **10**, 234–236.

SATO, R. and KUBO, H. (1964). The water pollution caused by organophosphorus insecticides in Japan. *Adv. Water Pollut. Res.* **1**, 95–99.

SETHUNATHAN, N. (1972). Diazinon degradation in submerged soil and rice paddy water. *Adv. Chem. Ser.* **111**, 244–255.

SETHUNATHAN, N. and MacRAE, I. C. (1969). Persistence and biodegradation of diazinon in submerged soils. *J. agric. Fd Chem.* **17**, 221–225.

SETHUNATHAN, N. and PATHAK, M. D. (1971). Development of a diazinon degrad-ing bacterium in paddy water after repeated applications of diazinon. *Can. J. Microbiol.* **17**, 699–702.

SETHUNATHAN, N. and PATHAK, M. D. (1972). Increased biological hydrolysis of diazinon after repeated application in rice paddies. *J. agric. Fd Chem.* **20**, 586–589.

SETHUNATHAN, N. and YOSHIDA, T. (1969). Fate of diazinon in submerged soil, accumulation of hydrolysis product. *J. agric. Fd Chem.* **17**, 1192–1195.

SETHUNATHAN, N. and YOSHIDA, T. (1972). Conversion of parathion to para-nitrophenol by diazinon-degrading bacterium, *Flavobacterium* sp. *Inst. Environ. Tech. Meet. Proc.* **18**, 255–257.

SETHUNATHAN, N. and YOSHIDA, T. (1973a). Degradation of chlorinated hydro-carbons by *Clostridium* sp. isolated from lindane-amended flooded soil. *Pl. Soil* **38**, 663–666.

SETHUNATHAN, N. and YOSHIDA, T. (1973b). A *Flavobacterium* sp. that degrades diazinon and parathion. *Can. J. Microbiol.* **19**, 873–875.

SETHUNATHAN, N. and YOSHIDA, T. (1973c). Parathion degradation in submerged rice soils in the Philippines. *J. agric. Fd Chem.* **21**, 504–506.

SETHUNATHAN, N., BAUTISTA, E. M. and YOSHIDA, T. (1969). Degradation of benzene hexachloride by a soil bacterium. *Can. J. Microbiol.* **15**, 1349–1354.

SKUJIŅŠ, J. J. (1967). Enzymes in soil. *In* "Soil Biochemistry" (A. D. McLaren and G. H. Peterson, Eds), vol. 1, pp. 371–414. Marcel Dekker, Inc., New York.

STENERSON, J. H. B. (1965). DDT metabolism in resistant and susceptible stable flies and in bacteria. *Nature, Lond.* **207**, 660–661.

SWEENEY, R. A. (1969). Metabolism of lindane by unicellular algae. *Proc. 12th Conf. G. Lakes. Res.* 98–102.

TABUCHI, T. (1955a). Microbial degradation of nicotine and nicotinic acid. *J. agric. Chem. Soc. Japan* **29**, 222–225.

TABUCHI, T. (1955b). Microbial degradation of nicotine and nicotinic acid. Part 2. Degradation of nicotine (1). *J. agric. Chem. Soc. Japan* **29**, 219–221.

TIEDJE, J. M. and ALEXANDER, M. (1967). Microbial degradation of organophosphorus insecticides and alkyl phosphates. *Agron. Abst.* 94.

TRELA, J. M., RALSTON, W. J. and GUNNER, H. B. (1968). Metabolism of diazinon by soil microflora. *Bact. Proc.* p. 6.

TSUKANO, Y. and KOBAYASHI, A. (1972). Formation of γ-3,4,5,6-tetrachloro 1-cyclohexene (BTC) in flooded rice field soils treated with γ-1,2,3,4,5,6-hexachlorocyclohexane (BHC). *Agr. Biol. Chem.* **36**, 166–167.

TU, C. M., MILES, J. R. W. and HARRIS, C. R. (1968). Soil microbial degradation of aldrin. *Life Sci.* **7**, 311–322.

TUCKER, B. V. and PACK, D. E. (1972). Bux insecticide soil metabolism. *J. agric. Fd Chem.* **20**, 412–416.

TWEEDY, B. G., LOEPPKY, C. and ROSS, J. A. (1970). Metobromuron: acetylation of the aniline moiety as a detoxification mechanism. *Science, N.Y.* **168**, 482–483.

VOCKEL, D. and KORTE, F. (1974). Beitraege zur oekologischen chemie LXXX. Versuche zum mikrobiellen Abbau von dieldrin und 2,2'-dichlorobiphenyl. *Chemosphere* **3**, 177–182.

WADA, E. (1957). Microbial degradation of the tobacco alkaloids, and some related compounds. *Archs. Biochem. Biophys.* **72**, 145–161.

WADA, E. and YAMASAKI, K. (1953). Mechanism of microbial degradation of nicotine. *Science, N.Y.* **117**, 152–153.

WADA, E. and YAMASAKI, K. (1954). Degradation of nicotine by soil bacteria. *J. Am. chem. Soc.* **76**, 155–157.

WALKER, W. W. and STOJANOVIC, B. J. (1973). Microbial versus chemical degradation of malathion in soil. *J. Environ. Quality* **2**, 229–232.

WEDEMEYER, G. (1966). Dechlorination of DDT by *Aerobacter aerogenes*. *Science, N.Y.* **152**, 647.

WEDEMEYER, G. (1968). Partial hydrolysis of dieldrin by *Aerobacter aerogenes*. *Appl. Microbiol.* **16**, 661–662.

WEISGERBER, I., DETERA, S. and KLEIN, W. (1974). Beitraege zur oekologischen Chemie LXXXVIII. Isoliering und identifizierung einiger heptachlor ^{14}C-metabolite aus Pflanzen und Boden. *Chemosphere* **3**, 221–226.

YASUNO, M., HIRAKOSO, S., SASA, M. and UCHIDA, M. (1965). Inactivation of some organophosphorus insecticides by bacteria in polluted water. *Jap. J. exp. Med.* **35**, 545–563.

YULE, W. N., CHIBA, M. and MORLEY, H. V. (1967). Fate of insecticide residues. Decomposition of lindane in soil. *J. agric. Fd Chem.* **15**, 1000–1004.

ZAYED, S. M. A. D., MOSTAFA, I. Y. and HASSAN, A. (1965). Organische p-haltige insecticide im stoffwechsel VII. Umwandlung von ^{32}P-markiertem dipterex durch mikroorganismen. *Arch. Mikrobiol.* **51**, 118–121.

ZUCKERMAN, B. M., DEUBERT, K., MACKIEWICZ, M. and GUNNER, H. (1970). Studies on the biodegradation of parathion. *Pl. Soil* **33**, 273–281.

Chapter 11

Microbial Degradation of Herbicides

ROGER E. CRIPPS AND TERRY R. ROBERTS

Shell Research Limited, Woodstock Laboratory,
Sittingbourne Research Centre,
Sittingbourne, Kent, England

I. Introduction

In recent years, the extensive literature on the microbial metabolism of naturally-occurring organic compounds has been supplemented by an ever increasing number of reports of studies on the degradation of synthetic organic chemicals. These studies, which have been carried out with pure cultures of isolated microorganisms, with mixed microbial populations such as those which occur in soil and, in some cases, with microbial enzymes, have revealed that many of these chemicals are susceptible to bio-degradation by microbes. The degree of degradation observed varies from compound to compound. Some molecules can be utilized as sole source of carbon and energy for the growth of a particular organism leading, in some but not all cases, to the complete metabolism of the substrate. Other compounds are only partially degraded to non-metabolizable products while some are apparently completely resistant to microbial attack. This variation in susceptibility to microbial metabolism is not surprising since any degradation of a compound of non-biosynthetic origin is presumably due to a low specificity of one or more catabolic enzymes present in the capable organisms.

Bio-degradation studies have, therefore, provided considerable insight into the nature of chemical substituents in a molecule which make that molecule unavailable for metabolism, that is, some relationships between chemical structure and bio-degradability can be identified. Such data are extremely valuable in environmental terms if molecules with a predetermined degree of bio-degradability are to be produced for use as, for example, pesticides or detergents.

It is the intention of this contribution to review the available evidence with regard to the microbial degradation of herbicides, taking results obtained from studies with soil systems and with isolated microorganisms. The methodology involved in the relevant types of study has been described in Chapter 5 of this volume. The large number of herbicides currently available commercially fall within several well-defined chemical groups. Each of these groups will be considered separately as far as possible and some individual compounds that fall outside these classes, will also be discussed in a separate section.

II. The Phenoxyalkanoates

The phenoxyalkanoic acids (Table 1) have been used for the past 30 years for the selective control of broad-leaved weeds in cereals and other crops and recently as total defoliants in chemical warfare. These compounds are primarily applied to the foliage of crops and weeds and their metabolism in plants has been extensively studied. The persistence and degradation of the phenoxyalkanoic acids in the soil environment has also been the subject of detailed study since these compounds reach the soil both during application and by transfer from plants to the soil by rainwater.

TABLE 1

Structures of phenoxyalkanoate herbicides

General structure (with OR, A, B, C substituents on benzene ring)

Substituents		Chemical name	Common name
$R = -CH_2CO_2H$	$A = CH_3$ $B = Cl$ $C = H$	2-methyl-4-chloro-phenoxyacetate	MCPA (MCP)
$R = -CH_2CO_2H$	$A = Cl$ $B = Cl$ $C = H$	2,4-dichlorophenoxy-acetate	2,4-D
$R = -CH_2CO_2H$	$A = Cl$ $B = Cl$ $C = Cl$	2,4,5-trichloro-phenoxyacetate	2,4,5-T
$R = -\overset{\overset{CH_3}{\mid}}{C}HCO_2H$	$A = CH_3$ $B = Cl$ $C = H$	2-(2-methyl-4-chlorophenoxy)-propionate	mecoprop (MCPP; CMPP)
$R = -\overset{\overset{CH_3}{\mid}}{C}HCO_2H$	$A = Cl$ $B = Cl$ $C = H$	2-(2,4-dichloro-phenoxy)propionate	dichlorprop (2,4-DP)
$R = -\overset{\overset{CH_3}{\mid}}{C}HCO_2H$	$A = Cl$ $B = Cl$ $C = Cl$	2-(2,4,5-trichloro-phenoxy)propionate	fenoprop (silvex; 2,4,5-TP)
$R = -CH_2CH_2CH_2CO_2H$	$A = CH_3$ $B = Cl$ $C = H$	4-(2-methyl-4-chloro-phenoxy)butyrate	MCPB
$R = -CH_2CH_2CH_2CO_2H$	$A = Cl$ $B = Cl$ $C - H$	4-(2,4-dichloro-phenoxy)butyrate	2,4-DB
$R = -CH_2CH_2CH_2CO_2H$	$A = Cl$ $B = Cl$ $C = Cl$	4-(2,4,5-trichloro-phenoxy)butyrate	2,4,5-TB

Early work, during 1945–1952, showed that 2,4-D underwent degradation in soil more rapidly than did MCPA or 2,4,5-T. The involvement of microorganisms in this process was demonstrated by several workers (e.g. Audus, 1952, 1964; Loos, 1969) using the soil perfusion technique originally described by Lees and Quastel (1946).

A lag phase was observed during which there was no appreciable decrease in herbicide concentration and this was followed by a rapid disappearance of the compound. This, and the fact that the addition of sodium azide to the system effectively stopped the degradation, suggested that microbes were involved. However, very little further work has been reported on the degradation of phenoxyalkanoates in soils and data on which, if any, degradation products are formed *in soil* are virtually non-existent.

In contrast, many studies have established that phenoxyalkanoates can be degraded by pure cultures of microorganisms. This capability has been identified in several genera of bacteria as well as in fungi and actinomycetes (see Evans *et al.*, 1971b; Loos, 1969). In addition, many phenoxyalkanoates, including those possessing plant growth regulating activity, have been shown to act as sole sources of carbon and energy for growth of some bacteria and in these cases much progress has been made in elucidating the metabolic pathways involved.

In this context, an *Arthrobacter* sp. was isolated (Loos *et al.*, 1967a) which could grow at the expense of 2,4-D. During growth all the chlorine in the molecule was released as chloride ion and the rest of the herbicide molecule was completely degraded to carbon dioxide, water and bacterial mass. Detailed investigations with this microorganism have revealed most features of the breakdown pathway of 2,4-D. The first metabolic reaction was shown to be removal of the side chain to form 2,4-dichlorophenol (Loos *et al.*, 1967b; Loos *et al.*, 1967c). Studies with 2,4-D radio-labelled in the side chain led to the suggestion that this C_2 unit was released as glyoxylate (Tiedje and Alexander, 1969), indicating that an oxidative cleavage occurs. The radiolabel was actually recovered from reaction mixtures as alanine which was postulated to have arisen from subsequent reactions of glyoxylate. In a study of the mechanism of the side chain cleavage reaction using 2,4-D containing ^{18}O in the ether oxygen (Helling *et al.*, 1968) it was shown that whole cells of the *Arthrobacter* sp. formed 2,4-dichlorophenol with complete retention of the ^{18}O. This indicated that cleavage occurred between the ether oxygen atom and the methylene group of the C_2 unit.

Whole cells and extracts of the *Arthrobacter* sp. could degrade not only 2,4-D but also MCPA and 2- and 4- chlorophenoxyacetates with the release of Cl^- ion (Loos *et al.*, 1967a, c). 2,4-Dichlorophenol and 2-methyl-4-chlorophenol were also metabolized. This phenol-oxidizing activity of the extracts was examined further and the enzyme involved was shown to be dependent on NADPH and molecular oxygen (Bollag *et al.*, 1968a). The product of the reaction was shown to be a catechol which in the case of 2,4-dichlorophenol was 3,5-

dichlorocatechol and with 2-methyl-4-chlorophenol was 3-methyl-5-chlorocatechol. The 3,5-dichlorocatechol could also be metabolized by the extracts by an "ortho" ring fission mechanism to form a muconic acid (Bollag *et al.*, 1968b; Tiedje *et al.*, 1969) which was further metabolized to succinate and (presumably) acetate by the series of reactions shown in Fig. 1 (Duxbury *et al.*, 1970; Sharpee *et al.*, 1973).

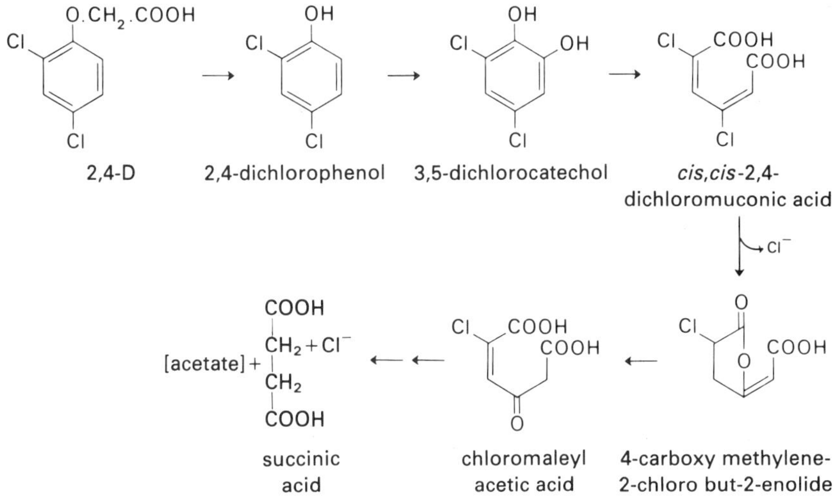

Fig. 1. Proposed degradative pathway of 2,4-D by *Arthrobacter* sp.

A slightly less clear picture of 2,4-D metabolism has been obtained with two *Pseudomonas* spp. by Evans *et al.* (1971a). During growth on this herbicide various related compounds were detected in the cultures, including 2,4-dichlorophenol, 3,5-dichlorocatechol and 2,4-dichloro-6-hydroxyphenoxyacetate. In addition, one species accumulated 2-chlorophenol together with large amounts of 2-chloromuconic acid. The pseudomonads were able to oxidize a variety of mono- and dichlorophenols and phenoxyacetates. The 6-hydroxy-2,4-D was not oxidized. Enzymatic studies confirmed that the organisms were capable of carrying out similar reactions to those shown for the *Arthrobacter* sp. (Fig. 1). However, the role of the monochlorinated derivatives found in, and oxidized by, the cultures is not clear, and a degradative pathway involving 2-chlorophenol and 2-chloromuconate, such as shown in Fig. 2, may also operate. Furthermore, the role of 6-hydroxy-2,4-D, which is formed but apparently not further metabolized, is not clear.

Other studies by Evans *et al.* (1971b) showed that a *Pseudomonas* sp. could degrade 4-chlorophenoxyacetate by a series of reactions initiated

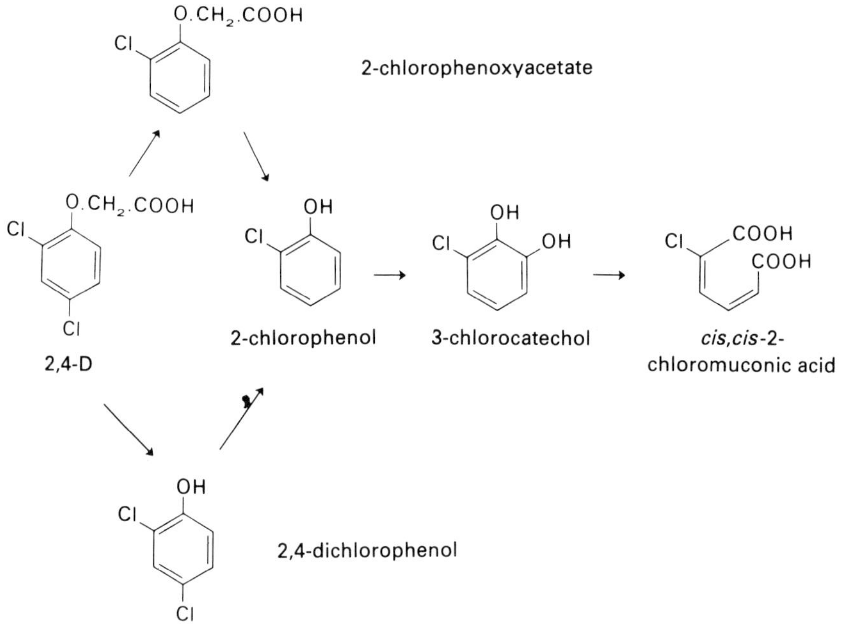
O.CH₂.COOH
Cl
2-chlorophenoxyacetate
O.CH₂.COOH
Cl
Cl
2,4-D
OH
Cl
2-chlorophenol
OH
Cl
OH
3-chlorocatechol
Cl
COOH
COOH
cis,cis-2-
chloromuconic acid
OH
Cl
Cl
2,4-dichlorophenol

Fig. 2. Possible reactions in 2,4-D metabolism by *Pseudomonas* sp. (after Evans *et al.*, 1971a).

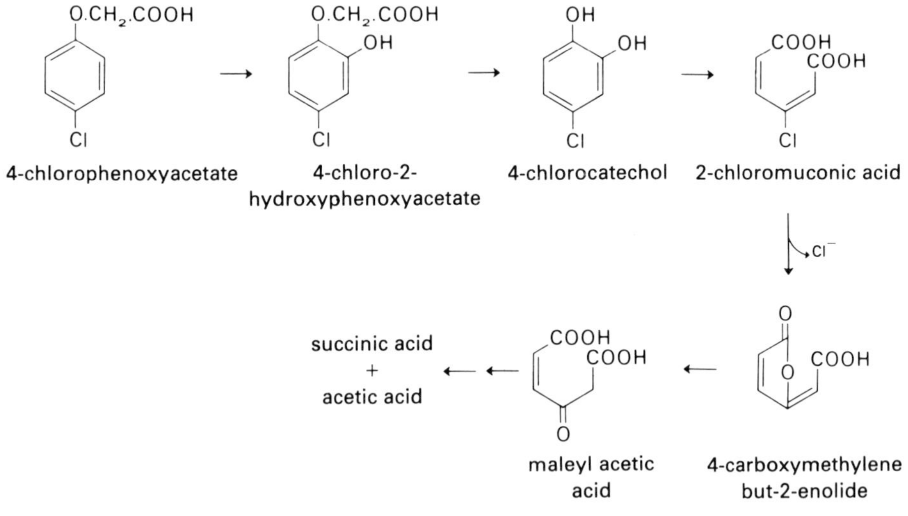
O.CH₂.COOH
Cl
4-chlorophenoxyacetate
O.CH₂.COOH
OH
Cl
4-chloro-2-
hydroxyphenoxyacetate
OH
OH
Cl
4-chlorocatechol
COOH
COOH
Cl
2-chloromuconic acid
Cl⁻
O
O
COOH
4-carboxymethylene
but-2-enolide
COOH
COOH
O
maleyl acetic
acid
succinic acid
+
acetic acid

Fig. 3. Degradation of 4-chlorophenoxyacetate by *Pseudomonas* sp. (after Evans *et al.*, 1971b).

by ring hydroxylation to form 2-hydroxy-4-chlorophenoxyacetate. Side-chain cleavage then resulted in 4-chlorocatechol formation and this was in turn further degraded (Fig. 3).

In contrast, the metabolism of MCPA by a *Pseudomonas* sp. may not involve ring hydroxylation before side-chain removal (Gaunt and Evans, 1971a, b). The isolation of compounds from cultures during growth and a detailed study of some of the enzymes involved support the degradation pathway shown in Fig. 4. However, since 6-hydroxy

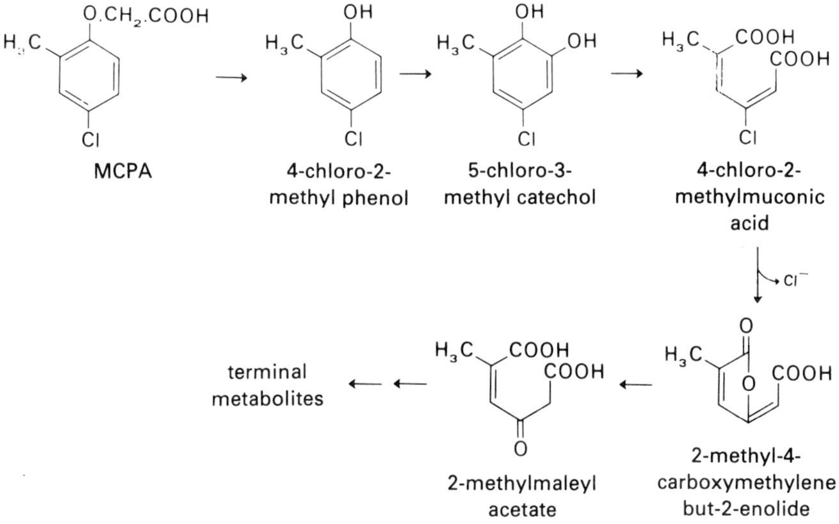

Fig. 4. Degradation of MCPA by a *Pseudomonas* sp. (after Gaunt and Evans, 1971a, b).

MCPA was also shown to be formed, Fig. 4 cannot represent the whole picture. The use of the above culture revealed the exact nature of the side-chain cleavage product. Using extracts of this organism, Gamar and Gaunt (1971) showed that, in the presence of NADPH, MCPA was converted into 5-chloro-o-cresol and glyoxylate (Fig. 5).

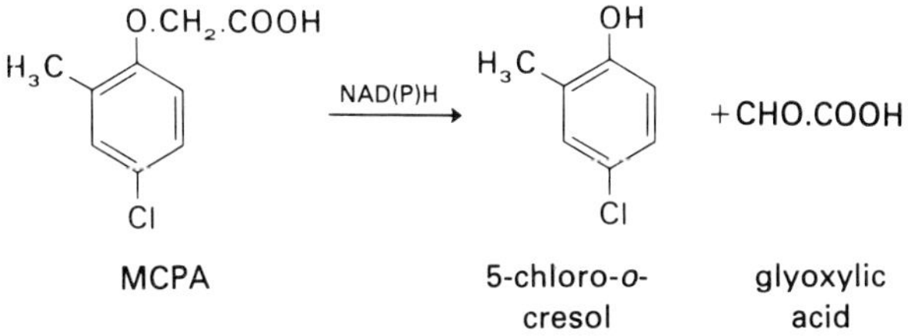

Fig. 5. Side chain cleavage of MCPA.

This confirmed that an oxidative cleavage reaction, probably a mixed function oxidase type reaction, operated at this point in phenoxyacetate degradation.

The production of ring-hydroxylated phenoxyacetates by bacteria may be a minor metabolic route, unimportant in comparison with the degradative pathways discussed above. However, in fungi, ring hydroxylation would seem to be the principal metabolic activity. In an extensive study by Woodcock and co-workers using the replacement culture technique, *Aspergillus niger* was shown to produce a variety of hydroxylated products from various phenoxyacetates. Thus phenoxyacetic acid itself was converted into a mixture of the 2-, 3- and 4-hydroxy isomers (Clifford and Woodcock, 1964) which indicated a lack of specificity of the system. Other studies (Faulkner and Woodcock, 1961) showed that 4-chlorophenoxyacetate was hydroxylated at the 2- and 3-positions by the fungus while 2-chlorophenoxyacetate was attacked primarily at positions 4- and 5- with smaller amounts of the 3- and 6-hydroxy derivatives being formed. Thus, hydroxylation occurred at all available sites on the aromatic ring. An additional product from 2-chlorophenoxyacetate was 2-hydroxyphenoxyacetate, indicating that *A. niger* was capable of hydrolytic dechlorination. Faulkner and Woodcock (1965) also studied the metabolism of 2,4-D and MCPA by this fungus. Two products were obtained from 2,4-D, namely the 5-hydroxy derivative and 2,5-dichloro-4-hydroxyphenoxyacetate. This latter product indicated that *A. niger* was capable of effecting a shift in position of a chlorine atom during ring hydroxylation. Such shifts have been shown to occur during the metabolism of several aromatic compounds (Daly *et al.*, 1968). MCPA was converted only into 2-methyl-4-chloro-5-hydroxyphenoxyacetic acid.

The microbial metabolism of 2,4,5-T has received little attention and the molecule is considered to be less easily degradable than the other herbicidal phenoxyacetates. However, Horvath (1970a) has reported that 2,4,5-T was oxidized by cells of a *Brevibacterium* sp. grown on benzoic acid. One chlorine atom was released as chloride and 3,5-dichlorocatechol was formed. Since this catechol is an intermediate of 2,4-D metabolism (Fig. 1), it would seem that the total microbial mineralization of 2,4,5-T may be possible.

Many other studies have been carried out on the microbial metabolism of phenoxyacetates (see Loos, 1969). No evidence has been obtained which would suggest degradation routes other than those described above.

Phenoxyalkanoates with side chains longer than two carbon atoms can also be metabolized by microorganisms by ring hydroxylation and side-chain cleavage. However, in these cases a further metabolic opportunity exists and has been widely demonstrated, namely β-oxidative attack on the side chain. In this mode of degradation, two-carbon fragments of the aliphatic side chain are successively removed as acetyl-CoA. Thus, species of *Nocardia*, *Pseudomonas* and *Micrococcus* were shown by Taylor and Wain (1962) to be able to effect β-oxidation of the side-chain of several phenoxyalkanoates, and the fungus *A. niger* (Byrde and Woodcock, 1957) was shown to have a similar capability. The latter organism also effected ring hydroxylation of these compounds. The β-oxidation of side chains with an even number of carbon atoms eventually leads to the formation of phenoxyacetates which can be degraded by the pathways already discussed. However, an alternative mode of metabolism has been reported, whereby the phenoxyalkanoate side-chain is removed in a single reaction. Thus a *Flavobacterium* sp. which can grow on 2,4-DB, metabolises this compound to 2,4-dichlorophenol and a C_4 acid indicating cleavage at the ether linkage (MacRae *et al.*, 1963).

III. The Phenylamides

There are many herbicides falling within this general class and these have been subdivided into the acylanilides (Table 2), the phenylureas (Table 3) and the phenylcarbamates (Table 4).

Amongst the acylanilides, which are a relatively new group of herbicides, propanil is the compound with the biggest sales, and is used mainly for the control of barnyard grass in rice (Oelke and Morse, 1968). Solan (pentanochlor) and dicryl (chloranocryl) have found use as post-emergence herbicides for tomatoes and cotton, respectively.

The phenylcarbamate herbicides, swep, propham and chlorpropham are applied pre-emergence for the control of annual grasses and broad-leaved weeds, and barban has been used for the control of wild oats.

Some phenylurea herbicides (e.g. diuron and monuron) have been in use for over 20 years and new members have been added to this general class of herbicide during this time. Linuron, chloroxuron and fluometuron were introduced in 1960 for pre- and post-emergence control of weeds in fruit, vegetables and cereals, although the main outlet for fluometuron is cotton. Metobromuron, buturon and siduron are more recent additions to the urea herbicide group.

TABLE 2

Structures of acylanilide and related herbicides

Structure	Chemical name	Common name
	N-(3,4-dichlorophenyl) propionamide	propanil (DPA)
	N-(3-chloro-4-methylphenyl)-2-methylpentanamide	pentanochlor (solan)
	N-(3,4-dichlorophenyl) methacrylamide	chloranocryl (dicryl)
	N-isopropyl-2-chloro-acetanilide	propachlor
	N-(3',4'-dichlorophenyl) 2-methylpentanamide	— (Karsil)
	ethyl-N-benzoyl-N-(3,4-dichlorophenyl)-2-amino-propionate	benzoylprop-ethyl

A common feature of all the phenylamides is that their structures are based on substituted anilines; a discussion of the fate of anilines derived from these herbicides is included in this section.

A. ACYLANILIDES

Soil microorganisms have been shown to be able to hydrolyse almost all acylanilide herbicides with the formation of the corresponding anilines. The major exception to this rule is the 2-chloroacetamides (e.g. propachlor) which are not readily hydrolysed (Matsunaka, 1971).

TABLE 3
Structures of phenylurea herbicides

Structure	Chemical name	Common name
Cl—⟨ring⟩—NH—CO—N(CH$_3$)(CH$_3$)	3-(4-chlorophenyl)-1,1-dimethylurea	monuron (CMU)
Cl, Cl—⟨ring⟩—NH—CO—N(CH$_3$)(CH$_3$)	3-(3,4-dichlorophenyl)-1,1-dimethylurea	diuron (DCMU; DMU)
Cl—⟨ring⟩—NH—CO—N(CH$_3$)(OCH$_3$)	3-(4-chlorophenyl)-1-methoxy-1-methylurea	monolinuron
Cl, Cl—⟨ring⟩—NH—CO—N(CH$_3$)(OCH$_3$)	3-(3,4-dichlorophenyl)-1-methoxy-1-methylurea	linuron
Br—⟨ring⟩—NH—CO—N(CH$_3$)(OCH$_3$)	3-(4-bromophenyl)-1-methoxy-1-methylurea	metobromuron
Cl, Br—⟨ring⟩—NH—CO—N(CH$_3$)(OCH$_3$)	3-(4-bromo-3-chlorophenyl)-1-methoxy-1-methylurea	chlorbromuron
F$_3$C—⟨ring⟩—NH—CO—N(CH$_3$)(CH$_3$)	1,1-dimethyl-3-(3-trifluoromethyl-phenyl) urea	fluometuron
Cl—⟨ring⟩—O—⟨ring⟩—NH—CO—N(CH$_3$)(CH$_3$)	3-[4-(4-chlorophenoxy)-phenyl]-1,1-dimethylurea	chloroxuron
⟨ring⟩—NH—CO—NH—⟨cyclohexyl, CH$_3$⟩	1-(2-methylcyclohexyl)-3-phenylurea	siduron
Cl—⟨ring⟩—NH—CO—N(CH$_3$)(CH—CH$_3$; C≡CH)	3-(4-chlorophenyl)-1-methyl-1-(1-methyl-prop-2-ynyl) urea	buturon

TABLE 4

Structures of phenylcarbamate herbicides

Structure	Chemical name	Common name
[structure] NH—COO—CH(CH₃)₂	isopropyl phenyl-carbamate	propham (IPC, IPPC)
[structure] Cl	NH—COO—CH(CH₃)₂	isopropyl 3-chloro-phenylcarbamate	chlorpropham (CIPC)
[structure] Cl	NH—COO—CH₂—CH₂—Cl	2-chloroethyl-3'-chloro-phenylcarbamate	CEPC
[structure] Cl, Cl	NHCOOCH₃	methyl-3,4-dichloro-carbanilate	swep
[structure] Cl	NH—COO—CH₂—C≡C—CH₂Cl	4-chlorobut-2-ynyl 3-chlorophenylcarbamate	barban

Several microbial species have been obtained which can use members of this class of herbicide as sole source of carbon and energy for growth, including *Pseudomonas striata, Fusarium solani* and species of *Penicillium* and *Pulullaria* (Bartha and Pramer, 1970). In all cases, these organisms possess an acylamidase activity which cleaves the substrate forming an aniline and an aliphatic moiety. The latter is used as a growth substrate by the organism and the former accumulates. This mechanism has been demonstrated for *F. solani* growing on propanil (Lanzilotta and Pramer, 1970a, b); the organism utilized the propionic acid formed by hydrolysis and released 3,4-dichloroaniline. Similarly, two species of *Penicillium* and one of *Pulullaria* have been shown to release 3,4-dichloroaniline from Karsil and to grow on 2-methylvaleric acid (Sharabi and Bordeleau, 1969). In neither case was the chlorinated aniline metabolized.

The acylamidases responsible have been obtained from extracts of *F. solani* (Lanzilotta and Pramer, 1970b) and a *Penicillium* sp. (Sharabi and Bordeleau, 1969) and their properties have been examined. The specificities of the enzymes from the two sources were found to differ markedly.

In contrast, *Rhizopus japonicus* metabolized acylanilides by a different mechanism (Wallnöfer *et al.*, 1972, 1973a). Instead of the hydrolytic reaction found with other fungi, this organism was able to hydroxylate the aliphatic portion of Karsil and chloranocryl (dicryl) to form the products shown in Fig. 6. These products, identified by mass spectroscopy and n.m.r., were shown to accumulate in cultures of *Rh. japonicus* growing in a glucose medium supplemented with the herbicides. With both Karsil and chloranocryl, hydroxylation in this way effected a detoxification of the herbicides.

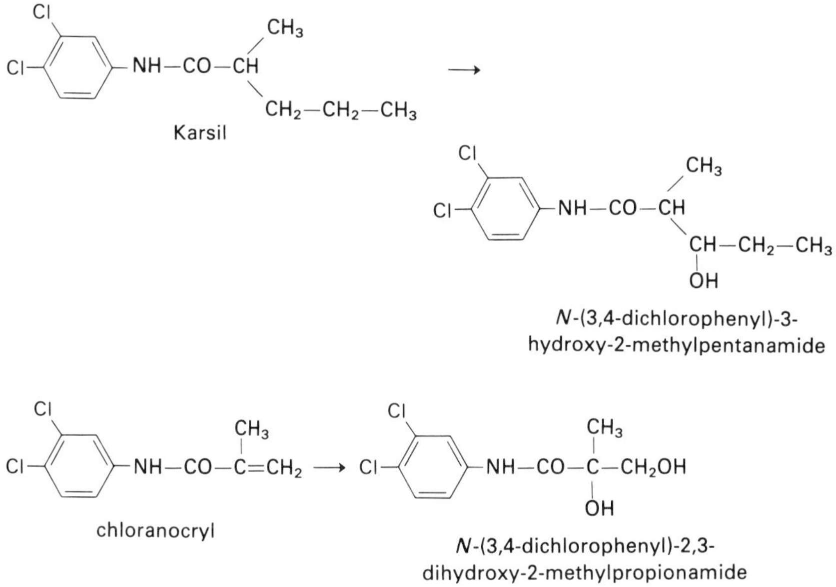

N-(3,4-dichlorophenyl)-3-
hydroxy-2-methylpentanamide

N-(3,4-dichlorophenyl)-2,3-
dihydroxy-2-methylpropionamide

Fig. 6. Metabolism of Karsil and chloranocryl (dicryl) by *Rhizopus japonicus* (after Wallnofer *et al.*, 1972, 1973a).

Recent work has demonstrated that the anilines formed during acylanilide degradation in soil can in some cases, react to form condensation products. Thus, Bartha and Pramer (1967) demonstrated that when propanil was added to soil at the 500 ppm level, 3,4-dichloroaniline and 3,3',4,4'-tetrachloroazobenzene (TCAB) were formed and the degradation of the side chain of the herbicide to CO_2 was inferred. This was the first of a number of reports of the formation of condensation products by aniline-based herbicides, particularly the acylanilides, when application rates to soil were high. Bartha (1969) showed that when propanil and solan were incubated with soil together, both at the 500 ppm level, the mixed azobenzene was formed

(3,3′,4-trichloro-4′-methylazobenzene) in addition to the expected symmetrical derivatives. In separate studies, the formation of 3-chloro-4-methylaniline from solan was demonstrated (Shirakawa, 1970).

A more extensive study of the metabolism of propanil in soils using radio-labelled materials was reported by Chisaka and Kearney (1970). The release of $^{14}CO_2$ from (carbonyl-^{14}C)- and (ring-^{14}C)-labelled propanil was measured when the compounds were added to a range of Japanese soils. More than 60% of the carbonyl radio-label appeared as $^{14}CO_2$ within five days compared with less than 3% from the ring position after 25 days. 3,4-Dichloroaniline and TCAB were also formed and several unidentified products were detected, indicating that the metabolic picture was not as simple as had appeared from earlier work. One of the unknown products was subsequently identified as 1,3-bis-(3,4-dichlorophenyl)triazene (Plimmer et al., 1970).

Kearney et al. (1970) supplied data on the analysis of field soils which had received repeated applications of propanil at a high rate. TCAB was found to be present at the 0·01–0·18 ppm level which showed that, although present, the concentrations were low considering the high application rates of propanil used. Nevertheless it was of interest to observe that condensation reactions could occur under field conditions. Other studies (Hughes and Corke, 1974) showed that the formation of TCAB from propanil varied considerably from soil to soil and in some cases no TCAB was formed at all.

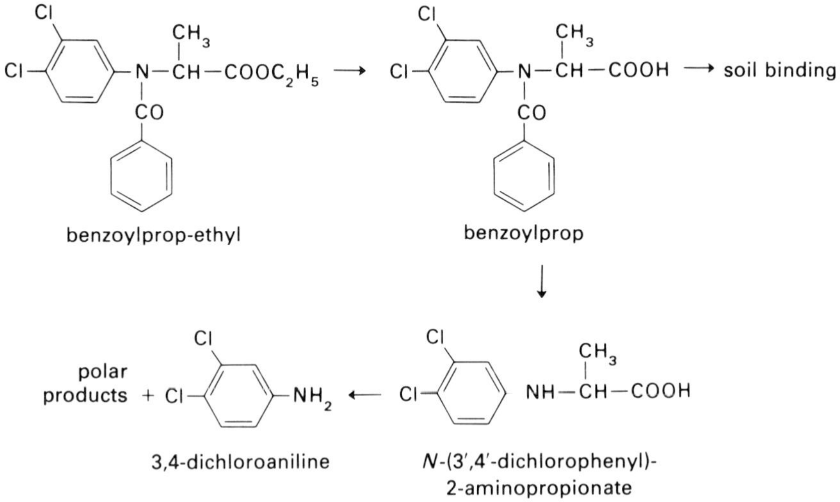

Fig. 7. Metabolism of benzoylprop-ethyl in soil (after Beynon et al., 1974).

The soil metabolism of propanil has been compared with that of chloranocryl, Karsil and propachlor (Bartha, 1968). Dicryl and Karsil degraded at a similar rate to that of propanil but the tri-substituted compound, propachlor, was less readily metabolized, giving rise to negligible amounts of aniline and no azobenzene. Benzoylprop-ethyl is also tri-substituted at the nitrogen atom and its rate of degradation to the parent aniline in soil was very slow. The major soil metabolite was the corresponding acid which underwent a slow binding process to become strongly bound to soil, as well as a slow debenzoylation to give an unstable acid which was metabolized to 3,4-dichloroaniline and polar products (Fig. 7) (Beynon *et al.*, 1974).

B. PHENYLCARBAMATES

Numerous studies with pure cultures, *in vitro*, have established that the major route of phenylcarbamate metabolism and detoxification is by hydrolysis to the free aniline, carbon dioxide and an alcohol. It is still not clear whether the amide or the ester bond is cleaved in the reaction since, in either case, the products would be the same. If the ester bond was hydrolysed, an unstable phenylcarbamic acid plus an alcohol would result; if the amide bond was attacked, an aniline plus an unstable alkyl carbonate would be formed. These possibilities are illustrated in Fig. 9 for chlorpropham (CIPC). Reaction mechanisms

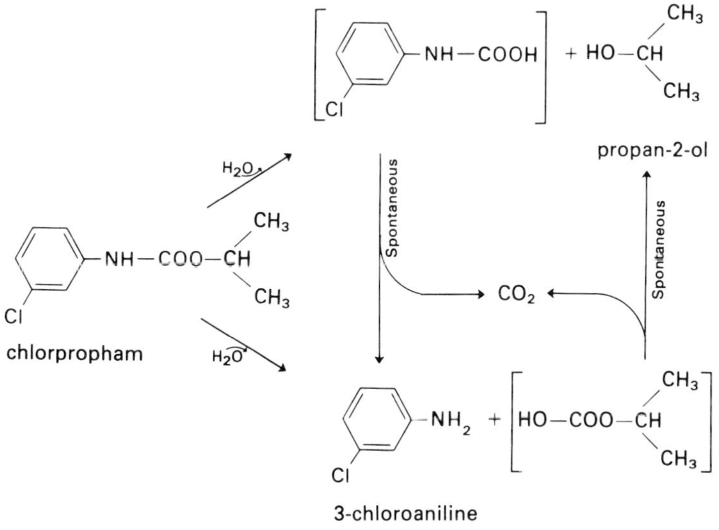

Fig. 8. Possible hydrolytic reactions of chlorpropham.

for these types of hydrolytic attack have been discussed by Herrett (1969).

Many microorganisms have been shown to possess the ability to hydrolyse phenylcarbamates. Thus, Kaufman and Kearney (1965) isolated species of *Pseudomonas, Agrobacterium, Flavobacterium* and *Achromobacter* which could use CIPC as sole carbon source and, also, a species of *Arthrobacter* and a further *Achromobacter* sp. which utilized CEPC. These isolates all converted their substrates into 3-chloroaniline and finally released all available chlorine atoms as chloride ion, indicating that mineralization was complete. The organisms isolated on CIPC also degraded CEPC but at a slower rate. In addition, they only released a maximum of 50% of the chlorine in the CEPC molecule. The herbicide IPC (propham) was degraded rapidly by all the isolates.

Similar results were obtained by Clark and Wright (1970a, b) with an *Arthrobacter* sp. and an *Achromobacter* sp., isolated by virtue of their ability to use IPC as sole carbon source. IPC was converted into aniline (and isopropanol) and both these compounds were used as carbon sources by the isolates. Respirometric data indicated that the aniline was degraded via catechol. The organisms were unable to grow on CIPC or barban, but formed 3-chloroaniline from both these herbicides. More recently, Wright and Forey (1972) have demonstrated the metabolism of barban by *Penicillium jenseni*. In this case, the metabolism of the 3-chloroaniline intermediate did not result in the release of chloride ion.

Several other studies have revealed that the ability to hydrolyse phenylcarbamates is very widespread amongst microorganisms. Thus, Kaufman and Blake (1973) and McClure (1974) have reported the isolation of several species of fungi and bacteria able to grow on IPC, CIPC and swep (as well as some acylanilide herbicides). The substrate specificity of each isolate towards various phenylcarbamates and the degree to which halide ion was released was found to vary considerably from organism to organism (Kaufman and Blake, 1973).

Some details of the hydrolytic reaction have been obtained from studies of an enzyme from *Pseudomonas striata* Chester (Kearney, 1965, 1967). This hydrolase has a very low specificity for phenylcarbamates and the cleavage reaction has been shown to be influenced by the physico-chemical properties of the substrate. The rate of reaction was dependent on the size of the alcohol moiety (steric effect) and was also influenced by inductive effects of ring substituents.

The soil metabolism of this class of herbicides has not been studied extensively. Since the subject was reviewed by Kaufman (1965) and Herrett (1969) there have been few further reports. Bartha and Pramer

(1969) made a brief study of the fate of swep in soil and found that, when application rates were high (500 ppm), it was converted into 3,4-dichloroaniline and TCAB, the formation of the latter compound showing a similarity to that of other aniline-based herbicides. The subject of azobenzene formation is discussed in depth in section III.D. Aniline formation was observed in earlier work on the degradation of CIPC and CEPC in soil perfusion experiments. In these cases, 3-chloroaniline was formed (Kaufman and Kearney, 1965).

C. PHENYLUREAS

The major mechanism of phenylurea detoxification is by N-dealkylation or N-dealkoxylation. Geissbühler et al. (1963) demonstrated the stepwise N-demethylation of chloroxuron in soils, and similar reactions have been shown to occur with diuron (Dalton et al., 1966) and fluometuron (Bozarth and Funderburk, 1971). In addition, Dalton et al. (1966) analysed field soils which had been treated with annual applications of 5 lb acre^{-1} of diuron for 4–7 years and these were found to contain $0 \cdot 2$–$0 \cdot 3$ ppm of metabolites in which one or both of the N-methyl groups had been lost ($0 \cdot 2$–$0 \cdot 3$ ppm of unchanged diuron was also present).

The N-demethoxylation of methoxy-substituted phenylureas has also been described. Schuphan (1974) followed the evolution of $^{14}CO_2$ from ^{14}C-methoxy-labelled monolinuron when added to sterile and non-sterile soil. Over a period of 12 months, 58% of the applied radio-label was released from the non-sterile soil compared with only 9% from the sterile soil, indicating a biological process was involved.

Similar reactions have been reported to be carried out by pure strains of fungi, in vitro. For example, Talaromyces wortmanii, when grown in the presence of metobromuron, converted this compound into the demethylated and demethoxylated derivatives (Tweedy et al., 1970). Also formed were p-bromophenylurea and p-bromoacetanilide. The identification of these metabolites in cultures of T. wortmanii allowed a metabolic pathway to be proposed as shown in Fig. 9. Thus metobromuron was converted into p-bromophenylurea by successive demethylation and demethoxylation, or vice versa, and the urea so formed was presumed to be metabolized to the corresponding aniline which was then acetylated. p-Bromoaniline was not isolated but it was shown that the fungus was able to rapidly acetylate this compound, a result which suggested that the accumulation of p-bromoaniline was unlikely. Similar reaction sequences to those shown in Fig. 9 have been suggested for the metabolism of chlorbromuron by *Rhizoctonia*

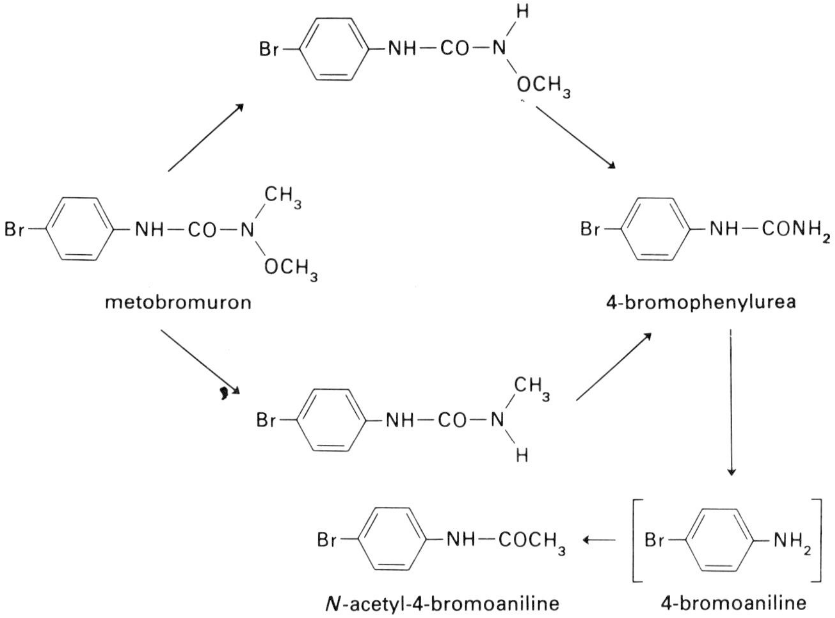

Br—⟨benzene⟩—NH—CO—N(H)(OCH₃)

Br—⟨benzene⟩—NH—CO—N(CH₃)(OCH₃)
metobromuron

Br—⟨benzene⟩—NH—CONH₂
4-bromophenylurea

Br—⟨benzene⟩—NH—CO—N(CH₃)(H)

Br—⟨benzene⟩—NH—COCH₃ ← [Br—⟨benzene⟩—NH₂]
N-acetyl-4-bromoaniline   4-bromoaniline

Fig. 9. Degradation of metobromuron by *Talaromyces wortmanii* (after Tweedy *et al.*, 1970).

solani (see Bollag, 1972) and for linuron and monolinuron by *Aspergillus niger* (Börner, 1967). Both organisms were able to catalyse either demethylation or demethoxylation of the parent substrate and further metabolize these products to the unsubstituted halogenophenylurea.

A slightly different reaction was obtained with *Rhizopus japonicus* (Wallnöfer *et al.*, 1973b) which effected the removal of only a single alkyl group from a range of phenylurea herbicides. Thus, monuron, monolinuron and fluometuron were mono-demethylated and buturon was metabolized by removal of the C_4 group. These products were not further degraded by the fungus.

The mechanism of the microbial demethylation of phenylurea herbicides is unknown but may involve an initial formation of the hydroxymethyl derivatives (Geissbühler, 1969; Bollag, 1972). This type of reaction has been demonstrated in plants (Tanaka *et al.*, 1972).

In contrast to the foregoing, siduron has been reported to be metabolized in soil by hydroxylation (Belasco and Langsdorf, 1969). (^{14}C) Siduron was applied to soil and, subsequently, the 4-hydroxyphenyl and the 4-hydroxycyclohexane derivatives were recovered. A product containing both hydroxyl substitutions in the same molecule was also formed. All three of these compounds had previously been encountered as animal metabolites (Belasco and Reiser, 1969) having been obtained from dog urine and rigorously identified by spectroscopic techniques.

A further mechanism of microbial degradation of phenylureas has been shown to be carried out by *Bacillus sphaericus*, ATCC 12123 (Wallnöfer and Bader, 1970; Engelhardt *et al.*, 1972). This organism, isolated from monolinuron-treated soil, was able to convert several phenylurea herbicides into the corresponding anilines. This reaction was carried out by an acylamidase which has been partially purified and shown to be specific for phenylureas substituted with an *N*-methoxy group. The reaction catalysed is shown for linuron and monolinuron in Fig. 10. The ureido carbon atom of the substrate is converted into CO_2 and *N,O*-dimethylhydroxylamine is also formed. This latter metabolite has recently been reported as a soil metabolite of monolinuron (Schuphan, 1974).

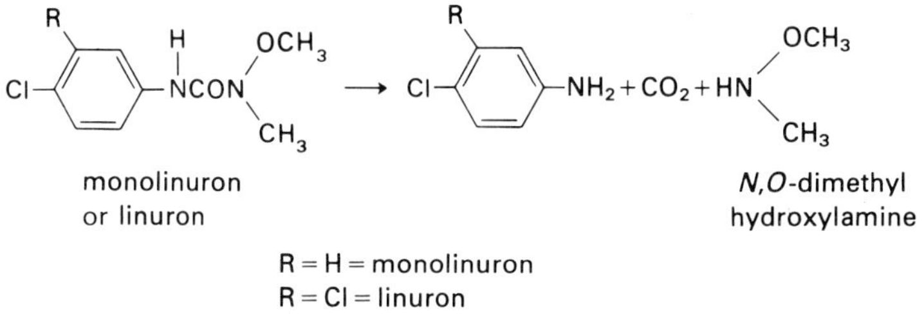

monolinuron
or linuron

N,O-dimethyl
hydroxylamine

R = H = monolinuron
R = Cl = linuron

Fig. 10. Degradation of linuron and monolinuron by an acylamidase from *Bacillus sphaericus* ATCC 12123 (after Engelhardt *et al.*, 1972).

The formation of the parent aniline during phenylurea degradation in soils is generally assumed to occur but the evidence has often been circumstantial (Geissbühler and Voss, 1971). One reason for this is that anilino-radiolabelled ureas were not used in early metabolism studies and the formation of labelled anilines may have been missed. Nevertheless, Dalton *et al.*, (1966) in their field analyses for diuron and its metabolites, detected 3,4-dichloroaniline in soils following repeated applications of that herbicide. Börner (1965) also found 3,4-dichloroaniline and 4-chloroaniline in soils treated with linuron and monolinuron, respectively.

In their work with fluometuron, Bozarth and Funderburk (1971) used ^{14}C-trifluoromethyl-labelled material and, in addition to demethylated products, tentatively identified 3-trifluoromethylaniline (TFMA) as a soil degradation product. Since TFMA was found in both autoclaved and non-autoclaved soils, they concluded that its formation may not have been the result of a biological reaction. The observation that the concentration of TFMA in non-sterile soil did not increase with time was consistent with the fact that $^{14}CO_2$ was released at a steady rate from these soils throughout the incubation period, indicating that further breakdown (presumably of TFMA) was occurring. Based on the available evidence, the proposed degradation of fluometuron in soil is shown in Fig. 11.

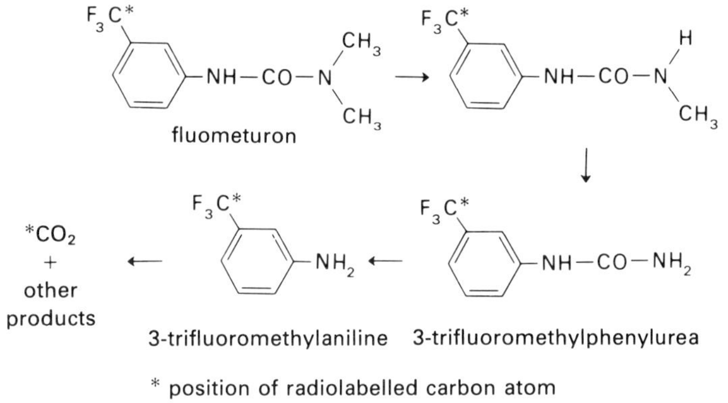

Fig. 11. Degradation of fluometuron in soil (after Bozarth and Funderburk, 1971).

D. HERBICIDE-DERIVED ANILINES

The formation of the parent aniline condensation products from phenylamide herbicides, as described above, has provided a great stimulus to investigations of the microbial metabolism of the herbicide-derived anilines themselves, both in soils and in pure cultures. In retrospect, it is unfortunate that many of the early studies carried out with soil involved the use of very high aniline concentrations when in practice, levels are likely to be much less than 1 ppm. The use of applications of 500–1000 ppm resulted in the formation of azobenzenes in high yield and overemphasized the importance of this aspect of the fate of anilines in soil. For example, Bartha et al. (1968) studied the extent of azobenzene formation from a range of substituted anilines added to soil at concentrations of 1000 ppm. No azobenzenes were formed from aniline itself or from trichloroanilines but all monochloroanilines and most, but not all, dichloroanilines produced symmetrical azobenzenes. Mixtures of chloroanilines led to the formation of the asymmetrical azobenzenes, in addition to the expected symmetrical compounds (Fig. 12) (Kearney et al., 1969a). These studies also revealed that several other unidentified products were formed during aniline metabolism.

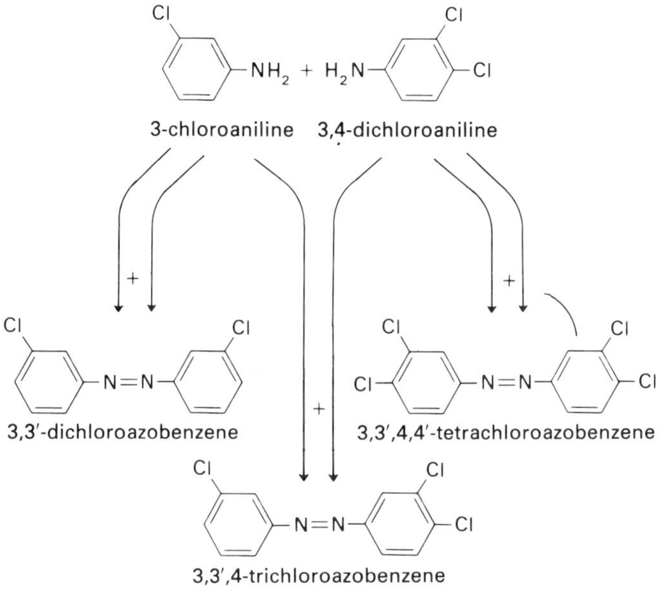

Fig. 12. Azobenzenes formed in soil from mixtures of chloroanilines (after Kearney *et al.*, 1969).

The formation of mixed azobenzenes was confirmed by Bordeleau and Bartha (1970) and these workers produced evidence for the participation of chlorophenylhydroxylamines in chloroazobenzene formation. The suggested mechanism was that the aniline was oxidized in a biological reaction to the hydroxylamine derivative which then could condense with excess aniline to form the azobenzene. Since the latter reaction took place in biologically inactive media, it was considered to be a chemical process.

Evidence that this type of amino-group oxidation reaction could occur in soils as a result of microbial activity has been provided by *in vitro* studies with the fungus, *Fusarium oxysporum* Schlect. This organism can metabolize anilines to a variety of products including some which arise by the condensation of two aromatic units. Thus, the addition of 3,4-dichloroaniline to cultures growing on sucrose, can, depending on the cultural conditions, lead to the formation of 3,3',4,4'-tetrachloroazoxybenzene, -azobenzene or -triazene (Kaufman *et al.*, 1972). 3,4-Dichloronitrobenzene was also formed. The accumulation of these compounds led to the postulation that aniline metabolism proceeds by oxidation of the amino-group forming, successively, the phenylhydroxylamine, nitrosobenzene and nitrobenzene (Fig. 13). Self-condensation of the hydroxylamine or condensation with the nitrosobenzene can result in azoxybenzene formation, whereas reaction of unchanged aniline with the hydroxylamine or nitrosobenzene could lead to the azobenzene. The mechanism by which the triazene was formed was not clear and may have involved a reaction of the 3,4-dichloroaniline with biologically produced nitrite.

Direct evidence that *F. oxysporum* Schlect oxidized the amino group of anilines has come from studies with 4-chloroaniline (Kaufman *et al.*, 1973). In similar experiments to those with 3,4-dichloroaniline, the chlorophenylhydroxylamine, nitroso- and nitrobenzenes were actually detected in cultures of the fungus growing in the presence of this monochloroaniline, as well as the expected azo- and azoxy-benzenes. These results strongly support a metabolic route similar to the scheme shown in Fig. 13. On prolonged incubation of this system, phenolic compounds were also detected and, ultimately, an almost quantitative release of chloride ion was observed. This suggests that the aniline condensation products may be, in this case at least, susceptible to degradation.

Further studies on the formation of azobenzenes from chlorinated anilines have been made by Bordeleau and Bartha (1972a, b, c). In soil, this reaction was correlated with soil peroxidase activity and, *in vitro*, incubation of 3,4-dichloroaniline with hydrogen peroxide and peroxidase had previously been shown to result in the formation of TCAB (Bordeleau *et al.*, 1972). Amino-group oxidation followed by condensation reactions was again postulated. At the microbiological level, a fungus, *Geotrichum candidum* was isolated from soil which possessed high concentrations of peroxidase and which could bring about TCAB formation from 3,4-dichloroaniline. This organism also possessed a second enzyme, termed aniline oxidase, which could catalyse azobenzene formation. The two enzymes were separated, partially purified and studied as a model system of aniline metabolism in soil. A large number of substituted anilines were exposed to the enzymes and the reaction products were identified. The results showed that with increasing electron density at the amino group, susceptibility to enzymic transformation increased. Yields of azobenzenes were reduced when electron-releasing substituents were present in the aniline molecule but in these cases, higher molecular weight polymers were formed.

The more recent work, providing data from studies with soil, microorganisms and enzymes, has revealed mechanisms by which aniline condensation products can be formed and have shown several of the factors involved in these reactions. These data could be useful in predicting the fate of herbicide-derived anilines in soil but other studies have revealed that other factors are involved and that conden-

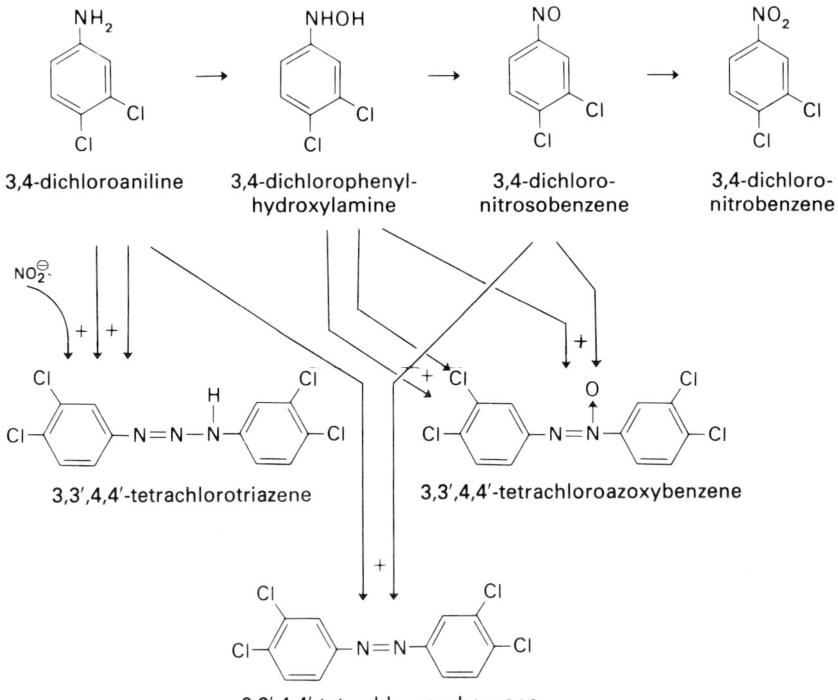

NH₂ → NHOH → NO → NO₂

3,4-dichloroaniline

3,4-dichlorophenyl-
hydroxylamine

3,4-dichloro-
nitrosobenzene

3,4-dichloro-
nitrobenzene

NO₂⁻

Cl—⟨⟩—N=N—N(H)—⟨⟩—Cl

3,3′,4,4′-tetrachlorotriazene

Cl—⟨⟩—N=N(→O)—⟨⟩—Cl

3,3′,4,4′-tetrachloroazoxybenzene

Cl—⟨⟩—N=N—⟨⟩—Cl

3,3′,4,4′-tetrachloroazobenzene

Fig. 13. Metabolism of 3,4-dichloroaniline by *Fusarium oxysporum* Schlect and the formation of condensation products (after Kaufman *et al.*, 1972).

sation reactions are not as important when concentrations more in keeping with normal pesticide application rates are involved. For example, Bartha (1971), using ring-^{14}C-labelled propanil and 4-chloroaniline, demonstrated that a proportion of the aniline became firmly bound to soil and could not be extracted with organic solvents. Furthermore, the proportion of soil-bound radiolabel increased from 53% to 77% as the concentration of the applied compounds was decreased from 50 to 5 ppm. It was also observed that ^{14}CO₂ was released from ring labelled (^{14}C) propanil and (^{14}C) 4-chloroaniline at a slow but positive rate (1·5–2·1% within 2–3 weeks).

Kearney and Plimmer (1972) applied (^{14}C) 3,4-dichloroaniline to soil at concentrations of 1000, 100, 10 and 1 ppm. When applied at 1 or 10 ppm to a silty clay loam soil, less than 1% was converted into TCAB. At 1000 ppm, a 14% conversion was observed. More recent work has shown that no TCAB was detected when 3,4-dichloroaniline was added at 17·5 ppm to a sandy loam soil (Beynon *et al.*, 1974) but that highly polar products were formed which could only be released from the soil with sodium hydroxide at 80°C. These reports emphasize, therefore, that although azobenzenes can be formed when chloroanilines are applied to soil, the major process, at concentrations likely to arise in use, is one of binding to the soil and the formation of highly polar products. The precise nature of these products remains as yet unknown although there is evidence for the presence of complexes of anilines with humic acids (Bartha, 1971; Beynon et al., 1974).

Other metabolic reactions of anilines have been reported with pure cultures of bacteria. These involve ring hydroxylation but not amino-group oxidation. Thus, Walker and Harris (1969) isolated a bacterium which, after growth on aniline, oxidized aniline and catechol but not phenylhydroxylamine. Similarly, Clark and Wright (1970b) with *Arthrobacter* sp. and *Achromobacter* sp. able to grow on IPC, showed that when these organisms were induced to oxidize aniline, catechol could also be metabolized. Although detailed studies on the reaction are lacking, the implication is that the bacterial metabolism of aniline involves conversion into catechol (with the concomitant formation of ammonia). Catechol is known to be further metabolized by many species of microorganisms to aliphatic compounds and, ultimately, to be completely mineralized. Hence the possibility arises that chlorinated anilines may also be susceptible to these reactions to some degree and that this may account for the observed release of CO_2 from the aniline ring of phenylamide herbicides.

IV. The *s*-Triazines

The herbicidal properties of the substituted *s*-triazines were discovered in the early 1950s and these herbicides have become widely used in agriculture. The history of the development of the *s*-triazines as herbicides has been summarized by Knüsli (1970) in an introduction to a symposium on the behaviour of *s*-triazines in soil. Since the 2-chloro-*s*-triazines were introduced, 2-methoxy- and 2-methyl-thio derivatives have been introduced commercially and the most important triazine herbicides are listed in Table 5.

Numerous workers have shown that *s*-triazines can be degraded by pure cultures of microorganisms and that, in many cases, these compounds can act as nutrient sources for the growth of microbial species. This capability has been demonstrated most commonly with species of fungi although there are some reports to show that actinomycetes and bacteria can also degrade the *s*-triazines. For example, Kaufman et al. (1965) showed that species of *Aspergillus*,

TABLE 5

Structures of s-triazine herbicides

General structure

Substituents			Chemical name	Common name
R_1	R_2	R_3		
Cl	NH.C$_2$H$_5$	NH.C$_2$H$_5$	2-chloro-4,6-bis-ethylamino-1,3,5-triazine	simazine
Cl	NH.C$_2$H$_5$	NH.isoC$_3$H$_7$	2-chloro-4-ethylamino-6-isopropylamino-1,3,5-triazine	atrazine
Cl	NH.isoC$_3$H$_7$	NH.isoC$_3$H$_7$	2-chloro-4,6-bis-isopropylamino-1,3,5-triazine	propazine
Cl	NH.C$_2$H$_5$	N.(C$_2$H$_5$)$_2$	2-chloro-4-diethylamino-6-ethylamino-1,3,5-triazine	trietazine
Cl	N.(C$_2$H$_5$)$_2$	N.(C$_2$H$_5$)$_2$	2-chloro-4,6-bis-diethylamino-1,3,5-triazine	chlorazine
Cl	N.(C$_2$H$_5$)$_2$	NH.isoC$_3$H$_7$	2-chloro-4-diethylamino-6-isopropylamino-1,3,5-triazine	ipazine
Cl	NH.C$_2$H$_5$	NH.C(CH$_3$)$_2$.CN	2-chloro-4-(1-cyano-1-methyl-ethylamino)-6-ethylamino-1,3,5-triazine	cyanazine
SCH$_3$	NH.C$_2$H$_5$	NH.C$_2$H$_5$	2-methylthio-4,6-bis-ethylamino-1,3,5-triazine	simetryne
SCH$_3$	NH.C$_2$H$_5$	NH.isoC$_3$H$_7$	2-methylthio-4-ethylamino-6-isopropylamino-1,3,5-triazine	ametryne
SCH$_3$	NH.isoC$_3$H$_7$	NH.isoC$_3$H$_7$	2-methylthio-4,6-bis-isopropyl-1,3,5-triazine	prometryne
SCH$_3$	NH.C$_2$H$_5$	NH.C(CH$_3$)$_3$	2-methylthio-4-ethylamino-6-tert.butylamino-1,3,5-triazine	terbutryne
OCH$_3$	NH.C$_2$H$_5$	NH.C$_2$H$_5$	2-methoxy-4,6-bis-ethylamino-1,3,5-triazine	simeton
OCH$_3$	NH.C$_2$H$_5$	NH.isoC$_3$H$_7$	2-methoxy-4-ethylamino-6-isopropylamino-1,3,5-triazine	atraton
OCH$_3$	NH.isoC$_3$H$_7$	NH.isoC$_3$H$_7$	2-methoxy-4,6-bis-isopropylamino-1,3,5-triazine	prometon

Fusarium, Penicillium, Rhizopus, Trichoderma and *Arthrobacter* could use simazine as a sole or supplemental source of carbon for growth, whereas Bortels *et al.* (1967) isolated several soil microorganisms (mainly fungi) which could use simazine only as a source of nitrogen. Similar results have been obtained with other herbicidal s-triazines (Kaufman and Kearney, 1970). In addition, the methylthio-s-triazine, prometryne, has been shown to act as a source of sulphur for *Aspergillus niger*, *Aspergillus tamarii* and *Aspergillus flavus* (Murray *et al.*, 1970).

The details of the ways in which *s*-triazines are degraded by microorganisms are not yet clear but the process has been elucidated in broad outline. It is apparent that the most common degradative mechanisms are dealkylation and deamination and that ring-cleavage has seldom been conclusively shown to occur. Most studies have been carried out with *Aspergillus fumigatus* Fres, an organism capable of degrading many *s*-triazines (Kaufman and Kearney, 1970). This fungus can utilize simeton as the sole source of either carbon or nitrogen or both but only the side chains of the molecule are degraded. Experiments with (^{14}C) simeton showed that substantial amounts of radioactivity were incorporated into cellular material from chain-labelled substrate but not from ring-labelled material (Kaufman *et al.*, 1965). This, and the observation that no $^{14}CO_2$ was released from ring-labelled simeton indicated that ring-cleavage was not effected by this fungus.

A. fumigatus also degraded simazine and atrazine without ring-cleavage (Kearney *et al.*, 1965a; Kaufman and Blake, 1970) and with these *s*-triazines some aspects of the degradative pathways have been revealed. Thus, when this organism was cultured in the presence of ^{14}C- or ^{36}Cl-labelled simazine, two products were detected (Kearney *et al.*, 1965a). One of these was conclusively shown to be the dealkylated product 2-amino-4-chloro-6-ethylamino-*s*-triazine. The other metabolite, which was not identified, retained the chlorine atom and the intact ring structure but had lost both ethyl groups. Ammelide (2-amino-4,6-dihydroxy-*s*-triazine) was also shown to be formed from simazine by this organism, indicating that deamination is also possible. Attempts to isolate enzymes from *A. fumigatus* Fres which could degrade simazine were not successful.

Atrazine has been shown to be degraded by *A. fumigatus* Fres in a similar way (Kaufman and Blake, 1970). There was evolution of $^{14}CO_2$ only from chain-^{14}C-labelled substrate and several dealkylated metabolites have been obtained from cultures, including 2-amino-4-chloro-6-isopropylamino-*s*-triazine and 2-amino-4-chloro-6-ethyl-amino-*s*-triazine. Other products, which as yet are unidentified, were also formed.

With *A. fumigatus* no release of the chlorine atom from the chloro-*s*-triazine molecule has been observed. However, when *Fusarium roseum* was cultured with sucrose and atrazine, hydroxyatrazine (2-ethyl-amino-4-hydroxy-6-isopropylamino-*s*-triazine) was formed (Couch *et al.*, 1965). This indicates that microbial hydrolytic dechlorination can occur.

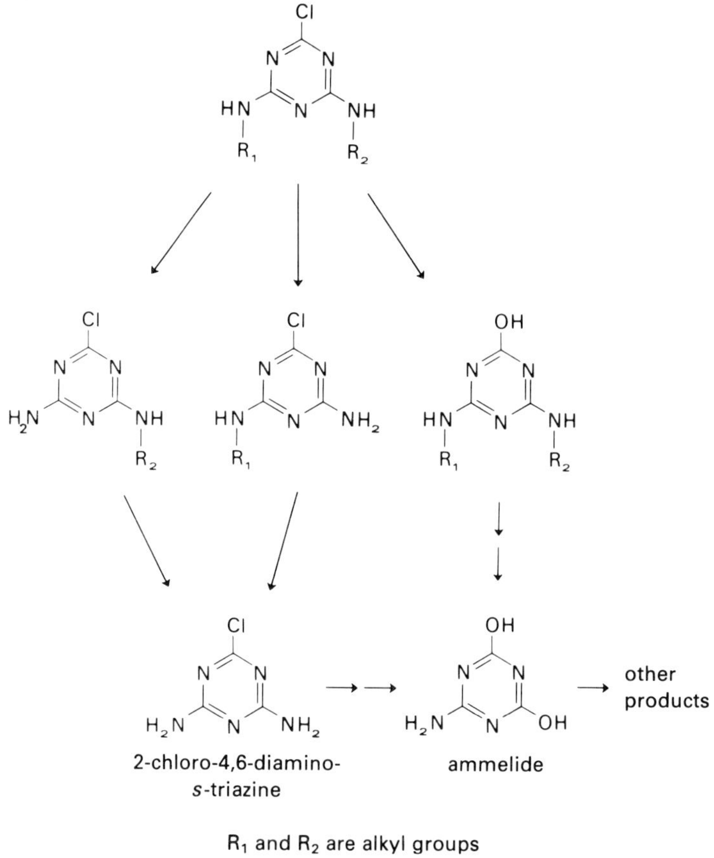

Fig. 14. Proposed pathways of chloro-*s*-triazine degradation by microorganisms.

These results allow a scheme for chloro-*s*-triazine bio-degradation to be postulated as shown in Fig. 14. Not all the reactions indicated have been demonstrated conclusively, particularly those occurring after the initial dealkylations. Furthermore, the scheme does not include any ring-cleavage reactions since the validity of most reports of ring-cleavage of *s*-triazines is difficult to assess. Many reports of the release of $^{14}CO_2$ from ring-labelled *s*-triazine herbicides from complex, mixed culture situations, including soil, have been published (see below), indicating that breakdown of the ring structure is possible under some conditions. However, convincing evidence that pure microbial isolates can bring about ring-cleavage of the herbicidal *s*-triazines is not forthcoming. Nevertheless, ammelide, the last metabolite shown in Fig. 14, is susceptible to microbial attack. Ammelide, as well as cyanuric acid (2,4,6-trihydroxy-*s*-triazine), and ammeline (2,4-bis-amino-6-hydroxy-s-triazine) have all been shown to undergo nitrification (Hauck and Stephenson, 1964) and cyanuric acid has been shown to act as a nitrogen source for some microorganisms (Jensen and Abdel-Ghaffar, 1969). Hence, although the details of the reaction are not known, it is evident that certain s-triazine compounds can undergo ring-cleavage by microbial action.

Most of the reactions demonstrated in pure culture studies have also been shown to occur in soil but, here, it has been demonstrated that the major degradative reaction of s-triazines is hydroxylation (Harris, 1967; Armstrong et al., 1967; Skipper et al., 1967; Obien and Green, 1969; Beynon et al., 1972a, b). Harris (1967) compared the fate of simazine, atrazine and propazine in four soils and found that, although the rate of hydroxylation varied somewhat with the soil type, the hydroxylation of atrazine and propazine was more rapid than that of simazine. For example, eight weeks after the treatment of soils with atrazine and propazine, on average, 40% of the applied herbicide had been hydroxylated. This compared with approximately 30% hydroxylation for simazine in the same time.

Although many cases of the biological hydroxylation of pesticides have been reported, and despite the reported hydroxylation of atrazine by F. roseum (Couch et al., 1965) there is evidence to show that, in soils, hydroxylation of the chloro-s-triazines is mainly the result of chemical hydrolysis. Armstrong et al. (1967) demonstrated that the hydrolysis of atrazine to 2-hydroxyatrazine followed first order kinetics in soil-free systems, in sterilized soil and in a perfusion system. Furthermore, no microbial degradation was detected when a soil-free atrazine medium was incubated with perfusate. Similarly, Skipper et al. (1967) concluded that the hydroxylation of atrazine in soils was a chemical process. In addition, these authors showed that $^{14}CO_2$ was released three times faster from ring-labelled 2-hydroxyatrazine than from ring-labelled atrazine itself.

Cyanazine differs from the aforementioned s-triazines in containing a nitrile group in the side chain. This is the initial site of attack on the compound when it is added to soil (Fig. 15) (Beynon et al., 1972a, b; Sirons et al. 1973). Nevertheless, hydroxylation at the 2-position subsequently occurs although it has not been confirmed that this is the result of chemical rather than microbial attack.

N-Dealkylation in soils, in contrast to hydroxylation, has, almost without exception, been attributed to microorganisms. Skipper and Volk (1972) studied the degradation of atrazine in three Oregon soils using samples of the herbicide radiolabelled, separately, with ^{14}C in the ring, the ethyl group and the isopropyl group. They observed that although hydroxylation was the major degradative reaction, attack on

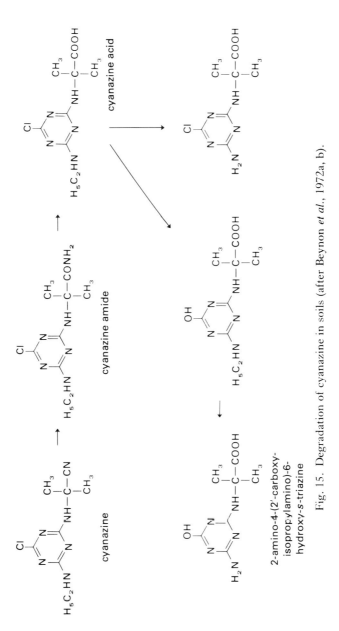

Fig. 15. Degradation of cyanazine in soils (after Beynon *et al.*, 1972a, b).

the ethyl group also occurred. The isopropyl group was attacked to a lesser extent. In one soil, evolution of $^{14}CO_2$ from the side chain amounted to only 4–5% over a 28 day period compared with the formation of 40% of hydroxyatrazine. In the other two soils, 9–10% of the ethyl-^{14}C label was released as $^{14}CO_2$ during the same period and hydroxyatrazine accounted for only 10% of the applied atrazine. Dealkylation of cyanazine at the 6-ethylamino group has also been demonstrated but was less important than attack at the cyano-methylethyl group which led to the formation of cyanazine amide and cyanazine acid (Beynon *et al.*, 1972a, b) (Fig. 15). Traces of the de-ethylated derivatives of cyanazine and cyanazine amide were also detected.

In the case of dialkylamino triazines, sequential dealkylation has been postulated as an explanation for the observed increase in phyto-toxicity of these compounds in soil (Kaufman and Kearney, 1970; Harris *et al.*, 1965). Two examples cited are chlorazine and ipazine; the former could give rise to simazine if one ethyl group were lost from the 4- and 6-positions and the latter would be converted into atrazine after mono-dealkylation (Fig. 16). These reactions have, however, not been shown to occur in soils and remain hypothetical.

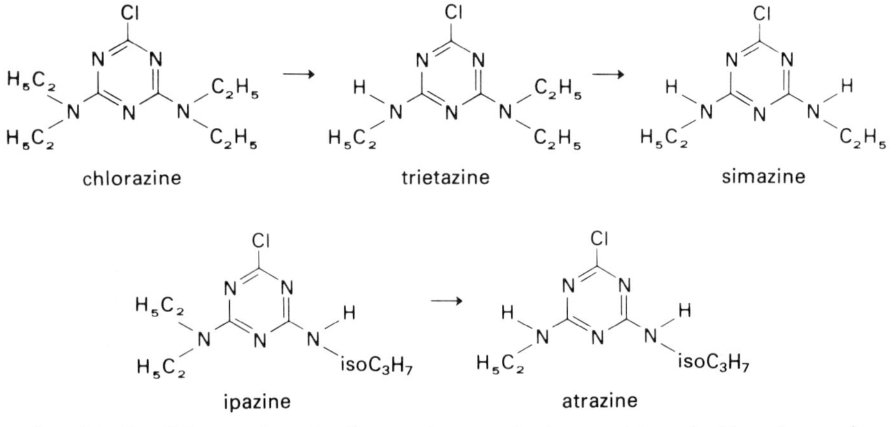

Fig. 16. Possible reactions leading to increased phytotoxicity of chlorazine and ipazine.

The result of dealkylation and dechlorination/hydroxylation of *s*-triazines is the formation of derivatives substituted with hydroxyl or amine groups. Further degradation (apart from deamination) requires the cleavage of the triazine ring but the available evidence points to this being a very minor reaction. In most cases, the evolution of $^{14}CO_2$ from ring-labelled triazines has been taken as evidence of ring cleavage and there are many varied reports of the extent to which this occurs in soil (see Knuesli *et al.*, 1969; Kaufman and Kearney, 1970). Wagner and Chahal (1966) incubated 10 mg litre^{-1} of ring-labelled (^{14}C) atrazine with soil for six months during which time less than 2% of the radio-activity was converted into $^{14}CO_2$. Similarly, Skipper and Volk (1972), also with atrazine, showed that only 0.1% of the compound was converted into $^{14}CO_2$ within two weeks. However, hydroxy-atrazine was degraded more readily than atrazine and $^{14}CO_2$ production amounted to 2.4–3.1%, depending on the soil type, over the two week period. Goswami and Green (1971) also found that the evolution of CO_2 was greater from hydroxyatrazine than from atrazine in Hawaiian soils, although in 90 days, the amounts evolved as CO_2 were only 1·67% and 0·005%, respectively. They also showed that neither compound evolved $^{14}CO_2$ (or $^{14}CH_4$) from the ring under anaerobic conditions.

V. The Benzoic Acid Herbicides

This class of herbicide, which includes not only substituted benzoic acids but also benzonitriles and thioamides (Table 6), are mainly used as post-emergence herbicides. For example, dicamba and bromoxynil octanoate are used for the control of weeds in cereals. Chlorthiamid and dichlobenil, however, are pre-emergence, soil-applied herbicides, which have been used effectively against a range of broad-leaved and grass weeds in top-fruit, soft fruit and vines.

The literature on the microbial metabolism of these compounds is not extensive although the degradation of several benzoic acids in soil has been studied in some detail.

The release of $^{14}CO_2$ from carboxyl-labelled dicamba by cultures of *Bacillus cereus* var. *mycoides* (Cain, 1968) and by an actinomycete (Würzer and Corbin, 1969) has been reported. Evidence for the microbial degradation of dicamba in soil was provided by Smith (1973a) who showed it to be degraded rapidly in Canadian prairie soils but not when the soil was autoclaved. When (*carboxy*-^{14}C) dicamba was added to a heavy clay soil only 10% remained unchanged after five weeks. 3,6-Dichlorosalicyclic acid was identified as a major degradation product and $^{14}CO_2$ was liberated. A bound form of dicamba was also present in soil; it was not extracted by calcium chloride solution but was released by sodium hydroxide (Smith, 1973b). Subsequent work with ring-labelled (^{14}C) dicamba confirmed these findings (Smith, 1974) and furthermore, it was observed that $^{14}CO_2$ was

TABLE 6

Structures of benzoic acid herbicides

Structure	Chemical name	Common name
	3-amino-2,5-dichlorobenzoic acid	chloramben (amiben)
	2,3,6-trichlorobenzoic acid	2,3,6-TBA
	3,6-dichloro-2-methoxy-benzoic acid	dicamba
	2,6-dichlorothiobenzamide	chlorthiamid
	2,6-dichlorobenzonitrile	dichlobenil
	3,5-dibromo-4-hydroxybenzonitrile	bromoxynil
	4-hydroxy-3,5-di-iodobenzonitrile	ioxynil

released which indicated that ring cleavage had occurred. 3,6-Dichloroanisole and 3,6-dichlorophenol were not detected as intermediates in the degradation.

Smith's work showed that the rate of degradation of dicamba was rapid in soils of pH values between 5·2 and 7·7. This is in constrast to observations using bioassay techniques that the rate of loss of dicamba in soil was rapid at pH 5·3 and very slow at pH 7·5 (Corbin and Upchurch, 1967; Parker and Hodgson, 1966) and emphasizes that soil pH is not the only factor which affects the rate of degradation.

The degradation of chlorthiamid and dichlobenil in soil has recently been reviewed by Beynon and Wright (1972). The major degradation product of these herbicides is 2,6-dichlorobenzamide (BAM) (Beynon and Wright, 1968) and the formation of BAM from dichlobenil is thought to be microbiological. BAM itself is resistant to microbial attack. Work in our own laboratory and by Fournier (1974) using the soil perfusion technique, has shown that BAM degraded very slowly but that $^{14}CO_2$ was released from (^{14}C) BAM labelled in the amide group. The rate of $^{14}CO_2$ evolution was reduced if unlabelled 3,5-dichlorobenzamide was added to the soil percolators and it was stopped completely if the 2,5-isomer was present (Fournier, 1974). This latter process was accompanied by the accumulation of radio-labelled 2,6-dichlorobenzoic acid. Previous work (Fournier and Catroux, 1972) had shown that some *Aspergillus* spp. could convert 2-, 3- and 4-chlorobenzamides and 2,5- and 3,5-dichlorobenzamides into the corresponding acids and utilize the amide nitrogen atom released. In this system, BAM did not act as a source of nitrogen for these fungi but the general scheme was established whereby chlorinated benzoic acids could be formed from these amides. Swanson (1969) has also claimed that several fungi can convert dichlobenil into the benzoic acid.

There is some controversy as to whether or not 2,6-dichlorobenzoic acid is formed in the soil from these herbicides. Although its formation might be expected, it is likely that steric hindrance from the 2,6-dichloro-substituents could prevent it. Beynon and Wright (1972) in their review concluded that since BAM was rather stable in soil, any conversion into the corresponding carboxylic acid would be a very slow process but that small amounts of it could be formed in some soils. The degradation pathway of dichlobenil and chlorthiamid is given in Fig. 17.

Similar reactions at the nitrile group of bromoxynil have been suggested (Smith and Cullimore, 1974) by studies *in vitro* using a *Flexibacterium* sp. With this compound, the nitrile group is not hindered sterically and both the amide and the acid accumulated in cultures of the *Flexibacterium* sp. containing bromoxynil. Both the amide and the acid subsequently disappeared from the cultures and cleavage of the aromatic ring was inferred by a corresponding decrease in the ultraviolet absorbance.

Bromoxynil also underwent rapid degradation in soil, with an initial half-life of two weeks (Smith, 1971). Again there was little degradation

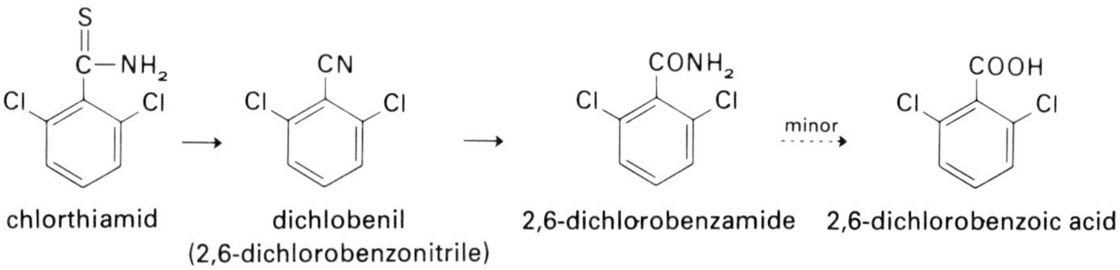

Fig. 17. The degradation of chlorthiamid and dichlobenil in soils (after Beynon and Wright, 1972).

of the herbicide in autoclaved soil. Although this work was executed without the use of radio-labelled bromoxynil, chromatographic evidence was obtained for the presence of small amounts of 3,5-dibromo-4-hydroxy-benzamide and 3,5-dibromo-4-hydroxybenzoic acid. Using the soil perfusion technique, Collins (1973) studied the degradation of bromoxynil octanoate in five soils using cyano-^{14}C and ring-^{14}C-labelled compounds. The herbicide was rapidly metabolized in all soils and up to 80% and 63%, respectively, of the radiolabel from the ^{14}CN group and the ^{14}C-ring was liberated as ^{14}CO$_2$. Collins also suggested that the cyano group was hydrolysed via the amide to the acid with subsequent decarboxylation leading to ring opening, as summarized in Fig. 18. In support of this proposed scheme, trace amounts of the amide and the acid were detected in the perfusates.

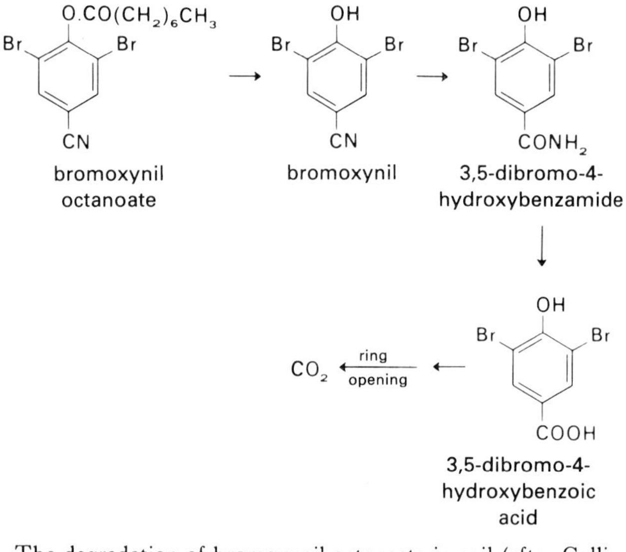

Fig. 18. The degradation of bromoyxnil octanoate in soil (after Collins, 1973).

There is little evidence to show that any of the commonly used benzoic acid herbicides can act as sole carbon sources for any species of microorganism, but since benzoic acid itself is readily bio-degradable many studies have been made on the degradation of substituted benzoic acids. Horvath (1971) described a *Brevibacterium* sp. which, after growth on unsubstituted benzoic acid, oxidized 2,3,6-trichlorobenzoic acid (2,3,6-TBA). One mole of oxygen was consumed and one mole each of CO_2 and Cl^- ion was released for each mole of herbicide metabolized by the cells while a product, identified as 3,5-dichlorocatechol, accumulated. The catechol was not further metabolized by the *Brevibacterium* sp. but an *Achromobacter* sp., also grown on benzoic acid, was shown to be able to cleave the aromatic ring to form 3,5-dichloro-2-hydroxymuconic semialdehyde (Horvath, 1970b). Since 3,5-dichlorocatechol is also a postulated intermediate in a metabolic pathway proposed for the mineralization of 2,4-D by an *Arthrobacter* sp. (Fig. 1) the further metabolism of 2,3,6-TBA is, at least in theory, possible. There is little detailed information on the metabolism of 2,3,6-TBA in soil although Dewey *et al.* (1962) reported that $^{36}Cl^-$ ions were released in up to 50% yield from uniformly labelled (^{36}Cl)2,3,6-TBA. No intermediates retaining the ^{36}Cl-label were detected.

VI. The Chlorinated Aliphatic Acids

The chlorinated aliphatic acids (Table 7) were amongst the first organic herbicides to be used and there are many references to studies of their microbial degradation. In particular, the fate of dalapon and of TCA in pure cultures has frequently been studied. However, few studies have been made of their degradation in soils.

The literature on the metabolism of the chlorinated aliphatic acids has been extensively reviewed (Kearney *et al.*, 1965b; Foy, 1969). In

TABLE 7

Structures of chlorinated aliphatic acid herbicides

Structure	Chemical name	Common name
$\begin{array}{c} \text{Cl} \\ \vert \\ \text{H}_3\text{C}-\text{C}-\text{COOH} \\ \vert \\ \text{Cl} \end{array}$	2,2-dichloropropionic acid	dalapon
$Cl_3C-COOH$	trichloroacetic acid	TCA

many cases these compounds act as carbon and energy sources for microorganisms and quantitative release of chloride ion has often been observed. This property is possessed by bacteria, fungi and actinomycetes. The ability of microorganisms to dissimilate the simpler chlorinated acetic and propionic acids is believed to be inducible in some of the bacterial species studied (Jensen, 1960).

It has been assumed that the microbial metabolism of compounds of this type occurs by a cleavage of the C–Cl bond to form chloride ion accompanied by the introduction of a hydroxyl group derived from a water molecule. This would mean, for example, that dalapon would ultimately be converted into pyruvate and TCA into oxalic acid. The former conversion has been demonstrated with cell-free extracts of an *Arthrobacter* sp. which was able to utilize dalapon as a growth substrate (Kearney *et al.* 1964). This enzymatic dehalogenation of dalapon to pyruvate probably involves 2-hydroxy-2-chloropropionate as an intermediate (Fig. 19) but, since this compound is unstable and would

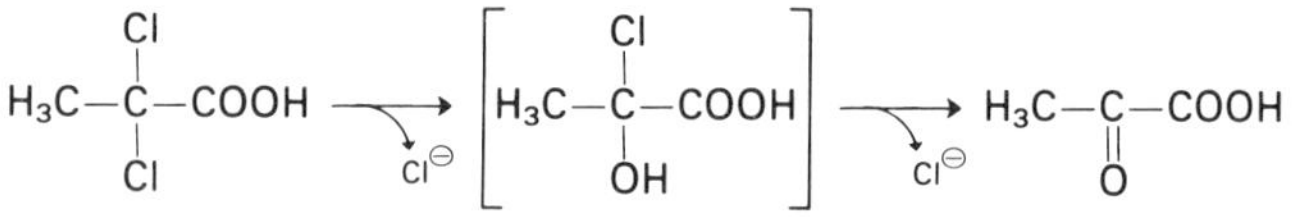

$$\begin{array}{c}\text{Cl}\\|\\\text{H}_3\text{C}-\text{C}-\text{COOH}\\|\\\text{Cl}\end{array} \xrightarrow{\text{Cl}^{\ominus}} \left[\begin{array}{c}\text{Cl}\\|\\\text{H}_3\text{C}-\text{C}-\text{COOH}\\|\\\text{OH}\end{array}\right] \xrightarrow{\text{Cl}^{\ominus}} \begin{array}{c}\\\text{H}_3\text{C}-\text{C}-\text{COOH}\\\|\\\text{O}\end{array}$$

| dalapon | 2-hydroxy-2-chloropropionate | pyruvate |

Fig. 19. Possible pathway of pyruvate formation from dalapon (after Kearney *et al.*, 1964).

decompose spontaneously to pyruvate and chloride ion, its formation has not been proved. Kearney (1966) has suggested two reaction mechanisms for the enzymatic formation of 2-hydroxy-2-chloropropionate from dalapon. The first involved substitution by hydroxyl ion by direct nucleophilic attack leading to the elimination of chloride ion (Fig. 20); the second mechanism involves the base-catalysed β-elimination of HCl from the molecule to form 2-chloroacrylic acid (Fig. 20) which could subsequently be hydrated to form the proposed intermediate. It is not known which of these mechanisms is actually involved.

The conversion of TCA into oxalate has not been demonstrated but some information on the degradation pathway of TCA is available. Kearney *et al.*, (1969b) reported the growth of a pseudomonad on TCA and established, using radiolabelled materials, that more CO_2 was produced from the carboxyl carbon atom than from the trichloromethyl carbon atom during growth. This indicated that free oxalic acid could not have been formed as an intermediate since, as it is

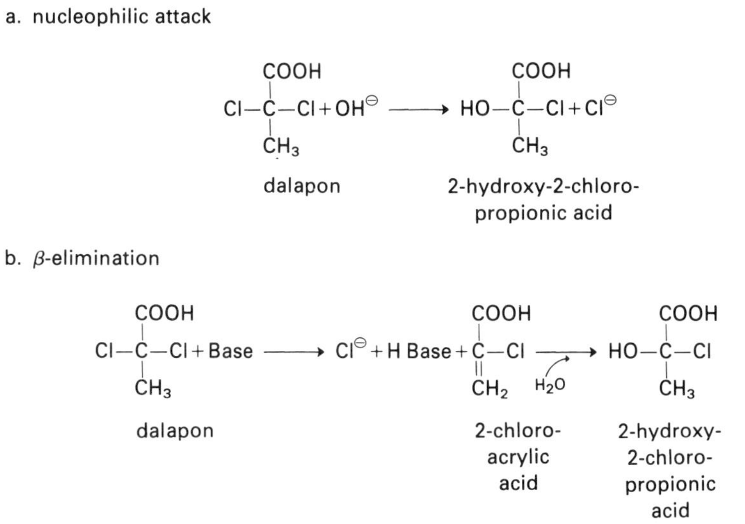

a. nucleophilic attack

$$Cl-\underset{\underset{CH_3}{|}}{\overset{\overset{COOH}{|}}{C}}-Cl + OH^\ominus \longrightarrow HO-\underset{\underset{CH_3}{|}}{\overset{\overset{COOH}{|}}{C}}-Cl + Cl^\ominus$$

dalapon   2-hydroxy-2-chloro-
propionic acid

b. β-elimination

$$Cl-\underset{\underset{CH_3}{|}}{\overset{\overset{COOH}{|}}{C}}-Cl + Base \longrightarrow Cl^\ominus + H\,Base + \underset{\underset{CH_2}{||}}{\overset{\overset{COOH}{|}}{C}}-Cl \xrightarrow[H_2O]{} HO-\underset{\underset{CH_3}{|}}{\overset{\overset{COOH}{|}}{C}}-Cl$$

dalapon   2-chloro-   2-hydroxy-
acrylic   2-chloro-
acid   propionic
acid

Fig. 20. Proposed mechanisms of dalapon dehalogenation (after Kearney, 1966).

a symmetrical molecule, its formation would allow the same amount of CO_2 to be formed from each carbon atom. Further studies showed that extracts of the organism could catalyse the formation of $^{14}CO_2$ from both $(1-^{14}C)$ and $(2-^{14}C)$ TCA. As with whole cells, $^{14}CO_2$ was derived from the carboxyl carbon atom at a greater rate. Reaction mixtures, using a partially purified enzyme preparation, were supplemented with ATP, coenzyme A, glutathione, thiamin pyrophosphate, Mg^{2+} and NADPH but only NADPH and coenzyme A were shown to be required in the reaction. This latter observation may provide an explanation of the position of oxalate in the breakdown of TCA since, if all the dehalogenation steps of the reaction sequence were to take place with the coenzyme A esters of the intermediates and not the free acids, the end product would be oxalyl-CoA (Fig. 21) not oxalate.

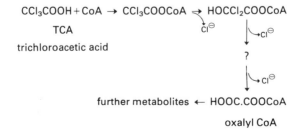

Fig. 21. Proposed formation of oxalyl-CoA from TCA (after Kearney et al., 1969b).

Hence, the two carbon atoms of the TCA molecule would not become metabolically equivalent during breakdown and the formation of unequal amounts of CO_2 from these two carbon atoms might be explained.

Detailed studies of the degradation of the chlorinated aliphatic acids in soil have not been reported. However, it is clear from the available evidence (Foy, 1969 and references cited therein) that their microbial decomposition does occur in soil. In one case, it was shown that $^{14}CO_2$ was released from both (1-^{14}C) and (2-^{14}C) dalapon when added to soil, which suggested that complete degradation of this compound had occurred.

VII. The Bipyridyls

The structures of diquat and paraquat are shown in Table 8. The herbicidal activity of diquat was discovered in 1955 and this led to the later discovery of paraquat.

TABLE 8

Structures of bipyridyl herbicides

Structure	Chemical name	Common name
	1,1'-ethylene-2,2'-bipyridilium dibromide	diquat
	1,1'-dimethyl-4,4'-bipyridilium dichloride	paraquat

The rapid defoliant action of these herbicides has been widely used in agriculture. Both paraquat and diquat have been used as desiccants and defoliants on potatoes, peas, sugar beet and other crops and they are also effective aquatic herbicides. The bipyridyls are inactivated rapidly in soils owing to their strong adsorption properties and their interaction with humic substances (Khan, 1974; Burns and Hayes, 1974). For this reason they are suitable for clearing weeds prior to sowing a crop. The observed inactivation of these compounds in soil raised the question of whether this was due to microbial action and whether these compounds were susceptible to microbial degradation.

A strain of *Corynebacterium fascians* was isolated (Baldwin *et al.*, 1966) which, in pure cultures degraded paraquat. Some 30–40% decomposition was obtained from a 10 ppm solution over a period of three weeks. Similar activity was found with some unidentified anaerobes and a strain of *Clostridium pasteurianum*. The same authors also studied *Lipomyces starkeyi* which degraded paraquat, using the compound as a source of nitrogen. The greatest rate of breakdown was obtained when sucrose was present as a source of carbon and under these conditions degradation was virtually complete. Radio-labelled CO_2 was released by this organism from the ^{14}C-labelled herbicide. More recent studies (Anderson and Drew, 1972) have confirmed the ability of *Lipomyces* sp. to degrade paraquat and have shown that this ability is an adaptive property of the organism. *Lipomyces starkeyi* can also degrade diquat (Funderburk, 1969).

The products of paraquat degradation by an unidentified bacterium have been reported to be 1-methyl-4,4'-dipyridinium ion and 1-methyl-4-carboxypyridinium ion (*N*-methylisonicotinic acid) (Funderburk and Bozarth, 1967). This has led to the proposal that the pathway of degradation involves an initial demethylation followed by cleavage of one of the pyridine rings (Fig. 22). Actinomycetes, including species of *Streptomyces* and *Nocardia*, have also been reported to degrade paraquat forming 1-methyl-4-carboxypyridinium ion (Namdeo, 1972). Identification of these products was by electrophoresis in the former study and by thin layer chromatography (t.l.c.) in the latter. *N*-Methylisonicotinate has also been shown to be formed by the photochemical degradation of the herbicide.

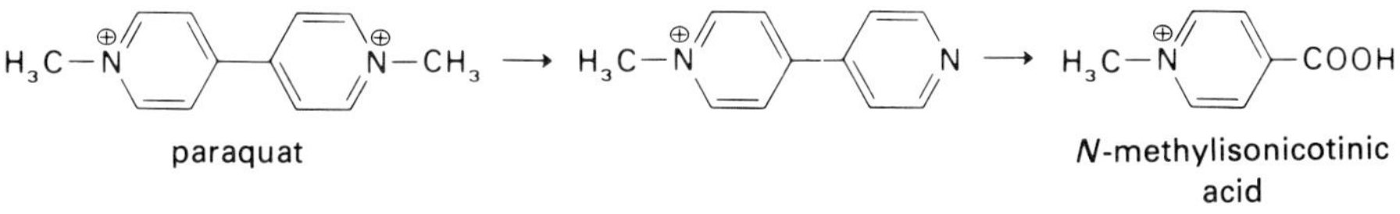

$$H_3C-\overset{\oplus}{N}\text{⟨⟩}\text{⟨⟩}\overset{\oplus}{N}-CH_3 \longrightarrow H_3C-\overset{\oplus}{N}\text{⟨⟩}\text{⟨⟩}N \longrightarrow H_3C-\overset{\oplus}{N}\text{⟨⟩}-COOH$$

paraquat   *N*-methylisonicotinic
                                                                                        acid

Fig. 22. Proposed pathway of paraquat degradation by an unidentified bacterium (after Funderburk and Bozarth, 1967).

Several studies have been carried out on the microbial metabolism of *N*-methylisonicotinate. A Gram-positive bacterium, designated 4Cl, was isolated by Orpin *et al.* (1972*a*) by virtue of its ability to grow on this compound. Cells of the organism were able to oxidize the 2-hydroxylated derivatives of the growth substrate, and cell-free extracts were able to convert 2-hydroxyisonicotinic acid into the 2,6-dihydroxy derivative by a hydroxylating reaction in which the oxygen atom of the hydroxyl group was apparently derived from water and not from molecular oxygen. Studies with specifically radio-labelled substrates showed that the N-methyl group of the molecule was liberated as formaldehyde and that the carboxyl group was released as CO_2. It was proposed that cleavage of the heterocyclic ring (by an unspecified mechanism) produced maleamate which was shown to be further metabolized by extracts to maleate and fumarate, giving the metabolic pathway shown in Fig. 23.

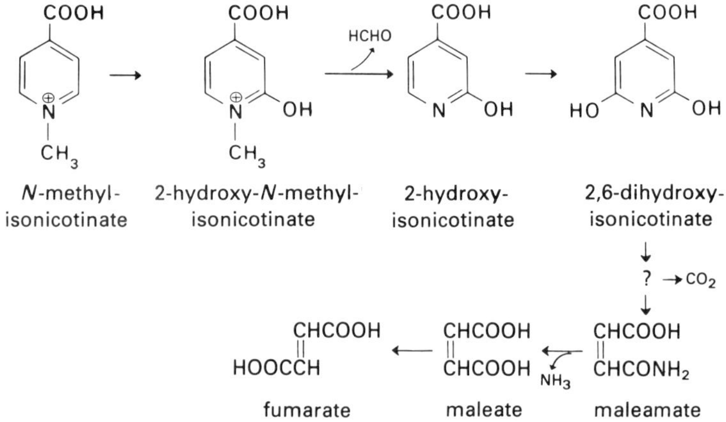

Fig. 23. Metabolism of N-methylisonicotinic acid by strain 4C1 (after Orpin *et al.*, 1972a).

A completely different degradative pathway has been proposed for an *Achromobacter* sp. also able to grow at the expense of N-methyl-isonicotinate (Cain *et al.*, 1970; Wright and Cain, 1972a, b). Cells of this organism, unlike those of strain 4C1, were unable to metabolize hydroxylated derivatives and analogues of the growth substrate, which suggested that hydroxy-intermediates were not involved in the degradative sequence. Also, the N-methyl group was liberated stoicheiometrically as methylamine by whole cells and by cell-free extracts supplemented with NADH or NAD^+. The metabolism of N-methyl-isonicotinate by cell extracts was totally dependent on the presence of NADH or NAD^+ and was accompanied by the uptake of 1 mole of oxygen and the release of 1 mole of CO_2 per mole of substrate utilized. The NADH was initially oxidized in the reaction with crude extracts but the NAD^+ formed was ultimately reduced again. These NADH-oxidizing and NAD^+-reducing activities could be separated by ammonium sulphate fractionation of the crude cell-free extracts. In addition to methylamine and CO_2, other products of the metabolism were shown to be formate and succinate. The origin of these products in the carbon atoms of the pyridine ring has been shown conclusively by the use of specifically radiolabelled substrate (Wright and Cain, 1972b). Formate was derived from the C_2 atom of the molecule while succinate arose from carbon atoms 3, 4, 5 and 6. The CO_2 was derived from the carboxyl carbon atom. This isotope distribution pattern is consistent with ring cleavage occurring between C_2 and C_3. Studies with carbonyl trapping reagents revealed that succinic semialdehyde was an intermediate of the reaction sequence and the presence of succinic semialdehyde dehydrogenase, which was induced by growth on N-methylisonicotinate, indicated that this compound was converted into succinate. These data led to the postulation of a scheme (Fig. 24) in which a partial reduction of the pyridinium ring occurs followed by direct oxidative cleavage. Several of the intermediates of this scheme have not yet been demonstrated conclusively.

Some studies have also been made on the metabolism of the photo-chemical degradation product of diquat, picolinamide (pyridine-2-carboxamide). A Gram-negative rod has been isolated which utilized this compound as sole source of carbon and nitrogen (Orpin *et al.* 1972b). Cells of this organism oxidized picolinic acid (pyridine-2-carboxylate), 6-hydroxypicolinate, maleamate and maleate, while

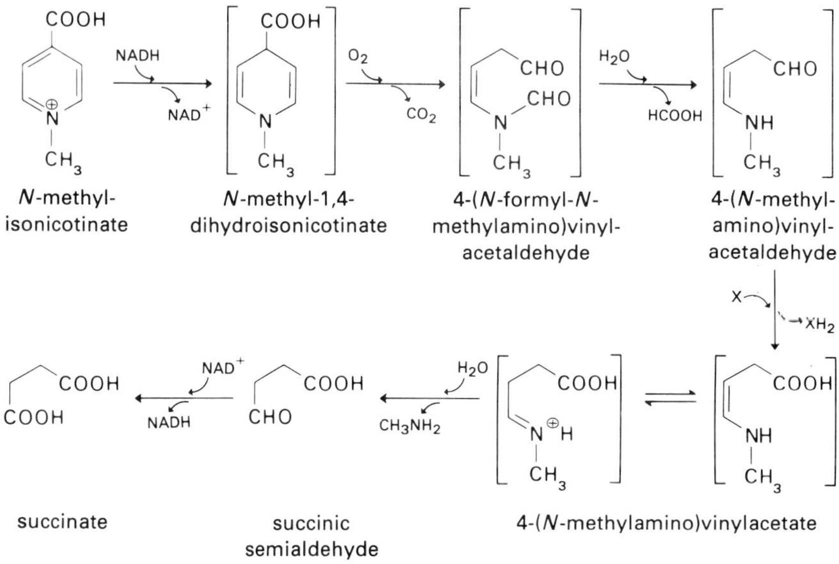

Fig. 24. Proposed pathway of N-methylisonicotinate degradation by *Achromobacter D.* (after Wright and Cain, 1972a, b).

arsenite inhibition of picolinamide metabolism allowed 6-hydroxy-picolinate to accumulate. Cultures of the organism, when grown under conditions of high picolinamide concentration, high pH and low oxygen tension formed 2,5-dihydroxypyridine. Cell-free extracts were able to convert picolinamide into picolinate and to hydroxylate picolinate to 6-hydroxypicolinate by a reaction not involving molecular oxygen. Extracts did, however, catalyse an oxygenase reaction which cleaved 2,5-dihydroxypyridine, leading to the formation of formate and maleamate and the latter could be converted into maleate and fumarate. These data are consistent with the metabolic pathway shown in Fig. 25, although details of the reactions involved in the formation and cleavage of the proposed intermediate, 2,5-dihydroxypyridine, are not clear.

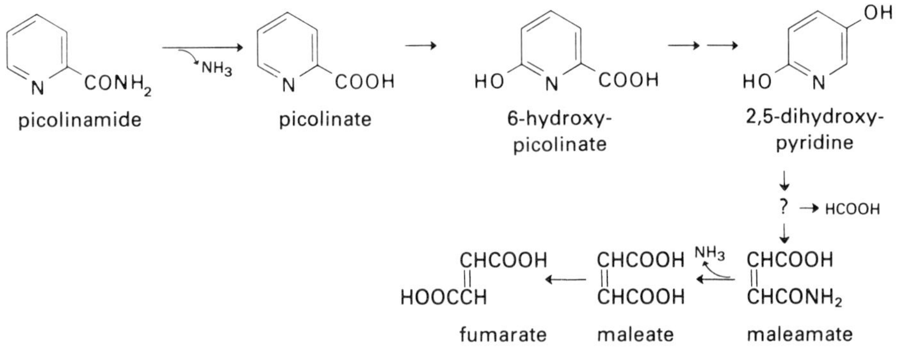

Fig. 25. Proposed microbial metabolism of picolinamide (after Orpin et al., 1972b).

In contrast to these studies with pure cultures, the predominant feature of the behaviour of paraquat and diquat in soils is their well known strong adsorption to soil. Those reports which have implied that paraquat and diquat undergo microbial degradation in soil are now believed to be erroneous (Boon, 1967; Calderbank and Tomlinson, 1968; Tu and Bollen, 1968). A recent paper by Giardina et al. (1973) suggested that about half of the paraquat applied to soil at dosage rates of 100–400 ppm was degraded after 20 days. However, since ^{14}C-labelled compound was not used, it is possible that poor recoveries in soil extracts gave rise to these results since no evidence of product formation was given.

Both paraquat and diquat eventually penetrate the clay lattices in soil to form adsorption complexes (Calderbank and Tomlinson, 1968) and so are not readily available for microbial attack. This effect was demonstrated neatly by Weber and Coble (1968) by the addition of montmorillonite clay to nutrient solutions containing soil microorganisms and (^{14}C) diquat. Microbial degradation, which was determined by monitoring $^{14}CO_2$ evolution, ceased when the montmorillonite was added. The clay was also shown to remove all of the unchanged (^{14}C) diquat from solution.

The ability of *Lipomyces starkeyi* to degrade paraquat, which has been discussed earlier, was used by Burns and Audus (1970) to study the microbial degradation of paraquat in soil. *Lipomyces* was cultured on paraquat until no paraquat remained, at which time soil and (^{14}C) paraquat were added to the culture. The release of $^{14}CO_2$ was measured and, of the four soils used, only one, a silt loam soil, gave rise to a significant amount of $^{14}CO_2$ above background and controls. This evolution of $^{14}CO_2$ ceased, however, after 96 hours and no further degradation of the herbicide was observed. The authors suggested the following explanation of these results. When paraquat is first added to soils it is possible that the initial adsorption by organic matter is more rapid than by clay and at this early stage some microbial degradation is possible with clay soils. As the paraquat penetrates the clay lattices microbial degradation stops. However, this approach of using mixtures of pure cultures and soils is fraught with experimental difficulties and poor repeatability. Considering the available evidence it is probable that if microbial degradation of paraquat and diquat occurs in soil at all it is only to a very limited extent.

VIII. The Dinitroanilines

The most prominent compound in the dinitroaniline series of herbicides (Table 9) is trifluralin which is a selective pre-emergence herbicide used for the control of grasses and broad-leaved weeds. Trifluralin has been used since the early 1960s for weed control in a range of crops including cotton, soybeans, carrots, tomatoes and peppers. Other dinitroanilines which have similar pre-emergence applications include benefin (benfluralin) (see Probst and Tepe, 1969), nitralin (Schieferstein and Hughes, 1966) and dinitramine.

The metabolism of trifluralin in soil and in pure cultures has been studied in some detail. Probst *et al.* (1967) reported a study of the metabolism of trifluralin in soil under normal aerobic conditions and under anaerobic conditions generated by waterlogging the soil. They found that under aerobic conditions several metabolites resulting from dealkylation and reduction reactions were formed in small amounts but the evidence suggested that they were short-lived and were converted into polar products. These polar products appeared to be the

TABLE 9
Structures of dinitroaniline herbicides

General structure

R₁, R₂ on N; O_2N and NO_2 on ring; R_4 and R_3 positions

R₁	R₂	R₃	R₄	Chemical name	Common name
n-C_3H_7	n-C_3H_7	CF_3	H	2,6-dinitro-NN-dipropyl-4-trifluoromethylaniline	trifluralin
C_2H_5	n-C_4H_9	CF_3	H	N-butyl-N-ethyl-2,6-dinitro-4-trifluoromethylaniline	benfluralin
n-C_3H_7	n-C_3H_7	SO_2CH_3	H	4-methylsulphonyl-2,6-dinitro-NN-dipropylaniline	nitralin
C_2H_5	C_2H_5	CF_3	NH_2	$N'N'$-diethyl-2,6-dinitro-4-trifluoromethyl-m-phenylenediamine	dinitramine
s-C_4H_9	H	t-C_4H_9	H	N-s-butyl-4-t-butyl-2,6-dinitroaniline	— (Amex)
n-C_3H_7	n-C_3H_7	SO_2NH_2	H	3,5-dinitro-N^4N^4-dipropyl-sulphanilamide	oryzalin
n-C_3H_7	n-C_3H_7	iso-C_3H_7	H	4-isopropyl-2,6-dinitro-NN-dipropylaniline	isopropalin

result of strong binding of anilines to the soil since an aromatic triamine was formed upon strong chemical reduction. Under anaerobic conditions, the rate of degradation was more rapid and reduction reactions predominated and preceded dealkylation. The proposed aerobic and anaerobic degradation pathways are shown in Fig. 26.

The above work was carried out with (^{14}C) trifluoromethyl-labelled trifluralin. Messersmith *et al.* (1971) followed the release of $^{14}CO_2$ from (1-*propyl* ^{14}C) trifluralin in soil and observed that only 3·5% of the radio-label was evolved as $^{14}CO_2$ in ten months. In this work, the higher rate of degradation of trifluralin in waterlogged soil was confirmed.

Using unlabelled trifluralin, Párr and Smith (1973) published evidence to show that the initial degradation of trifluralin in soil was microbiological under anaerobic conditions. This evidence included the enhanced degradation of trifluralin in the presence of an organic substrate and the compound's stability in moist, anaerobic soil after autoclaving.

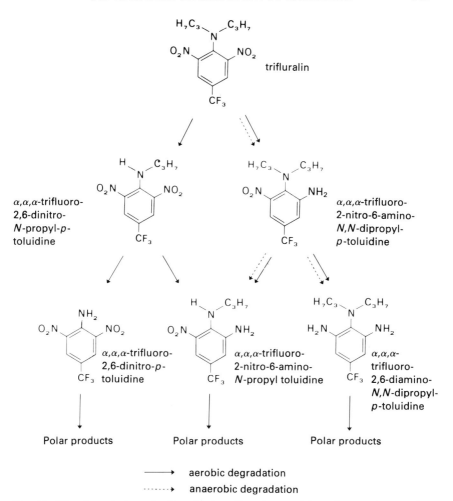

Fig. 26. The degradation of trifluralin in soil under aerobic and anaerobic conditions.

The procedure of producing anaerobic conditions by flooding soil has been criticized (Willis *et al.*, 1974). These workers used redox potential measurements to characterize the intensity of reduction in an anaerobic soil suspension system. The degradation of trifluralin was studied at different redox potentials and it was shown that the exclusion of oxygen by soil flooding caused the rapid degradation of the herbicide only when the potential decreased below a critical value which was not accurately determined but was shown to lie between $+150$ and $+50 \, \text{mV}$.

Benfluralin has been shown to degrade in soil in a similar way to trifluralin under normal and flooded soil conditions (Probst and Tepe,

1969). One exception was that α,α,α-trifluoro-2,6-dinitro-p-cresol was formed from this compound but this has not been detected as a soil degradation product of trifluralin.

There are few reports of the metabolism of dinitroaniline herbicides *in vitro*. Laanio *et al.* (1973) have reported the metabolism of several compounds of this class, including trifluralin and dinitramine, by cultures of the fungi *Aspergillus fumigatus, Fusarium oxysporum* and *Paecilomyces* sp. It was shown that *A. fumigatus* degraded dinitramine with the formation of three products, two of which were identified by t.l.c. and mass spectroscopy as the ethylamino and amino derivatives. The third product was identified as a benzimidazole derivative (Fig. 27) which has also been shown to be a soil metabolite of dinitramine (Smith *et al.*, 1973). The mechanism of formation of this compound is not clear but it may represent a general metabolic route for these herbicides since benzimidazoles have also been shown to be formed in

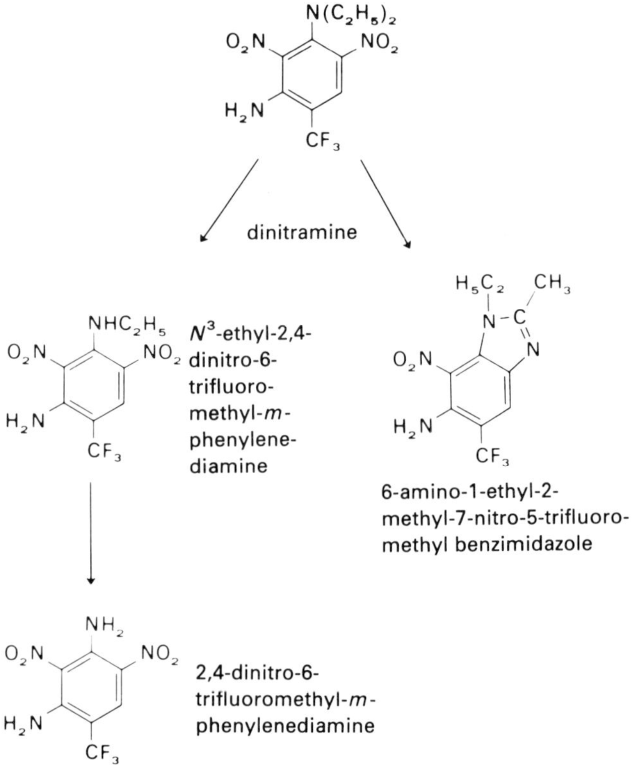

Fig. 27. Proposed metabolism of dinitramine by *Aspergillus fumigatus* (after Laanio *et al.*, 1973).

soil under field conditions from trifluralin, oryzalin and isopropalin (Golab and Amundson, 1974).

Cell-free extracts of *A. fumigatus*, when supplemented with NADPH and ferrous ions, catalysed the mono-dealkylation of dinitramine. The activity was, however, very low.

Trifluralin has been shown to be degraded, under anaerobic conditions, by the rumen bacteria *Bacteroides ruminicola* and *Lachnospira multiparus* (Williams and Feil, 1971). The products which were identified included many which are shown as intermediates in the degradative sequence in Fig. 26, indicating that these bacteria can reduce one or both nitro-groups to an amino function and can bring about dealkylation.

The metabolism of another new dinitroaniline herbicide, *N*-s-butyl-4-t-butyl-2,6-dinitroaniline (Amex), has been studied in soil and in a culture solution of the soil fungus, *Paecilomyces* sp. (Kearney *et al.*, 1974). The major product formed in soil was 4-t-butyl-2,6-dinitroaniline. Other soil metabolites were found but these have not yet been characterized. The major metabolite isolated from fungal extracts was identified as 3-(4-t-butyl-2,6-dinitroanilino)-2-butanol using n.m.r. spectroscopy. These structures are given in Fig. 28.

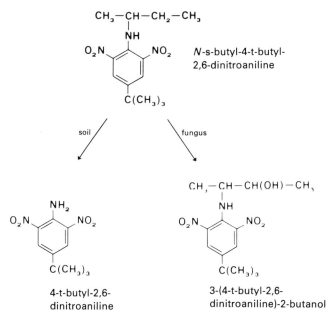

Fig. 28. Degradation products of *N*-s-butyl-4-t-butyl 2,6-dinitroaniline (Amex) in soil and in cultures of *Paecilomyces* sp. (after Kearney *et al.*, 1974).

IX. Miscellaneous Herbicides

Several herbicides that do not fall within the general classes discussed so far have also been studied with respect to their susceptibility to microbial degradation and a few of these are considered in this section. With the exception of pyrazon, little detailed information on the breakdown mechanisms is available. The relevant chemical structures are given in Table 10.

TABLE 10

Structures of pyrazon, endothal, pentachlorophenol and propyzamide

Structure	Chemical name	Common name
	5-amino-4-chloro-2-phenyl-3-pyridazinone	pyrazon (PCA)
	3,6-endoxohexahydrophthalic acid	endothal
	pentachlorophenol	— (PCP)
	3,5-dichloro-N-(1,1-dimethylpropynyl)-benzamide	pronamide (propyzamide)

A. PYRAZON

Pyrazon can be utilized as a source of carbon by soil bacteria (Engvild and Jensen, 1969; Froehner *et al.*, 1970; Lingens, 1973). The details of the metabolic pathway involved have largely been elucidated and have shown that the phenyl group is the only part of the molecule that is

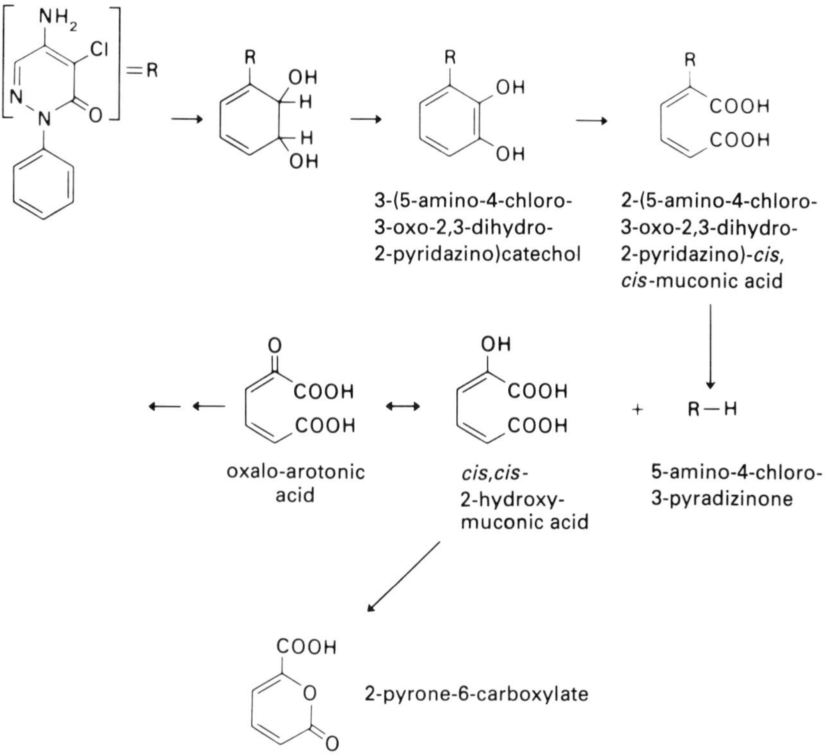

3-(5-amino-4-chloro-
3-oxo-2,3-dihydro-
2-pyridazino)catechol

2-(5-amino-4-chloro-
3-oxo-2,3-dihydro-
2-pyridazino)-*cis*,
cis-muconic acid

oxalo-arotonic
acid

cis,cis-
2-hydroxy-
muconic acid

5-amino-4-chloro-
3-pyradizinone

2-pyrone-6-carboxylate

Fig. 29. Proposed degradation of pyrazon by a soil bacterium (after Lingens, 1973).

metabolized. The proposed route of degradation is shown in Fig. 29. Of the intermediates shown, the dihydrodiol (shown to be in the *cis*-configuration), the muconic acid, 5-amino-4-chloropyridazin-3(2H)-one and 2-pyrone-6-carboxylate were isolated from cultures of the bacterium and rigorously identified (de Frenne *et al.*, 1973). The pyridazinone derivative accumulated in stoicheiometric amounts and has not been shown to be further degraded by any microorganism. Further evidence for the pathway was obtained by the observation that extracts of the organism catalysed an NAD-dependent dehydrogenation of the dihydrodiol to the corresponding catechol (Haug *et al.*, 1973). The subsequent metabolism of the postulated intermediate 2-hydroxymuconate (4-oxalocrotonate) was not definitely established, although extracts could metabolize this compound, but it was suggested that pyruvate and acetaldehyde might be formed by a reaction sequence similar to that proposed for 4-oxalocrotonate metabolism by an *Azotobacter* sp. (Sala-Trepat and Evans, 1971).

B. ENDOTHAL

Endothal can be used as a sole carbon source by some *Arthrobacter* spp. (Jensen, 1964; Sikka and Saxena, 1973). Sikka and Saxena (1973) have studied the incorporation of ^{14}C-substrate into cellular components of the organism and have shown that citrate, aspartate and glutamate are among the earliest compounds radiolabelled but the pathway of assimilation still remains obscure.

C. PENTACHLOROPHENOL

Pentachlorophenol has been reported to be converted into pentacoloroanisole and tetrachlorohydroquinone dimethyl ether by a *Bacillus* sp. (Suzuki and Nose, 1971). In addition, this herbicide apparently degraded rapidly in soil under reducing (anaerobic) conditions forming, within three weeks, all four possible isomeric tetrachlorophenols, 2,4,5- and 2,3,5-trichlorophenols, 3,4- and 3,5-dichlorophenols and 3-chlorophenol. Kirsch and Etzel (1973) found that an adapted mixed microbial population attacked pentachlorophenol under non-proliferating conditions forming $^{14}CO_2$ from radiolabelled substrate, while Chu and Kirsch (1972) isolated an organism by a continuous flow enrichment technique which, in pure culture, could also mineralize this compound from solutions containing up to 200 mg litre^{-1}.

D. PRONAMIDE

Pronamide (propyzamide), a broad spectrum herbicide effective in the control of annual grasses and several broad-leaved weeds, has been

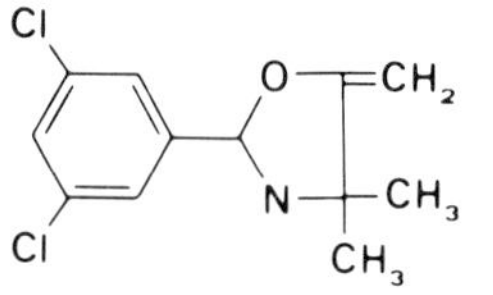

.2-(3,5-dichlorophenyl)-
4,4-dimethyl-5-
methylenoxazoline

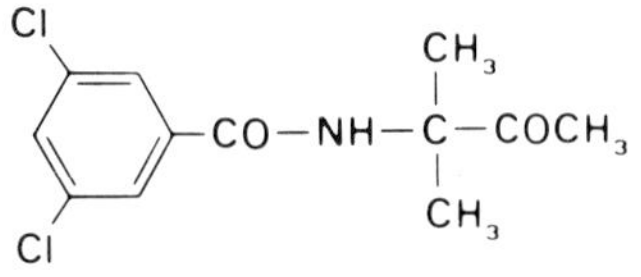

N-(1,1-dimethylacetonyl)-
3,5-dichlorobenzamide

Fig. 30. Metabolites of pronamide (propyzamide) in soil (after Yih *et al.* 1970, and Yih and Swithenbank, 1971).

shown to degrade in soils (Yih *et al.*, 1970; Yih and Swithenbank, 1971; Fisher, 1974). $^{14}CO_2$ was released from material radio-labelled in the carbonyl group and incubated in non-sterilized soil (Fisher, 1974), but autoclaved soil caused no mineralization. Several metabolites have been identified in soil, including the cyclic compound 2-(3,5-dichlorophenyl)-4,4'-dimethyl-5-methylenoxazoline and *N*-(1,1-dimethylacetonyl)-3,5-dichlorobenzamide (Fig. 30).

X. Conclusions

It can be seen from the foregoing that there is a considerable amount of information on the microbial metabolism of herbicidal compounds. The degree to which any particular compound is degraded varies but in many instances pure cultures of effective microorganisms have been isolated and in some cases the detailed biochemistry of the degradative processes has been elucidated. The question arises as to the relevance of these academic studies to the persistence and fate of these herbicides under field conditions. This is often difficult to evaluate because detailed biochemical data are most easily obtained with readily metabolizable compounds which are often relatively non-persistent in the environment.

In former years, a lack of awareness of the possible ecological effects of synthetic organic chemicals meant that no detailed studies in soil were effected. This is exemplified by the phenoxyalkanoate herbicides. Although the persistence in soil of the parent molecules was studied, the identification of possible degradation products in soil was largely ignored. Considerable work with pure cultures, however, was carried out as discussed above. A possible reason for this is that at the time these herbicides were first used, radio-tracer methodology was still in its infancy. Later experience has shown how essential is the use of radioisotopes in the study of the soil metabolism of pesticides. The fate of more recently developed herbicides has, in general, been studied much more extensively in both soil and pure culture systems and, furthermore, some of the well established compounds have been re-examined (although this is not true for the phenoxyalkanoates). The result of this is that a better balanced viewpoint can be obtained of the behaviour of these compounds as regards their susceptibility to bio-degradation and whether they will actually be bio-degraded under field conditions. Perhaps the major criticism of pure culture studies is that they may not reveal the true fate of herbicides in natural environments, particularly where, for example, the compounds or their metabolites react strongly with soils. That is to say, in effect, that a culture flask can never reproduce *in vivo* conditions. This is obviously true but, nevertheless, it is clear that in many cases pure culture and soil studies of herbicide degradation have complemented each other and that both types of study have been shown to be necessary and desirable.

This chapter was submitted in March 1975.

References

ANDERSON, J. R. and DREW, E. A. (1972). Growth characteristics of a species of *Lipomyces* and its degradation of paraquat. *J. gen. Microbiol.* **70**, 43–58.

ARMSTRONG, D. E., CHESTERS, G. and HARRIS, R. F. (1967). Atrazine hydrolysis in soil. *Soil Sci. Soc. Am. Proc.* **31**, 61–66.

AUDUS, L. J. (1952). The decomposition of 2,4-dichlorophenoxyacetic acid and 2-methyl-4-chlorophenoxyacetic acid in the soil. *J. Sci. Fd Agric.* **3**, 268-274.

AUDUS, L. J. (1964). Herbicide behaviour in the soil. *In* "The Physiology and Biochemistry of Herbicides", (L. J. Audus, Ed.) pp. 163–206. Academic Press, London and New York.

BALDWIN, B. C., BRAY, M. F. and GEOGHEGAN, M. J. (1966). The microbial decomposition of paraquat. *Biochem. J.* **101**, 15.

BARTHA, R. (1968). Biochemical transformations of anilide herbicides in soil. *J. agric. Fd Chem.* **16**, 602–604.

BARTHA, R. (1969). Pesticide interaction creates hybrid residue. *Science, N.Y.* **166**, 1299–1300.

BARTHA, R. (1971). Fate of herbicide-derived chloroanilines in soil. *J. agric. Fd Chem.* **19**, 385–387.

BARTHA, R. and PRAMER, D. (1967). Pesticide transformation to aniline and azo compounds in soil. *Science, N.Y.* **156**, 1617–1618.

BARTHA, R. and PRAMER, D. (1969). Transformation of the herbicide methyl-*N*-(3,4-dichlorophenyl)-carbamate (Swep) in soil. *Bull. Environ. Contam. Toxicol.* **4**, 240–245.

BARTHA, R. and PRAMER, D. (1970). Metabolism of acylanilide herbicides. *Adv. appl. Microbiol.* **13**, 317–341.

BARTHA, R., LINKE, H. A. B. and PRAMER, D. (1968). Pesticide transformations: Production of chloroazobenzenes from chloroanilines. *Science, N.Y.* **161**, 582–583.

BELASCO, I. J. and LANGSDORF, W. P. (1969). Synthesis of ^{14}C-labelled Siduron and its fate in soil. *J. agric. Fd Chem.* **17**, 1004–1007.

BELASCO, I. J. and REISER, R. W. (1969). Metabolic fate of Siduron in the animal. *J. agric. Fd Chem.* **17**, 1000–1003.

BEYNON, K. I. and WRIGHT, A. N. (1968). Breakdown of the herbicide ^{14}C-chlorthiamid. I. Laboratory studies of the breakdown in soils. *J. Sci. Fd Agric.* **19**, 723–726.

BEYNON, K. I. and WRIGHT, A. N. (1972). The fates of the herbicides chlorthiamid and dichlobenil in relation to residues in crops, soils and animals. *Residue Rev.* **43**, 23–53.

BEYNON, K. I., STOYDIN, G. and WRIGHT, A. N. (1972a). The breakdown of the triazine herbicide cyanazine in soils and maize. *Pestic. Sci.* **3**, 293–305.

BEYNON, K. I., STOYDIN, G. and WRIGHT, A. N. (1972b). The breakdown of the triazine herbicide cyanazine in wheat and potatoes grown under indoor conditions in treated soils. *Pestic. Sci.* **3**, 379–387.

BEYNON, K. I., ROBERTS, T. R. and WRIGHT, A. N. (1974). Degradation of the herbicide benzoylprop-ethyl in soil. *Pestic. Sci.* **5**, 451–463.

BOLLAG, J. M. (1972). Biochemical transformations of pesticides by soil fungi. *Critical Rev. Microbiol.* **2**, 35–38.

BOLLAG, J. M., HELLING, C. S. and ALEXANDER, M. (1968a). 2,4-D metabolism. Enzymic hydroxylation of chlorinated phenols. *J. agric. Fd Chem.* **16**, 826–828.

BOLLAG, J. M., BRIGGS, G. G., DAWSON, J. E. and ALEXANDER, M. (1968b). 2,4-D metabolism. Enzymic degradation of chlorocatechols. *J. agric. Fd Chem.* **16**, 829–833.

BOON, W. R. (1967). The quaternary salts of bipyridyl – a new agricultural tool. *Endeavour* **26**, 27–32.

BORDELEAU, L. M. and BARTHA, R. (1970). Azobenzene residues from aniline-based herbicides; evidence for labile intermediates. *Bull. Environ. Contam. Toxicol.* **5**, 34–37.

BORDELEAU, L. M. and BARTHA, R. (1972a). Biochemical transformation of herbicide-derived anilines in culture medium and in soil. *Can. J. Microbiol.* **18**, 1857–1864.

BORDELEAU, L. M. and BARTHA, R. (1972b). Biochemical transformations of herbicide-derived anilines: purification and characterisation of causative enzymes. *Can. J. Microbiol.* **18**, 1865–1871.

BORDELEAU, L. M. and BARTHA, R. (1972c). Biochemical transformations of herbicide-derived anilines: requirements of molecular configuration. *Can. J. Microbiol.* **18**, 1872–1882.

BORDELEAU, L. M., ROSEN, J. D. and BARTHA, R. (1972). Herbicide-derived chloroazobenzene residues: pathway of formation. *J. agric. Fd Chem.* **20**, 573–578.

BÖRNER, H. (1965). Untersuchungen uber den Abbau von Afalon *N*-(3,4-dichlorophenyl)-*N'*-methoxy-*N'*-methylharnstoff und Aresin *N*-(4-chlorophenyl)-*N'*-methylharnstoff im Boden. *Z. PflKrankh. PflPath. PflSchutz.* **72**, 517–531.

BÖRNER, H. (1967). Der Abbau von Harnstoffherbiziden im Boden. *Z. PflKrankh. PflPath. PflSchutz.* **74**, 135–143.

BORTELS, H., FRICKE, E. and SCHNEIDER, R. (1967). Simazine decomposition by micro-organisms in various soils. *NachrBl. dt. PflSchutzdienst.* **19**, 101–105.

BOZARTH, G. A. and FUNDERBURK, H. H., Jr. (1971). Degradation of fluometuron in sandy loam soil. *Weed Sci.* **19**, 691–695.

BURNS, I. G. and HAYES, M. H. B. (1974). Some physico-chemical principles involved in the adsorption of the organic cation paraquat by soil humic materials. *Residue Rev.* **52**, 117–146.

BURNS, R. G. and AUDUS, L. J. (1970). Distribution and breakdown of paraquat in soil. *Weed Res.* **10**, 49–58.

BYRDE, R. J. W. and WOODCOCK, D. (1957). Fungal detoxication 2. The metabolism of some phenoxyalkylcarboxylic acids by *Aspergillus niger*. *Biochem. J.* **65**, 682–686.

CAIN, P. S. (1968). An investigation of the herbicidal activity of 2-methoxy-3,6-dichlorobenzoic acid. *Weed Abstr.* **17**, 223.

CAIN, R. B., WRIGHT, K. A. and HOUGHTON, C. (1970). Microbial metabolism of the pyridine ring: bacterial degradation of 1-methyl-4-carboxypyridinium chloride, a photolytic product of paraquat. *Meded. Fac. Landb. Rijksuniv. Gent* **35**, 785–798.

CALDERBANK, A. and TOMLINSON, T. E. (1968). The fate of paraquat in soils. *Outlook on Agriculture* **5**, 252–257.

CHISAKA, H. and KEARNEY, P. C. (1970). Metabolism of propanil in soils. *J. agric. Fd Chem.* **18**, 854–858.

CHU, J. P. and KIRSCH, E. J. (1972). Metabolism of pentachlorophenol by an axenic bacterial culture. *Appl. Microbiol.* **23**, 1033–1035.

CLARK, C. G. and WRIGHT, S. J. L. (1970a). Detoxication of isopropyl *N*-phenylcarbamate (IPC) and isopropyl *N*-3-chlorophenylcarbamate (CIPC) in soil, and isolation of IPC-metabolizing bacteria. *Soil Biol. Biochem.* **2**, 19–26.

CLARK, C. G. and WRIGHT, S. J. L. (1970b). Degradation of the herbicide isopropyl-*N*-phenylcarbamate by *Arthrobacter* and *Achromobacter* spp. from soil. *Soil. Biol. Biochem.* **2**, 217–226.

CLIFFORD, D. R. and WOODCOCK, D. (1964). Metabolism of phenoxyacetic acid by *Aspergillus niger* van Tiegh. *Nature, Lond.* **203**, 763.

COLLINS, R. F. (1973). Perfusion studies with bromoxynil octanoate in soil. *Pestic. Sci.* **4**, 181–192.

CORBIN, F. T. and UPCHURCH, R. P. (1967). Influence of pH on detoxication of herbicides in soil. *Weeds* **15**, 370–377.

COUCH, R. W., GRAMLICH, J. V., DAVIS, D. E. and FUNDERBURK, H. H. Jr. (1965). The metabolism of atrazine and simazine by soil fungi. *Proc. Sth. Weed Control Conf.* **18**, 623–631.

DALTON, R. L., EVANS, A. W. and RHODES, R. C. (1966). Disappearance of diuron from cotton field soils. *Weeds* **14**, 31–33.

DALY, J., GUROFF, G., JERINA, D., UDENFRIEND, S. and WITKOP, B. (1968). Intramolecular migrations during hydroxylation of aromatic compounds. The NIH shift. *Adv. Chem. Ser.* **77**, 279–289.

DEWEY, O. R., LYNDSAY, R. V. and HARTLEY, G. S. (1962). Biological destruction of 2,3,6-trichlorobenzoic acid in soil. *Nature, Lond.* **195**, 1232.

DUXBURY, J. M., TIEDJE, J. M., ALEXANDER, M. and DAWSON, J. E. (1970). 2,4-D metabolism. Enzymatic conversion of chloromaleylacetic acid to succinic acid. *J. agric. Fd Chem.* **18**, 199–201.

ENGELHARDT, G., WALLNOFER, P. R. and PLAPP, R. (1972). Identification of N,O-dimethylhydroxylamine as a microbial degradation product of the herbicide linuron. *Appl. Microbiol.* **23**, 664–666.

ENGVILD, K. C. and JENSEN, H. L. (1969). Microbiological decomposition of the herbicide pyrazon. *Soil Biol. Biochem.* **1**, 295–300.

EVANS, W. C., SMITH, B. S. W., FERNLEY, H. N. and DAVIES, J. I. (1971a). Bacterial metabolism of 2,4-dichlorophenoxyacetate. *Biochem. J.* **122**, 543–551.

EVANS, W. C., SMITH, B. S. W., MOSS, P. and FERNLEY, H, N. (1971b). Bacterial metabolism of 4-chlorophenoxyacetate. *Biochem. J.* **122**, 509–517.

FAULKNER, J. K. and WOODCOCK, D. (1961). Fungal detoxication. Pt V. Metabolism of o- and p-chlorophenoxyacetic acids by *Aspergillus niger*. *J. Chem. Soc.* 1961, Part IV, 5397–5400.

FAULKNER, J. K. and WOODCOCK, D. (1965). Fungal detoxication. Pt. VII. Metabolism of 2,4-dichlorophenoxyacetic and 4-chloro-2-methylphenoxyacetic acids by *Aspergillus niger*. *J. Chem. Soc.* 1965, Part I, 1187–1191.

FISHER, J. D. (1974). Metabolism of the herbicide pronamide in soil. *J. agric. Fd Chem.* **22**, 605–606.

FOURNIER, J. C. (1974). Degradation microbienne de la 2,6-dichlorobenzamide dans des modeles de laboratoire Pt. I. *Chemosphere* **3**, 77–82.

FOURNIER, J. C. and CATROUX, G. (1972). Etude du metabolisme de la benzamide et de ses derives mono et dichlores par des souches d' *Aspergillus* sp. *Compt. Rend.* **275**, 1723–1726.

FOY, C. L. (1969). The chlorinated aliphatic acids. *In* "Degradation of Herbicides", (P. C. Kearney and D. D. Kaufman, Eds), pp. 207–253. Marcel Dekker Inc, New York.

de FRENNE, E., EBERSPACHER, J. and LINGENS, F. (1973). The bacterial degradation of 5-amino-4-chloro-2-phenyl-3(2H)-pyridazinone. *Eur. J. Biochem.* **33**, 357–363.

FROEHNER., C., OLTMANNS, O. and LINGENS, F. (1970). Isolierung und Charakterisierung Pyrazon-abbauender Bakterien. *Arch. Mikrobiol.* **74**, 82–89.

FUNDERBURK, H. H. (1969). Diquat and paraquat. *In* "Degradation of Herbicides" (P. C. Kearney and D. D. Kaufman, Eds), pp. 283–297. Marcel Dekker Inc, New York.

FUNDERBURK, H. H., Jr. and BOZARTH, G. A. (1967). Review of the metabolism and decomposition of diquat and paraquat. *J. agric. Fd Chem.* **15**, 563–567.

GAMAR, Y. and GAUNT, J. K. (1971). Bacterial metabolism of 4-chloro-2-methylphenoxyacetate. Formation of glyoxylate by side-chain cleavage. *Biochem. J.* **122**, 527–531.

GAUNT, J. K. and EVANS, W. C. (1971a). Metabolism of 4-chloro-2-methylphenoxyacetate by a soil pseudomonad. Preliminary evidence for the metabolic pathway. *Biochem. J.* **122**, 519–526.

GAUNT, J. K. and EVANS, W. C. (1971b). Metabolism of 4-chloro-2-methyl-phenoxyacetate by a soil pseudomonad. Ring-fission, lactonising and delactonising enzymes. *Biochem. J.* **122**, 533–542.

GEISSBÜHLER, H. (1969). The substituted ureas. *In* "Degradation of Herbicides", (P. C. Kearney and D. D. Kaufman, Eds), pp. 79–111. Marcel Dekker Inc, New York.

GEISSBÜHLER, H. and VOSS, G. (1971). Metabolism of substituted urea herbicides. *In* "Pesticide Terminal Residues" (A. S. Tahori, Ed.), pp. 305–332. Butterworths, London.

GEISSBÜHLER, H., HASELBACH, C., AEBI, H. and EBNER, L. (1963). The fate of *N'*-(4-chlorophenoxy)-phenyl-*N,N*-dimethyl urea (C-1983) in soils and plants III. Breakdown in soils and plants. *Weed Res.* **3**, 277–297.

GIARDINA, M. C., TOMATI, U. and PIETROSANTE, W. (1973). Effects of paraquat on some soil bacteria responsible for hydrolytic activity. *Nuovi Annali Ig. Microbiol.* **24**, 191–196.

GOLAB, T. and AMUNDSON, M. E. (1974). Degradation of trifluralin, oryzalin and isopropalin in soil. Presented at the Third Int. Congress Pest. Chem., Helsinki, July, 1974.

GOSWAMI, K. P. and GREEN, R. E. (1971). Microbial degradation of the herbicide atrazine and its 2-hydroxy analogue in submerged soils. *Environ. Sci. Technol.* **5**, 426–429.

HARRIS, C. I. (1967). Fate of 2-chloro-*s*-triazine herbicides in soil. *J. agric. Fd Chem.* **15**, 157–162.

HARRIS, C. I., KAUFMAN, D. D., SHEETS, T. J., NASH, R. G. and KEARNEY, P. C. (1965). Behaviour and fate of *s*-triazines in soils. *Adv. Pest Control Res.* **8**, 1–55.

HAUCK, R. D. and STEPHENSON, H. F. (1964). Nitrification of triazine nitrogen. *J. agric. Fd Chem.* **12**, 147–151.

HAUG, S., EBERSPACHER, J. and LINGENS, F. (1973). Enzymatic and chemical preparation of 5-amino-4-chloro-2-(2',3'-dihydroxyphenyl)-3-(2H)-pyridazinone. *Biochem. Biophys. Res. Commun.* **54**, 760–763.

HELLING, C. S., BOLLAG, J. M. and DAWSON, J. E. (1968). Cleavage of ether oxygen bond in phenoxyacetic acid by an *Arthrobacter* species. *J. agric. Fd. Chem.* **16**, 538–539.

HERRETT, R. A. (1969). Methyl- and phenylcarbamates. *In* "Degradation of Herbicides" (P. C. Kearney and D. D. Kaufman, Eds), pp. 113–145. Marcel Dekker Inc., New York.

HORVATH, R. S. (1970a). Microbial cometabolism of 2,4,5-trichlorophenoxyacetic acid. *Bull. Environ. Contam. Toxicol.* **5**, 537–541.

HORVATH, R. S. (1970b). Cometabolism of methyl- and chloro-substituted catechols by an *Achromobacter sp.* possessing a new meta-cleaving oxygenase. *Biochem. J.* **119**, 871–876.

HORVATH, R. S. (1971). Cometabolism of the herbicide 2,3,6-trichlorobenzoate. *J. agric. Fd Chem.* **19**, 291–293.

HUGHES, A. F. and CORKE, C. T. (1974). Formation of tetrachloroazobenzene in some Canadian soils treated with propanil and 3,4-dichloroaniline. *Can. J. Microbiol.* **20**, 35–39.

JENSEN, H. L. (1960). Decomposition of chloroacetates and chloropropionates by bacteria. *Acta Agric. Scand.* **10**, 83–103.

JENSEN, H. L. (1964). Studies of soil bacteria (*Arthrobacter globiformis*) capable of decomposing the herbicide endothal. *Acta Agric. Scand.* **14**, 193–207.

JENSEN, H. L. and ABDEL-GHAFFAR, A. S. (1969). Cyanuric acid as nitrogen source for micro-organisms. *Arch. Microbiol.* **67**, 1–5.

KAUFMAN, D. D. (1965). Degradation of carbamate herbicides in soil. *J. agric. Fd Chem.* **15**, 582–591.

KAUFMAN, D. D. and BLAKE, J. (1970). Degradation of atrazine by soil fungi. *Soil Biol. Biochem.* **2**, 73–80.

KAUFMAN, D. D. and BLAKE, J. (1973). Microbial degradation of several acetamide, acylanilide, carbamate, toluidine and urea pesticides. *Soil Biol. Biochem.* **5**, 297–308.

KAUFMAN, D. D. and KEARNEY, P. C. (1965). Microbial degradation of isopropyl-*N*-3-chlorophenylcarbamate and 2-chloroethyl-*N*-3-chlorophenylcarbamate. *Appl. Microbiol.* **13**, 443–446.

KAUFMAN, D. D. and KEARNEY, P. C. (1970). Microbial degradation of *s*-triazine herbicides. *Residue Rev.* **32**, 235–265.

KAUFMAN, D. D., KEARNEY, P. C. and SHEETS, T. J. (1965). Microbial degradation of simazine. *J. agric. Fd Chem.* **13**, 238–242.

KAUFMAN, D. D., PLIMMER, J. R., IWAN, J. and KLINGEBIEL, U. I. (1972). 3,3′4,4′-tetrachloroazoxybenzene from 3,4-dichloroaniline in microbial culture. *J. agric. Fd Chem.* **20**, 916–919.

KAUFMAN, D. D., PLIMMER, J. R. and KLINGEBIEL, U. I. (1973). Microbial oxidation of 4-chloroaniline. *J. agric. Fd Chem.* **21**, 127–132.

KEARNEY, P. C. (1965). Purification and properties of an enzyme responsible for hydrolysing phenylcarbamates. *J. agric. Fd Chem.* **13**, 561–564.

KEARNEY, P. C. (1966). Metabolism of herbicides in soils. *Adv. Chem. Ser.* **60**, 250–262.

KEARNEY, P. C. (1967). Influence of physicochemical properties on biodegradability of phenylcarbamate herbicides. *J. agric. Fd Chem.* **15**, 568–571.

KEARNEY, P. C. and PLIMMER, J. R. (1972). Metabolism of 3,4-dichloroaniline in soils. *J. agric. Fd Chem.* **20**, 584–585.

KEARNEY, P. C., KAUFMAN, D. D. and BEALL, M. L. (1964). Enzymatic dehalogenation of 2,2-dichloropropionate. *Biochem. Biophys. Res. Commun.* **14**, 29–33.

KEARNEY, P. C., KAUFMAN, D. D. and SHEETS, T. J. (1965a). Metabolites of simazine by *Aspergillus fumigatus*. *J. agric. Fd Chem.* **13**, 369–372.

KEARNEY, P. C., HARRIS, C. I., KAUFMAN, D. D. and SHEETS, T. J. (1965b). Behaviour and fate of chlorinated aliphatic acids in soils. *Adv. Pest Control Res.* **6**, 1–30.

KEARNEY, P. C. PLIMMER, J. R. and GUARDIA, F. B. (1969a). Mixed chloro-azobenzene formation in soil. *J. agric. Fd Chem.* **17**, 1418–1419.

KEARNEY, P. C., KAUFMAN, D. D., von ENDT, D. W. and GUARDIA, F. S. (1969b). TCA metabolism by soil micro-organisms. *J. agric. Fd Chem.* **17**, 581–584.

KEARNEY, P. C., SMITH, R. J., PLIMMER, J. R. and GUARDIA, F. S. (1970). Propanil and TCAB residues in rice soils. *Weed Sci.* **18**, 464–466.

KEARNEY, P. C., PLIMMER, J. R., WILLIAMS, V. P., KLINGEBIEL, U. I., ISENSEE, A. R., LAANIO, T. L., STOLZENBERG, G. E. and ZAYLSKIE, R. G. (1974). Soil persistence and metabolism of *N-sec* butyl-4-*tert* butyl-2,6-dinitroaniline. *J. agric. Fd Chem.* **22**, 856–859.

KHAN, S. U. (1974). Humic substances reactions involving bipyridylium herbicides in soil and aquatic environments. *Residue Rev.* **52**, 1–26.

KIRSCH, E. J. and ETZEL, J. E. (1973). Microbial decomposition of penta-chlorophenol. *J. Wat. Pollut. Control Fed.* **45**, 359–364.

KNÜSLI, E. (1970). History of the development of triazine herbicides. *Residue Rev.* **32**, 1–9.

KNUESLI, E., BERRER, D., DUPIUS, G. and ESSER, H. (1969). 'S-triazines'. I*n* "Degradation of Herbicides" (P. C. Kearney and D. D. Kaufman, Eds), pp. 51–78. Marcel Dekker Inc., New York.

LAANIO, T. L., KEARNEY, P. C. and KAUFMAN, D. D. (1973). Microbial metabolism of dinitramine. *Pestic. Biochem. Physiol.* **3**, 271–277.

LANZILOTTA, R. P. and PRAMER, D. (1970a). Herbicide transformation 1. Studies with whole cells of *Fusarium solani. Appl. Microbiol.* **19**, 301–306.

LANZILOTTA, R. P. and PRAMER, D. (1970b). Herbicide transformation II. Studies with an acylamidase of *Fusarium solani Appl. Microbiol.* **19**, 307–313.

LEES, H. and QUASTEL, J. H. (1946). Biochemistry of nitrification in soil. I. Kinetics of, and the effects of poisons on, soil nitrification, as studied by a soil perfusion technique. *Biochem. J.* **40**, 803–815.

LINGENS, F. (1973). Abbau von Herbiziden Fungiziden durch Mikroorganismen des Bodens. *Chimia* **27**, 38–635.

LOOS, M. A. (1969). Phenoxyalkanoic acids. *In* "Degradation of Herbicides" (P. C. Kearney and D. D. Kaufman, Eds), pp. 1–49. Marcel Dekker Inc., New York.

LOOS, M. A., ROBERTS, R. N. and ALEXANDER, M. (1967a). Phenols as inter-mediates in the decomposition of phenoxyacetates by an *Arthrobacter species. Can. J. Microbiol.* **13**, 679–690.

LOOS, M. A., ROBERTS, R. N. and ALEXANDER, M. (1967b). Formation of 2,4-dichlorophenol and 2,4-dichloroanisole from 2,4-dichlorophenoxyacetate by *Arthrobacter sp. Can. J. Microbiol.* **13**, 691–699.

LOOS, M. A., BOLLAG, J. M. and ALEXANDER, M. (1967c). Phenoxyacetate herbicide detoxication by bacterial enzymes. *J. agric. Fd Chem.* **15**, 858–860.

MacRAE, I. C., ALEXANDER, M. and ROVIRA, A. D. (1963). The decomposition of 4-(2,4-dichlorophenoxy) butyric acid by *Flavobacter sp. J. gen. Microbiol.* **32**, 69–76.

MATSUNAKA, S. (1971). Metabolism of the acylanilide herbicides. *In* "Pesticide Terminal Residues" (A. S. Tahori, Ed.), pp. 343–365. Butterworths, London.

McCLURE, G. W. (1974). Degradation of anilide herbicides by propham-adapted micro-organisms. *Weed Sci.* **22**, 323–329.

MESSERSMITH, C. G., BURNSIDE, O. C. and LAVY, T. L. (1971). Biological and non-biological dissipation of trifluralin from soil. *Weed Sci.* **19**, 285–290.

MURRAY, D. S., RIECK, W. L. and LYND, J. Q. (1970). Utilisation of methylthio-*s*-triazine for growth of soil fungi. *Appl. Microbiol.* **19**, 11–13.

NAMDEO, K. N. (1972). Biodegradation of paraquat dichloride. *Ind. J. Exp. Biol.* **10**, 133–135.

OBIEN, S. R. and GREEN, R. E. (1969). Degradation of atrazine in four Hawaiian soils. *Weed Sci.* **17**, 509–514.

OELKE, E. A. and MORSE, M. D. (1968). Propanil and molinate for control of barnyard grass in water-seeded rice. *Weed Sci.* **16**, 235–239.

ORPIN, C. G., KNIGHT, M. and EVANS, W. C. (1972a). The bacterial oxidation of *N*-methylisonicotinate, a photolytic product of paraquat. *Biochem. J.* **127**, 833–844.

ORPIN, C. G., KNIGHT, M. and EVANS, W. C. (1972b). The bacterial oxidation of picolinamide, a photolytic product of diquat. *Biochem. J.* **127**, 819–831.

PARKER, C. and HODGSON, G. L. (1966). Some studies on the fate of picloram and dicamba in soils underlying bracken. *Proc. 8th Brit. Weed Cont. Conf.* **2**, 614–615.

PARR, J. F. and SMITH, S. (1973). Degradation of trifluralin under laboratory conditions and soil anaerobiosis. *Soil Sci.* **155**, 55–63.

PLIMMER, J. R., KEARNEY, P. C., CHISAKA, H., YOUNT, J. B. and KLINGEBIEL, U. I. (1970). 1,3-bis (3,4-dichlorophenyl) triazene from propanil in soils. *J. agric. Fd Chem.* **18**, 859–861.

PROBST, G. W. and TEPE, J. B. (1969). Trifluralin and related compounds. *In* "Degradation of Herbicides" (P. C. Kearney and D. D. Kaufman, Eds), pp. 255–282. Marcel Dekker Inc., New York.

PROBST, G. W., GOLAB, T., HERBERG, R. J., HOLZER, F. J., PARKA, S. J., van der SCHANS, C. and TEPE, J. B. (1967). Fate of trifluralin in soils and plants. *J. agric. Fd Chem.* **15**, 592–599.

SALA-TREPAT, J. M. and EVANS, W. C. (1971). The meta cleavage of catechol by *Azotobacter* species. *Eur. J. Biochem.* **20**, 400–413.

SCHIEFERSTEIN, R. H. and HUGHES, W. J. (1966). SD 11831—A new herbicide from Shell. *Proc. 8th Brit. Weed Cont. Conf.* **2**, 1966, 377–381.

SCHUPHAN, I. (1974). Zum metabolismus von Phenylharnstoffen II. Abbau und metabolismus von Monolinuron-O-methyl ^{14}C im boden. *Chemosphere* **3**, 127–130.

SHARABI, N. E. and BORDELEAU, L. M. (1969). Biochemical decomposition of the herbicide *N*-(3,4-dichlorophenyl)-2-methylpentanamide and related compounds. *Appl. Microbiol.* **18**, 369–375.

SHARPEE, K. W., DUXBURY, J. M. and ALEXANDER, M. (1973). 2,4-Dichloro-phenoxyacetate metabolism by *Arthrobacter sp*. Accumulation of a chloro-butenolide. *Appl. Microbiol.* **26**, 445–447.

SHIRAKAWA, N. (1970). Selective herbicide, solan (CMMP, 3'-chloro-2-methyl-*p*-valerotoluidide). IV. Decomposition in soil and its inhibition by some chemicals. *Zasso Kenkyu* **10**, 32–36.

SIKKA, H. C. and SAXENA, J. (1973). Metabolism of endothal by aquatic micro-organisms. *J. agric. Fd Chem.* **21**, 402–406.

SIRONS, G. J., FRANK, R. and SAWYER, T. (1973). Residues of atrazine, cyanazine and their phytotoxic metabolites in a clay loam soil. *J. agric. Fd Chem.* **21**, 1016–1020.

SKIPPER, H. D. and VOLK, V. V. (1972). Biological and chemical degradation of atrazine in three Oregon soils. *Weed Sci.* **20**, 344–347.

SKIPPER, H. D., GILMOUR, C. M. and FURTICK, W. R. (1967). Microbial versus chemical degradation of atrazine in soils. *Soil Sci. Soc. Am. Proc.* **31**, 653–656.

SMITH, A. E. (1971). Degradation of bromoxynil in Regina heavy clay. *Weed Res.* **11**, 276–282.

SMITH, A. E. (1973a). Degradation of dicamba in prairie soils. *Weed Res.* **13**, 373–378.

SMITH, A. E. (1973b). Transformation of dicamba in Regina heavy clay. *J. agric. Fd Chem.* **21**, 708–710.

SMITH, A. E. (1974). Breakdown of the herbicide dicamba and its degradation product 3,6-dichloro-salicylic acid in prairie soils. *J. agric. Fd Chem.* **22**, 601–605.

SMITH, A. E. and CULLIMORE, D. R. (1974). The *in vitro* degradation of the herbicide bromoxynil. *Can. J. Microbiol.* **20**, 773–776.

SMITH, A. E., BELLES, W. S., SHEN, K. W. and WOODS, W. G. (1973). The degradation of dinitramine in soil. *Pestic. Biochem. Physiol.* **3**, 278–288.

SUZUKI, T. and NOSE, K. (1971). Decomposition of pentachlorophenol in farm soil. II: PCP metabolism by a microorganism isolated from soil. *Noyaku Seisan Gijutsu*, **26**, 21–24, via *Chem. Abstr.* **77**, 4024g (1972).

SWANSON, C. R. (1969). The benzoic acid herbicides. *In* "Degradation of Herbicides" (P. C. Kearney and D. D. Kaufman, Eds), pp. 299–320. Marcel Dekker Inc., New York.

TANAKA, F. S., SWANSON, H. R. and FREAR, D. S. (1972). An unstable hydroxy-methyl intermediate formed in the metabolism of 3-(4-chlorophenyl)-1-methyl urea in cotton. *Phytochem.* **11**, 2701–2708.

TAYLOR, H. F. and WAIN, R. L. (1962). Side-chain degradation of certain w-phenoxyalkanecarboxylic acids by *Nocardia coeliaca* and other micro-organisms isolated from soil. *Proc. R. Soc. Ser. B.* **156**, 172–186.

TIEDJE, J. M. and ALEXANDER, M. (1969). Enzymatic cleavage of the ether bond of 2,4-dichlorophenoxyacetate. *J. agric. Fd Chem.* **17**, 1080–1084.

TIEDJE, J. M., DUXBURY, J. M., ALEXANDER, M. and DAWSON, J. W. (1969). 2,4-D metabolism: Pathway of degradation of chlorocatechols by *Arthrobacter sp. J. agric. Fd Chem.* **17**, 1021–1026.

TU, C. M. and BOLLEN, W. B. (1968). Interactions between paraquat and microbes in soils. *Weed Res.* **8**, 38–45.

TWEEDY, B. G., LOEPPKY, C. and ROSS, J. A. (1970). Metabolism of 3-(*p*-bromophenyl)-1-methoxy-1-methyl urea (metobromuron) by selected soil microorganisms. *J. agric. Fd Chem.* **18**, 851–853.

WAGNER, G. H. and CHAHAL, K. S. (1966). Decomposition of carbon-14-labelled atrazine in soil samples from Sanborn field. *Soil Sci. Soc. Am. Proc.* **40**, 752–754.

WALKER, N. and HARRIS, O. (1969). Aniline utilisation by a soil pseudomonad. *J. appl. Bact.* **32**, 457–462.

WALLNÖFER, P. R. (1969). The decomposition of urea herbicides by *Bacillus sphaericus* isolated from soil. *Weed Res.* **9**, 333–339.

WALLNÖFER, P. R. and BADER, J. (1970). Degradation of urea herbicides by cell-free extracts of *Bacillus sphaericus. Appl. Microbiol.* **19**, 714–717.

WALLNÖFER, P. R., SAFE, S. and HUTZINGER, O. (1972). Die Hydroxylation des Herbizids Karsil (*N*-(3,4-dichlorophenyl)-2-methylpentamid) durch *Rhizopus japonicus. Chemosphere* **1**, 155–158.

WALLNÖFER, P. R., SAFE, S. and HUTZINGER, O. (1973a). Microbial hydroxylation of the herbicide *N*-(3,4-dichlorophenyl)-methacrylamide (dicryl). *J. agric. Fd Chem.* **21**, 502–504.

WALLNÖFER, P. R., SAFE, S. and HUTZINGER, O. (1973b). Microbial demethylation and debutynylation of four phenylurea herbicides. *Pestic. Biochem. Physiol.* **3**, 253–258.

WEBER, J. B. and COBLE, H. D. (1968). Microbial decomposition of diquat adsorbed on montmorillonite and kaolinite clays. *J. agric. Fd Chem.* **16**, 475–478.

WILLIAMS, P. P. and FEIL, V. J. (1971). Identification of trifluralin metabolites from rumen microbial cultures. Effect of trifluralin on bacteria and protozoa. *J. agric. Fd Chem.* **19**, 1198–1204.

WILLIS, G. H., WANDER, R. C. and SOUTHWICK, L. M. (1974). Degradation of trifluralin in soil suspensions as related to redox potential. *J. Environ. Qual.* **3**, 262–265.

WRIGHT, K. A. and CAIN, R. B. (1972a). Microbial metabolism of pyridinium compounds. Metabolism of 4-carboxy-1-methylpyridinium chloride, a photolytic product of paraquat. *Biochem. J.* **128**, 543–559.

WRIGHT, K. A. and CAIN, R. B. (1972b). Microbial metabolism of pyridinium compounds. Radioisotope studies of the metabolic fate of 4-carboxy-1-methylpyridinium chloride. *Biochem. J.* **128**, 561–568.

WRIGHT, S. J. L. and FOREY, A. (1972). Metabolism of the herbicide barban by a soil *Penicillium. Soil Biol. Biochem.* **4**, 207–213.

WURZER, B. and CORBIN, F. T. (1969). Studies on the microbial detoxification of the herbicides dicamba and amiben. *Weed Abstr.* **18**, 70.

YIH, R. Y. and SWITHENBANK, C. (1971). Identification of metabolites of *N*-(1,1-dimethylpropynyl)-3,5-dichlorobenzamide in soil and alfalfa. *J. agric. Fd Chem.* **19**, 314–319.

YIH, R. Y., SWITHENBANK, C. and MacRAE, D. H. (1970). Transformations of the herbicide *N*-(1,1-dimethylpropynyl)-3,5-dichlorobenzamide in soil. *Weed Sci.* **18**, 604–607.

Chapter 12

Microbial Degradation of Fungicides, Fumigants and Nematocides

DAVID WOODCOCK

*University of Bristol, Research Station,
Long Ashton, Bristol, England*

I. Introduction

On the basis of their mode of application fungicides may be divided into three classes: eradicant, protectant and systemic (or chemotherapeutant). Eradicant fungicides, usually compounds of low specificity which are applied during dormancy to minimize phytotoxic effects, comprise both seed sterilants and soil chemicals; the latter include fumigants also possessing nematocidal activity. Protectant or foliage fungicides are applied to healthy plant surfaces to provide protection against attack by fungal spores, but will also incidentally fall on the surrounding soil or sward. Systemic fungicides may be applied as foliage sprays or as soil drenches and in both cases there is a surface residue. Thus the metabolic activity of soil microorganisms is of paramount importance and much of present day research activity is devoted to the detection and estimation of fungicide metabolites and terminal residues, not only in the soil but in cereal, vegetable and fruit crops.

II. Soil-applied Chemicals and Seed Protectants

The importance of soil as an almost ideal reaction medium for the breakdown of "foreign" chemicals can hardly be overemphasized. With its moisture and oxygen contents, its organic constituents and its siliceous minerals, which provide potential catalytic surfaces for a variety of reactions, to say nothing of a huge and varied microbial population, it is a matrix concomitantly promoting both biological and non-biological transformation processes. It is with the former that most of this chapter is concerned, although accurate delineation is sometimes rather difficult.

A. ALIPHATIC COMPOUNDS (EXCLUDING ORGANOSULPHUR)

Successful soil fumigants are in general highly toxic to a wide range of soil microflora, but their application is often followed by a marked stimulation of selected microorganisms and this phenomenon appears to be independent of soil type (Chandra and Bollen, 1961). There is, however, growing evidence that some resistant bacteria, actinomycetes and fungi can metabolize even these simple compounds. Thus carbon disulphide, which has been used both as a seed treatment and as a fumigant against *Phylloxera*, gradually disappears from soil and the corresponding appearance of SO_4^{2-} seems highly significant. Likewise the formation of Br^- from ethylene dibromide, despite the fact that although Hanson and Nex (1953) found a very slow decomposition of ethylene dibromide in stored soil, about 91% being destroyed in 172 days, neither they nor Wade (1954) were able to demonstrate microbial degradation conclusively.

Methyl bromide is a broad spectrum soil toxicant and both Wensley (1953) and McKeen (1954) have noted its differential effect on soil microflora. Munnecke and Lindgren (1954) found that whilst methyl bromide controlled non-microsclerotic fungi it failed to kill micro-sclerotium-forming *Verticillium alboatrum*. Some saprophytic *Ascomycetes* (e.g. *Thielavia* and *Chaetomium* spp., *Penicillium vermiculatum* and *Trichoderma viride*) have also been shown to be resistant (Wensley, 1953) and whilst selective penetration and permeation could provide the answer it is possible that an inherent utilization mechanism is involved.

Such an explanation is unlikely in the case of chloropicrin, however, for in addition to its nematocidal properties it is virtually non-selective towards fungi. Nevertheless, it is perhaps significant that usage at fungicidal levels does cause stimulation of some soil bacteria (Klemmer, 1957), though no evidence of microbial degradation of the molecule seems to have been reported, despite the well known vulnerability of the nitro group.

By contrast, allyl alcohol, which would be expected to be even more reactive, is indeed rapidly degraded in soil. This compound became prominent as a treatment for the control of damping-off pathogens which attack pine seedlings (Lindgren and Henry, 1949), and the first suggestion of possible microbial breakdown came from Overman and Burgis (1956). They found that although allyl alcohol controlled *Pythium* and *Rhizoctonia* spp. it caused a marked stimulation of the growth of *Trichoderma viride* when used as a soil surface drench. Several strains of *T. viride* were able to use allyl alcohol as an energy source and its rapid detoxification in soil was fairly conclusively demonstrated by Jensen (1961). Using barley seedlings as indicator plants to monitor its herbicidal effect, he showed that allyl alcohol, added at the rate of $0 \cdot 2–1$ ml kg^{-1}, disappeared within 4–8 days at 25°C in a normal soil, an interval that could be decreased by the addition of *Pseudomonas fluorescens*. At 5°C, or with restricted oxygen access, the inhibitory effect on test plants could be extended to four weeks or even longer, but significantly this was also the case with sterile soil which suggests that a non-microbial oxidative mechanism can also operate. No chemical evidence of breakdown was offered but the findings of Legator and Racusen (1959) are significant. They showed that the toxicity of allyl alcohol to *Rhizoctonia solani* is attributable to its oxidation *in vivo*, to acrolein which is a potent reagent for the thiol group.

The formation of Cl$^-$ in soil from the nematocidal "DD mixture" (1,2- and 1,3-dichloropropenes) has been known for some time though without reports of any intermediate degradation products. The disappearance of *cis*- and *trans*-1,3-dichloropropenes likewise from soil was noted by Hannon *et al*. (1963) but although no metabolites could be detected by them, Castro and Belser (1966) demonstrated non-biological hydrolysis to *cis*- and *trans*-3-chloroallyl alcohols respectively. More recently the same workers found further degradation of these alcohols to CO_2 and Cl^- by a *Pseudomonas* sp. isolated from soil (Belser and Castro, 1971). Using 1,2,3-^{14}C-labelled *cis*-3-chloroallyl alcohol *(1)* and a g.l.c. technique, the breakdown sequence was shown to involve oxidation to *cis*-3-chloroacrylic acid *(2)*, dehalogenation to formylacetic acid *(3)*, followed by rapid decarboxylation and total oxidation (Fig. 1). The *trans*-alcohol also followed a similar pattern.

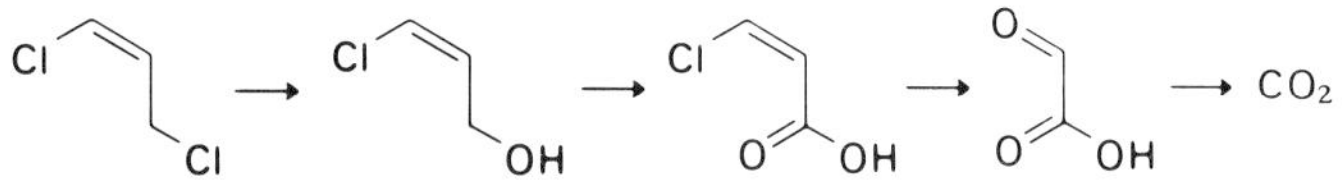

Cl⌐⟍⟍⟍⟍Cl → Cl⌐⟍⟍⟍⟍OH → Cl⌐⟍⟍O⟍OH → O⟍⟍O⟍OH → CO_2

cis-1,3-   *(1)*   *(2)*   *(3)*
dichloropropene

Fig. 1. Breakdown pathway of *cis*-1,3-dichloropropene.

2,2-Dibromo-3-nitrilopropionamide, originally used as an effective seed sterilant (Nolan and Hechenbleikner, 1947) has recently become very useful in the control of microorganisms in cooling tower and paper mill waters (Wolf and Sterner, 1972) and its environmental decomposition has been studied by Exner *et al* (1973). Its degradation or disappearance in the presence of soil which might conceivably be due, *inter alia*, to adsorption, hydrolysis or to chemical or microbial attack, seems unlikely to be caused by the two former processes, especially since hydrolysis in aqueous solution at pH 7 was found to be negligible. Whereas the half life for hydrolysis at pH 6 was 155 hours, its disappearance from soil slurries at pH 5·8 took only 6–15 hours; had control experiments with sterile soil been performed it might have been possible to implicate soil organisms. Confirmation of at least concomitant microbial degradation, however, has been provided by Wolf (1971) who showed that the 2-^{14}C-labelled compound yielded 40% [14]CO_2 after two weeks in the presence of unaerated waste treatment sludge.

B. ORGANOMERCURIALS

The breakdown of both aliphatic and aromatic compounds of mercury in the soil was originally ascribed to chemical breakdown (Daines, 1936) but microbial involvement has become increasingly apparent.

Thus a mercury-resistant strain of *Penicillium roqueforti* Thom. able to absorb a large amount of the metal from nutrient media containing phenylmercury acetate, was reported by Russell (1955). Further, in a study of the effect of methoxyethylmercury chloride, phenylmercury urate and ethylmercury phosphate on *Helicobasidium mompa* in different soils, although the marked decrease in fungicidal effect found in most environments except sand (Gondo and Kubo, 1958) could have been due to adsorption of the toxicant on soil particles, microbial action cannot be ruled out.

More positive evidence of microbial degradation has been provided by the work of Munnecke and his collaborators, who have investigated the breakdown of semesan(2-chloro-4-hydroxymercuriphenol), panodrench(cyanomethylmercuriguanidine) and panogen(methylmercury dicyandiamide). When semesan was applied as an aqueous suspension to non-sterile soil, it was found to be inactivated after 2–3 weeks, whereas its activity persisted for some months in steam-sterilized soil (Munnecke and Solberg, 1958). It was also found that several *Penicillium, Aspergillus* and *Trichoderma* spp. isolated from semesan-treated soil could grow satisfactorily on potato-dextrose agar containing up to 10 ppm of the fungicide, and more significantly that semesan was rapidly inactivated on soil that had been incubated with *Penicillium* spp. and one *T. viride* isolate. The two other mercury compounds showed interesting and somewhat contrasting behaviour (Spanis and Munnecke, 1962; Spanis *et al.*, 1962). Whereas a *Bacillus* sp. isolated from soil was found to inactivate successive additions of panodrench at accelerated rates, both in soil and in broth culture, panogen and semesan were completely inhibitory at comparable levels. A mutant of this organism, however, appeared incapable of inactivating panodrench, although tolerant of it. Other interesting examples of specificity among these organomercurials included several *Penicillium* and *Aspergillus* spp. which although conditioned to survive normally lethal doses of semesan and able to inactivate it, were completely ineffective against the cyano compound, though a few *Penicillium* spp. were able to tolerate low concentrations in soil. In contrast, several *Bacillus* spp. that could neither inactivate nor even tolerate semesan could be induced to tolerate and inactivate panodrench quite quickly. No information regarding metabolic pathways or degradation products was offered in these studies but it was suggested that fungicide inactivation took place in the following stages: initial uptake of toxicant, metabolic breakdown and finally, possible utilization of suitable fragments for energy purposes.

The metabolism of panogen by microorganisms has also been studied by Czerwinska *et al.* (1969). They found that Gram-negative bacteria were more resistant than Gram-positive bacteria, and they considered that the inactivation of the fungicide was due to formation of an (unspecified) inorganic mercury compound.

The degradation of organomercury fungicides in the soil has also been investigated by Kimura and Miller (1964). They detected both metallic mercury and volatile organomercury compounds in the air above soil which had been treated with ethyl or phenylmercury acetate or methylmercury derivatives. They concluded that the degradation process was microbial, though a large proportion of the mercury compounds persisted intact in the soil over periods of 30–50 days. More recently, however, the microbial degradation of phenylmercury acetate has been the subject of various studies. Using the ^{203}Hg-labelled compound, Tonomura et al. (1968) found that a mercury-resistant organism K62, isolated from soil, seemed to be able to bring about a conversion to a substance more volatile than phenylmercury acetate. A definite advance was made when Furakawa et al. (1969) using the ^{203}Hg- or ^{14}C-labelled compound showed that some 70% of ^{203}Hg or 80% of ^{14}C disappeared in two hours on aerobic incubation with washed cells previously grown in the presence of 20 ppm of phenylmercury acetate. Fission of the carbon–mercury bond yielded metallic mercury, together with benzene detected by g.l.c.; no benzene was detected in an abiotic control experiment. Confirmation of this breakdown was obtained when it was shown that ethylmercury phosphate similarly yielded ethane and methyl mercury choride formed methane, in addition to the metal.

A further advance was made by Matsumura et al. (1971) who incubated ^{203}Hg-labelled phenylmercury acetate (2×10^{-6} M) at 30°C for 10 days with cultures of some 35 microorganisms isolated from natural lake bottom sediments or soils. The major metabolite produced by all of the microbial cultures was shown by analysis and infrared spectroscopy to be diphenylmercury, but a second metabolite was not identified; furthermore, there was no evidence of any direct conversion of phenylmercury to methylmercury (MeHg$^+$). The possibility that inorganic mercury might be converted to MeHg$^+$ by living organisms was first confirmed by Jensen and Jernelov (1969) in model experiments using mercury(II) chloride and lake bottom sediments, for no methylation took place using sterile controls. Wood et al. (1968) suggested that anaerobic methane-producing bacteria were responsible, the modus operandi being the transfer of methyl groups from the Co^{3+} of cobolamin to Hg^{2+}, a process which they showed could also be non-enzymatic. Microbial involvement was confirmed, however, by Kaars Sijpesteijn and Vonk (1973). Using three fungal and five bacterial species in pure culture experiments, they demonstrated aerobic as well as anaerobic production of $MeHg^+$, which was monitored by a g.l.c. technique using an electron capture detector (Westöö, 1968). They also inclined to the view that the biological methylation of mercury may in the case of certain organisms be carried out by methylcobolamine, but sounded a cautious note on generalizing about methylation under natural conditions from the results of pure culture studies.

C. ORGANOSULPHUR COMPOUNDS

1. *Dithiocarbamates*

The degree of involvement of microorganisms in the breakdown of dithiocarbamates in soil appears to be surprisingly small and there are some cases which are far from unequivocal. Nabam (disodium ethyl-enebisdithiocarbamate (4)) was one of the first of this class of compound to be used as a soil fungicide, and although its breakdown in this medium is rapid, it has been shown by Munnecke (1958) to be abiotic. Under neutral conditions nabam forms ethylenethiourea (5) and ethylenethiuram monosulphide, the latter being slowly converted to the former (Thorn and Ludwig, 1954); other compounds such as sulphur and carbonyl sulphide may also be formed (Moje *et al.*, 1964).

The decay curves for ferbam (ferric dimethyldithiocarbamate) in soil were found to be comparable with those for nabam, though less steep, and once again the available evidence suggested that a physicochemical rather than a microbial degradation process was operative

$$(4) \qquad (5) \qquad (6) \qquad (7)$$

(Munnecke, 1958). On the other hand, ethylenethiourea (5) which occurs both as an impurity in technical ethylenebisdithiocarbamates and as a degradation product, does seem susceptible to biological conversion. Thus Kaufman and Fletcher (1973) using the ^{14}C-labelled compound showed that ethylenethiourea was rapidly converted, through ethyleneurea (6), to $^{14}CO_2$ in non-sterile soils, and they also isolated four other degradation products, of which two were identified by co-chromatography as hydantoin (7) and 1-(2'-imidazoline-2'yl)-2-imidazolinethione (8). Despite this, Kaars Sijpesteijn and Vonk (1974) were unable to isolate any ethylenethiourea-degrading organisms from ethylenethiourea-treated soil.

(8)   (9)   (10)   (11)

The persistence of thiram (tetramethylthiuram disulphide (9)), which is used as a seed treatment against damping-off diseases, has been examined by Richardson (1954) who found that whilst it rapidly disappeared from compost soil, it persisted for over two months in sandy soil. The rate of disappearance suggested that more than one factor was involved and microbial involvement appeared likely. The fact that seedling protection was observed after the disappearance of the fungicide, pointed to the possibility of an active degradation product, though no attempts at isolation were made.

Little appears to be known of the fate of the dimethyldithiocarbamate ion in soil, although its uptake by cucumber seedlings was first demonstrated by Dekhuisen (1961). When taken up by these plants, it is rapidly converted into the β-glucoside, to a lesser extent into the alanine derivative and also into an unknown fungitoxic compound, presumably also a conjugate of some kind (Kaslander et al., 1961, 1962; Dekhuisen, 1964; Kaars Sijpesteijn and van der Kerk, 1965). Thiram can be reduced easily to the dithiocarbamate ion and can thus yield the same products. Massaux (1963) has in fact applied (^{35}S)thiram to begonia and cucumber leaves and detected only the same fungitoxic compounds. However, in an investigation of the delayed phytotoxicity of dimethyldithiocarbamate fungicides on papaya, Hylin and Chin (1967) used ^{35}S- and ^{14}C-labelled compounds and noted the formation of carbon disulphide and dimethylamine together with four unidentified metabolites, in addition to the β-glucoside and alanine conjugates.

Microorganisms, however, have been shown to behave differently. Thus sodium dimethyldithiocarbamate (10) was converted to α-aminobutyric acid, not only by washed cell suspensions of *Saccharomyces cerevisiae*, *Hansenula anomala* and *Bacterium coli*, but also by mycelial pellets of *Glomerella cingulata*, *Aspergillus niger* and *Cladosporium cucumerinum* (Kaars Sijpesteijn et al., 1962). Rather surprisingly similar products were not obtained using the related compound, nabam (*4*) or,. metham–sodium(sodium methyldithio-carbamate (*11*)).

2. Captan

Captan(N-(trichloromethylthio)-3a,4,7,7a-tetrahydrophthalimide (*12*)) has been used extensively in soil applications, largely because of its effectiveness against *Pythium* spp. and especially against many seedling diseases of beans, and for the control of the cotton seedling disease complex, especially in conjunction with quintozene(penta-chloronitrobenzene) (Bird *et al.*, 1957). Despite the ease with which it is hydrolysed, there are conflicting reports of its stability in soil, Munnecke (1958) mentioning a survival period as long as 65 days, whereas Burchfield (1959) reported a half life of only 3–4 days at pH 7. This latter finding seems more in keeping with a fungicide which reacts so readily with thiols, witness the antagonism of cysteine, homocysteine, glutathione, thioglycerol and coenzyme A to its fungi-toxicity against *Saccharomyces pastorianus* (Lukens and Sisler, 1958). These workers also reported the formation of significant amounts of carbon disulphide and thiophosgene from captan in the presence of yeast cells, and in studies with cysteine *in vitro*, the primary reaction products included cystine, tetrahydrophthalimide, hydrogen sulphide, carbon disulphide and hydrochloric acid. Two schemes for the breakdown of captan are both supported by available evidence. In the first (Fig. 2) an unstable intermediate (*13*) formed by an S_N2 reaction,

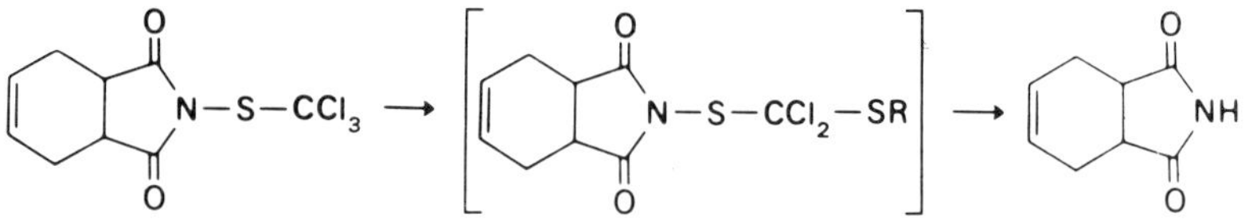

captan
(*12*)

(*13*)

$$2RS\cdot \rightarrow R-S-S-R$$
(*14*)

$$CSCl_2 + 2RSH \rightarrow S{=}C(SR)_2 + 2HCl$$
(*15*)

Fig. 2. Captan metabolism.

liberates thiophosgene and a thiol free radical which generates the disulphide (*14*). Interaction of thiophosgene with further thiols forms the trithiolcarbonate (*15*) which may be stable and isolatable as is that from 4-nitrothiophenol (Owens and Blaak, 1960), or which may decompose to yield CS_2 and the corresponding monosulphide (Lukens, 1959). The weakness of this scheme would seem to be that no intermediates such as (*13*) have so far been isolated and there is no direct evidence for free-radical participation. The second, and more likely explanation involves the N—S bond, in which the imide nucleophile is exchanged for the thiol residue (Lukens and Sisler, 1958; Owens and Blaak, 1960), the resultant mixed disulphide reacting with a further molecule of thiol to form a symmetrical disulphide, together with the unstable trichloromethylmercaptan which decomposes spontaneously to thiophosgene and HCl.

3. *Prothiocarb*

S-ethyl-*N*-(3-dimethylaminopropyl)thiocarbamate hydrochloride (*16*) (common name, prothiocarb) was recently introduced as a systemic soil fungicide, capable of being taken up and translocated in plants and of exerting a fungistatic effect on various *Phycomycetes* such as *Pythium*, *Phytophthora* and *Peronospora* spp. (Bastiaansen *et al.*, 1974; Kaars Sijpesteijn *et al.*, 1974). Its mobility and fate in biological systems have been examined by Iwan and Goller (1974) using [3]H- and ^{14}C-compounds labelled in two different positions of the molecule. Oxidative decomposition of (^{14}C)prothiocarb in all non-sterile soil samples kept in the dark at 20°C was immediate, no distinct lag phase being observed. Two likely degradation products were detected: *N,N*-dimethyl-1,3-diaminopropane (*17*) and another, judged by its chromatographic behaviour, was believed to be ethyl *N*-(3-aminopropyl)thiocarbamate (*18*) (Fig. 3). Apart from carbon dioxide, the only

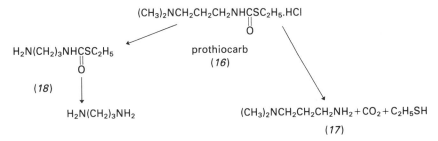

Fig. 3. Microbial degradation of prothiocarb.

other product observed also occurred in the sterile control experiment and was thus believed to be artefactual. These two degradation products ranged in extent from trace amounts to 10% after six months, though the amount of N,N-dimethyl-1,3-diaminopropane isolated after day 30 never exceeded 1%.

D. QUINONES

Chloranil(tetrachloro-1,4-benzoquinone (*19*)) and dichlone, (2,3-dichloro-1,4-naphthoquinone (*20*)) have both been used as effective seed treatments. Both compounds can participate in substitution

(*19*)   (*20*)   (*21*)

(*22*)   (*23*)

reactions involving a reactive chlorine atom and functional groups present in amino acids, peptides and proteins, and they may also interfere with electron transport systems. The work of Gause *et al.* (1967) would seem to lend support to this view. They showed fairly conclusively using an electron spin resonance technique that the initial step in the reaction of thiols with chloranil was a redox-reaction involving the formation of a semiquinone free radical of type (*23*) which can then react further.

Owens (1953) was one of the first to show that dichlone could inhibit certain enzymes containing free thiol groups. With simple thiols both chlorine atoms are replaced forming compounds of type (*21*), while with higher molecular weight thiols such as glutathione and some proteins, steric hindrance precludes distribution, and the reaction forming compounds of this type (*22*) is reversible. Evidence for this latter type of reaction was supplied by Owens and Novotny

(1958) working with intact conidia of *Neurospora sitophila*, by way of the coenzyme A derivative (*22*, R = CoA), though little if anything appears to be known of the fate of the modified dichlone molecule. In this context the findings of Owens and Miller (1957) are interesting. They showed that when conidia of *N. sitophila* which had been treated with dichlone were sonically disintegrated and the homogenate centrifuged, about 75% of the fungicide was present in the super-natant solution in a bound form, much of it associated with soluble proteins. This accorded with results obtained in another investiga-tion in which a large proportion of dichlone which had been in-cubated aerobically with *N. sitophila* spores, was found to have reacted rapidly to form at least five water-soluble products which were not identified.

Similar reactions may well influence the stability of quinones in a soil environment, often rich in organic matter. Indeed, Burchfield (1959) found that dichlone decomposed much more rapidly in a silt loam soil than in an aqueous medium, and since its stability could not be correlated with the hydrolysis rate in aqueous buffer, it seems that this represents true microbial degradation; it is unfortunate that control experiments in sterile soil were not carried out. There is also circumstantial evidence for metabolic breakdown in connection with *Rhizobia* spp. While seed protectants based on heavy metals are extremely toxic to such organisms, chloranil and dichlone appear considerably less so (Ruhloff and Burton, 1951) and indeed dichlone had only moderate activity against all species except *Rhizobium phaseoli*. The ability of an isolate of *Rhizoctonia solani* from seedling cotton to become more tolerant through serial transfer (Elsaid and Sinclair, 1962; Elsaid, 1963) may also point to metabolic breakdown.

Microbial detoxication might also provide an explanation for the fact that post-infectional rather than pre-infectional applications are most effective when the pectolytic enzyme inhibitory property of

(*24*)

rufianic acid (*24*) is used in the systemic protection of tomato plants from *Fusarium oxysporum* (Grossmann, 1963).

E. AROMATIC COMPOUNDS

1. *Halogenonitrohydrocarbons*

The fact that bacteria exist which are capable of metabolizing aromatic hydrocarbons has long been recognized (e.g. Tausson, 1928). While degradation of an aromatic compound by soil microorganisms usually takes place by a pathway involving a dihydric phenol (Rogoff, 1961; Dagley, 1967), the position of substituents is often critical. Thus *meta*-isomers tended to resist degradation by various *Pseudomonas* spp. (Kameda *et al.*, 1957; Cartwright and Cain, 1959), and the deactivating effect of nitro- and sulphonic acid groups is just as evident *in vivo* as *in vitro*. Increased nuclear substitution, especially by halogens, also tends to retard or even prevent degradation.

Thus although a soil bacterium capable of reducing *p*-dinitrobenzene to *p*-nitraniline is known (Villanueva, 1959), no evidence for the breakdown of 1,2,4-trichloro-3,6-dinitrobenzene or quintozene(pentachloronitrobenzene) by a number of microbial isolates from a compost soil was found by Teuteberg (1964). The increased tolerance to this fungicide shown by isolates of *Rhizoctonia solani* from plots treated with quintozene (Shatla and Sinclair, 1963) might seem to reflect the ability on the part of the pathogen to degrade the compound, but Vredeveld (1965) later suggested that the differential uptake of quintozene by *Rhizoctonia solani*, compared to the more resistant *Pythium ultimum*, may be the reason for the selectivity. However, the first report of the actual microbial degradation of quintozene came with the work of Chacko *et al.* (1966) who showed that eight fungal species and eight actinomycetes could reduce quintozene to pentachloroaniline. Thus *Streptomyces aureofaciens* grown in liquid nutrient medium, with quintozene (10–20 μg ml^{-1}) added, was able to degrade the fungicide to the extent of 36% in six days. This degradation only occurred during the active growth phase and the organism was not able to use quintozene as sole carbon source. The first mention of metabolites had been made by Gorbach and Wagner (1967) who used a g.l.c. detection method to show the presence of pentachloroaniline, together with an unidentified second metabolite in quintozene-treated potatoes. This reduction of quintozene is enhanced by anaerobic conditions (Ko and Farley, 1969) and 56% could be accounted for as amine after three weeks. The remainder appeared to be lost by non-biological processes, since a similar amount was unaccounted for after the same incubation period in sterilized soil; volatilization seems the most likely explanation (Caseley, 1968).

Nakanishi and Oku (1969) confirmed the greater uptake of toxicant by sensitive fungi such as *Rhizoctonia solani* and showed that quintozene added at 25 ppm to a liquid culture of *Fusarium oxysporum* was not only reduced to pentachloraniline but was also converted to pentachloromethylthiobenzene (*25*).

(*25*)

Recently, Beck and Hansen (1974) have examined the disappearance of quintozene and its technical impurities penta- and hexachlorobenzenes as well as the known metabolite pentachloroaniline in controlled laboratory experiments at 18–20°C and over a period of 600 days. Pentachlorobenzene, pentachloroaniline and methylthiopentachlorobenzene were all shown to be metabolites and can thus be expected in quintozene-treated field soils.

The stability of 1-fluoro-2,4-dinitrobenzene in both moist and dry silt loam soils has been investigated by Burchfield (1959). Despite its rapid disappearance from the former (half life, 12 hours) nucleophilic replacement of the reactive halogen seems to be precluded both on the grounds that application rate makes little difference and by the lack of correlation of survival rate with hydrolysis rate in aqueous buffer. No mention appears to have been made of any reduction taking place and, without studies in sterile soil, the possibility of microbial involvement must remain unresolved.

2. Phenols

The microbial degradation of phenols is important not only because of the use of chloro- and nitrophenols and their derivatives as pesticides, but also because dihydric phenols can represent key intermediates in the breakdown of aromatic compounds by soil microorganisms (Dagley, 1967). Members of the following genera are able to grow on phenolic materials as the sole carbon source: *Bacillus*, *Micrococcus*, *Mycobacterium*, *Pseudomonas* and *Spirillum*. The ability to degrade phenols is not confined to bacteria, however, for certain soil fungi, wood-rotting and other fungal species also possess this ability e.g. *Aspergillus*, *Penicillium*, *Oospora* and *Neurospora* spp. Yeasts have also been shown to utilize phenols (Zimmerman, 1958; Mills *et al.*, 1971).

The efficacy of pentachlorophenol as a herbicide and as an eradicant fungicide (particularly on inert materials) is well known and, although complete chlorination of the benzene nucleus might have been expected to have enhanced stability as well as fungitoxicity, there are some indications of microbial degradation. Thus Loustalot and Ferrer (1950) found that its toxicity had almost completely disappeared at all levels of application after one month in a saturated soil, whereas there was no appreciable inactivation in an air-dry soil. Later Lyr (1963) reported on the detoxication of pentachlorophenol *in vitro* by certain wood-rotting basidomycetes such as *Trametes versicolor*, which operate by secreting laccase into the medium. Although no evidence of metabolite identification was given, changes in the ultraviolet absorption spectrum, together with the liberation of Cl^- and the appearance of coloured products, all point to microbial degradation. In studies with various chlorinated phenols Lyr estimated the enzymatically released Cl^- and showed that stability to laccase and tyrosinase activity increased with increasing degree of chlorination of the nucleus. To what extent light energy is also involved in such degradations is not easy to assess but it is likely to be considerable (Munakata and Kuwahara, 1969). More recent reports on the metabolism of pentachlorophenol by fungi and bacteria have also been made (Chu and Kirsch, 1972; Ide *et al.* 1972; Alliot, 1973; Kirsch and Etzel, 1973).

TABLE 1

Decomposition of chloro- and bromophenols in soil suspensions[a]

Substituent	Days for complete disappearance
None	2
2-Chloro	14
3-Chloro	>72
4-Chloro	9
2-Bromo	14
3-Bromo	>72
4-Bromo	16
2,4-Dichloro	9
2,5-Dichloro	>72
2,4,5-Trichloro	>72
2,4,6-Trichloro	5
2,3,4,6-Tetrachloro	>72
Pentachloro	>72

[a] From Alexander and Aleem (1961) by courtesy of the American Chemical Society.

That the number and position of substituent chlorine atoms present in the phenolic nucleus is important is abundantly clear from the results of Alexander and his co-workers, some of which are given in Table 1 (Alexander and Aleem, 1961; MacRae and Alexander, 1965). In their investigations, the candidate phenol was added to a previously sterilized mineral salts medium, together with a fresh soil sample to supply the microbial inoculum, and during incubation at 30°C on a rotary shaker, disappearance of the phenol was monitored by regular spectrophotometric assay (Whiteside and Alexander, 1960). In general, the more highly halogenated phenols were more stable, but it was significant that no phenol with a substituent in the *meta* position (3 or 5) was degraded to any appreciable extent. The use of sodium azide in control experiments confirmed that the observed disappearance of the phenol was attributable to microbiological causes, no change in ultraviolet absorbance being observed in the azide-treated flasks. Other examples of the blocking effect of a *meta*-halogen are the failure of a *Mycoplana* sp. which, while able to grow satisfactorily on 4-chlorophenol, was unable to utilize the corresponding 3-chloro-isomer (Walker and Newman, 1956), and the stability of several substituted phenoxyalkylcarboxylic acids which has been similarly explained (Steenson and Walker, 1957; Bell, 1957).

More recently, however, the metabolism of both 4- and 3-chloro-phenols by phenol-grown cells of a strain of *Rhodotorula glutinis* has been reported by Walker (1973), 4-chlorocatechol being identified as a product in both cases. Such cells also consumed oxygen in the presence of 2-chloro-, 4-bromo-, 2,4-dichloro- and 2,4-dibromo-phenols. Similar co-metabolic findings were also made with benzoate-grown cells and the 2-, 3- and 4-chlorobenzoates. The subject of microbial co-metabolism has been reviewed by Horvath (1972).

The metabolism of 2,3,4,6-tetrachlorophenol by microorganisms also recently came under close scrutiny because of musty taint in chickens (Curtis *et al.* 1972, 1974). Ninety nine out of 116 isolates from broiler house litter, comprising 26 fungal species, metabolized the phenol and of these 68 produced 2,3,4,6-tetrachloroanisole (Gee and Peel, 1974). Most efficient methylation was noted with *Penicillium corylophilum*, whilst other spp., notably *Penicillium brevicompactum* metabolized the phenol without methylation. Progress studies with such species suggested that more than one metabolic degradation route operated.

Nitrophenols are interesting from the point of view of microbial degradation, for although the nucleus has been deactivated by the introduction of the nitro group, this substituent provides a new focal point for microbial attack. The earliest observation appears to be that of Erikson (1941) who found that 2,4,6-trinitrophenol and tri-nitroresorcinol could be utilized for growth by certain strains of *Micromonospora chalceae* isolated from lake mud, although no investigation of metabolites was made. The liberation of nitrite ion from 2- and 4-nitrophenols by two *Pseudomonas* spp. isolated from soil was reported by Simpson and Evans (1953). Using adaptive criteria they found, rather surprisingly, that the corresponding aminophenols did not feature as metabolic steps and they suggested that detoxication involved oxidative elimination of the nitro group with subsequent formation of a dihydric phenol, which could then be degraded by the known metabolic pathway (Evans *et al.*, 1951). A more recent paper (Raymond and Alexander, 1971) once again illustrated the uniqueness of *meta* substituents, for while a bacterium isolated from soil could use *p*-nitrophenol as both carbon and energy sources, producing nitrite stoicheiometrically, it could not use *m*-nitrophenol, although this compound was found to be oxidized to nitrohydroquinone.

Gundersen and Jensen (1956) used a strain of *Corynebacterium simplex* to obtain information about the degradation of nitrophenols. Where compounds of this class were satisfactory as the sole carbon and nitrogen source for the growth of the organism, the focal point of attack was often the *p*-nitro group, although 2-butyl-4,6-dinitrophenol (alkyl group not fully specified) and 3,5-dinitrosalicylic acid were notable exceptions. Since the organisms were still viable after three weeks in the presence of the two latter compounds, it is possible that some adverse steric factor was preventing their utilization. In contrast to these results, however, Douros and Reid (1956) showed that strains of both *Pseudomonas aeruginosa* and *Pseudomonas putida* were able to grow readily in a mineral salts medium containing 2,4-dinitrophenol, 2,4-dinitrotoluene or 2-s-butyl-4,6-dinitrophenol as sole carbon source.

The use of 2,4-dinitrophenol in timber preservatives has led to the recognition of another interesting detoxification system, the metabolic activity of *Fusarium oxysporum* making possible the growth of the wood-rotting fungus *Coprinus micaceus* which is normally much less tolerant of the un-attacked molecule. In contrast to the microbial reduction of one nitro group, both nitro groups of 2,4-dinitrophenol provided the foci for detoxification by *F. oxysporum* (Madhosingh, 1961). Both 2-amino-4-nitro- and 4-amino 2-nitrophenol were formed together with an unidentified metabolite which might have resulted from further breakdown of one or other of these aminophenols. The formation of 2,4-dinitrophenol in cultures with added 2-amino-4-nitrophenol suggested the presence of a reversible oxidation–reduction system in the fungus and the suggested metabolic pathway is

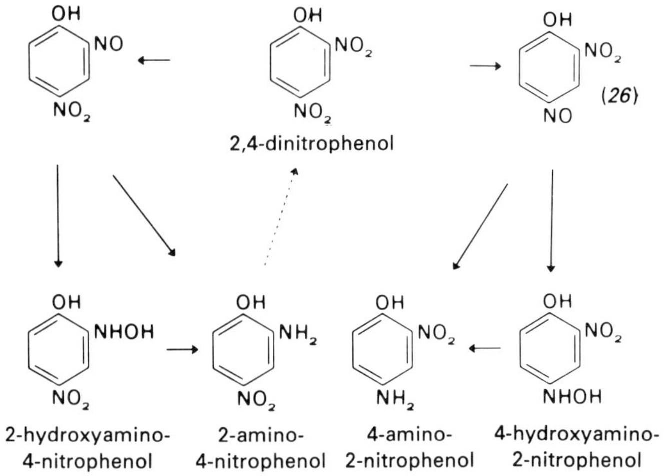

Fig. 4. Suggested degradation of 2,4-dinitrophenol.

shown in Fig. 4. Data suggesting that with at least certain bacteria the nitroso compound (26) is the immediate precursor of the amine are already known (Yamashima et al., 1954).

3. Fenaminosulf

The photosensitivity of fenaminosulf(sodium *p*-dimethylaminoben-zenediazosulphonate) a seed dressing and soil fungicide for the control of *Pythium*, *Aphanomyces* and *Phytophthora* spp. is well known and has recently been reviewed (Woodcock, 1977). When protected from light however, Hills and Leach (1962) found that it was about ten times more effective against *Pythium ultimum* than against *Rhizoctonia solani*. While this may be due to preferential uptake, it is possible that microbial detoxification is involved, particularly as Hills (1962) demonstrated a wide distribution of the ^{14}C-labelled fungicide in sugar beet seedlings together with the presence of four unidentified metabolites. This is supported by the findings of Tolmsoff (1962) who showed that cells from sugar beet and *Rhizoctonia* spp. but not from *Pythium ultimum*, contained a mitochondrial system that decomposed fenaminosulf in the presence of DPNH. The unstable intermediate produced might well be *NN'*-dimethyl-*p*-phenylenediamine, for this compound has recently been identified as a metabolic product formed by the soil bacterium *Pseudomonas fragi* Bk9 (Karanth et al., 1974). This organism, whilst unable to use the fungicide as a sole carbon source, did produce the diamine as a co-metabolic product and the presence of a specific reductase in cell-free extracts was demonstrated.

Another sulphonate shown to undergo microbial degradation is dodecylbenzenesulphonate which was the sole carbon source added to continuously perfusing columns of mixed soils (Bernarde *et al.*, 1965). There was a stoicheiometric relationship between the reduction of sulphonate ion concentration and the appearance of inorganic sulphur compounds.

F. HETEROCYCLIC COMPOUNDS

1. *Anilazine*

The fungitoxicity of anilazine(2,4-dichloro-6-(*o*-chloroanilino)-1,3,5-triazine (*27*)) like that of the quinones chloranil and dichlone, is also

(*27*)

dependent upon the presence of reactive halogen atoms in the molecule. By the same token, this renders it capable of reaction with metabolites and protoplasmic constituents containing amino and sulfhydryl groups (Burchfield and Storrs, 1956). These same workers showed that much of the fungicide which is rapidly taken up by *Neurospora sitophila* spores evidently combines irreversibly with protoplasmic constituents because the amount recoverable on immediate extraction was only 20–25% (Burchfield and Storrs, 1957). The stability of anilazine in a moist silt loam soil has also been studied and a half life of 12 hours found (Burchfield, 1959). Breakdown of the fungicide was much slower in air-dried soil, but since no control experiments were carried out with sterile soil it is difficult to assess the significance of the fact that there was no correlation between stability in moist soil and the hydrolysis rate in aqueous buffer solution. Microbial involvement in the disappearance of anilazine in soil is thus an open question.

2. *Hymexazol*

Hymexazol(3-hydroxy-5-methylisooxazole (*28*)) exhibits remarkable activity against soil-borne diseases of many crops caused by

Aphanomyces, *Pythium*, *Fusarium* and *Corticium* spp. and some isolates of *Rhizoctonia solani*. Its characteristics as a soil fungicide and its metabolism in plants and soil have recently been reviewed (Tomita *et al.*, 1973). Disease-controlling activity persisted and its breakdown in unsterilized soil, which was studied using the ^{14}C-labelled compound, appeared to be microbial in origin since only a trace of $^{14}CO_2$ was evolved from sterilized soil.

By extraction of treated unsterilized soil with 1 M hydrochloric acid, two metabolites were obtained and shown to be identical with acetoacetamide (*29*) and 5-methyl-2(3H)-oxazolone (*30*) respectively, using t.l.c. with seven different solvent systems. The breakdown pathway of hymexazol in soil was proposed (Fig. 5).

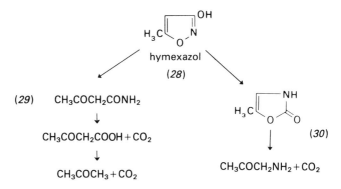

Fig. 5. Breakdown pathway of hymexazol in soil.

5-Methyl-2(3H)-oxazolone has also been reported to be formed from hymexazol by ultraviolet radiation (Göth *et al.*, 1967) but this factor is not critical since the Japanese workers showed that the compound was also produced from hymexazol-treated soil which had been kept in the dark.

III. Protectant Fungicides

A. ORGANOTIN COMPOUNDS

The results of a critical study of the persistence of fentin acetate(triphenyltin acetate) in soil have been reported by Barnes *et al.* (1973). Using the ^{14}C-labelled compound at levels of 5 and 10 ppm in light-shielded columns of soil, the evolution of $^{14}CO_2$ was monitored continuously and showed that although the rate decreased with time, the half life was approximately 140 days. Significantly there was no lag period, suggesting a very rapid adaptation of soil microflora to the metabolism of fentin acetate. Very little $^{14}CO_2$ evolution (0·47% in 60 days) took place in steam-sterilized soil proving that degradation was indeed microbial in origin. It was noted however, that a preliminary abiotic cleavage of the Sn—C bonds is possible, producing aromatic moieties which are then rapidly subject to microbial attack. The decreased rate of $^{14}CO_2$ evolution which occurred after 66% of the aromatic carbon had been accounted for, could indicate resistance of a monophenyltin compound to further attack. Failure to achieve complete extraction of radio-labelled metabolites is almost certainly due to strong adsorption to soil particles. Among the various micro-organisms that were isolated from soil enriched with fentin acetate after 150 days incubation were *Aspergillus* spp. and a Gram-negative rod-shaped species, all of which had the ability to degrade fentin acetate. The relatively small recovery of $^{14}CO_2$ in these experiments probably reflects, at least in part, the incorporation of fentin acetate carbon into the growing cultures.

B. AROMATIC COMPOUNDS

1. *Dicloran*

Dicloran(2,6-dichloro-4-nitroaniline *(31)*) is effective against a wide range of fungal pathogens, particularly *Botrytis, Rhizopus* and *Sclerotinia* spp. causing hyphal distortion, but having little effect on spore

(31)   (32)   (33)

germination. The first hint of microbial degradation came from the work of Groves and Chough (1970) who used the ^{35}Cl-labelled compound to examine its fate in plants and soils. They showed that after incubation of soils with added dicloran, the rate of decomposition into CO_2 and other (unidentified) products increased rapidly; a rod-shaped bacterium was isolated, but no mention was made of specific metabolites. Bacterial cultures to which (^{14}C)dicloran had been added yielded 25–50% of the radioactivity as $^{14}CO_2$; residual activity was not extractable with benzene, indicating complete alteration of the original fungicide. A major study was completed by van Alfen and Kosuge (1974) who used (^{14}C)dicloran to show that both *Escherichia coli* and *Pseudomonas cepacia* in liquid culture converted it into 2,6-dichloro-*p*-phenylenediamine *(32)* and 4-amino-3,5-dichloroacetanilide *(33)*, together with at least four unidentified metabolites. With the former organism, the diamine *(32)* accumulated in the culture medium, whereas the *Pseudomonas* sp. completely metabolized the fungicide to the anilide *(33)* in 48 hours. An unidentified bacterium, isolated from the surface of a peach fruit, was also very efficient in converting dicloran to 4-amino-3,5-dichloroacetanilide and none of the inter-mediate diamine could be detected. In all cases the rate of conversion of dicloran to the diamine was found to be significantly greater under nitrogen than under normal aerobic conditions, thus substantiating the earlier results of Wang (1972) who found more rapid degradation of the fungicide under flooded (anaerobic) conditions.

2. *Chlorothalonil*

The fungicidal properties of chlorothalonil(2,4,5,6-tetrachloro-1,3-dicyanobenzene *(34)*) were first described by Turner *et al.* (1964) and

CN
Cl   Cl

Cl   CN
Cl

(34)

its activity, together with that of related isophthalonitriles has been correlated with their rate of reaction with a model thiol derivative, 4-nitrothiophenoxide (Turner and Battershell, 1969). By virtue of the cyano-activated halogen atoms, chlorothalonil reacts with functional cellular groups by nucleophilic substitution and its reaction with cells and sub-cellular components of *Saccharomyces pastorianus* has been investigated by Tillman *et al.* (1973) using the ^{14}C-labelled fungicide and Na$_2$^{35}SO$_4$-labelled cells. The initial uptake of fungicide was rapid with an accumulation of a large amount in the cells and the rapid formation of reduced glutathione (GSH) derivatives, until all GSH had reacted. The fungicide also reacted concurrently with proteins, resulting in an eventual inhibition of specific NAD thiol-dependent glycolytic and respiratory enzymes. Chlorothalonil thus has similar reaction characteristics to captan but there is one major difference between the two compounds in the stability of the complex formed with high molecular weight cell constituents. ^{35}S from captan remained bound to cellular proteins during precipitation with various reagents (Siegel and Sisler, 1968) whereas a large proportion of the radioactivity of the (^{14}C)chlorothalonil protein complex did not remain attached to protein in similar circumstances.

IV. Systemic Fungicides

A. ORGANOPHOSPHORUS COMPOUNDS

Despite relatively high mammalian toxicities, the ease of breakdown of fungicidal compounds containing organophosphorus moieties makes them attractive substitutes for fungicides based on heavy metals. The comprehensive reviews of Grapov and Melnikov (1973) and of Fest and Schmidt (1973) make interesting reading. Ever since the introduction of the first fungicide of this class, triamiphos(5-amino-1-(bisdimethylaminophosphinyl)-3-phenyl-1,2,4-triazole), on which there appears to be a complete lack of metabolic information, exploitation of this class of compound has gained momentum, particularly in the replacement of mercury fungicides in the control of rice blast, caused by *Pyricularia oryzae*. The impressive results obtained by various Japanese research workers in this connection are worthy of note.

1. *Edifenphos*

The metabolism of edifenphos(O-ethyl-S,S-diphenyl phosphoro-dithioate (*35*)) by mycelial cells of *Pyricularia oryzae*, the causal agent of rice blast, has been studied by Uesugi and Tomizawa (1971a) using ^{32}P and ^{35}S-labelled as well as the unlabelled compound. The fungicide was shaken at 27°C with pregrown fungal mycelia in M/30 phosphate buffer, pH 7, containing 1% glucose, then centrifuged at appropriate intervals to separate cellular material. Water-soluble fractions were separated by ion-exchange on Dowex 1-X8 resin, the main metabolite, ethyl-S-phenyl hydrogen phosphorothioate (*36*), being formed by des-S-phenylation involving one P–S linkage. The formation of this metabolite gradually increased up to 24 hours and thereafter declined, with corresponding increase of the proportions of ethyl dihydrogenphosphate and orthophosphoric acid. The major toluene-soluble metabolite was found to be O-ethyl-S-(p-hydroxyphenyl)-S-phenyl phosphorodithioate (*37*) and this also underwent des-S-phenylation to yield the same product as edifenphos. No difference was found in the rate and mode of degradation of the fungicide by resistant and susceptible strains of the fungus. The various metabolic products are summarized in Fig. 6.

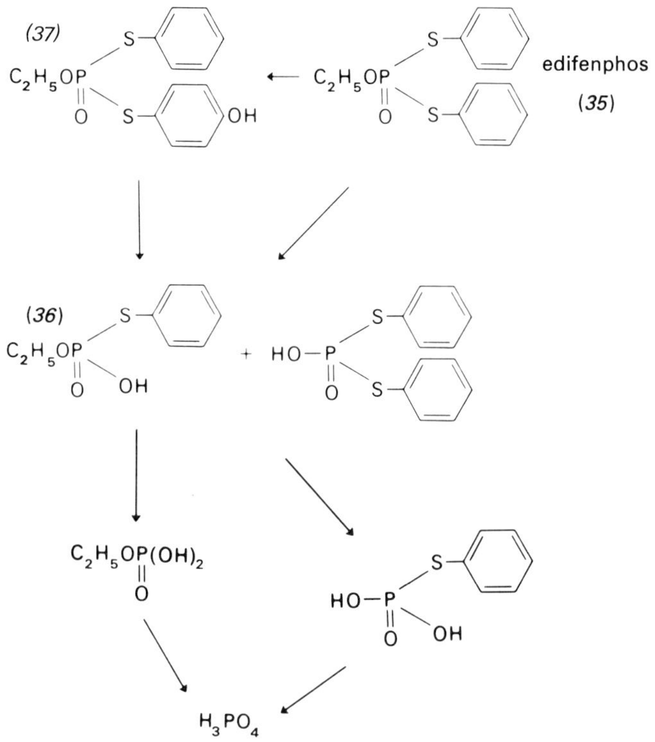

Fig. 6. Metabolism of edifenphos by mycelia of *Pyricularia oryzae.*

2. *Inezin*

Uesugi and Tomizawa (1971b) have also examined the metabolism of Inezin(S-benzyl-O-ethyl phenylphosphonothioate (38)) by mycelial cells of *P. oryzae* (Fig. 7). A g.l.c. study revealed O-ethyl hydrogen phenylphosphonothioate (39) and ethyl hydrogen phenylphosphonate (40) as metabolites, formed by des-benzylation. The fate of the latter group was followed using the [14]C (benzyl)-labelled fungicide, the water-soluble metabolites being separated as before on a Dowex column, while toluene-soluble materials were separated by t.l.c. Examination of the methylated derivatives of the latter by g.l.c. revealed the presence of a *m*-hydroxyphenol derivative (41) but no trace of any *o*- or *p*-hydroxylation. Once again there appears to be no

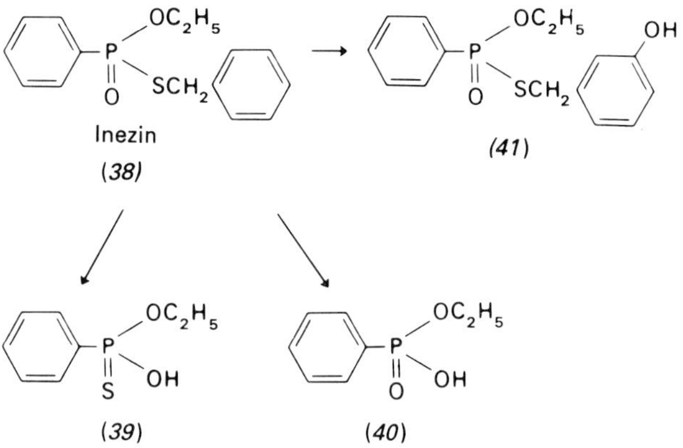

Fig. 7. Metabolism of Inezin by mycelial cells of *Pyricularia oryzae*.

significant difference in the extent or pattern of metabolism of Inezin by susceptible and resistant strains of the fungus.

3. *Kitazin P*

The same workers (Tomizawa and Uesugi, 1972) have carried out a similar study with Kitazin P(S-benzyl-O,O-diisopropyl phosphorothioate (42)). Ion exchange separation of the metabolic products from the ^{32}P- and ^{35}S-labelled fungicide after incubation with mycelium of *P. oryzae* revealed the main metabolite as O,O-diisopropyl hydrogen phosphorothioate (43) (Fig. 8).

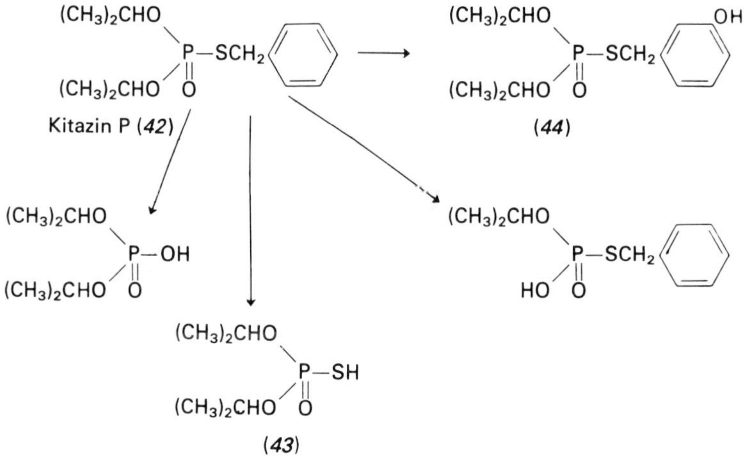

Fig. 8. Metabolism of Kitazin P by mycelial cells of *Pyricularia oryzae*.

Among several metabolites obtained from the toluene-soluble fraction, S-m-hydroxybenzyl-O,O-diisopropyl phosphorothioate (*44*) was identified by t.l.c. and by a comparison of its methylated and acetylated derivatives with authentic compounds using g.l.c. As with Inezin there was no evidence of o- and p-hydroxylation and a differential metabolic rate did not appear to play a critical role in resistant strains.

4. *Pyrazophos*

Another group of fungicidal organophosphorus compounds comprise phosphoric esters of pyrazolopyrimidines. Pyrazophos (O-6-ethoxy-carbonyl-5-methylpyrazolo-(1,5-a)-pyrimidin-2-yl O,O-diethyl phosphorothioate (*45*)), despite poor root uptake which precludes its use in soil application or as a seed dressing, provides excellent control of powdery mildews (Hay, 1971).

(*45*)   (*46*)

A comprehensive study of pyrazophos, including metabolism, has recently been made by de Waard (1974). Using the ^{14}C-labelled fungicide at a concentration of 10^{-5} M he demonstrated a rapid breakdown by *Pyricularia oryzae*, forming the oxygen analogue (PO-pyrazophos) in addition to the des-phosphorylated pyrazolopyrimidine (*46*) and various unidentified metabolites. Since incubation in fungal culture filtrates did not result in the formation of PO-pyrazophos, this conversion probably takes place intra-cellularly. A microsomal oxidation reaction, hitherto unknown in fungi, was proposed. Microsomal mixed-function oxidases might also be responsible for the direct breakdown of pyrazophos to the des-phosphorylated compound (*46*) but enzymatic hydrolysis of PO-pyrazophos was thought to be more probable. *Pythium debaryanum* and *Saccharomyces cerevisiae*, which are relatively insensitive, did not metabolize pyrazophos and this is possibly related to poor uptake.

Fission of the O—P bond has also been reported in cucumber plants (Gorbach *et al.*, 1974), where the fungicide has a half life of 3–4 days, the β-glucoside of 6-ethoxycarbonyl-2-hydroxy-5-methyl-(1,5-a)-pyrazolopyrimidine being formed.

B. ANTIBIOTICS

Of the many substances synthesized by microorganisms, antibiotics are of increasing importance as a new class of antimicrobial agents, particularly useful in the control of bacterial diseases of horticultural crops even though they undergo both physical sequestration and microbial degradation. One of the earliest investigations was made by Waksman and Woodruff (1940) who attributed the inactivation of actinomycin in soil, both to adsorption on soil particles and also to microbial degradation. Siminoff and Gottlieb (1951) stressed the importance of the chemical nature particularly for the former process, and noted the strong adsorption and consequent inactivation of the basic antibiotics streptomycin, streptothricin and subtilin. Later, Katz and Pienta (1957) demonstrated a direct relationship between the size of the inoculum and the extent of decomposition of actinomycin D by an *Achromobacter* sp. and made preliminary investigations into the nature of the degradation products. Streptomycin was similarly inactivated, not only by soil adsorption (Jefferys, 1952) but also by the action of soil microorganisms in non-sterile soil (Pramer and Starkey, 1951). There can be little doubt about the latter results, for not only was there no loss of activity in heat-sterilized soil over a period of three weeks, but all antibiotic action had completely disappeared from a non-sterile soil within two weeks. The main organism involved was believed to be a *Pseudomonas* sp. In contrast to these results, Nissen (1954) failed to demonstrate the breakdown of either aureomycin or streptomycin using a carbon dioxide measurement technique.

The neutral antibiotic chloromycetin (*47*), a product of the actinomycete *Streptomyces venezuelae*, does not appear to be appreciably inactivated by adsorption but soil microflora, notably *Bacillus subtilis*, possessed a marked ability to metabolize it (Gottlieb and Siminoff, 1952). After an initial lag period of 1 day, disappearance was rapid, 70% having disappeared within 14 days. Despite this fairly rapid breakdown, detection of metabolites was rather inconclusive, no evidence being found for the expected arylamino moiety, or for *p*-nitrophenylserinol (*48*) and dichloracetic acid, which had been isolated and characterized earlier from actively growing broth cultures notably of *Proteus vulgaris* and *B. subtilis* to which chloromycetin had been added before incubation (Smith *et al.*, 1949).

$$O_2N-\langle\ \rangle-\underset{\underset{OH}{|}}{CH}-\underset{\underset{CH_2OH}{|}}{CH}-NHCOCHCl_2$$

(*47*)

$$O_2N-\langle\ \rangle-\underset{\underset{OH}{|}}{CH}-\underset{\underset{CH_2OH}{|}}{CH}-NH_2$$

(*48*)

The acidic antibiotic clavacin, which inhibits both bacteria and fungi, and the neutral cycloheximide have been investigated by Gottlieb et al. (1952). Clavacin lost activity only slowly in a sterile black prairie loam soil but was rapidly degraded under non-sterile conditions. Cycloheximide was also degraded in non-sterile soil and lost practically all its activity after 11 days, although some 70% was lost in the same period in sterile soil, possibly through incomplete sterilization.

Griseofulvin (49, R = Cl, R' = OCH₃), produced by *Penicillium nigricans*, has perhaps received most attention, not least because of its translocatability and activity against many important plant pathogens. Applied to root systems, it protects foliage from infection, although with decreasing effect, because of slow degradation in plant tissue (Crowdy et al., 1955, 1956). Abbott and Grove (1959) used a micro-spectrophotometric technique and bioassay to follow the disappearance of griseofulvin from media in which *Botrytis allii* and *Mucor ramannianus* were growing; apart from an unidentified phenolic compound, no metabolites were detected.

(49)

The first evidence for microbial breakdown in soil appears to have been obtained by Jefferys (1952) who observed rapid inactivation of the compound in fresh garden soil. Similar observations were made by Wright (1955), and this breakdown was studied in more detail by Wright and Grove (1957) who found that following an initial lag phase, griseofulvin rapidly disappeared from a fresh garden loam, further additions of the antibiotic resulting in an increased rate of inactivation. There was a corresponding increase in the number of *Pseudomonas* spp. and the same organisms were isolated after the 5-day incubation of a mineral salts medium, containing griseofulvin as sole carbon source, with an inoculum of the soil. No metabolites containing an intact aromatic ring could be detected using the ultraviolet absorption between 220–400 nm and gravimetric estimation of Cl⁻ confirmed the ring breakdown. In an analysis of metabolic products from griseofulvin in a similar experiment terminated at an intermediate stage, the acidic fraction showed an ultraviolet absorption curve very similar to that of the trione (49, R = Cl, R' = OH), a logical metabolic intermediate. The same *Pseudomonas* sp. thought to be responsible for the breakdown of griseofulvin also degraded two analogues, dechlorogriseofulvin (*49*, R = H, R' = OCH₃) and the amine (*49*, R = Cl, R' = NH₂), when these compounds were supplied as sole carbon source.

Nowhere has there been greater development of antibiotics for agricultural use than in Japan. Blasticidin-S (*50*, R = NH₂) one of the first successes, stimulated the research which led to other excellent compounds such as kasugamycin and the polyoxins. It is active against both bacteria and fungi but is particularly selective in its antifungal activity. At concentrations as low as 5–10 μg ml^{-1}, it is up to 100 times as active as organomercury fungicides, thus making control of rice paddy blast even more effective. Its degradation in the rice plant, by sunlight and in the soil has been the subject of study by Misato *et al.* (1972). Various microorganisms usually found in a paddy field environment, such as *Pseudomonas marginalis*, *Pseudomonas ovalis* and

(*50*)

(*51*)

Fusarium oxysporum, depressed activity when the fungicide was added to growing cultures, although mycelia or washed cell suspensions were less effective; the major metabolite was cytomycin (*51*, R = NH₂). Mycelia of an unnamed fungal strain isolated from soil, however, showed the greatest ability to inactivate blasticidin-S, the metabolites in this case, identified by t.l.c. and ultraviolet absorption spectra, being deaminohydroxyblasticidin (*50*, R = OH) and deaminohydroxycytomycin (*51*, R = OH) in addition to cytomycin.

C. AROMATIC COMPOUNDS

1. *Chloroneb*

Chloroneb(1,4-dichloro-2,5-dimethoxybenzene (*52*, R = R' = CH₃)) is a somewhat unlikely systemic fungicide which is useful in reducing soil-borne seedling diseases in certain crops, being effective against some *Rhizoctonia* spp. (Fielding and Rhodes, 1967), *Pythium* spp. (Littrell *et al.*, 1969) and *Ustilago* spp. (Lukens, 1967). It has also been

(*52*)

successfully used against Typhula blight on turf (Vargas and Beard, 1970). (^{14}C)Chloroneb has been used to follow its uptake by roots, its accumulation therein and also in lower portions of stems (Fielding and Rhodes, 1967; Darrag and Sinclair, 1969; Kirk *et al.*, 1969; Rhodes *et al.*, 1971). Whereas Kirk *et al.* (1969) found only unchanged chloroneb in the stems of cotton seedlings, Rhodes *et al.* (1971) found DCMP(2,5-dichloro-4-methoxyphenol (*52*, R = H, R' = CH$_3$)) to be a major metabolite in both cotton and bean plants grown for 12 days in soil containing the fungicide. They also reported the presence of an unidentified polar metabolite which accounted for 2% of the extracted radioactivity. Thorn (1973) identified the latter as 2,5-dichloro-4-methoxyphenol β-glucoside and found that it amounted to some 30% of extractable radioactivity, the free phenol being present to the extent of only 7%; he also noted trace amounts of the completely demethyl-ated phenol 2,5-dichlorohydroquinone (*52*, R = R' = H) first reported by Rhodes *et al.* (1971). As a result of bioassays using *Rhizoctonia solani* and *Pythium ultimum* he concluded that the chloroneb metabol-ites do not play an important role in the protection afforded to seed-lings by the fungicide.

Hock and Sisler (1969) found that *Rhizoctonia solani* demethylated chloroneb to the extent of 50% in 24 hours. The whole question of its stability in the presence of pure cultures of soil microorganisms has been examined by Wiese and Vargas (1972, 1973). Both chloroneb and DCMP were unaltered in liquid shake cultures when growth was minimal or static, but in nutrient broth where active growth was supported, several organisms not only demethylated chloroneb but also re-methylated the resultant DCMP. Of 23 organisms grown in the presence of 5 μg ml^{-1} chloroneb, 13 caused demethylation, *Fusarium* spp. being most active in this respect, up to 50% conversion being possible in five days. When grown in the presence of 5 μg ml^{-1} of DCMP, eight organisms, particularly *Trichoderma viride* and *Mucor*

TABLE 2

Degradation and synthesis of chloroneb by microorganisms grown 10 days in basal medium amended with chloroneb or DCMP (5 μg ml^{-1})[c]

Microorganism	DCMP recovered from chloroneb medium (μg ml^{-1})	Chloroneb recovered from DCMP medium (μg ml^{-1})
Sterile medium control	0	0
Fusarium solani f. pisi[a]	3·0	0·2
Fusarium solani f. phaseoli[a]	2·5	<0·1
Sclerotinia sclerotiorum[a]	1·5	0
Mucor ramannianus	1·1	0·6
Helminthosporium victoriae[a]	1·0	0
Corynebacterium fascians	1·0	0
Stemphyllium sarcinaeforme	0·5	0
Cephalosporium gramineum	0·5	0·1
Aspergillus fumigatus	0·4	0·6
Cladosporium cucumerinum	0·3	0
Verticillium albo-atrum[a]	0·3	<0·1
Cephalosporium gregatum	<0·1	0·4
Chaetomium globosum	<0·1	<0·1
Helminthosporium sativum[a]	0·1	0
Trichoderma viride	0	0·9
Penicillium frequentans	0	0·1
Rhizoctonia solani[b]	0	<0·1
Pythium aphanidermatum[a]	0	0
Thielaviopis basicola	0	0
Agrobacterium tumifaciens	0	0
Bacillus subtilis	0	0
Saccharomyces cerevisiae	0	0
Alternaria tenuis[a]	0	0

[a] Growth (dry wt) reduced 10–50% in presence of chloroneb (5 μg ml^{-1}).

[b] No growth in presence of chloroneb (5 μg ml^{-1}).

[c] From Wiese and Vargas (1973) by courtesy of Academic Press Inc. and the authors.

ramannianus, reconverted 20% to the original fungicide. *Cephalosporium gramineum*, *Rhizoctonia solani*, *M. ramannianus* and various *Fusarium* spp. were capable of both methylation and demethylation, and such microbial processes may account for the relative stability of chloroneb in the soil. Because the microbial specificities for chloroneb degradation and synthesis differ (Wiese and Vargas, 1973) (Table 2), methylation is not a simple reversal of the demethylation reaction but may involve a microbial methyl transferase (Axelrod, 1955).

2. Carboxylic acid anilides

The metabolism of mebenil (2-methylbenzanilide) and 2-chloro-benzanilide by pure cultures of *Rhizopus japonicus* has been investigated by Wallnöfer *et al.* (1971). The identities of the hydroxy metabolites, determined by n.m.r. and mass spectral analysis, were 2-methyl-4'-hydroxy- and 2-chloro-4'-hydroxybenzanilide respectively. These metabolites appeared with the beginning of the exponential growth phase, and highest concentrations were reached when fungal growth ended.

3. Thiophanates

Thiophanate (1,2-bis-(3-ethoxycarbonyl-2-thioureido)benzene (*53*, R = C_2H_5)) and the corresponding methoxy compound thiophanate-methyl (*53*, R = CH_3) control several important phytopathogenic fungi, including apple powdery mildew, apple scab, pre-harvest rot and mould of citrus, pear scab, blast and sheath blight of rice, and *Cercospora* leaf spot of sugar beet. Their relatively facile cyclization to ethyl and methyl benzimidazol-2-yl-carbamates (*54*, R = C_2H_5 or CH_3) respectively, takes place not only in refluxing 50% aqueous acetic acid in the presenceof copper(II) acetate (Selling *et al.*, 1970) but also more slowly in tap water, buffer or sterile saline. This conversion has been

NHCSNHCOOR
NHCSNHCOOR

(*53*)

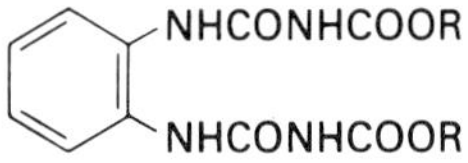

NHCONHCOOR
NHCONHCOOR

(*55*)

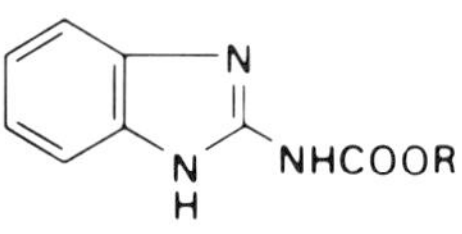

(*54*) benzimidazole-NHCOOR

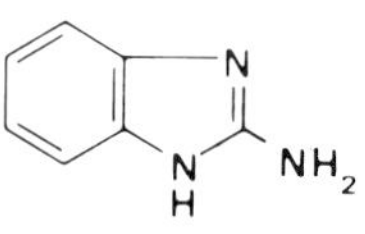

(*56*) benzimidazole-NH_2

reported to occur also in bean and cucumber plants by Noguchi *et al.* (1971) who used [14]C- and ^{35}S-labelled compounds and co-t.l.c.-autoradiography to demonstrate cleavage of the C—S bond and subsequent benzimidazole ring formation. More recently Buchenauer *et al.* (1973) found almost complete conversion of thiophanate-methyl to carbendazim (*54*, R = CH_3) in cotton plants in six days.

A detailed study of the behaviour of thiophanate-methyl in soil has been made by Noguchi (1972). [14]C- or ^{35}S-labelled compounds were again used and incubated for 60 days in two types of soil, though in the case of the silty loam soil complete data was not obtainable because of unextractable radioactivity. Thiophanate-methyl was found to be transformed almost entirely within seven days in both types of soil, the amount of carbendazim increasing up to that time; no evidence was found for the formation of dimethyl o-phenylene-4,4-bis-allophanate (*55*, R = CH$_3$), which is a very minor metabolite in plants. Thereafter carbendazim began to disappear, the rate being slightly faster in the silty loam than in the sandy loam soil, falling to about 20% after 60 days. It was suggested that the activity of soil microorganisms might be an important factor in this degradation.

The persistence and metabolism of thiophanate-methyl in soil has also been examined by Flecker et al. (1974). They confirmed the rapid conversion to carbendazim and also noted that the rate was four times faster at pH 7·4 than at pH 5·6, though steam-sterilization slowed down, but did not altogether halt the transformation, which again points to some measure of microbial involvement.

D. HETEROCYCLIC COMPOUNDS

1. *Substituted benzimidazoles*

Benomyl (methyl(1-butylcarbamoyl)benzimidazol-2-ylcarbamate (*57*)) has also been shown to be converted, in varying degrees depending on the particular environment, to carbendazim (*54*, R = CH$_3$). Clemons and Sisler (1969), who first reported this, demonstrated a 50% breakdown of benomyl in 1 hour on incubation at 30°C of a nutrient medium to which it had been added. Sims et al. (1969) were able to detect carbendazim, but not benomyl, in extracts of cotton plants four weeks after treatment with the latter, while Peterson and Edgington (1969) found that benomyl was completely transformed to carbendazim in five days by bean plants. Further metabolism of carbendazim has also been reported by Siegel (1973) who examined its distribution in strawberry plants using the ^{14}C-labelled material. After 88 days he found that (^{14}C)carbendazim was present to only 32%, together with 10–18% 2-aminobenzimidazole (*56*); the remaining radioactivity was presumed to be in unidentified metabolites either water-soluble or still bound to residual plant material after extraction.

(*57*)   (*58*)

The first hint of microbial degradation was mentioned by Worthing (1974). He found a rapid loss of benomyl in soils, which could not be attributed to leaching or to plant uptake. In laboratory experiments a rapid loss of fungicide occurred when benomyl or carbendazim was added to soil from plots previously drenched with benomyl; such a loss did not occur in soil samples which has been γ-irradiated (at 10 Mrad), or in soil taken from plots not previously treated with benomyl.

The degradation of benomyl in soil has also been examined by Baude *et al.* (1974) using the ^{14}C-labelled fungicide. They found half lives of 3–6 and 6–12 months in turf and soil respectively, the major product being carbendazim, with 2-aminobenzimidazole as a minor metabolite. Flecker *et al.* (1974) used both *ring*-2-^{14}C- and *methyl*-^{14}C-labelled carbendazim to examine its stability in soil. These compounds released < 1 and 16%, respectively, of the applied radioactivity as ^{14}CO$_2$. The recovery of radioactivity from acetone extracts of soil which had been treated with carbendazim was 79–91% in samples extracted immediately and 53–78% in samples taken 43 days after treatment, depending on the nature of the soil and the rate of application (10–100 ppm). The slow rate of ^{14}CO$_2$ production suggested stability of the benzimidazole ring to microbial oxidation but an enigmatic feature of this work is the proportion of ^{14}C not recoverable, which appears to increase with time. The explanation may lie in fission of the benzimidazole nucleus which has recently been demonstrated in melon plants grown in a liquid nutrient solution containing benomyl (20 μg ml^{-1}) (Rouchaud *et al.*, 1974). The seven metabolites identified by g.l.c. included not only carbendazim (*54*, R = CH$_3$) and 2-amino-benzimidazole (*56*), together with their respective β-glucosides, but also benzimidazole itself, *o*-aminobenzonitrile and aniline.

In contrast, thiabendazole (2-(thiazol-4-yl)benzimidazole) (*58*) appears remarkably stable. Apart from the observation that 75% of the fungicide disappeared in ten days from the leaves of pepper plants and that "rapid degradation . . . probably takes place outside the living cell" there are no reports of breakdown (Ben-Aziz and Aharonson, 1974).

Parbendazole (methyl 5(6)-butyl-benzimidazol-2-ylcarbamate, (*59*, R = CH$_3$)) an anthelmintic like thiabendazole, has yielded a variety of metabolites (Dunn *et al.*, 1973), including three produced by fungi. *Cunninghamella bainieri* produced the alcohol (*59*, R = CH$_2$OH), a likely precursor of the acid (*59*, R = CO$_2$H) also formed, while a *Paecilomyces* sp. was responsible for γ-oxidation of the butyl substituent with the formation of the ketone (*60*).

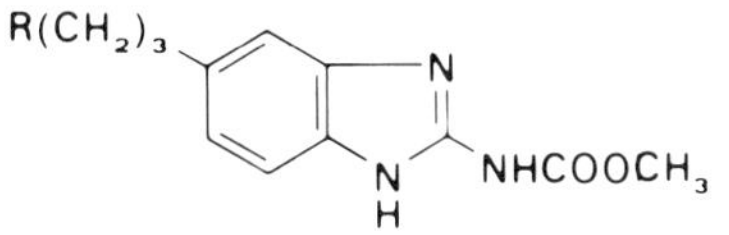

(59) 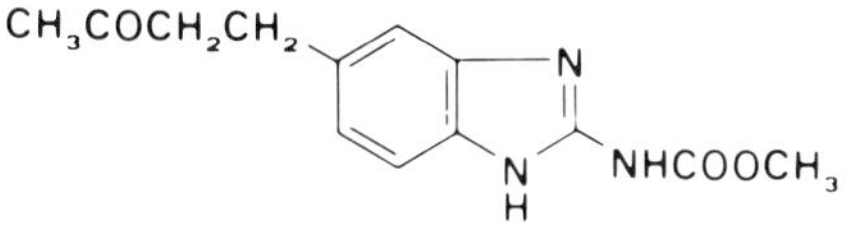   (60)

2. Carboxylic acid anilides

The stability of carboxin (2,3-dihydro-6-methyl-5-phenylcarbamoyl-1,4-oxathiin (61, R = H)) has been examined by Chin et al. (1969) using *hetero ring*-^{14}C and *benzene ring*-^{14}C labelled materials. They found that whilst oxidation to the 4-oxide (62) took place not only in

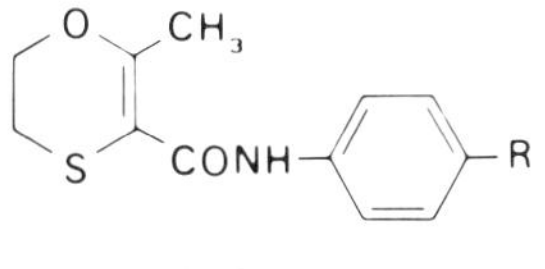

(61) 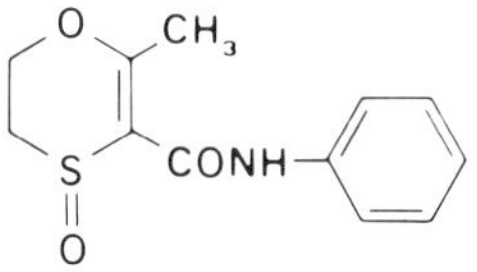   (62) 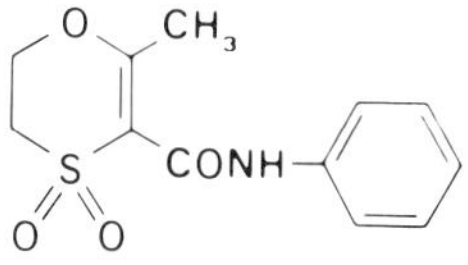   (63)

water and soil but also in barley, wheat and cotton plants, hydrolysis was not detected. The change was facilitated by low pH but there was not evidence of further oxidation to the corresponding sulphone (oxycarboxin, 63). Snel and Edgington (1970) investigated uptake, translocation and metabolism of carboxin in cotton plants using the ^{14}C-labelled compound and confirmed conversion to the non-fungitoxic sulphoxide (62) only. Analysis of root tissue, however, also showed that the —CONH-linkage had been hydrolysed with the formation of aniline, which subsequently became bound to plant polymers and also converted to water-soluble conjugates. More recently Briggs et al. (1974) have shown the formation of the phenol (61, R = OH) as the major soluble metabolite formed in growing barley seedlings.

Lyr et al. (1973) noted that mitochondria from both carboxin-sensitive and insensitive fungi could oxidize carboxin, and found that *Ustilago maydis* generated the 4-oxide (62) more readily than any other species. Further oxidation to the sulphone (63) was very limited and only occurred under certain conditions, though it had been suggested earlier that the energy requirement for sulphoxide-sulphone conversion was not available under biological conditions.

The metabolism of furcarbanil (2,5-dimethyl-3-phenylcarbamoyl-furan (64) by *Rhizopus japonicus* and related fungi (Wallnöfer et al., 1972) also involves oxidation, on this occasion of methyl groups. The isomeric hydroxymethyl compounds (65, 66) which accumulated after 1 week's growth of the fungus were isolated and identified by examination of their n.m.r. and mass spectra.

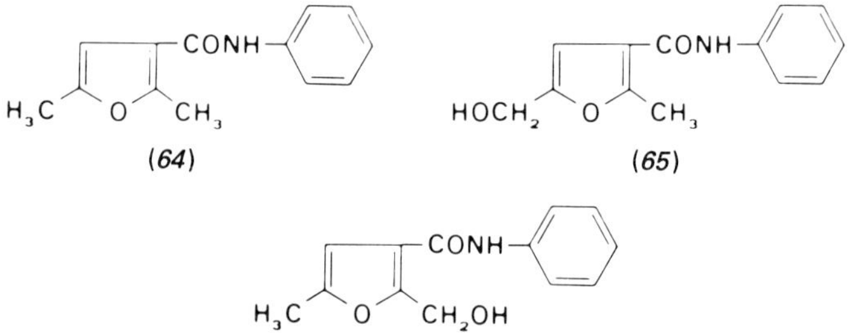

(64)

(65)

(66)

3. 3-(3',5'-Dichlorophenyl)-5,5-dimethyloxazolidine-2,4-dione (DDOD)

DDOD (67) which is active against *Sclerotinia sclerotiorum* and *Botrytis cinerea*, has been the subject of uptake and distribution studies by Sumida *et al.*(1973), who used *carbonyl*-^{14}C- or *phenyl*-^{3}H-labelled compounds for studies in bean plants and vines and in the soil. This is yet another case where enzymatic and non-enzymatic degradations both appear to operate, DDOD being considerably degraded in 14 days, not only in sterile or non-sterile nutrient solution but also at low concentrations (0·22–2·5 ppm) in distilled water. Opening of the

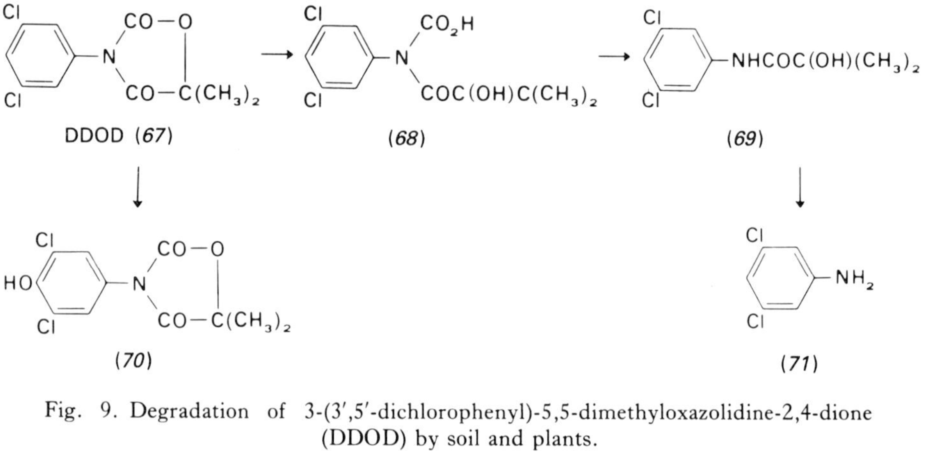

DDOD (67)

(68)

(69)

(70)

(71)

Fig. 9. Degradation of 3-(3',5'-dichlorophenyl)-5,5-dimethyloxazolidine-2,4-dione (DDOD) by soil and plants.

heterocyclic ring (Fig. 9) gave *N*-(3,5-dichlorophenyl)-*N*-α-hy-droxyisobutyrylcarbamic acid (68) from which α-hydroxyisobutyryl-3,5-dichloranilide (69) was formed by decarboxylation. Comparable metabolism of DDOD in bean plants was minimal, but yet another metabolite, the dione (70) was tentatively identified in the leaves of vines which had been injected with (^{14}C) DDOD. Degradation to the carbamic acid (68) and anilide (69) took place to a lesser extent in sterilized than in non-sterile soil, presumably because of microbial action in the latter, although reduced water content and/or alteration in soil structure in the former case, have been suggested as possible causes for reduced breakdown. In experiments to study the fate of the aromatic ring, (^{3}H)DDOD was used and a terminal residue found to be 3,5-dichloroaniline (71).

4. Triadimefon

Triadimefon (1-(4-chlorophenoxy)-3,3-dimethyl-1-(1,2,4-triazol-1-yl)-2-butanone (72)), a new systemic fungicide which also has protective and curative actions, has recently been shown to undergo an interesting transformation by *Aspergillus niger* (Clifford *et al.*, personal communication). While there was a 40% conversion to the corresponding carbinol (73) in marrow plants, this rarely exceeded 3% in

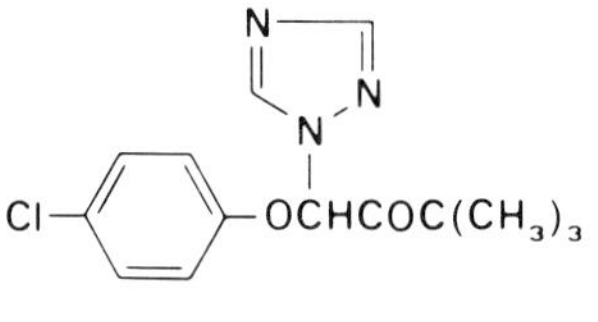

(72)

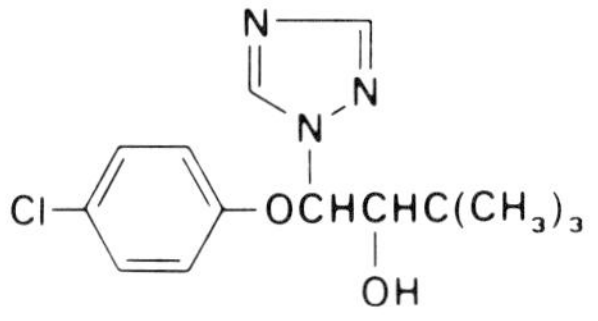

(73)

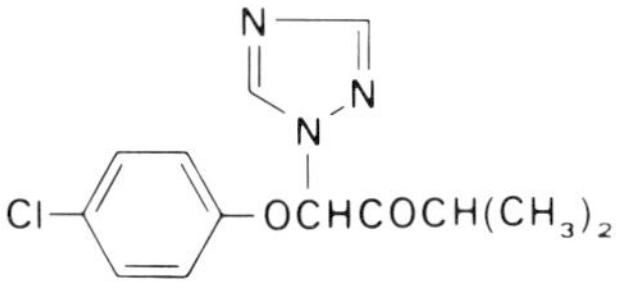

(74)

the fungal mycelium. Here desmethylation occurred to the extent of 35% with the formation of the isopropyl analogue of triadimefon (74). No such desmethylation took place when the carbinol (73) was incubated with *A. niger*.

5. S-n-*butyl*-S'-p-*t*-*butylbenzyl*-N-*3*-*pyridyldithiocarbonimidate*

A study of the degradation of S-*n*-butyl-S'-*p*-t-butylbenzyl-*N*-3-pyridyldithiocarbonimidate (75), a fungicide highly effective in controlling various powdery mildew diseases, by light, plants and soils has revealed several interesting transformations in the latter media (Fig. 10) which are almost certainly microbial in origin (Ohkawa *et al.*,

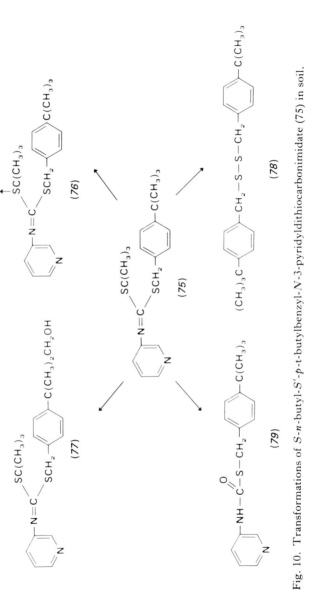

Fig. 10. Transformations of S-n-butyl-S'-p-t-butylbenzyl-N-3-pyridyldithiocarbonimidate (75) in soil.

1976). Products identified included the S-oxide (76), the hydroxylated compound (77), the disulphide (78) and S-p-t-butylbenzyl-N-3-pyridylthiolocarbamate (79).

E. ALIPHATIC COMPOUNDS

In a search for naturally occurring substances suitable for use as agricultural chemicals Misato $et\ al.$ (1972) found that N-lauroyl-L-valine (80) showed excellent activity against rice blast. Since hydrolysis of the amide linkage would only release the already naturally-occurring compounds lauric acid and L-valine, residue problems do not exist and the compound appears to be a very attractive proposition. Possible utilization of N-lauroyl-L-valine by microorganisms

$$CH_3(CH_2)_{10}CONHCHCH(CH_3)_2$$
$$CO_2H$$

(80)

was examined by incorporation into the culture medium as a source of carbon and nitrogen, and of nine aerobic soil bacteria included, $Pseudomonas\ aeruginosa$ proved to be most active. This organism was accordingly used for degradation studies with the $amide$-^{14}C-labelled compound. Release of $^{14}CO_2$ indicated hydrolytic cleavage and also, presumably, oxidation of the lauric acid formed through the normal TCA cycle.

Prothiocarb (S-ethyl-(N,N-dimethylaminopropyl)thiocarbamate (81)) is a systemic fungicide which is applied to soil as the hydrochloride. Two metabolites have been tentatively identified as N,N-dimethylpropylene diamine and the demethylated compound S-ethyl-(N-methylaminopropyl) thiocarbamate (82) (Iwan and Goller,

$$(CH_3)_2N(CH_2)_3COSC_2H_5 \qquad CH_3NH(CH_2)_3COSC_2H_5$$

(81)   (82)

unpublished data, cited by Vonk and Kaars Sijpesteijn, 1977). It seems likely that the latter compound is of microbial origin (Pieroh $et\ al.$, 1975).

V. Conclusions

Investigations concerned with the metabolism of fungicides, as with other pesticides, are necessary for several reasons, not least of which is to assess possible hazards arising from their widespread use in the environment. The identification and toxicological assessment of metabolic products is thus of paramount importance and knowledge of such compounds may also be useful in establishing modes of action. Such studies may lead to the development of more sophisticated and safer fungicides possessing greater biological activity and improved specificity.

While the soil has always been recognized as an active degradative environment, past workers have often been content to measure rates of disappearance of organic compounds from soils by bioassay methods and have not attempted to isolate, much less identify, possible metabolites or delineate metabolic pathways. Much of the work that has been done has concerned insecticides and herbicides – presumably because of the much greater usage of these agrochemicals – but there is a growing interest in the fate of fungicides. It is also becoming increasingly apparent that in environments such as the soil and plant surfaces, the effects of air, light, moisture and enzymes (both microbial and plant) are completely integrated and it is often difficult to distinguish between enzymatic and non-enzymatic reactions. The magnitude of transformations photochemically initiated or promoted has become clearly recognized and has been discussed elsewhere (Woodcock, 1977).

It is clear from the foregoing account that the fungicide scene has changed rapidly in the past decade and that increased environmental and toxicological demands have been met only by increased use of, and improvements in labelling, isolation and scanning techniques.

References

ABBOTT, M. J. J. and GROVE, J. F. (1959). Uptake and translocation of organic compounds by fungi. II. Griseofulvin. *Expl. Cell Res.* **17**, 105–113.

ALEXANDER, M. and ALEEM, M. I. H. (1961). Effect of chemical structure on microbial decomposition of aromatic herbicides. *J. agric. Fd Chem.* **9**, 44–49.

ALLIOT, H. (1973). A bibliography of organic solvent-based wood preservatives. *Brit. Wood Preserving Assoc. News Sheet* No. 127.

AXELROD, J. (1965). The formation and metabolism of physiologically active compounds by *N*- and *O*-methyltransferases. *In* "Transmethylation and Methionine Biosynthesis" (S. K. Shapiro and F. Schlenk, Eds), pp. 71–83. Univ. of Chicago Press, Chicago.

BARNES, R. D., BULL, A. T. and POLLER, R. C. (1973). Studies on the persistence of the organotin fungicide fentin acetate (triphenyltin acetate) in the soil and on surfaces exposed to light. *Pestic. Sci.* **4**, 305–317.

BASTIAANSEN, M. G., PIEROH, E. A. and AELBERS, E. (1974). Prothiocarb, a new fungicide to control *Phytophthora fragariae* in strawberries and *Pythium ultimum* in flower bulbs. *Meded. Fac. Llandb. Rijksuniv. Gent* **39**, 1019–1026.

BAUDE, F. J., PEASE, H. L. and HOLT, R. F. (1974). Fate of benomyl in field soil and turf. *J. agric. Fd Chem.* **22**, 414–418.

BECK, J. and HANSEN, K. E. (1974). The degradation of quintozene pentachlorobenzene, hexachlorobenzene and pentachloroaniline in soil. *Pestic. Sci.* **5**, 41–48.

BELL, G. R. (1957). Some morphological and biochemical characteristics of a soil bacterium which decomposes 2,4-dichlorophenoxyacetic acid. *Can. J. Microbiol.* **3**, 821–840.

BELSER, N. O. and CASTRO, C. E. (1971). Biodehalogenation – The metabolism of the nematocides *cis*- and *trans*-3-chloroallyl alcohols by a bacterium isolated from soil. *J. agric. Fd Chem.* **19**, 23–26.

BEN-AZIZ, A. and AHARONSON, N. (1974). Dynamics of uptake, translocation and disappearance of thiabendazole and methyl 2-benzimidazole carbamate in pepper and tomato plants. *Pestic. Biochem. Physiol.* **4**, 120–126.

BERNARDE, M. A., KOFT, B. W., HORVATH, R. and SHANLIS, L. (1965). Microbial degradation of the sulphonate of dodecylbenzenesulphonate. *Appl. Microbiol.* **13**, 103–105.

BIRD, L. S., RANNEY, C. D. and WATKINS, G. M. (1957). Evaluation of fungicides mixed with the covering soil at planting as a control measure for the cotton seedling-disease complex. *Pl. Dis. Reptr* **41**, 165–173.

BRIGGS, D. E., WARING, R. H. and HACKETT, A. M. (1974). The metabolism of carboxin in growing barley. *Pestic. Sci.* **5**, 599–607.

BUCHENAUER, H., ERWIN, D. C. and KEEN, N. T. (1973). Systemic fungicidal effect of thiophanate-methyl on *Verticillium* wilt of cotton and its transformation to methyl 2-benzimidazole carbamate. *Phytopathology* **63**, 1091–1095.

BURCHFIELD, H. P. (1959). Comparative stabilities of dyrene, 1-fluoro-2,4-dinitrobenzene, dichlone and captan in silt loam soil. *Contrib. Boyce Thompson Inst.* **20**, 205–215.

BURCHFIELD, H. P. and STORRS, E. E. (1956). Chemical structures and dissociation constants of amino acids, peptides and proteins in relation to their reaction rates with 2,4-dichloro-6-(*o*-chloroaniline)-*s*-triazine. *Contrib. Boyce Thompson Inst.* **18**, 395–418.

BURCHFIELD, H. P. and STORRS, E. E. (1957). Effect of chlorine substitution and isomerism on the interactions of *s*-triazine derivatives with conidia of *Neurospora sitophila*. *Contrib. Boyce Thompson Inst.* **18**, 429–462.

CARTWRIGHT, N. J. and CAIN, R. B. (1959). Bacterial degradation of the nitrobenzoic acids. *Biochem. J.* **71**, 248–261.

CASELEY, J. C. (1968). The loss of three chloronitrobenzene fungicides from the soil. *Bull. Environ. Contam. Toxicol.* **3**, 180–193.

CASTRO, C. E. and BELSER, N. O. (1966). Hydrolysis of *cis*- and *trans*-1,3-dichloropropene in wet soil. *J. agric. Fd Chem.* **14**, 69–70.

CHACKO, C. I., LOCKWOOD, J. L. and ZABIK, M. (1966). Chlorinated hydrocarbon pesticides: degradation by microbes. *Science N.Y.* **154**, 893–895.

CHANDRA, P. and BOLLEN, W. (1961). Effects of nabam and mylone on nitrification, soil respiration and microbial numbers in four Oregon soils. *Soil Sci.* **92**, 387–393.

CHIN, W. T., STONE, G. M. and SMITH, A. E. (1969). The fate of carboxin in soil, plants and animals. *Proc. 5th Br. Insectic. Fungic. Conf.* **2**, 332–339.

CHU, J. P. and KIRSCH, E. J. (1972). Metabolism of pentachlorophenol by axenic bacterial culture. *Appl. Microbiol.* **23**, 1033–1035.

CLEMONS, G. P. and SISLER, H. D. (1969). Formation of a fungitoxic derivative from Benlate. *Phytopathology* **59**, 705–706.

CROWDY, S. H., GARDNER, D., GROVE, J. J. and PRAMER, D. (1955). The translocation of antibiotics in higher plants. I. Isolation of griseofulvin and chloramphenicol from plant tissue. *J. exp. Bot.* **6**, 371–383.

CROWDY, S. H., GROVE, J. F., HEMMING, H. G. and ROBINSON, K. C. (1956). The translocation of antibiotics in higher plants. II. The movement of griseofulvin in broad bean and tomato. *J. exp. Bot.* **7**, 42–64.

CURTIS, R. F., LAND, D. G., GRIFFITHS, N. M., GEE, M. G., ROBINSON, D., PEEL, J. L., DENNIS, C. and GEE, J. M. (1972). 2,3,4,6-Tetrachloroanisole: association with musty taint in chickens and microbiological formation. *Nature, Lond.* **235**, 223–224.

CURTIS, R. F., DENNIS, G., GEE, J. M., GEE, M. G., GRIFFITHS, N. M., LAND, D. G., PEEL, J. L. and ROBINSON, D. (1974). Chloroanisoles as a cause of musty taint in chickens and their microbiological formation from chlorophenols in broiler house litters. *J. Sci. Fd Agric.* **25**, 811–828.

CZERWINSKA, E., JANKO, Z., KOWALIK, R. and STEFANIAK, H. (1969). Microbial decomposition of methyl mercury dicyanidiamide. *Roczn. Naukro ln.* **95**, 219–229.

DAGLEY, S. (1967). The microbial metabolism of phenolics. *In* "Soil Biochemistry" (A. D. McLaren and G. H. Peterson, Eds), Vol. 1, pp. 287–317. Marcel Dekker, New York.

DAINES, R. H. (1936). Some principles underlying the action of mercury in soils. *Phytopathology* **26**, 90.

DARRAG, I. E. M. and SINCLAIR, J. B. (1969). Evidence for systemic protection against *Rhizoctonia solani* with chloroneb in cotton seedlings. *Phytopathology* **59**, 1102–1105.

DEKHUISEN, H. M. (1961). Transformation in plants of sodium dimethyldithiocarbamate into other fungitoxic compounds. *Nature, Lond.* **191**, 198–199.

DEKHUISEN, H. M. (1964). The systemic action of dimethyldithiocarbamates on cucumber scab by *Cladosporium cucumerium* and the conversion of these compounds by plants. *Neth. J. Pl. Path.* **70**, Suppl. 1 (75 pp.).

DOUROS, J. D. and REID, J. J. (1956). Decomposition of certain herbicides by soil microflora. *Bact. Proc.* **56**, 23–24.

DUNN, G. L., GALLAGHER, G., DAVIS, L. D., HOOVER, J. R. E. and STEDMAN, R. J. (1973). Metabolites of methyl 5 (6)-butyl-2-benzimidazolecarbamate (parbendazole). Structure and synthesis. *J. Mednl Chem.* **16**, 996–1002.

ELSAID, H. M. (1963). Adapted tolerance of *Rhizoctonia solani* to increased concentrations of three soil fungicides. *Phytopathology* **53**, 875.

ELSAID, H. M. and SINCLAIR, J. B. (1962). Adapted tolerance to organic fungicides by an isolate of *Rhizoctonia solani*. *Phytopathology* **52**, 731.

ERIKSON, D. (1941). Studies on some lake-mud strains of *Micromonospora*. *J. Bact.* **41**, 277–300.

EVANS, W. C., SMITH, B. S. W., LINSTEAD, R. P. and ELVIDGE, J. A. (1951). Chemistry of the oxidative metabolism of certain aromatic compounds by microorganisms. *Nature, Lond.* **168**, 772–775.

EXNER, J. H., BURK, G. A. and KYRIACOU, D. (1973). Rates and products of decomposition of 2,2-dibromo-3-nitrilopropionamide. *J. agric. Fd Chem.* **21**, 838–842.

FEST, C. and SCHMIDT, K.-J. (1973). "The Chemistry of Organophosphorus Pesticides". Springer Verlag, New York.

FIELDING, M. J. and RHODES, R. C. (1967). Studies with ^{14}C-labelled chloroneb fungicide. *Proc. Cotton Dis. Council* **27**, 56.

FLECKER, J. R., LACY, H. M., SCHUTZ, I. R. and HOUKOM, E. C. (1974). Persistence and metabolism of thiophanate-methyl in soil. *J. agric. Fd Chem.* **22**, 592–595.

FURAKAWA, K., SUSUKI, T. and TONOMURA, K. (1969). Decomposition of organic mercurial compounds by mercury-resistant bacteria. *Agric. Biol. Chem.* **33**, 128–130.

GAUSE, E. M., MONTALVO, D. A. and ROWLANDS, J. R. (1967). Electron spin resonance studies of the chloranil–cysteine, dichlone–cysteine and dichlone–glutathione reactions. *Biochim. biophys. Acta* **141**, 217.

GEE, J. M. and PEEL, J. L. (1974). Metabolism of 2,3,4,6-tetrachlorophenol by microorganisms from broiler house litter. *J. gen. Microbiol.* **85**, 237–243.

GONDO, M. and KUBO, T. (1958). Effect of some fungicides on *Helicobasidium mompa* Tanaka in soil (abs.). *Rev. appl. Mycol.* **37**, 677.

GORBACH, S. and WAGNER, U. (1967). Pentachloronitrobenzene residues in potatoes. *J. agric. Fd Chem.* **15**, 654–656.

GORBACH, S., THIER, W., KELLNER, H. M., SCHULZE, E. F., KUNSLER, K. and FISHER, H. (1974). Environmental impact of pyrazophos. 1. Degradation in plants and in the rat. *3rd Int. Congress of Pesticide Chemistry* abs. 524.

GÖTH, H., GAGNEUX, A. R., EUGSTER, C. H. and SCHMID, H. (1967). 2(3H)-Oxazolone durch photoumlagerung von 3-hydroxyisoxazolen. Synthese von muscazon. *Helv. Chim. Acta* **50**, 137–142.

GOTTLIEB, D. and SIMINOFF, P. (1952). The effect of metabolites on antimicrobial agents. *Phytopathology* **42**, 91–97.

GOTTLIEB, D., SIMINOFF, P. and MARTIN, M. M. (1952). The production and role of antibiotics in soil. IV. Chloromycetin. *Phytopathology* **42**, 493–496.

GRAPOV, A. F. and MELNIKOV, N. N. (1973). Organophosphorus fungicides. *Russ. chem. Revs.* **42**, 772–780.

GROSSMANN, F. (1963). Untersuchungen über die Hemmung pektolytischer Enzyme von *Fusarium oxysporum* f. lycopersici III Wirkung einiger Hemmstoff *in vivo*. *Phytopathol. Z.* **45**, 139–159.

GROVES, K. and CHOUGH, K. S. (1970). Fate of fungicide 2,6-dichloro-4-nitroaniline (DCNA) in plants and soils. *J. agric. Fd Chem.* **18**, 1127–1128.

GUNDERSEN, K. and JENSEN, H. L. (1956). A soil bacterium decomposing organic nitro compounds. *Acta agric. Scand.* **6**, 100.

HANNON, C. I., ANGELINI, J. and WOLFORD, R. (1963). Detection of dichloropropene-dichloropropane in soil by gas chromatography. *J. Gas Chromat.* **1**, 27.

HANSON, W. J. and NEX, R. W. (1953). Diffusion of ethylene bromide in soils. *Soil Sci.* **76**, 209–214.

HAY, S. J. B. (1971). The control of apple mildew with HOE 2873. *Proc. 6th Br. Insectic. Fungic. Conf.* **1**, 134–140.

HILLS, F. J. (1962). Uptake, translocation and chemotherapeutic effect of *p*-dimethylaminobenzenediazo sodium sulphonate (Dexon) in sugar beet seedlings. *Phytopathology* **52**, 389–392.

HILLS, F. J. and LEACH, L. D. (1962). Photochemical decomposition and biological activity of *p*-dimethylaminobenzenediazo sodium sulphonate (Dexon). *Phytopathology* **52**, 51–56.

HOCK, W. K. and SISLER, H. D. (1969). Metabolism of chloroneb by *Rhizoctonia solani* and other fungi. *J. agric. Fd Chem.* **17**, 123–128.

HORVATH, R. S. (1972). Microbial co-metabolism and the degradation of organic compounds in nature. *Bact. Rev.* **36**, 146–155.

HYLIN, J. W. and CHIN, B. H. (1967). Delayed phytotoxicities produced by dithiocarbamate fungicide residues. *6th Int. Congress of Plant Protection*, 614 (abstr.).

IDE, A., NIKI, Y., SAKAMOTO, F., WATANABE, I. and WATANABE, H. (1972). Decomposition of pentachlorophenol in paddy soil. *Agric. biol. Chem.* **36**, 1937–1944.

IWAN, J. and GOLLER, D. (1974). Behaviour of SN41703 (Prothiocarb) in soils and plants. *3rd Int. Congress of Pesticide Chemistry*. abs. 135.

JEFFERYS, E. G. (1952). The stability of antibiotics in soils. *J. gen. Microbiol.* **7**, 295–312.

JENSEN, H. L. (1961). Biologisk soderdeling af ukrudtsmidler i jorbunden II Allyl alcohol. *Tidskr. PlAvl.* **65**, 185.

JENSEN, S. and JERNELÖV, A. (1969). Biological methylation of mercury by aquatic organisms. *Nature, Lond.* **223**, 753–754.

KAARS SIJPESTEIJN, A. and VAN DER KERK, G. J. M. (1965). Fate of fungicides in plants. *A. Rev. Phytopath.* **3**, 127–152.

KAARS SIJPESTEIJN, A. and VONK, J. W. (1973). Methylation of inorganic mercury by bacteria and fungi. *Meded. Fac. Llandbouwet. Rijksuniv. Gent* **38**, 759–768.

KAARS SIJPESTEIJN, A. and VONK, J. W. (1974). Decomposition of bisdithiocarbamates and metabolism by plants and microorganisms. *3rd Int. Congress of Pesticide Chemistry*. abs. 033.

KAARS SIJPESTEIJN, A., KASLANDER, J. and VAN DER KERK, G. J. M. (1962). On the conversion of sodium dimethyldithiocarbamate to its α-aminobutyric acid derivative by microorganisms. *Biochim. biophys. Acta* **62**, 587–589.

KAARS SIJPESTEIJN, A., KERKENAAR, A. and OVEREEM, J. C. (1974). Observations on selectivity and mode of action of prothiocarb (SN 41703). *Meded. Fac. Llandbouwet. Rijksuniv. Gent.* **39**, 1027–1034.

KASLANDER, J., KAARS SIJPESTEIJN, A. and VAN DER KERK, G. J. M. (1961). On the transformation of dimethyldithiocarbamate into its β-glucoside by plant tissues. *Biochim. biophys. Acta* **52**, 396–397.

KASLANDER, J., KAARS SIJPESTEIJN, A. and VAN DER KERK, G. J. M. (1962). On the transformation of the fungicide sodium dimethyldithiocarbamate into its alanine derivative by plant tissue. *Biochim. biophys. Acta* **60**, 417–419.

KAMEDA, Y., TOYOURA, E. and KIMURA, Y. (1957). Metabolic activities of soil bacteria towards derivatives of benzoic acid, amino acids and acylamino acids. *Kanazawa Daigaku Yakugakubu Kenkyu Nempo* **7**, 37: *Chem. Abstr.* **52**, 4081.

KARANTH, N. G. K., BHAT, S. G., VAIDYANATHAN, C. S. and VASANTHARAJAN, V. N. (1974). Conversion of Dexon (*p*-dimethylaminobenzenediazo sodium sulphonate) to *N,N*-dimethyl-*p*-phenylenediamine by *Pseudomonas fragi* Bk9. *Appl. Microbiol.* **27**, 43–46.

KATZ, E. and PIENTA, P. (1957). Decomposition of actinomycin by a soil organism. *Science, N.Y.* **126**, 402–403.

KAUFMAN, D. D. and FLETCHER, C. L. (1973). Ethylenethiourea degradation in soil. *2nd Int. Congress of Plant Pathology.* abstr. 1018.

KIMURA, Y. and MILLER, V. L. (1964). Degradation of organomercury fungicides in soil. *J. agric. Fd Chem.* **12**, 253–257.

KIRK, B. T., SINCLAIR, J. B. and LAMBREMONT, E. N. (1969). Translocation of ^{14}C-labelled chloroneb and DMOC in cotton seedlings. *Phytopathology* **59**, 1473–1476.

KIRSCH, E. J. and ETZEL, J. E. (1973). Microbial decomposition of pentachlorophenol. *J. Water Pollution Control Federation* **45**, 359–364.

KLEMMER, H. W. (1957). Response of bacterial, fungal and nematode populations of Hawaiian soils to fumigation and liming. *Bact. Proc.* **57**, 12.

KO, W. H. and FARLEY, J. D. (1969). Conversion of pentachloronitrobenzene to pentachloroaniline in soil and the effect of these compounds on soil microflora. *Phytopathology* **59**, 64–67.

LEGATOR, M. and RACUSEN, D. (1959). Mechanism of allyl alcohol toxicity. *J. Bact.* **77**, 120–121.

LINDGREN, R. M. and HENRY, B. W. (1949). Promising treatments for controlling root disease and weeds in a southern pine nursery. *Pl. Dis. Reptr* **33**, 228–231.

LITTRELL, R. H., GAY, J. D. and WELLS, H. D. (1969). Chloroneb fungicide for control of *Pythium aphanidermatum* on several crop plants. *Pl. Dis. Reptr* **53**, 913–915.

LOUSTALOT, A. J. and FERRER, R. (1950). The effect of some environmental factors on the persistence of sodium pentachlorophenate in soil. *Proc. Am. Soc. Hort. Sci.* **56**, 294-298.

LUKENS, R. J. (1959). Chemical and biological studies on a reaction between captan and the dialkyldithiocarbamates. *Phytopathology* **49**, 339–343.

LUKENS, R. J. (1967). Chemical control of stripe smut. *Pl. Dis. Reptr* **51**, 355–356.

LUKENS, R. J. and SISLER, H. D. (1958). Chemical reactions involved in the fungitoxicity of captan. *Phytopathology* **48**, 235–244.

LYR, H. (1963). Enzymatische detoxication chorierter phenole. *Phytopathol. Z.* **47**, 73–83.

LYR, H., RITTER, G. and BANASIAK, L. (1973). Detoxication of carboxin. *Z. allg. Mikrobiol.* **14**, 313–320; *Chem. Abstr.* **81**, 59127.

MacRAE, I. C. and ALEXANDER, M. (1965). Microbial degradation of selected herbicides in soil. *J. agric. Fd Chem.* **13**, 72–76.

MADHOSINGH, C. (1961). The metabolic detoxication of 2,4-dinitrophenol by *Fusarium oxysporum*. *Can. J. Microbiol.* **7**, 553–567.

MASSAUX, F. (1963). Contribution à l'etude de l'action systémique du thiram (TMTD) chez les plantes à l'aide du thiram marqué au ^{35}S. *Meded. LandbHogesch. Gent* **28**, 590–596.

MATSUMURA, F., GOTOH, Y. and BOUSH, G. M. (1971). Phenyl mercuric acetate: metabolic conversion by microorganisms. *Science, N.Y.* **173**, 49–51.

McKEEN, C. D. (1954). Methyl bromide as a soil fumigant for controlling soil-borne pathogens and certain other organisms in vegetable seeds. *Can. J. Bot.* **32**, 101–113.

MILLS, C., CHILD, J. J. and SPENCER, J. F. T. (1971). The utilization of aromatic compounds by yeast. *Antonie van Leeuwenhoek* **37**, 281–287.

MISATO, T., YAMAGUCHI, I. and HOMMA, Y. (1972). A new approach in the development of biodegradable pesticides. *In* "Environmental Toxicology of Pesticides" (F. Matsumura, G. M. Boush and T. Misato, Eds), pp. 587–606. Academic Press, New York.

MOJE, W., MUNNECKE, D. E. and RICHARDSON, L. T. (1964). Carbonyl sulphide, a volatile fungitoxicant from nabam in soil. *Nature, Lond.* **202**, 831–832.

MUNAKATA, K. and KUWAHARA, M. (1969). Photochemical degradation products of pentachlorophenol. *Residue Rev.* **25**, 13–23.

MUNNECKE, D. E. (1958). The persistence of non volatile diffusible fungicides in soil. *Phytopathology* **48**, 581–585.

MUNNECKE, D. E. and LINDGREN, D. L. (1954). Chemical measurements of methyl bromide concentration in relation to kill of fungi and nematodes in nursery soil. *Phytopathology* **44**, 605–606.

MUNNECKE, D. E. and SOLBERG, R. A. (1958). Inactivation of semesan in soil by fungi. *Phytopathology* **48**, 396.

NAKANISHI, T. and OKU, H. (1969). Metabolism and accumulation of PCNB by phytopathogenic fungi in relation to selective toxicity. *Phytopathology* **59**, 1761–1762.

NISSEN, T. V. (1954). Effects of antibiotics on carbon dioxide production in soil. *Nature, Lond.* **174**, 226–227.

NOGUCHI, T. (1972). Environmental evaluation of systemic fungicides. *In* "Environmental Toxicology of Pesticides" (F. Matsumura, G. M. Boush and T. Misato, Eds), pp. 607–632. Academic Press, New York.

NOGUCHI, T., OHKUMA, K. and KOSAKA, S. (1971). Relation of structure and activity in thiophanates. *In* "Pesticide Chemistry" (A. S. Tahori, Ed.), Vol. 5, pp. 263–280. Gordon and Breach, London.

NOLAN, H. K. and HECHENBLEIKNER, I. (1947). Seed and plant disinfectants (to American Cyanamid Co.). U.S. Pat. 2419888 (April 29).

OHKAWA, H., SHIBAIKE, R., OKIHARA, Y., MORIKAWA, M. and MIYAMOTO, J. (1976). Degradation of the fungicide Denmert® (S-*n*-butyl S′-*p-tert*-butylbenzyl N-3-pyridyldithiocarbonimidate, S-1358) by plants, soils and light. *Agric. biol. Chem.* **40**, 943–951.

OVERMAN, A. J. and BURGIS, D. S. (1956). Allyl alcohol as a soil fungicide. *Phytopathology* **46**, 532–535.

OWENS, R. G. (1953). Studies on the nature of fungicide action. 1. Inhibition of sulphydryl-amino, iron and copper-dependent enzymes *in vitro* by fungicides and related compounds. *Contrib. Boyce Thompson Inst.* **17**, 221–242.

OWENS, R. G. and BLAAK, G. (1960). Chemistry of the reactions of dichlone and captan with thiols. *Contrib. Boyce Thompson Inst.* **20**, 475–498.

OWENS, R. G. and MILLER, L. P. (1957). Intracellular distribution of metal ions and organic fungicides in fungal spores. *Contrib. Boyce Thompson Inst.* **19**, 177–178.

OWENS, R. G. and NOVOTNY, H. M. (1958). Mechanism of action of the fungicide dichlone (2,3-dichloro-1,4-naphthoquinone). *Contrib. Boyce Thompson Inst.* **19**, 463–482.

PETERSON, C. A. and EDGINGTON, L. V. (1969). Translocation of the fungicide benomyl in bean plants. *Phytopathology* **59**, 1044.

PIEROH, E. A., AHRENS, C. H. and SCHUMACHER, J. (1975). Prothiocarb, a new systemic fungicide against phycomycetous fungi. *VIIIth International Plant Protection Congress*, Moscow, 1975, Section II, p. 575.

PRAMER, D. and STARKEY, R. L. (1951). Decomposition of streptomycin. *Science, N.Y.* **113**, 127.

RAYMOND, D. G. M. and ALEXANDER, M. (1971). Microbial metabolism and co-metabolism of nitrophenols. *Pestic. Biochem. Physiol.* **1**, 123–130.

RHODES, R. C., PEASE, H. L. and BRANTLEY, R. K. (1971). Fate of ^{14}C-labelled chloroneb in plants and soils. *J. agric. Fd Chem.* **19**, 745–749.

RICHARDSON, L. T. (1954). The persistence of thiram in soil and its relationship to the microbiological balance and damping-off control. *Can. J. Bot.* **32**, 335–346.

ROGOFF, M. H. (1961). Oxidation of aromatic compounds by bacteria. *Adv. appl. Microbiol.* **3**, 193–221.

ROUCHAUD, J. P., DECALLONNE, J. R. and MEYER, J. A. (1974). Metabolic fate of methyl 2-benzimidazole carbamate in melon plants. *Phytopathology* **64**, 1513–1517.

RUHLOFF, M. and BURTON, J. C. (1951). Compatibility of rhizobia with seed protectants. *Soil Sci.* **72**, 283–290.

RUSSELL, P. (1955). Inactivation of phenyl mercuric acetate in groundwood pulp by a mercury-resistant strain of *Pencillium roqueforti* Thom. *Nature, Lond.* **176**, 1123–1124.

SELLING, H. A., VONK, J. W. and KAARS SIJPESTEIJN, A. (1970). Transformation of the systemic fungicide methyl thiophanate into 2-benzimidazole carbamic acid methyl ester. *Chemy Ind.* 1625.

SHATLA, M. N. and SINCLAIR, J. B. (1963). Tolerance to PCNB and pathogenicity correlated in naturally occurring isolates of *Rhizoctonia solani. Phytopathology* **52**, 752.

SIEGEL, M. R. (1973). Distribution and metabolism of methyl-2-benzimidazole carbamate, the fungitoxic derivative of benomyl in strawberry plants. *Phytopathology* **63**, 890–896.

SIEGEL, M. R. and SISLER, H. D. (1968). Fate of the phthalimide and trichloromethylthio ($SCCl_3$) moieties of folpet in toxic action on cells of *Saccharomyces pastorianus. Phytopathology* **58**, 1123–1128.

SIMINOFF, P. and GOTTLIEB, D. (1951). The production and role of antibiotics in the soil. I. The fate of streptomycin. *Phytopathology* **41**, 420–429.

SIMPSON, J. R. and EVANS, W. C. (1953). The metabolism of nitrophenols by certain bacteria. *Biochem. J.* **55**, xxiv.

SIMS, J. J., MEE, H. and ERWIN, D. C. (1969). Methyl 2-benzimidazole carbamate, a fungitoxic compound isolated from cotton plants treated with methyl 1-(butylcarbamoyl)-2-benzimidazole carbamate (benomyl). *Phytopathology* **59**, 1775–1776.

SMITH, G. N., WORREL, C. S. and LILLIGREN, B. L. (1949). The enzymic hydrolysis of chloroamphenicol. *Science, N.Y.* **110**, 297–298.

SNEL, M. and EDGINGTON, L. V. (1970). Uptake, translocation and decomposition of oxathiin fungicides in bean. *Phytopathology* **60**, 1708–1716.

SPANIS, W. and MUNNECKE, D. E. (1962). Bacterial inactivation of organic fungicides. *Phytopathology* **52**, 365.

SPANIS, W., MUNNECKE, D. E. and SOLBERG, R. A. (1962). Biological breakdown of two organic mercurial fungicides. *Phytopathology* **52**, 455–462.

STEENSON, T. I. and WALKER, N. (1957). The pathway of breakdown of 2,4-dichloro- and 4-chloro-2-methylphenoxyacetic acids by bacteria. *J. gen. Microbiol.* **16**, 146–155.

SUMIDA, S., YOSHIHARA, R. and MUJAMOTO, J. (1973). Degradation of 3-(3',5'-dichlorophenyl)-5,5-dimethyloxazolidine-2,4-dione by plants, soil and light. *Agric. biol. Chem.* **37**, 2781–2790.

TAUSSON, W. O. (1928). Die oxydation des phenanthrens durch bakterien. *Planta* **5**, 239–272.

TEUTEBERG, A. (1964). Breakdown of halonitrobenzenes by soil bacteria. *Arch. Microbiol.* **48**, 21–49.

THORN, G. D. (1973). Uptake and metabolism of chloroneb by *Phaseolus vulgaris. Pestic. Biochem. Physiol.* **3**, 137–140.

THORN, G. D. and LUDWIG, R. A. (1954). The aeration products of disodium ethylenebisdithiocarbamate. *Can. J. Chem.* **32**, 872.

TILLMAN, R. W., SIEGEL, M. R. and LONG, J. W. (1973). Mechanism of action and fate of the fungicide chlorothalonil (2,4,5,6-tetrachloroisophthalonitrile) in biological systems 1. Reactions with cells and sub-cellular components of *Saccharomyces pastorianus*. *Pestic. Biochem. Physiol.* **3**, 160–167.

TOLMSOFF, W. J. (1962). Respiratory activities of cell-free particles from *Pythium ultimum, Rhizoctonia solani* and sugar beet seedlings. *Phytopathology* **52**, 755.

TOMITA, K., TAKAHI, Y., ISHIZUKA, R., KAMIMURA, S., NAKAGAWA, M., ANDO, M., NAKANISHI, T. and OKUDAIRA, H. (1973). Hymexazol, a new plant protecting agent. *Ann. Sankyo Res. Lab.* **25**, 1–51.

TOMIZAWA, C. and UESUGI, Y. (1972). Metabolism of *S*-benzyl *O*,*O*-diisopropyl phosphorothioate (Kitazin P) by mycelial cells of *Pyricularia oryzae*. *Agric. biol. Chem.* **36**, 294–300.

TONOMURA, K., MAEDA, K., FUTAI, F., NAKAGAMI, T. and YAMADA, M. (1968). Stimulative vaporisation of phenylmercuric acetate by mercury-resistant bacteria. *Nature, Lond.* **217**, 644.

TURNER, N. J. and BATTERSHELL, R. D. (1969). The relative influence of chemical and physical properties on the fungitoxicity of tetrachloroisophthalonitrile and some analogues. *Contrib. Boyce Thompson Inst.* **24**, 139–148.

TURNER, N. J., LIMPEL, L. E., BATTERSHELL, R. D., BLUESTONE, H. and LAMONT, D. (1964). A new foliage protectant fungicide, tetrachloro-isophthalonitrile. *Contrib. Boyce Thompson Inst.* **22**, 303–310.

UESUGI, Y. and TOMIZAWA, C. (1971a). Metabolism of *O*-ethyl-*S*,*S*-diphenyl-phosphorodithioate (Hinosan) by mycelial cells of *Pyricularia oryzae*. *Agric. biol. Chem.* **35**, 941–949.

UESUGI, Y. and TOMIZAWA, C. (1971b). Metabolism of *S*-benzyl-*O*-ethyl phenyl-phosphorothioate (Inezin) by mycelial cells of *Pyricularia oryzae*. *Agric. biol. Chem.* **36**, 313–317.

VAN ALFEN, N. K. and KOSUGE, T. (1974). Microbial metabolism of the fungicide 2,6-dichloro-4-nitroaniline. *J. agric. Fd Chem.* **22**, 221–224.

VARGAS, J. M. and BEARD, J. B. (1970). Chloroneb, a new fungicide for control of Typhula blight. *Pl. Dis. Reptr* **54**, 1075.

VILLANUEVA, J. R. (1959). Nitroreductase systems of a *Nocardia* species. *J. gen. Microbiol.* **20**, vi.

VONK, J. W. and KAARS SIJPESTEIJN, A. (1977). Metabolism. *In* "Systemic Fungicides" 2nd edn. (edited by R. W. Marsh assisted by R. J. W. Byrde and D. Woodcock), pp. 160–175. Longman, London and New York.

VREDEVELD, N. G. (1965). Selective action of pentachloronitrobenzene on soil fungi. PhD thesis, Michigan State Univ.

de WAARD, M. A. (1974). Mechanisms of action of the organophosphorus fungicide pyrazophos. Ph.D. thesis, Agricultural University, Wageningen, The Netherlands.

WADE, P. (1954). Soil fumigation. II. The stability of ethylene dibromide in soil. *J. Sci. Fd Agric.* **5**, 288–291.

WAKSMAN, S. A. and WOODRUFF, H. B. (1940). The soil as a source of micro-organisms antagonistic to disease-producing bacteria. *J. Bact.* **40**, 581–600.

WALKER, N. (1973). Metabolism of chlorophenols by *Rhodotorula glutinis. Soil Biol. Biochem.* **5**, 525–530.

WALKER, R. L. and NEWMAN, A. S. (1956). Microbial decomposition of 2,4-dichlorophenoxyacetic acid. *Appl. Microbiol.* **4**, 201–206.

WALLNÖFER, P. R., SAFE, S. and HUTZINGER, O. (1971). Metabolism of the systemic fungicides 2-methylbenzanilide and 2-chlorobenzanilide by *Rhizopus japonicus. Pestic. Biochem. Physiol.* **1**, 458–463.

WALLNÖFER, P. R., KÖNIGER, M., SAFE, S. and HUTZINGER, O. (1972). Metabolism of the systemic fungicide 2,5-dimethyl-3-furane carboxylic acid anilide (BAS 3191) by *Rhizopus japonicus* and related fungi. *J. agric. Fd Chem.* **20**, 20–22.

WANG, C. H. (1972). The effect of soil properties on losses of 2-chloronitrobenzene fungicide from soil. Ph.D. Dissertation, Dept. of Plant Pathology, University of California, Davis, California.

WENSLEY, R. N. (1953). Microbiological studies of the action of some selected soil fumigants. *Can. J. Bot.* **31**, 277–308.

WESTÖÖ, G. (1968). Determination of methylmercury salts in various kinds of biological material. *Acta Chem. Scand.* **22**, 2277–2280.

WHITESIDE, J. S. and ALEXANDER, M. (1960). Measurement of microbial effects of herbicides. *Weeds* **8**, 204–213.

WIESE, M. V. and VARGAS, J. M. (1972). Degradation and synthesis of chloroneb by soil microorganisms. *Phytopathology* **62**, 1121 (abstr.).

WIESE, M. V. and VARGAS, J. M. (1973). Interconversion of chloroneb and 2,5-dichloro-4-methoxyphenol by soil microorganisms. *Pestic. Biochem. Physiol.* **3**, 214–222.

WOLF, P. A. (1971) cited by Exner *et al.* (1973). *J. agric. Fd Chem.* **21**, 838–842.

WOLF, P. A. and STERNER, P. W. (1972). 2,2-Dibromo-3-nitrilopropionamide, a compound with slimicidal activity. *Appl. Microbiol.* **24**, 581–584.

WOOD, J. M., SCOTT KENNEDY, F. and ROSEN, C. G. (1968). Synthesis of methylmercury compounds by extracts of a methanogenic bacterium. *Nature, Lond.* **220**, 173–174.

WOODCOCK, D. (1977). Non-biological conversions of fungicides. *In* "Antifungal Compounds" (M. R. Siegel and H. D. Sisler, Eds), Vol. 2, pp. 209–249. Marcel Dekker, New York.

WORTHING, C. R. (1974). Soil applied benomyl and carbendazim for control of tomato diseases. *Rep. Glasshouse Crops Res. Inst for 1973*, p. 69.

WRIGHT, J. M. (1955). The production of antibiotics in soil. II. Production of griseofulvin by *Penicillium nigricans. Ann. appl. Biol.* **43**, 288–296.

WRIGHT, J. M. and GROVE, J. F. (1957). The production of antibiotics in soil. V. Breakdown of griseofulvin in soil. *Ann. appl. Biol.* **45**, 36–43.

YAMASHINA, I., SHIKATA, S. and EGAMI, F. (1954). Enzymic reduction of aromatic nitro, nitroso and hydroxyl amino compounds. *Bull. Soc. Chem. Japan* **27**, 42–45.

ZIMMERMAN, R. (1958). Über phenolspaltende Hefen. *Naturwissenschaften* **45**, 165.

PESTICIDE LIST

Chemical designation of pesticides referred to in the text by common or trade names

The accepted common name of the British Standards Institution (BSI) or the International Standards Organization (ISO) is used wherever possible. If no common name has been approved by the BSI or ISO the common name accepted by an alternative organization is given. If no common name is available, the chemical name, trivial name or trade name (with an initial capital letter) is used.

In compiling this list reference has been made to the following sources:

Crop Chemicals Index (1977), 8th edition. ICI Plant Protection Division, Bracknell, England.

PANS Pesticide Index (1977). Centre for Overseas Research, London, England.

Pesticide Manual (1977), 5th edition. British Crop Protection Council.

Recommended Common Names for Pesticides, BS1831 (1969) and supplements no. 1 (1970), no. 2 (1970) and no. 3 (1974). British Standards Institution.

KEY TO ABBREVIATIONS USED IN PESTICIDE LIST

Authorities for common names (given in brackets)	Classification of pesticide use
B British Standards Institution	A acaricide
O International Standards Organization	B bactericide
J Japanese Ministry of Agriculture and Forestry	F fungicide
W Weed Science Society of America	H herbicide
E Entomological Society of America	I insecticide
U USA Standards Institute	M molluscicide
C Canadian Standards Association	N nematocide
* organic compounds not requiring common names (BSI)	P plant growth regulator
	R rodenticide
	V avicide

Common, trivial, or trade name	Chemical name	Use	Structure on page
Accothion, *see* fenitrothion			
acrolein (W)	acrylaldehyde (or 2-propenal)	HA	—
Agelon; mixture, *see* atrazine and prometryne			
Agrosan GN	phenylmercury acetate + ethylmercury chloride	F	—
alachlor (BOWC)	2'-chloro-2,6-diethyl-*N*-methoxymethylacetanilide	H	—
aldicarb (BO)	2-methyl-2-(methylthio)propionaldehyde *O*-(methylcarbamoyl)oxime	I	—
aldrin (BOW)	contains 95% HHDN (qv.) (100% HHDN in Canada, USSR and Denmark)	I	655
Alipur; mixture, *see* chlorbufam and cycluron			
allethrin (BOC)	(±)-3-allyl-2-methyl-4-oxocyclopent-2-enyl (±)-*cis*, *trans*-chrysanthemate	I	—
allidochlor (BOC)	*NN*-diallylchloroacetamide	H	—
allyl alcohol*	—	F	—
ametryne (B)	2-ethylamino-4-isopropylamino-6-methylthio-1,3,5-triazine	H	693
Amiben, *see* chloramben			
aminotriazole, *see* amitrole			
amitrole (OW)	3-amino-1,2,4-triazole	H	167
anilazine (BO)	2,4-dichloro-6 (2-chloroanilino)-1,3,5-triazine	F	749
Aresin, *see* monolinuron			
asulam (BOWC)	methyl (4-aminobenzenesulphonyl)-carbamate	H	—
atraton (BO)	2-ethylamino-4-isopropylamino-6-methoxy-1,3,5-triazine	H	693
atrazine (BOW)	2-chloro-4-ethylamino-6-isopropylamino-1,3,5-triazine	H	693
azinphos-methyl (BOC)	*S*-(3,4-dihydro-4-oxobenzo[*d*]-[1,2,3]-triazin-3-ylmethyl) dimethyl phosphorothiolothionate	AI	—
barban (BOWUC)	4-chlorobut-2-ynyl 3-chlorophenylcarbamate	H	680
Bayluscide, *see* niclosamide			
benfluralin (BO) (benefin; W)	*N*-butyl-*N*-ethyl-2,6-dinitro-4-trifluoromethylaniline	H	712
Benlate, *see* benomyl			
benomyl (BO)	methyl 1-(butylcarbamoyl) benzimidazol-2-ylcarbamate	F	763

Common, trivial, or trade name	Chemical name	Use	Structure on page
bensulide (BOW)	N-2-(OO-di-isopropylphosphoro-thiolothionyl) ethyl benzenesulphonamide	H	—
bentazone (B)	3-isopropyl-(1H)-benzo-2,1,3-thiadiazin-4-one 2,2-dioxide	H	—
benzoylprop-ethyl (B)	ethyl 2-(N-benzoyl-3,4-dichloroanilino)-propionate	H	678
benzthiazuron (BO)	1-(2-benzothiazolyl)-3-methylurea	H	167
BHC (BO)	mixed isomers of 1,2,3,4,5,6-hexachlorocyclohexane	IR	655
blasticidin-S (J)	blasticidin-S benzyl aminobenzene-sulphonate	F	759
Bordeaux mixture	copper sulphate + hydrated lime solution	F	—
Botran, see dicloran			
bromacil (BOWC)	5-bromo-6-methyl-3-s-butyluracil	H	—
bromophos (BOC)	O-(4-bromo-2,5-dichlorophenyl) OO-diethyl phosphorothioate	AI	—
bromoxynil (BOWC)	3,5-dibromo-4-hydroxybenzonitrile	H	700
bufencarb (metalkamate)	3-ethylpropylphenyl methyl carbamate and 3-methylbutylphenyl methylcarbamate	I	658
butralin (BW)	4-(1,1-dimethylethyl)-N(1-methylpropyl)-2,6-dinitrobenzene amine	H	—
buturon (BOWC)	3-(4-chlorophenyl)-1-methyl-1-(1-methylprop-2-ynyl)urea	H	679
—	S-n-butyl-S'-p-t-butylbenzyl-N-3-pyridyldithiocarbonimidate	F	768
—	N-s-butyl-4-t-butyl-2,6-dinitroaniline	H	715
Bux, see bufencarb			
cacodylic acid (W)	hydroxydimethyl arsenine oxide	H	—
—	calcium cyanamide	H	—
—	calcium cyanide	I	—
camphechlor (B)	a reaction mixture of chlorinated camphenes containing 67–69% chlorine	AI	—

Common, trivial, or trade name	Chemical name	Use	Structure on page
captafol (BOU)	N-(1,1,2,2-tetrachloroethylthio)-3a,4,7,7a-tetrahydrophthalimide	F	—
captan (BOC)	N-(trichloromethylthio)-3a,4,7,7a-tetrahydrophthalimide	F	739
carbaryl (BOUC)	1-naphthyl methylcarbamate	I	656
carbathion, ?see carbophenothion			
carbendazim (BO)	methyl benzimidazol-2-ylcarbamate	F	762
carbetamide (BOW)	D-1-(ethylcarbamoyl)ethyl phenylcarbamate	H	—
carbofuran (BOU)	2,3-dihydro-2,2-dimethylbenzofuran-7-yl methylcarbamate	I	—
carbon disulphide*	—	IN	—
carbopheno-thion (BOC)	S-(4-chlorophenylthiomethyl)OO-diethyl phosphorodithioate	AI	656
carboxin (BO)	2,3-dihydro-6-methyl-5-phenylcarbamoyl-1,4-oxathiin	F	765
Carbyne, see barban			
CCC(chlorocholine chloride), see chlormequat			
CEPC	2-chloroethyl-3'-chloro-phenylcarbamate	H	680
chloramben (BOWC)	3-amino-2,5-dichlorobenzoic acid	H	700
chloramphen-icol (J)	D($-$)-threo-2,2-dichloro-N-(β-hydroxy-α(hydroxymethyl)-p-nitrophenyl)acetamide	B	—
chloranil	2,3,5,6-tetrachloro-1,4-benzoquinone	F	741
chloranocryl (B)	N-(3,4-dichlorophenyl)methacrylamide	H	678
chlorazine (B)	2-chloro-4,6-bisdiethylamino-1,3,5-triazine	H	693
chlorbromuron (BOW)	3-(4-bromo-3-chlorophenyl)-1-methoxy-1-methylurea	H	679
chlorbufam (BO)	1-methylprop-2-ynyl 3-chlorophenyl-carbamate	H	—
chordane (BO)	1,2,4,5,6,7,8,8-octachloro-3a,4,7,7a-tetrahydro-4,7-methanoindane (mixed active isomers)	I	656
chlordecone (BO)	decachloropentacyclo[3,3,2,O^{2,6},O^{3,9},O^{7,10}]-decan-4-one	I	—
chlordimeform (BO)	N'-(4-chloro-o-tolyl)-N,N-dimethylformamidine	AI	656
chlorfenac (BO)	2,3,6-trichlorophenylacetic acid	H	—

Common, trivial, or trade name	Chemical name	Use	Structure on page
chlorfenvinphos (BOC)	2-chloro-1-(2,4-dichlorophenyl)vinyl diethyl phosphate	I	—
chlorflurazole (B)	4,5-dichloro-2-trifluoromethyl-benzimidazole	H	—
chlormequat (BOC)	2-chloroethyltrimethylammonium ion	HP	—
chlorobenzilate (BOCJ)	ethyl 4,4'-dichlorobenzilate	A	656
chloroform*	—	BF	—
—	p-chloroformanilide	H	—
chloroneb (BOU)	1,4-dichloro-2,5-dimethoxybenzene	F	760
chloronitropropane	1-chloro-2-nitropropane	F	—
chloropicrin*	trichloronitromethane	IN	—
chloropropylate (BOC)	isopropyl 4,4'-dichlorobenzilate	A	656
chlorothalonil (B)	tetrachloroisophthalonitrile	F	752
chloroxuron (BOW)	3,4-(4-chlorophenoxy)phenyl-1,1-dimethylurea	H	679
chlorpropham (BOWC)	isopropyl 3-chlorophenylcarbamate	H	680
chlorpyrifos (BOU)	OO-diethyl O-3,5,6-trichloro-2-pyridyl phosphorothioate	AI	656
chlorthal dimethyl (BO)	dimethyl 2,3,5,6-tetrachloroterephthalic acid	H	—
chlorthiamid (BO)	2,6-dichlorothiobenzamide	H	700
chlortoluron (BO)	3-(3-chloro-p-tolyl)-1,1-dimethylurea	H	—
CIPC, see chlorpropham			
copper oxychloride(O)	—	F	—
copper sulphate*	—	F	—
Cotoran, see fluometuron			
Counter, see terbufos			
4-CPA (BOWC)	4-chlorophenoxyacetic acid	H	674
cyanatryn (B)	4-(1-cyano-1-methylethylamino)-6-ethylamino-2-methylthio-1,3,5-triazine	H	—
cyanazine (BW)	2((4-chloro-o(ethylamino)-s-triazin-2-yl)amino)-2-methylpropionitrile	H	693

Common, trivial, or trade name	Chemical name	Use	Structure on page
cycloate (BW)	*S*-ethylcyclohexylethylthiocarbamate	H	—
cycloheximide (BCJ)	3-(2-(3,5-dimethyl-2-oxocyclohexyl)-2-hydroxyethyl)glutarimide	F	—
cycluron (BOWC)	3-cyclo-octyl-1,1-dimethylurea	H	—
2,4-D (BOC)	2,4-dichlorophenoxyacetic acid	HP	671
2,4-DA = 2,4-D amine, *see* 2,4-D			
dalapon (BOWC)	2,2-dichloropropionic acid	H	703
dazomet (BOWC)	tetrahydro-3,5-dimethyl-2*H*-1,3,5-thiadiazine-2-thione	FHN	—
2,4-DB (BW)	4-(2,4-dichlorophenoxy)butyric acid	H	671
DCA	dichloroacetic acid	H	—
DCMU (J), *see* diuron			
DCU	dichloral urea	H	—
D-D	1,2-dichloropropane-1,3-dichloropropene mixture	N	734
DDOD, *see* dichlozoline			
DDT (BC)	technical dichlorodiphenyltrichloroethane (mixture in which *pp'*-DDT predominates)	IR	—
pp'-DDT (B)	1,1,1-trichloro-2,2-di-(4-chlorophenyl)-ethane	IR	656
demeton (BO); mixture, *see* demeton-O and demeton-S			
demeton-O (BO)	diethyl 2-(ethylthio)-ethyl phosphorothionate	AI	—
demeton-S (BOC)	diethyl *S*-[2-(ethylthio)ethyl] phosphorothiolate	AI	—
2,4-DEP (W)	contains at least 90% of tris-(2,4-dichlorophenoxyethyl)phosphite and bis(2,4-dichlorophenoxyethyl) phosphite	H	—
2,4-DES, *see* disul			
desmetryne (B) (desmetryn; OW)	2-isopropylamino-4-methylamino-6-methylthio-1,3,5-triazine	H	—
Dexon, *see* fenaminosulf			
DFP	*OO*-diisopropyl phosphorofluoridate	I	657
diazinon (BOC)	*OO*-diethyl *O*-(2-isopropyl-6-methyl-4-pyrimidinyl) phosphorothioate	AI	657
—	dibromochloropropane (1,2-dibromo-3-chloropropane)	N	—
—	2,2-dibromo-3-nitropropionamide	FB	—

Common, trivial, or trade name	Chemical name	Use	Structure on page
dicamba (BOWC)	3,6-dichloro-2-methoxybenzoic acid	H	700
dichlobenil (BOWC)	2,6-dichlorobenzonitrile	H	700
dichlone (BOC)	2,3-dichloro-1,4-naphthoquinone	F	741
dichlorophen (BO)	di-(5-chloro-2-hydroxyphenyl)methane	F	—
1,3-dichloropropene, *see* Telone			
dichlorprop (BOWC)	(±)-2-(2,4-dichlorophenoxy)propionic acid	H	671
dichlorvos (BOC)	2,2-dichlorovinyl dimethyl phosphate	I	657
dichlozoline (BJ)	3(3,5-dichlorophenyl)-5,5-dimethyloxazolidine-2,4-dione	F	766
dicloran (B)	2,6-dichloro-4-nitroaniline	F	751
dicryl(O), *see* chloranocryl			
dieldrin (BOC)	contains not less than 85% HEOD	I	657
diflubenzuron	1-(4-chlorophenyl)-3-(2,6-difluoro-benzoyl)urea	I	—
difluronyl, *see* diflubenzuron			
dimethirimol (BO)	5-butyl-2-dimethylamino-4-hydroxy-6-methylpyrimidine	F	—
dimethoate (BOC)	dimethyl S-(N-methylcarbamoylmethyl) phosphorothiolothionate	AI	—
dinitramine (BW)	N'N'-diethyl-2,6-dinitro-4-trifluoromethyl-m-phenylenediamine	H	712
—	2,4-dinitrophenol	F	748
dinoseb (BOWC)	2,4-dinitro-6-s-butylphenol	FHI	—
diphenamid (BOWC)	NN-dimethyldiphenylacetamide	H	—
diquat (BOWC)	1,1'-ethylene-2,2'-bipyridylium ion	H	706
disul (O)	2-(2,4-dichlorophenoxy)ethyl hydrogen sulphate	H	—
disulfoton (BOC)	diethyl S-(2-(ethylthio)ethyl) phosphorothiolothionate	I	—
diuron (BOW)	3-(3,4-dichlorophenyl)-1,1-dimethylurea	H	679
DMPA	O-(2,4-dichlorophenyl)O'-methyl N-isopropylphosphoroamidothioate	H	—
DNBP (JO), *see* dinoseb			
DNOC (BOWC)	2-methyl-4,6-dinitrophenol	HI	—
dodine (BOC)	dodecylguanidine acetate	F	—
drazoxolon (BO)	4-(2-chlorophenylhydrazono)-3-methyl-5-isoxazolone	F	167
Dursban, *see* chlorpyrifos			
Dyfonate, *see* fonofos			
Dyrene, *see* anilazine			

Common, trivial, or trade name	Chemical name	Use	Structure on page
EDB (J), *see* ethylene dibromide			
EDCT	mixture of ethylene dichloride and carbon tetrachloride	N	—
edifenphos (BO)	*O*-ethyl *SS*-diphenyl phosphorodithioate	F	754
EMC, *see* ethylmercury chloride			
EMP (J)	ethylmercury phosphate	F	—
endosulfan (BOC)	6,7,8,9,10,10-hexachloro-1,5,5a,6,9,9a-hexahydro-6,9-methano-2,4,3-benzo(e)dioxathiepin 3-oxide	AI	657
endothal (B)	7-oxabicyclo(2,2,1)heptane-2,3-dicarboxylic acid	HP	716
endothal–sodium, *see* endothal			
endrin (BOC)	1,2,3,4,10,10-hexachloro-6,7-epoxy-1,4,4a,5,6,7,8,8a-octahydro-*exo*-1,4-*exo*-5,8-dimethanonaphthalene	IV	657
EPN (EJ)	*O*-ethyl *O*-4-nitrophenyl phenylphosphonothioate	AI	657
Eptam, *see* EPTC			
EPTC (BOW)	*S*-ethyl dipropylthiocarbamate	H	—
ESBP (J)	*S*-benzyl *O*-ethyl phenylphosphonothioate	F	755
ethirimol (B)	5-butyl-2-ethylamino-4-hydroxy-6-methylpyrimidine	F	167
ethoprophos (B)	*O*-ethyl *SS*-dipropyl phosphorodithioate	IN	—
Ethylan CP	condensate of octylphenol/ethylene oxide	F	—
ethylene dibromide*	1,2-dibromoethane	IN	—
—	ethylmercury acetate	F	—
ethylmercury chloride*	—	F	—
ethylmercury phosphate, *see* EMP			
fenac (WC), *see* chlorfenac			
fenaminosulf (B)	sodium 4-dimethylaminobenzene-diazosulphonate	FB	—
fenamiphos (B)	*O*-methyl *O*-(3-methyl-4-methylthiophenyl)isopropyl-phosphoramidate	N	—
fenitrothion (BO)	*OO*-dimethyl *O*-(3-methyl-4-nitrophenyl)phosphorothioate	I	657

Common, trivial, or trade name	Chemical name	Use	Structure on page
fenoprop (BO)	(±)-2-(2,4,5-trichlorophenoxy)propionic acid	H	671
fensulfothion (BOC)	diethyl 4-(methylsulphinyl)phenyl phosphorothionate	IN	—
fenthion (BOC)	OO-dimethyl O-(3-methyl-4-methylthiophenyl)phosphorothioate	I	658
fentin acetate (BO)	triphenyltin acetate	FIM	—
fenuron (BOWC)	1,1-dimethyl-3-phenylurea	H	—
ferbam (BOC)	ferric dimethyldithiocarbamate	F	—
fluometuron (BOWC)	1,1-dimethyl-3-(3-trifluoromethylphenyl)urea	H	679
fluorodifen (BOC)	4-nitrophenyl 2-nitro-4-trifluoromethyl-phenyl ether	H	—
—	1-fluoro-2,4-dinitrobenzene	F	—
folpet (BC)	N-(trichloromethylthio)phthalimide	F	—
fonofos (BO)	ethyl S-phenyl ethylphosphonodithioate	I	658
formalin	40% aqueous solution of formaldehyde*	FB	
fuberidazole (BO)	2-(2-furyl)benzimidazole	F	167
fumarin (BC)	3-(α-acetonylfurfuryl)-4-hydroxycoumarin	R	167
Fungex	copper ammonium carbamate	F	—
Fungicidin = nystatin			
furcarbanil (B)	2,5-dimethyl-3-furanilide	F	766
gamma-BHC (B) gamma isomer of BHC			
Gesaprim, *see* atrazine			
Gesaran, *see* methoprotryne			
glyphosate (BUW)	N-(phosphonomethyl)glycine	H	—
Gramoxone, *see* paraquat			
griseofulvin (BOC)	7-chloro-4,6,2'-trimethoxy-6'-methylgris-2'-en-3,4'-dione	F	154
HEOD (BO)	1,2,3,4,10,10-hexachloro-6,7-epoxy-1,4,4a,5,6,7,8,8a-octahydro-*exo*-1,4-*endo*-5,8-dimethanonaphthalene	I	657
heptachlor (BOC)	1,4,5,6,7,8,8-heptachloro-3a,4,7,7a-tetrahydro-4,7-methanoindene	I	658
hexachlordane, *see* chlordane			
hexachloro-phene	2,2'-methylene bis(3,4,6-tri-chlorophenol)	F	—

Common, trivial, or trade name	Chemical name	Use	Structure on page
HHDN (BO)	1,2,3,4,10,10-hexachloro-1,4,4a,5,8,8a-hexahydro-*exo*-1,4-*endo*-5,8-dimethanonaphthalene	I	655
HPMTS	2-hydroxypropyl methanethio-sulphonate	F	—
hymexazol	3-hydroxy-5-methylisoxazole	F	750
IAA	indole-3-acetic acid	P	167
IBP (J)	S-benzyl-OO-diisopropyl phosphorothioate	F	755
Inezin, *see* ESBP			
ioxynil (BOWC)	4-hydroxy-3,5-di-iodobenzonitrile	H	700
ipazine (B)	2-chloro-4-diethylamino-6-isopropylamino-1,3,5-triazine	H	693
IPC, *see* propham			
Isobac, *see* hexachlorophene			
isobenzan (BOC)	1,3,4,5,6,7,8,8-octachloro-1,3,3a,4,7,7a-hexahydro-4,7-methanoisobenzofuran	I	—
isocil (BOWC)	5-bromo-3-isopropyl-6-methyluracil	H	—
isopropalin (BW)	4-isopropyl-2,6-dinitro-NN-dipropylaniline	H	712
Karsil	N-(3'4'-dichlorophenyl)2-methylpentanamide	H	678
kasugamycin (J)	(5-amino-2-methyl-6-(2,3,4,5,6-pentanhydroxycyclohexyloxy)pyran-3-yl) amino-α-iminoacetic acid	F	—
Kitazin P, *see* IBP			
KOCN (W)	potassium cyanate	H	—
Kuron, *see* fenoprop			
Kurosal SL, *see* fenoprop			
Lanstan, *see* chloronitropropane			
—	N-lauroyl-L-valine	F	769
lenacil (BOW)	3-cyclohexyl-6,7-dihydro-1H-cyclopentapyrimidine-2,4(3H, 5H)-dione	H	—
lindane (O), *see* gamma–BHC			
linuron (BOWC)	3-(3,4-dichlorophenyl)-1-methoxy-1-methylurea	H	679
Lumeton; mixture, *see* methoprotryne, simazine and mecoprop			
malathion (BOC)	S-[1,2-di(ethoxycarbonyl)ethyl] dimethyl phosphorothiolothionate	AI	658

Common, trivial, or trade name	Chemical name	Use	Structure on page
maleic hydrazide*	6-hydroxy-3-(2H)-pyridazinone	P	—
mancozeb (BO)	complex of zinc and maneb containing 20% manganese and 2·5% zinc	F	—
maneb (BOC)	manganese ethylenebisdithiocarbamate (polymeric)	F	—
MCA	monochloroacetic acid*	H	—
MCPA (BOWC)	2-methyl-4-chlorophenoxyacetic acid	H	671
MCPB (BOWC)	4-(4-chloro-2-methylphenoxy)butyric acid	H	671
mebenil	2-methylbenzanilide	F	—
mecoprop (BOWC)	(±)-2-(4-chloro-2-methylphenoxy)propionic acid	H	671
MEMC	2-methoxyethylmercury chloride	F	—
menazon (BOC)	S-(4,6-diamino-1,3,5-triazin-2-ylmethyl) dimethyl phosphorothiolothionate	I	—
methabenzthiazuron (BO)	1,2'-benzothiazolyl-1,3-dimethylurea	H	—
metham (BW)	methyldithiocarbamic acid	FHIN	—
metham-sodium (B)	sodium N-methyldithiocarbamate dihydrate	FHIN	738
methazole (BOUW)	2-(3,4-dichlorophenyl)-4-methyl-1,2,4-oxadiazolidine-3,5-dione	H	—
methoprotryne (BOC)	2-isopropylamino-4-(3-methoxypropylamino)-6-methylthio-1,3,5-triazine	H	—
methoxychlor (BOC)	1,1,1-trichloro-2,2-di-(4-methoxyphenyl)ethane	I	658
methoxyethylmercury chloride, *see* MEMC			
—	methylarsenic sulphide	F	—
methyl bromide*	bromomethane	AFHIN	—
methylmercury dicyandiamide*	N-cyano-N'-(methylmercury)guanidine	FN	—
methyl parathion, *see* parathion methyl			
metobromuron (BOWC)	3-(4-bromophenyl)-1-methoxy-1-methylurea	H	679
metoxuron (BO)	3-(3-chloro-4-methoxyphenyl)-1,1-dimethylurea	H	—
metribuzin (BW)	4-amino-4,5-dihydro-3-methythio-6-t-butyl-1,2,4-triazin-5-one	H	—
mevinphos (BOC)	2-methoxycarbonyl-1-methylvinyl dimethyl phosphate	IN	—

Common, trivial, or trade name	Chemical name	Use	Structure on page
mexacarbate (BU)	4-dimethylamino-3,5-xylyl methylcarbamate	AIM	658
mirex (E)	dodecachlorooctahydro-1,3,4-methano-2H-cyclobuta[cd]pentalene	I	659
MMDD, *see* methylmercury dicyandiamide			
Mobam	4-benzothienyl methylcarbamate	I	167
Mocap, *see* ethoprophos			
molinate (BW)	S-ethyl NN-hexamethylenethiolocarbamate	H	—
monolinuron (BOWC)	3-(4-chlorophenyl)-1-methoxy-1-methylurea	H	679
monuron (BOWC)	3-(4-chlorophenyl)-1,1-dimethylurea	H	679
morphothion (BOC)	OO-dimethyl S-(morpholino-carbonylmethyl)phosphorodithioate	I	167
MSMA (W)	monosodium methanearsonate	H	—
Murbetol; mixture, *see* endothal–sodium and propham			
nabam (BOC)	disodium ethylenebisdithiocarbamate	F	737
naptalam (BOWC)	N-1-naphthylphthalamic acid	H	—
neburon (BOWC)	1-butyl-3-(3,4-dichlorophenyl)-1-methylurea	H	—
Neo-Voronite	mixture of a benzimidazole and carbamylic acid	F	—
niclosamide (BO)	5-chloro-N-(2-chloro-4-nitrophenyl) salicylamide	M	—
nicotine*	1-methyl-2-(3-pyridyl) pyrrolidine	I	659
nitralin (BW)	4-methylsulphonyl-2,6-dinitro-NN-dipropylaniline	H	712
nitrapyrin (B)	2-chloro-6-trichloromethylpyridine	B	—
nitrofen (BOW)	2,4-dichloro-4'-nitrodiphenyl ether	H	—
noruron (BO)	3-(hexahydro-4,7-methanoindan-5-yl)-1,1-dimethylurea	H	—
N-Serve, *see* nitrapyrin			
Ordram, *see* molinate			
oryzalin (BW)	3,5-dinitro-N^4N^4-dipropyl-sulphanilamide	H	712
oxycarboxin (BOJ)	2,3-dihydro-6-methyl-5-phenylcarbamoyl-1,4-oxathiin 4,4-dioxide	F	—

Common, trivial, or trade name	Chemical name	Use	Structure on page
Panacide, *see* dichlorophen			
Panodrench, *see* methylmercurydicyandiamide			
Panogen, *see* methylmercurydicyandiamide			
para-oxon	*OO*-diethyl *O*-4-nitrophenyl phosphate	I	—
paraquat (BOWC)	1,1'-dimethyl-4,4'-bipyridylium ion	H	706
parathion (BOC)	*OO*-diethyl *O*-*p*-nitrophenyl phosphorothioate	AIV	659
parathion-methyl (BOC)	*OO*-dimethyl *O*-*p*-nitrophenyl phosphorothioate	AI	—
parbendazole	methyl 5(6)-butyl-benzimidazol-2-yl carbamate	F	763
PCNB (J), *see* quintozene			
PCP (W), *see* pentachlorophenol			
pebulate (BOW)	*S*-propyl butylethylthiocarbamate	H	—
pentachlorophenol*	—	FHI	716
pentanochlor (BO)	*N*-(3-chloro-4-methylphenyl)-2-methyl-pentanamide	H	678
permethrin	3-phenoxybenzyl(±)-*cis,trans*-3-(2,2-dichlorovinyl)-2,2-dimethylcyclo-propanecarboxylate	I	—
phenmedipham (BOW)	3-methoxycarbonylaminophenyl *N*-(3'-methylphenyl)carbamate	H	—
phenobenzuron (B)	1-benzoyl-1-(3,4-dichlorophenyl)-3,3-dimethylurea	H	—
phentriazophos, *see* triazophos			
phenylmercuriurea*	—	F	—
phenylmercury acetate*	—	FH	—
phenylmercury urate, *see* phenylmercuriurea			
phorate (BOC)	*OO*-diethyl *S*-(ethylthiomethyl) phosphorodithioate	I	659
phosmet (B)	*OO*-dimethyl phthalimidomethyl phosphorodithioate	AI	659
Phosphene, *see* mevinphos			
phoxim (BO)	α-cyanobenzylideneamino diethyl phosphorothionate	AI	—
Phytar, *see* cacodylic acid			
picloram (BOWC)	4-amino-3,5,6-trichloropicolinic acid	H	167
pirimicarb (BO)	2-dimethylamino-5,6-dimethylpyrimidin-4-yl dimethylcarbamate	I	—
pirimiphos-ethyl (B)	*O*-(2-diethylamino-6-methylpyrimidin-4-yl) *OO*-diethyl phosphorothioate	I	—

Common, trivial, or trade name	Chemical name	Use	Structure on page
pirimiphos-methyl (B)	O-(2-diethylamino-6-methylpyrimidin-4-yl)OO-dimethyl	AI	—
PMA, see phenylmercury acetate			
potassium cyanate, see KOCN			
Preforan, see fluorodifen			
Printazol; mixture, see MCPA and 2,4-D			
prometon (BOW)	2,4-bisisopropylamino-6-methoxy-1,3,5-triazine	H	693
prometryne (B)	2,4-bisisopropylamino-6-methylthio-1,3,5-triazine	H	693
pronamide (W), see propyzamide			
propachlor (BOWC)	2'-chloro-N-isopropylacetanilide	H	678
propanil (BOWC)	N-(3,4-dichlorophenyl)propionamide	H	678
propazine (BOW)	2-chloro-4,6-bisisopropylamino-1,3,5-triazine	H	693
propham (BOWC)	isopropyl phenylcarbamate	H	680
Prophos, see ethoprophos			
propoxur (BO)	2-isopropoxyphenyl methylcarbamate	I	—
propyzamide (BOJ)	3,5-dichloro-N-(1,1-dimethyl-propynyl)benzamide	H	716
prothiocarb (B)	S-ethyl-N-(3-dimethylaminopropyl)thiocarbamate hydrochloride	F	740
pyrazon (B)	5-amino-4-chloro-2-phenyl-3-pyridazinone	H	716
pyrazophos (B)	O-6-ethoxycarbonyl-5-methyl-pyrazolo[1,5-a]pyrimidin-2-yl diethylphosphorothionate	F	756
pyriclor	2,3,5-trichloropyridin-4-ol	H	—
quinomethion-ate (B)	6-methyl-2-oxo-1,3-dithiolo(4,5-b)quinoxaline	AFI	167
—	quinonoxime benzoylhydrazine	F	—
quintozene (BOC)	pentachloronitrobenzene	F	—

Reglone, see diquat
Rhizoctol, see methylarsenic sulphide

Common, trivial, or trade name	Chemical name	Use	Structure on page
rotenone*	1,2,12,12a-tetrahydro-8,9-dimethoxy-2-(1-methylenyl)-(1)benzopyrano(3,4,*b*)furo(2,3,*h*)(1)-benzopyran-6(6H)-one	I	—
—	rufianic acid	F	742
schradan (BOC)	bis-*NNN'N'*-tetramethylphosphorodiamidic anhydride	AI	—
Semesan	2-chloro-4-hydroxymercuriphenol	F	—
Sevidol; mixture, *see* carbaryl and gamma-BHC			
Sevin, *see* carbaryl			
siduron (BOWC)	1-(2-methylcyclohexyl)-3-phenylurea	H	679
silvex (UW), *see* fenoprop			
simazine (BOWC)	2-chloro-4,6-bisethylamino-1,3,5-triazine	H	693
simeton (B)	2,4-bisethylamino-6-methoxy-1,3,5-triazine	H	693
simetryne (BO)	2,4-bisethylamino-6-methylthio-1,3,5-triazine	H	693
—	sodium arsenite	A	—
sodium cacodylate, *see* cacodylic acid			
—	sodium chlorate	H	—
—	sodium hexachlorophene	F	—
sodium pentachlorophenate, *see* pentachlorophenol			
Stam, *see* propanil			
streptomycin (BOC)	2,4-diguanidino-3,5,6-trihydroxycyclohexyl 5-deoxy-2-*O*-(2-deoxy-2-methylamino-α-L-glucopyranosyl)-3-*C*-formyl-β-L-*lyxo* pentanofuranoside	BF	—
Subamycin, *see* tetracycline			
sulfallate (BOC)	2-chloroallyl diethyldithiocarbamate	H	—
—	sulphur	AF	—
swep (WC)	methyl-3,4-dichlorocarbanilate	H	680
2,4,5-T (BOWC)	2,4,5-trichlorophenoxyacetic acid	H	671
2,4,5-TB (BOC)	4-(2,4,5-trichlorophenoxy)butyric acid	H	671
2,3,6-TBA (BWOC)	2,3,6-trichlorobenzoic acid	H	700
TCA (BOW)	sodium trichloroacetate	H	703
TCMTB	2-(thiocyanomethylthio)benzothiazole	F	—
TCNB, *see* tecnazene			

Common, trivial, or trade name	Chemical name	Use	Structure on page
TDE (BOC)	1,1-dichloro-2,2-di-(4-chlorophenyl)ethane	I	640
tebuthiuron (BW)	1-(5-t-butyl-1,3,4-thiadiazol-2-yl)-1,3-dimethylurea	H	167
tecnazene(BOC)	1,2,4,5-tetrachloro-3-nitrobenzene	F	—
Telone	1,3-dichloropropene	N	734
TEPP (BOC)	tetraethyl pyrophosphate	I	659
terbacil (BW)	3-t-butyl-5-chloro-6-methyluracil	H	—
terbufos	S-t-butylthiomethyl-OO-diethyl phosphorodithioate	I	—
terbutryne (B) (terbutryn; OW)	2-ethylamino-6-methylthio-4-t-butylamino-1,3,5-triazine	H	693
tetracycline	4-(dimethylamino)1,4,4a,5,5a,6,11,12a-octahydro-3,6,10,12,12a-pentahydroxy-6-methyl-1,11-dioxo-2-naphthacenecarboxamide	B	—
thiabendazole (B)	2-(thiazol-4-yl) benzimidazole	F	763
thiazon = ?Thiazone, see dazomet			
Thimet, see phorate			
thionazin (B)	diethyl 2-pyrazinyl phosphorothionate	IN	—
thiophanate (B)	1,2-di-(3-ethoxycarbonyl-2-thioureido)benzene	F	762
thiophanate-methyl (BO)	1,2-di-(3-methoxycarbonyl-2-thioureido)benzene	F	762
thiram (BOC)	tetramethylthiuram disulphide	F	738
Tillam, see pebulate			
Tordon, see picloram			
toxaphene (OC), see camphechlor			
TPA, ?see fenoprop			
triadimefon	1-(4-chlorophenoxy)-3,3-dimethyl-1-(1,2,4-triazol-1-yl)-2-butanone	F	767
triamiphos (BOC)	5-amino-1-(bisdimethylaminophosphinyl)-3-phenyl-1,2,4-triazole	F	—
triarimol (B)	2,4-dichloro-α-pyrimidin-5-ylbenzhydrol	F	—
triazophos (BO)	OO-diethyl O-1-phenyl-1,2,4-triazol-3-yl phosphorothioate	AI	167
tricamba (BOWC)	3,5,6-trichloro-2-methoxybenzoic acid	H	—

Common, trivial, or trade name	Chemical name	Use	Structure on page
trichloronate (B)	ethyl 2,4,5-trichlorophenyl ethylphosphonothionate	I	—
trichlorphon (B) (trichlorfon; OC)	dimethyl 2,2,2-trichloro-1-hydroxyethyl phosphonate	I	659
tridemorph (BO)	2,6-dimethyl-4-tridecylmorpholine	F	—
trietazine (BOWJ)	2-chloro-4-diethylamino-6-ethylamino-1,3,5-triazine	H	693
trifluralin (BOWC)	2,6-dinitro-NN-dipropyl-4-trifluoromethylaniline	H	712
triforine (BO)	1,4-di-(2,2,2-trichloro-1-formanidoethyl)piperazole	F	—
Ustinex GL; mixture, see diuron, bromacil, aminotriazole, and 2,4-D			
VCS-438, see methazole			
Verdasan, see phenylmercury acetate			
vernolate (W)	S-propyl NN-dipropyl thiolcarbamate	H	—
Vorlex	mixture of dichloropropane-dichloropropene (see D-D) and methyl isothiocyanate	F	—
—	xylene	H	—
Zectran, see mexacarbate			
zineb (BOC)	zinc ethylenebisdithiocarbamate (polymeric)	F	—
ziram (BOC)	zinc dimethyldithiocarbamate	F	—

Subject Index

A

Acanthamoeba castellanii
pesticide effects on, 622
Acarina, 604
pesticide effects on, 609
Accothion, effect on ureolytic microbes, 364
Acetomonas, transformation of DDT, 641
Acetylation
by microorganisms, 119, 182, 631, 752
in soil, 719
Acetylene reduction test, for nitrogen-fixation, 282–283
Achromobacter, 41, 62
aromatic ring cleavage of chlorinated biphenyls, 636, 637
fungicide action on, 353
transformation of pesticides, 684, 692, 703, 708–709, 757
Achromobacter eurydice, nitro-reduction of parathion, 648
Achromobacter nephridii, herbicide action on, 324
Acid production
fungal, herbicide influence, 447
Acinetobacter, 62
transformation of DDT, 641
Acridine orange
staining microbes in soil, 261
Acrolein
from oxidation of allyl alcohol, 733
volatilization from water, 99
Acrophialophora, insecticide action on, 362

Actinomyces griseus, hydrolysis of trichlorphon, 634
Actinomycetes, *see also* individual organisms
decarboxylation of dicamba, 699
fumigant effects on, 371–372
fungicide effects on, 348–349, 487–489
herbicide effects on, 332–333, 428–441, 446
insecticide effects on, 360–362, 366, 496–497
transformation of insecticides 141, 647
Actinomycin
transformation by microorganisms, 757
transformation in soil, 757
Activation of pesticides, 7, 140, 150
Acylanilide herbicides, *see also* individual compounds
effects on micro-algae, 566–567
transformation of, 678–683
Adenosine triphosphate, ATP
in algae, pesticide effects on, 550, 552, 555, 556, 563, 564
determination of, 266–267
as index of soil microbial population, 266–268
in sea water, 41
Adsorption of pesticides to soil, 82, 86, 87–88, 89, 95, 99–100, 101, 170, 706, 710, *see also* Desorption of pesticides from soil, and Bound pesticide residues in soil
Aerobacter, herbicide action on, 381